THE STATE OF THE ART OF NEUTRINO PHYSICS

A Tutorial for Graduate Students and Young Researchers

ADVANCED SERIES ON DIRECTIONS IN HIGH ENERGY PHYSICS

ISSN: 1793-1339

Published

Vol. 1 – High Energy Electron–Positron Physics (*eds. A. Ali and P. Söding*)
Vol. 2 – Hadronic Multiparticle Production (*ed. P. Carruthers*)
Vol. 3 – CP Violation (*ed. C. Jarlskog*)
Vol. 4 – Proton–Antiproton Collider Physics (*eds. G. Altarelli and L. Di Lella*)
Vol. 5 – Perturbative QCD (*ed. A. Mueller*)
Vol. 6 – Quark–Gluon Plasma (*ed. R. C. Hwa*)
Vol. 7 – Quantum Electrodynamics (*ed. T. Kinoshita*)
Vol. 9 – Instrumentation in High Energy Physics (*ed. F. Sauli*)
Vol. 10 – Heavy Flavours (*eds. A. J. Buras and M. Lindner*)
Vol. 11 – Quantum Fields on the Computer (*ed. M. Creutz*)
Vol. 12 – Advances of Accelerator Physics and Technologies (*ed. H. Schopper*)
Vol. 13 – Perspectives on Higgs Physics (*ed. G. L. Kane*)
Vol. 14 – Precision Tests of the Standard Electroweak Model (*ed. P. Langacker*)
Vol. 15 – Heavy Flavours II (*eds. A. J. Buras and M. Lindner*)
Vol. 16 – Electroweak Symmetry Breaking and New Physics at the TeV Scale (*eds. T. L. Barklow, S. Dawson, H. E. Haber and J. L. Siegrist*)
Vol. 17 – Perspectives on Higgs Physics II (*ed. G. L. Kane*)
Vol. 18 – Perspectives on Supersymmetry (*ed. G. L. Kane*)
Vol. 19 – Linear Collider Physics in the New Millennium (*eds. K. Fujii, D. J. Miller and A. Soni*)
Vol. 20 – Lepton Dipole Moments (*eds. B. Lee Roberts and William J. Marciano*)
Vol. 21 – Perspectives on Supersymmetry II (*ed. G. L. Kane*)
Vol. 22 – Perspectives on String Phenomenology (*eds. B. Acharya, G. L. Kane and P. Kumar*)
Vol. 23 – 60 Years of CERN Experiments and Discoveries (*eds. H. Schopper and L. Di Lella*)
Vol. 24 – The High Luminosity Large Hadron Collider: The New Machine for Illuminating the Mysteries of Universe (*eds. O. Brüning and L. Rossi*)
Vol. 25 – Massive Neutrinos: Flavor Mixing of Leptons and Neutrino Oscillations (*ed. H. Fritzsch*)
Vol. 26 – The Standard Theory of Particle Physics: Essays to Celebrate CERN's 60th Anniversary (*eds. L. Maiani and L. Rolandi*)
Vol. 27 – Technology Meets Research 60 Years of CERN Technology: Selected Highlights (*eds. C. Fabjan, T. Taylor, D. Treille and H. Wenninger*)
Vol. 28 – The State of the Art of Neutrino Physics: A Tutorial for Graduate Students and Young Researchers (*ed. A. Ereditato*)

Forthcoming

Vol. 8 – Standard Model, Hadron Phenomenology and Weak Decays on the Lattice (*ed. G. Martinelli*)

Advanced Series on
Directions in High Energy Physics — Vol. 28

THE STATE OF THE ART OF NEUTRINO PHYSICS

A Tutorial for Graduate Students and Young Researchers

Editor

Antonio Ereditato

University of Bern, Switzerland

World Scientific

NEW JERSEY • LONDON • SINGAPORE • BEIJING • SHANGHAI • HONG KONG • TAIPEI • CHENNAI • TOKYO

Published by

World Scientific Publishing Co. Pte. Ltd.
5 Toh Tuck Link, Singapore 596224
USA office: 27 Warren Street, Suite 401-402, Hackensack, NJ 07601
UK office: 57 Shelton Street, Covent Garden, London WC2H 9HE

Library of Congress Cataloging-in-Publication Data
Names: Ereditato, Antonio, editor.
Title: The state of the art of neutrino physics : a tutorial for graduate students and young researchers / editor, Antonio Ereditato (University of Bern, Switzerland).
Other titles: Advanced series on directions in high energy physics ; v. 28.
Description: Singapore ; Hackensack, NJ : World Scientific, [2018] | Series: Advanced series on directions in high energy physics ; vol. 28 | Includes bibliographical references and index.
Identifiers: LCCN 2017033409| ISBN 9789813226081 (hardcover ; alk. paper) | ISBN 9813226080 (hardcover ; alk. paper)
Subjects: LCSH: Neutrinos.
Classification: LCC QC793.5.N42 S73 2018 | DDC 539.7/215--dc23
LC record available at https://lccn.loc.gov/2017033409

British Library Cataloguing-in-Publication Data
A catalogue record for this book is available from the British Library.

For any available supplementary material, please visit
http://www.worldscientific.com/worldscibooks/10.1142/10600#t=suppl

Desk Editor: Ng Kah Fee

Typeset by Stallion Press
Email: enquiries@stallionpress.com

Preface

The neutrino is amongst the most fascinating elementary particles due to its very elusive nature and outstanding properties that have attracted the interest of generations of physicists since 1930, when it was first postulated by Wolfgang Pauli as a "desperate remedy" to explain the apparent energy violation in the beta decay. Many fundamental discoveries in particle physics had the neutrino involved in one way or another.

To date, neutrino physics is still one of the hottest topics of modern particle physics. The discovery of neutrino oscillations in 1998 by the Super-Kamiokande collaboration and the Nobel Prize for 2015 jointly awarded to Takahaki Kajita and Art McDonald for the discovery are just two milestones of a long series of events and scientific achievements that have marked the last 25 years of research. Key experiments and significant theoretical developments have contributed in building up what we can now call the Standard Model of Neutrino Physics.

Neutrino flavor and mass eigenstates mix and oscillate as described by the Pontecorvo–Maki–Nakagawa–Sakata (PMNS) matrix. The occurrence of oscillations hence implies that the neutrino is a massive particle like all other fermions, although with a much smaller mass. The mechanism that generates such low neutrino masses is still unknown and this leads to strong hints for the existence of new physics beyond the Standard Model of particles and interactions. Moreover, there is still a long list of unknowns in neutrino physics that demand the continued effort of physicists: the Dirac or Majorana nature of the particle, the possible presence of a CP violating phase in the mixing matrix, the reason of the non-diagonal nature of the matrix, the ordering of the three neutrino mass eigenstates and their values, the question of the existence of additional neutrino states beyond the three known flavors — the so-called sterile neutrinos, the relevance of neutrinos in the early and current Universe, the neutrino emission from point astrophysical sources, the role of neutrinos in Supernova explosions, the effect of neutrinos in new physics paradigms etc. This shopping list will likely keep neutrino physicists occupied for several more decades.

The aim of this book is to provide a tutorial summarizing the present scenario and the future challenges to those students and young researchers who decide to work in neutrino physics with the objective of solving for the unknowns mentioned above. Several distinguished colleagues contributed to the work, paying much attention to educational and didactic aspects. I am sincerely grateful to them for their effort and dedication. My colleagues and I hope that future researchers will find here answers to their questions and, hopefully, incentives to progress further in this fascinating research field.

Contents

Chapter 1

A Brief History of Neutrino

A. Bettini

University of Padua. INFN. Padua Division (Italy)
bettini@pd.infn.it

Abstract

The history of neutrino physics is an instructive example of how the physics knowledge advances in interplay between theoretical developments and progress of the experimental art. Neutrinos, in particular, have always surprised us, behaving differently from the expectations and suggesting us something that we did not initially understand. We shall discuss the most relevant chapters of neutrino history from the original proposal of its existence in 1930 to when the phenomena proving that neutrino have a mass were firmly established about the year 2000. In doing that, we shall only briefly recall the well-known discoveries.

1. Introduction

Neutrinos are very difficult to study because they have only weak interactions (gravitational apart) and because their masses are very small, even on the elementary particle scale, less than one millionth of the lightest charged particle, the electron.

In the 85 years since the Pauli hypothesis, neutrinos gave us a series of surprises, quite often behaving differently from the expectations of the current theory. Indeed, even today, the BEH mechanism giving mass to elementary fermions in the Standard Model (SM) is not implemented for neutrinos, which are assumed to be massless. However, they have a mass. In addition, the flavor eigenstates do not have a definite mass. Neutrinos cannot be detected directly. Rather we detect the charged lepton produced via W exchange (charged-current weak interaction, CC in the following) together with the neutrino or produced by the neutrino when it interacts. For example, $\bar{\nu}_e$ is by definition the neutral particle produced together with an electron in a β^- decay, ν_e the neutral particle produced together with a positron in a β^+ decay, and similarly for the other flavors. But $\bar{\nu}_e$ is also the neutral particle producing a positron, which we detect when it interacts (by W exchange). The question arises: are the flavors at production and detection the same? The experimental answer is "yes" if the proper time elapsed is small, it may be "no" if this is large enough or if the neutrino travels in matter of slowly varying density.

We recall for completeness that neutrino interact also via Z^0 exchange, namely through weak neutral currents (NC), which are flavour independent.

The SM assumes neutrinos to be helicity eigenstates with a negative eigenvalue and antineutrinos with a positive eigenvalue. Again, this is false. The SM assumes neutrinos and antineutrinos to be distinguished by the lepton number, but the assumption is not experimentally proven. As a matter of fact, there is no experimental proof that a neutrino and antineutrino are different particles. They might well be two helicity states of the same completely neutral particle, similar to what the photon is.

The SM assumes the existence of three neutrino flavors. The assumption has been confirmed by direct experiments so far, in particular by the LEP experiments giving the value $N_\nu = 2.984 \pm 0.008$, and by cosmology giving $N_{\nu-eff} = 3.15 \pm 0.23$. As a matter of fact, claims are discussed in the literature pointing to the possible existence of some type of sterile neutrinos, but all the claimed so-called anomalies have systematic uncertainties that are larger than the effect itself or have been disproven by subsequent experiments. We shall not consider the issue any further.

Neutrinos are shy, but have often appeared to be kind with us. Some relevant unknown parameters, when finally measured after many efforts, turned out to had a value large enough to make our life less difficult in the subsequent steps. Examples are the large mixing angle (LMA) solution of the solar neutrino puzzle and the quite large value of the smallest of the mixing angles, θ_{13}. Hopefully, still unknown quantities might be around the corner again.

A variety of neutrino sources are used for the experimental study. Natural sources include nuclear collisions of cosmic rays in the atmosphere (ν_e and ν_μ and their antiparticles at production), the Sun (ν_e at the origin), the supernova explosions (all flavors), and possibly other cosmic phenomena. Neutrino beams are built at the proton accelerator facilities (mainly ν_μ or $\bar{\nu}_\mu$ at production) and power reactors ($\bar{\nu}_e$ at production).

In this chapter, I shall briefly summarise the historical steps that led to the picture we have presently gained of neutrino physics from the very beginning till the final establishing of neutrino nonzero mass and mixing around the year 2000. I shall try to discuss, in particular, the pieces of history that are important but less well known.

2. Prehistory

In the 1920s, three elementary particles were known: one negative, the electron; one positive, the proton and one neutral, the photon. The models of atomic nuclei were built out of these particles. A nucleus of atomic number A and charge Z was assumed to be made of A positive protons and A–Z negative electrons. The model presented three very severe problems, namely the order of magnitude of the energy of the electrons measured in beta decay, the continuous beta decay energy spectra and the nuclear statistics.

First, the electrons emitted in a beta decay were thought to be present in the nucleus even before the decay. Under these conditions, their wave function had to be confined in the nuclear size, which was known to be on the order of the femtometer. Then, the uncertainty principle predicts a typical electron momentum (and kinetic energy) of about 100 MeV. But the observed maximum energies in beta decays were one order of magnitude smaller.

Second, if the beta decay of nucleus (A, Z) into $(A, Z+1)$ is $(A, Z) \rightarrow (A, Z+1) + e^-$, the electron energy spectrum should be a line at the mass difference $m\,(A, Z) - m\,(A, Z+1)$, but experiments showed a continuous spectrum ending at that mass difference.

Third, between 1928 and 1929, Rasetti[1] performed a series of beautiful experiments measuring the Raman rotation spectra of a number of diatomic gases, in particular of homonuclear molecules like H_2, N_2 and O_2. The rotational bands he observed showed a sequence of lines of alternate intensity. Let us see why, following what Fermi wrote in a beautiful book.[2] These systems of two identical particles should obey Bose or Fermi statistics, depending on their nuclear spin I being integer or half-integer. It is easy to find that of the total $(2I+1)^2$ linearly independent functions of the spins of the two nuclei, $I(2I+1)$ are antisymmetric and $(I+1)(2I+1)$ are symmetric under the exchange of the spin coordinates. If the nuclei obey Bose statistics, the product of the spin and space wave functions must be symmetric. Hence, the weight of space antisymmetric states is $I(2I+1)$ and the weight of space symmetric states is $(I+1)(2I+1)$. Then, in the case of Bose statistics, the ratio between the spin weights of space symmetric over space antisymmetric states is $(I+1)/I$ in the case of Bose statistics, and for the same argument, it is $I/(I+1)$ in the case of Fermi statistics. For example, for the H_2 molecule, the two ratios are 3 and 1/3, respectively. Molecular transitions happen between two space-symmetric or two space-antisymmetric states. Depending on the statistics, Bose or Fermi, the former have then weights greater or smaller than the latter. Therefore, the *symmetric lines* in the Bose case are more intense than the *antisymmetric lines* and vice versa in the Fermi case. Since symmetric and antisymmetric lines are in alternate positions, one observes lines of alternate intensities. From the spectra of Rasetti, it was learned, for example, that the N_2 nuclei obey Bose statistics. However, in the current model, the N_2 nucleus contained 14 protons and 7 electrons, namely 21 spin 1/2 particles and its spin should have been half-integer.

3. From theory to reality, 1930–1956

One should consider that in that period quantum mechanics was quite young and not yet completely developed. In addition, in the preceding years, many assumptions that had been considered well established for centuries had been contradicted by experiments in atomic physics. One could not exclude that more of them could fail inside nuclei, so much that Bohr had advanced the hypothesis that even the most "sacred" law of physics, energy conservation, might hold only statistically and not on an event by event basis.

The first step in the right direction was a guess of Pauli, who assumed that an unknown neutral light particle, which he called neutron, might exist in the nucleus and be emitted together with the electron in beta decay. On the occasion of a nuclear physics meeting in Tübingen on 4 December 1930, Pauli wrote the famous one-page letter on his *desperate remedy* to his *Liebe Radioaktive Damen und Herren.* He justified his absence by writing that "unfortunately I cannot appear in Tübingen since I am indispensable here in Zurich because of a ball on the night of 6–7 December" (Pauli was 30-years old and apparently, he had something more interesting to do). The substance is that "because of the wrong statistics of the ^{14}N and ^{6}Li nuclei and the continuous beta spectrum, I have hit upon a desperate remedy to save the 'exchange theorem' of statistics and the law of conservation of energy. Namely, the possibility that there could exist in the nuclei electrically neutral particles, that I wish to call neutrons, which have spin 1/2 and obey the exclusion principle and which further differ from light quanta in that they do not travel with the velocity of light. The mass of the neutrons should be of the same order of magnitude as the electron mass and in any event not larger than 0.01 proton masses. The continuous beta spectrum would then become understandable by the assumption that in beta decay a neutron is emitted together with the electron such that the sum of the energies of the neutron and the electron is constant."

Clearly, what Pauli liked to call neutron is what we call neutrino, a name soon later given by Fermi. Note that the desperate remedy would solve two of the above-mentioned problems. The problem of the energy cost of the wave function confinement remained unsolved. It was rather aggravated, because now it regarded both the electron and the "neutron."

The next step was experimental. Two years later, in 1932, J. Chadwick[3] discovered the neutron, a neutral particle of mass similar to the proton. This could not be the light particle guessed by Pauli, but its presence in the nucleus together with protons could solve the problem of the wrong statistics.

The three problems were solved in a single shot by Fermi the next year. It was the leap of the genius, amounting not only to the discovery of a new fundamental interaction (beyond gravity and electromagnetism) but also to a fundamental step in the development of the second quantization. Fermi published his article titled "Tentativo di una teoria dell'emissione di raggi beta" (Attempt for a theory of the beta rays emission) in *La Ricerca Scientifica* in December 1933[4] and, in an extended version, in Il Nuovo Cimento[5] and, translated in German, in Zeitschrift für Physik in January 1934. The abstract speaks of: "Theory of β rays emission of radioactive substances, built on the hypothesis that the electrons emitted by the nuclei do not exist before the decay. On the contrary they are created together with a neutrino, similarly to the creation of a light quantum that goes with a quantum jump of an atom. Comparison of theory and experiment." I underline that electron and neutrino do not exist before the decay. The problem of the energy cost of wave-function confinement disappears. The two particles are created at the moment of the decay. Fermi assumes the number of electrons and neutrinos not to be necessarily constant and that proton and neutron can be considered as two states of the same particle (according to a proposal by Heisenberg) and writes down the Hamiltonian using the mathematical formalism developed by Dirac, Jordan and Klein called the second quantization (which is not what we mean today). Fermi assumed the space–time structure of the new interaction to be analogous to the electromagnetic one but without a mediator. More specifically he assumed a contact interaction between the vector current of heavy particles (n,p) and the vector current of light particles (e, ν), in the form $g\left(\bar{\psi}_p\gamma_\mu\psi_n\right)\left(\bar{\psi}_e\gamma^\mu\psi_\nu\right)$. From the available data, he calculated the value of the coupling constant $g = 4 \times 10^{-50}\,\text{cm}^3$ erg, meaning in SI units $g = 0.4 \times 10^{-63}\,\text{m}^3$ J. This is clearly what we call the Fermi constant G_F. This fundamental constant is known today with a relative uncertainty of 5.1×10^{-7} and is usually given in natural units. Its value in SI units, is about $1.4 \times 10^{-63}\,\mathrm{m}^3$ J. We see that the Fermi's evaluation was not too bad, considering that the assumed interaction was V rather than $V - A$, which amounts to a factor 2.

Notice that "neutrino" is still one. There is not yet any difference between neutrino and antineutrino. By the way, we still do not know what such difference really is. From the available data, Fermi could conclude that the neutrino mass is much smaller than the electron mass. He suggested to determine it by measuring the beta spectrum near to the end point E_0. In particular, the slope of the spectrum is vertical for nonzero mass, horizontal if it is zero. Figure 1 is the sketch he drew in his paper for large (grande), small (piccola) and zero neutrino mass μ.

In the same year, Bethe and Peierls published a beautiful short paper on Nature titled "The neutrino",[6] in which they compared: (a) the Pauli hypothesis (nuclei contain neutrinos) with (b) the Fermi model (neutrinos are created at the decay). They observed that in the second case *one would expect that positive electrons could sometime be created together with a neutrino in nuclear transformations. Therefore the experiments of Curie and Joliot*[7] *on artificial positive β-decay give strong support to model (b), as one can scarcely assume the existence of positive electrons in the nucleus.* If Fermi is right, their argument continues, *the possibility of creating neutrinos necessarily implies the existence of annihilation processes. The most interesting amongst them would be the following: a neutrino hits a nucleus and a positive or negative electron is created while the neutrino disappears and the charge of the nucleus changes by 1.* We call the process, namely $\nu + {}^{Z}_{A}N \to e^{\mp} + {}^{Z\pm1}_{A}N$, inverse β decay. Bethe and Peierls then wrote down the dimensional connection between the lifetime of a β decay and the cross section of its inverse, in terms of the fundamental constants (h and c) and of the mass of the nucleus. Having found a limit to the cross section $\sigma < 10^{-44}\,\mathrm{cm}^2$

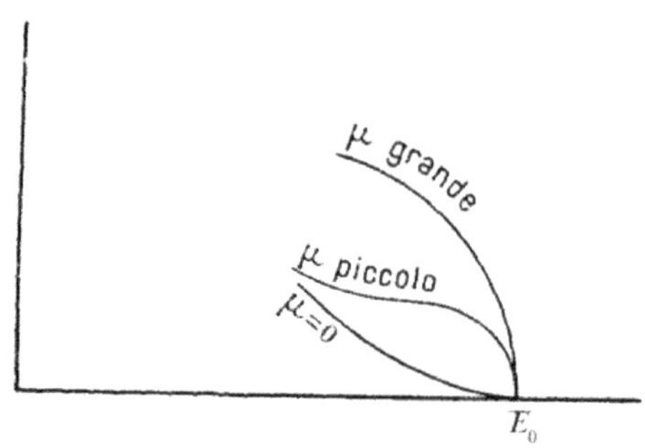

Fig. 1. Beta decay energy spectrum near the end-point as predicted by Fermi for different neutrino mass μ.

at a few MeV energy, *corresponding to a penetrating power of 10^{16} km in solid matter* (1000 light years), the authors concluded that "it is therefore absolutely impossible to observe processes of this kind with the neutrinos created in nuclear transformations."

And Pauli is told to have said:[8] "I did a horrible thing, which no theorist should do, I postulated a particle that cannot be detected."

Their estimate of the cross section was astonishingly close to the presently known values. However, fortunately, their conclusion was premature. Indeed, the problem can be solved with a sufficiently large neutrino flux and with a sufficiently massive detector. Twenty years had to pass before such fluxes would become available from nuclear power reactors of the GW scale. Such a reactor became critical in the mid 1950s. Indeed, the neutrino flux at 10 m from the core of 1 GW reactor is about $\Phi = 10^{17}\,\mathrm{m}^{-2}\mathrm{s}^{-1}$. The corresponding interaction rate on an Avogadro number of nuclei (N_A) is $R = N_A\sigma\Phi = 6\times10^{-8}\,\mathrm{mol}^{-1}\mathrm{s}^{-1}$. The expected rate in, for example, 1 t detector with nuclei of say $A = 100$ is $R = 2/\mathrm{h}$. This is not so small. The real problem to control, as we shall see, is due to the backgrounds.

However, and notwithstanding the extreme difficulty, the problem of neutrino detection has been challenging experimental physicists since the 1930s. Someone was thinking to use for neutrinos the radiochemical method that had been used since decades with neutrons. The target material under study was exposed to neutrons for several weeks and then the induced radioactivity was measured. Could the same method be used with neutrinos? In 1939, Crane[9] was brave enough to perform a bold attempt of this type in his salt bag experiment. Consider the inverse beta reaction (no difference between neutrino and antineutrino yet)

$$\nu + {}^{35}\mathrm{Cl} \to {}^{35}\mathrm{S} + e^{+}. \tag{1}$$

The final isotope, namely ^{35}S is radioactive with a half-life of about 3 months, which is neither too long nor too short. After having exposed for a period comparable with the half-life a quantity of chlorine to neutrinos, one could try to detect the decays of any ^{35}S that might have been produced. Crane exposed a three-pound bag of common salt (NaCl) for 90 days to a 1 mC capsule of radioactive material inside the bag. He then attempted to extract any S atoms with classical chemical methods and counted what he had extracted in an ionization counter. No activity was detected corresponding to the limit of $\sigma < 10^{-30}\,\mathrm{cm}^2$.

Crane was well aware that his limit was some 12 orders of magnitude larger than the expectations. However, he had made the first step toward the radiochemical neutrino detection. The classical chemical methods were very far from allowing the separation of chemical species at the level of one in 10^{30}, which was necessary for exposures to neutrinos. As a matter of fact, from the next decennia an up to the present, the problem of this level of separation has been solved for only two species, ^{37}Ar and ^{71}Ge.

The next steps are credited to Pontecorvo. In those years (1945–1946), he was working in the nuclear British programme at the Chalk River facility in Canada, where a heavy water nuclear reactor was under development. Indeed, at the end of the Second World War and at the start of the Cold War, the programme had a very high strategic military interest. The programme was called, as a coverage, "Tube Alloys". Obviously, the internal reports were classified. Two of these, PD-141 and PD-205, are on neutrino detection with the radiochemical method. While the latter is well known, the former came to the attention of the community only recently. Let us see what they contain.

The first report, PD-141[10] written in 1945, remained classified till 1964 (sic!). Here Pontecorvo states that "The object of this note is to show that the experimental observation via inverse β process is not out of the question and to suggest a method that might make the experiments feasible." He thanks Guèron for having suggested him to use CCl_4, an inexpensive liquid that can be assembled in large masses, as a target. Pontecorvo considered exactly the same reaction as Crane had done 16 years before him but omitted to quote him.

In a footnote, Pontecorvo credits Maurice Pryce, a member of Tube Alloys, having pointed out *that the flux of neutrinos from the sun is quite considerable*. Pontecorvo concluded that the expected flux of $10^{14}\,\mathrm{m}^{-2}\,\mathrm{s}^{-1}$ *is too low for the experiment suggested* considering a minimum flux of $10^{20}\,\mathrm{m}^{-2}\,\mathrm{s}^{-1}$ needed.

The second Pontecorvo report, PD-205[11] of 1946, was declassified already in 1948. The report contains the famous the Cl–Ar method proposal, namely to exploit the reaction

$$\nu + {}^{37}\mathrm{Cl} \rightarrow {}^{37}\mathrm{Ar} + e^{-}. \tag{2}$$

Notice that now the Cl isotope is ^{37}Cl rather than ^{35}Cl. Indeed, in 1946 but not in 1945, the half-life of ^{37}Ar was known to be 35 days, sufficiently long to be a candidate. The Pontecorvo's argument for having to extract Ar is that it is a volatile and inert gas. Consequently, a separation at the $1/10^{30}$ level might have been feasible. Time will show that Pontecorvo was right. However, unlike Bethe and Peierls, Pontecorvo overestimated the cross section by two orders of magnitude. He only briefly considered the crucial background issue, namely processes different from Eq. (2) capable of producing ^{37}Ar, but did not discuss this point at any depth, simply stating that (p,n) effects induced by cosmic rays *are estimated to be very small* (but they are not). He stated that probably the most convenient neutrino source is a power reactor.

Also notice that, now we know but it was then unknown, the Cl–Ar reaction can be induced by neutrinos, not by antineutrinos. On the contrary, the Cl–S reaction of Eq. (1) is induced by antineutrinos and might be thought to be usable (in theory but not in practice) to monitor nuclear reactors. May this be the reason for which the classification of PD-141 was confirmed in its periodic audits till 1964?

The merit of transforming a brilliant idea into a fully developed experimental proposal in 1949 must be credited to Louis Alvarez.[12] The motivation of Alvarez is one of the curious episodes of neutrino physics history. Alvarez recalls that Majorana had proposed in 1937 *a modified Fermi theory which involves the equivalence of neutrino and antineutrino,* and adds that very recently *an experiment*[13] *has given very strong evidence that the Majorana modification is correct.* This would have been the discovery of the neutrino-less double beta decay (see Section 4) but unfortunately the Fireman experiment was later found to be wrong.

Alvarez studied all the details in an experiment based on the Cl–Ar reaction with a nuclear reactor as the source and a detector of 40 t of CCl_4. The results were published in the Berkeley unclassified internal report UCRL-238.[12] Alvarez correctly evaluated the cross section to $2 \times 10^{-45}\,\text{cm}^2$ and the neutrino flux, addressed the Ar separation problem and that of the counter for the ^{37}Ar decays. He developed models for the different background sources from cosmic rays, radioactive impurities in the counter and the pile. He found out that the former background was too high for an experiment on the surface and proposed to perform it in a tunnel 12 m under the pile. Indeed, underground experiments will be crucial in neutrino history, as we shall see.

Alvarez, however, finally did not do the experiment he had designed. It was Raymond Davis[14] who did that in 1958 with the most powerful (0.8 GW) existing power reaction, at Savannah River, searching whether the Cl–Ar reaction was induced by reactor neutrinos. In modern notation, the reaction would be $\bar{\nu}_e + {}^{37}\text{Cl} \rightarrow {}^{37}\text{Ar} + e^-$.

Davis used a target of 11 400 l (18.5 t) of C_2Cl_4. The calculated sensitivity was twenty times larger than the minimum needed for detection if neutrino and antineutrino are equivalent. The experiment was sensitive enough, but no event was detected.

We can conclude that the neutral particle produced in β^- decay does not induce the Cl–Ar reaction. The further step, assumed in the SM, that neutrino and antineutrino are different particles, is however arbitrary. Indeed, the Cl–Ar reaction does not happen for a combination of two different reasons, namely the $V - A$ structure of the interaction and the smallness of neutrino mass.

The detector technique that made the discovery of the neutrino possible and that subsequently dominated neutrino experiments employs a detection medium that can be assembled in large volumes, scalable to larger ones when needed, and purified from radioactive contaminant up to an extreme purity. These are the liquid scintillators in which charged particles from the reaction under study produce scintillation light that can be detected by photomultipliers (PMs). The PMs can be lodged on the walls surrounding the volume thanks to the good transparency of the liquids. Solid scintillator detectors had been known since long times, but they could be used only in limited sizes due, mainly, to the lack of transparency. In 1949–1950, three different groups[15–17] discovered that suitable ternary liquid mixtures scintillate too, with very long attenuation lengths.

Using the reactor used by Davis as the neutrino source, in 1956, Reines and Cowan[18] discovered the neutrino, to be precise the electron antineutrino, in their famous experiment. The reaction was

$$\bar{\nu}_e + p \to n + e^+. \tag{3}$$

We just recall that the target was Cd doped water, providing the proton target, in two tanks sandwiched between liquid scintillator detectors. The positron immediately annihilates giving two gammas in opposite directions, both with 511 keV energy. The Compton electrons they produce in two opposite scintillator tanks are detected in coincidence as a prompt signal. A further powerful discrimination against backgrounds is given by the detection of the neutron, after thermalisation. Cadmium is used because its nucleus has a high capture cross section for thermal neutrons. The photons emitted in the excited Cd decay are detected as a delayed signal.

On 14 June 1956, Frederick Reines and Clyde Cowan sent a telegram to Pauli: "We are happy to inform you that we have definitely detected neutrinos from fission fragments by observing inverse beta decay of protons. Observed cross section agrees well with expected six time ten to the minus forty four square centimeters." The reply of Pauli is reproduced in Fig. 2.

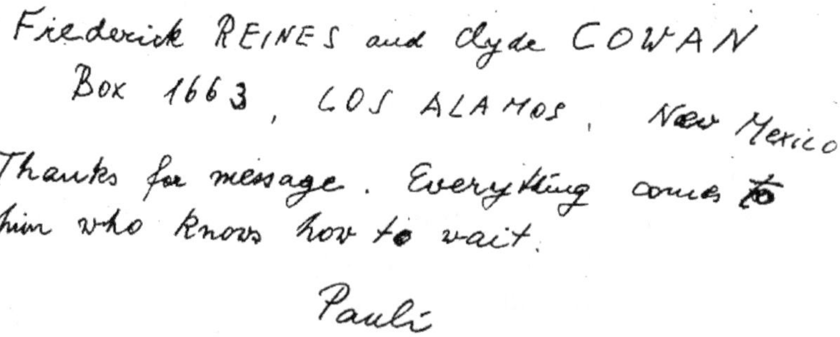

Friederick REINES and Clyde COWAN
Box 1663, LOS ALAMOS, New Mexico
Thanks for message. Everything comes to him who knows how to wait.
Pauli

Fig. 2. The message of Pauli to Reines and Cowan on 15 June 1956. CERN Pauli Archives.

4. Is neutrino a completely neutral particle?

Let us now go back to 1937, when Ettore Majorana,[19] who was not happy with the Dirac electron theory due to the asymmetry, in the derivation process, between particle and antiparticle, wrote a fundamental paper titled "Symmetric positron electron theory". The abstract reads: "We show that it is possible to achieve complete formal symmetrisation in the electron and positron quantum theory by means of a new quantization process. The meaning of Dirac equations is somewhat modified and it is no more necessary to speak of negative-energy states; nor to assume, for any other type of particles, especially neutral ones, the existence of antiparticles, corresponding to the 'holes' of negative energy". But, much more important, the new approach allows "also to build a substantially novel theory for the particles deprived of electric charge (... hypothetical neutrinos)".

The Dirac solution of the Dirac equation is a bi-spinor ψ, with four degrees of freedom, corresponding to the two helicity states of electron and positron. Its components are complex.

It is well known that only the anticommutators, the Clifford algebra, of the γ matrices are defined. Majorana's question is: can the γ matrices be chosen to have all the components of ψ real? The answer is positive, namely the matrices

$$\tilde{\gamma}_0 = \sigma_2 \otimes \sigma_1, \quad \tilde{\gamma}_1 = i\sigma_1 \otimes 1, \quad \tilde{\gamma}_2 = i\sigma_3 \otimes 1, \quad \tilde{\gamma}_3 = i\sigma_2 \otimes \sigma_2, \tag{4}$$

satisfy the Clifford algebra and are purely imaginary. Consequently, all the $i\tilde{\gamma}$ being real, the Dirac equation

$$(i\tilde{\gamma}^\mu \partial_\mu - m)\,\tilde{\psi}(x) = 0 \tag{5}$$

can govern a $\tilde{\psi}$ with all real components. The Majorana solution has only two degrees of freedom corresponding to the two helicity states of a completely neutral particle of spin 1/2. Completely neutral means that it has no charge of any type, including lepton number. We know of several completely neutral bosons, like the photon, the Z, the Higgs boson and two of the gluons. Notice that a completely neutral boson can be massless or massive. A completely neutral fermion must have a mass.

In his article Majorana did not address the question of how to experimentally distinguish between the two theories. He simply stated that it was probably "not yet possible to ask to the experience to decide between this new theory and the simple extension of the Dirac equations to the neutral particles".

After three quarters of a century and an impressive experimental progress, we do not have the answer yet, but we have a way to follow. Let us see its historical origins.

In 1935, Goeppert-Mayer,[20] theoretically proposed the existence of a very rare decay that should exist when β decay is energetically forbidden, namely the double beta disintegration accompanied by two neutrinos ($2\nu 2\beta$), which is

$$ {}^{Z}_{A}N \rightarrow {}^{Z+2}_{A}N + 2e^{-} + 2\bar{\nu}_{e}. \tag{6}$$

She evaluated, using the Fermi theory, that the half-lives had to be in excess of 10^{17} yr. Several half-lives $2\nu 2\beta$ have been measured since then and found typically on the order of 10^{21} yr. Recent results are, for example, for ^{136}Xe two measured values of $2.165 \pm 0.016 \pm 0.059$[21] and $2.11 \pm 0.04 \pm 0.21$,[22] and for ^{76}Ge $1.84^{+0.14}_{-0.10}$[23] all in 10^{21} yr units.

In 1937, soon after Majorana's paper, Giulio Racah[24] gave the first proposal for a possible experimental distinction between Dirac and Majorana neutrinos. In the former case, the neutrinos (that we call antineutrinos) emitted in a β^{-} decay can produce only positrons and vice versa those emitted in a β^{+} decay can produce only electrons. If they are Majorana particles, any neutrino can produce both electrons and positrons. This was a very nice idea but unfortunately cannot work, for two independent reasons that Racah did not know, namely the $V - A$ space–time structure of the neutrino coupling to W and the smallness of neutrino mass.

The next theoretical step is credited to W. H. Furry[25] in 1939. He observed that *In the older theory, double-β disintegration involves emission of four particles, two electrons (or positrons) and two antineutrinos (or neutrinos), and the probability is extremely small. In the Majorana theory only two particles — the electrons or positrons — have to be emitted, and transition probability is much larger.* This is the neutrino-less double beta decay ($0\nu 2\beta$), a process that, in modern terms, violates lepton number by two units and is forbidden by the standard model.

Furry was completely right in principle, but, once more, he did not know that neutrino coupling to W is $V - A$ and that neutrino masses m are very small compared to neutrino energies E, namely $m \ll E$. Unfortunately, "the transition probability" of $0\nu 2\beta$ is not "much larger" but much smaller than that of $2\nu 2\beta$. Let us see why.

As we already stated, electron neutrino, ν_e, is by definition the neutral particle produced with a positron in the β^+ decay. It has two helicity components. Due to the $V - A$ coupling, the negative helicity component is largely dominant (and is the only one assumed to exist in the SM), the positive helicity component has amplitude proportional to m/E (where m is the neutrino mass and E its energy). Similarly, the antineutrino, namely $\bar{\nu}_e$, is the neutral particle produced with an electron in the β^- decay.

If neutrinos are Dirac particles, distinguished by the lepton number, both helicity components of ν_e produce electrons and both helicity components of $\bar{\nu}_e$ produce positrons in their CC interactions. If neutrinos are Majorana particles, the dominant negative helicity component of ν_e produces electrons, exactly as in the Dirac case, but the positive helicity component produces positrons and vice versa happens for the $\bar{\nu}_e$.

In conclusion, the difference between Dirac and Majorana is hidden in the small m/E component. In the $0\nu 2\beta$ the typical neutrino energy is 10 MeV, while neutrino masses are 100 meV (or less). The disintegration probability has to pay the factor $(m/E)^2 \sim 10^{-14}$. Furry was right; without this factor the $0\nu 2\beta$ would have been much faster than $2\nu 2\beta$. Calculations show that the $0\nu 2\beta$ and $2\nu 2\beta$ would have been comparable for neutrino masses around 10 eV. But now we know that they are much smaller. Indeed, several experiments have been done, are running and are planned in underground laboratories worldwide. From time to time, positive evidence was reported, starting from the above-mentioned experiment by Fireman, who claimed a half-life $T_{1/2}\left({}^{124}\mathrm{Sn}\right) = (0.4 - 0.9) \times 10^{16}$ yr, an extremely short value seen from our present knowledge. None of them was ever confirmed by more sensitive experiments in underground laboratories. The best lower limits at present are $T_{1/2}\left({}^{136}\mathrm{Xe}\right) > 1.1 \times 10^{26}$ yr from KZ in the Kamioka Laboratory in Japan[26] and $T_{\frac{1}{2}}\left({}^{76}\mathrm{Ge}\right) > 5.3 \times 10^{25}$ yr from GERDA in the INFN LNGS in Italy[27] both at 90% C.L.

5. The proton accelerators

Proton accelerators of energies of 10 GeV and larger entered in operation in the main high-energy laboratories worldwide at the end of the 1950s. In historical order, these have been the 10 GeV Synchro-phasatron at the

JINR at Dubna USSR, in Spring 1957; the 28 GeV Proton Synchrotron (CPS) at CERN, Geneva Switzerland, in the Autumn 1959 and the 33 GeV Alternate Gradient Synchrotron (AGS) at the Brookhaven National Laboratory, USA, in July 1960. These machines played a fundamental role for advancing the frontiers in all the sectors of particle physics. It became possible, in particular, to design and implement high-energy and high-intensity neutrino beams for the first time. The subsequent generations of proton accelerators, till the present, with their higher energy or luminosity (the latter being particularly important for neutrinos) always included neutrino beams facilities of performance increasing with time. In parallel, neutrino detectors evolved as functions of the open problems in neutrino physics.

The first idea of a neutrino beam from pion decays *in flight* is credited to M. A. Markov, who in 1957 proposed to Fakirov, one of his students in Moscow, a thesis "On the Possibility to Investigate the Interaction of High Energy Neutrinos with Matter at Accelerators."

In his thesis defended in 1958,[28] Fakirov considered proton energies of 6.2, 10, 25 and 50 GeV and a beam providing 10^{10} protons/s on target. He first calculated the expected pion energy spectrum and angular distribution from the proton collisions on an internal target. For this is used a model by Landau,[29] because experimental data were not available at the time. He then calculated neutrino fluxes at 50 m from the primary target. Fakirov also calculated the interaction rate induced by 1 GeV neutrinos in a Pb target of 1 m^3 volume at 100 m from the target to be approximately two to three events per day. He also considered the possibility to focus the pions and estimated a possible gain by a factor larger than 10 in neutrino beam intensity. The final sentences of the Fakirov thesis are: "At present, it becomes worthwhile to discuss a real physical experiment with neutrinos of high energy at the contemporary accelerators. The preliminary estimates show that with an adequate shield against the neutrons from the set-up and the cosmic ray neutrinos, the background can be significantly lower than the effect under discussion."[a]

His results can be checked on the real values of the 1962 experiment at BNL (scaling distances, energies and detector mass as needed). The comparison shows that Fakirov had missed the neutrino fluxes by one order of magnitude only, which is not at all a bad result considering the many unknowns that he had to cope with. Twenty-six years later, at the 17th International Conference of History of Science,[30] Markov recalled to have

[a]I am grateful to S. Petcov for all the information on the Fakirov thesis.

proposed in 1959 a report on the topic "On the High-Energy Neutrino Physics" to the International Conference on High-Energy Physics in Kiev. But, after negative reactions of a few colleagues and of the convenor of the weak interactions section, who asked "Are you serious?", Markov withdrew the report. In his introductory talk at the "Neutrino 1977", conference Markov[31] further stated on purpose that "Suggestions of carrying out experiments at high-energy neutrinos at accelerators contained in this work (the Fakirov's thesis) as well as a suggestion of Prof. B. M. Pontecorvo on neutrino experiments with low energy neutrinos could not be realized in our country due to the absence of proper accelerators". Indeed, neutrino beams require much more than just the availability of protons accelerated at the GeV scale, as we shall see.

Several physics issues to be addressed with a high-energy neutrino beam were under discussion. The principal ones were: (1) Does the neutrino interaction cross-section grow as E^2 with energy? (2) Is the Fermi interaction point-like or is it mediated by a $W^{\pm}$ boson? (3) Are the neutral particles produced in the μ decay (ν_μ) and in β decay (ν_e) different or equal? Note that the question of the possible existence of neutral weak currents was not given a high priority till 1971 when G. 't Hooft and M. Veltman demonstrated that interactions mediated by massive vector bosons are renormalizable.

The experimental answer to question 1 was completely unexpected. Even more, it was giving us revolutionary messages that nobody initially understood. Consider CC neutrino interactions on a target of mass m_t, possibly mediated by a vector boson W of mass M_W as a function of the laboratory neutrino energy E_ν and let $s = 2m_tE_\nu$ be the centre of mass energy squared. The cross-section dependence on energy can be expressed on the basis of dimensional arguments in three different energy ranges. Remember on purpose that the cross section is proportional to the square of the Fermi constant, which has physical dimensions $[G_F^2] = [E^{-4}]$. Also, in natural units, the dimensions of the cross section are $[\sigma] = [L^2] = [E^{-2}]$.

a. Neutrino energy is quite smaller than the target mass, $\sqrt{s} \ll m_t$. This is like a ball hitting on a brick wall. The only energy of the game is neutrino energy and we conclude that

$$\sigma \propto G_F^2 E_\nu^2. \tag{7}$$

This was supposed to be the case for neutrino beams of a few GeV interacting with nuclei of dozens of GeV masses.

b. Neutrino energy is larger than the target mass but much smaller than the mediator mass, namely $m_t \ll \sqrt{s} \ll M_W$. The kinematic quantity is now

s and we can write

$$\sigma \propto G_F^2 s \propto 2m_t G_F^2 E_\nu \propto G_F^2 E_\nu, \tag{8}$$

namely the cross section rises linearly with energy.

c. At energies on the order of the mediator mass, the cross section goes as

$$\sigma \propto G_F^2 \frac{M_W^2}{s + M_W^2} s. \tag{9}$$

And, finally for very large neutrino energies $\sqrt{s} \gg M_W \gg m_t$, the cross section becomes constant, namely

$$\sigma \propto G_F^2 M_W^2. \tag{10}$$

We see that an answer to question (2) could not be given with neutrino beams of any feasible energy, as a consequence of the large value of the W mass. We leave as an exercise to the reader the calculation of the minimum neutrino energy needed for that.

The first neutrino beam at CERN became available in 1962. Heavy liquid bubble chambers were used as targets and detectors at the same time. Heavy liquids, such as freons and propane, were used, rather than hydrogen, both in order to have larger target masses, on the order of the ton, and to make electromagnetic showers, due to electrons and photons, well recognizable (Fig. 3).

In 2001, Perkins, one of the protagonists of neutrino physics in the second half of the past century, gave us his recollection of accelerator neutrino physics at CERN in the paper "An early neutrino experiment: how we missed quark substructure in 1963".[32] Here he recalls how it was soon found that the inelastic cross section in a few GeV region was very large, almost an order

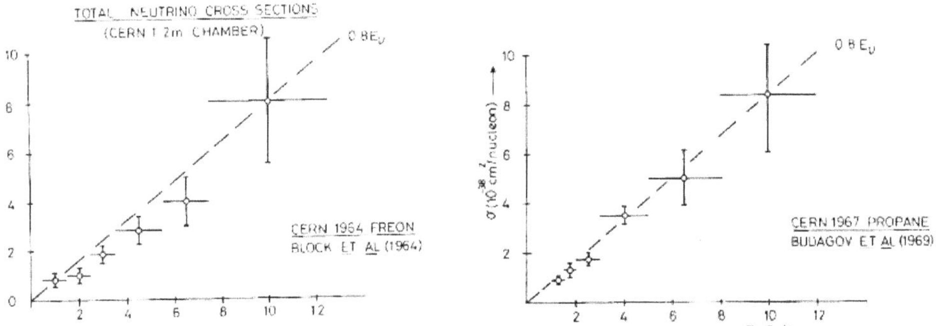

(a) (b)

Fig. 3. Total neutrino cross sections vs energy as measured at CERN in (a) a freon bubble chamber and (b) a propane bubble chamber.

of magnitude larger that the elastic cross section, and rapidly increasing with energy. "They were clearly trying to tell us a very simple thing, but unfortunately, we were just not listening!"

What neutrinos were trying to tell us became clear after the discovery in the deep inelastic experiment of electrons at SLAC of the presence of point-like structures inside the nucleons. In Perkins's words: "I re-plotted the old freon data and the more recent propane data. It was indeed linear with neutrino energy (see Figure 3). This just shows you that, when you know what the answer is, it's easy." However, the linear dependence was showing up at very low energy, already around a GeV, an effect that became known as "precocious scaling". Once more, neutrinos were trying to tell us something; once more we did not listen. The message is clear today. The mass of the targets hit by neutrinos is very small, namely $m_t \ll 1\,\text{GeV}$. The mass of the quarks is much smaller than the nucleon mass. These were just the first contributions of neutrinos to the establishment of the quark and gluon structure of the nucleons and to the measurement of their properties.[32]

Coming to point 3, in a recollection of 1982, Pontecorvo wrote:[33] "I have to come back a long way (1947–1950). Several groups, among which J. Steinberger, E. Hincks and I, and others were investigating the (cosmic) muon decay. The result of the investigations was that the decaying muon emits 3 particles: one electron ... and two neutral particles, which were called by different people in different ways: two neutrinos, neutrino and neutretto, ν and ν', etc. I am saying this to make clear that for people working on muons in the old times, the question about different types of neutrinos was always present."

Here we just recall that, as the reader knows well, 1957 was a crucial year for weak interaction, with the discovery of parity violation by Wu and collaborators,[34] and the introduction of the $V-A$ structure of the charge currents weak interactions by Sudarsham and Marshak[35] and by R. Feynman and M. Gell-Mann.[36]

The first proposal for a possible experiment to search whether ν_μ and ν_e are different particles or not was made in 1959 by B. Pontecorvo,[37] then working in Dubna. The suggestion was to look at e^+ in interactions of neutrinos from decays of pions *at rest* in a target hit by the proton beam. The π^- are captured by nuclei and do not decay. If $\nu_\mu \neq \nu_e$, the decay chain of π^+ is

$$\pi^+ \to \mu^+ + \nu_\mu; \quad \mu^+ \to e^+ + \nu_e + \bar{\nu}_\mu. \tag{11}$$

Notice that there are no $\bar{\nu}_e$. The produced neutrinos would interact in a separate target, where the observation of any positron would prove that $\bar{\nu}_\mu = \bar{\nu}_e$.

Observe, however, that the energy of the ν_μ produced by π^+ at rest is below threshold for being able to interact and produce an observable muon through the reaction $\nu_\mu + N \to \mu + N'$. Consequently, if $\nu_\mu \neq \nu_e$ nothing is observed. But a nonobservation cannot prove something positive. Even more seriously, neutrinos from the decay at rest go in all the directions uniformly filling the 4π solid angle. The fraction reaching the detector would be very small, on the order of 10^{-3} or less at 10 m distance. This is certainly a price that cannot be paid in any neutrino experiment.

In 1959, Mel Schwartz,[38] unaware of the activity in Russia, proposed to produce neutrinos from decay *in flight* of the high-energy pions (and kaons) produced from proton interactions on an internal target of a proton synchrotron, exploiting the natural collimation of the neutrinos from the fact that the parent pions would be extremely relativistic. Preliminary calculations showed that an interaction rate on a 10 t detector on the order of one event per hour was feasible, a result similar to that of Fakirov. In June 1960, with the AGS operational at BNL and the full support of the Director General, M. Goldhaber, preliminary measurements of the pion fluxes started.

The full experiment leading to the discovery of the second neutrino was done in 1962 by Schwartz *et al.*[39] In brief, the experiment was as follows. When the proton beam reached 15 GeV, a fast magnet drove the proton on an internal Be target. Secondary particles, mainly pions in a solid angle centred at 7° were aimed at the detector. The internal target had been lodged in a (3 m) *long* straight section to avoid interference for the produced hadrons with the fringing field of the nearest dipole. Hadrons and muons were absorbed in 13.5 m Fe shield. The iron, which was procured at a low cost by using an old ship, was not enough to absorb completely the muon that could be produced at the full energy of the proton beam (28 GeV). A further paraffin shield absorbed the neutrons.

Of fundamental importance for the discovery was the choice of the detector. It had to be massive (10 t), because it is also the target for the neutrinos, and had to provide strong μ/π and e/π discrimination power. Fortunately, such a detector, the spark chamber, had been recently invented by Fukui and Miyamoto in 1959[40] (see Fig. 4). In its total exposure, the experiment observed 34 "single muon" events, namely events consisting of a single straight track crossing a large number of the 2.5 cm thick plates of the chamber. No electron candidate event was observed. In conclusion, the neutrino born together with a μ in the π decays produces a μ, not e, when it interacts.

Fig. 4. Melvin Schwartz and the 10 t spark chamber in a multiple exposure of cosmic-ray μ tracks.

Following the Schwartz ideas, a proposal for the two-neutrino experiment was presented by Steinberger in 1961 at CERN, where he was in a sabbatical. A *short* straight section of the CERN proton synchrotron (CPS), 1.5 m long only, was assigned to the project. Longer straight sections were given to different priorities. However, von Dardel computed that, because of neglect of the defocusing effect of the fringing field of the CPS magnets, the neutrino flux had been overestimated by an order of magnitude. In addition, the mass of the chosen detector, the "Ramm" bubble chamber was quite modest (1/2 t). von Dardel calculated that the total statistics would be only three or four events after several months of running. CERN management decided to cancel the experiment. Schwartz is told to have commented: "Early in 1961 it looked like Jack and his associates at CERN would certainly have the first neutrino events. Then we received a piece of fabulous news. CERN had cancelled the neutrino experiment."

We recall for completeness that searches for the third sequential lepton family, a replica of the first two, was started by Zichichi at CERN in 1963,[41] on the basis of pure experimental intuition. He named them "Heavy Lepton and its neutrino." The basic idea was to search for an $e\mu$ pair noncoplanar with the beam. The search started at CERN continued at the ADONE electron–positron collider,[42] without success because of the too low energy.

The third neutrino, ν_τ, was finally discovered in year 2000 by K. Niwa[43] and collaborators with the emulsion technique on a neutrino beam of the Fermilab.

Going back to our history, we have seen how CERN had lost the first battle. However, CERN will win the second one and more, thanks to three major developments: the fast ejection of the proton beam, the focussing horn and the giant bubble chambers. Let us touch on each of them.

The fast ejection system was proposed in 1959, built and installed in 1963 by Beread Kuiper and Günther Plass.[44] Extraction of a bunch was performed in two stages. First a "kicker" magnet was raised to the full field in less than 100 ns, i.e., between the passage of two proton bunches in the CPS. The angular deflection of 1 mrad given by the kicker resulted in a betatron oscillation of maximum amplitude of about 20 mm. A septum magnet was placed with its septum a few millimitres from the unperturbed beam downstream at 1/4 of the betatron wavelength. The magnetic field of the second magnet deflected the protons out through a window in the beam pipe.

The idea of magnetic horn focussing was developed by van der Meer starting in 1961.[45] The device collects the secondary pions and kaons from the external target and directs them toward the experiments. It is made of two axially symmetric conductors. Figure 5 shows a possible geometry. An intense current sheet runs down the inner and returns on the outer one. The Lorentz force $q\mathbf{v} \times \mathbf{B}$ focuses forward the mesons of one sign, while it deflects outside those of the opposite sign. This allows choosing to run with ν_μ or with $\bar{\nu}_\mu$. In practice, two horns of somewhat different shapes in series are used to further improve the neutrino flux.

The first "gigantic" bubble chamber was Gragamelle, named after the mother of Gargantua in the giant novels of the 1530s by Rebelais. André Lagarrigue started to conceive the device in 1963, led its design and construction at Saclay and then the subsequent physics programme at CERN. The chamber was 4.8 m long and 2 m in diameter. It weighed 1000 t and held nearly 12 m^3 of heavy liquid freon (CF_3Br). The instrument

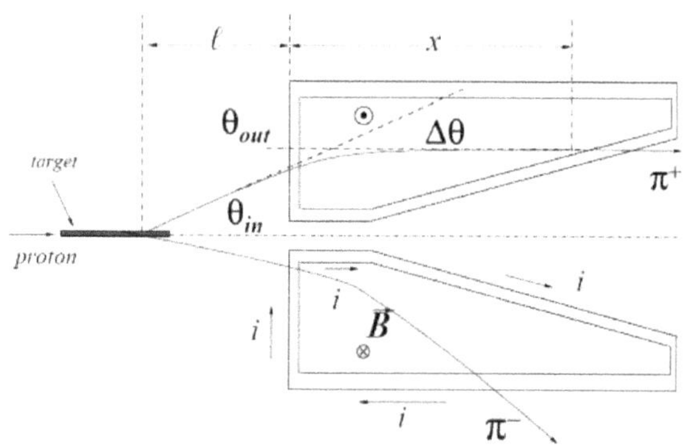

Fig. 5. Principle of the magnetic horn.[44]

operated between 1970 and 1978 on neutrino beams performing a rich experimental programme, culminated with the discovery of the weak neutral currents (NC in the following) in 1973.[46] The premature death of Lagarrigue in 1975 is presumably the reason for such a momentous discovery not having been awarded the Nobel prize. For a complete pedagogical discussion on this discovery by D. Heidt and A. Pullia, see Ref. 47.

6. Mixing, oscillations, adiabatic flavor conversion in matter

We have seen how a claim of the discovery of $0\nu 2\beta$ decay, which later was found wrong, stimulated L. Alvarez to transform the Pontecorvo idea of the Cl–Ar reaction in a fully developed experimental proposal. A similar episode happened in 1957 when R. Davis was running his experiment, based on that technique, to try to detect reactor antineutrinos. In that period, rumours of positive evidence reached Pontecorvo in Dubna. Again, they were wrong, but stimulated the idea of neutrino oscillations.[48] He wrote: "Recently there came to our attention a paper of Davis (still in press), who investigated the production of ^{37}Ar from ^{37}Cl under bombardment of neutral leptons produced by a powerful reactor. Davis result — a measurable probability of the investigated process — if confirmed, definitely indicates that neutrino charge is not strictly conserved". If the process is observed, Pontecorvo reasoned, it might be due to antineutrinos transforming into neutrinos. At the time, only the "first" neutrino was known and Pontecorvo thought of an oscillation between neutrino and antineutrino similar to those between K^0 and $\bar{K}^0$.

For this to happen, in analogy with K^0_S and K^0_L, two completely neutral neutrino states should exist, namely neutrinos should be Majorana particles Pontecorvo wrote:

"... neutrino and antineutrino are ... symmetrical and antisymmetrical combinations of two truly neutral Majorana particles ν_1 and ν_2

$$\nu_1 = \frac{1}{\sqrt{2}}\,(\nu + \bar{\nu})\,; \quad \nu_2 = \frac{1}{\sqrt{2}}\,(\nu - \bar{\nu})\,. \tag{12}$$

It follows that neutrinos in vacuum can transform themselves into antineutrinos and vice versa.... So, a beam of neutral leptons from a reactor which first consists mainly of antineutrinos will change in composition and at a certain distance... will be composed of neutrinos and antineutrinos in equal quantities."

He considered the Sun as a possible source, but concluded that, unfortunately, the solar physics is not enough under control. The theoretical work to develop a reliable solar model was done by J. Bahcall a few years later, as we shall see.

The Pontecorvo idea was that also fermions, as mesons, can be completely neutral. This was a revolutionary idea, even if neutrino mixing and oscillations were found to be completely different from those of the neutral mesons.

For the next episode, we move to Japan, where in 1955, Shoichi Sakata had proposed at the annual meeting of the Physical Society of Japan a composite model of the hadrons. The paper was published the next year.[49] In the Sakata model, hadrons were composite particles constituted of the "fundamental baryons" p, n and Λ and their antiparticles. The Sakata model was developed by A. Gamba, R. E. Marshak and S. Okubo[50] who introduced the concept of a "fundamental" lepton–baryon (L–B) symmetry between (ν, e, μ) and (p, n, Λ). A further evolution, known as Nagoya model, was developed in 1960 by Z. Maki *et al.*[51] who assumed the fundamental baryons to be made of "B^+ matter" (which carries baryonic number and positive electric charge) and leptons, respectively as

$$p = \nu B^+, \;\; n = eB^+, \;\; \Lambda = \mu B^+. \tag{13}$$

In 1962, with the discovery of the second neutrino, the L–B symmetry was lost. To try to recover it, two different groups of theorists, first Katayama *et al.*[52] and then Maki *et al.*[53] assumed that the "true" neutrinos are linear combinations of the "weak" neutrinos ν_e and ν_μ, say

$$\nu_1 = \nu_e \cos\delta + \nu_\mu \sin\delta \quad \nu_2 = -\nu_e \sin\delta + \nu_\mu \cos\delta, \tag{14}$$

and that, for unknown reasons, only one "true" neutrino combines with the B^+ matter and enters the L–B symmetry. With two flavors, mixing is described by a single angle, while for three flavors we need a matrix. There are two differences between the two papers. Katayama *et al.* not only assumed neutrino to be massless but also advanced the hypothesis that both true neutrinos might couple to B^+, with a consequence that a fourth fundamental baryon (it will be charm) should exist. Maki *et al.* gave neutrino a mass and foresaw that "transmutations" between flavors might happen. Notice that the predicted oscillation is now between flavors.

Ironically, as another coincidence, the Sakata model was found to contradict the experimental facts just in the same year by Lipkin and collaborators.[54] They showed that the reaction $\bar{p}p \to K_L^0 K_S^0$, which proceeds through an odd parity state and had been observed, is forbidden by the model.[b]

A few years later, when the Homestake experiment on solar neutrinos had started, two theoretical contributions to neutrino oscillations were given by Pontecorvo and by Gribov and Pontecorvo, the first when the experiment had started but not yet released data and the second when the first results had been published. We shall come back to that after discussing the Homestake experiment in the next section.

As we already mentioned, neutrinos change flavor not only through oscillations but also via matter effects. In 1978, Wolfenstein[55] noticed that the neutrino eigenstates in matter are different from those in a vacuum. In 1985, Mikheyev and Smirnov[56] showed that two are the possible effects of matter on neutrinos, the first is the resonant enhancement of the oscillation amplitude, the second is the adiabatic flavor conversion, called MSW effect. The latter takes place in a medium of slowly varying density and happens, in particular, in the Sun for neutrinos in the higher part of the energy spectrum. Electron neutrinos are produced in the very high density core of the sun, where the ν_2 mass eigenstate in matter almost coincides with ν_e. Then this mass eigenstate propagates as such in the decreasing density medium and exits as a vacuum-ν_2, which is a coherent superposition of different flavors. The phenomenon is very different from the oscillation, which needs an interference, and no interference can happen in presence of a single eigenstate. Unfortunately, however, the adiabatic conversion is often called oscillation in the literature, a misleading terminology that should be avoided. The point is thoroughly discussed in the chapter by Smirnov.

7. The first great surprise — solar neutrinos

Our history now moves to the USA, where in 1964, John Bahcall[57] and Raymond Davis[58] published two epochal face-to-face articles. The theoretical one, by Bahcall, contained the first complete development of the solar model and the prediction of the electron neutrino flux and energy spectrum. Bahcall

---

[b]The argument goes as follows. We start by observing that isospin and parity conservation forbid the $\Lambda\bar{\Lambda} \to \pi^+\pi^-$ annihilation for odd parity states because the initial state has $I = 0$ and the $I = 0$ states of two pions are symmetric under the interchange of the two pions. Consequently parity is even. Now, in the Sakata model there is an SU(3) symmetry that interchanges p with Λ everywhere. Under this transformation, charged pions, namely $\bar{p}n$ and $\bar{n}p$, become neutral kaons, namely $\bar{\Lambda}n$ and $\bar{n}\Lambda$, and the selection rule forbids the $p\bar{p} \to K^0\bar{K}^0$ annihilation in odd parity. In the quark model, there is no such selection rule, because the analogous transformation on pions and kaons interchanging u and s quarks, mixes Λ with Σ rather than Λ and p.

had been working on the Sun and its neutrinos since 1962. The initial results had been discouraging showing a rate too small to be detectable. However, already in 1963, Bahcall noticed the presence of a super-allowed transition to an analogue state of ^{37}Ar at about 5 MeV excitation energy, which he had not considered before. This lucky circumstance increased the estimated capture rate of the higher energy neutrinos, those from ^{8}B decay (see Eq. (19)), by almost a factor of 20. The experiment was feasible. By 1964 R. Davis had completed the experimental design.

In the following paragraphs, we shall give the essential elements on neutrino production in the Sun; more detailed discussion is to be found in other chapters of this book. The "solar standard model" (SSM) we have today is due mainly to 40 years of work by Bahcall and collaborators. For a historical overview see reference Ref. 59.

The Sun is a main-sequence star in the stable hydrogen-burning stage. Its density is about $10^5\,\mathrm{kg\,m^{-3}}$ in the centre gradually diminishing towards the surface. The overall reaction that produces 95% of the energy is the fusion of four protons into a helium nucleus, namely

$$4p \to \mathrm{He}^{++} + 2\nu_e + 2e^+. \tag{15}$$

The two positrons immediately annihilate with two electrons. Therefore the energy generation process is

$$4p + 2e^- \to \mathrm{He}^{++} + 2\nu_e + 26.7\,\mathrm{MeV}. \tag{16}$$

The basic elementary reaction called the "pp fusion" is

$$p + p \to {}^2\mathrm{H} + e^+ + \nu_e. \tag{17}$$

The thermonuclear reactions take place in the central part, the core, of the star where the thermal energy is on the order of dozens of keV. These energies are much smaller than the Coulomb barriers of the interacting nuclei and consequently the cross sections are very small, large enough for the above-mentioned reactions to proceed but not for the sun to explode like a bomb.

Only a small part of the energy released by Eq. (16) is taken by neutrinos, while the largest fraction, is transported by photons. The original MeV-energy photons interact with the solar medium producing other photons of decreasing energies and in increasing number. While the light of the Sun and of the stars is a surface phenomenon, neutrinos reach us directly from the centre of the Sun, without any absorption. The propagation of a neutrino wave in matter is characterized by a refraction index, as is the case for light.

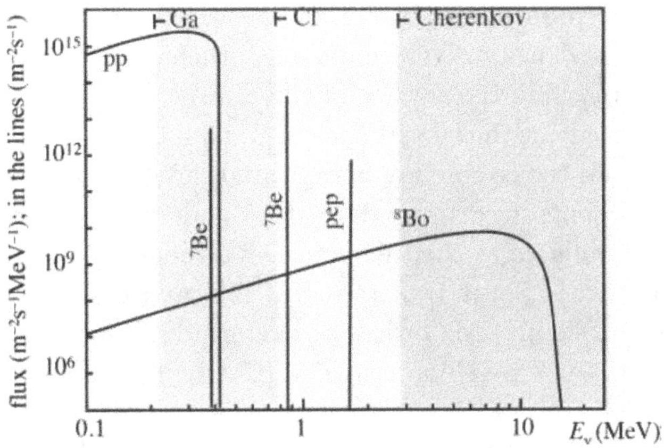

Fig. 6. Principal components of the SSM solar neutrino spectrum and sensitivity ranges of the experimental techniques. Notice the logarithmic scales.

No absorption means that the imaginary part of the index is zero, but its real part is different from one. This fact is at the origin of the MSW effect.

The largest fraction of the energy flux is due to the elementary reaction (17), namely *pp* fusion. This component of the flux can be evaluated from the measured luminous flux and is almost independent of the details of the solar model. However, it is the most difficult experimentally due to the very low neutrino energies, which are below 420 keV. Figure 6 shows the principal components of the calculated solar neutrino spectrum together with the sensitivity ranges of the different detector techniques.

Two processes produce higher-energy neutrinos. The first one gives the so-called "beryllium neutrinos" through the reaction

$$^7\text{Be} + e^- \rightarrow {}^7\text{Li} + \gamma + \nu_e. \tag{18}$$

The flux is dichromatic, with the principal line at 0.86 MeV. The second process gives the "boron neutrinos"

$$^8\text{B} \rightarrow 2\alpha + e^+ + \nu_e. \tag{19}$$

This is the highest-energy component with a spectrum reaching 14 MeV (neglecting much weaker components). As such, it is the least difficult to detect. Notice that the boron is produced by the beryllium via the reaction

$$^7\text{Be} + p \rightarrow {}^8\text{B} + \gamma. \tag{20}$$

The latter processes contribute marginally to the light energy flux and, as a consequence, our knowledge of the corresponding neutrino fluxes strongly depends on the solar model and its parameters.

Neutrinos are produced by the above reactions in the central region of the Sun, where the density is very high. They then cross a decreasing density medium before reaching the surface of the Sun.

The total flux at Earth is $6\times10^{14}\,\mathrm{m}^{-2}\mathrm{s}^{-1}$, namely much smaller than that from a reactor, and the cosmic ray backgrounds are consequently much more relevant. Experiments need to be done deep underground. No underground laboratory was available at the time and Davis lodged his detector in a gold mine, at Homestake in South Dakota, USA, 1480 m deep. The cosmic ray flux is reduced to $\sim4\mu/(\mathrm{d\,m}^2)$, six orders of magnitude less that on the surface.

Davis used 615 t of perchloroethylene (C_2Cl_4) (containing 133 t di ^{37}Cl) as the target-detector medium. About one Ar nucleus per day produced by the Cl–Ar reaction (2) was expected. Every few weeks, Davis extracted the metastable Ar nuclei with the atoms they had formed using a helium stream. After an especially developed chemical processing, he introduced the extracted gas into a counter to detect the ^{37}Ar decays. Since the signal rate is only of a few counts per month, it is mandatory not only to work deep underground but also to use only materials extremely free of radioactive components. The energy threshold of 814 keV allowed the detection of the highest energy neutrinos, namely the beryllium and boron neutrinos produced by reactions (18) and (19). On 11 August 1967, Davis wrote to Fowler: "Dear Willy, I do have a preliminary result from our first good run. The sample was taken June 22 and counting has continued until today. I am now removing the sample and will rerun background. So we do have a result and during the last few weeks I have told a few people that are interested." The first results were published in 1968 again in a back-to-back comparison between measurements[60] and solar model predictions.[61] No signal had been detected yet, but the upper limit on the rate was a factor 2.5 smaller than the prediction of the SSM. In the next three decades, improvements led first to the detection of the signal and then to a slow reduction of the uncertainties in both the experiment and the theory. The efficiency of the extraction process was tested with exposures to radioactive sources. The best estimates for the observational and for the theoretical prediction remained essentially constant. The discrepancy between the SSM prediction and the chlorine observation became known as the solar neutrino puzzle.

Let us now come back to the two above-mentioned Russian theoretical papers. As another curious coincidence, the first one, by Pontecorvo,[62] was received by the Soviet journal ZETF on 9 June 1967, just 2 weeks before 22 June when the first Homestake extraction took place. In that paper, Pontecorvo, came back to the idea of neutrino oscillations, still between neutrino and antineutrino but now between flavors as well. Having analysed the existent experimental data, he reached the conclusion that ample room was left for lepton flavor violation. Here he, following Bahcall and Davis, reconsidered the Sun as a neutrino source he had excluded in 1945 and, very importantly, added the possibility to use it to observe oscillations. He writes that "From an observational point of view the ideal object is the sun. If the oscillation length is smaller than the radius of the sun region effectively producing neutrinos (let us say, one tenth of the sun radius $R_\odot$ or one million km for the ^{8}B neutrinos, which will give the main contribution in the experiments being planned now) direct oscillations will be smeared out and unobservable. The only effect on the earth's surface would be that the flux of observable sun neutrinos must be two times smaller than the total (active and sterile) neutrino flux." Note that with "sterile", Pontecorvo means neutrino species different from ν_e and not detectable by the Cl–Ar process and that the factor 1/2 comes from him still thinking of oscillations similar to those of the K^0, namely with maximum mixing. Pontecorvo concludes on a pessimistic note stating that Sun physics was "not known well enough to permit a comparison of the expected and observable neutrino intensities." Neither the Japanese articles on neutrino mixing and "transmutations" nor those of 1964 of Bahcall and Davis are quoted in the references. While it is very likely that the former had gone unnoticed, it looks strange that Pontecorvo discussed as planned a running experiment (and used the plural) and did not recognize and credit the work of Bahcall.

One year after the first Davis results, Gribov and Pontecorvo[63] proposed that the explanation of the solar neutrino puzzle was that neutrinos were produced as ν_e oscillated to a state not observable through the Cl–Ar reaction, in the same frame as the 1967 Pontecorvo paper. We now know that the boron neutrino deficit was not due to oscillations but due to adiabatic flavor conversion in matter (MSW).

8. The underground laboratories

As we have just seen, a cavity in a gold mine was the site of the fundamental experiment by R. Davis, which paved the way to neutrino oscillations. However, the next generation experiments have been possible, thanks to the development of novel experimental infrastructures, namely the underground physics laboratories. Indeed, a hall in a mine is not a laboratory. A real laboratory must provide services underground and on the surface. For example, the electric power must be of high quality and for a fraction of it, continuous operation must be guaranteed. Proper air quality (dust free, with low Radon level, stable in temperature and humidity, etc.) must be provided at a sufficient rate. The same applies for water, taking into account that different experiments may require different levels of radio purity. Screening facilities must be provided to help the experiments in the selection of the needed extreme radio-purity components. Safe and easy to use access tunnels or shafts must be guaranteed. A laboratory needs facilities on the surface such as a mechanical shops, electronics and chemicals laboratories, storage rooms, offices, meeting and conference rooms, sleeping rooms, IT services, etc.[64]

The first full-fledged underground laboratory is the Baksan Neutrino Observatory (BNO).[65] In 1966, under the action of Markov, head of the Physics Division, the Academy of Sciences of the USSR obtained a Decree of the Soviet Government for the construction of the underground and surface facilities (the "Neutrino village"). Scientific activity started under the leadership of Zatzepin and Chudakov. The underground laboratory, including a horizontal access tunnel, was excavated and built under Mount Andyrchi in the Caucasus Mountains.

In 1979, a double tunnel was under construction for the freeway between Rome and the Adriatic Sea under the Gran Sasso Mountain. A. Zichichi, then President of the INFN, saw the unique opportunity of building a world-class underground laboratory with a broad spectrum of potential scientific programmes. After having checked the geological and radio cleanness properties of the rocks, he proposed to the Italian Parliament to build such a large laboratory. In 1982, the Parliament approved and funded the construction, which was completed in 1987 at a very low cost.[66]

In 1983, M. Koshiba established the Kamioka Underground Observatory, in a modern working mine with horizontal access, to host the KamiokaNDE water-Cherenkov detector.[67] Later on, its bigger successor was built, Super-Kamiokande (SK), which in 1998 discovered neutrino oscillations in the muon neutrinos from the atmosphere as we shall see, complementing the Davis and Bahcall discovery.

Several other sites of different sizes, different levels of support, different degrees of opening to the international community and different rigor in the selection of the experiments were built after these.

Coming back to solar neutrinos, the first confirmation of the deficit came from the Kamiokande experiment in 1987. Like its larger successor Super-Kamiokande, it was a water Cherenkov detector. This technique, allowing the assembling of very large volumes of extra-clean liquid, that contributes to neutrino physics enormously. The instrument detects the charged lepton produced by neutrinos, in particular in an elastic scattering

$$\nu_x + e^- \to \nu_x + e^-. \tag{21}$$

All neutrino flavors contribute. However, while electron neutrinos scatter via both NC and CC, the other two flavors scatter only via NC and, consequently, with a cross section about 1/6 of the former. Having a high-energy threshold, the experiment is sensitive to the boron neutrinos only. The rate measured by Kamiokande[68] was about one half of that expected from the SSM, a value that was later confirmed by SK.[69]

By measuring the direction of the electron hit by the neutrino in the elastic scattering, SK also established that neutrinos were coming from the Sun (the first neutrino telescope). At this point the existence of a problem was well established. However, was the solution of the puzzle to be found in some flaw in the solar model or in an anomalous behaviour of the neutrinos? Indeed, the high-energy neutrino flux is very sensitive to the values of the parameters of the model. For example, it depends on the temperature of the core as T^{18}.

The answer could come from the measurement of the *pp* neutrino flux, which can be calculated from the solar luminosity with a 2% uncertainty. Two radiochemical experiments were built for this purpose, one, GALLEX in the LNGS in Italy and one, SAGE, in the BNO in Russia. Both employed gallium as the target, 30 t and 60 t, respectively. They are sensitive to electron neutrinos via the inverse beta decay reaction

$$\nu_e + {}^{71}\text{Ga} \to e^- + {}^{71}\text{Ge} \tag{22}$$

with 233 keV energy threshold. GALLEX published the first results in 1992[70] finding that the measured rate was about 60% of the prediction of the SSM. For a detailed history, see Ref. 71. SAGE, which had previously published a rate compatible with zero, soon confirmed this value. GALLEX ended in 1997, becoming, in an improved version, GNO, which in turn ended in 2003. SAGE is still running. The measured values of the electron neutrinos capture rates[72,73] are in mutual agreement, and again a fraction of the SSM predictions.

Actually, already in 1995, GALLEX[74] had reached such a precision to exclude the "solar solution" of the puzzle by the following argument. Further controls included an exposure to a known neutrino flux from a radioactive source of known activity, to check the efficiency of the extraction and counting procedures (that is different from a calibration). The rate measured by GALLEX is the sum of three main contributions: from *pp*,

Eq. (17); from boron, Eq. (18) and from beryllium, Eq. (19). However, the sum of the first, as evaluated from the solar luminosity, and the second, as measured by SK, was already larger than the total rate measured by GALLEX. Consequently, no space was left in the budget for beryllium neutrinos. However, these must be present, because the boron, which exists as observed by SK, is a daughter of beryllium, Eq. (20). In conclusion, the neutrino deficit could not be explained by any modification of the solar model.

However, a "nuclear solution" of the puzzle still existed, because the SSM needed values of the nuclear cross section at very low energies, in the so-called Gamow peak, where they are so small that they had not been measured. Rather they had been extrapolated, over several orders of magnitude, from higher energy. However, if any resonance was present in the unexplored energy region, the extrapolation might have been wrong by large factors. The relative contributions to the solar spectrum of the ^{7}Be and ^{8}B neutrinos on one side, the *pp* ones on the other, is determined by the $^3\mathrm{He} + {}^3\mathrm{He} \rightarrow {}^4\mathrm{He} + 2p$ cross section. If this would be larger than assumed in the SSM, the flux at high energy would be smaller. This was the "nuclear solution" of the solar neutrino puzzle (but not of the "^{7}Be–^{8}B puzzle"). To clarify the issue, a low-energy accelerator facility, the "Laboratory for Underground Nuclear Astrophysics" (LUNA), was built in the necessary low background environment of the LNGS. In 1998, LUNA published the measurement of that cross section down to the solar energies.[75] No resonance was observed; the value used in the SSM had been correct; the nuclear solution was excluded.

In conclusion, by 1998, it had become clear that the solution of the puzzle had to be credited to an anomalous behaviour of neutrinos. The experiments had been sensitive only (or almost so in the case of the Cherenkov experiments) to electron neutrinos. We had learnt that a large fraction of them disappeared on their way from the solar centre to the Earth. The most probable hypotheses were neutrino either oscillation or adiabatic flavor conversion in slowly varying density matter (MSW effect) or both. However, this was not yet proven.

The final proof for the adiabatic transition came from the Sudbury Neutrino Observatory (SNO) in 2002 in a mine near Sudbury, Canada. The experiment, a heavy water-Cherenkov detector, was sensitive to electron neutrinos via CC. In addition, it detected NC interactions, which are induced by all the neutrino flavors. In other words, the experiment detected both the neutrinos that did not change flavour and those produced by the adiabatic conversion, even if not distinguishing their flavors. The firm conclusion was that the measured flux of all the neutrino flavors[76] agrees, within the about 10% uncertainties, with the SSM predictions.

Subsequently, the different components of the solar neutrino spectrum have been (and are being) measured by the BOREXINO experiment at LNGS. When a neutrino scatters elastically on an electron, the latter is detected by the light flash it delivers in a liquid scintillator of 100 t fiducial mass. This also gives the energy to the electron. The unprecedented radio-purity level reached by the experiment allows working with a low energy threshold and measuring, in particular, the different components of the neutrino flux. Electron neutrinos behave in the Sun as they behave in vacuum at low energies ($E \ll 2\,\text{MeV}$), while adiabatic transition dominates for $E \gg 2\,\text{MeV}$. The ν_e survival probability in the transition between the two regimes is sensitive to physics beyond the Standard Solar Model. Once more, BOREXINO showed that its predictions are correct.[77]

9. The second great surprise — atmospheric neutrinos

Neutrino oscillations were discovered by the SK experiment in 1998[78] in the Kamioka underground observatory in Japan. The detector is a 22.5 kt fiducial mass water-Cherenkov detector. Atmospheric neutrinos are detected by their CC scattering processes

$$\nu_\mu + N \to \mu + N'; \quad \nu_e + N \to e + N'. \tag{23}$$

In both cases, a unique ring signalling the charged lepton is observed. The ring is sharp in the case of the μ corresponding to its straight track, while it is fuzzier in the case of the electron, whose track scatters due to bremsstrahlung. This allows the single-ring events to be classified as "e-type" or "μ-type". Neutrino energies range from a few hundred MeV to many GeV. In this range, the differential cross sections are strongly forward peaked and consequently the measured charged lepton direction is almost the same as the neutrino one. Knowing the incident neutrino direction, we know the distance it has travelled from its production point in the atmosphere, as illustrated in Fig. 7. Keep in mind that neutrinos pass through the Earth without absorption.

Calling θ the zenith angle of the neutrino direction with the zenith, the flight length varies from about 10 km for $\theta = 0$ to more than 12 000 km for $\theta = \pi$.

The detector gives a rough measurement of the charged lepton energy, which is statistically correlated to the incident neutrino energy. Both e-type

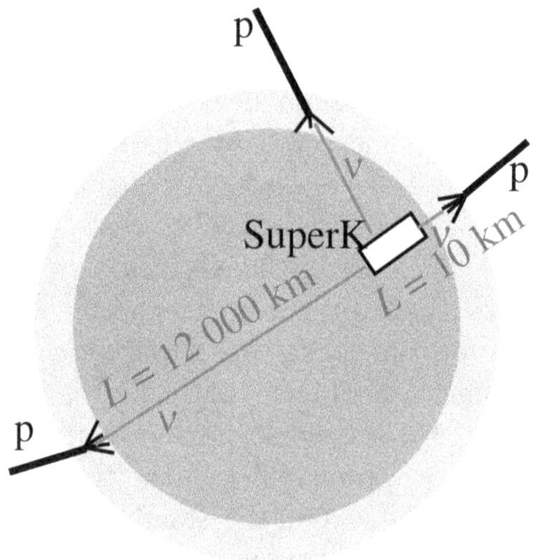

Fig. 7. Correlation between direction and path length of neutrinos.

and μ-type events are then divided into a low-energy sample, less than about one GeV, and a high-energy sample, up to several GeV.

An essential component of the experiment is the calculation, based on a number of measurements, of the muon and electron neutrino fluxes as expected in the absence of oscillations, as functions of the energy and of the zenith angle. The calculated fluxes are quite reliable, being based on data from cosmic rays experiments performed worldwide over several decennia.

Having defined four categories of events, "e-type" and "μ-type", low energy and high energy, Fig. 8 shows the distributions of the cosine of the zenith angle for each of them. The dotted curves are the predictions in absence of oscillations. We will discuss the formalism of neutrino oscillations in other chapters. Here we just mention that the continuous curves are obtained assuming oscillations and fitting to the data with relevant square mass difference Δm^2 and mixing angle θ_{23} as free parameters. The fitting procedure determines their best values.

The ν_e data do not show signs of oscillations, which implies that θ_{13} is small. All the ν_μs coming in at small zenith angles, corresponding to short flight lengths, reach the detector. However, above a certain distance the muon neutrino flux is one half of the expected flux, showing the oscillation with $\theta_{23} \approx \pi/4$. Finally, the low-energy ν_μs oscillate even at shorter flight lengths as expected from the dependence of oscillations on the energy to flight length ratio.

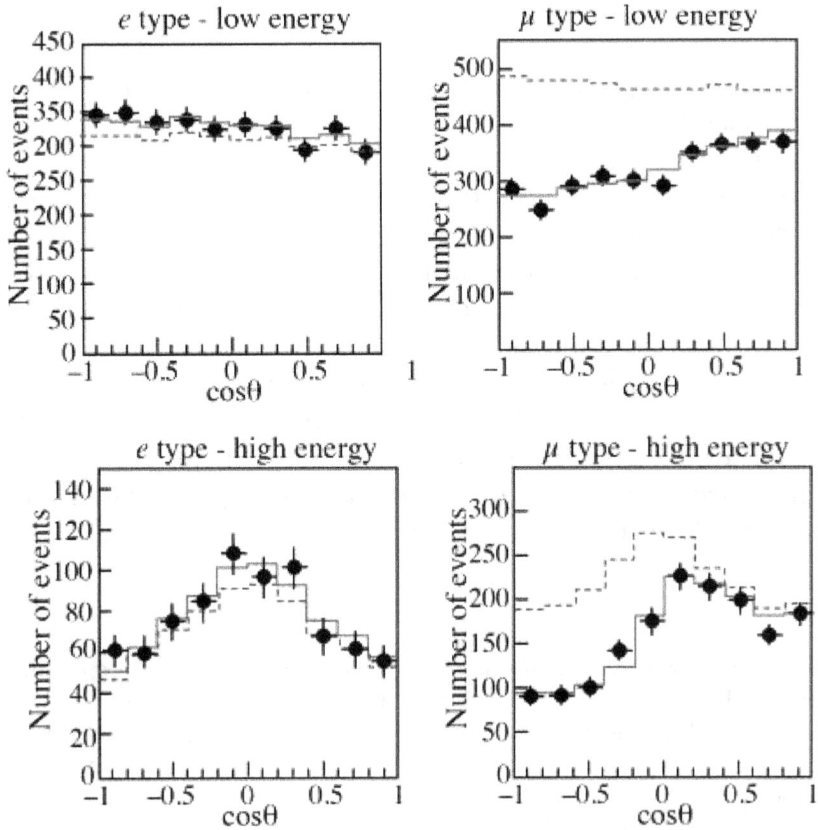

Fig. 8. Electron and muon neutrino fluxes vs. zenith angle.

The ν_μ disappearance phenomenon was contemporarily confirmed by other two experiments on atmospheric neutrinos, MACRO at the INFN LNGS and SOUDAN2 in the USA.

10. Conclusions and outlook

We have finished our history at about the year 2000, when two different phenomena proving that neutrinos are massive, namely oscillations and the adiabatic flavor conversion in matter (MSW effect), had been firmly established. Today, the neutrino system is described by nine quantities, the masses m_1, m_2 and m_3 of the three stationary states, three mixing angles θ_{12}, θ_{23}, and θ_{13}, three phases, one "Dirac" and two "Majorana" ones. The latter do not exist if neutrinos are Dirac particles. As discussed in other chapters of this book, we know within a few percept accuracy the three mixing angles and the absolute values of two square mass differences, which are one much smaller than the other. We also know, via MSW effect in the Sun, the sign of the smaller one. We do not know the sign of the other, the absolute value of the mass spectrum and the phases. Very remarkably, eighty years after the Majorana theory, we do not know yet if neutrinos and antineutrinos are different particles or not.

Neutrinos might be kind with us again; we already see clear hints that the Dirac phase might be large, possibly as much as $-\pi/2$. This would make the determination of the sign of the larger mass difference easier, and indeed we have preliminary evidence that m_3 is the largest. These hints may well be solid evidence by the time this book is distributed. If it will be so, this knowledge will be exploited for optimizing the scope and the use of resources in the design of the next generation of experiments.

Double beta decay experiments are making strong progress too, pushing forward the extreme low-background frontier. Nobody knows were the signal is, if any, but it might well be around the corner.

References

1. F. Rasetti, *Nuovo Cimento* **6**, 356 (1929); F. Rasetti, *Phys. Rev.* **34**, (1929) 367; F. Rasetti, *Z. Physik* **61**, 598 (1930).
2. E. Fermi, *Molecole e cristalli*, Zanichelli Bologna, 1934. English translation: *Molecules Crystals and Quantum Statistics*, W. A. Benjamin, inc. New York 1966.
3. J. Chadwick, F. R. S, *Proc. Royal Soc. of London* **A136**, 692 (1932), *Nature* **192**, 312 (1932).
4. E. Fermi, La Ricerca Scientifica, Vol II, n. 12 (1933).
5. E. Fermi, *Nuovo Cimento* **11**, (1934) 1; English transl. Fred L. Wilson *Am. J. Phys.* **36**, 1150 (1968).
6. H. Bethe and R. Peierls, *Nature* **133**, 532 (1934).
7. I. Curie and F. Joliot, *Nature* **133**, 201 (1934).
8. Quoted by Reines, in the preface to *Spaceship Neutrino* by C. Sutton (1992), p. XI.
9. H. R. Crane, *Phys. Rev.* **55**, 501 (1934); H. R. Crane, *Rev. Mod. Phys.* **20**, 278 (1948).
10. B. Pontecorvo, *On a method of detection of free neutrinos. Report PD-141.* Chalk River (1946). Canada. Available at The National Archives, Kew Gardens, London, UK.
11. B. Pontecorvo, *On a method of detection of free neutrinos. Report PD-205.* Chalk River (1946). Canada. See B. Pontecorvo *Selected scientific works* p. 21, Società Italiana di Fisica, Bologna, Italia.
12. L. Alvarez, *The proposed experimental test of neutrino theory.* Report UCRL-238. http://escholarship.org/uc/item/1sh4k6s2#page-4.
13. E. L. Fireman, *Phys. Rev.* **75**, 323 (1949).

14. R. Davis, Unesco Conference 1958. Paris. Vol. 1 p. 728.
15. M. Ageno *et al.*, Atti Acc. Naz. Lincei **6**, 626 (1949); *Phys. Rev.* **79**, 720 (1950).
16. H. Kallmann, *Phys. Rev.* **78**, 621 (1950).
17. G. I. Reynolds, F. B. Harrison, G. Salvini, *Phys. Rev.* **78**, 488 (1950).
18. C. L. Cowan *et al.*, *Science* **124**, 124 (1956).
19. E. Majorana, *Nuovo Cimento* **14**, 171 (1937).
20. M. Goeppert-Mayer "Double beta disintegration" *Phys. Rev.* **48**.
21. J. B. Albert *et al.*, *Phys. Rev.* **C89**, (2014) 015502.
22. A. Gando *et al.*, *Phys. Rev.* **C85**, (2012) 045504.
23. M. Agostini *et al.*, *J. Phys.* **G40**, (2013) 035110.
24. G. Racah, *Nuovo Cimento* **14**, 323 (1937).
25. W. H. Furry, "On the transition probabilities in double beta-decay" *Phys. Rev.* **56**, 1184.
26. A. Gando *et al.*, arXiv:1695.02889.
27. M. Agostini *et al.*, Neutrino 2016.
28. D. Fakirov, Diploma thesis (in Bulgarian), Department of Physics, Moscow State University (1958).
29. L. Landau, Uspekhi Fizicheskih Nauk v. LII, p. 185, 1954.
30. M. A. Markov, "Early development of weak interactions in the USSR" Nauka Publisher, Central Depatment of Oriental Literature, Moscow (1985).
31. M. A. Markov, Proceedings of Neutrino'77, Nauka, Moscow (1978).
32. D. H. Perkins, An early neutrino experiment: how we missed quark substructure in 1963. *Eur. Phys. J. H* **38**, 713–726 (2013).
33. B. Pontecorvo, The infancy and youth of neutrino physics. Some recollections. *Jour. De Phys.* **43**(12), C8–221 (1982).
34. C. S. Wu *et al.*, *Phys. Rev.* **197**, 324 (1957).
35. R. E. Marshak and C. G. Sudarsham, Proceedings of the Padua-Venice Conference on "Mesons and recently discovered particles" Società Italiana di Fisica (1958) and *Phys. Rev.* **109**, 1860 (1958).
36. R. P. Feynman and M. Gell-Mann, *Phys. Rev.* **109**, 193 (1958).
37. B. Pontecorvo, Electron and Muon neutrinos. *Sov. J. Phys.* JETP 10: 1236–1240 (1960).
38. M. Schwartz, *Phys. Rev. Lett.* **4**, 306 (1960).
39. G. Danby *et al.*, *Phys. Rev. Lett.* **9**, 36 (1962).
40. S. Fukui and S. Myiamoto, *Nuovo Cimento* **11**, 113 (1959).
41. M. Bernardini *et al.*, INFN/AE-67/3, 20 March 1967.
42. A. Zichichi *et al.*, CERN-63-26. N P Div/ 27:6:1963; T. Massam *et al.*, *Nuovo Cimento* **39**, 464 (1965).
43. K. Kodama *et al.*, *Phys. Lett.* **B504**, 218 (2001).
44. B. Kuiper and G. Plass, *Operational experience with the CPS fast ejection system* CERN-NPA/Int-65-26 (1965). https://cds.cern.ch/record/2029480/files/CERN-NPA-Int-65-26.pdf.
45. S. van der Meer, "A directive device for charged particles and its use in an enhanced neutrino beam," CERN-61-07 (1961).
46. F. J. Hasert *et al.*, *Phys. Lett.* **B46**, 121 (1973) and *ibid.* 138 (1973); *Nucl. Phys.* **B73**, 1 (1974).

47. D. Haidt and A. Pullia, The weak neutral current-discovery and impact. *Riv. Nuov. Cim.* **36**, 335 (2013).
48. B. M. Pontecorvo, JINR Preprint P-95, Dubna, 1957.
49. S. Sakata, *Progr. Theor. Phys.* **16**, 686 (1956).
50. A. Gamba, R. E. Marshak, and S. Okubo, *Proc. Nat. Acad. Sci.* **45**, 881 (1959).
51. Z. Maki, M. Nakagawa, Y. Ohnuki, and S. Sakata, *Progr. Theor. Phys.* **23**, 1270.
52. Y. Katayama *et al.*, *Progr. Theor. Phys.* **28**, 675 (1962).
53. Z. Maki *et al.*, *Progr. Theor. Phys.* **28**, 870 (1962).
54. C. A. Levinson *et al.*, *Phys. Lett.* **1**, 125 (1962).
55. L. Wolfenstein, *Phys. Rev.* **D17**, 2369 (1978); ibid **D20**, 2634 (1979).
56. S. P. Mikheyev and A. Smirnov, *Yad. Fiz.* **42**, 1441 (1985) [*Sov. J. Nucl. Phys.* **42**, 913 (1985)].
57. J. Bahcall, *Phys. Rev. Lett.* **12**, 300 (1964).
58. R. Davis Jr, *Phys. Rev. Lett.* **12**, 303 (1964).
59. J. Bhacall, *Nucl. Phys.* **B** (Proc. Suppl.) **118**, 77 (2003).
60. R. Davis Jr *et al.*, *Phys. Rev. Lett.* **20**, 1205 (1968).
61. J. Bhcall *et al.*, *Phys. Rev. Lett.* **20**, 1209 (1968).
62. B. Pontecorvo, *Zh. Eksp. Teor. Fiz.* **53**, 1717 (1967) [Sov. Phys. JETP **26**, 984 (1968)].
63. V. Gribov and B. Pontecorvo, *Phys. Lett.* **B28**, 493 (1969).
64. A. Bettini, *The World Underground Laboratories.* EPJplus **127**, 114 (2012).
65. V. V. Kuzminov, EPJplus **127**, 113 (2012).
66. L. Votano, EPJplus **127**, 127 (2012).
67. Y. Suzuki and K. Inoue, EPJplus **127**, 111 (2012).
68. K. S. Hirata, K. S. *et al.*, *Phys. Rev. Lett.* **63**, 16 (1989).
69. J. Hosaka *et al.*, *Phys. Rev.* **D 73**, 112001 (2006).
70. P. Anselmann *et al.*, *Phys. Lett.* **B285**, 375 and 390 (1992).
71. T. Kirsten, Il Nuovo Saggiatore, **31**, (2015) n. 1-2, pag. 46 open access at http://prometeo.sif.it/papers/online/sag/031/01-02/pdf/07-percorsi.pdf.
72. M. Altman *et al.*, *Phys. Lett.* **B616**, 174 (2005).
73. J. N. Abdurashitov *et al.*, *JETP* **95**, 181 (2002).
74. P. Anselmann *et al.*, *Phys. Lett.* **B357**, 237 (1995).
75. M. Junker *et al.*, *Phys. Rev.* **C57**, 2700 (1998).
76. B. Aharmin *et al.*, *Phys. Rev.* **C72**, 055502 (2005).
77. G. Bellini *et al.*, *Riv. Nuov. Cim.* **35**, 481 (2012).
78. Y. Fukuda *et al.*, *Phys. Rev. Lett.* **81**, 1562 (1998).

Chapter 2

Introduction to the Formalism of Neutrino Oscillations

G. Fantini*,‡, A. Gallo Rosso*,§, V. Zema*,¶ and F. Vissani†,||

**Gran Sasso Science Institute, L'Aquila (AQ), Italy*
†INFN, Laboratori Nazionali del Gran Sasso, Assergi (AQ), Italy and Gran Sasso Science Institute, L'Aquila (AQ), Italy
‡guido.fantini@gssi.it, §andrea.gallorosso@gssi.infn.it, ¶vanessa.zema@gssi.it and ||francesco.vissani@lngs.infn.it

Abstract

The recent wide recognition of the existence of neutrino oscillations concludes the pioneer stage of these studies and poses the problem of how to communicate effectively the basic aspects of this branch of science. In fact, the phenomenon of neutrino oscillations has peculiar features and requires to master some specific idea and some amount of formalism. The main aim of these introductory notes is exactly to cover these aspects, in order to allow the interested students to appreciate the modern developments and possibly to begin to do research in neutrino oscillations.

The structure of these notes: The structure of these notes is the following. In the first section, we describe the context of the discussion. Then we will introduce the concept of neutrino mixing and analyze its implications. Next, we will examine the basic formalism of neutrino oscillations, recalling a few interesting applications. Subsequently, we discuss the modifications to neutrino oscillations that occur when these particles propagate in the matter. Finally, we offer a brief summary of the results and outline the perspectives. Several Appendix sections supplement the discussion and collect various technical details.

We strive to describe all relevant details of the calculations in order to allow the reader to understand thoroughly and to appreciate the physics of neutrino oscillation. We do not aim to achieve completeness and/or to collect the most recent results. We limit the reference list to a minimum: We cite the seminal papers of this field in the next section, mention some few books and review papers in the last section, and occasionally make reference to certain works that are needed to learn more or on which we relied to some large extent for an aspect or another. These choices are dictated not only by the existence of a huge amount of research work on neutrinos, but also and most simply in view of the introductory character of these notes.

We assume that the reader knows special relativity and quantum mechanics and some basic aspects of particle physics. As a rule we will adopt the system of "natural units" of particle physics, defined by the choices

$$\hbar = c = 1. \tag{1}$$

In the equations, the repeated indices are summed, whenever this is not reason of confusion. Our metric is defined by $xp = x_\mu p^\mu = x_0 p_0 - \vec{x} \cdot \vec{p}$ where $x = (x_0, \vec{x})$ and $p = (p_0, \vec{p})$ are two quadrivectors. Unless stated otherwise, we will use the Dirac (or nonrelativistic) representation of the Dirac matrices; see the Appendix for technical details.

1. Introduction

In this section, the main aspects and features of the neutrinos are recalled (Section 1.1) and an introduction to the concept of neutrino oscillations is offered (Section 1.2). In this manner, the interested reader can review the basic concepts and can diagnose or retrieve, when necessary, the missing information. This material, along with the Appendix, is aimed to introduce to the discussion of the main content of this work, exposed in the subsequent three sections.

In view of the introductory character of the present section, we do not list much of the works of historical interest. However, there are a few papers that should be read by whoever is really interested in understanding the roots of the formalism. These include the seminal papers on neutrino oscillations by Pontecorvo,[1] the one on neutrino mixing by Maki, Nakagawa and Sakata[2] the papers of Wolfenstein[3] and of Mikheyev and Smirnov[4] on the matter effect.

1.1. *Overview of neutrinos*

We begin with a brief historical outline in Section 1.1.1, focussed on the basic properties of neutrinos and on their characteristic interactions, called (charged current) weak interactions or, formerly, β interaction.[a] Then, we offer in Section 1.1.2 a slightly more formal overview of some important

[a]Recall that the term β ray was introduced by Rutherford to describe a type of nuclear radiation that we know to be just high-energy electrons or positrons.

aspects, introducing the hypothesis of nonzero neutrino mass and showing that neutrino masses play a rather peculiar role. Finally, we discuss in Section 1.1.3 the reasons why neutrinos require us to master the relativistic formalism and, in particular, require a full description of relativistic spin 1/2 particles, i.e., Dirac equation. See the appendix for a reminder of the main formal aspects of the Dirac equation, and note incidentally that Pauli "invented" the neutrino just after Dirac's relativistic theory of the electron (1928) was proposed and before it was fully accepted.

1.1.1. *A brief history of the major achievements*

The main aim of this section is just to introduce some concepts and terms that are essential for the subsequent discussion; in other words, we use this historical excursion mostly as a convenient excuse. For accurate historical accounts with references, the reader is invited to consult the tables of Ref. 5, chapter 1 of Ref. 6, Ref. 7 and the essay by Bettini in chapter 1.

Existence of the neutrino: The first idea of the existence of neutrinos was conceived by Pauli in 1930, who imagined them as components of the nucleus.[b] The modern theory is due to Fermi (1933), in which (anti)neutrinos are created in association with β rays in certain nuclear decays. From this theory, Wick (1934) predicted the existence of electron capture; the nuclear recoil observed by Allen (1942) with $^7\mathrm{Be} + e \to \nu + {}^7\mathrm{Li}$ provided evidence of the neutrino. The first attempt to detect the final states produced by neutrino interactions was by Davis (1955) following a method outlined by Pontecorvo (1948). The first successful measurement was by Reines and Cowan (1956), using a reaction discussed by Bethe and Peierls (1934). For this reason, Reines received the Nobel prize (1995).

The three families (=copies) of neutrinos: Pontecorvo argued that the e and μ capture rates are the same (1947). Then Puppi (1948) suggested the existence of a new neutrino corresponding to the muon; see also Klein (1948), Tiomno and Wheeler (1949), and Lee, Rosenbluth, Yang (1949). The fact that the ν_μ is different from the ν_e was demonstrated by Lederman, Schwartz, Steinberger in 1962 (Nobel prize in 1988). Evidences of the τ lepton, a third

[b] Before 1930, the prevailing theory of the nucleus was that it is formed by protons and electrons tightly bound, e.g., D $= (2p + e)$. Then the spin of $^6\mathrm{Li} = (6p + 3e)$ and $^{14}\mathrm{N} = (14p + 7e)$ was predicted to be wrong and the β-decaying nuclei were predicted to have monochromatic decay spectra, which is, once again, wrong. Pauli improved this model assuming, e.g., that $^6\mathrm{Li} = (6p + 3e + 3\nu)$; this assumption was proposed before knowing the existence of the neutron (funnily enough, Pauli called "neutron" the light particle ν that we call today "neutrino").

type of lepton after e and μ, have been collected since 1974: these are the reasons for awarding the Nobel to Perl (1995). The corresponding tau neutrino ν_τ was first seen by DONUT experiment (2000), but the number of neutrinos undergoing weak interactions, $N_\nu = 3$, was known since 1990, thanks to LEP measurements of the Z width.

Nature of weak interaction and of neutrinos: A turning point in the understanding of weak interactions is the hypothesis that they violate parity, formulated by Lee and Yang (1956), a fact confirmed by the experiment of Wu (1957) and recognized by the Nobel committee in 1957. This was the key to understand the structure of weak interactions and it allowed Landau, Lee and Yang and Salam to conclude that, for neutrinos, the spin and the momentum have opposite directions while, for antineutrinos, the direction is the same one. One talks also of negative helicity of neutrinos and positive helicity of antineutrinos. The final proof of this picture was obtained by the impressive experiment of Goldhaber *et al.* (1958). Eventually, the theoretical picture was completed arguing for an universal vector-minus-axial (V − A) nature of the charged-current weak interactions (Sudarshan and Marshak (1958); Feynman and Gell-Mann (1958)).

Neutrino mixing and oscillations: The first idea of neutrino oscillations was introduced by Pontecorvo (1957). The limitations of the first proposal were overcome by the same author, who developed the modern theory of neutrino transformation in vacuum (1967). The new ingredient is the mixing of different families of neutrinos, introduced by Katayama, Matumoto, Tanaka, Yamada and independently and more generally by Maki, Nakagawa, Sakata in 1962. The connection of neutrino mixing with neutrino mass was outlined by Nakagawa, Okonogi, Sakata, Toyoda (1963). Wolfenstein (1978) pointed out a new effect that concerns neutrinos propagating in ordinary matter, called nowadays matter effect; its physical meaning and relevance was clarified by Mikheyev and Smirnov (1986). The evidence of oscillations accumulated from the observation of solar and atmospheric neutrinos over many years. The decisive role of the results of Sudbury Neutrino Observatory (SNO) and Super-Kamiokande (SK) as a proof of oscillations was recognized by the Nobel committee (2015); however, the number of experiments that have contributed significantly to this discovery is quite large.

For the above reasons, a couple of acronyms are currently used in the physics of neutrinos and in particular in neutrino oscillations:

(1) PMNS mixing, after Pontecorvo, Maki, Nagakawa, Sakata, to indicate the *neutrino (or leptonic) mixing* discussed in the next section;

(2) MSW effect, after Mikheyev, Smirnov, Wolfenstein, to indicate the *matter effect* described later.

1.1.2. *Neutrino properties*

In this section, we offer an introductory discussion of some important neutrino properties. In particular, we will discuss the difference between neutrinos and antineutrinos and introduce the masses of the neutrinos. Although we use the formalism of quantum field theory, we illustrate the results with a pair of pictures that we hope will make the access to the concepts easier. See also Sections 2.1.2 and A.2 for more formal details.

Neutrinos, antineutrinos, their interactions, lepton number: When considering an electrically charged particle, say an electron, the difference between this particle and its antiparticle is evident: one has charge $-e$, the other $+e$. What happens when the particle is neutral? There is no general answer, e.g., the photon or the π^0 coincides with the their own antiparticle, whereas the neutron or the neutral kaon K^0 does not.

The case in which we are interested is the one of neutrinos. The charged-current weak interactions allow us to tag neutrinos and antineutrinos, thanks to the associated charged lepton. In fact, the relativistic quantum field theory predicts the existence of several processes with the same amplitude; this feature is called *crossing symmetry*. A rather important case concerns the six processes listed in Table 1.

The kinematics of these reactions, however, is not the same; moreover, in some of these cases, the nucleon should be inside a nucleus to trigger the decay and/or the initial lepton should have enough kinetic energy to trigger the reaction. The fact that neutrinos and antineutrinos are different means, e.g., that the basic neutrinos from the Sun, from $pp \to De^+\nu$, will never trigger the inverse beta decay (IBD) reaction $p\bar{\nu} \to ne^+$.

Table 1. Most common charged-current weak interaction processes, characterized by the same amplitude: $\beta^\pm$ decay, electron (positron) capture, inverse beta decay (IBD).

β^- decay	$n \to pe^-\bar{\nu}$	β^+ decay	$p \to ne^+\nu$
β^+ capture	$ne^+ \to p\bar{\nu}$	β^- capture	$pe^- \to n\nu$
IBD on n	$n\nu \to pe^-$	IBD	$p\bar{\nu} \to ne^+$

It is easy to see that the above set of reactions is compatible with the conservation of the *lepton number*; namely, the net number of leptons (charged or neutral) in the initial and in the final states does not change.

In the Fermi theory (or generally in quantum field theory), the leptonic charged current describes the transition from one neutral lepton to a charged one and the other reactions connected by the crossing symmetry. The V – A structure of weak interactions means that this current has the form,

$$V_\mu^{\text{lept}} - A_\mu^{\text{lept}} = \bar{e}\gamma_\mu \nu - \bar{e}\gamma_\mu\gamma_5\nu = 2\bar{e}\gamma_\mu P_{\text{L}}\nu, \tag{2}$$

where we introduced the chirality projector $P_{\text{L}} = (1 - \gamma_5)/2$. This structure implies that the wave functions of the neutrinos and of the antineutrinos appear necessarily in the combinations,[c]

$$\psi_\nu(x) = e^{-ipx} P_{\text{L}}\, u \quad \text{and} \quad \psi_{\bar{\nu}}(x) = e^{-ipx} P_{\text{R}}\, u, \tag{3}$$

where the Four-spinors u obey the Dirac equation and where we have considered plane waves for definiteness; thus $p\ x = p_\mu x^\mu = Et - \vec{p}\cdot\vec{x}$; of course, the energy is $E = \sqrt{p^2 + m^2}$ where $p = |\vec{p}|$ is the momentum and m is the mass. In the ultrarelativistic limit $p \gg m$, we have $E \approx p$ and the Dirac Hamiltonian that rules the propagation of a massive fermion can be written as

$$H_{\text{D}} = \vec{\alpha}\cdot\vec{p} + \beta m \approx \vec{\alpha}\cdot\vec{p} = \vec{\Sigma}\cdot\vec{p}\,\gamma_5 \approx E \times \vec{\Sigma}\cdot\vec{n}\,\gamma_5, \tag{4}$$

where $\vec{n} = \vec{p}/p$ is the direction of the momentum. The projection of the spin in the direction of the momentum, $\vec{\Sigma}\cdot\vec{n}$, is called the *helicity*. Thus, when the kinetic energy is much larger than the mass, we find that the energy eigenstates given in Eq. (3) satisfy

$$\vec{\Sigma}\cdot\vec{n}\,\psi_\nu(x) \approx -\psi_\nu(x) \quad \text{and} \quad \vec{\Sigma}\cdot\vec{n}\,\psi_{\bar{\nu}}(x) \approx +\psi_{\bar{\nu}}(x). \tag{5}$$

Stated in plain words, we see that, in the ultrarelativistic limit, neutrinos have negative helicity whereas antineutrinos have positive helicity. In other terms, we have another way to identify what a neutrino is and what an antineutrino is, as illustrated in Fig. 1. However, in this manner the question arises: What happens if the mass is nonzero and we invert the direction of the momentum? Is it possible to retain a distinction between neutrinos and antineutrinos?

[c] The wave functions ψ that appear in the following equation are the same as those used in the field definition Eq. (31) and the following properties can be derived using the results on charge conjugation described in Section A.2.

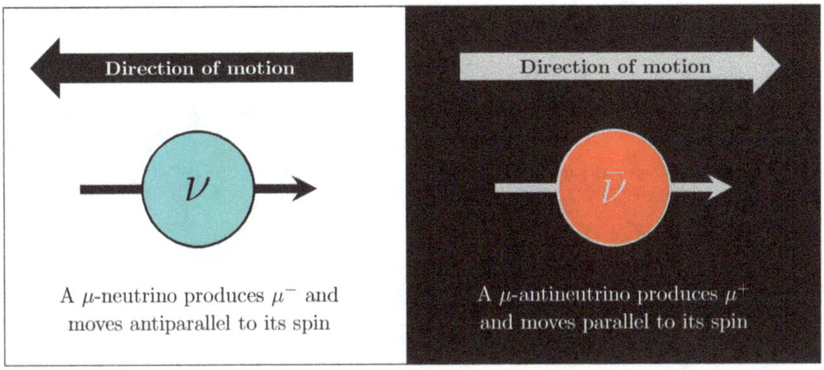

Fig. 1. The chiral (or V − A) nature of the charged-current interactions allows us to define what a neutrino is and what an antineutrino is *in the ultrarelativistic limit*, when chirality coincides with helicity and the value of the mass plays only a minor role.

Neutrino masses: The answers to the questions raised just above are not known and to date we have to rely on theoretical considerations. The main hypotheses that are debated are the following two:

(1) The mass of neutrinos has the same character as the mass of any other charged spin 1/2 particles; in more formal terms, we assume the same type of mass originally hypothesized by Dirac for the electron. A closer example is the neutron, which is a neutral particle just as the neutrino. In a relativistic quantum field theory, this type of mass entails a strict separation between particle and antiparticle states. In more detail, it means that in the rest frame there are four distinct states, as for the neutron or the electron, namely, two spin states for the neutrino and two spin states for the antineutrino. This is illustrated in Fig. 2a. If we accelerate the two neutrino states (depicted in blue in Fig. 2a) at ultrarelativistic velocity and in the direction of the spin, one of these states will be allowed to react with the matter through weak interactions, whereas the other state will not; the same is true for antineutrinos.

(2) The second hypothesis is the one put forward by Majorana. In this case, there are just two spin states in the rest frame; the symmetry under rotations implies that these are two states of the same particle, or in other words, particle and antiparticle coincide. This is illustrated in Fig. 2b. This hypothesis can be reconciled with the property of weak interactions, summarized in Fig. 1, simply remarking that helicity is *not* an invariant quantity for a massive particle. Therefore, the distinction between neutrinos and antineutrinos is just a feature of ultrarelativistic motions and not a fundamental one. This hypothesis is more economical than the previous one, being based on a smaller number of states, and

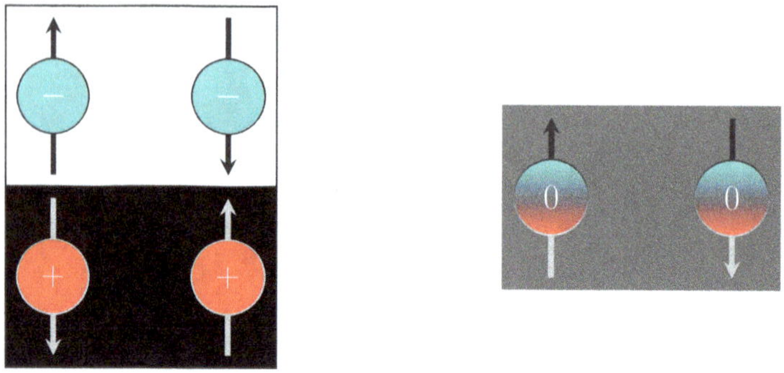

(a) (b)

Fig. 2. Massive fields in their rest frames. The arrows show the possible directions of the spin. (a) The four states of a Dirac massive field. The signs indicate the charge that distinguishes particles and antiparticles, e.g., the electric charge of an electron. (b) The two states of a Majorana massive field. The symbol "zero" indicates the absence of any charge (not only of the electric charge): particles and antiparticles coincide.

it is considered plausible by many theorists, for various reasons that we will not examine here in detail.

Arguably, the question of settling which of these hypotheses is correct is the most important open question in neutrino physics to date. In principle, it would be possible to observe the difference between the two types of masses in some experiments[d] even if we know that the effects we are searching for experimentally are quite small.

The difference between the two types of masses is however irrelevant for the description of other important phenomena, including neutrino oscillations. In fact, the main experimental evidence arising from the Majorana mass would be the total lepton number violation. On the contrary, neutrino oscillations concern $\nu_\ell \to \nu'_\ell$ and $\bar{\nu}_\ell \to \bar{\nu}'_\ell$ transitions, i.e., a transformation from one lepton to another lepton, and not a violation of the total lepton number. This shows that the distinctive feature of Majorana mass is not probed. This was first remarked and proved in Ref. 8 and will be discussed in the following.

1.1.3. *The major role of relativity for neutrinos*

From the standpoint of the standard model of elementary particle, all fermions have a similar status. However, electrons and neutrinos behave

[d]Most plausibly, this can be done through the search of lepton number violating phenomena, such as the decay $(A, Z) \to (A, Z + 2) + 2e^-$, called *neutrinoless double beta decay* and discussed elsewhere in this volume.

very differently in many situations. One reason for this difference is just the velocity of the motion. The external electrons of an atom revolve with a slow velocity $v \sim c \times \alpha$, where $\alpha \approx 1/137$ is the fine-structure constant; thus, the role of relativistic considerations is not very central. By contrast, relativity is of paramount relevance for neutrinos, for several (more or less evident) reasons that will be recalled here.

Smallness of neutrino mass: Since the beginning, it has been believed that neutrinos have a small mass (Pauli, Fermi, Perrin). Its existence was demonstrated only recently and, despite the fact we do not know its precise value yet, we are sure that it is more than one million times smaller than the mass of the electron. The main experiment that will investigate neutrino mass in laboratory is the KArlsruhe TRItium Neutrino (KATRIN): It aims to study the endpoint region of the tritium decay hoping to improve by 10 times the current upper limit of 2.0 eV, combined result of Mainz and Troitsk experiments. In the context of three neutrino oscillations, these results can be compared directly with other ones: Those from SN1987A (or those from pion and tau decay) are (much) weaker than those described above, while the most recent combined cosmological analyses claim much tighter limits, that in fact have the potential to discriminate the neutrino mass spectrum. All these aspects are discussed in details elsewhere in this volume. As discussed above, not only the "absolute" value of the neutrino masses, but also the nature of the mass is at present unknown.

Features of the main phenomena of neutrino emission: Neutrinos are emitted in nuclear transitions where the typical kinetic energy ranges from few keV to few MeV; the lowest energy neutrinos observed by Borexino (solar neutrinos) have few hundred keV. In many high-energy processes, neutrinos are emitted with much larger kinetic energies: the highest-energy events attributed to neutrinos are those seen by IceCube with energy of few PeV. In all practical cases in which we will be interested here, the kinetic energy is much more than the mass, and neutrinos propagate in the *ultrarelativistic* regime.

Cross sections growth with energy: Neutrino interactions, as a rule, increase with their energy E_ν. In fact, there is a characteristic constant named after Fermi, G_F, that appears in the amplitude of neutrino interactions. This has dimensions of an area (or equivalently an inverse energy squared); thus any cross section behaves typically as $\sigma \sim G_\mathrm{F}^2 E_\nu^2$ or $G_\mathrm{F}^2 m E_\nu$, where m is some characteristic mass. Incidentally, the main feature of neutrinos, namely the smallness of their interaction cross sections (weak interactions) is evident from the numerical value of

$$G_{\mathrm{F}}^2 = 5.297 \times 10^{-44} (\mathrm{cm/MeV})^2. \qquad (6)$$

This is why these particles have been observed only at relatively high energies. Moreover, as discussed above in most cases (and all cases of interest for the study of oscillations), we can assume that neutrinos are produced in ultrarelativistic conditions.

Helicity (spin-momentum correlation): In weak interactions, ultrarelativistic neutrinos (resp., antineutrinos) have as a rule the spin antialigned (resp., aligned) with the momentum. This is completely different from what happens to the electrons in atomic physics, where the spin can be both up or down. Various aspects of this connection have been discussed in Section 1.1.2, stressing the assumption $p \gg m$. We add here only one formal remark. When we consider ultrarelativistic motions, the wave equations for neutrinos and antineutrinos can be written as a pair of Weyl equations, namely,

$$i\, \partial_t\, \phi = \mp \vec{\sigma} \cdot \vec{p}\, \phi, \qquad (7)$$

where σ_i are the Pauli matrices and $\vec{p} = -i\vec{\nabla}$. In this formalism, the role of helicity is very transparent. (Formally, this remark concerns the structure of the Dirac equation and becomes evident in the representation of the Dirac matrices in that the chirality matrix γ_5 is diagonal: See Sections A.1 and A.3.)

Role of antineutrinos: Finally, unlike atomic physics where the existence of positrons and antibaryons can be neglected, in most practical applications neutrinos and antineutrinos have the same importance: Consider, for instance, big bang or supernovae events that have energies at which the particle–antiparticle production cannot be neglected. Note that the possibility to create (or to destroy) particles is a specific feature of relativistic phenomena. Indeed, the model of Fermi was proposed since the start as a relativistic quantum field theory.

To summarize, we have described the main reasons why neutrinos are, as a rule, ultrarelativistic and the theory of neutrinos has to be a relativistic theory. This implies that the formalism to describe neutrinos differs to large extent from the one commonly used for the physics of the electrons in the atoms, i.e., of ordinary matter.

1.2. *Introduction to oscillations*

As mentioned above, three different types of neutrinos (also called flavors) exist. It is common usage to call neutrino oscillation the observable transformation of a neutrino from one type to another, from the moment when it was produced to the moment when it is detected.[e] In his Nobel lecture (1995), Reines depicts this phenomenon by using the vivid allegory of the "dog which turns into a cat" which, however, causes a rather disconcerting feeling, being quite far from what we experience in reality.

As remarked immediately by its discoverers, neutrino oscillation is a typical quantum mechanics phenomenon that can be easily described resting on the wave nature of neutrinos. In order to introduce it as effectively as possible, we consider three different physical systems in the following. The first is just the propagation of polarized light in a birefringent crystal, the second is the spin states of an electron in a magnetic field, the third one is the neutral kaon system. All these systems can be considered as sources of precious analogies for us; the last one was originally invoked to introduce neutrino oscillations (Pontecorvo, 1957).

Transformation of the polarization states of the light: Consider a wave guide (or a transparent crystal) on a table, oriented along the (horizontal) x axis and assume that the two orthogonal directions y and z have different refractive indices n_y and n_z, e.g., birefringent crystal. Suppose that a plane wave that propagates along the x axis enters the crystal and is linearly polarized at 45° in the y–z plane, bisecting the first and third quadrants. This polarization is realized when the two oscillating components of the electric field on the y and z axes have a phase difference $\delta = 0$. Inside the crystal, the first component propagates according to

$$\exp\left[2\pi i\left(\frac{x}{\lambda_y} - \frac{t}{T}\right)\right] \quad \text{with} \quad \lambda_y = \frac{\lambda_0}{n_y} \tag{8}$$

and similarly for the z component. Owing to the fact that the two refractive indices are different, the relative phase between the y and the z components changes. In particular, if the length of the crystal L satisfies the condition,

$$\delta(L) = 2\pi(n_y - n_z)\frac{L}{\lambda_0} = \pi, \tag{9}$$

the wave will exit with a polarization at 135° that bisects the second and fourth quadrants. Thus, the wave will be orthogonal to the initial wave, the

[e]In a more restrictive and precise sense, this term refers to the sinusoidal/cosinusoidal character of some connected phenomena.

one that entered the crystal. (This device and arrangement is called in optics *waveplate.*)

Another description of the same phenomenon is as follows. The definition $n = \lambda_0/\lambda$ used above[f] implies

$$n = \frac{c}{v_{\rm p}}, \tag{10}$$

where $v_{\rm p} = \lambda/T = \omega/k$ is the phase velocity; thus, the phase velocities are different in the y and z directions. We can then say that the components of the wave in the y and z directions propagate with different velocities and this causes the relative change of phase that modifies the polarization of the propagating wave (and/or of the photons).

An important remark concerns the interpretation of this situation in terms of photons, i.e., the impinging photons get transformed into their orthogonal states. This conclusion is quite dramatic indeed and we are entitled to talk of transmutation or transformation of photons.

Transformation of the spin states of an electron: Consider a region where there is a magnetic field aligned along the z axis with intensity B and an electron whose spin lies in the $x-y$ plane. It is evident to any physicist that, in this situation, the spin of the electron will remain in the plane and simply revolve, in a motion of precession around the z axis.

Let us examine this situation from the point of view of quantum mechanics. The (matrix elements of the spin) wave function that describes an electron, polarized in the x direction at time $t = 0$, is

$$\psi_{x+}(0) = \frac{1}{\sqrt{2}}\begin{pmatrix}1\\1\end{pmatrix}. \tag{11}$$

The (matrix elements of the spin) Hamiltonian that describes the coupling of $\vec{B}$ to the magnetic moment $\vec{\mu}$, is

$$H = -\vec{B}\cdot\vec{\mu} = \begin{pmatrix}-B\mu & 0\\ 0 & B\mu\end{pmatrix}. \tag{12}$$

The state of the electron will be not stationary and it is easy to find the solution of the Schrödinger equation as

$$\psi_{x+}(t) = \frac{1}{\sqrt{2}}\begin{pmatrix}e^{i\Phi}\\ e^{-i\Phi}\end{pmatrix}, \quad \text{where} \quad \Phi = \frac{B\,\mu\,t}{\hbar}. \tag{13}$$

---

[f]We recall that $\lambda = 2\pi/k$, $\lambda_0 = 2\pi c/\omega$ and $\omega = 2\pi/T$ since the wave behaves as $\exp[i\,(kz - \omega t)]$.

This implies for instance that the electron will not remain polarized in the direction of the x axis in the course of the time and in fact, it can eventually turn into an orthogonal state. In order to verify this statement, we evaluate the probability to find it in the state with opposite polarization, namely aligned along

$$\psi_{x-}(0) = \frac{1}{\sqrt{2}} \begin{pmatrix} 1 \\ -1 \end{pmatrix}. \tag{14}$$

It is straightforward to verify that

$$P_{x+\to x-}(t) \equiv |\, \langle \psi_{x-}(0)|\psi_{x+}(t)\rangle \,|^2 = \left| \frac{e^{i\Phi} - e^{-i\Phi}}{2} \right|^2 = \sin^2\left[\frac{B\,\mu\,t}{\hbar}\right], \tag{15}$$

which is nonzero; it is between 0 and 1 as any respectable probability, it has an oscillatory character and in fact it becomes 1 (signaling a full transformation) when $\Phi = \pi/2$.

Transformation of neutral kaons: Shortly after the discovery of K mesons (kaons) and Λ particles, it was realized that a new quantum number, strangeness, is conserved by strong interactions. The neutral kaon K^0 produced, e.g., in $\pi^- + p \to K^0 + \Lambda$, having an internal nonvanishing strangeness quantum number, is different from its own antiparticle $\bar{K}^0$. Gell-Mann and Pais in 1955 remarked that the common decay channel K^0 or $\bar{K}^0 \to \pi^+\pi^-$ implies necessarily the existence of nonzero transition amplitudes $\langle \pi^+\pi^-|H_{\text{weak}}|K^0\rangle$ and $\langle \pi^+\pi^-|H_{\text{weak}}|\bar{K}^0\rangle$ at the order G_{F}. Thus, strangeness is not respected in weak interactions. The effective Hamiltonian of the two neutral kaons has nonzero transition element $\langle K^0|H_{\text{eff}}|\bar{K}^0\rangle = \delta m - i\,\delta\Gamma/2$, namely,

$$H_{\text{eff}} = \begin{pmatrix} M & \delta m \\ \delta m^* & M \end{pmatrix} - \frac{i}{2}\begin{pmatrix} \Gamma & \delta\Gamma \\ \delta\Gamma^* & \Gamma \end{pmatrix} \quad \text{with} \quad \begin{cases} M = \mathcal{O}(1) \\ \Gamma, \delta M, \delta\Gamma = \mathcal{O}(G_{\text{F}}^2), \end{cases} \tag{16}$$

where the non-Hermitian part accounts for the weak decays. This Hamiltonian is nondiagonal; thus the propagation eigenstates differ from the strangeness eigenstates K^0 and $\bar{K}^0$. These are indicated with K_{L} and K_{S}, from "long" and "short" (with reference to their very different lifetimes) and are occasionally also called weak eigenstates.[g] Even if their mass difference is so small that it cannot be measured directly, it entails the occurrence of

[g]In first approximation, they coincide with the CP eigenstates K_1 and K_2.

virtual transitions $K^0 \rightleftarrows \bar{K}^0$, quoting the words of Gell-Mann and Pais. In more detail, a K^0 produced at the time $t = 0$ is a combination of $K_{\rm L}$ and $K_{\rm S}$, so it is possible that it will be detected as a $\bar{K}^0$ at a subsequent time t. This phenomenon is called kaon transformation or transmutation or (in the modern language) *kaon oscillation*.[h]

Another interesting behavior of the neutral kaon system was predicted in 1955 by Pais and Piccioni: since K^0 and $\bar{K}^0$ interact differently with nuclei, the eigenstates of propagation in the ordinary matter are not $K_{\rm L}$ and $K_{\rm S}$. Thus, when a beam of $K_{\rm L}$ traverses a slab of matter, we will also have $K_{\rm S}$ at the exit. This phenomenon is called *kaon regeneration*.

The inception of neutrino oscillations: After the experimental observation of the kaon transformation phenomenon, Pontecorvo (1957) asked whether something similar could occur to other systems such as neutrino–antineutrino, neutron–antineutron, atoms–antiatoms. While this does not correspond to the current physical picture of the phenomenon, it is its first specific description in the scientific literature. Later, Maki *et al.*[2] mentioned the possible occurrence of *virtual transmutation* of neutrinos, again without elaborating the details. This was emended, once again, by Pontecorvo who described the modern formalism in 1967. It was the phenomenon of kaons regeneration that inspired Zavattini to ask Wolfenstein about the possible occurrence of something similar in neutrino physics. (Note the interesting fact: the former was an experimentalist and the latter a theorist, just as in the case of Mikheyev and Smirnov. Similarly, Pontecorvo belonged to a school of physicists where the distinction between theorists and experimentalists was quite vague.)

2. Leptonic mixing

2.1. *General considerations*

2.1.1. *Definition and context*

Neutrino flavor states: The concept of *neutrino with given flavor* identifies the neutral particle associated by weak interactions to the charged leptons with given flavor, namely the electron, the muon or the tau. For what we know to date, the association works in such a manner that the flavor is conserved in the interaction point. This allows us to define, e.g., a ν_e as the

---

[h] The first proof was as follows: a beam of kaons originally derived from $p+n \to p+\Lambda+K^0$ and therefore composed by mesons of positive strangeness, K^0, was able upon propagation to produce hyperons of negative strangeness, Σ^-.

particle associated to the e^+ in the β^+ decay of ^{30}P, or equivalently, the state emitted in electron capture processes, say, $e^- + {}^{37}\mathrm{Ar} \to \nu_e + {}^{37}\mathrm{Cl}$. A similar definition holds true for antineutrinos.

At the basis of this definition is the assumption, consistent with all known facts, that charged-current weak interactions are described by a relativistic quantum field theory, that is, Fermi theory at low energies and the standard electroweak model (based on $\mathrm{SU(3)_C \times SU(2)_L \times U(1)_Y}$) at higher ones. For this reason, many different processes evolve to the same type of neutrino, just as in the previous example with ν_e. Another hidden assumption is that neutrino masses play a negligible role in these interactions, as we will discuss here.

Neutrino mixing: Let us assume that the quantized neutrino fields that have given flavor ν_ℓ *do not coincide* with quantized neutrino fields that have given mass but rather they coincide with linear combinations of fields ν_i that have given mass m_i, namely,

$$\nu_\ell = \sum_{i=1}^{N} U_{\ell i}\, \nu_i \quad \text{with} \quad \begin{cases} \ell = e, \mu, \tau & \text{[flavor]} \\ i = 1, 2, 3 & \text{[mass]}, \end{cases} \tag{17}$$

where $U_{\ell i}$ are the elements of the *leptonic mixing matrix*. In the following, we will emphasize the case of three light neutrinos, $N = 3$, and we will assume that $U_{\ell i}$ are the elements of a 3×3 unitary matrix. These assumptions are consistent: (1) with the measured width of Z^0 that receives a contribution from the three light neutrinos, (2) with the fact that the neutrinos with given flavors have all the same interactions — they are *universal* and (3) with cosmological observations. The existence of a sizable admixture with other neutrinos would imply new phenomena (if they are heavy, it would result into nonunitarity of the 3×3 part of the mixing matrix and it could lead to observable violations of flavor universality) that currently are not observed.

Connection with Lagrangian densities: The above situation holds true, for instance, if we assume that the mass of neutrinos are described by a Dirac Lagrangian density,

$$\mathcal{L}_\mathrm{D} = -\sum_{\ell,\ell'} \left(\overline{\nu_{\ell\mathrm{R}}} \mathcal{M}^\mathrm{D}_{\ell\ell'} \nu_{\ell'\mathrm{L}} + \overline{\nu_{\ell'\mathrm{L}}} (\mathcal{M}^\mathrm{D}_{\ell\ell'})^\dagger \nu_{\ell\mathrm{R}} \right), \tag{18}$$

with

$$\mathcal{M}^\mathrm{D} = V\, \mathrm{diag}(m_1, m_2, m_3)\, U^\dagger, \tag{19}$$

where $\nu_{\ell'\mathrm{L}} = P_\mathrm{L}\nu_{\ell'}$ and $\nu_{\ell\mathrm{R}} = P_\mathrm{R}\nu_\ell$ are the projections of the neutrino field — later expanded in oscillators, see Eqs. (26) and (31) — in the left-handed and right-handed subspaces, with $P_{\rm R,L} = (1 \pm \gamma_5)/2$, and V is another mixing matrix that concerns only right neutrino fields; it can be set equal to U when we are interested only in neutrino oscillations and if we assume that only the known forces are present. The same mixing matrix U stems from a Majorana Lagrangian density,

$$\mathcal{L}_{\rm M} = -\frac{1}{2} \sum_{\ell,\ell'} \left(\overline{\nu^{\rm c}_{\ell {\rm R}}} \mathcal{M}^{\rm M}_{\ell\ell'} \nu_{\ell' {\rm L}} + \overline{\nu_{\ell' {\rm L}}} (\mathcal{M}^{\rm M}_{\ell\ell'})^\dagger \nu^{\rm c}_{\ell {\rm R}} \right), \tag{20}$$

with

$$\mathcal{M}^{\rm M} = U^* {\rm diag}(m_1, m_2, m_3) U^\dagger, \tag{21}$$

where we can define one right spinor in terms of the left spinor by using the charge conjugation matrix as follows: $\nu^{\rm c}_{\rm R} \equiv C \overline{\nu_{\rm L}}^{\rm t}$. From this definition, we derive $\overline{\nu^{\rm c}_R} = -\nu^{\rm t}_{\rm L} C^{-1}$; see Section A.2 for further discussion on the charge conjugation matrix and on its properties. Let us clarify that the fields ν_ℓ with $\ell = e, \mu, \tau$ that we consider in the rest of our discussion are the left chiral fields implied by ordinary weak interactions and the presence of the chiral projector $P_{\rm L}$ is not indicated but only to simplify the notations.

2.1.2. *Relation between flavor and mass states*

Plane waves: Let us assume that the free neutrinos are confined in a cube of volume $V = L^3$ subject to periodicity conditions so that the momenta are "quantized", i.e., are given in terms of integer numbers n_x, n_y, n_z as $p_x = n_x \times 2\pi/L$, $p_y = n_y \times 2\pi/L$, $p_z = n_z \times 2\pi/L$. We identify the states fully by means of helicity and use bispinors with given helicity, namely, $(\vec{\sigma} \cdot \vec{p}/p)\varphi_\lambda = \lambda \varphi_\lambda$, where $\lambda = \pm 1$. We group the momentum and the helicity in the collective label,

$$\alpha = \vec{p}, \lambda \tag{22}$$

to shorten a bit the notation.

In the "standard" representation of the Dirac matrices[i] the *plane waves* of free neutrinos of mass m_i are the eigenvectors of the Dirac Hamiltonian

[i] Also named after Dirac, Pauli, or both and also called "nonrelativistic" representation. Recall that in this representation, the correspondence with ordinary quantum mechanics is more transparent; however, it can be used for any particle, including the ultrarelativistic ones (see Section A.1).

$H_{\rm D}$ given by

$$\psi_i(\vec{x},\alpha) = \frac{\exp(i\,\vec{p}\,\vec{x})}{\sqrt{2V}} \begin{pmatrix} \sqrt{1+\varepsilon_i}\;\varphi_\lambda \\ \lambda\sqrt{1-\varepsilon_i}\;\varphi_\lambda \end{pmatrix}, \quad \text{where} \quad \varepsilon_i = \frac{m_i}{E_i}, \tag{23}$$

namely $H_{\rm D}\psi_i = E_i\psi_i$ with $E_i = \sqrt{p^2+m_i^2}$. We recall incidentally that the bispinors are usually grouped into four-spinors,

$$u_i(\alpha) = \frac{1}{\sqrt{2}} \begin{pmatrix} \sqrt{1+\varepsilon_i}\;\varphi_\lambda \\ \lambda\sqrt{1-\varepsilon_i}\;\varphi_\lambda \end{pmatrix} \tag{24}$$

that are normalized according to $u^\dagger u = 1$. We use this notation occasionally; see, e.g., Eq. (104). Note the single-particle (nonrelativistic) normalization conditions for the $\int \mathrm{d}^3x\;\psi^\dagger\psi = 1$ for the plane waves and $\varphi_\lambda^\dagger\varphi_\lambda = 1$ for the bidimensional spinors hold.

In the typical situation when the neutrinos are ultrarelativistic, these functions take a mass-independent universal form,

$$\psi_i(\vec{x},\alpha) \approx \psi(\vec{x},\alpha) \quad \text{with} \quad \psi(\vec{x},\alpha) = \frac{\exp(i\,\vec{p}\,\vec{x})}{\sqrt{2V}} \begin{pmatrix} \varphi_\lambda \\ \lambda\,\varphi_\lambda \end{pmatrix}. \tag{25}$$

The reason why we do not show the dependence of the time is that, immediately below, we introduce quantized fields, given in the interaction representation.

Fields and oscillators: We begin from the quantized field of a neutrino with mass m_i,

$$\nu_i(x) = \sum_\alpha \left(a_i(t,\alpha)\,\psi_i(\vec{x},\alpha) + b_i^\dagger(t,\alpha)\,\psi_i^{\rm c}(\vec{x},\alpha) \right), \tag{26}$$

where

$$\begin{cases} a_i(t,\alpha) = a_i(\alpha)e^{-iE_it}, \\ b_i(t,\alpha) = b_i(\alpha)e^{-iE_it}. \end{cases} \tag{27}$$

Once again, the index i corresponds to the mass of the neutrino. Above, we introduced the charge-conjugate spinor,

$$\psi_i^{\rm c} = C\overline{\psi_i}^{\rm t}. \tag{28}$$

See Section A.2 for a reminder. The oscillators a_i and b_i are dimensionless operators that obey the condition

$$\{a^\dagger(\alpha), a(\beta)\} = \delta_{\alpha\beta} \tag{29}$$

namely there is one independent fermionic oscillator for each value of α.[j] The above field describes a Dirac neutrino or also a Majorana neutrino, simply replacing $b_i = a_i$.

We are interested in the neutrino field,

$$\nu_\ell(x) = U_{\ell i}\ \nu_i(x), \tag{30}$$

where the repeated index i is summed over. If we consider ultrarelativistic neutrinos $p \gg m_i$, and if we do not measure the energy too precisely (so to identify the various mass components), we see from Eq. (25) that the field with given flavor, which is associated to the corresponding charged lepton, can be approximated to

$$\nu_\ell(x) \approx \sum_\alpha \Big(a_\ell(t,\alpha)\, \psi(\vec{x},\alpha) + b_\ell^\dagger(t,\alpha)\, \psi^{\rm c}(\vec{x},\alpha) \Big) \tag{31}$$

with the very important identification,

$$\begin{cases} a_\ell(t,\alpha) = U_{\ell i}\ a_i(t,\alpha), \\ b_\ell(t,\alpha) = U_{\ell i}^*\ b_i(t,\alpha). \end{cases} \tag{32}$$

It is important to note that, in this approximate expression:

- the plane waves have the same "universal form" ψ valid in the ultrarelativistic limit;
- the operators with given flavor ℓ, namely a_ℓ and b_ℓ, are weighted sums of operators with given mass;
- the particle (a_ℓ-type) operators are summed with the matrix $U_{\ell i}$; whereas the antiparticle (b_ℓ-type) operators with its conjugate $U_{\ell i}^*$.

Momentum eigenstates: We can simplify the formulae even further when we consider the fact that the neutrino fields are always multiplied by the left chiral projector $P_{\rm L} = (1-\gamma_5)/2$ in the known weak interactions. For this reason, as it is well known, we will have only neutrinos with negative helicity $\lambda = -1$ and antineutrinos with positive helicity $\lambda = +1$ in the

[j] In other words, $\delta_{\alpha\alpha'} = \delta_{\vec{p}\vec{p}'}\ \delta_{\lambda\lambda'}$ is a product of Kronecker deltas; recall that in our formalism the momenta are quantized in order to provide a transparent physical interpretation.

ultrarelativistic (UR) limit. Therefore, we can write

$$\begin{cases} a_\ell(t,\vec{p},-) = U_{\ell i}\ a_i(\vec{p},-)\ e^{-iE_it} & \text{for UR neutrinos,} \\ b_\ell(t,\vec{p},+) = U^*_{\ell i}\ b_i(\vec{p},+)\ e^{-iE_it} & \text{for UR antineutrinos.} \end{cases} \tag{33}$$

In this manner, it is possible to define formally the states of ultrarelativistic neutrino and antineutrinos as

$$\begin{cases} |\nu_\ell,\vec{p}\,\rangle \equiv a^\dagger_\ell(0,\vec{p},-)|0\rangle = U^*_{\ell i}|\nu_i,\vec{p}\,\rangle & \text{for UR neutrinos,} \\ |\bar{\nu}_\ell,\vec{p}\,\rangle \equiv b^\dagger_\ell(0,\vec{p},+)|0\rangle = U_{\ell i}|\nu_i,\vec{p}\,\rangle & \text{for UR antineutrinos} \end{cases} \tag{34}$$

and also the evolved vectors in the representation of Schrodinger,

$$\begin{cases} |\nu_\ell,\vec{p},t\rangle \equiv a^\dagger_\ell(t,\vec{p},-)|0\rangle = U^*_{\ell i}|\nu_i,\vec{p}\,\rangle e^{-iE_it} & \text{for UR neutrinos,} \\ |\bar{\nu}_\ell,\vec{p}.t\rangle \equiv b^\dagger_\ell(t,\vec{p},+)|0\rangle = U_{\ell i}|\nu_i,\vec{p}\,\rangle e^{-iE_it} & \text{for UR antineutrinos.} \end{cases} \tag{35}$$

A few remarks are in order:

- The mixing matrix enters differently in the relation for neutrinos and antineutrinos states (unless this matrix is real).
- It should be noted that the previous very important equations are derived directly from quantum field theoretical relations.
- Note that we have omitted the helicity labels in the states; or more precisely, we have replaced them with an explicit indication of the character of the state, either neutrino or antineutrino, which is well defined in the ultrarelativistic limit.

Again on the ultrarelativistic limit: In order to complete the discussion in Section 1.1.3 making it more specific, we collect here some final remarks on the ultrarelativistic approximation.

- Consider the lowest-energy neutrinos that can be detected to date, namely the solar neutrinos above ~300 keV that can be seen in Borexino, along with the (very conservative) bound of 2 eV on neutrino masses. In this case, the value of the parameter that quantifies the deviation from the UR limit Eq. (23) is $\varepsilon_i < 10^{-5}$.
- It is possible to think of various situations when the above assumptions do not apply. For example, very near the endpoint of the β spectrum, only the lightest neutrino mass state ν_1 is emitted together with the electron, and not all three of them, simply to conserve energy. However, it is not easy to imagine how to form a usable neutrino beam with this composition.
- More interestingly, the neutrinos produced in the big bang now have momenta $p \sim kT \approx 0.2\,\text{meV}$; the two heavier states have masses larger

than $\sqrt{|\Delta m^2_{31}|} \approx 50\,\text{meV}$ and $\sqrt{\Delta m^2_{21}} \approx 8.6\,\text{meV}$ and are now nonrelativistic. However, a discussion of oscillations is hardly needed, as we can simply treat these states as mass states.

Therefore, for all practical applications, we are interested to discuss neutrino oscillations of the ordinary neutrinos only when the ultrarelativistic limit applies.

2.2. *The parameters relevant to oscillations*

2.2.1. *General considerations*

Number of parameters in a unitary matrix: The number of free parameters of an $n \times n$ mixing matrix U satisfying unitarity $UU^\dagger = U^\dagger U = \mathbb{1}$ is easily found taking into account that there are n constraints $\sum_{i=1}^{n} |U^2_{\ell i}| = 1$ and two times $n(n-1)/2$ further constraints $\sum_{i=1}^{n} U^*_{\ell i} U_{\ell' i} = 0$, with $\ell \neq \ell'$.[k] Thus we have n^2 real constraints and n^2 real parameters in the matrix U. The determinant is a phase factor that can be explicitly factored out, corresponding to the factorization of the group $\text{U}(n) = \text{U}(1) \otimes \text{SU}(n)$. (The same counting can be done even more simply, by considering that a unitary matrix can be written as $U = \exp(iH)$ where H is Hermitian; the subgroup $\text{SU}(n)$ corresponds to the subset of the traceless Hermitian matrices.) In summary, in the 2×2 case we have four parameters and in the 3×3 case we have nine of them.

Overall phases do not matter: As we have mentioned just above, the states with given flavor are superpositions of states with given mass and the former evolve in time in a nontrivial manner, acquiring different phase factors. For this reason, flavor transformation occurs: This is the conceptual core of the neutrino oscillation phenomena.

Note that in quantum mechanics there is the freedom to determine at our will the overall phase factors of the states, and in particular, the phase factors of the states with given flavor and those with given mass. To be sure, we note that when these phases are changed, the mixing matrix *does change*. In fact, if we change the phases of the flavor and of the mass states, according to

$$|\nu_\ell\rangle' = e^{i\zeta_\ell}|\nu_\ell\rangle \quad \text{and} \quad |\nu_i\rangle' = e^{i\zeta_i}|\nu_i\rangle, \tag{36}$$

[k] In fact, the number of complex out-of-diagonal entries of the Hermitian matrix $M = U^\dagger U$ is $n(n-1)$.

we will have

$$|\nu_\ell\rangle' = (U'_{\ell i})^* |\nu_i\rangle' \tag{37}$$

with a mixing matrix where we have changed the phases in all rows and columns,

$$U'_{\ell i} = e^{-i\zeta_\ell}\, U_{\ell i}\, e^{i\zeta_i}. \tag{38}$$

On the other hand, it is evident that this redefinition will not change the absolute values,

$$|U'_{\ell i}| = |U_{\ell i}|, \tag{39}$$

and, most importantly, it will not change the probabilities of transition in which we are interested, such as

$$|\langle \nu_{\ell_2}|\nu_{\ell_1}, t\rangle|^2 = |{}'\langle \nu_{\ell_2}|\nu_{\ell_1}, t\rangle'|^2, \tag{40}$$

where ${}'\langle\nu|$ is the bra corresponding to the ket $|\nu\rangle'$; recall that we have $|\nu_\ell\rangle \equiv |\nu_\ell, t=0\rangle$. The possibility to redefine the overall phase factors has, however, a prominent consequence: Not all the n^2 parameters of a unitary matrix are relevant for the phenomenon that we are interested in discussing. Next, we proceed to count the parameters that, instead, do matter.

Number of parameters relevant to oscillations: The parameters that are relevant to neutrino and antineutrino oscillations are those that are invariant under a redefinition of the phases of the flavor fields and of the mass neutrino fields. Evidently these are less than n^2, the parameters of a unitary matrix. We present three different ways to count them, in view of the importance of this counting.[1]

(1) The standard way to count these parameters is to consider that for each flavor or mass field, we can impose a constraint on $U_{\ell i}$ exploiting the redefinition of the phases of the fields. For instance we can choose to make real a full row and a full column of the mixing matrix, but the two have one parameter in common; thus we can arrange $2n-1$ constraints only.

(2) Another consideration that leads to the same conclusion is that we have n phase factors for the flavor fields and n of them for the mass fields,

[1]The same counting applies to the quark sector; for this reason, it is common usage to call this subset of parameters as *physical parameters*. However, this terminology is misleading since for neutrinos other phases are potentially measurable — even if not by means of (ultrarelativistic) neutrino oscillations.

but the global phase factors of the flavor fields and mass fields amount to a single effective parameter, not two.

(3) As already noted in Eq. (39), when using different phases of the fields, the modulus of the elements of the mixing matrix stays unchanged. Thus, let us count the number of independent parameters $|U^2_{\ell i}|$. The unitarity relations $\sum_{i=1}^{n} |U^2_{\ell i}| = 1$ fix one parameter for each row and similarly for each column. This consideration implies that the real matrix with elements $|U^2_{\ell i}|$ has only $(n-1)^2$ independent parameters. For example, in the 3×3 example, we have four of them,

$$\begin{pmatrix} \text{free} & \text{free} & \text{fixed} \\ \text{free} & \text{free} & \text{fixed} \\ \text{fixed} & \text{fixed} & \text{fixed} \end{pmatrix}. \tag{41}$$

All in all, the result is that the number of free (or independent) parameters is

$$\text{Number of free parameters} = n^2 - (2n-1) = (n-1)^2. \tag{42}$$

This means four in the 3×3 case while in the 2×2 case we have only one relevant parameter.

Angles and phases in the mixing matrix: It is not difficult to identify these free parameters more precisely. First, consider the subclass of unitary matrices that are also real, $U = U^*$. It is easy to verify that this corresponds to the orthogonal real matrices that have $n(n-1)/2$ parameters. These parameters are *angles*; e.g., in the 2×2 case, we are dealing with a single angle, whereas in the 3×3, case we have the three Euler angles. The remaining physical parameters are *phases* relevant to oscillations,

$$\begin{aligned} \text{Number of phases} &= (n-1)^2 - n(n-1)/2 \\ &= (n-1)(n-2)/2 \quad \text{for} \quad n \geq 2. \end{aligned} \tag{43}$$

In other words, a 2×2 matrix can be made real by a suitable choice of phase factors so that no phases have relevance to oscillations, whereas a 3×3 matrix has one physical phase factor relevant to neutrino oscillations.

Quartets: As we will see later, a complex combination of mixing elements, sometimes called *quartet*, enters the explicit expressions of the oscillation probabilities,

$$\mathcal{Q}_{\ell\ell',ij} = U_{\ell i}\, U^*_{\ell' i}\, U^*_{\ell j}\, U_{\ell' j}, \tag{44}$$

and indeed it has the same value for any choice of the phases of the ν_ℓ and ν_i fields, i.e., it is not convention dependent. Moreover, it satisfies various interesting properties, such as

$$\begin{cases} \mathcal{Q}_{\ell\ell',ij} = \mathcal{Q}^*_{\ell'\ell,ij} = \mathcal{Q}^*_{\ell\ell',ji}, \\ \sum_i \mathcal{Q}_{\ell\ell',ij} = \delta_{\ell\ell'}|U^2_{\ell j}|, \\ \mathcal{Q}_{\ell\ell,ij} = |U^2_{\ell i}U^2_{\ell j}|, \\ \dots \end{cases} \tag{45}$$

We can usefully define the real and imaginary parts,

$$\mathcal{Q}_{\ell\ell',ij} = R_{\ell\ell',ij} + i\, I_{\ell\ell',ij}, \quad \text{namely} \quad \begin{cases} R_{\ell\ell',ij} = \text{Re}[\mathcal{Q}_{\ell\ell',ij}], \\ I_{\ell\ell',ij} = \text{Im}[\mathcal{Q}_{\ell\ell',ij}], \end{cases} \tag{46}$$

which are of interest since the expression of the oscillation probabilities depend upon these two real quantities, as will be discussed in Section 3.1.1. These quantities can be always expressed in terms of the mixing angles and of the physical phase factor; however, they are interesting on their own, being parameterization independent. Therefore, we examine here the real part of the quartets R and their imaginary part I. We will show (focussing on the three-flavor case) that it is possible to calculate both of them in terms of the simplest phase-independent quantities, namely $|U^2_{\ell i}|$, up to a sign that remains undetermined.

Expression of R and I in terms of $|U_{\ell i}|$: The symmetry properties are as follows: The real part is even in the exchange of $\ell \leftrightarrow \ell'$ and $i \leftrightarrow j$, whereas the imaginary one is odd in both exchanges. Thus, in the case $\ell = \ell'$, we have

$$R_{\ell\ell,ij} = |U^2_{\ell i}U^2_{\ell j}| \quad \text{and} \quad I_{\ell\ell,ij} = 0. \tag{47}$$

We are also very interested in the cases when $\ell \neq \ell'$ and $i \neq j$. The expressions are as follows:

$$R_{\ell\ell',ij} = \frac{1}{2}\left(a_k^2 - a_i^2 - a_j^2\right), \tag{48}$$

$$|I_{\ell\ell',ij}| = 2\sqrt{p(p-a_k)(p-a_j)(p-a_i)} \tag{49}$$

with the following definitions:

$$a_i = |U_{\ell i}U_{\ell' i}|, \quad a_j = |U_{\ell j}U_{\ell' j}|, \quad a_k = |U_{\ell k}U_{\ell' k}|, \quad p = (a_i + a_j + a_k)/2,$$

where $\ell \neq \ell'$ and i, j and k are a permutation of 1, 2 and 3. The first expression is immediately obtained by considering the relation $U_{\ell k}U^*_{\ell' k} = -U_{\ell i}U^*_{\ell' i} - U_{\ell j}U^*_{\ell' j}$, taking its absolute value and then recalling the definition of $R_{\ell\ell',ij}$. In order to derive the second expression, one can proceed in following steps:

(1) Let us define the following complex numbers: $z_i = U_{\ell i}U^*_{\ell' i}$ with $i = 1, 2, 3$, and let us express them as $z_i = a_i \exp(i\theta_i)$, where of course $a_i = |z_i|$.
(2) The orthogonality relation $U_{\ell k}U^*_{\ell' k} + U_{\ell i}U^*_{\ell' i} + U_{\ell j}U^*_{\ell' j} = 0$ reads $z_k + z_i + z_j = 0$ that can be seen as a triangle in the complex plane.
(3) Its area is $A = |a_i a_j \sin\theta_{ij}/2|$ where we introduced the angle $\theta_{ij} = \theta_i - \theta_j$ between two sides.
(4) If we rewrite $I_{\ell\ell',ij} = \text{Im}[z_i z_j^*] = \text{Im}[a_i e^{i\theta_i} a_j e^{-i\theta_j}] = a_i a_j \sin\theta_{ij} = 2 \times A$, we realize that the absolute value of this quantity is twice the area of the above triangle.
(5) But the area can also be evaluated with Heron's formula, since length of the sides is $|z_i| = a_i$ with $i = 1, 2, 3$.

It is remarkable that just as we have a single phase, we have a single imaginary quantity in the three-neutrino case, e.g.,

$$I_{e\mu,12} = -I_{e\tau,12} = +I_{e\tau,13} = \ldots \tag{50}$$

The proof of the first relation goes as follows: let us consider $I_{e\mu,12} = \text{Im}[U_{e1}U^*_{\mu 1}U^*_{e2}U_{\mu 2}]$. From the orthogonality relation, we have $I_{e\mu,12} = \text{Im}[U_{e1}$ $U^*_{e2}(-U^*_{e1}U_{e2} - U^*_{\tau 1}U_{\tau 2})]$ which is also equal to $-\text{Im}[U_{e1}U^*_{e2}U^*_{\tau 1}U_{\tau 2}] = -I_{e\tau,12}$ which is the desired relation. Similarly, we can obtain the other ones. An interesting consequence of this result is that, if some element of the mixing matrix is zero, this quantity will be zero. Conversely, a necessary condition for this quantity to be large is that all the elements of the mixing matrix are large (more on that in the following pages — see Eqs. (58) and (59) — and then in Section 3.2.2).

2.2.2. *The standard parameterization*

Explicit expression: The leptonic (PMNS) mixing matrix U is conventionally written in terms of the mixing angles θ_{12}, θ_{13} and θ_{23} and the CP-violating phase δ that plays a role in neutrino oscillations. The most common convention is

$$U = \begin{pmatrix} c_{12}c_{13} & s_{12}c_{13} & s_{13}e^{-i\delta} \\ -s_{12}c_{23} - c_{12}s_{13}s_{23}e^{i\delta} & c_{12}c_{23} - s_{12}s_{13}s_{23}e^{i\delta} & c_{13}s_{23} \\ s_{12}s_{23} - c_{12}s_{13}c_{23}e^{i\delta} & -c_{12}s_{23} - s_{12}s_{13}c_{23}e^{i\delta} & c_{13}c_{23} \end{pmatrix}, \tag{51}$$

where $s_{ij}, c_{ij} \equiv \sin\theta_{ij}, \cos\theta_{ij}$ and where the angles lie in the first quadrant, whereas the phase δ is generic, $\delta \in [0, 2\pi)$. Note the usage of the same phase convention and parameterization of the quark (CKM) mixing matrix even if, of course, the values of the parameters are different. This convention makes evident that the CP phase is unphysical if $\theta_{13} \to 0$. Indeed this limit holds if *any* mixing angle is zero.

Expression in terms of a product of matrices: The same matrix can be written as a product of three Euler rotations and a matrix of phases as follows:

$$U = R_{23}\ \Delta\ R_{13}\ \Delta^*\ R_{12} \quad \text{with} \quad \Delta = \begin{pmatrix} 1 & 0 & 0 \\ 0 & 1 & 0 \\ 0 & 0 & e^{i\delta} \end{pmatrix}, \tag{52}$$

where

$$R_{23} = \begin{pmatrix} 1 & 0 & 0 \\ 0 & c_{23} & s_{23} \\ 0 & -s_{23} & c_{23} \end{pmatrix}, \quad R_{13} = \begin{pmatrix} c_{13} & 0 & s_{13} \\ 0 & 1 & 0 \\ -s_{13} & 0 & c_{13} \end{pmatrix},$$
$$R_{12} = \begin{pmatrix} c_{12} & s_{12} & 0 \\ -s_{12} & c_{12} & 0 \\ 0 & 0 & 1 \end{pmatrix}. \tag{53}$$

An alternative expression: By changing the phases, one can obtain apparently different matrices whose parameters however maintain the same meaning and therefore describe exactly the same oscillation phenomena. An example of such a choice of some interest is,

$$U' = \begin{pmatrix} c_{12}c_{13} & s_{12}c_{13} & s_{13} \\ -c_{12}s_{13}s_{23} - s_{12}c_{23}e^{-i\delta} & -s_{12}s_{13}s_{23} + c_{12}c_{23}e^{-i\delta} & c_{13}s_{23} \\ -c_{12}s_{13}c_{23} + s_{12}s_{23}e^{-i\delta} & -s_{12}s_{13}c_{23} - c_{12}s_{23}e^{-i\delta} & c_{13}c_{23} \end{pmatrix}. \tag{54}$$

This new choice is somewhat advantageous for the discussion of neutrinoless double beta decay, since the phase δ does not play any role for the parameter that summarizes the effect of the light Majorana neutrino masses, i.e., $m_{\beta\beta} \equiv \sum_i U_{ei}^2 m_i$.[m] Below, we exhibit a direct construction of the unitary mixing matrix where the single physical phase factor is singled out, which yields the above form of the matrix.

[m]The parameter $m_{\beta\beta}$ depends only upon the phases of the states with given masses, named "Majorana phases", that could be simply and usefully included in m_i, namely $m_{\beta\beta} = \sum_i |U_{ei}^2|(m_i e^{i2\phi_i})$.

It is quite natural to choose the phases of the fields in order to make real and positive one row and one column. Choosing the ones that cross in $U_{e3} \equiv \sin\theta \equiv s$, we get

$$U'' = \left(\begin{array}{c|c} c\,\vec{n} & s \\ \hline u & c\,\vec{m} \end{array} \right), \tag{55}$$

where of course $c = \cos\theta$. Note that in order to have a simpler notation, we assume that the row and column vectors are evident from the context and do not indicate the transpose sign on the vectors $\vec{m}$ and $\vec{n}$. Here, we introduced two real unit vectors that we can decompose as

$$\vec{n} = (n_1, n_2) = (c_{12}, s_{12}) \quad \text{and} \quad \vec{m} = (m_1, m_2) = (s_{23}, c_{23}). \tag{56}$$

The 2×2 matrix u is still to be determined. Now, consider the second row r_2 of the full matrix U'', with components $r_2 = (\alpha\,\vec{n} + \beta\,\vec{n}_\perp,\ cm_1)$, where α and β are in general two complex numbers and $\vec{n}_\perp = (-s_{12}, c_{12})$ is a real vector orthogonal to $\vec{n}$. When we impose the orthogonality with the first row, we find that $\alpha = -s\,m_1$, and when we impose that $||r_2||^2 = \sum_i |U''_{\mu i}|^2 = 1$, we find that $|\beta^2| = m_2^2$. Thus we conclude that the second row is $r_2 = (-sm_1\,\vec{n} + e^{i\phi}\,m_2\,\vec{n}_\perp,\ cm_1)$. For similar reasons, the third row is $r_3 = (-sm_2\,\vec{n} - e^{i\psi}\,m_1\,\vec{n}_\perp,\ cm_2)$. Finally, imposing the orthogonality with the third row, one realizes that $\phi = \psi$. In short, the matrix u defined above is

$$u = -s\,\vec{m} \otimes \vec{n} + e^{i\phi}\,\vec{m}_\perp \otimes \vec{n}_\perp. \tag{57}$$

Thus, by choosing the parameters $\phi = -\delta$, $\theta = \theta_{13}$ and with the phase choice for $\vec{m}_\perp = (c_{23}, -s_{23})$, we conclude that $U'' = U'$.

Measure of CP violation: As we have already discussed, for three neutrinos there is a single imaginary quantity that signals the presence of a complex mixing matrix, and that, therefore, rules the leptonic CP-violation phenomena in neutrino oscillations: See Eq. (50). In the standard parameterization, the expression of this quantity is

$$J_{\text{CP}} \equiv I_{e\mu,12} = s_{13}c_{13}^2\ s_{12}c_{12}\ s_{23}c_{23}\ s_\delta. \tag{58}$$

Notice that (consistently with the above discussion) this quantity is bound to vanish if anyone of the angles is zero or if $\delta = 0$. It is easy to show that the maximum value is $J_{\text{CP}}^{\text{max}} = 1/(6\sqrt{3})$, obtained when $\theta_{12} = \theta_{23} = \pi/4$ and $\sin\theta_{13} = 1/\sqrt{3}$ and $\delta = \pi/2$; the minimum value is just the opposite one. The symbol J is used to honor C. Jarlskog, who originally introduced such a quantity to describe CP violation in quark systems.

2.2.3. *What we know on the parameters of neutrino oscillations*

As we will see better below, neutrino oscillations depend upon the difference of neutrino masses squared, and this is why the relevant massive parameters are sometimes indicated symbolically as delta-m-squared. The analysis of oscillation data have allowed us to discover and measure two different values of delta-m-squared, in a manner that will be recalled shortly later and apart from an important remaining ambiguity. The results of these experiments and analyses are illustrated graphically in Fig. 3; note that two different types of neutrino mass hierarchies (or orderings, or spectra) are compatible with the existing data.

The values of the parameters of the leptonic mixing matrix, obtained from a global analysis of all oscillation data available in 2016,[9] are presented in Table 2. We present the best fit values and an estimation of the accuracy, obtained from the two sigma ranges as follows: For any parameter x, at a given confidence level corresponding to n sigma in the gaussian approximation, the uncertainty is given[9] as an asymmetric interval $x_{\rm min} < x_{\rm bestfit} < x_{\rm max}$. In order to compute a single number able to give, at first glance, an overview of the order of magnitude of the relative uncertainty $\Delta x/x_{\rm bestfit}$, we set $\Delta x = (x_{\rm max} - x_{\rm min})/2n$. Note that these results do not depend strongly upon the type of mass hierarchy, except for the parameter θ_{23}. Just for reference, the corresponding angles, in the case of normal mass hierarchy, are $\theta_{13} = 8.5°$, $\theta_{12} = 33°$, $\theta_{23} = 42°$, $\delta = 250°$. For what concerns the value of δ, it was determined by the combined data from T2K and NOvA that are experiments with appearance channels (T2K was the first to constrain the CP-violation phase).

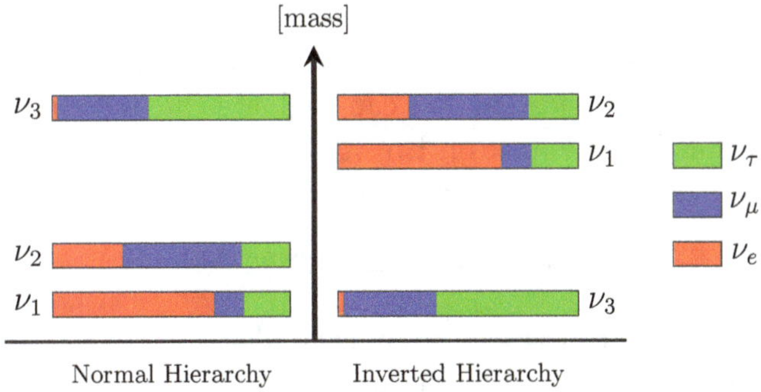

Fig. 3. Illustration of the mass spectra compatible with the data from neutrino oscillations; left, normal hierarchy; right, inverted hierarchy. The length of the colored bars are proportional to the flavor content of each mass state, $|U_{\ell i}|^2$.

Table 2. Results of the global analysis of oscillation data of the Bari group (2016).[9] The precise meaning of the parameters and the error estimate are discussed in the text.

	Normal (Inverted)	Error	Units
Δm^2	2.50 (2.46)	18 %	10^{-3} eV^2
δm^2	7.37 (7.37)	2.4 %	10^{-5} eV^2
$\sin^2\theta_{13}$	2.17 (2.19)	4.8 %	10^{-2}
$\sin^2\theta_{12}$	2.97 (2.97)	6.2 %	10^{-1}
$\sin^2\theta_{23}$	4.43 (5.75)	16 %	10^{-1}
δ	1.39 (1.39)	19%	π

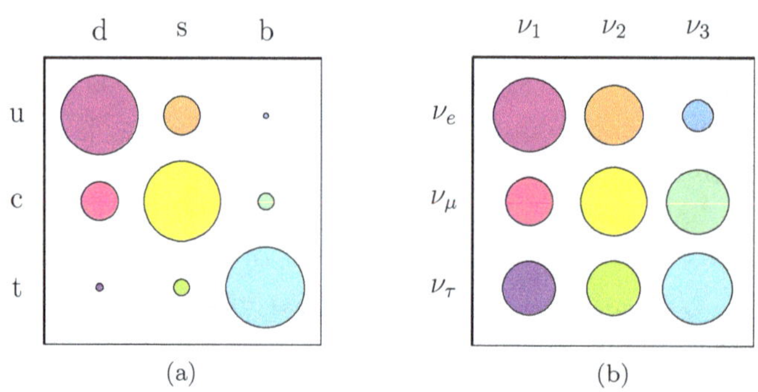

Fig. 4. The surfaces of the circles represent the size of the mixing elements, assuming normal mass hierarchy. From top to bottom, from left to right: (a) Quark mixing (CKM) elements $|V_{ud}|, |V_{us}|, |V_{ub}|, |V_{cd}|, \ldots$; (b) Lepton mixing (PMNS) elements $|U_{e1}|, |U_{e2}|, |U_{e3}|, |U_{\mu 1}|, \ldots$ In both panels, the mixing matrices are supposed to be unitary. The hierarchical structure of quark mixing elements contrasts with the structure of lepton mixing elements.

For further illustration, we compare in Fig. 4 the absolute values of the lepton mixing elements (i.e., the PMNS mixing matrix) with the corresponding values of the quark mixing elements (i.e., the CKM mixing matrix). As a direct application, we can evaluate the universal CP-violating quantity that has been defined in Eq. (58). With the present central values and assuming normal mass hierarchy, we have

$$J_{\text{CP}} = -0.03 = -32\%\ J_{\text{CP}}^{\text{max}}. \tag{59}$$

The result for inverted mass hierarchy is similar.

The specific choices of the delta-m-squared parameters used in this analysis are

$$\delta m^2 = m_2^2 - m_1^2 \quad \text{and} \quad \Delta m^2 = |m_3^2 - (m_2^2 + m_1^2)/2| \tag{60}$$

that, denoting the lightest neutrino mass by m, are equivalent to the following set of relations:

$$\begin{cases} m_1 &= m \qquad\qquad\qquad\qquad (= \sqrt{m^2 + \Delta m^2 - \delta m^2/2}), \\ m_2 &= \sqrt{m^2 + \delta m^2} \qquad\qquad (= \sqrt{m^2 + \Delta m^2 + \delta m^2/2}), \\ m_3 &= \sqrt{m^2 + \Delta m^2 + \delta m^2/2} \ (= m), \end{cases} \tag{61}$$

where the expression outside (inside) the brackets applies for normal (inverse) mass hierarchy.

Note incidentally that the minimum mass is not probed by oscillations. It should be stressed that the case of normal mass hierarchy is slightly favored from the present experimental information at $\Delta\chi^2 = 2.8$ (namely, about 1.7σ) from the same analysis.[9]

3. Vacuum neutrino oscillations

In the first section (Section 3.1), we derive the formulae of the oscillation probability, namely the transition probability of neutrino of a given flavor ℓ to a flavor ℓ', and discuss the standard manipulations that are needed to understand thoroughly the underlying physics. In the next section (Section 3.2), we will consider various applications of these formulae that will allow us to appreciate their usefulness and explore the flexibility of this formalism.

3.1. *General formalism*

Using the results on the mixing matrix U obtained in the previous section, we discuss how to describe the oscillation probabilities in the situations when the flavor is unchanged (survival or disappearance probability) or instead when the flavor of the neutrino changes (appearance probability). We consider the general case (Section 3.1.1) and also the special case when we have two neutrinos only (Section 3.1.2). Then we consider a formal development, i.e., we discuss neutrino oscillations in the context of the field theoretical formalism (Section 3.1.3), introduce the effective Hamiltonians (Section 3.1.4), and finally discuss what new effects are expected describing the neutrinos as wave packets (Section 3.1.5).

3.1.1. *Oscillations with n-flavors and n-mass states*

Time dependence and transition amplitude: The most general case we can consider describes n neutrino flavor eigenstates $|\nu_\ell\rangle$ as linear combinations of the n neutrino mass eigenstates $|\nu_j\rangle$. The flavor states change in time as the energy eigenstates evolve,

$$|\nu_\ell, t\rangle = \sum_{j=1}^{n} U_{\ell j}^* |\nu_j, t\rangle, \quad \text{where} \quad |\nu_j, t\rangle = |\nu_j, 0\rangle\, e^{-iE_j t}, \tag{62}$$

where we use a somewhat redundant but transparent notation for $|\nu_j, 0\rangle = |\nu_j\rangle$ and similarly for $|\nu_\ell, 0\rangle = |\nu_\ell\rangle$. Therefore, we easily find

$$|\nu_\ell, t\rangle = \sum_{j=1}^{n} U_{\ell j}^*\, e^{-iE_j t} |\nu_j, 0\rangle = \sum_{j=1}^{n} U_{\ell j}^*\, e^{-iE_j t} \left(\sum_{\ell'} U_{\ell' j} |\nu_{\ell'}, 0\rangle \right) \tag{63}$$

and eventually

$$|\nu_\ell, t\rangle = \sum_{\ell'} \left(\sum_{j=1}^{n} U_{\ell j}^*\, e^{-iE_j t}\, U_{\ell' j} \right) |\nu_{\ell'}, 0\rangle, \tag{64}$$

where we used the basic property of unitary matrices $U^\dagger = U^{-1}$. From this expression, the formula for the *amplitude of transition* follows immediately,

$$\mathcal{U}_{\ell\ell'}(t) \equiv \langle \nu_{\ell'}, 0 | \nu_\ell, t\rangle = \langle \nu_{\ell'}, 0| \sum_{\ell''} \left(\sum_{j=1}^{n} U_{\ell j}^*\, e^{-iE_j t}\, U_{\ell'' j} \right) |\nu_{\ell''}, 0\rangle, \tag{65}$$

leading to

$$\mathcal{U}_{\ell\ell'}(t) = \sum_{j=1}^{n} U_{\ell j}^*\, e^{-iE_j t}\, U_{\ell' j}. \tag{66}$$

The reason why we prefer to indicate this amplitude with the symbol $\mathcal{U}$, rather than with the common symbol $\mathcal{A}$ used by other authors, will be discussed in Sections 3.1.3 and 3.1.4.

Probabilities: With few more straightforward manipulation,[n] we obtain the probability

$$P_{\nu_\ell \to \nu_{\ell'}}(t) = |\mathcal{U}_{\ell\ell'}(t)|^2 = \sum_{i=1}^{n} |U_{\ell i}|^2 |U_{\ell' i}|^2 + \sum_{i \neq j}^{n} \mathcal{Q}_{\ell\ell', ij}^*\, e^{-i(E_i - E_j)t} \tag{67}$$

[n] In particular,

$$\begin{aligned} P_{\nu_\ell \to \nu_{\ell'}} &= \mathcal{U}_{\ell\ell'} \mathcal{U}_{\ell\ell'}^* = \left(\sum_i U_{\ell i}^* U_{\ell' i} e^{-iE_i t} \right) \left(\sum_j U_{\ell j}^* U_{\ell' j} e^{-iE_j t} \right)^* = \\ &= \sum_{ij} U_{\ell i}^*\, U_{\ell' i}\, U_{\ell' j}^*\, U_{\ell j}\, e^{-i(E_i - E_j)t} = \sum_{ij} \mathcal{Q}_{\ell\ell', ij}^* e^{-i(E_i - E_j)t}. \end{aligned}$$

Finally, we note that $\mathcal{Q}_{\ell\ell', ii} = |U_{\ell i}^2||U_{\ell' i}^2|$.

that has a compact expression using the quartet $\mathcal{Q}_{\ell\ell',ij}$, defined and discussed in Section 2.2.1 — see Eq. (44). Noting that $\mathcal{Q}_{\ell\ell',ij} = \mathcal{Q}^*_{\ell\ell',ji}$, we see that the second term in the last expression can be rewritten as

$$P_{\nu_\ell \to \nu_{\ell'}}(t) = \sum_{i=1}^{n} |U_{\ell i}|^2 |U_{\ell' i}|^2 + \sum_{i>j}^{n} 2\,\mathrm{Re}[\mathcal{Q}^*_{\ell\ell',ij}\, e^{-i(E_i - E_j)t}\,]. \tag{68}$$

At this point, it is useful to consider two opposite limits. The first one is when the time is very small so that the phases are negligible. From $|\mathcal{U}_{\ell\ell'}(0)| = \delta_{\ell\ell'}$, we get the useful identity,

$$\delta_{\ell\ell'} = \sum_{i=1}^{n} |U_{\ell i}|^2 |U_{\ell' i}|^2 + \sum_{i>j}^{n} 2\,\mathrm{Re}[\mathcal{Q}^*_{\ell\ell',ij}]. \tag{69}$$

The second limit is when the time is very large so that all the phases $(E_i - E_j)t \gg 1$; we suppose that there is no degeneracy in the energy level (i.e., none of the masses are equal). In this case, we can consider the average over time,

$$\langle P_{\nu_\ell \to \nu_{\ell'}} \rangle \equiv \lim_{T\to\infty} \frac{1}{T} \int_0^T P_{\nu_\ell \to \nu_{\ell'}}(t)\,\mathrm{d}t = \sum_{i=1}^{n} |U_{\ell i}|^2 |U_{\ell' i}|^2. \tag{70}$$

The *averaged* value of the probability is an important quantity whose physical meaning will be discussed in Section 3.2.5. Adopting the definition of the phase,

$$\varphi_{ij} = \frac{E_i - E_j}{2}\, t, \tag{71}$$

along with the identity given in Eq. (69), we find the final, general expression,[10]

$$P_{\nu_\ell \to \nu_{\ell'}} = \delta_{\ell\ell'} - \sum_{i>j} (4\,\mathrm{Re}[\mathcal{Q}_{\ell\ell',ij}]\,\sin^2\varphi_{ij} + 2\,\mathrm{Im}[\mathcal{Q}_{\ell\ell',ij}]\,\sin 2\varphi_{ij}). \tag{72}$$

A further simplification of the expression is possible in the 3×3 case, since the imaginary part of the quartet, $I_{\ell\ell',ij}$, can be rewritten using Eqs. (50) and (58),

$$I_{\ell\ell',ij} = J_{\mathrm{CP}} \sum_{\ell'',k} \epsilon_{\ell\ell'\ell''}\,\epsilon_{ijk} = \begin{cases} J_{\mathrm{CP}} & \text{for } \ell,\ell',i,j = e,\mu,1,2, \text{ etc.,} \\ -J_{\mathrm{CP}} & \text{for } \ell,\ell',i,j = e,\mu,2,1, \text{ etc.,} \\ 0 & \text{for } \ell = \ell' \text{ or } i = j, \end{cases} \tag{73}$$

where J_{CP} is the single (Jarlskog) invariant that quantifies CP violation. Thus, with the definitions of Eq. (46), the expression for the probabilities of disappearance (or survival, $\ell = \ell'$) and of appearance ($\ell \neq \ell'$), valid in the three flavor case, read

$$\begin{cases} P_{\nu_\ell \to \nu_\ell} = 1 - \sum_{i>j} 4\, |U_{\ell i}^2 U_{\ell j}^2|\, \sin^2 \varphi_{ij} & [\ell = \ell'], \\ P_{\nu_\ell \to \nu_{\ell'}} = - \sum_{i>j} 4\, R_{\ell\ell',ij}\, \sin^2 \varphi_{ij} \pm 8\, J_{\rm CP} \prod_{i>j} \sin \varphi_{ij} & [\ell \neq \ell'], \end{cases} \tag{74}$$

where for the second formula we used a few standard trigonometrical manipulations and where the sign in front of the CP-violating part is $\pm = \sum_{\ell''} \epsilon_{\ell\ell'\ell''}$. These equations can be possibly expressed in terms of the elements of the mixing matrix squared by using Eqs. (48) and (49) if one wishes so, but recall that the sign of $J_{\rm CP}$ is not fixed by those relations.

3.1.2. *The case with two flavors and two masses*

The simplest (but very useful!) model one can think of in order to explain neutrino oscillations is the one with two flavor states. We consider for definiteness the two mass eigenstates $|\nu_i\rangle$ $i = 2, 3$ and the flavor eigenstates $|\nu_\ell\rangle$ $\ell = \mu, \tau$, which are a reasonable approximation of a real situation (as we discuss later in Section 3.2.1). There is just a single parameter that fully describes the relevant part of the mixing matrix. Thus, we can assume the following natural parametrization:

$$\begin{pmatrix} \nu_\mu \\ \nu_\tau \end{pmatrix} = \begin{pmatrix} U_{\mu 2} & U_{\mu 3} \\ U_{\tau 2} & U_{\tau 3} \end{pmatrix} \begin{pmatrix} \nu_2 \\ \nu_3 \end{pmatrix}, \quad \text{where} \quad U = \begin{pmatrix} c & s \\ -s & c \end{pmatrix}, \tag{75}$$

where we indicate with $c \equiv \cos(\theta)$ and $s \equiv \sin(\theta)$ the cosine and sine of the single θ parameter: the mixing angle. We have

$$|\nu_\mu\rangle = U_{\mu i} |\nu_i\rangle \quad \text{and} \quad |\bar{\nu}_\mu\rangle = U_{\mu i} |\bar{\nu}_i\rangle \tag{76}$$

since the unitary matrix is real in the 2×2 case: $U_{\ell i}^* = U_{\ell i}$. Now we evaluate the so-called *survival probability*, namely the probability to detect a $|\nu_\mu\rangle$ after the neutrino propagation in vacuum for a time t. The amplitude is

$$\mathcal{U}_{\mu\to\mu}(t) \equiv \langle \nu_\mu | \nu_\mu, t\rangle = \sum_{j,k} \langle \nu_k | U_{\mu k} U_{\mu j}^*\, e^{-iE_j t} | \nu_j \rangle = \sum_k |U_{\mu k}|^2\, e^{-iE_k t}. \tag{77}$$

It is worth to stress that this formula is valid both for models with two or n mass states if we allow the sum to run on all the possible values of k and keep the values E_k distinct. We are just one step from calculating the survival probability,

$$P_{\nu_\mu \to \nu_\mu} = |\mathcal{U}_{\mu\to\mu}(t)|^2 = \left| (1 - |U_{\mu 3}|^2) + |U_{\mu 3}|^2 \, e^{-i(E_3 - E_2)t} \right|^2, \tag{78}$$

where we have used the unitarity condition $\sum_b |U_{ab}|^2 = 1$ valid $\forall a$. If we define $\varphi \equiv (E_3 - E_2)t/2$, we can rewrite such a probability in a much more useful manner,

$$\begin{aligned} P_{\bar\nu_\mu \to \bar\nu_\mu} &= |1 - |U_{\mu 3}|^2 (1 - e^{i2\varphi})|^2 = |1 - e^{i\varphi}|U_{\mu 3}|^2 (e^{-i\varphi} - e^{i\varphi})|^2 \\ &= |1 + 2ie^{i\varphi}|U_{\mu 3}|^2 \sin\varphi|^2 = |e^{-i\varphi} + 2i|U_{\mu 3}|^2 \sin\varphi|^2 \\ &= |\cos\varphi + i\sin\varphi(2|U_{\mu 3}|^2 - 1)|^2 = 1 - 4|U_{\mu 3}|^2(1 - |U_{\mu 3}|^2)\sin^2(\varphi) \end{aligned} \tag{79}$$

The dependence of this expression on time (length) and mass spectrum is completely embedded in the φ parameter and the energy has the standard expression $E_i = \sqrt{m_i^2 + p^2}$, where p is the momentum of the incoming neutrino. It is quite useful to insert the numerical value of the physical constants in the formulae in order to make their meaning transparent. In the ultrarelativistic limit, the energy difference between two mass eigenstates of given momentum can be approximated to

$$E_3 - E_2 = \frac{E_3^2 - E_2^2}{E_3 + E_2} = \frac{m_3^2 - m_2^2}{E_3 + E_2} \simeq \frac{m_3^2 - m_2^2}{2E} \equiv \frac{\Delta m^2}{2E}, \tag{80}$$

where $E_2 \approx E_3 \approx p \approx E$. If we plug this into the φ definition, we get

$$\sin^2(\varphi) = \sin^2\left(\frac{(E_3 - E_2)t}{2E}\right) \approx \sin^2\left(\frac{\Delta m^2 t}{4E}\right). \tag{81}$$

Since we are dealing with ultrarelativistic neutrinos (namely $p \gg m_i$), using the definition of natural units where $\hbar = c = 1$ we can rewrite the argument of the sine in previous expression, i.e., the phases, as

$$\frac{\Delta m^2 c^4 L}{4E\,\hbar c} = \frac{\Delta m^2 c^4}{\mathrm{eV}^2}\frac{L}{\mathrm{km}}\frac{\mathrm{GeV}}{E}\left[\frac{\mathrm{eV}^2\ \mathrm{km}}{4\mathrm{GeV}(\hbar c)}\right], \tag{82}$$

and then recalling $\hbar c \sim 197.396\,\mathrm{MeV\ fm}$, we get

$$\frac{\mathrm{eV}^2 \times \mathrm{km}}{4\,\mathrm{GeV} \times (\hbar c)} \simeq \frac{\mathrm{eV}^2 \times 10^3\,\mathrm{m}}{4 \times 10^9\,\mathrm{eV} \times (197.396 \times 10^6\,\mathrm{eV} \times 10^{-15}\,\mathrm{m})} \simeq 1.267, \tag{83}$$

the notorious number that appears in almost every paper on neutrino oscillation.

A situation between two and three flavors: It might be interesting to stress analogies and differences with a slightly more complex model. Let us consider a world (a theory) not very different from the one we live in, where three mass eigenstates exist but two of them are degenerate. This translates into the following assumption:

$$m_2 - m_1 = 0. \tag{84}$$

We are interested in computing the survival probability in such a system,

$$\langle \nu_\ell | \nu_\ell(t) \rangle = e^{-iE_1 t}[(|U_{\ell 1}|^2 + |U_{\ell 2}|^2 \, e^{-i(E_2-E_1)t}) + |U_{\ell 3}|^2 \, e^{-i(E_3-E_1)t}], \tag{85}$$

where we factored out the first phase contribution. Under the assumption of degeneracy of $2 \leftrightarrow 1$ states, one can drop the phase arising from the second term of the sum. Using the unitarity of the mixing matrix, we have

$$P_{\ell\ell} = \left| (1 - |U_{\ell 3}|^2) + |U_{\ell 3}|^2 \, e^{-i(E_3-E_1)t} \right|^2 \tag{86}$$

that coincides with Eq. (78), obtained in the two-flavor case. Note that the condition of degeneracy should not hold exactly; for our purpose, it is sufficient that the milder condition $(E_2 - E_1)t \ll 1$ holds true, which means that the distance between production and detection is small enough. In this sense, we can say that the two-flavor formula can be used to describe (under suitable conditions) three-flavor situations. It is worth noting that in this approximation, the oscillation probabilities of neutrinos and those of antineutrinos are the same; thus CP-violation effects are not visible. This conclusion can be usefully presented in another manner also: The probability in Eq. (86) depends only on the three parameters $U_{\ell 3}$, but one of them is fixed by the unitarity condition $\sum_\ell |U_{\ell 3}|^2 = 1$. Thus, the oscillation probabilities depend only upon two physical mixing angles, as it is clear from the standard parameterization. This implies that the third mixing angle can be put to zero without changing the physics, and the Jarlskog invariant J_{CP} vanishes, as evident from Eq. (58).

3.1.3. *Oscillations in field theoretical formalism*

It is useful to develop in some detail the connection with the field theoretical formalism. Let us consider the case when a neutrino is produced or detected by charged-current weak interactions; in this case, the key quantity is just the matrix elements of the neutrino field between an initial (or final) neutrino (or antineutrino) and the vacuum.

Let us begin from the simple case of a neutrino mass state. In this case, indicating explicitly the indices of mass i, j and the indices of the four-spinor

a, b, the matrix element of interest is

$$\langle 0|(P_{\rm L})_{ab}\hat{\nu}_{aj}(x)|\nu_i,\vec{p}\rangle \stackrel{\rm UR}{=} \psi_a(\vec{x},\vec{p},-)\times\delta_{ij}e^{-iE_it}, \tag{87}$$

where as usual we consider the ultrarelativistic limit so that the one-to-one connection between helicity and chirality holds true; note that the presence of the chirality projector leads to a negative helicity -1 for neutrinos. The universal function $\psi(\vec{x},\vec{p},-)$ is given in Eq. (25). The matrix element shown in Eq. (87) is diagonal in the mass indices, as it is *a priori* evident from the fact that the states with given mass and momentum are stationary states; furthermore, it has the characteristic space–time dependence of de Broglie wave, namely, $\sim e^{i(\vec{p}\cdot\vec{x}-E_it)}$.

Now, we compare this result with the one that we obtain when we have a state with given flavor and a field with another flavor. Using Eqs. (17) and (34), we find

$$\begin{aligned}\langle 0|(P_{\rm L})_{ab}\hat{\nu}_{a\ell'}(x)|\nu_\ell,\vec{p}\rangle &\stackrel{\rm UR}{=} U_{\ell' j}U^*_{\ell i}\langle 0|(P_{\rm L})_{ab}\,\hat{\nu}_{bj}(x)|\nu_i,\vec{p}\rangle\\ &\stackrel{\rm UR}{=} \psi_a(\vec{x},\vec{p},-)\times U_{\ell' i}U^*_{\ell i}\,e^{-iE_it}\\ &\equiv \psi_a(\vec{x},\vec{p},-)\times\mathcal{U}_{\ell\ell'}(t).\end{aligned} \tag{88}$$

In this matrix element, the spinorial function ψ_a is multiplied by transition amplitude $\mathcal{U}_{\ell\ell'}(t)$ that we have obtained previously. Comparing Eqs. (87) and (88), we see that, for flavor states, the transition amplitude $\mathcal{U}_{\ell\ell'}(t)$ plays the same role as the ordinary phase factor e^{-iE_it} for mass states.

Of course, the above field theoretical formalism leads to the same result of Section 3.1.2. Moreover, it suggests the fact that the amplitude is a unitary matrix, as we can easily check by a direct computation. Using the definition of the amplitude in Eq. (66), we have

$$\begin{aligned}\left[\mathcal{U}\mathcal{U}^\dagger\right]_{\ell_1\ell_2} &= \sum_\ell \mathcal{U}_{\ell_1\ell}\mathcal{U}^*_{\ell_2\ell} = \sum_{\ell,j,k}\left[U^*_{\ell_1 j}\,e^{-iE_jt}\,U_{\ell j}\right]\left[U_{\ell_2 k}\,e^{iE_kt}\,U^*_{\ell k}\right]\\ &= \sum_{j,k}U^*_{\ell_1 j}\,e^{-iE_jt}\,U_{\ell_2 k}\,e^{iE_kt}\,\delta_{jk} = \sum_j U^*_{\ell_1 j}\,U_{\ell_2 j} = \delta_{\ell_1\ell_2},\end{aligned} \tag{89}$$

where we have repeatedly used the unitarity of the mixing matrix.[o] From a point of view of the physical interpretation, it is interesting to consider the

[o]From a mathematical point of view, we can reach the same conclusion even more simply, noticing that $\mathcal{U}(t)$ is a product of three unitary matrices: $\mathcal{U}(t) = U^*\mathrm{diag}[\exp(-iE_it)]U^{\rm t}$.

diagonal term,

$$1 = [\mathcal{U}\, \mathcal{U}^\dagger]_{\ell\ell} = \sum_{\ell'} |\mathcal{U}_{\ell\ell'}|^2 = \sum_{\ell'} P_{\nu_\ell \to \nu'_\ell}, \tag{90}$$

namely, the condition that the oscillation probabilities summed over all possible final states gives 1, that is, the neutrino does not disappear; it only changes flavor.

3.1.4. *The vacuum Hamiltonians*

The neutrinos with given mass propagate according with the standard free Hamiltonian. In quantum field theory and with the notations of Eq. (26), this Hamiltonian for the ith mass state is just

$$\begin{aligned} \mathbf{H}_i^{\text{free}} &= \int \mathrm{d}^3x \, :\mathcal{H}_i^{\text{free}}: \\ &= \sum_{\vec{p}} \sqrt{p^2 + m_i^2} \sum_{\lambda=\pm 1} [a_i^\dagger(\vec{p},\lambda)\, a_i(\vec{p},\lambda) + b_i^\dagger(\vec{p},\lambda)\, b_i(\vec{p},\lambda)]. \end{aligned} \tag{91}$$

Two remarks are in order: (1) we use the boldface to stress the operatorial character; (2) the term with the b oscillators should be omitted for Majorana neutrinos. Therefore, we obtain the obvious solutions for the evolution of the energy eigenstates $|\nu_i, t\rangle$ in the Schrodinger representation, those used, e.g., in Eq. (62).

The corresponding solutions for the flavor states, given again in Eq. (62), require the assumption of Eqs. (33) and (34): These equations, as discussed in Section 2.1.2, are valid in the ultrarelativistic limit. Under these conditions, we can usefully introduce a matrix on flavor space that describes the evolution of the state. This is obtained as follows:

$$i\frac{\mathrm{d}}{\mathrm{d}t}|\nu_\ell, t\rangle = U_{\ell j}^* \, i\frac{\mathrm{d}}{\mathrm{d}t}|\nu_j, t\rangle = U_{\ell j}^* E_j |\nu_j, t\rangle = U_{\ell j}^* E_j U_{\ell' j} |\nu_{\ell'}, t\rangle, \tag{92}$$

which leads to

$$i\frac{\mathrm{d}}{\mathrm{d}t}|\nu_\ell, t\rangle \equiv [H_0]_{\ell\ell'}\, |\nu_{\ell'}, t\rangle, \tag{93}$$

where of course the repeated indices are summed. In matricial notation, we have

$$H_0 = U^* \,\mathrm{diag}(E)\, U^{\mathrm{t}}, \quad \text{where} \quad \mathrm{diag}(E)_{ij} = \begin{cases} E_i & \text{if } i = j, \\ 0 & \text{if } i \neq j. \end{cases} \tag{94}$$

With this matrix, the results of Eq. (64) or Eq. (66) can be presented[p] as

$$|\nu_\ell, t\rangle = \left[e^{-iH_0 t}\right]_{\ell\ell'} |\nu_{\ell'}, 0\rangle \quad \text{or} \quad \mathcal{U}(t) = e^{-iH_0 t}, \tag{95}$$

respectively. Therefore, we see that the transition amplitude can be regarded as an evolutor, in the quantum mechanical sense; also compare with Eq. (89). For the antineutrinos, we need to replace $U_{\ell i}$ with $U^*_{\ell i}$; otherwise, the results are identical.

In summary, we can describe the flavor transformation of neutrinos, caused by the effect of the relative phases of the neutrinos with given mass and by a nontrivial mixing matrix, by introducing the "effective Hamiltonians" valid in the ultrarelativistic limit,

$$H_{0,\nu} \equiv U^* \text{diag}(E) U^{\text{t}} \quad \text{and} \quad H_{0,\bar\nu} \equiv U \text{diag}(E) U^\dagger, \tag{96}$$

for the propagation of free neutrinos and antineutrinos respectively, where we adopted the matricial notation and defined the diagonal matrix,

$$\text{diag}(E)_{ii} = E_i. \tag{97}$$

The "effective Hamiltonians" $H_{0,\nu}$ and $H_{0,\bar\nu}$ are commonly called *vacuum Hamiltonians* for two reasons: (1) to emphasize that the flavor transformation occurs for neutrinos and antineutrinos that propagate in vacuum and (2) to remark the difference with the (additional) matter Hamiltonian term that will be introduced and discussed later on. Note that, again in the ultrarelativistic approximation ($p \sim E$), one can write

$$E_i = \sqrt{m_i^2 + p^2} \simeq p + \frac{1}{2}\frac{m_i^2}{E}. \tag{98}$$

The first term gives rise to the diagonal matrix $p \times \mathbb{1}$ and this can be dropped, because it gives rise to an overall phase factor, common to all the flavor states and thus irrelevant for oscillations. Thus, one can equivalently write

$$H_{0,\nu} = \frac{1}{2E} U^* \text{diag}(m^2) U^{\text{t}} \quad \text{or also} \quad H_{0,\nu} = \frac{1}{2E} U^* \text{diag}(\Delta m^2) U^{\text{t}}, \tag{99}$$

where, e.g., $\text{diag}(\Delta m^2)_{ii} = m_i^2 - m_1^2$.

Note that the "effective Hamiltonians" depend explicitly upon the value of the momentum; thus, they should be thought more properly as sets of

[p]The calculation of $e^{-iH_0 t}$ can be performed by the Taylor expansion, $e^{-iH_0 t} = 1 + (-iH_0 t) + 1/2(-iH_0 t)^2 + \ldots.$, noticing that $H_0^n = U^* \text{ diag}(E)^n \, U^{\text{t}}$.

matrix elements of a true Hamiltonian between states with given momentum, i.e., plane waves. This is why the symbol H_0 is not written in boldface, e.g., in Eq. (91).

3.1.5. *Oscillations and wave packets*

The usage of the plane waves to derive oscillation formulae sometimes generates confusion. In fact, it is not evident how to define the transit time or the distance between production and detection with plane waves that are not localized. We discuss here how to improve the description.

Scalar wave packet: Let us consider a packet of scalar waves in one dimension, propagating along the z axis,

$$f(z,t) = \sum_q e^{i(qz-E_q t)} f(q). \tag{100}$$

We recall that we assume that the particle is in a box so that its momenta are quantized. Suppose that this packet has typical momentum p (namely, $|f(q)^2|$ is localized around the point $q = p$, where it attains its maximum) and then let us expand the energy around this point,

$$E_{p+q} \approx E_p + q v_p, \quad \text{where} \quad v_p = \frac{\mathrm{d}E_p}{\mathrm{d}p}. \tag{101}$$

We can change variable $q = p + k$ and rewrite

$$f(z,t) \approx e^{i(pz-E_p t)} \sum_k e^{ik(z-v_p t)} f(p+k) \equiv e^{i(pz-E_p t)} F(z - v_p t), \tag{102}$$

where we introduced the auxiliary function, $F(z) \equiv \sum_k e^{ikz} f(p+k)$. The two factors of Eq. (102) are amenable to the following interpretation: The first one describes a (de Broglie) plane wave, the second one describes the position of the particle in the space. Two remarks are in order:

(1) The function $f(z,t)$ is approximately an eigenstate of the momentum $-i\,\mathrm{d}/\mathrm{d}z$ if $p \times |F(z-v_p t)| \gg |\mathrm{d}F(z-v_p t)/\mathrm{d}z|$, i.e., if the function $F(z - v_p t)$ does not vary much when it is nonzero. The same condition implies that $f(z,t)$ is also an approximate eigenstate of the energy, assuming the dispersion relation dictated by relativity.[q]

[q]First we multiply both sides by $v_p = v = p/E$ and note that the left-hand side $p^2/E\,|F(z-vt)|$ is smaller than $E|F(z-vt)|$. In the right-hand side, instead, we have $v \times |\mathrm{d}F(z-vt)/\mathrm{d}z| = |\mathrm{d}F(z-vt)/\mathrm{d}t|$. Thus we obtain $E|F(z-vt)| \gg |\mathrm{d}F(z-vt)/\mathrm{d}t|$ that implies that $f(z,t)$ is an approximate eigenvalue of the energy $i\,\mathrm{d}/\mathrm{d}t$.

(2) The modulus of the function is $|f(z,t)|^2 \approx |F(z-v_p t)|^2$. Note that in the ultrarelativistic regime in which we are particularly interested, we have $E \approx pc$ in good approximation and the velocity is just $v_p \approx c$. The wave packet becomes nondispersive, in good approximation i.e., it propagates maintaining its shape.

Neutrino wave packet: Let us consider the state of a neutrino with flavor ℓ propagating with momentum along the z axis. Summing over the component of the momentum along z, call it q, we have

$$|\nu_\ell, g\rangle = \sum_q g(q)|\nu_\ell, q\rangle, \tag{103}$$

where $g(q)$ is a adimensional factor such that $\sum_q |g(q)|^2 = 1$. Now we calculate the transition from this state to the vacuum, following the same manipulations described in Section 3.1.3. We obtain

$$\begin{aligned} \langle 0|P_\mathrm{L}\hat{\nu}_{\ell'}(x)|\nu_\ell, g\rangle &\overset{\mathrm{UR}}{=} U_{\ell' j}U^*_{\ell j}\sum_q g(q)\psi(z,q,-)e^{-iE_j t} \\ &= U_{\ell' j}U^*_{\ell j}\sum_q g(q)\frac{e^{i(qz-E_j t)}}{\sqrt{V}}\, u_-, \end{aligned} \tag{104}$$

where $E_j = \sqrt{q^2+m_j^2}$. The ultrarelativistic four-spinor of negative helicity[r] u_- factors out; so it does the dependence upon the helicity. Thus, the considerations concerning the scalar wave packet can be applied quite directly. Suppose that the typical momentum of the wave packet is p. Then, define the auxiliary wave function concerning a massive neutrino,

$$\mathcal{G}(z) \equiv \sum_q \frac{e^{iqz}}{\sqrt{V}}\, g(p+q)\, u_-. \tag{106}$$

Finally, as in Eq. (102), we conclude

$$\langle 0|P_\mathrm{L}\hat{\nu}_{\ell'}(x)|\nu_\ell, g\rangle \overset{\mathrm{UR}}{=} U_{\ell' j}\; e^{i(pz-E_j t)}\; \mathcal{G}(z-v_j t)\; U^*_{\ell j}. \tag{107}$$

This is our main formula, where the formalism previously described has been enhanced to include a description of the wave packet. In fact, when we set

---

[r] In the "standard" representation of the Dirac matrices, this is just

$$u_- = \frac{1}{\sqrt{2}}\begin{pmatrix} 0 \\ +1 \\ 0 \\ -1 \end{pmatrix}. \tag{105}$$

$g(p+q) = \delta_{q,0}$, we get $\mathcal{G} \to \psi$, i.e., we have plane waves. If, furthermore, we replace $v_j \to c$, we end up with Eq. (88). Note that the right-hand side of Eq. (107) can be equated to the ℓ' component of the wave function,

$$[\Psi^\ell(t)]_{\ell'} = U_{\ell' j} e^{i(pz - E_j t)} \mathcal{G}(z - v_j t) U_{\ell j}^*, \tag{108}$$

that describes a neutrino ℓ at $t = 0$ that evolves in the course of the time. Let us proceed with a detailed discussion of Eq. (107).

Discussion of the formula: Neutrino masses enter Eq. (107) or equivalently Eq. (108) through the energy $E_j = \sqrt{p^2 + m_j^2}$ and the velocity $v_j = p/E_j$. Thus, they produce two kinds of effects: the relative phases get different after a time $(E_1 - E_2)t_{\text{osc}} \sim 1$ and the wave packets separate after a time $(v_1 - v_2)t_{\text{dec}} \sim \sigma_z$, where σ_z is the size of the wave packet. These are two different effects: interferences connected to the phase change, namely neutrino oscillations, and separation of the wave packets, namely decoherence. It is important to stress that oscillations occur when the components of the wave packet associated to the different mass states overlap. Using $\sigma_E \approx c\,\sigma_p \sim c/\sigma_z$, we have

$$t_{\text{osc}} \sim t_{\text{dec}} \times \frac{\sigma_E}{E}. \tag{109}$$

Therefore, if the energy uncertainty σ_E is much less than the value of the energy $E \approx pc$, neutrino oscillations will happen earlier and decoherence will happen later.

Equivalently, we can carry on the discussion by introducing and comparing three lenghts: propagation length (i.e., distance between production and detection) $L = ct$, oscillation length $L_{\text{osc}} = ct_{\text{osc}}$, coherence length $L_{\text{dec}} = ct_{\text{dec}}$. Similarly, it is useful to note the two phenomena mentioned above correspond to different velocities among the individual components: oscillations are caused by the difference among phase velocities E_j/p; decoherence is caused by the difference among group velocities $\mathrm{d}E_j/\mathrm{d}p = p/E_j$ instead.

Summary and illustration: At this point, we have a general picture of what happens to ultrarelativistic neutrinos during their propagation:

(1) The light neutrinos are produced according to the mixing matrix, since we assume that their kinetic energy is much larger than their mass.
(2) Initially, we can assume in excellent approximation that the components with different masses, and thus also their wave packets, travel with velocity c. Thus it is appropriate to use $\mathcal{G}(z - ct)$ for all mass components

in Eq. (107) or Eq. (108). The usual transition amplitude $\mathcal{U}_{\ell\ell'}$, multiplied by a flavor-independent factor, e.g., Eq. (108), reads

$$[\Psi^{\ell}]_{\ell'} = \mathcal{U}_{\ell\ell'}(t)\ \mathcal{G}(z - ct). \tag{110}$$

The differences of phases $(E_i - E_j)t$ deviate from zero and oscillations in proper sense occur.

(2') Subsequently, the differences of phases of the components $(E_i - E_j)t$ appearing in $\mathcal{U}_{\ell\ell'}$ become very large. Thus, the phase factors oscillate very rapidly when we vary, even slightly, the distance or the energy at the source or at the detection. For this reason, the effect of the oscillatory terms is not measurable in practice anymore.

(3) Eventually, the wave packets separate and the description becomes even simpler conceptually. When the different components do not overlap anymore, oscillations are completely lost (i.e., not only for practical purposes). Interestingly, neutrinos with different masses could be in principle detected separately in this stage even if, in practice, this is extremely difficult. If they are not measured, the probabilities of transmutation that follow from step (2′) and step (3) are the same.

The steps (1), (2) and (3) are illustrated graphically in Fig. 5.

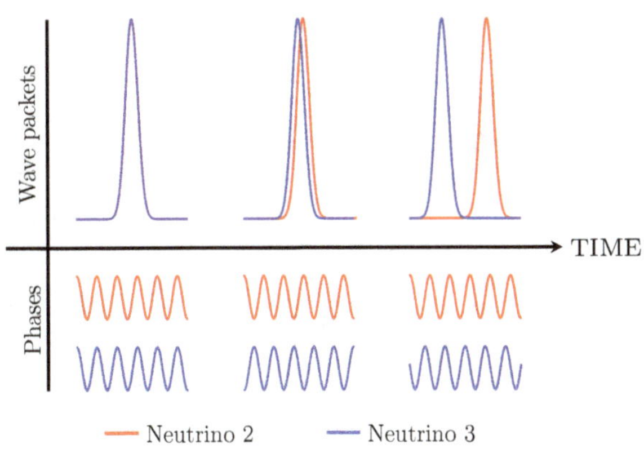

Fig. 5. Physics of oscillations illustrated with the ν_μ–ν_τ system. We assume the existence of two mass components ν_2 and ν_3, indicated in red and blue, respectively. The arrow shows the time. We display the wave packets — plots over the arrow — and the phases of the two mass components — plots below the arrow — at three representative times: (1) Initially (left), the two wavepackets are overlapped; until the two mass components are in phase, the state is $|\nu_\mu\rangle$. (2) Then (middle), the two wavepackets are still overlapped; when the two mass components get counterphased, the state becomes $|\nu_\tau\rangle$. (3) After a long time (right), when the mass components are fully separated, we have a 50/50 flavor mixture.

3.2. *Applications and examples*

We will proceed to illustrate the formulae given above by considering several specific applications and particular cases. However, we would like to begin simply by justifying better the name of "oscillations" given to this phenomenon, especially in the modern literature. We have seen the appearance of a several oscillatory functions; however, this point is illustrated much better by a plot. In Fig. 6, we show the probabilities that an electron neutrino remains so or changes in the course of its propagation, with the parameters given in Table 2.

3.2.1. *Why two-flavor formulae are so useful*

Two-flavor oscillation formulae in vacuum allow us to discuss the main facts concerning the observed neutrino oscillations. In order to see how, recall that (1) electron antineutrinos are produced in nuclear fission reactors and then detected; (2) similarly muon neutrinos are produced by charged pion decays (either naturally or artificially) and then detected.

In both cases, we are interested in the probability of survival. Let us consider

$$|\bar{\nu}_e\rangle = U_{ei}|\bar{\nu}_i\rangle \quad \text{and} \quad |\nu_\mu\rangle = U^*_{\mu i}|\nu_i\rangle. \tag{111}$$

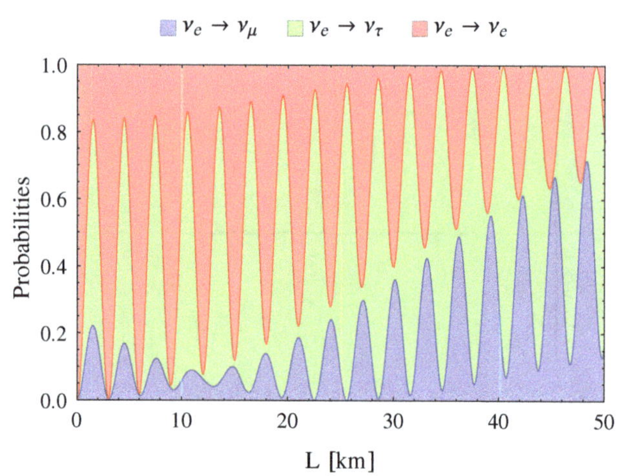

Fig. 6. In this plot, we consider an electron neutrino of 3 MeV that propagates for several kilometers. For each fixed distance, the probability that the electron neutrino becomes a muon (or tau) neutrino in the course of the propagation is given by the purple (or green) region. The probability that it remains an electron neutrino is given by the purple region. The plot emphasizes that the sum of the three probabilities is one, i.e., there is no loss of probability.

When the distances are not too large, as quantified below, only the third neutrino, the one that has the larger mass difference with the other two states, causes oscillations. The other two neutrinos' mass states have effectively the same mass. In other words, the discussion of Section 3.1.2 applies to the real situations, due to the value of the oscillation parameters given in Table 2. Let us recall the amplitudes of oscillations of interest, namely,

$$\langle \bar{\nu}_e | \bar{\nu}_e, t \rangle = \sum_{i=1}^{3} |U_{ei}|^2 e^{-iE_i t} \approx e^{-iE_1 t} \left[1 - |U_{e3}|^2 + |U_{e3}|^2 e^{i(E_3 - E_1)t} \right], \tag{112}$$

where we used the above assumption, along with unitarity. The two-flavor formulae follow,

$$P_{\bar{\nu}_e \to \bar{\nu}_e} = 1 - 4|U_{e3}^2| \left(1 - |U_{e3}^2|\right) \sin^2 \varphi, \tag{113}$$

$$P_{\nu_\mu \to \nu_\mu} = 1 - 4|U_{\mu 3}^2| \left(1 - |U_{\mu 3}^2|\right) \sin^2 \varphi, \tag{114}$$

and adopting the standard parameterization (Eq. (51) and Table 2), we can replace

$$|U_{e3}| = \sin\theta_{13} \quad \text{and} \quad |U_{\mu 3}| = \cos\theta_{13} \sin\theta_{23}. \tag{115}$$

Using these simple formulae, Daya Bay and Reno collaborations (reactor experiments) measured $\theta_{13} \sim 9°$; SK and MACRO (atmospheric neutrino experiments[s]) and subsequently K2K and MINOS (accelerator experiments) measured $\theta_{23} \sim 45°$. Due to the fact that $\Delta m^2_{31} \approx 2.5 \times 10^{-3}\,\mathrm{eV}^2$, oscillations manifested in these experiments, and indeed the phases in the conditions of these experiments are large enough,

$$\varphi = 1.267\, \frac{\Delta m^2_{31} L}{E_\nu} \approx 1 \quad \text{if} \quad \begin{cases} L = 1\,\mathrm{km},\ E_\nu = 3\,\mathrm{MeV} & [\text{reactor } \nu], \\ L = 700\,\mathrm{km},\ E_\nu = 2\,\mathrm{GeV} & [\text{long baseline}]. \end{cases} \tag{116}$$

The average value of the electron survival probability for solar neutrinos that follows from the above formula including the effect of θ_{13} is $\langle P_{\nu_e \to \nu_e} \rangle = \langle P_{\bar{\nu}_e \to \bar{\nu}_e} \rangle = 1 - (\sin^2 2\theta_{13})/2 = 0.95$: This is too small to account for solar neutrino disappearance. In fact, it is the effect of the splitting between the other two mass states that explains solar neutrinos! This hypothesis was tested by KamLAND reactor experiments at distances of the order of

[s]Neutrinos are produced at a height of some $H = 15\text{–}20\,\mathrm{km}$ height in the atmosphere. Those coming from the horizontal direction travel $L \approx (2HR_\oplus)^{1/2} = 450\text{–}500\,\mathrm{km}$, while the vertical ones travel much less.

$L \sim 100\,\text{km}$ that determined precisely $\Delta m^2_{21} \approx 7.4 \times 10^{-5}\,\text{eV}^2$ using the formula,[t]

$$P_{\bar{\nu}_e \to \bar{\nu}_e} = \sin^4\theta_{13} + \cos^4\theta_{13}(1 - \sin^2 2\theta_{12} \sin^2\phi), \tag{117}$$

where this time $\phi = 1.267\,\Delta m^2_{21} L/E_\nu$. The value of the last angle, which describes the composition of ν_e in terms of the last two mass states, is $\theta_{12} \approx 33°$. The ambiguity $\theta_{12} \leftrightarrow 90° - \theta_{12}$ due to the occurrence of $\sin^2 2\theta_{12}$ in the formula is resolved by high-energy solar neutrinos that are affected by matter (MSW) effect and will be discussed in Section 3.2.2.

The above considerations show that, despite their simplicity, the two-flavor formulae (or their direct extensions) allow us to understand a great deal of facts concerning the evidences of neutrinos oscillations. See Table 2 for an updated summary of the values of the oscillation parameters.

3.2.2. *A special case: Maximal mixing*

Definitions: The cases of maximal mixing are of special interest. We begin the discussion with the two-flavor case. One assumes that a pair of neutrinos, say μ and τ, is described by

$$|\nu_\mu\rangle = \frac{|\nu_1\rangle + |\nu_2\rangle}{\sqrt{2}} \quad \text{and} \quad |\nu_\tau\rangle = \frac{-|\nu_1\rangle + |\nu_2\rangle}{\sqrt{2}}, \tag{118}$$

namely,

$$|\nu_\ell\rangle = U^*_{\ell i}|\nu_i\rangle \quad \text{with} \quad U = \frac{1}{\sqrt{2}}\begin{pmatrix} +1 & +1 \\ -1 & +1 \end{pmatrix} \tag{119}$$

so that the mixing elements satisfy $|U_{\ell i}|^2 = 1/2$. This simple case is a good approximation of some situations of physical interest. The maximal mixing in the 3×3 case is a bit more intricate. It is based on the assumption that the mixing matrix is

$$U = \frac{1}{\sqrt{3}}\begin{pmatrix} 1 & 1 & 1 \\ 1 & \omega & \omega^2 \\ 1 & \omega^2 & \omega \end{pmatrix}, \quad \text{where} \quad \omega = e^{i\,2\,\pi/3} \tag{120}$$

so that $\omega^3 = 1$ and all matrix elements satisfy $|U_{\ell i}|^2 = 1/3$. This case is to be thought as a toy model since it does not correspond to

[t]This equation follows by averaging to zero the fast oscillation due to Δm^2_{31}, namely, adopting the following approximation, $P_{\bar{\nu}_e \to \bar{\nu}_e} = |\,|U^2_{e1}| + |U^2_{e2}|e^{-i2\,\varphi_{21}} + |U^2_{e3}|e^{-i2\,\varphi_{31}}\,|^2 \approx |\,|U^2_{e1}| + |U^2_{e2}|e^{-i2\,\varphi_{21}}\,|^2 + |U^4_{e3}|$.

any situation of physical reality but it is anyway useful as test bed, in particular to better understand CP-violating phenomena. In fact, we find $J_{\rm CP} = {\rm Im}[U_{e1}U^*_{\mu 1}U^*_{e2}U_{\mu 2}] = 1/(6\sqrt{3})$, namely, the maximal amount of CP violation, as discussed after Eq. (58). In fact, it can be checked by direct calculation that, by adopting a suitable redefinition of phases as in Eq. (38), the mixing matrix given in Eq. (120) is *equivalent* to the standard form given in Eq. (51) with the values $\theta_{12} = \theta_{23} = \pi/4$, $\sin\theta_{13} = 1/\sqrt{3}$ and $\delta = +\pi/2$.

Two-flavor case: Using the oscillation probability of Eq. (67) and the definition of maximal mixing, it is easy to show that

$$P_{\mu\tau} = P_{\tau\mu} = \sin^2\phi = 1 - P_{\mu\mu} = 1 - P_{\tau\tau} \quad \text{where} \quad \phi = \frac{\Delta m^2 L}{4E}. \tag{121}$$

Thus:

(1) the probabilities of appearance and disappearance are all connected;
(2) the averaged values are $\langle P_{\ell\ell'}\rangle = 1/2$ for $\ell, \ell' = \mu, \tau$;
(3) it is possible that the original type of neutrino fully disappears when $\phi = \pi/2$.

Three-flavor case: We recall that (by definition) all matrix elements satisfy $|U_{\ell i}|^2 = 1/3$; moreover, it is interesting to also impose the additional condition

$$\phi = \phi_{21} = \phi_{32} = \phi_{31}/2 \tag{122}$$

or equivalently $\Delta m^2 = m_2^2 - m_1^2 = m_3^2 - m_2^2 = (m_3^2 - m_1^2)/2$. This is a rather particular case that is not realized in nature but has some didactic interest for what concerns CP (or better T) violation. We have,

$$P_{\ell\ell} = \frac{(1 + 2\cos\phi)^2}{9} \quad \text{and} \quad P_{\ell\ell'} = \frac{1 - P_{\ell\ell}}{2} + \frac{2\,\xi}{3\sqrt{3}}\sin\phi\,(1 - \cos\phi), \tag{123}$$

where $\ell \neq \ell'$. The last term describes CP violation and the sign ξ is

$$\xi = -1 \text{ for } \ell\ell' = e\mu, \mu\tau, \tau e \quad \text{and} \quad \xi = +1 \text{ for } \ell\ell' = \mu e, \tau\mu, e\tau. \tag{124}$$

It is noticeable that when $\phi = 2\pi/3$ we have

$$\begin{pmatrix} P_{ee} & P_{e\mu} & P_{e\tau} \\ P_{\mu e} & P_{\mu\mu} & P_{\mu\tau} \\ P_{\tau e} & P_{\tau\mu} & P_{\tau\tau} \end{pmatrix} = \begin{pmatrix} 0 & 0 & 1 \\ 1 & 0 & 0 \\ 0 & 1 & 0 \end{pmatrix}. \tag{125}$$

Therefore, the chosen combination of mixing angles and of phases of oscillations produces the maximum possible deviation from the case when the oscillation probability is symmetric, $P_{\ell\ell'} = P_{\ell'\ell}$. In other words, this is the maximum amount of T violation, that is, the maximum difference between the cases when $\nu_\ell \to \nu_{\ell'}$ and $\nu_{\ell'} \to \nu_\ell$ with $\ell \neq \ell'$.

3.2.3. *A two-flavor case of oscillations with neutrino interactions*

When we have muon and tau neutrinos of high energies, we should consider the fact that they oscillate, approximatively with maximal mixing $\theta_{\text{max}} = 45°$, and they also interact (differently) with matter. (This is relevant, for example, for the neutrinos produced in the center of the Sun due to the annihilation or decay of hypothetical dark matter particles.[13]) After charged-current interactions, neutrinos get converted into the corresponding charged leptons that are eventually slowed down and absorbed. This can be formally described including an imaginary term in the Hamiltonian that accounts for their disappearance at the original energy.[u] Thus,

$$H = U(\theta_{\text{max}}) \begin{pmatrix} -\kappa & 0 \\ 0 & \kappa \end{pmatrix} U^{-1}(\theta_{\text{max}}) - \frac{i}{2} \begin{pmatrix} \Gamma_\mu & 0 \\ 0 & \Gamma_\tau \end{pmatrix} \tag{126}$$

with

$$\kappa = \frac{\Delta m^2}{4E}, \tag{127}$$

where $\Gamma_{\mu,\tau}$ are the interaction rates with matter. Setting

$$\Gamma = \frac{\Gamma_\mu + \Gamma_\tau}{2} \quad \gamma = \frac{\Gamma_\mu - \Gamma_\tau}{4}, \tag{128}$$

we get the convenient form of the Hamiltonian,

$$H = \begin{pmatrix} -i\gamma & \kappa \\ \kappa & +i\gamma \end{pmatrix} - \frac{i\Gamma}{2} \begin{pmatrix} 1 & 0 \\ 0 & 1 \end{pmatrix} \tag{129}$$

In the case of constant density, we can easily calculate the evolutor $\mathcal{U} = \exp(-iHr)$, where r is the distance travelled by the neutrino, proceeding

[u] Considering again the analogy with optics, this term is, in principle, similar to the complex term that is introduced in the refractive index definition, $n \equiv n_1 + in_2$, when a light signal propagates in a dispersive medium; n_2 accounts for the possible absorption of the signal. There is also a stringent analogy with the kaon system, where the complex term in the Hamiltonian allows the disappearance of the particle, e.g., for a decay process (see Eq. (16)). Note that we consider charged-current interactions; the inclusion of neutral-current interactions is more complicated, because a term that describes the appearance of a neutrino in the final state should be introduced.

as follows. The second part, proportional to the unit matrix, is trivially exponentiated. For the rest, let us begin by assuming the parameter γ as imaginary. The resulting matrix is Hermitian and can be easily diagonalized. Its eigenvalues $\lambda_\pm$ and mixing angles are

$$\lambda_\pm = \pm\alpha \ , \ \cos 2\theta = \frac{i\gamma}{\alpha}, \ \sin 2\theta = \frac{\kappa}{\alpha}, \quad \text{where} \quad \alpha = \sqrt{\kappa^2 - \gamma^2}. \tag{130}$$

At this point, it is easy to exponentiate the full Hamiltonian. The result is

$$\mathcal{U} = e^{-\Gamma r/2} \begin{pmatrix} \cos\varphi - \frac{\gamma}{\alpha}\sin\varphi & -i\frac{\kappa}{\alpha}\sin\varphi \\ -i\frac{\kappa}{\alpha}\sin\varphi & \cos\varphi + \frac{\gamma}{\alpha}\sin\varphi \end{pmatrix} \quad \text{with} \quad \varphi = \alpha r. \tag{131}$$

Finally, we use analytical continuation to come back to the case when γ is real, i.e., the case in which we are actually interested. We find the explicit formula for $\nu_\nu \to \nu_\tau$ conversion,

$$P_{\nu_\mu \to \nu_\tau} = e^{-r\,\Gamma} \, \frac{\sin^2\left(\dfrac{\Delta m^2 \, r}{4\,E_\nu}\sqrt{1-\varepsilon^2}\right)}{1-\varepsilon^2} \quad \text{with} \quad \varepsilon = \frac{\Gamma_\mu - \Gamma_\tau}{\Delta m^2/E_\nu}. \tag{132}$$

It is possible to find in Ref. 13 the numerical values of the interaction rates Γ and more discussion of this and similar situations.

3.2.4. *Electron survival probability in three flavors*

The electron survival probability in the full three-flavor regime contains an imprint of the type of mass hierarchy. The goal of the future experiment JUNO is just to probe this effect; see Ref. 11 for a very detailed discussion. Here we derive the oscillation probability and demonstrate this effect. The formal manipulations are slightly more demanding than the previous ones and for this reason we use a brief and special notation. The amplitude of oscillation for $\bar{\nu}_e \to \bar{\nu}_e$ in three flavors is

$$\begin{aligned} \mathcal{U}_3 &= e^{-iE_3 t}\left[C_{12}C_{13}\, e^{if_{31}} + S_{12}C_{13}\, e^{if_{32}} + S_{13}\right] \\ &= e^{-iE_3 t}\left[C_{13}\,\mathcal{U}_2 + S_{13}\right] \end{aligned} \tag{133}$$

where the corresponding expression in two flavors (or if $\theta_{13} = 0$) is

$$\mathcal{U}_2 = C_{12}\, e^{if_{31}} + S_{12}\, e^{if_{32}}, \tag{134}$$

and where we simplified the notation by using the standard parameterization and setting,

$$C_{ij} = \cos^2\theta_{ij},\ S_{ij} = \sin^2\theta_{ij} \quad \text{and} \quad f_{ij} = (E_i - E_j)t \approx \frac{m_i^2 - m_j^2}{2E} L. \tag{135}$$

In this manner we get

$$\begin{aligned} P_3 &= C_{13}^2 |\mathcal{U}_2|^2 + S_{13}^2 + 2\,C_{13}S_{13}\,\mathrm{Re}[\mathcal{U}_2] \\ &= C_{13}^2 P_2 + S_{13}^2 + 2\,C_{13}S_{13}\,\mathrm{Re}[\mathcal{U}_2], \end{aligned} \tag{136}$$

where we set by definition

$$P_2 = |\mathcal{U}_2|^2 \quad \text{and} \quad P_3 = |\mathcal{U}_3|^2. \tag{137}$$

The piece in which we are mostly interested is

$$\mathrm{Re}[\mathcal{U}_2] = C_{12}\cos(f_{31}) + S_{12}\cos(f_{32}). \tag{138}$$

The two large phases of oscillations, that is, f_{31} and f_{32}, differ but only by a small amount. This is given by the phase of the "solar" (or better KamLAND's) neutrino oscillations,

$$f_{31} - f_{32} = f_{21} \equiv f. \tag{139}$$

It is convenient to introduce a special choice of the *common part* of the large phase factor Φ such that the slow and the fast oscillations get neatly separated. This choice is[12]

$$f_{31} = \alpha\,\Phi + S_{12}\,f \quad \text{and} \quad f_{32} = \alpha\,\Phi - C_{12}\,f, \tag{140}$$

where $\alpha = \pm 1$ corresponds to normal/inverse hierarchy, respectively. In fact, we have

$$\begin{aligned} \mathrm{Re}[\mathcal{U}_2] =& C_{12}\cos(\Phi + \alpha S_{12}\,f) + S_{12}\cos(\Phi - \alpha C_{12}\,f) \\ =& \cos\Phi\ [\ C_{12}\cos(S_{12}\,f) + S_{12}\cos(C_{12}\,f)] \\ &- \alpha\sin\Phi\ [\ C_{12}\sin(S_{12}\,f) - S_{12}\sin(C_{12}\,f)]\,. \end{aligned} \tag{141}$$

This expression is consistent with the following definition of φ:

$$\begin{aligned} \cos\varphi\ |\mathcal{U}_2| &\equiv C_{12}\cos(S_{12}\,f) + S_{12}\cos(C_{12}\,f), \\ \sin\varphi\ |\mathcal{U}_2| &\equiv C_{12}\sin(S_{12}\,f) - S_{12}\sin(C_{12}\,f). \end{aligned} \tag{142}$$

This special definition allows us to recast the term of interest in a very neat manner,

$$\mathrm{Re}[\mathcal{U}_2] = |\mathcal{U}_2| \times \cos(\Phi + \alpha\varphi). \tag{143}$$

In this manner, we have separated the effects of the fast and the slow oscillations in the terms $\mathrm{Re}[\mathcal{U}_2]$: The phase Φ is large and produces the fast oscillations, whereas the phase φ is small and varies only with the solar phase f, just as the term $|\mathcal{U}_2| = \sqrt{P_2}$.

To summarize, our expression for the electron antineutrino survival probability is[v]

$$\begin{aligned} P_3 &= C_{13}^2 |\mathcal{U}_2|^2 + S_{13}^2 + 2\, C_{13} S_{13}\, \mathrm{Re}[\mathcal{U}_2] \\ &= C_{13}^2 P_2 + S_{13}^2 + 2\, C_{13} S_{13} \sqrt{P_2} \cos(\Phi + \alpha\varphi). \end{aligned} \tag{144}$$

The last term shows that in the region of Φ-driven fast oscillations, the prediction of the two mass hierarchies is slightly "out of tune" between them; the detection of this effect requires a very good energy resolution. Moreover, the overall factor $S_{13} = \sin^2\theta_{13}$ suppresses this effect, which contributes to make the task of JUNO demanding.

3.2.5. *The limit of fully averaged oscillations*

An important limiting case is when we have vacuum oscillations and the phases of oscillation are very large. In this case, the oscillatory (cosinusoidal) terms get averaged to zero in any practical situation, e.g., when the production or the detection points are not perfectly known, when there is a distribution of initial energies and this is not perfectly measured in the detector, etc. The case is known as *averaged oscillations* and sometimes also as Gribov–Pontecorvo regime.[14] Strictly speaking, one cannot speak of "oscillations" in proper sense. However, this case is of wide physical interest: For instance, it is important for low-energy solar neutrinos (when "matter effects" are negligible) or also for high-energy neutrino from cosmic sources, as those that have been plausibly measured with IceCube.

The formulae for averaged oscillations are very simple and are given in Ref. 15,

$$P_{\nu_\ell \to \nu_{\ell'}} = P_{\bar{\nu}_\ell \to \bar{\nu}_{\ell'}} = \sum_{i=1}^{3} |U_{\ell i}|^2 |U_{\ell' i}|^2. \tag{145}$$

---

[v] In order to compare with Ref. 9, the following replacements are needed: $\Phi \to 2\Delta_{ee}$, $f \to 2\delta$, $C_{ij} \to c_{ij}^2$, $S_{ij} \to s_{ij}^2$.

Of course, the averaged probabilities are subject to unitarity constraints as any other set of oscillation probabilities, e.g., $\sum_{\ell'=e,\mu,\tau} P_{\nu_\ell \to \nu_{\ell'}} = 1$, or just as the elements of the mixing matrix squared $|U_{\ell i}|^2$. However, owing to the symmetry in the exchange $\ell \leftrightarrow \ell'$, the averaged probabilities depend only upon three parameters, and not four as $|U_{\ell i}|^2$. The probabilities can be arranged in a symmetric matrix,

$$\mathcal{P} = \begin{pmatrix} \frac{1}{3} + 2P_0 & \frac{1}{3} - P_0 + P_1 & \frac{1}{3} - P_0 - P_1 \\ & \frac{1}{3} + \frac{P_0}{2} - P_1 + P_2 & \frac{1}{3} + \frac{P_0}{2} - P_2 \\ & & \frac{1}{3} + \frac{P_0}{2} + P_1 + P_2 \end{pmatrix}, \tag{146}$$

where we introduced three new parameters that can be expressed in terms of the usual ones as follows:

$$P_0 = \frac{1}{2}\left\{(1-\epsilon)^2\left[1 - \frac{\sin^2 2\theta_{12}}{2}\right] + \epsilon^2 - \frac{1}{3}\right\}, \tag{147}$$

$$P_1 = \frac{1-\epsilon}{2}\left\{\gamma \cos 2\theta_{12} + \beta\frac{1-3\epsilon}{2}\right\}, \tag{148}$$

$$P_2 = \frac{1}{2}\left\{\gamma^2 + \frac{3}{4}\beta^2(1-\epsilon)^2\right\}, \tag{149}$$

and the small quantities $\alpha, \beta, \gamma, \epsilon$ (only the last one is known to date) are

$$\begin{cases} \epsilon = \sin^2\theta_{13}, \\ \alpha = \sin\theta_{13}\cos\delta\sin 2\theta_{12}\sin 2\theta_{23}, \\ \beta = \cos 2\theta_{23}, \\ \gamma = \alpha - \dfrac{\beta}{2}\cos 2\theta_{12}(1+\epsilon). \end{cases} \tag{150}$$

It is interesting that the averaged oscillation probabilities depend upon the CP-violating phase and also upon the departure of θ_{23} from 45°, through $\cos 2\theta_{23}$. However, there are only three free parameters; thus one combination of the usual parameters is degenerate. More precisely, P_0 is well known; P_1 contains most uncertainties; P_2 is positive and very small. The above results are taken from Ref. 16, where further discussion, references, applications and numerical results can be found.

4. Neutrino oscillations in matter

In this section, we introduce and discuss the matter term of Wolfenstein. Then we present a few analytical solutions of the equation of propagation, including the one discussed by Mikheyev and Smirnov. The aim is to illustrate the characteristic features of oscillations due to the matter term. Finally, we consider some physical applications.

4.1. *The matter (or Wolfenstein) term*

Origin: Vacuum oscillations are caused by the relative *phases* that the components of a neutrino with given flavor obtain in the course of the time t. This is a typical undulatory or quantum phenomenon, as it is evident from the connection between energies E_i of each component and the frequencies of oscillation ν_i of each component: $E_i = h\nu_i$. Indeed, as we have shown in Section 3.1.4, vacuum oscillations can be described by an effective Hamiltonian. In this section, we discuss a new point: electron neutrinos that propagate in a material medium receive a special, additional phase of scattering, as first remarked by Wolfenstein. We can think of this new phase as an additional term in the Hamiltonian of propagation or also as a flavor-specific refraction term.

Consider the amplitude of scattering of neutrinos onto electrons at rest and more specifically consider the case when the neutrino momentum is unchanged (=forward scattering),

$$\nu_\ell(\vec{p})\, e^- \to \nu_\ell(\vec{p})\, e^-. \tag{151}$$

This amplitude of scattering is maximum when $\ell = e$ due to charged-current interactions. Thus, electronic neutrinos receive a special phase that contributes to the effective Hamiltonian and eventually also to oscillations. Also neutral currents give an additional phase due to the forward scattering of neutrinos with any particle X of the medium (excepting neutrinos themselves) $\nu_\ell(\vec{p})X \to \nu_\ell(\vec{p})X$; however, this contribution is the same for all types of neutrinos and therefore it contributes only to an irrelevant overall phase factor for all neutrinos.[w]

The most important case when this new phase is relevant is when the matter, where neutrinos propagate, contains electrons at rest or almost at rest. We will focus the subsequent discussion on this case. However, in principle, one can also have effects due to polarization, collective motions, etc.

[w]Two remarks are in order: (1) If we also have new (sterile) neutrinos that do not interact, these phases become relevant. (2) If the density of neutrinos is very high, the scattering of neutrinos onto neutrinos can cause additional effects; however, their complete description still eludes our understanding.

Formal derivation of the matter term: We are interested in describing the contribution of Fermi interactions to the forward scattering of an electron neutrino. The idea is to evaluate the amplitude of forward scattering in terms of the matrix element of the Hamiltonian. Thus, we need a formal description of the Hamiltonian, state of the neutrino and state of the medium. We show how to derive, under reasonable assumptions, the standard result (namely, the Wolfenstein or matter term).

Fermi Hamiltonian for neutrino–electron interaction: The Hamiltonian density of the Fermi interactions contains the following interaction between electrons and electron neutrinos:

$$\mathcal{H}^{\rm cc}_{\nu_{ee}} = +\frac{G_{\rm F}}{\sqrt{2}}\, \bar{\nu}_e \gamma_a (1-\gamma_5) e\; \bar{e}\gamma^a (1-\gamma_5)\nu_e, \tag{152}$$

where $G_{\rm F} > 0$ whose sign is predicted by the standard electroweak model. This implies the following Hamiltonian, written in terms of relativistic quantum fields:

$$\mathbf{H}^{\rm cc}_{\nu_{ee}} = \sqrt{2}\, G_{\rm F} \int {\rm d}^3x\; \bar{\nu}_{e{\rm L}} \gamma_a \nu_{e{\rm L}}\; \bar{e}\gamma^a (1-\gamma_5) e, \tag{153}$$

where:

(1) we have integrated over the space coordinates to pass from the Hamiltonian density to the Hamiltonian operator;
(2) we have introduced the chiral projector $P_{\rm L} = (1-\gamma_5)/2$ and defined $\nu_{e{\rm L}} = P_{\rm L}\nu_e$;
(3) we have applied the Fierz transformation described in Section A.3.

Matrix element for forward scattering: Let us consider the previous Hamiltonian at a given time, say, $t = 0$. Then, we consider its matrix element between the *same* initial and final state $|s\rangle$, namely,

$$\langle s|\mathbf{H}^{\rm cc}_{\nu_{ee}}(t=0)|s\rangle \quad \text{with} \quad |s\rangle = |\nu_e, e_1, e_2, e_3 \ldots.\rangle. \tag{154}$$

The state $|s\rangle$ describes a single electron neutrino, localized around $\vec{x} = \vec{x}_0$, and a large number of electrons almost at rest, localized around the positions $\vec{x} = \vec{x}_i$ with $i = 1, 2, 3, \ldots$ (we have in mind a neutrino propagating in the matter, e.g., of the Sun or of the Earth). The only operator containing the electron fields that has large matrix elements is the time component of the vectorial part $\bar{e}\gamma_0 e = e^\dagger e$, that is the electron density operator.[x]

[x]The other operator is $e^\dagger \vec{\alpha}\, \hat{e}$ that is proportional to the current and is thus suppressed with $\beta = v/c$, $e^\dagger \vec{\alpha}\gamma_5 e$, which is proportional to the polarization and assumed to be small, and finally $e^\dagger \gamma_5 e$ that is even smaller.

Thus, the relevant part of the matrix element is

$$\langle s|\mathbf{H}^{\rm cc}_{\nu_{ee}}(t=0)|s\rangle = \sqrt{2}G_{\rm F}\int {\rm d}^3x\ \langle s|\ \mathbf{n}_e(\vec{x})\ \mathbf{n}_{\nu_{e\rm L}}(\vec{x})\ |s\rangle, \tag{155}$$

where we indicate the density operators of the electron and of the neutrino with $\mathbf{n}_e(\vec{x}) = e^{\dagger}(0,\vec{x})e(0,\vec{x})$ and $\mathbf{n}_{\nu_{e\rm L}}(\vec{x}) = \nu_{e\rm L}{}^{\dagger}(0,\vec{x})\nu_{e\rm L}(0,\vec{x})$. We factorize the state $|s\rangle$ and consider separately the matrix elements of the two operators. We have

$$\langle\nu_{e\rm L}|\mathbf{n}_{\nu_{e\rm L}}(\vec{x})|\nu_{e\rm L}\rangle = |\Psi_{\nu_{e\rm L}}(\vec{x},\vec{x}_0)|^2, \tag{156}$$

$$\langle e_1, e_2, e_3 \ldots .|\mathbf{n}_e(\vec{x})|e_1, e_2, e_3 \ldots .\rangle = \sum_i |\Psi_e(\vec{x},\vec{x}_i)|^2, \tag{157}$$

where we have introduced the wave functions of the neutrino and electrons, emphasizing the position of each particle. We supposed for simplicity that the electrons are distributed similarly, but around different positions.

In the case of interest, the individual electrons are distributed in small cells, e.g., atomic sizes of $\sigma_e \sim 10^{-8}$ cm, whereas the neutrino is distributed in a much larger scale σ_ν. Another important quantity is the *macroscopic* electron density that we denote by $n_e(\vec{x})$. We suppose that the scale of variation of the macroscopic electron density, Δr_e, is much larger than the size of the neutrino wave function. For the Sun, e.g., this is $\Delta r_e \sim R_\odot/10$, and also in the Earth, it is typically km size or more. To summarize, we suppose that

$$\Delta r_e \gg \sigma_\nu \gg \sigma_e. \tag{158}$$

In the above assumptions, we can evaluate the integral in Eq. (155). We divide the region where the neutrino wave function is nonzero in many cubes $a = 1, 2, 3....$ Each cube has size Δx such that $\sigma_\nu \gg \Delta x \gg \sigma_e$ and volume Δx^3. In each cube, there are a lot of electrons, while the neutrino density is almost constant with value $|\Psi_{\nu_{e\rm L}}(\vec{x}_a,\vec{x}_0)|^2$. If we integrate the electron density in the cube with label a, we have $\int_a {\rm d}^3x \sum_i |\Psi_e(\vec{x},\vec{x}_i)|^2 = N_e(\vec{x}_a) \gg 1$, namely the number of electrons inside the cube. This is almost the same for all a, because we suppose that $\Delta r_e \gg \Delta x$. We can then approximate: $N_e(\vec{x}_a) \approx n_e(\vec{x}_0)\Delta x^3$, where we indicate the position $\vec{x}_0$ of the neutrino since we are interested in the value of the electronic density where the neutrino is. Thus, the integral is

$$I(\vec{x}_0) = \int {\rm d}^3x \sum_i |\Psi_e(\vec{x},\vec{x}_i)|^2 \times |\Psi_{\nu_{e\rm L}}(\vec{x},\vec{x}_0)|^2, \tag{159}$$

which can be solved as

$$\begin{aligned} I(\vec{x}_0) &\approx \sum_a N_e(\vec{x}_a)|\psi_{\nu_{e\mathrm{L}}}(\vec{x}_a,\vec{x}_0)|^2 \\ &\approx n_e(\vec{x}_0)\sum_a \Delta x^3\ |\Psi_{\nu_{e\mathrm{L}}}(\vec{x}_a,\vec{x}_0)|^2 \\ &\approx n_e(\vec{x}_0)\times\int \mathrm{d}^3x\ |\Psi_{\nu_{e\mathrm{L}}}(\vec{x},\vec{x}_0)|^2 = n_e(\vec{x}_0). \end{aligned} \tag{160}$$

Therefore, the matrix element of interest, for a single neutrino located around $\vec{x}=\vec{x}_0$, is just

$$\langle \mathbf{H}^{\mathrm{cc}}_{\nu_{ee}}\rangle = \sqrt{2}G_{\mathrm{F}}\ n_e(\vec{x}_0). \tag{161}$$

Following the above derivation, it is easy to see that an electron antineutrino has the same matrix element but with the opposite sign, due to the fact that the operator $\mathbf{n}_{\nu_{e\mathrm{L}}}$ counts the net number of electron neutrinos; therefore, an antineutrino counts -1.

4.1.1. *Hamiltonians of propagation including the matter term*

Thus, we should add to the 3×3 vacuum Hamiltonians that describe free neutrino/antineutrino propagation the matter Hamitonian that acts on flavor states,

$$H_{\mathrm{MSW}} = \pm\sqrt{2}\,G_{\mathrm{F}}n_e(\vec{x})\begin{pmatrix}1&0&0\\0&0&0\\0&0&0\end{pmatrix}, \quad \text{where} \quad \begin{cases} + & \text{for } \nu, \\ - & \text{for } \bar{\nu}, \end{cases} \tag{162}$$

where $\vec{x}$ is the position of the neutrino and $n_e(\vec{x})$ the number density of electrons. To be sure, note that this term has the right dimension, namely "energy", since the Fermi constant has 1/energy^2 and the electron density is 1/volume = energy^3 in natural units. Note that we consider a *linear* contribution in the Fermi constant (whereas cross sections or decay widths depend upon G_{F}^2). This fact shows that we are considering amplitudes and/or wave functions rather than probabilities and is a witness to the quantum nature of the phenomenon. A last remark is in order: This new term can be regarded as a contribution to the refraction of the electronic neutrino wave.

To summarize, the flavor of ultrarelativistic neutrinos of given energy E that propagate in a medium with electronic density $n_e(x)$ changes according to neutrino and antineutrino Hamiltonians that have one part due to vacuum oscillations and another one due to the Wolfenstein (or matter) term. For neutrinos, this is given by

$$H_\nu = U\,\mathrm{diag}(k)\,U^\dagger + V(x)\,\mathrm{diag}(1,0,0) \tag{163}$$

with

$$k_i = m_i^2/(2E), \tag{164}$$

$$V(x) \equiv \sqrt{2}G_\mathrm{F} n_e(x). \tag{165}$$

For antineutrinos we need to replace

$$U \to U^* \quad \text{and} \quad V \to -V. \tag{166}$$

We have adopted a shorthand k for the quantity that describes the effect that is purely due to neutrino masses (i.e., of vacuum oscillations).

Before discussing some specific example and some solutions, we note that these Hamiltonians contain the energy $E = pc$ (and/or the momentum) of the neutrino and its position x, too. This description might seem to be in contradiction with the Heisenberg indetermination principle, i.e., with some basic facts of quantum mechanics. However, as we have discussed above, neither of these observables are exactly measured. The idea is that, in actual circumstances, neutrinos travel in wave packets, which are supposed to be nondispersive.

4.1.2. *Remarks*

Numerical value: The numerical value of the matter term and vacuum term, given in the same units of inverse meter, is

$$V = \sqrt{2}\,G_\mathrm{F} n_e = \frac{3.868\times 10^{-7}}{\mathrm{m}} \times \frac{n_e}{\mathrm{mol/cm^3}}, \tag{167}$$

$$k = \frac{\Delta m^2}{2E} = \frac{2.533}{\mathrm{m}} \times \frac{\Delta m^2}{\mathrm{eV^2}} \times \frac{\mathrm{MeV}}{E}, \tag{168}$$

Let us remark that, when the matter term is small in comparison to the other one, we reduce to the case of vacuum oscillations. In order to clarify when this happens, let us examine the ratio of these two terms. We find

$$\frac{V}{k} \equiv \frac{\sqrt{2}\,G_\mathrm{F} n_e}{\Delta m^2/(2E)} \approx \left(\frac{\rho\, Y_e}{100\,\mathrm{mol/cm^3}}\right)\left(\frac{8\times 10^{-5}\,\mathrm{eV^2}}{\Delta m^2}\right)\left(\frac{E}{5\,\mathrm{MeV}}\right), \tag{169}$$

where ρ is the mass density and Y_e is the fraction of electrons and/or protons (since ordinary matter is neutral). The values of the parameters are those needed to interpret solar neutrino data: (1) the central density of the Sun is expected to be about 150 g/cm^3, with some fraction of Helium; (2) the numerical value of Δm^2 is close to the one given in Table 2; (3) the energy thresholds of SK and SNO are close to 5 MeV. The comparison in Eq. (169) shows that at low neutrino energies $E \ll 5\,\text{MeV}$, the matter (Wolfenstein) term has a negligible effect whereas at high energy it is relevant.

A comment on CP and T violation: Let us compare the dependence of certain probabilities of conversion in various cases,

$$\begin{aligned} P_{\nu_\mu \to \nu_e} &= F(U, V; E_\nu, L), \\ P_{\bar{\nu}_\mu \to \bar{\nu}_e} &= F(U^*, -V; E_\nu, L), \\ P_{\nu_e \to \nu_\mu} &= F(U^*, V; E_\nu, L), \\ P_{\bar{\nu}_e \to \bar{\nu}_\mu} &= F(U, -V; E_\nu, L), \end{aligned} \tag{170}$$

where F is a certain function. When we exchange the initial and final states, we consider T reversal which implies $U \neq U^*$. Furthermore, we see that there are also C- and CP-violating effects when we exchange particles and antiparticles that should be attributed to the exchange $V \to -V$. In fact, ordinary matter contains electrons, and therefore, it induces C- and CP-violating effects in the neutrino propagation that are of "environmental nature", on top of the CP (or T) violating effects of fundamental nature, due to the fact that the leptonic mixing matrix is complex.

Amplitude of three-flavor neutrino oscillations: In order to obtain a general expression for the oscillation amplitude in matter in the three-flavor case, we can consider the mixing matrix $U = R_{23}\ \Delta\ R_{13}\ \Delta^*\ R_{12}$, given in Eq. (52), such that the propagation Hamiltonian in Eq. (163) is

$$H_\nu = R_{23}\ \Delta\ \tilde{H}\ \Delta^*\ R_{23}^{\mathrm{t}} \tag{171}$$

with

$$\tilde{H} = R_{13}\ R_{12}\ \mathrm{diag}(k)\ R_{12}^{\mathrm{t}}\ R_{13}^{\mathrm{t}} + V(x)\mathrm{diag}(1, 0, 0), \tag{172}$$

where $\tilde{H}$ is independent[y] on θ_{23} and δ. The formal expression of the amplitude of the three-flavor neutrino oscillations, when the matter effect

[y]The definition of $\tilde{H}$ in Eq. (172) is obtained from $[\Delta,\ R_{12}] = 0$ and $[V,\ R_{23}] = [V,\ \Delta] = 0$.

is included, can hence be written as,[z]

$$\mathcal{U} = \text{Texp}\left[-i\int \mathrm{d}t\ H_\nu(t)\right] \equiv R_{23}\Delta \begin{pmatrix} u_{11} & u_{12} & u_{13} \\ u_{21} & u_{22} & u_{23} \\ u_{31} & u_{32} & u_{33} \end{pmatrix} \Delta^* R_{23}^{\mathrm{t}}, \qquad (173)$$

where u_{ij} depend upon Δm_{23}^2, θ_{13}, Δm_{12}^2, θ_{12} and on the type of mass hierarchy (normal/inverted) that can be easily fixed changing a discrete parameter $h = \pm 1$. When we solve the evolution equations with a certain assumption on the matter density, we calculate the complex numbers u_{ij} and therefore the amplitudes and the probabilities of oscillations. For example, at the web address

http://pcbat1.mi.infn.it/-battist/cgi-bin/oscil/index.r,

the interface of the "Neutrino Oscillations Simulator" allows the user to plot and download the probabilities for neutrinos or antineutrinos that propagate from two points on the surface of the Earth. The solutions are obtained numerically, as described in Ref. 17 (see in particular Eqs. (1), (3), (5), (6) there) and using the preliminary reference Earth model (PREM) of the electronic Earth density illustrated in Fig. 7.

4.2. *Analytical solutions of the equations of propagation in the matter*

For a generic profile of electronic density, it is always possible to resort to numerical methods to solve the equations of propagations. However, the solutions have an oscillatory character and this makes the problem difficult; we need to use a sufficiently small step that can be eventually time consuming.

However, there are various cases when the propagation equations have exact or approximate analytical solutions that are useful to describe the physics. We will discuss two of them in the following. The first is simply the case when we have a constant density; this can be applied in sequence to layers of constant density, finding eventually the overall amplitude. The second case instead is when the density varies slowly, which is called the adiabatic case; this admits a simple solution, which moreover is physically important.

[z]If we divide the interval $[0, t]$ into n equal intervals of size $dt = t/n$ and indicate with $t_j = jdt$ the discrete values of the time, where j is an integer with $n \geq j > 0$, we can define the time-ordered product as

$$\text{Texp}\left[-i\int_0^t d\tau H(\tau)\right] = \lim_{n\to\infty} e^{-i\,dt\,H(t_n)} e^{-i\,dt\,H(t_{n-1})} \ldots . e^{-i\,dt\,H(t_1)}.$$

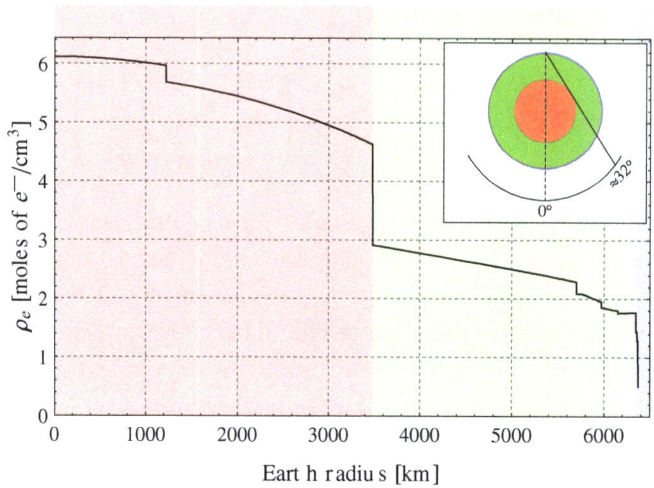

Fig. 7. Density profile of the Earth according to the PREM model.[18] Different colors correspond to different Earth layers: (1) inner and outer cores in red, (2) mantle in green and (3) crust in blue. Visualizing the shell structure in scale, we notice that a neutrino can cross the core only if its angle is smaller than ≈32°.

4.2.1. *Constant matter density and "resonance"*

In the case when we have two flavors (2×2 matrices) and when matter with constant density is particularly instructive, we have

$$U(\theta) = \begin{pmatrix} c & s \\ -s & c \end{pmatrix} \quad \text{and} \quad k \equiv k_2 - k_1 = \frac{\Delta m^2}{2E}, \tag{174}$$

where we can assume $\Delta m^2 > 0$ and $0 < \theta < \pi/2$.

Note that, for the description of oscillations, it is harmless if the Hamiltonian of propagation is modified by adding a real constant: In fact, this constant yields the same phase for all flavor states that is irrelevant for the flavor transformations. Thus, let us choose the constant in such a manner that our Hamiltonian is traceless, i.e., let us subtract the term $-\mathrm{tr}(H)/2 \cdot \mathbb{1}$, getting,[aa]

$$H = \frac{k}{2} U \,\mathrm{diag}(-1,1)\, U^{\mathrm{t}} + \frac{V}{2} \mathrm{diag}(1,-1) = \frac{1}{2} \begin{pmatrix} V - kc_2 & ks_2 \\ ks_2 & kc_2 - V \end{pmatrix}, \tag{175}$$

where we use the obvious shorthands $c_2 = \cos 2\theta$ and $s_2 = \sin 2\theta$.

[aa] This procedure relies on the same principle that allows us to change the overall factor in the vacuum Hamiltonian — see Eq. (99), for example

At this point, we can simply *redefine* the angles and the parameters k in such a manner that this hamiltonian coincides formally with a vacuum hamiltonian. This is obtained by setting,

$$\begin{cases} kc_2 - V \equiv k_m c_{2m} \\ ks_2 \equiv k_m s_{2m} \end{cases} \quad \Longleftrightarrow \quad \begin{cases} t_{2m} = \dfrac{s_2}{c_2 - V/k} \\ k_m = k\sqrt{s_2^2 + (c_2 - V/k)^2} \end{cases} \tag{176}$$

In this manner, we can just reinterpret the solutions, that have been already obtained for the case of vacuum oscillations. We need only to examine the so-called mixing angle in matter θ_m and the parameter $k_m \equiv \Delta m_m^2/(2E)$.

Let us discuss the most interesting feature. Suppose that $V > 0$ (as it is usually the case) and kc_2 has the same sign (namely, suppose $\theta < \pi/4$). By varying the energy of the neutrino E, we vary k from 0 to ∞. Thus, it is always possible to realize the case $c_{2m} = 0$, namely, the case when $\theta_m = \pi/4$ (maximal mixing). This case is called *resonance* and the corresponding energy is called the resonance energy. The resonance point corresponds to the minimum splitting between the energy levels of the mass eigenstates. The possibile transition between the two energy levels (let us call them $\lambda_\pm$) is usually defined *level crossing*. From Eq. (176), it is evident that the minimum distance between $\lambda_\pm = \pm k_m/2$ occurs for $V = kc_2$. Note that (1) this is possible for any value of the mixing angle in vacuum; (2) at the resonance energy, k_m and thus Δm_m^2 get the minimum value.

To summarize, we will have three limiting situations for neutrinos,

$$\begin{cases} V \ll k & \Rightarrow \theta_m \approx \theta, \\ V = kc_2 & \Rightarrow \theta_m \approx 45^\circ, \\ V \gg k & \Rightarrow \theta_m \approx 90^\circ, \end{cases} \tag{177}$$

while for the antineutrinos instead, the sign of V will be opposite and the mixing angle in matter will be smaller than the one in vacuum. We leave to the interested reader a complete study of Δm_m^2, θ_m and $P_{e\mu}(k_m, \theta_m)$, as the parameter V is varied.

4.2.2. *Adiabatic propagation/MSW effect*

Generally, as the neutrino proceeds along its path, the electron density varies. The case when the variation is slow (in a sense to be made precise later) is particularly important and in this case it is possible to obtain a solution of the equations of propagation. This is called the adiabatic solution. Let us proceed step by step:

(1) First of all, we rewrite the Hamiltonian by introducing the local eigenvalues. We rewrite

$$i\partial_t \nu = [\, U_m \,\mathrm{diag}(k_m)\, U_m^\dagger \,]\; \nu, \tag{178}$$

where ν is the vector of the flavor states, $\nu = (\nu_e, \nu_\mu, \nu_\tau)^{\mathrm{t}}$, while U_m is the mixing matrix that depends upon the position and/or the time along the path.

(2) Now we introduce the local mass eigenstates $n = (n_1, n_2, n_3)^{\mathrm{t}}$,

$$\nu \equiv U_m n \;\Rightarrow\; i\partial_t n = [\mathrm{diag}(k_m) - i(U_m^\dagger \partial_t U_m)]n. \tag{179}$$

If the second term was not there, it would be easy to solve the problem.

(3) In order to understand the physics, we focus from here on the 2×2 case, when all matrices are real and say $\nu = (\nu_e, \nu_\mu)^{\mathrm{t}}$ and $n = (n_1, n_2)^{\mathrm{t}}$. After a few direct calculations, we find that the propagation equation is,[bb]

$$i\partial_t \begin{pmatrix} n_1 \\ n_2 \end{pmatrix} = \begin{pmatrix} -k_m/2 & -i\partial_t\theta_m \\ i\partial_t\theta_m & k_m/2 \end{pmatrix} \begin{pmatrix} n_1 \\ n_2 \end{pmatrix}, \tag{180}$$

where the effect of the change in the electron density is contained in $\partial_t\theta_m$.

(4) Now we introduce the definition,

$$\text{Adiabaticity parameter:} \quad \gamma = \left| \frac{2\dot{\theta}_m}{k_m} \right|, \tag{181}$$

and we call "adiabatic condition" the one when γ is always small so that the new term is negligible and we can easily read the solution. In order to have a more precise idea, we note that the case when k_m is minimum, i.e., at the "resonance", the adiabaticity parameter is

$$V(x_{\mathrm{res}}) = kc_2 \;\Rightarrow\; \gamma(x_{\mathrm{res}}) = \left| \frac{\dot{V}}{V} \frac{c_2}{k\, s_2^2} \right|. \tag{182}$$

The case when the condition of adiabaticity is violated significantly is a bit more complex. However, it can still be treated analytically,[cc] and the effects are described by a negative exponential in the parameter $1/\gamma$.

[bb]We have added the term $-\mathrm{tr}[\mathrm{diag}(k_m)]/2 \cdot \mathbb{1}$ that modifies only the overall phase irrelevant for neutrino oscillations.

[cc]This was obtained in Ref. 19 and it is reviewed, e.g., in Ref. 20 We will not need its expression explicitly but will discuss it very briefly later on in Section 4.3.3.

(5) Suppose that γ is small. The local mass eigenstates n_1 and n_2 will not change and will only evolve getting a phase

$$n_1(t) = e^{i\xi}\, n_1(0) \quad n_2(t) = e^{-i\xi}\, n_2(0), \quad \text{where} \quad \xi = \int \frac{k_m}{2} \mathrm{d}t. \tag{183}$$

This is the formal statement that describes the case of "adiabatic propagation" in our two-flavor system. At this point, it is easy to find the propagation amplitude, e.g., $\mathcal{U}_{e\to e} = cc_m e^{i\xi} + ss_m e^{-i\xi}$. (In order to derive this equation, we begin from $\nu = U_m n$, we find that $\nu_e = c_m n_1 + s_m n_2$ and then we calculate $\nu_e(t)$ using $n_1(t)$ ed $n_2(t)$, thanks to the linearity of the equation; finally, we use the definition $\mathcal{U}_{e\to e} = \langle \nu_e, \nu_e(t)\rangle$, or in other words we take the scalar product).

(6) Since the phases ξ are large, we can average to zero the interference terms in the formulae of the probability of electron survival, getting

$$P_{ee} = c^2 c_m^2 + s^2 s_m^2. \tag{184}$$

This is the formula that is usually employed and in which we are interested. We recall that the case when the interference terms drop out is a typical classical average. In the three-flavor case, the obvious generalization is $P_{ee} = |U_{ei}^2||U_{ei}^m|^2$ and will be discussed in the last part of this section.

(7) For practical purposes, it is useful to note that when $V \gg k$, the electron neutrino is $\nu_e \approx n_2$, since, as just explained, it propagates in matter with the combination $\nu_e(t) = c_m n_1(t) + s_m n_2(t)$ and for $V \gg k$, $\theta_m \approx 90°$, as discussed in Section 4.2.1. Since the central density in the Sun is about 150 g/cm^3, this is a good approximation for the propagation of solar neutrinos above about 5 MeV.

Let us note that the above description is strictly analogous to the situation of a two-level Hamiltonian that depends upon one parameter. This analogy can help to understand the propagation in the adiabatic regime, as discussed in the following.

Time-dependent Hamiltonian in adiabatic approximation: Let us consider a time-dependent Hamiltonian and introduce the eigenstates and eigenvalues

$$H(t)|n(t)\rangle = E_n(t)|n(t)\rangle \tag{185}$$

and suppose the Hamiltonian is not degenerate.

Let us derive w.r.t. time

$$\dot{H}|n\rangle + H|\dot{n}\rangle = \dot{E}_n|n\rangle + E_n|\dot{n}\rangle. \tag{186}$$

Then we multiply by $\langle m|$ with $n \neq m$,

$$\begin{aligned} l.h.s &= \langle m|\dot{H}|n\rangle + \langle m|H|\dot{n}\rangle = \langle m|\dot{H}|n\rangle + E_m\langle m|\dot{n}\rangle, \\ r.h.s &= \dot{E}_n\langle m|n\rangle + E_n\langle m|\dot{n}\rangle = E_n\langle m|\dot{n}\rangle. \end{aligned} \tag{187}$$

So we find

$$\langle m|\dot{n}\rangle = \frac{\langle m|\dot{H}|n\rangle}{E_n - E_m} \quad \text{if } n \neq m. \tag{188}$$

Thus the eigenstate acquires overlap with other eigenstates if the derivative of H is large in comparison to the energy differences.

Instead in the case $n = m$, we have

$$\langle n|n\rangle = 1 \Rightarrow \langle\dot{n}|n\rangle + \langle n|\dot{n}\rangle = 0 \Rightarrow \langle n|\dot{n}\rangle = -\langle n|\dot{n}\rangle^*. \tag{189}$$

In other words, this number is imaginary. We define

$$\eta_n = i\langle n|\dot{n}\rangle \in \boldsymbol{R}. \tag{190}$$

Now we can write the generic states as

$$|\psi\rangle = \sum_n e^{-iS_n} c_n|n\rangle, \text{ where } S_n(t) = \int_0^t \mathrm{d}\tau\ E_n(\tau). \tag{191}$$

Its time evolution is dictated by the Schrödinger equation $i|\dot{\psi}\rangle = H|\psi\rangle$, this implies

$$\sum_n \dot{S}_n e^{-iS_n} c_n|n\rangle + i\sum_n \left(e^{-iS_n}\dot{c_n}|n\rangle + e^{-iS_n}c_n|\dot{n}\rangle\right) = \sum_n e^{-iS_n} c_n H\,|n\rangle. \tag{192}$$

The first and last term cancel out. Multiplying by $\langle m|$, we have

$$\dot{c}_m = -\sum_n e^{i(S_m - S_n)} c_n\langle m|\dot{n}\rangle = -c_m\langle m|\dot{m}\rangle - \sum_{n\neq m} e^{i(S_m - S_n)} c_n\langle m|\dot{n}\rangle. \tag{193}$$

Using the above result,

$$\dot{c}_m = -c_m\langle m|\dot{m}\rangle - \sum_{n\neq m} c_n \frac{\langle m|\dot{H}|n\rangle}{E_n - E_m} e^{i(S_m - S_n)}. \tag{194}$$

If the second term is negligible, namely in the adiabatic case, we have

$$c_m(t) = c_m(0) \exp\left[i \int_0^t d\tau \; \eta_m(\tau)\right], \tag{195}$$

namely, $|c_m|^2$ does not change with time.

From the time evolution of the mass eigenstates in Eq. (180), we have

$$i\partial_x n_2 = \frac{k_m}{2}\left(i\frac{2\dot{\theta}_m}{k_m} n_1 + n_2\right) = \frac{k_m}{2}(i\gamma n_1 + n_2), \tag{196}$$

where γ is the adiabatic parameter in Eq. (181). Since in the adiabatic approximation the limit $\gamma \ll 1$ holds, the mass eigenstate time evolution is

$$\begin{cases} i\,\partial_x n_1 = -\frac{k_m}{2}\, n_1, \\ i\,\partial_x n_2 = \frac{k_m}{2}\, n_2, \end{cases} \tag{197}$$

i.e., the time evolutions (time and spatial evolutions are equivalent in the relativistic regime) of the two eigenstates are independent of one another. As a consequence, in the adiabatic approximation, if $\nu_e \approx n_2$ in the production point (e.g., for solar neutrinos), ν_e remains in that eigenstate during the whole adiabatic evolution.

To summarize, the case of adiabatic propagation applies if the probability of transition between the two eigenstates is always negligible, even when the two levels get as close as possible without crossing each other (a sub-case called "avoided crossing" in quantum mechanics).

Effect of the Earth on neutrino propagation: As we have just seen, in the case of solar ν_e, we expect to receive on Earth just the mass eigenstate ν_2 above a certain energy (adiabatic conversion). However, it is possible that neutrinos reach the detector passing through the Earth, i.e., in night time. Therefore, it is interesting to calculate the probability of conversions from ν_2 to ν_e in two-flavor approximation. This has a very simple and useful expression in the case of constant matter density that we discuss here.

In the two-flavor basis, for the time-independent hamiltonian in Eq. (175), the evolutor that describes the effect of the propagation in the Earth is

$$\begin{aligned} \mathcal{U} &= U_m \operatorname{diag}[e^{i\varphi_m}, e^{-i\varphi_m}]\; U_m^{\mathrm{t}} \\ &= \begin{pmatrix} c_m + i s_m \cos 2\theta_m & -i s_m \sin 2\theta_m \\ -i s_m \sin 2\theta_m & c_m - i s_m \cos 2\theta_m \end{pmatrix}, \end{aligned} \tag{198}$$

where $\varphi_m = \Delta m_m^2 L/(4E)$, $c_m = \cos\varphi_m$, $s_m = \sin\varphi_m$. Consider the case of solar neutrinos; the initial state is ν_2, which in flavor space is represented by $(\sin\theta, \cos\theta)$, and the final state is ν_e, which in flavor space is represented by $(1, 0)$. The relevant matrix element of the evolutor is

$$\begin{aligned}\mathcal{U}_{2\to e} &= (c_m + i s_m \cos 2\theta_m)\sin\theta - i s_m \sin 2\theta_m \cos\theta \\ &= c_m \sin\theta - i s_m \sin(2\theta_m - \theta).\end{aligned} \tag{199}$$

The last factor can be rewritten using the formulae for the mixing angles in matter,

$$\sin(2\theta_m - \theta) = \sin\theta \, \frac{1+\varepsilon}{\sqrt{(1+\varepsilon)^2 - 4\varepsilon\cos^2\theta}}, \tag{200}$$

where as usual $\varepsilon = V/k = 2EV/\Delta m^2$. With a bit of algebra, we obtain the desired result,

$$P_{\nu_2\to\nu_e} = |\mathcal{U}_{2\to e}|^2 = \sin^2\theta \left[1 + 4\varepsilon\ \cos^2\theta \sin^2\left(\frac{\Delta m^2 L}{4E}\xi\right) / \xi^2\right] \tag{201}$$

with

$$\xi = \sqrt{(1+\varepsilon)^2 - 4\varepsilon\cos^2\theta}, \tag{202}$$

where of course we have in mind the so-called "solar neutrino mass splitting" (the lowest one) and θ is in good approximation, $\theta_{12} \approx 33°$.

4.3. *Applications and examples*

4.3.1. *High-energy atmospheric neutrinos*

In this section, taken from Ref. 21, we examine the oscillations of high-energy atmospheric neutrinos. This is interesting, since one can discriminate the neutrino mass hierarchy by observing the details of the matter effect.

We begin by using some simplifying assumptions, postponing derivations and refinements. Let us consider oscillations with a single scale, and let us consider oscillations in constant matter density. In fact, the second hypothesis is rather inaccurate in the conditions which we are interested in for the experiments, and a more accurate evaluation of the oscillation probabilities for the actual calculations is required. However, a qualitative discussion based on simple-minded analytical results is sufficient to correctly understand the relevant features.

Under the above assumptions, the probability that a muon neutrino gets converted into an electron neutrino is simply

$$P_{\mu e} = \sin^2\theta_{23}\ \sin^2 2\widetilde{\theta_{13}}\ \sin^2\widetilde{\varphi} \quad \text{with} \quad \widetilde{\varphi} = \frac{\widetilde{\Delta m^2} L}{4E}. \tag{203}$$

Most of the results which we are interested in follow from the above simple formula. In Eq. (203), we introduced the usual matter-modified mixing angle and squared-mass difference,

$$\begin{cases} \sin 2\widetilde{\theta_{13}} = \sin 2\theta_{13}\ /\Delta, \\ \cos 2\widetilde{\theta_{13}} = (\cos 2\theta_{13} - \varepsilon)/\Delta, \\ \widetilde{\Delta m^2} = \Delta m^2 \times \Delta, \end{cases} \tag{204}$$

where

$$\Delta = \pm\sqrt{(\cos 2\theta_{13} - \varepsilon)^2 + \sin^2 2\theta_{13}}. \tag{205}$$

The sign of Δ is a matter of convention; the ratio between matter and the vacuum term is,

$$\varepsilon \equiv \pm\frac{\sqrt{2}G_{\rm F}n_e}{\Delta m^2/(2E)} \approx \pm\frac{\rho}{5.5\,\text{g/cm}^3} \times \frac{Y_e}{1/2} \times \frac{2.4\times 10^{-3}\,\text{eV}^2}{\Delta m^2} \times \frac{E}{5.5\,\text{GeV}}, \tag{206}$$

where $G_{\rm F}$ is the Fermi coupling and we identify Δm^2 with Δm^2_{23}. Now, instead, the sign is important: It is "plus" for normal hierarchy and "minus" for inverted hierarchy. Considering the average matter density of the Earth $\rho = 5.5\,\text{g/cm}^3$ and $Y_e = 1/2$, we get $n_e = 1.7\times 10^{24} e^-/\text{cm}^3$ for the electronic density. Thus, the characteristic length of MSW theory is

$$L_* \equiv \frac{1}{\sqrt{2}G_{\rm F}n_e} \sim 1000\,\text{km}. \tag{207}$$

We see that, for normal hierarchy, the maximum of $P_{\mu e}$ is obtained when (1) Δ is as small as possible, in order to maximize $\sin 2\widetilde{\theta_{13}}$; moreover (2) the phase of propagation is $\widetilde{\varphi} \sim \pi/2$. These conditions are met when the neutrino energy and the propagation distance are

$$E_{\max} = \frac{\Delta m^2 L_*}{2}\cos 2\theta_{13} \quad \text{and} \quad L_{\max} = \frac{\pi L_*}{\tan 2\theta_{13}} \tag{208}$$

or, numerically,

$$E_{\max} \sim 5.5\,\text{GeV} \quad \text{and} \quad L_{\max} \sim 9000\,\text{km}. \tag{209}$$

In the case of inverted hierarchy, the matter effect depresses $P_{\mu e}$, which becomes negligible.

As we have told, one would like to observe the matter effect. From the point of view of the experiments, one direct approach is to check the simple prediction concerning $P_{\mu e}$. However, there are some practical advantages to detect muons rather than electrons. Then, let us consider the survival probability $P_{\mu\mu}$, focussing again on the normal hierarchy case. Let us discuss the case when the first local maximum of $P_{\mu\mu}$ is as small as possible. This results from $P_{\mu\tau}$ and from $P_{\mu e}$, since $P_{\mu\mu} = 1 - P_{\mu\tau} - P_{\mu e}$. We are interested in the case when the minimum of $P_{\mu\tau}$ happens in the vicinity of the energy identified in Eq. (209). When the phase of oscillation of $P_{\mu\tau}$ is close to the vacuum phase, the condition $\Delta m^2 L/(2E_{\max}) = 2\pi$ gives $L \sim 6000\,\text{km}$. This suggests that the distance that amplifies the matter effect on $P_{\mu\mu}$ is between 6000 and 9000 km, as confirmed by more complete analyses. These points that were obtained using the web interface mentioned in Section 4.1.2 are illustrated in Fig. 8.

Description of atmospheric neutrino oscillations at various levels of accuracy: The general case requires a numerical solution based on Eq. (173). However, when the "solar" Δm^2_{12} is set to zero, i.e., when its effects are negligible — the only nonzero out-of-diagonal elements u_{ij} in Eq. (173) are u_{13} and u_{31}. The CP-violating phase δ drops out from the probabilities

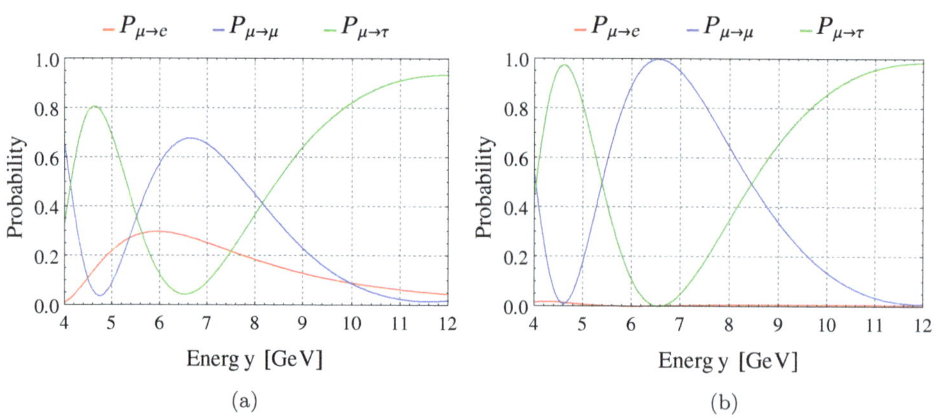

Fig. 8. Oscillation probabilities in the Earth for a path of $L = 7000$ km: the blue line is $P_{\nu_\mu\to\nu_\mu}$; the red line, $P_{\nu_\mu\to\nu_e}$; the green one, $P_{\nu_\mu\to\nu_\tau}$. The two panels refer to normal and inverted hierarchies, respectively. The increase of $P_{\nu_\mu\to\nu_e}$ for normal hierarchy, due to matter effect, causes the decrease of $P_{\nu_\mu\to\nu_\mu}$. (a) Normal hierarchy. (b) Inverted hierarchy.

$P_{\ell\ell'} = |\mathcal{U}_{\ell'\ell}|^2$, which moreover becomes symmetric, $P_{\ell\ell'} = P_{\ell'\ell}$ for each $\ell, \ell' = e, \mu, \tau$. Therefore, in this approximation, we have three independent probabilities and all the other ones are fixed. We can choose, e.g.,

$$\begin{cases} P_{e\mu} = \sin^2\theta_{23}|u_{13}|^2 \\ P_{e\tau} = \cos^2\theta_{23}|u_{13}|^2 \\ P_{\mu\tau} = \sin^2\theta_{23}\cos^2\theta_{23}|u_{33} - u_{22}|^2 \end{cases} \tag{210}$$

so that, e.g., $P_{ee} = 1 - P_{\mu e} - P_{\tau e} = |u_{11}|^2$. From these formulae, we obtain

$$P_{\mu e} = \sin^2\theta_{23}\,(1 - P_{ee}) \quad \text{and} \quad P_{\mu\tau} = \frac{1}{4}\sin^2 2\theta_{23}\left|1 - \sqrt{P_{ee}}\,e^{i\hat{\varphi}}\right|^2, \tag{211}$$

where $\hat{\varphi}$ is a (rapidly varying) phase factor. Two important remarks are in order:

(1) The last equation shows that $P_{\mu e}$ is large in the region where P_{ee} is small, and that $P_{\mu\tau}$ remains close to zero in the first nontrivial minimum near $\hat{\varphi} = 2\pi$, even when $P_{ee} \approx 0.3$–0.4 due to matter effect.
(2) The sign of Δm^2 controls the sign of the vacuum Hamiltonian; therefore, switching between the two mass hierarchies or switching between neutrinos and antineutrinos has the same effect, e.g., $P_{e\mu}(\text{IH}) = P_{\bar{e}\bar{\mu}}(\text{NH})$.

The first remark is consistent with our numerical findings that $P_{\mu e}$ is amplified and $P_{\mu\tau}$ does not deviate strongly from its behavior in vacuum in the conditions that are relevant for our discussion.

Proceeding further with the approximations, and considering at this point the case of constant matter density, we obtain simple and closed expressions. For the case of normal mass hierarchy, they read

$$\begin{cases} u_{13} = u_{31} = -i\sin\widetilde{\varphi}\sin\widetilde{2\theta_{13}}, \\ u_{11} = \cos\widetilde{\varphi} + i\sin\widetilde{\varphi}\cos\widetilde{2\theta_{13}} = u_{33}^*, \\ u_{22} = \cos\widetilde{\varphi}' + i\sin\widetilde{\varphi}', \end{cases} \tag{212}$$

where

$$\widetilde{\varphi}' = \frac{\Delta m^2 L}{4E}(1 + \varepsilon). \tag{213}$$

From Eqs. (210) and (212), we recover the expression of Eq. (203), used in the above discussion. In the approximation of constant matter density, the phase $\hat{\varphi}$ entering the expression of the probability $P_{\mu\tau}$ is given by $\sqrt{P_{ee}}\cos\hat{\varphi} \equiv \cos\widetilde{\varphi}\cos\widetilde{\varphi}' - \sin\widetilde{\varphi}\sin\widetilde{\varphi}'\cos\widetilde{2\theta_{13}}$. This is close to the vacuum phase when

ε is large or small in comparison to 1: in fact, we have $\cos 2\widetilde{\theta_{13}} \sim \pm 1$ and $\widetilde{\varphi} \sim \pm \Delta m^2 L/(4E)(1-\varepsilon)$ from Eq. (204) so that $\cos\hat{\varphi} \sim \cos[\Delta m^2 L/(2E)]$.

4.3.2. *Solar neutrinos*

Let us discuss how the adiabatic solution applies to solar neutrinos. First of all, we note that the expected matter density profile is approximately exponential, $n_e \propto \exp(-10x/R_\odot)$, where $R_\odot = 7 \times 10^8\,\text{m}$ is the solar radius. Considering the mass difference squared, $\Delta m^2_{12} = 8 \times 10^{-5}\,\text{eV}^2$ and $\sin^2 2\theta_{12} \sim 0.8$, Eq. (182) reads

$$\gamma \approx \left| \frac{10\ c_2}{kR_\odot\ s_2^2} \right| \ll 1 \tag{214}$$

for any possible relevant energy $E = 0.1\text{–}20\,\text{MeV}$. Thus, the local mass eigenstates n_1 and n_2 remain so always, just as described in Eq. (183); therefore, the result of Eq. (184) follows.

At this point, we can interpret solar neutrinos. Let us begin by noting that we expect, *a priori*, that matter effects are important only for solar neutrinos of high energy, see Eq. (169). Given the density of electrons in the center of the Sun, $100\,\text{mol/cm}^3$, the mixing angle in matter is about $90°$, see Eq. (177). Thus, since $s_m \approx 1$ and $c_m \approx 0$, the adiabatic formula for P_{ee} simply becomes

$$P_{ee} = \sin^2\theta_{12}, \tag{215}$$

and there is no problem of explaining the solar neutrino observations that indicate, at high energies, that the suppression factor of electron neutrinos is about 1/3. At this point, we know that the right solution is the one with $\theta_{12} \sim 30°$ (because if we choose the other one that is allowed by vacuum neutrino oscillations, with $\theta_{12} \sim 60°$, we would get 2/3 instead). Evidently, at low energies, we recover the vacuum oscillation formula,

$$P_{ee} = 1 - \frac{\sin^2 2\theta_{12}}{2}. \tag{216}$$

In fact, making reference to the cases of Eq. (177), the limit $\theta \sim \theta_m$ applies.

4.3.3. *Supernova neutrinos*

Neutrinos from the gravitational collapse have energies from a few MeV to almost 100 MeV, propagate at distances of the order of kpc and are produced in nuclear density regions, billion times larger than solar ones. They certainly undergo some oscillation effects. However, it is not clear, to date, to what extent it is possible to separate the effects of the oscillations from the astrophysical uncertainties since the latter are very large. We will limit ourselves to show here (without deriving them) a few basic formulas in view of the fact that neutrinos of this type do exist and are detectable.

First of all, we calculate the neutrino masses from the Hamiltonian of propagation given in Eq. (163) as a function of the electron density. These masses depend upon the neutrino masses in vacuum, energy of the neutrino, electron density and mixing angles θ_{12} and θ_{13} (see discussion in Eq. (173)). The result for the energy of a typical supernova neutrino is given in Fig. 9, assuming the case of normal mass hierarchy. We see that if the density of electrons is large enough, we will have two "level crossings" or "resonances". While in the case of solar neutrinos we had only one (of course, the one occurring for lowest densities), in the case of supernova, we will have both of them, since in the center of the star (where the neutrinos are produced) we reach nuclear densities.

Then, we consider a generalization of the two flavors discussion of neutrino oscillations given in Section 4.2.2. For generic values of the mixing angle θ and the corresponding Δm^2, the matrix containing the probabilities of survival or of conversion, averaged over the large phases, is

$$P = \begin{pmatrix} c^2 & s^2 \\ s^2 & c^2 \end{pmatrix} \begin{pmatrix} 1 - P_{\rm f} & P_{\rm f} \\ P_{\rm f} & 1 - P_{\rm f} \end{pmatrix} \begin{pmatrix} c_m^2 & s_m^2 \\ s_m^2 & c_m^2 \end{pmatrix}. \tag{217}$$

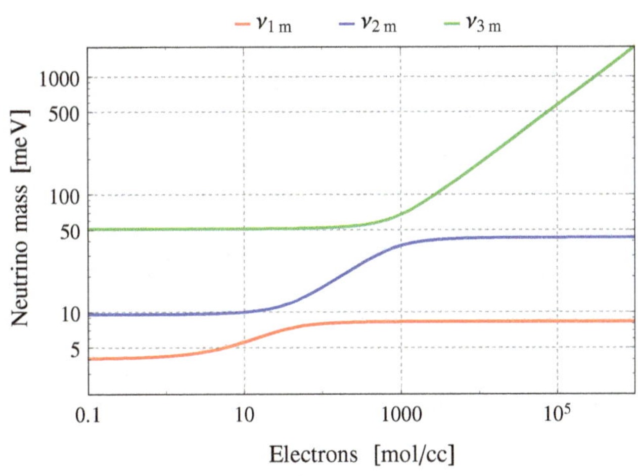

Fig. 9. Neutrino masses calculated from the Hamiltonian of propagation given in Eq. (163) and given as a function of the electron density for a typical energy of supernova neutrinos, $E_\nu = 20\,\text{MeV}$. We assumed normal mass hierarchy and fixed the lightest neutrino mass to be $m_1 = 4\,\text{meV}$ just for illustration purposes.

The probability $P_{\rm f}$ is called *flip probability* and describes the possibility that there is a deviation from the relations given in Eq. (183). It is easy to verify that, in the limit $P_{\rm f} \to 0$, the above formula yields P_{ee} as in Eq. (184). In fact, as we have checked in the case of solar neutrinos, the adiabaticity parameter γ is very small, and this boils down to the conclusion that $P_{\rm f} \to 0$.

The generalization to the three-flavor case is quite direct but requires knowledge of the type of neutrino mass hierarchy. In fact, we know from the solar neutrino oscillations in matter that ν_e is mostly in ν_1, which corresponds to the choice of the angle θ_{12} discussed in the previous paragraph. Conversely, we still do not know for sure whether the neutrino ν_3, which is responsible for atmospheric oscillations, is the heaviest (=direct or normal mass hierarchy) or the lightest (=inverse mass hierarchy), see Fig. 3. Assuming the first case, which is slightly favored at present and also the most plausible theoretically, one finds[22]

$$\begin{aligned} P = &\begin{pmatrix} |U_{e1}^2| & |U_{e2}^2| & |U_{e3}^2| \\ |U_{\mu1}^2| & |U_{\mu2}^2| & |U_{\mu3}^2| \\ |U_{\tau1}^2| & |U_{\tau3}^2| & |U_{\tau3}^2| \end{pmatrix} \cdot \begin{pmatrix} 1-P_{\rm S} & P_{\rm S} & 0 \\ P_{\rm S} & 1-P_{\rm S} & 0 \\ 0 & 0 & 1 \end{pmatrix} \cdot \\ &\cdot \begin{pmatrix} 1 & 0 & 0 \\ 0 & 1-P_{\rm A} & P_{\rm A} \\ 0 & P_{\rm A} & 1-P_{\rm A} \end{pmatrix} \cdot \begin{pmatrix} |U_{e1,m}^2| & |U_{e2,m}^2| & |U_{e3,m}^2| \\ |U_{\mu1,m}^2| & |U_{\mu2,m}^2| & |U_{\mu3,m}^2| \\ |U_{\tau1,m}^2| & |U_{\tau3,m}^2| & |U_{\tau3,m}^2| \end{pmatrix} . \end{aligned} \tag{218}$$

We see that now there are two flip probabilities. The first one is $P_{\rm A}$ connected to the "atmospheric" parameters (the delta mass squared Δm_{23}^2 and θ_{13}); the second is $P_{\rm S}$ connected to the "solar" parameters (the delta mass squared Δm_{12}^2 and θ_{12}). From this expression, we derive the electron survival probability,

$$P_{ee} = (1-P_{\rm A})|U_{e3}^2| + P_{\rm A}\left[\,(1-P_{\rm S})|U_{e2}^2| + P_{\rm S}|U_{e1}^2|\,\right]. \tag{219}$$

As in the case for solar neutrinos, we also expect that $P_{\rm S} \sim 0$ in the case of the supernova. The same is expected to be true for the other flip probability, $P_{\rm A} \sim 0$.[dd] Therefore, we conclude

$$P_{ee} \sim |U_{e3}^2| \sim 0. \tag{220}$$

[dd]This is expected to be true in all reasonable cases; the only possible (unlikely) exception is if the electron density changes very sharply.

This implies that the electron neutrinos that we detect at Earth were produced at the source as muon or tau neutrinos.

It is possible to recap this conclusion with the help of Fig. 9. Let us focus on the continuous (green) line that shows the evolution of the heaviest mass eigenstate ν_{3m} from the vacuum (lower electronic densities) to the innermost part of the star (highest electronic densities). We note that after a certain density ($n_e \sim 1000\,\mathrm{mol/cm^3}$), the mass of ν_{3m} increases linearly with the electron density (see Fig. 9). In fact, the matter term in Eq. (163) is the largest part of the Hamiltonian of propagation in the center of the star; it is linear in n_e, and evidently the heaviest eigenstate of the matter term is just the electron neutrino. This means that an electron neutrino, produced in the center of the star, is just the heaviest mass eigenstate of the full Hamiltonian in very good approximation, $|\nu_e\rangle \approx |\nu_{3m}\rangle$. The adiabatic evolution implies that a neutrino follows the evolution of the mass eigenstate; in our case, the electron neutrino produced in the center of the star always remains the heaviest mass eigenstate, $|\nu_e\rangle \approx |\nu_{3m}(t)\rangle$, for all times (up to an irrelevant overall phase) and exits from the star as $|\nu_3\rangle$. Thus, its survival probability is $|\langle\nu_e|\nu_3\rangle|^2 = |U_{e3}^2|$.

4.4. *Earth matter effect at solar and supernova energies*

At solar neutrino energies, the probability $P_{\nu_2\to\nu_e}$ given in Eq. (201) admits the simple numerical approximation

$$P_{\nu_2\to\nu_e} \approx 0.3 + 1.7\%\ \kappa\ \sin^2\left(\frac{\Delta m^2 L}{4E}\right) \tag{221}$$

with

$$\kappa = \left(\frac{7.37\times 10^{-5}\,\mathrm{eV}^2}{\Delta m^2}\right)\left(\frac{n_e}{2\,\mathrm{mol/cm^3}}\right)\left(\frac{E}{5\,\mathrm{MeV}}\right). \tag{222}$$

Therefore, we see that the detectors receive (slightly) more electron neutrinos in night than in day, which is the reason why this is called *day–night* effect or *regeneration* (adopting neutral kaon's terminology). This feature remains true on average when we consider the full description of the Earth density.

Something similar is expected to happen in the case of supernova neutrinos. In this case the typical energy is 20–30 MeV, larger than the one of solar neutrinos; thus, if neutrinos pass through the Earth, the effect is larger. In the approximation described above (Section 4.3.3), when we start from ν_e, we expect to receive on Earth ν_3 (or ν_2) for normal (or inverse) mass hierarchy; instead, $\bar{\nu}_e$ becomes $\bar{\nu}_1$ (or $\bar{\nu}_3$). If we receive ν_3

or $\bar{\nu}_3$, the relevant vacuum term (related to the larger value of Δm^2) is too large at these energies and no effect is expected. In the other cases, we can use almost immediately the above formula: In fact, we have the relation $P_{\nu_1 \to \nu_e} = 1 - P_{\nu_2 \to \nu_e}$, and we can obtain $P_{\bar{\nu}_1 \to \bar{\nu}_e}$ from $P_{\nu_1 \to \nu_e}$ by flipping the sign of the matter term; therefore it is sufficient to calculate just one of these probabilities and we can immediately use the above result. Note the curious fact that $P_{\bar{\nu}_1 \to \bar{\nu}_e}$ can be readily obtained from $P_{\nu_2 \to \nu_e}$ given above by simply exchanging $\theta \to \pi/2 - \theta$. This allows us to describe $\bar{\nu}_e$ in normal hierarchy and to understand that $\bar{\nu}_e$ are expected to be "regenerated" if they reach the detector after passing through the Earth.

For an exhaustive discussion in connection with solar neutrinos, see Ref. 33; for a study concerning supernova neutrinos, see Ref. 34. An eloquent animation is available at

http://www.bo.infn.it/selvi/aniMSWearth.gif.

5. Summary and discussion

Neutrino oscillations are an important topic of high-energy particle physics. With these notes, we have introduced the reader to the main points and results. The usefulness of these results has been illustrated by examining specific cases, applications, solutions of the equations of propagation and by discussing various formal developments. Moreover, we have thoroughly discussed the conceptual context and collected in the appendix various considerations concerning the description of relativistic fermions that complement a more or less standard knowledge on Dirac matrices from a course on high-energy/particle physics. In this brief summary, we list the most important basic equations concerning neutrino oscillations and related facts; we overview the status of the field and we remark once again the introductory character of these notes in the hope of encouraging the reader to join the research on neutrino physics.

Survival kit: For the purpose of recollecting the main facts and of allowing one to test the overall understanding of the expounded material, we offer our selection of the basic formulae discussed in the text that a reader is supposed to master at this point:

(1) Definition of the mixing matrix, Eq. (17), and its role in the expressions of neutrino and antineutrino states, Eq. (34).
(2) Standard expression of the mixing matrix, Eq. (51), values of the parameters, Table 2, explanation of the part that matters for oscillations, Eq. (38).

(3) Amplitude of transition for vacuum oscillations, Eq. (66).
(4) General expression of the probabilities of transition, Eq. (74).
(5) Numerical factor, 1.267, Eq. (83).
(6) Matter term, Eq. (162), and its numerical value, Eqs. (169) and (206).
(7) Vacuum Hamiltonians for the evolution of the states, Eq. (96) and also Eq. (99), matter Hamiltonians, Eq. (163).

Neutrino oscillations today: A detailed knowledge of this phenomenon is crucial to correctly interpret several results concerning neutrino experiments, either in laboratory or in the context of neutrino astronomy. In similar cases, it is necessary to master the formulae expounded here (or possibly their extensions), to dispose of a detailed model of the source and of the detector, to adopt suitable statistical procedures for data analysis. Of course, these formulae are necessary to investigate the value of the neutrino parameters: The hottest topics are the discrimination of neutrino mass hierarchy, a precise measurement of CP violation, the assessment of the deviation of θ_{23} from 45°. Moreover, they are needed to describe the impact of flavor transformations on the expectations, search for *other* neutrinos besides the usual ones,[ee] search for new "matter effects", hunt for (speculative) effects of neutrino magnetic moments, etc. The axiomatic aspects of neutrino oscillations are still considered a matter of debate, and a detailed description of the non-relativistic limit on neutrino oscillations is regarded with interest, even if the contribution of these discussions to the field does not have crucial importance. Moreover, in general, the resemblance of oscillations with other phenomena concerning the light suggests the interest in investigating the possibilities of analogous phenomena, such as neutrino reflection. The understanding of the flavor transformation of neutrinos in supernovae, where the neutrino density also plays a role (and the magnetic field is very high), is a very active area of research.

What else? There are several topics that we have not treated. For example, there are several useful analytical works to make more efficient the calculation of the oscillation effects; there are a lot of solutions that have been worked out and discussed in neutrino oscillations; furthermore, there are a lot of useful points of view and formalisms, concerning not only the advanced topics but also the basic ones. In this connection, an important and general question is: Where to get more information on neutrino oscillations? In view of the introductive character of these notes, we list here only a few

[ee]These are often called "sterile" neutrinos. We have various hints to date but no significant evidences or compelling theoretical indications for them.[23,24]

books[25–28] and review papers.[10,20,29–32] We offer them to the reader since we have found them useful, but we are deeply aware of the incompleteness of this list and of the subjective character of this choice. To conclude, we note that neutrino oscillations are, to date, the only successful probe we have of neutrino masses. However, they do not allow us to probe the lightest neutrino mass or to investigate the nature of the mass (either Dirac or Majorana). Likewise, the evidence of neutrino mass raises the question of their meaning in extensions of the standard model of the electroweak particles. All these important aspects are discussed elsewhere in this book.

Appendix. Special results on Dirac matrices

A.1. *All Dirac matrices are equivalent* (*Pauli theorem*)

Gamma matrices in physics: In physics, the Dirac γ matrices are labeled by $\mu = 0, 1, 2, 3$: γ_0 is Hermitian while the other three are anti-Hermitian. They obey $\gamma_\mu\gamma_\nu + \gamma_\nu\gamma_\mu = 2\eta_{\mu\nu}\mathbb{1}$, where $\eta = \text{diag}(1, -1, -1, -1)$. Thus, a one-to-one connection with a set of Hermitian γ_a matrices (with $a = 1, 2, 3, 4$) is possible, by setting $\gamma_4 \equiv \gamma_0$ and multiplying $\gamma_{1,2,3}$ by a factor i (or $-i$). They obey the same algebra where η_{ab} is replaced by the Kronecker δ_{ab}.

Theorem 1. *Any set of four Hermitian matrices that satisfy the algebra,*

$$\gamma_a\gamma_b + \gamma_b\gamma_a = 2\delta_{ab}\mathbb{1}, \quad \textit{where} \quad a, b = 1, 2, 3, 4, \tag{A.1}$$

is unitary equivalent, in the sense that γ_a *and* γ'_a *being two different sets, it exists as a unitary matrix* U *with*

$$\gamma'_a = U\gamma_a U^\dagger. \tag{A.2}$$

Proof. There are various proofs[ff] and we choose an inelegant one that, however, has the merit of being very direct and of using elementary and important considerations. Before starting, let us stress that if γ_a are a set of such matrices, any unitary matrix U defines another valid set of matrices $\gamma' \equiv U\gamma U^\dagger$. Then it is enough to prove that such a set of matrices exists and can be written in a universal form by a choice of the basis, i.e., by a unitary transformation.

[ff] For example, use a result of finite group theory: *given a group of order* N_G, *the dimensions of all its irreducible representations* d_i *satisfy* $\sum_{i\geq 1} d_i^2 = N_G$. Consider the group of order $N_G = 32$ with elements $1, \gamma_1, \gamma_2, \ldots, \gamma_1\gamma_2, \ldots, -1, -\gamma_1, \ldots, -\gamma_1\gamma_2\gamma_3\gamma_4$. Note that we have one four-dimensional representation and also the trivial one. Thus, $32 = 4^2 + 1 + \sum_{i\geq 3} d_i^2$, and there is no room for other representations with $d_i \geq 4$.

(1) The first step is to look for a set of $\gamma_a^\dagger = \gamma_a$ matrices and note that $\gamma_a^2 = \gamma_a^\dagger \gamma_a = \mathbb{1}$ implies that the matrices we are looking for are also unitary. Moreover, their eigenvalues are ± 1. Now,

$$\gamma_1 = -\gamma_2\gamma_1\gamma_2 \Rightarrow \operatorname{tr}(\gamma_1) = -\operatorname{tr}\left(\gamma_1\gamma_2^2\right) \Rightarrow \operatorname{tr}(\gamma_1) = 0. \tag{A.3}$$

Thus, we see that the dimension of the space should be even.[gg] Now, it is impossible to realize this algebra in $d = 2$ dimensions: three matrices could be the Pauli ones, but the fourth one cannot be realized. So, we need at least $d = 4$ dimensions.

(2) We assume $d = 4$ and choose the first matrix to be diagonal,

$$\gamma_1 = \operatorname{diag}(1, 1, -1, -1). \tag{A.4}$$

By imposing the anticommutation condition, the other three Hermitian matrices are of the form

$$\gamma_i = \begin{pmatrix} 0 & a_i \\ a_i^\dagger & 0 \end{pmatrix}, \quad i = 2, 3, 4, \tag{A.5}$$

and from $\gamma_i^2 = 1$, we find that the 2×2 matrices a_i are unitary. Of course, these are three different a_i; in the rest of the proof, we will fix the matrices a_i for $\gamma_{2,3,4}$.

(3) We are free to change basis, without changing the form of γ_1, using the subset of transformations,

$$\begin{pmatrix} u & 0 \\ 0 & v \end{pmatrix}. \tag{A.6}$$

where u and v are unitary and 2×2. Under this assumption, the a_i matrices transform as

$$a_i \to u\, a_i\, v^\dagger. \tag{A.7}$$

If we choose $u = v$, we can make one a_i diagonal, say, $a_2 = \operatorname{diag}(e^{i\alpha}, e^{i\beta})$. Using at this point $u = \mathbb{1}$ and $v = \operatorname{diag}(e^{i\alpha}, e^{i\beta}) = a_2$, we obtain

$$\gamma_2 = \begin{pmatrix} 0 & \mathbb{1} \\ \mathbb{1} & 0 \end{pmatrix}. \tag{A.8}$$

Thus, we have set the matrix γ_2 in a standard form. Note that with this position we retain the invariance of γ_1 and γ_2 under any matrices $u = v$ that we use from here on.

[gg]Please recall that $\operatorname{tr}(abc) = \operatorname{tr}(bca)$ and $h^\dagger = h \Rightarrow \operatorname{tr}(h) = \sum_i \lambda_i$, where λ_i are the eigenvalues of h.

(4) When we impose the anticommutation conditions of $\gamma_{3,4}$ with γ_2, we find

$$a_i + a_i^\dagger = 0, \quad i = 3, 4. \tag{A.9}$$

Therefore, we have that the last two a_i matrices are anti-Hermitian and satisfy

$$a_i = i\,\vec{m}_i \cdot \vec{\sigma} = i\sum_{j=1}^{3}(\vec{m}_i)_j\,\sigma_j, \text{ where } \vec{m}_i \in \mathbb{R}^3 \quad \text{and} \quad \vec{m}_i^2 = 1, \tag{A.10}$$

where σ_i are the Pauli matrices and the last condition follows from unitarity. A change of basis,

$$u = e^{i\frac{\vec{\alpha}\cdot\vec{\sigma}}{2}} = \cos(\alpha/2) + i\,\vec{n}\cdot\vec{\sigma}\,\sin(\alpha/2), \tag{A.11}$$

where

$$\alpha = |\vec{\alpha}| \quad \text{and} \quad \vec{n} = \vec{\alpha}/\alpha, \tag{A.12}$$

works as a rotation acting on $\vec{m}$ and allows us to choose $a_3 = i\sigma_1$ that defines γ_3. At this point, we still have freedom to change the basis without changing $\gamma_{1,2,3}$, by choosing $u = \exp[i\,\beta\sigma_1/2]$. In this manner, we can set $a_4 = i\sigma_2$, fixing the form of γ_4. □

To summarize, we have shown that given a set of Hermitian matrices γ that obey the anticommutative algebra, it is always possible to choose a suitable basis (i.e., to perform a unitary transformation) so that they are cast in a *universal* form: $U\gamma\,U^\dagger = \gamma_{\text{univ}}$. This implies that such sets of matrices are equivalent to each other.

Majorana representation: When we consider the above universal set of matrices and when we reinsert the factors i, we are free to set $\gamma_0' = \gamma_3$, $\gamma_1' = i\gamma_1$, $\gamma_2' = i\gamma_2$, $\gamma_3' = i\gamma_4$, which are all imaginary $(\gamma')^* = -\gamma'$. This set of matrices was introduced by Majorana and simplifies several equations. Note that the Hermiticity property is generic and invariant under a unitary change of basis, while the feature $(\gamma')^* = -\gamma'$ is a characteristic of Majorana basis and rather specific.

A.2. *Charge conjugation matrix*

An application of the Pauli theorem of great practical importance is the following one. Consider a given set of Dirac matrices γ_a. Since the opposite of their transposes, $-\gamma_a^{\rm t}$, satisfies the same algebra, we know that there is a unitary matrix C such that

$$-C^{-1}\gamma_a^{\rm t}C = \gamma_a. \tag{A.13}$$

From this property, we can deduce several useful forms, such as

$$\gamma^{\rm t}C = -C\gamma \quad \text{or} \quad C^{-1}\gamma^{\rm t} = -\gamma C^{-1}. \tag{A.14}$$

The unitary matrix C is called *charge conjugation matrix*, since it transforms particle into antiparticle and viceversa, without affecting the polarization, according to

$$\psi^{\rm c} = C(\bar{\psi})^{\rm t} \tag{A.15}$$

(see also the discussion after Eqs. (A.23) and (A.24)). Here we remark an important relation that involves C and the chiral projectors that allows us to handle the weak leptonic charged currents. This is

$$P_{\rm L}\psi^{\rm c} = C(\overline{\psi_{\rm R}})^{\rm t}, \tag{A.16}$$

where $P_{\rm L} = (1-\gamma_5)/2$ and $\gamma^5 = i\gamma^0\gamma^1\gamma^2\gamma^3$. The proof of this equality[hh] follows when one observes that

(1) $P_{\rm L}C = CP_{\rm L}^{\rm t}$ that is due to the definition of γ^5;
(2) $\gamma^0 P_{\rm L} = P_{\rm R}\gamma^0$, since $\gamma_\mu\gamma_5 = -\gamma_5\gamma_\mu$; and
(3) γ_5 is Hermitian that implies $P_{\rm L}^\dagger = P_{\rm L}$.

Note that this relation, in a more readable form, is

$$(\psi^{\rm c})_{\rm L} = (\psi_{\rm R})^{\rm c}. \tag{A.17}$$

Properties of the C matrix: By using the identities in Eq. (A.14), one can show that the matrix C is antisymmetric. This requires a bit of effort:

(1) The first step is to note that $\gamma C^{-1}C^{\rm t} = -C^{-1}\gamma^{\rm t}C^{\rm t} = C^{-1}C^{\rm t}\gamma$, namely $[C^{-1}C^{\rm t},\ \gamma] = 0$. As the γ matrices form a group under the anticommutation operation, for the Shur lemma, this implies that $C^{-1}C^{\rm t}$ is proportional to the identity $C^{\rm t} = \kappa C$.
(2) The next step is to again take the transpose, getting $C = (C^{\rm t})^{\rm t} = \kappa^2 C$, thus $\kappa = \pm 1$.

[hh]The complete proof is $P_{\rm L}C(\bar{\psi})^{\rm t} = CP_{\rm L}^{\rm t}(\bar{\psi})^{\rm t} = C(\bar{\psi}P_{\rm L})^{\rm t} = C(\psi^\dagger\gamma_0 P_{\rm L})^{\rm t} = C(\psi^\dagger P_{\rm R}\gamma_0)^{\rm t} = C((P_{\rm R}^\dagger\psi)^\dagger\gamma^0)^{\rm t} = C((P_{\rm R}\psi)^\dagger\gamma^0)^{\rm t} = C(\psi_{\rm R}^\dagger\gamma^0)^{\rm t} = C(\overline{\psi_{\rm R}})^{\rm t}$.

(3) The last step is the boring one. Considering the symmetry properties of 16 different combinations, we have $C^{\rm t} = \kappa C$, $(C\gamma_a)^{\rm t} = -\kappa C\gamma_a$, $(C[\gamma_a, \gamma_b])^{\rm t} = -\kappa C[\gamma_a, \gamma_b]$, $(C\gamma_5\gamma_a)^{\rm t} = \kappa C\gamma_5\gamma_a$, $(C\gamma_5)^{\rm t} = \kappa C\gamma_5$: There are 10 matrices with symmetry property $-\kappa$ and six matrices with symmetry property $+\kappa$. Now, a generic complex 4×4 matrix has 16 parameters. The 16 combinations introduced above are a complete basis for this space. Since, as known, a complex 4×4 symmetric matrix has 10 parameters, while the antisymmetric one has only six, we conclude that the case that occurs is $\kappa = -1$.

To summarize,

$$\begin{cases} C,\ C\gamma_5,\ C\gamma_5\gamma_a & \text{are antisymmetric matrices,} \\ C\gamma_a,\ C[\gamma_a, \gamma_b] & \text{are symmetric matrices.} \end{cases} \tag{A.18}$$

Notice the interesting consequence that, given a single (anticommuting) field χ, we can form only a few nonzero covariant quantities, namely, $\chi^{\rm t} C\chi$, $\chi^{\rm t} C\gamma_5\chi$, $\chi^{\rm t} C\gamma_5\gamma_a\chi$, since if M is a symmetric matrix, $\chi^{\rm t} M\chi \equiv \chi_a M_{ab}\chi_b = \chi_a\chi_b M_{ab} = -\chi_b\chi_a M_{ab} = -\chi_b M_{ba}\chi_a = -\chi^{\rm t} M\chi = 0$. This implies, for instance, that it is not possible to have the ordinary couplings of the field χ to the photon field A^a and to the electromagnetic tensor F^{ab}, namely,

$$\chi^{\rm t} C\gamma_a\chi \times A^a = \chi^{\rm t} CC^{-1}\gamma_a^{\rm t} C\chi \times A^a = -\chi^{\rm t} C\gamma_a\chi \times A^a = 0, \tag{A.19}$$

$$\chi^{\rm t} C\gamma_5[\gamma_a, \gamma_b]\chi \times F^{ab} = 0, \tag{A.20}$$

again using $CC^{-1} = \mathbb{1}$.

In other words, the field χ has neither charge nor magnetic moment of this type, while we can have nonzero operators using two different anticommuting fields χ_1, χ_2, since the symmetry properties would imply $\chi_1^{\rm t} M\chi_2 \equiv \chi_{1,a} M_{ab}\chi_{2,b} = -\chi_{2,b} M_{ba}\chi_{1,a} = -\chi_2^{\rm t} M\chi_1$ that in general is nonvanishing.

Discrete symmetries of the Dirac equation: Let us consider the Dirac equation in an external electromagnetic field $A_\mu(x)$,

$$\left[\hat{p} - e\hat{A}(x) - m\right]\psi(x) = 0, \tag{A.21}$$

where of course $p^\mu = i\partial/\partial x_\mu = (i\partial_t, -i\vec{\nabla})$, m is the mass (a real parameter), and the contractions with the gamma matrices are indicated by $\hat{A} = \gamma_\mu A^\mu$. This equation is covariant not only under proper Lorentz transformation but also under various discrete symmetries. In other words, by transforming the

Table A.1. Discrete symmetries of the Dirac equation. P: Spatial parity. C: Charge conjugation. T: Time reversal.

Coordinates		EM field	Fermionic field
P	$x_{\rm P}=(t,-\vec{x})$	$A_{\rm P}(x)=(A^0(x_{\rm P}),-A^i(x_{\rm P}))$	$\psi_{\rm P}(x)=\gamma_0\psi(x_{\rm P})$
C	$x_{\rm C}=(t,\vec{x})$	$A_{\rm C}(x)=(-A^0(x_{\rm C}),-A^i(x_{\rm C}))$	$\psi_{\rm C}(x)=C\gamma_0^t\psi(x_{\rm C})$
T	$x_{\rm T}=(-t,\vec{x})$	$A_{\rm T}(x)=(A^0(x_{\rm T}),-A^i(x_{\rm T}))$	$\psi_{\rm T}(x)=\gamma_5 C\psi^*(x_{\rm T})$

spacetime and the fields appropriately, as listed in Table A.1, we map valid solutions onto valid solutions.

We note that

(1) the transformation of the space–time coordinates justifies the names of P and T (whereas for C this is trivial, namely no transformation is required);
(2) the transformations of the electromagnetic field under P and T imply that the vector potential is a polar vector, and it transforms like the velocity thereby changing sign under T;
(3) for C symmetry, the change of the EM potential is equivalent to flip the sign of the charge $e\to -e$.

Let us discuss now the properties of the transformed fields. We begin from the **spatial parity** P. Applying the Dirac differential operator, we have

$$\left[\hat{p}-e\hat{A}_{\rm P}(x)-m\right]\psi_{\rm P}(x)=\gamma_0\left[\hat{p}_{\rm P}-e\hat{A}(x_{\rm P})-m\right]\psi(x_{\rm P}),\qquad\text{(A.22)}$$

where $p_{\rm P}=i\partial/\partial x_{\rm P}=(i\partial_t,i\vec{\nabla})$. Changing the names of the variables and more precisely replacing $-\vec{x}\to\vec{x}$, we recognize that the differential operator in the r.h.s. is just the Dirac one. Therefore, if the field $\psi(x)$ satisfies Dirac equation, the P-transformed field also does it.

Now we repeat the same steps for the **charge conjugation** C. Using the well-known property $\gamma_\mu\gamma_0=\gamma_0\gamma_\mu^\dagger$, we have

$$\begin{aligned}\left[\hat{p}-e\hat{A}_{\rm C}(x)-m\right]\psi_{\rm C}(x)&=-\left[\hat{p}+e\hat{A}(x)-m\right]\gamma_0 C\psi^*(x)\\&=-\gamma_0\left[\hat{p}^\dagger+e\hat{A}^\dagger(x)-m\right]C\psi^*(x).\end{aligned}\qquad\text{(A.23)}$$

We remember the defining property $C\gamma^{\rm t}=C\gamma^{\rm t}CC^{-1}=-\gamma C$ that implies $\gamma^{\rm t}C^\dagger=-C^\dagger\gamma$ and therefore $C\gamma^*=C\gamma^*C^{-1}C=-\gamma^\dagger C$. We also notice that

$p_a^* = (i\partial_a)^* = -p_a$, getting

$$\begin{aligned}\left[\hat{p} - e\hat{A}_{\rm C}(x) - m\right]\psi_{\rm C}(x) &= -\gamma_0\left[\hat{p}^\dagger + e\hat{A}^\dagger(x) - m\right]C\psi^*(x)\\ &= -\gamma_0 C\left(\left[\hat{p} - e\hat{A}(x) - m\right]\psi(x)\right)^*.\end{aligned} \tag{A.24}$$

Again, we conclude that if the field $\psi(x)$ satisfies the Dirac equation, the transformed configuration also does it. Note that the first equation can be interpreted as if the charge of the field e is formally inverted to $-e$; this justifies the name *charge conjugation* given to the matrix C.

Finally, we perform the same steps for **time reversal** T. Consider the above-defined spinor,

$$\left[\hat{p} - e\hat{A}_{\rm T}(x) - m\right]\psi_{\rm T}(x) = \gamma_5 C\left[\hat{p}^{\rm t} - e\hat{A}_{\rm T}^{\rm t}(x_{\rm T}) - m\right]\psi^*(x_{\rm T}). \tag{A.25}$$

Using the fact that the matrix γ_0 is Hermitian while γ_i are anti-Hermitian, we have $\hat{A}_{\rm T}^{\rm t} = \hat{A}^*$ and $\hat{p}^{\rm t} = \hat{p}_{\rm T}^*$, where $p_{\rm T} = (-i\partial_t, -i\vec{\nabla}) = (i\partial_{-t}, -i\vec{\nabla})$; thus,

$$\begin{aligned}\left[\hat{p} - e\hat{A}_{\rm T}(x) - m\right]\psi_{\rm T}(x) &= \gamma_5 C\left[\hat{p}^{\rm t} - e\hat{A}_{\rm T}^{\rm t}(x_{\rm T}) - m\right]\psi^*(x)\\ &= \gamma_5 C\left(\left[\hat{p}_{\rm T} - e\hat{A}(x_{\rm T}) - m\right]\psi(x_{\rm T})\right)^*,\end{aligned} \tag{A.26}$$

and once again we see that modulo the redefinition of the time coordinate, $-t \to t$, the fact that $\psi(x)$ is a solution of the Dirac equation implies that its time reversed conjugate is also a solution.

A.3. *Fierz identity for* $(V - A)$ *current–current operator*

Let us extend the set of the three Hermitian Pauli matrices by also introducing

$$\sigma_{ab}^0 \equiv \delta_{ab}. \tag{A.27}$$

The four matrices σ^α, with $\alpha = 0, 1, 2, 3$, form a basis of the 2×2 complex matrices. Now, consider the tensor $\sigma_{ab}^0\, \sigma_{cd}^0$; we would like to consider it as a set of matrices with indices c and b. Thus, each of them can be rewritten in terms of the basis of the 2×2 matrices that we have introduced by using four complex coefficients, namely,

$$\sigma_{ab}^0\, \sigma_{cd}^0 = \sum_{\alpha=0}^{3} \kappa_{ad}^\alpha\, \sigma_{cb}^\alpha. \tag{A.28}$$

In order to determine the coefficients κ, we multiply by σ^{β}_{bc} and sum over the repeated coefficients, finding

$$\sigma^{0}_{ab}\,\sigma^{\beta}_{bc}\,\sigma^{0}_{cd} = \sigma^{\beta}_{ad} = \sum_{\alpha=0}^{3} \kappa^{\alpha}_{ad}\ \mathrm{tr}[\sigma^{\alpha}\,\sigma^{\beta}] = 2\kappa^{\beta}_{ad}. \tag{A.29}$$

Therefore, we can replace k^{β}_{ad} with $\sigma^{\beta}_{ad}/2$ and thus we have proven the first identity,

$$\sigma^{0}_{ab}\,\sigma^{0}_{cd} = \frac{1}{2}\left[\sigma^{0}_{ad}\,\sigma^{0}_{cb} + \sigma^{i}_{ad}\,\sigma^{i}_{cb}\right], \tag{A.30}$$

where $i = 1, 2, 3$ is summed over. Similarly, we can deduce an identity for the combination, $\sigma^{i}_{ab}\,\sigma^{i}_{cd}$ and finally write the useful identity,

$$\eta_{\alpha\beta}\,\sigma^{\alpha}_{ab}\,\sigma^{\beta}_{cd} = -\eta_{\alpha\beta}\,\sigma^{\alpha}_{ad}\,\sigma^{\beta}_{cb}, \tag{A.31}$$

where α and $\beta = 0, 1, 2, 3$ are summed over and η is the (flat) metric tensor,

$$\eta = \mathrm{diag}(+1, -1, -1, -1). \tag{A.32}$$

Let us now consider four *anticommuting* bispinors named $\psi_{\mathrm{L}}, \chi_{\mathrm{L}}, \lambda_{\mathrm{L}}, \phi_{\mathrm{L}}$, namely quantized fermionic fields. The previous identity implies

$$\psi^{\dagger}_{\mathrm{L}}\bar{\sigma}^{\alpha}\chi_{\mathrm{L}}\ \lambda^{\dagger}_{\mathrm{L}}\bar{\sigma}_{\alpha}\phi_{\mathrm{L}} = +\psi^{\dagger}_{\mathrm{L}}\bar{\sigma}^{\alpha}\phi_{\mathrm{L}}\ \lambda^{\dagger}_{\mathrm{L}}\bar{\sigma}_{\alpha}\chi_{\mathrm{L}}, \tag{A.33}$$

where we have introduced the matrices,

$$\bar{\sigma}^{\alpha} = (\sigma^{0}, -\sigma^{i}). \tag{A.34}$$

The proof is based on the previous identity and on the anticommutativity that flips the "minus" sign into a "plus".

The last remark is that in the chiral (or Weyl) representation of the Dirac matrices, where $\gamma_5 = \mathrm{diag}(-1, -1, +1, +1)$ and $\gamma_0 = \mathrm{antidiag}(1, 1, 1, 1)$,

$$\psi = \begin{pmatrix} \psi_{\mathrm{L}} \\ \psi_{\mathrm{R}} \end{pmatrix} \quad \text{and also} \quad \gamma^{\alpha} P_{\mathrm{L}} = \begin{pmatrix} 0 & 0 \\ \bar{\sigma}^{\alpha} & 0 \end{pmatrix} \tag{A.35}$$

and similarly for any other four-spinor. Therefore, the previous identity concerning bispinors can be translated into an identity concerning (anticommuting) four-spinors,

$$\bar{\psi}\gamma^{\alpha}P_{\mathrm{L}}\chi\ \bar{\lambda}\gamma_{\alpha}P_{\mathrm{L}}\phi = \bar{\psi}\gamma^{\alpha}P_{\mathrm{L}}\phi\ \bar{\lambda}\gamma_{\alpha}P_{\mathrm{L}}\chi. \tag{A.36}$$

Namely, in this combination of $V-A$ currents, we can simply exchange the second and the fourth four-spinors. Owing to the fact that this is an identity between scalars, we conclude that this is a general fact, valid in *any* representation of Dirac matrices.

This Fierz identity Eq. (A.36) is useful in various situations, in particular:

(1) In the study of neutrinos, it can be used for the derivation of the matter terms of MSW theory.
(2) In the theory of weak interactions, it allows us to cast in the same form the two contributions from CC and NC.

References

1. B. Pontecorvo, "Mesonium and anti-mesonium," *Sov. Phys. JETP* **6**, 429 (1957) [Zh. Eksp. Teor. Fiz. **33**, 549 (1957)] and "Neutrino experiments and the problem of conservation of leptonic charge," *Sov. Phys. JETP* **26**, 984 (1968) [Zh. Eksp. Teor. Fiz. **53**, 1717 (1967)].
2. Z. Maki, M. Nakagawa, and S. Sakata, "Remarks on the unified model of elementary particles," *Prog. Theor. Phys.* **28**, 870 (1962). doi:10.1143/PTP.28.870.
3. L. Wolfenstein, "Neutrino oscillations in matter," *Phys. Rev. D* **17**, 2369 (1978). doi:10.1103/PhysRevD.17.2369.
4. S. P. Mikheyev and A. Y. Smirnov, "Resonance amplification of oscillations in matter and spectroscopy of solar neutrinos," *Sov. J. Nucl. Phys.* **42**, 913 (1985) [Yad. Fiz. **42**, 1441 (1985)].
5. B. M. Pontecorvo, "Pages in the development of neutrino physics," *Sov. Phys. Usp.* **26**, 1087 (1983) [Usp. Fiz. Nauk **141**, 675 (1983)].
6. S. Bilenky, "Introduction to the physics of massive and mixed neutrinos," *Lecture Notes in Physics 817* (2010), Springer-Verlag, Berlin.
7. M. Riordan, "Pauli's Ghost: The conception and discovery of neutrinos," In Caldwell, D. O. (ed.): *Current Aspects of Neutrino Physics*, 1–16. Springer, Berlin.
8. S. M. Bilenky, J. Hosek, and S. T. Petcov, "On oscillations of neutrinos with Dirac and Majorana masses," *Phys. Lett.* **94B**, 495 (1980).
9. F. Capozzi, E. Lisi, A. Marrone, D. Montanino, and A. Palazzo, "Neutrino masses and mixings: status of known and unknown 3ν parameters," *Nucl. Phys. B* **908**, 218 (2016). [arXiv:1601.07777 [hep-ph]].
10. A. Strumia, and F. Vissani, "Neutrino masses and mixings and...," (2010) [arXiv:hep-ph/0606054].
11. F. Capozzi, E. Lisi, and A. Marrone, "Neutrino mass hierarchy and electron neutrino oscillation parameters with one hundred thousand reactor events," *Phys. Rev. D* **89**, no. 1, 013001 (2014) [arXiv:1309.1638 [hep-ph]].
12. H. Nunokawa, S. J. Parke, and R. Zukanovich Funchal, "Another possible way to determine the neutrino mass hierarchy," *Phys. Rev. D* **72**, 013009 (2005) [hep-ph/0503283].
13. M. Cirelli, N. Fornengo, T. Montaruli, I. Sokalski, A. Strumia, F. Vissani, "Spectra of neutrinos from dark matter annihilation," *Nucl. Phys. B* **727**, 99 (2005) [E-*ibid.*B790 (2008) 338.].
14. V. N. Gribov, and B. Pontecorvo, "Neutrino astronomy and lepton charge," *Phys. Lett.* **28B**, 493 (1969).

15. S. M. Bilenky, and B. Pontecorvo, "Lepton mixing and neutrino oscillations," *Phys. Rept.* **41**, 225 (1978).
16. A. Palladino, and F. Vissani, "The natural parameterization of cosmic neutrino oscillations," *Eur. Phys. J. C* **75**, 433 (2015) [arXiv:1504.05238 [hep-ph]].
17. G. Battistoni, A. Ferrari, C. Rubbia, P. R. Sala, and F. Vissani, "Atmospheric neutrinos in a large liquid argon detector," hep-ph/0604182.
18. A. M. Dziewonski, and D. L. Anderson, "Preliminary reference Earth model," *Phys. Earth Planet. Interiors* **25**, 297 (1981).
19. S. J. Parke, "Nonadiabatic level crossing in resonant neutrino oscillations," *Phys. Rev. Lett.* **57**, 1275 (1986). doi:10.1103/PhysRevLett.57.1275.
20. T. K. Kuo and J. T. Pantaleone, "Neutrino oscillations in matter," *Rev. Mod. Phys.* **61**, 937 (1989).
21. C. Lujan-Peschard, G. Pagliaroli, and F. Vissani, "Counting muons to probe the neutrino mass spectrum," *Eur. Phys. J. C* **73**, 2439 (2013). doi:10.1140/epjc/s10052-013-2439-1 [arXiv:1301.4577 [hep-ph]].
22. A. S. Dighe, and A. Y. Smirnov, "Identifying the neutrino mass spectrum from the neutrino burst from a supernova," *Phys. Rev. D* **62**, 033007 (2000) [hep-ph/9907423].
23. M. Cirelli, G. Marandella, A. Strumia, and F. Vissani, "Probing oscillations into sterile neutrinos with cosmology, astrophysics and experiments," *Nucl. Phys. B* **708**, 215 (2005) [hep-ph/0403158].
24. S. Gariazzo, C. Giunti, M. Laveder, Y. F. Li, and E. M. Zavanin, "Light sterile neutrinos," *J. Phys. G* **43**, 033001 (2016) [arXiv:1507.08204 [hep-ph]].
25. B. Kayser, F. Gibrat-Debu, and F. Perrier, "The physics of massive neutrinos," *World Sci. Lect. Notes Phys.* **25**, 1 (1989).
26. M. Fukugita, and T. Yanagida, "Physics of neutrinos and applications to astrophysics," Berlin: Springer (2003).
27. C. Giunti, and C. W. Kim, *Fundamentals of Neutrino Physics and Astrophysics* Oxford: Oxford University Press (2007).
28. E. Canovi, G. Cò, D. Montanino, F. Vissani *Neutrino: The Mutant Particle,* Rome, Italy: Aracne editrice (2016).
29. S. M. Bilenky, and S. T. Petcov, "Massive neutrinos and neutrino oscillations," *Rev. Mod. Phys.* **59**, 671 (1987) [E ibid **61**, 169 (1989) and **60**, 575 (1988)].
30. E. K. Akhmedov, "Neutrino physics," hep-ph/0001264.
31. E. K. Akhmedov, and A. Y. Smirnov, "Paradoxes of neutrino oscillations," *Phys. Atom. Nucl.* **72**, 1363 (2009). doi:10.1134/S1063778809080122 [arXiv:0905.1903 [hep-ph]].
32. M. Blennow, and A. Y. Smirnov, "Neutrino propagation in matter," *Adv. High Energy Phys.* **2013**, 972485 (2013). doi:10.1155/2013/972485 [arXiv:1306.2903 [hep-ph]].
33. E. Lisi, and D. Montanino, "Earth regeneration effect in solar neutrino oscillations: An Analytic approach," *Phys. Rev. D* **56**, 1792 (1997).
34. N. Y. Agafonova *et al.*, "Study of the effect of neutrino oscillations on the supernova neutrino signal in the LVD detector," *Astropart. Phys.* **27**, 254 (2007).

Chapter 3

Neutrino Oscillation Detectors and Methods

D. Autiero

Institut de Physique Nucléaire de Lyon,
4, Rue E. Fermi, Villeurbanne 69622, France
autiero@cern.ch

Abstract

The study of neutrino oscillations requires a variety of challenging experimental techniques in order to reconcile, with the realization of massive fine-grained target detectors, the tininess of cross-sections with the need of performing sophisticated precision measurements. These studies often include the identification of final state charged leptons, required to detect the flavor of the interacting neutrino in appearance searches. The measurement of the energy of the final state leptons and of the energy of the system of recoiling hadrons allows reconstructing the impinging neutrino energy. This spectral information is fundamental in order to measure the L/E oscillatory pattern. Some detector techniques, representative of the basic aspects, will be discussed by stressing their respective merits with respect to different oscillation channels and neutrino energy ranges.

1. Basic principles

Given the tininess of interaction cross-sections, neutrino detection is a constant struggle. In addition, fluxes can be dramatically decreased by the distances required in long-baseline neutrino oscillation experiments. In order to collect samples of neutrino interactions of significant statistics, massive detectors must be employed, with targets weighting from a few *ton* to tens of *kton*. Detectors should, at the same time, act as neutrino targets and reconstruct the secondary particles generated in neutrino interactions. These two functions can be implemented by embedding heavy passive materials in an active detector structure. In other fields of high energy physics, particle detectors are usually built around a collision point or a target. In neutrino detectors, interactions may occur uniformly along the beam axis in all the materials constituting the target. This aspect imposes a uniformity requirement: the detector should provide similar response independently on the position of the neutrino interaction point. The capability of reconstructing individually secondary particles generated in neutrino interactions requires quite high readout granularity, going from cm to even μm, in special cases such as the τ lepton tagging. Practical and economical challenges arise when such a fine granularity has to be implemented on large size detectors. Furthermore, accurate tracking of particles is obviously badly compatible with the presence of heavy materials, which induce multiple scattering, showers and secondary interactions. This is yet another contradiction which must be solved in neutrino experiments. A first solution may be consisting in only reconstructing the tracks of muons, which are penetrating particles. By exploiting heavy materials, which make electrons and hadrons showering, an inclusive calorimetric approach can be then adopted for the measurement of the energy and average direction of all the remaining particles. Longitudinal segmentation, alternating passive and active materials along the neutrino beam axis, generally allows implementing an equivalent response independently on the neutrino interaction depth in the detector. This is a typical case in neutrino detectors with a *sampling calorimeter* structure. However, the best response and highest energy resolution can be often achieved only in *fully active detectors*, where the interaction and detection functionalities are joined in a single homogeneous medium (water, liquid scintillator, liquid argon (LAr)). The highest level of sophistication is reached by using fully active detectors with an active medium readout so finely segmented to allow tracking individually secondary particles. A full detailed reconstruction of the final state can be then performed. The basic design of neutrino detectors is generally dictated by a few fundamental requirements such as the energy range of the neutrinos to be studied, the kind of final state charged lepton identification to be achieved, the resolution needed on the measurement of the neutrino energy and on other kinematical quantities. In this chapter, for practical reasons, we will mainly focus on detectors which could be used for the study of neutrino oscillations at accelerators. We will discuss a few historical examples of detector systems working at neutrino energies going from a few hundred MeV to tens of GeV. The detection of lower energy neutrinos, in the MeV energy range, coming from the Sun or from nuclear reactors requires specific detector designs[7–9] which we will not discuss in this chapter. In accelerator neutrino interactions final states with several particles can be produced. The level of sophistication and of difficulty in building a detector for accelerator neutrinos depends on the capability of providing the following measurements, which, in addition, have often to be implemented at the scale of very large detectors:

- Tagging the charged lepton in the final state: μ, e, τ (in order of increasing difficulty)
- Measurement of the sign of the charged lepton (needs magnetic field)
- Capability of providing spectral information for the L/E measurement and related energy resolution. Obviously, achieving high energy resolution is a more challenging task
- Measurement of the energy of the hadronic system. More demanding: individual hadrons tracking, energy measurement and identification

Charged-current (CC) neutrino interactions allow tagging the impinging neutrino flavor by identifying the corresponding final state lepton: muon, electron or τ. This capability is fully exploited in measurements performed by detecting the *appearance* of a new neutrino flavor, originally absent from the neutrino source. This new flavor can then be generated by the neutrino oscillations mechanism, after a certain travel path. Typical examples are the detection of ν_τ and ν_e appearance in ν_μ oscillations occurring in the atmospheric neutrinos sector. Detectors tagging the final state lepton can be also used in *disappearance mode.* In this case two identical detectors are normally deployed to measure the flux of the neutrino flavor mostly composing the beam. A *near detector* is installed close to the neutrino source in order to characterize the nonoscillated flux. A *far detector*, located at a distance from the source possibly corresponding to the oscillation maximum, measures the flux, which is affected in a significant way by the oscillations mechanism. The effect of the oscillations in disappearance mode is visible as a depletion of the flux measured in a well-defined spectral region. Typical examples are the study of $\nu_\mu \to \nu_\tau$ oscillations in the atmospheric neutrinos sector performed at accelerators by looking at ν_μ disappearance or accomplished at nuclear reactors by measuring the $\bar{\nu}_e$ disappearance. Near detectors are almost mandatory for disappearance measurements. They can be also helpful in appearance measurements by reducing systematic uncertainties on background contaminations, for example, by measuring the intrinsic ν_e contamination in a ν_μ beam produced by an accelerator. This contamination, originally present at the source, will be competing with the ν_e signal in the $\nu_\mu \to \nu_e$ appearance channel. The measurement of the L/E oscillatory pattern is an important signature allowing to extract information on the oscillation parameters. In order to measure this pattern the detector must have the capability of reconstructing the impinging neutrino energy by measuring the energy of the final state charged lepton and, more demanding, by also measuring the energy of the system of recoiling hadrons. Neutrino oscillation physics has been evolving in the last decades to precision measurements. These measurements are becoming more and more challenging by requiring large far detectors without compromising on their capabilities and performance.

2. Bubble chambers: a reference in neutrino interactions imaging

Bubble chambers were used at the beginning of neutrino physics and contributed to important discoveries such as the detection of neutral-current (NC) interactions with the Gargamelle experiment.[1] They were as well exploited in first short-baseline searches for neutrino oscillations.[6] Bubble chambers were fully active and homogeneous detectors in which the ionization losses from charged particles, present in the final states of neutrino interactions, were producing visible tracks composed of aligned sets of small bubbles in superheated cryogenic liquids, such as hydrogen, deuterium or heavier noncryogenic liquids like heavy freon (CF_3Br).[2–5] These detectors were inserted in a magnetic field in order to measure momenta and charges from tracks curvature. Electromagnetic calorimetry could be provided as well by using the liquid as a showering medium. In general a bubble chamber could provide a full imaging of the final state particles and the identification and the energy measurement of part of them, very useful for the study of exclusive topologies (see Fig. 1). The data taking was first performed by taking pictures of the events which could be further analyzed and digitized. On the other hand, bubble chambers needed to operate with expansion/compression cycles in order to get to the critical conditions of the liquid at the moment of particles detection and in order to reset the detector for a new detection. The detector was then not fully active during all time. A fast bubble chamber could go through 30 cycles per second. Cycles had to be synchronized with the beam extraction from the accelerator in order not to introduce dead time. Events were not directly recorded in an electronic format and this detector design could not be extended to larger sizes or high intensity applications.

3. NOMAD: kinematical measurements and a near detector reference

The neutrino oscillation magnetic detector (NOMAD) detector was built at the beginning of the 90s in order to perform a short-baseline search of $\nu_\mu \to \nu_\tau$ and $\nu_\mu \to \nu_e$ oscillations in appearance mode.[12] It was operating,

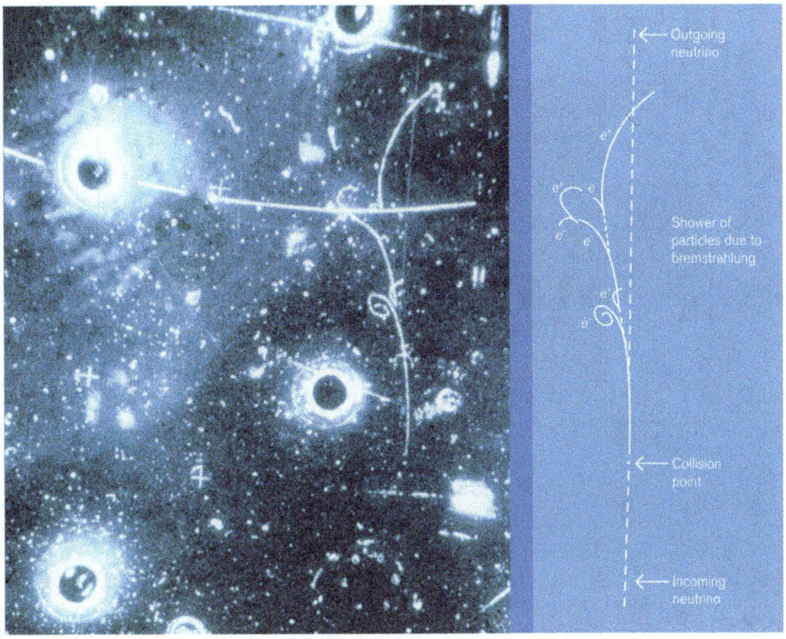

Fig. 1. A first leptonic NC interaction candidate with a single electron showering in the final state observed in the Gargamelle bubble chamber.

together with the CHORUS[13] detector, at CERN, in the West Area Neutrino Facility, at about 1 km distance from the target generating the neutrino beam. The beam was mainly composed of ν_μ , with an average energy of 24 GeV. The CHORUS experiment has, as well, an historical interest, since it pioneered the use of nuclear emulsions in ν_τ appearance searches based on topological criteria. This technique[14] will be described in more details in Section 6. It is interesting to discuss NOMAD, since it was a first attempt to achieve a *bubble chamber like* event reconstruction quality in a fully electronic detector inserted in a magnetic spectrometer. NOMAD pioneered ν_τ appearance searches based on kinematical criteria[15] applied to the final state of neutrino interactions. Previous neutrino detectors had never achieved the resolution needed by this technique. NOMAD can still represent a valid conceptual example for near detectors. A near detector inspired by the NOMAD design is for instance being considered in the framework of the DUNE experiment.[77] The LAr time projection chamber (TPC), which will be described more extensively in Section 7, can be also considered as an "electronic bubble chamber" and it is even closer to the original concept of bubble chambers. NOMAD was a completely electronic detector,[16] composed of a set of drift chambers embedded in a magnetic spectrometer operating at 0.4 T (Fig. 2). Neutrinos were interacting in the

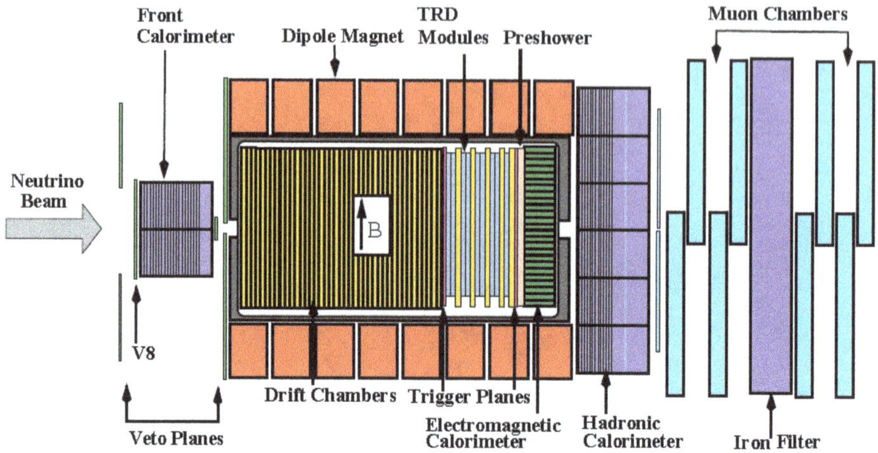

Fig. 2. A top view of the NOMAD detector.

material, mostly carbon, constituting the skin of the drift chambers. The target had a fiducial mass of 2.7 ton and an average density of $0.1\,\text{g/cm}^3$. The measurement of the curvature of tracks of final state particles allowed for charge identification and momentum measurement with a typical resolution of 3.5% below $10\,\text{GeV}/c$. The magnet used for NOMAD was recycled from the former UA1 CERN experiment.[10] This magnet has been used also to host the T2K near detector,[36] another interesting detector example that we will not discuss here.

NOMAD was designed with particular care for the identification of electrons in the final state. Good electron identification was motivated by the interest in searching for $\nu_\mu \to \nu_\tau$ oscillations by selecting events where τ decay into an electron and two neutrinos. The potential background to this channel was coming from the small ($\sim$1%) ν_e contamination in the neutrino beam. This background could be further suppressed by performing a kinematical analysis of the final state. In case of τ decay events the final state is characterized by two neutrinos. The presence of the undetected neutrinos can be inferred by measuring the missing momentum on the plane perpendicular to the incoming ν_τ direction. The missing transverse momentum vector has also specific angular correlations that can be exploited to select τ events. These angular correlations, defined on the plane transverse to the incoming neutrino, concern the direction of the missing transverse momentum with respect to the momentum of the visible tau daughter particle and with respect to the direction of the total momentum of the hadronic system produced by the recoiling quarks. In order to perform the ν_τ search in this channel, a detector with high electron identification sensitivity capable of reconstructing individual particles in the final state, as well as of precisely measuring their kinematics, was then needed. NOMAD had a few subdetectors devoted to electron identification, such as: a transition radiation detector,[18] a pre-shower and a lead glass electromagnetic calorimeter.[17] They were overall providing, in association with the spectrometer, a pion to electron rejection power exceeding 10^6. The target made of drift chambers was required to be transparent to electrons and photons as much as possible, in order not to perturb the measurement of their energy to be performed by the downstream electromagnetic calorimeter. The total target length was equivalent to 1 radiation length (X_0). Muon identification was performed with a set of chambers located after an iron filter, which was acting as a hadronic calorimeter. The iron filter was equivalent to eight interaction lengths. The muon identification efficiency was of the order of 98% for muons with momenta larger than 4 GeV/c.[11,16] The NOMAD detector acquired about 1.5 million neutrino interactions with "bubble chamber like" reconstruction quality. This represents the largest sample benefiting of such highly detailed reconstruction of the final states, collected so far in the deep inelastic region. Two typical events are shown in Fig. 3.

A $\nu_\mu \to \nu_e$ search[19] was motivated by the liquid scintillator neutrino detector (LSND) anomaly.[20] A sample of 5600 ν_e CC events was selected for this analysis with a remarkable purity of 98% corresponding to 44% selection efficiency. The NOMAD design is a good example of a neutrino detector with excellent electron identification and a detailed reconstruction of the neutrino interaction final state and of its kinematics. This concept, although still interesting as an example of compact near detector, cannot be extrapolated, for practical and economical reasons, to the dimensions needed to build a far detector. As we will see, a key element for the construction

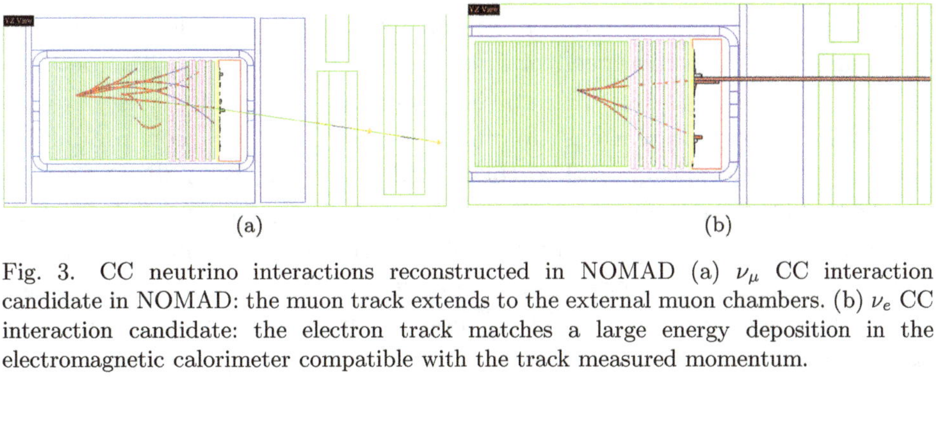

(a) (b)

Fig. 3. CC neutrino interactions reconstructed in NOMAD (a) ν_μ CC interaction candidate in NOMAD: the muon track extends to the external muon chambers. (b) ν_e CC interaction candidate: the electron track matches a large energy deposition in the electromagnetic calorimeter compatible with the track measured momentum.

of large far detectors, without compromising on the channels granularity, consists in performing the signals readout from the surface of the detector. This breaks the proportionality law of the number of channels with the detector volume and sensibly reduces technical complications and costs. The large scale application of this readout concept is allowed by the achievement of particular conditions in the active medium, such as good water transparency in water Čerenkov detectors or long lifetime of electrons in LAr detectors.

4. Super-Kamiokande and massive water Čerenkov detectors

The water Čerenkov technology played a fundamental role in neutrino physics by making possible the realization of very massive experiments capable of discriminating between ν_μ and ν_e. This technique was pioneered by the Kamiokande[22,23] and IMB[21] detectors, motivated by proton decay searches. It then allowed for the observation of neutrinos from the collapse of the supernova SN1987A[24–26] and for other important measurements on solar and atmospheric neutrinos, which confirmed the solar neutrinos deficit and opened the problem of the atmospheric neutrinos anomaly.[27] These pioneering developments in neutrino astrophysics were awarded by the 2002 Nobel Prize in Physics.[28] The water Čerenkov technique reached full maturity with the Super-Kamiokande (SK)[29] detector, making possible the discovery of neutrino oscillations in the atmospheric neutrinos sector (2015 Nobel Prize in Physics).[30] SK has also been exposed to man-made neutrinos with the K2K[31] and T2K[33] accelerator beams, in order to study ν_μ disappearance[32] and, in the T2K experiment, the ν_e appearance effect[34] related to the θ_{13} mixing angle. SK is a 50 kton water Čerenkov detector with a cylindrical shape of about 39 m diameter and 41 m height. The inner walls of the cylinder containing the active volume for neutrino interactions were covered by 11000 photomultipliers of 20″ diameter, used to reconstruct the rings of Čerenkov photons produced by charged particles traveling at a speed larger than the speed of light in water (see Fig. 4). This detector is particularly effective in the sub-GeV neutrino energy range. Here quasi-elastic scattering interactions dominate. Given the little pion production cross-section, the background to the e/μ identification is small. Depending on the fact that the interacting neutrino is a ν_e or a ν_μ, the quasi-elastic final state contains an electron or a muon plus a proton. The proton is generally below threshold for Čerenkov light emission. The electron and the muon will instead emit Čerenkov light all along their paths in water, projecting a ring of photons on the walls of the detector. This ring is sampled by the

Fig. 4. Inside view of the Super-Kamiokande water Čerenkov detector at the Kamioka underground observatory.

array of photomultipliers that cover the walls, providing a pixelized image of the light. Rings produced by electrons have a more fuzzy shape due to the presence of electromagnetic showers (Fig. 5). At low energy, the analysis of the neutrino interactions typically deals with events containing a single Čerenkov ring. The shape of the ring allows identifying electrons and muons with $\sim$1% misidentification probability.

Under the assumption that the neutrino interaction occurred in the quasi-elastic channel and by neglecting the Fermi momentum of the target nucleon, the incoming neutrino energy can be reconstructed by measuring the energy and the direction of the charged lepton with respect to the neutrino beam direction. For not-beam-generated events, such as atmospheric neutrinos interactions, this calculation is not possible. However, the direction and energy of the neutrino can be approximated by the ones of the charged lepton. SK proved the oscillations interpretation of the atmospheric neutrinos anomaly by measuring ν_μ and ν_e fluxes for upward going (oscillated neutrinos) and downward going (almost unoscillated) neutrinos. The zenith angle dependence is a self-normalizing (directly providing an internal crosscheck, given the isotropy of atmospheric neutrino fluxes) and clear signature of neutrino oscillations in disappearance mode.[35] The detector also allowed studying solar neutrinos down to a threshold of 5 MeV. The key point of the water Čerenkov technique is to make very massive homogeneous detectors affordable, with precise ν_μ and ν_e tagging

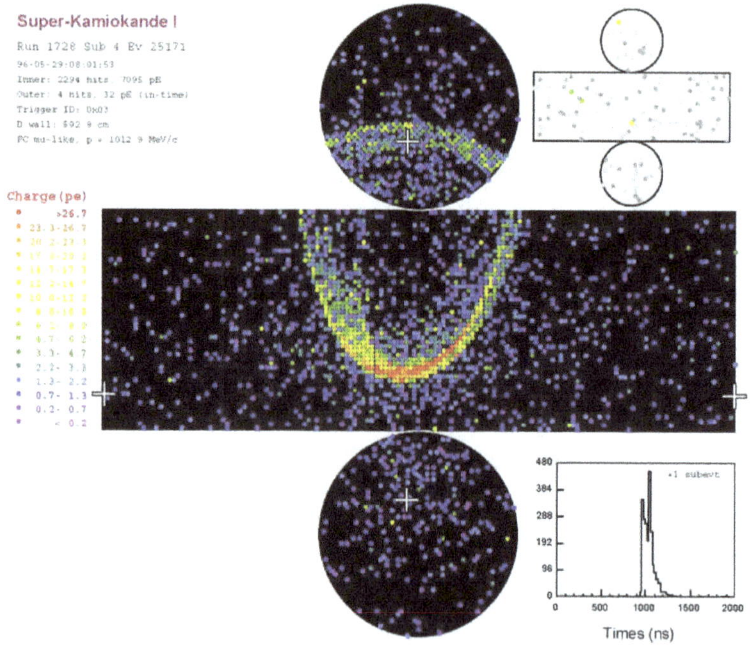

(a)

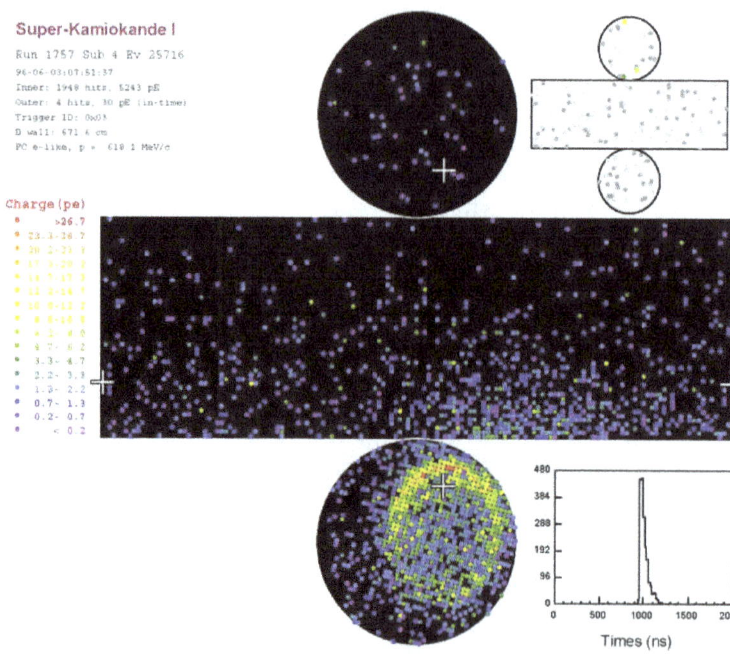

(b)

Fig. 5. e/μ identification in the SK detector. Pixels represent the light intensity seen by single photomultipliers. The edge of the Čerenkov ring produced by an electron is more fuzzy with respect to the ring of a muon, due to the fact that the electron showers in water. (a) ν_μ single ring CC candidate. (b) ν_e single ring CC candidate.

coupled with the capability of providing an adequate measurement of the neutrino energy. This technique is mostly suitable for low energy events in the region of few hundred MeV. At this energy, the final state reconstruction based on single rings and the energy reconstruction based on the charged lepton measurement provide the best performance. Water is a cheap target/active detector medium. Its transparency makes possible the readout of signals at the detector surface, which has to be equipped with an adequate photodetector coverage. This principle allows building very massive detectors. The Hyper-Kamiokande (HK) project[37] is developing an application of the water Čerenkov technology at the 0.5 Mton scale, in view of the search for CP violation in the neutrino sector. This search is based on the study of the subleading $\nu_\mu \rightarrow \nu_e$ oscillation at the $\Delta m^2{}_{23}$ of atmospheric neutrinos by comparing the measurements taken with ν_μ and anti-ν_μ beams. Given the considered baseline and neutrino energy, the water Čerenkov technology in HK allows studying at high statistics the populations of ν_μ and ν_e at the first oscillation maximum. However, it provides limited spectral information and it does not give access to the study of the second oscillation maximum. The low beam energy and the short baseline also make disentangling matter effects and studying the mass hierarchy independently on CP violation difficult. The mass hierarchy has to be determined from other measurements as on atmospheric neutrinos or provided by other experiments. HK is mostly designed as a high statistics e/μ counting experiment at the first oscillation maximum for the CP violation search. In a counting experiment, with limited spectral information, the assessment of normalization systematics is a crucial aspect of the measurements. A complementary approach for the CP search, working at higher energy and relying more on spectral information, will be introduced later when discussing LAr TPCs in Section 7.

5. MINOS and magnetized iron calorimeters: the spectral information in disappearance mode

Magnetized iron calorimeters are clear examples of detectors characterized by a longitudinal segmentation obtained by alternating instrumented planes with passive heavy materials acting as neutrino target. These detectors feature long-standing tradition in neutrino physics, having been used for precision electroweak measurements such as the measurement of the Weinberg angle by the CHDS and CCFR experiments.[44,45] The MINOS experiment was mainly built in order to check the atmospheric neutrino anomaly in disappearance mode with a man-made, energy tunable, ν_μ neutrino beam

Fig. 6. The MINOS magnetized iron/scintillator far detector.

mostly operated at 3 GeV average energy over a baseline of 735 km.[48] The experiment was based on a two detector setup, a near and a far detector of 980 ton and 5.4 kton, respectively (Fig. 6), with structure and response as similar as possible.[49] The near detector was located at 1 km distance from the target. The detectors were made by a longitudinal stack of magnetized iron plates (1" thickness, 1.5 X_0) with a toroidal magnetization scheme and scintillator strips of 4 cm width and 1 cm thickness.

MINOS introduced the large scale use of cheap extruded scintillator strips with wavelength shifting (WLS) fibers readout, in order to reduce the number of readout channels and instrument a very large detector size by minimizing costs. In the far detector the iron plates had an octagonal shape of 8 m size. The detector also included 486 active scintillator planes with a total of 95000 strips. CC interactions of muon neutrinos are characterized by the presence of a long muon track which is clearly sticking out from the hadronic shower, which is just extending nearby the neutrino vertex. ν_e CC events may be identified by looking at the shower profile that is more compact and regular for electrons than the one of hadronic showers in NC events (Fig. 7). The neutrino energy can be reconstructed by combining

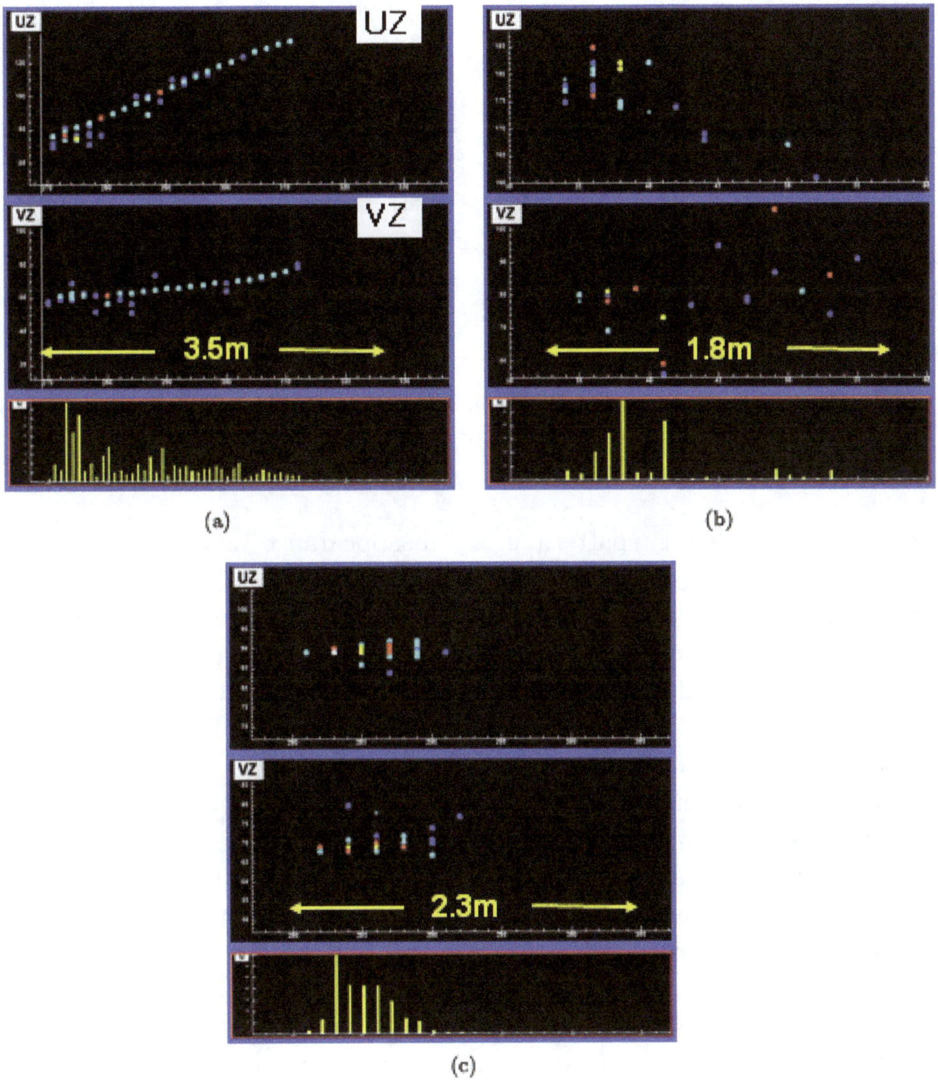

Fig. 7. Simulated neutrino interactions in MINOS: ν_μ CC events are characterized by the long penetrating muon track and and an hadronic shower around the vertex; NC interactions just have this hadronic shower at the vertex. NC may be distinguished from ν_e CC interactions, which are characterized by a more compact and regular electromagnetic shower. (a) ν_μ CC Event. (b) NC Event. (c) ν_e CC Event.

the measurement of the muon energy (typical resolution: ~6% by using range information, and ~13% by reconstructing the track curvature in the magnetic field) with hadronic calorimetry, which has a resolution of $55\%/\sqrt{E}$. This spectral information was a key issue for the measurement

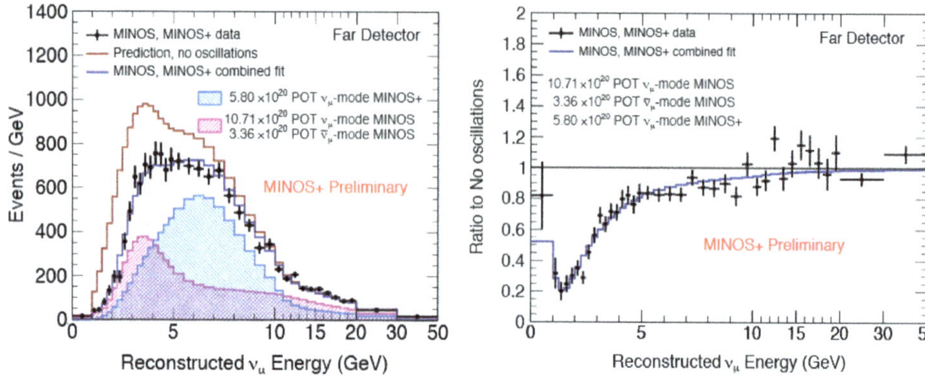

Fig. 8. Measured ν_μ CC energy spectrum from the MINOS and MINOS+ data and reconstruction of the L/E oscillatory pattern in disappearance mode.[50]

of the L/E oscillatory pattern in ν_μ disappearance and for the precise determination of the oscillation parameters ($\Delta m^2{}_{23}$ and θ_{23}) (Fig. 8).

Magnetized iron calorimeters have been also considered as detectors for future neutrino factories. The study of $\nu_\mu \to \nu_e$ oscillations in neutrino factories can be performed by looking at wrong sign muons from neutrino interactions in the far detector.[51] For this application, magnetized iron calorimeters are natural candidates, allowing for the implementation of muon tagging and charge measurement in massive detectors. Electron identification requires instead very fine granularity and longitudinal segmentation. Magnetized iron calorimeters are not optimized for the selection of ν_e samples. Calorimetric detectors with a longitudinal segmentation based on other passive materials, such as CHARM and CHARM II,[46,47] have been also operating in the past for electroweak measurements with neutrinos. An evolution of this detector concept, in the direction of improving electron identification capabilities, aims at implementing an almost fully active fine-grained detector with very little dead material. An example is given by the NOνA 14 kton far detector that is realized with independent cells filled with liquid scintillator and read out by WLS fibers (Fig. 9). This detector is currently employed, similarly as in the T2K experiment, in the study of the subleading atmospheric $\nu_\mu \to \nu_e$ oscillation which is determined at the first order by the θ_{13} angle and then by tinier effects such as CP violation and the mass hierarchy.[52,54] Another example of fully active fine-grained detector is given by the Minerva high statistics neutrino scattering experiment for the study of neutrino cross-sections[53] based on plastic scintillator.

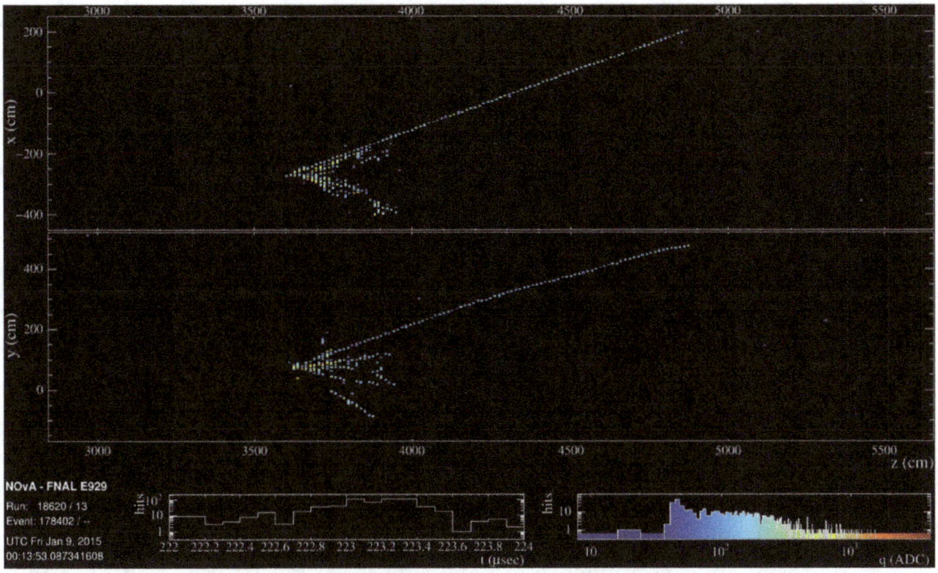

Fig. 9. ν_μ CC candidate in the NOνA fully active far detector.

6. OPERA: the challenge of long-baseline ν_τ appearance

We have seen that the study of ν_e CC interactions with the identification of the final state electron and the reconstruction of its energy is a challenging task. We will now discuss a special case of flavor tagging which is even more demanding for its implementation on very large detectors: the identification of the τ lepton. The OPERA experiment[39] was designed to perform a direct observation of ν_τ appearance in the $\nu_\mu \to \nu_\tau$ oscillations advocated to explain the ν_μ disappearance in the atmospheric neutrinos anomaly. The τ is a short living particle decaying at the energies of OPERA after 1 mm flight path and requiring quite sophisticated efforts for its detection. These may be based on kinematical cuts as discussed for NOMAD. This kinematical analysis can be also exploited in LAr TPCs such as ICARUS or future detectors based on the same technology. The complementary approach consists in visualizing the τ decay topology with a high-resolution tracking device. When the τ decays to a single charged particle + neutrino(s), this topology appears as a sharp angular discontinuity (*kink*) present on the reconstructed track trajectory that includes the short τ track followed by the one of its daughter particle. Searches based on this topological signature may further benefit from the combination with kinematical handles in order to suppress the background, coming for instance from hadronic interactions with a kink-like topology. These τ tagging techniques have to be implemented in a practically accessible manner on very large detectors having target masses sufficient in order to achieve enough sensitivity for the oscillations appearance searches. OPERA was operating in the Gran Sasso underground laboratory at a distance of 730 km from the target producing the neutrino beam at CERN. The CNGS[43] beam with an average energy of 17 GeV was optimized in order to produce the maximal number of ν_τ CC interactions in the detector. This was designed with a target mass of 1.5 kton in order to collect about 10 candidate events over 5 years of beam exposure.[40] The observation of the τ decay topology occurring at these energies over about 1 mm distance requires micrometric tracking accuracy, which is very challenging to be implemented on such a massive detector. In OPERA these two requirements were reconciled thanks to the concept of the emulsion cloud chamber (ECC) already used in the year 2000 by the DONUT experiment[38] for the first direct detection of the ν_τ. The ECC combines the high-accuracy tracking capabilities of nuclear emulsions with passive heavy material in a sandwich-like cells where single neutrino interactions can be isolated and studied on almost real-time. The basic ECC module (*brick*) in OPERA was weighing 8.3 kg. It was made of 57 ($12.5 \times 10.2\,\mathrm{cm}^2$) emulsion films interleaved with 56 lead sheets (1 mm thickness). Each film was actually made of two $42\,\mu$m emulsion layers separated by a $205\,\mu$m thick plastic base. The neutrino target and decay topology detector was constituted by a succession of walls filled with these bricks and interleaved by plastic scintillator hodoscopes used to locate the bricks containing neutrino interactions.[41] The target was fractioned in two identical detector modules (Fig. 10). A muon spectrometer at the end of each module was allowing for muon identification and the measurement of its charge and momentum.

The brick (Fig. 11) is an almost complete standalone detector. It allows for precise reconstruction of the neutrino vertex and of the τ decay topology. Each brick can act as well as a finely segmented electromagnetic calorimeter (total thickness $\sim$10 X_0) providing electron and gamma identification and their energy reconstruction. Momenta of secondary charged particles can be reconstructed as well by measuring their multiple scattering through the lead plates. The calorimetric measurement of electromagnetic showers and the momentum measurement of not-showering charged particles can ensure the reconstruction of the neutrino interaction kinematics. Particle identification (muons, pions, protons) can be provided by measuring the specific ionization losses of tracks in the emulsions close to their stopping points.[55] This measurement could be performed by including the bricks downstream the one containing the neutrino vertex down to each track

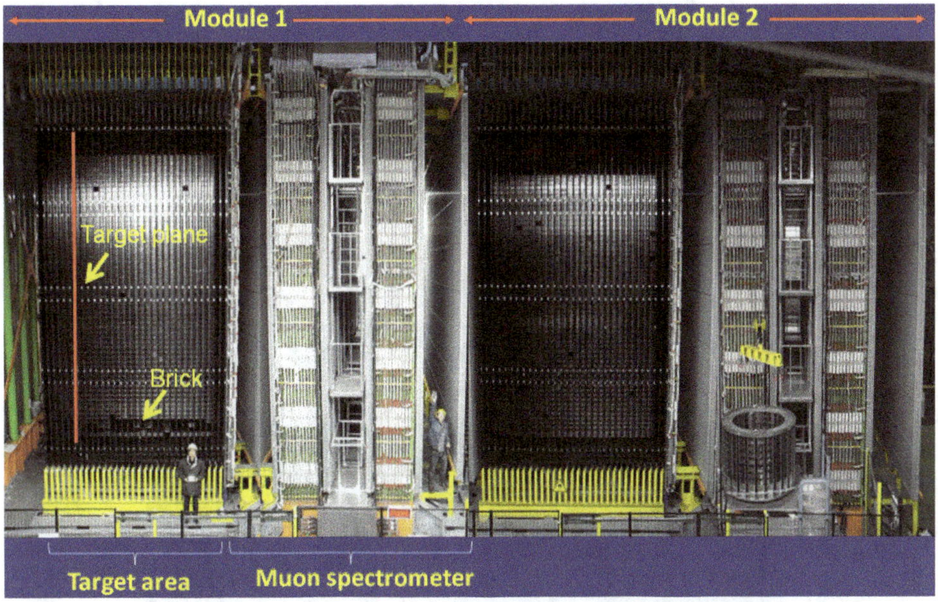

Fig. 10. Picture of the two OPERA detector modules.

stopping point. The bricks with neutrino interactions were localized with the scintillator hodoscopes and extracted from the target by a robotic system. The analysis of the emulsion films in the bricks was performed by scanning them with high speed automated microscopes. Figure 12 shows the first τ candidate observed in OPERA in 2010.[42] The event is interpreted as a ν_τ CC interaction followed by a $\tau \to \rho + \nu$ decay chain. From the event display one can appreciate the richness of details and the accuracy in the reconstruction of the primary vertex, the secondary tracks and the kink topology. The event includes two electromagnetic showers pointing to the secondary vertex presenting the kink topology. For a series of practical reasons it is difficult to think that the detection technology used by OPERA could be further extrapolated to larger detectors.

7. The liquid argon time projection chamber: precision measurements in neutrino physics

The LAr TPC[58] uses LAr as a fully active, homogeneous, interaction and detection medium. LAr is a relatively cheap medium which can be used to fill very large volumes. LAr is obtained from the distillation of air, where argon is present at the 1% level; at atmospheric pressure it boils at 87.3 K. The LAr TPC provides 3D imaging and reconstruction of neutrino interactions[63] as well as high-resolution measurement of the energy losses

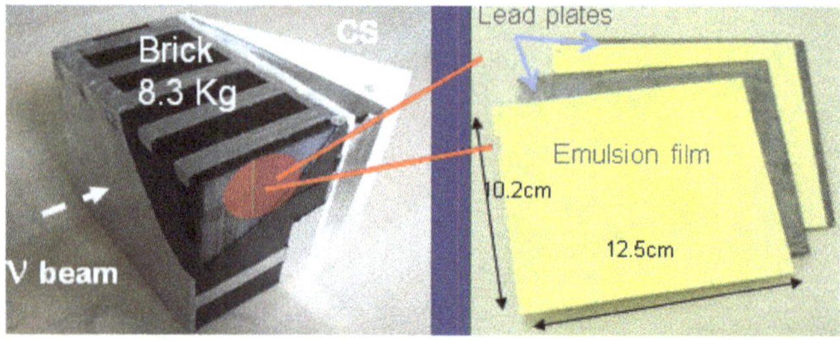

(a)

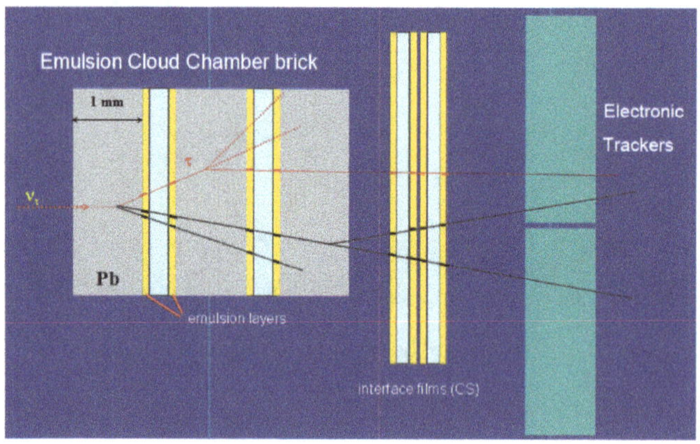

(b)

Fig. 11. The ECC bricks used in the OPERA hybrid detector for the ν_τ appearance tagging. (a) Picture of an ECC brick: sandwich of lead sheets and emulsion films. (b) τ decay topology reconstruction in the ECC and link to the tracks sampled in the electronic detectors.

Fig. 12. The first ν_τ CC candidate reconstructed in a ECC brick of the OPERA detector.

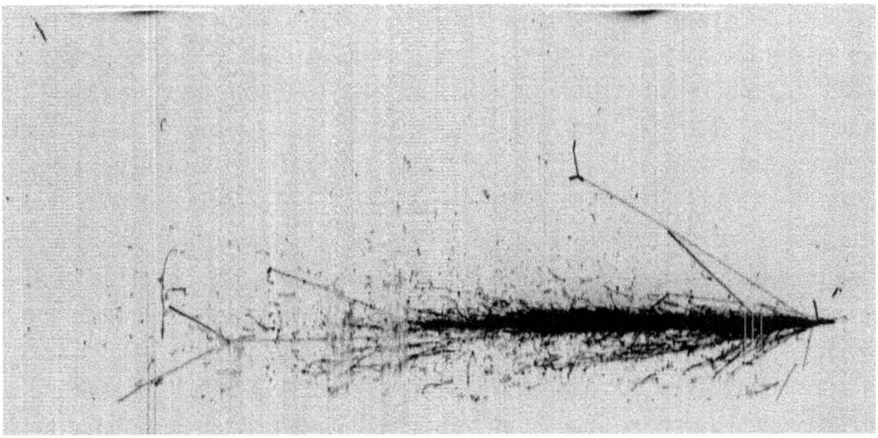

Fig. 13. Image of a ν_e CC neutrino interaction candidate in the ICARUS LAr TPC with a high energy electron shower.

of each charged particle produced in the event. An important aspect of this technology, crucial for the construction of large size detectors, is the possibility of performing the readout from a side of the detector, where the anode, collecting the electrons from the ionization made by secondary particles, is located. These electrons reach the anode after a drift path traveled at constant speed. The drift of the charges happens, similarly as in gas TPC, by enclosing the active LAr volume in a field shaping cage defining a uniform electric field of typically 0.5–1 kV/cm. This projective geometry makes the number of needed readout channels scaling proportionally to the surface of the detector instead than to its volume. LAr is also an excellent medium for electromagnetic calorimetry[64] ($1\ X_0 = 14$ cm) and it can be used for hadronic calorimetry as well ($1\ \lambda_I = 85.7$ cm). The original detector concept was pioneered by C. Rubbia in 1977. LAr does not attach the electrons produced by the energy losses of particles crossing its volume. Other important properties are related to the high electron mobility and low diffusion.[56] Residual electronegative impurities which would absorb the electrons, such as O_2, must be removed to ensure long enough drift paths of the electrons from their production point to the anode. A key point for the practical feasibility of the detector was then to demonstrate the capability of removing these impurities by purifying the LAr obtained from the distillation of air. The clouds of electrons produced by the ionization along charged particles trajectories conserve, during the drift, their original shape, slightly smeared by diffusion effects. The detector can then provide an accurate imaging of the original distribution of the ionization similarly as in bubble chambers.

After particles have ionized the LAr, part of the created electron–ion pairs will immediately recombine by emitting prompt UV scintillation light at 128 nm. The scintillation process has also a slower light emission component ($\tau = 1.6\,\mu$s). About 5000 photons/mm are produced for a track at the ionization minimum. This light can be detected by using cryogenic photodetectors coated with a wavelength-shifter material used to convert from UV to visible light. The detection of the prompt light component provides information about the time the interaction occurred and the ionization was produced in LAr (event T_0). When electrons reach the end of their drift, they can be collected by an anode made a plane of wires spaced with a typical readout pitch of 3 mm. The readout system may provide more than one view by including planes of induction wires upstream the collection wires. By setting at appropriate potentials on the readout planes, the induction wires will be almost transparent to the electrons transit. The electrons will end their drift path on the collection wires generating a bipolar signal while transiting nearby the induction wires. The coordinate perpendicular to the wires can be reconstructed by measuring the drift time with respect to the event T_0. The drift speed of the electrons is 1.5 m/ms for an electric field of 0.5 kV/cm. The readout of the electric signals from the wires implies the use of continuously sampling analog to digital converters (ADCs). Typically the signal waveforms are sampled by the ADC at 2.5 MHz. Reading just the collection wires would provide only one 2D view of the events. It is possible to reconstruct additional views by adding, upstream the collection plane along the drift direction, additional planes of induction wires. The electric potentials biasing the wire planes can be adjusted in order to get progressively stronger fields in the direction of the collection wires, thus ensuring high transparency of the induction planes for the drifting electrons. Under these conditions, most of the electrons will be then reaching the collection plane. By implementing at least two orthogonal views, the detector can perform a 3D reconstruction of the events. Unlike in gas detectors, it is difficult to obtain multiplication in proximity of wires immersed in the liquid due to the electrons mean free path in LAr and the need of very strong electric fields in order to develop avalanches. For this reason, immersed wires cannot be used to increase the original ionization. After local recombination at the ionization place, for a particle at the ionization minimum, the typical quantity of charge which is available to be drifted towards the anode is about 20000 electrons every 3 mm of track length. Even neglecting the electron losses during the drift, this is a tiny amount of charge which needs to be amplified in order to become exploitable. Low noise amplifiers must be used for that. The noise introduced by the amplifiers will be proportional to the

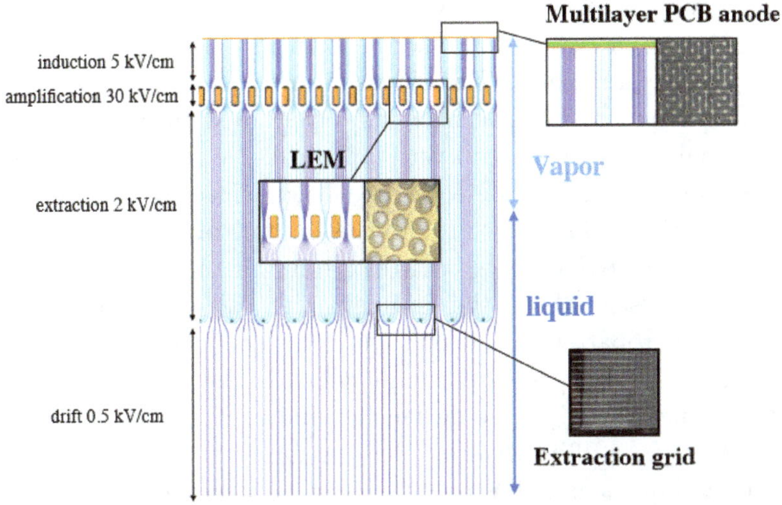

Fig. 14. Principle of charge amplification and collection in a dual-phase LAr TPC.

capacitance of the long wires (~a few meters) deployed for the charge readout plus the capacitance of the cables used to bring the electrical signals to the amplifiers. The typical S/N ratio for a minimum ionizing particle on the collection view is of the order of 10:1. The LAr purity typically needed to drift the electrons over distances of the order of meters corresponds to an equivalent contamination of O_2 at the level of ~0.1 ppb. This level of contamination results in 3 ms free electron lifetime before attachment to an electronegative impurity. This lifetime has to be compared to the drift time needed to reach the anode, which for a field of 0.5 kV/cm is of the order of 0.67 ms for each traveled meter of drift path. A dedicated purification/recirculation system of the LAr is necessary in order to achieve and preserve with time such a high level of purity.[62] ICARUS[60] was the first large-scale LAr TPC detector built for neutrino physics. Each detector module of 600 ton was made of two TPC of 300 ton each with a drift path of 1.5 m. ICARUS was part of the ν_τ appearance program based on the CERN neutrino beam to Gran Sasso.[59] The experiment was originally designed to reach a total detector mass of 5 kton. Due to boundary constraints, the target mass eventually ended up to be limited to the first 600 ton module only. The operation of this module and its exposure to the CNGS neutrino beam were however a remarkable milestone in the large-scale demonstration of this detector technology.[61]

As we have seen in Section 4, future precision measurements in neutrino oscillations will mostly rely on the study of $\nu_\mu \to \nu_e$ oscillations. They will then require very large detectors with excellent electron identification capabilities and good energy resolution. The LAr TPC is a natural technology fulfilling these requirements. The extension of the LAr technology to huge underground detectors[65] has been extensively studied during the last decade. In order to build a detector with very long drifts (∼10 m), it is necessary to introduce some gain in the detector itself compensating for the electrons' losses due to the residual electronegative impurities. This is possible by amplifying the drifted charges with electron avalanches occurring in the gas naturally present above the LAr level.[66–71] These avalanches can be generated in micropattern gas detectors. However, in normal gas detectors, some quenching components are usually added to the gas in order to absorb the UV photons created in the avalanches and prevent diffuse sparking. In the dual-phase LAr TPC the gas phase must be ultrapure and it is not possible to add quenchers. UV photons must be then mechanically confined in the gas detector structure itself. These detectors must also be operated at much lower gain (∼20–100) than standard gas detectors ($\sim 10^5$–10^6). The extraction of the electrons from the liquid to the gas phase[57] is achieved with full efficiency by applying across the gas–liquid interface an electric field stronger than 2 kV/cm defined by an immersed grid. Avalanches occur in a micropattern structure, made of 500 μm holes drilled with 800 μm pitch in a printed circuit board, the large electron multiplier (LEM). The LEM has a voltage difference of ∼3 kV applied on the two sides which produces a field of ∼30 kV/cm within the holes where the avalanches are produced. A second

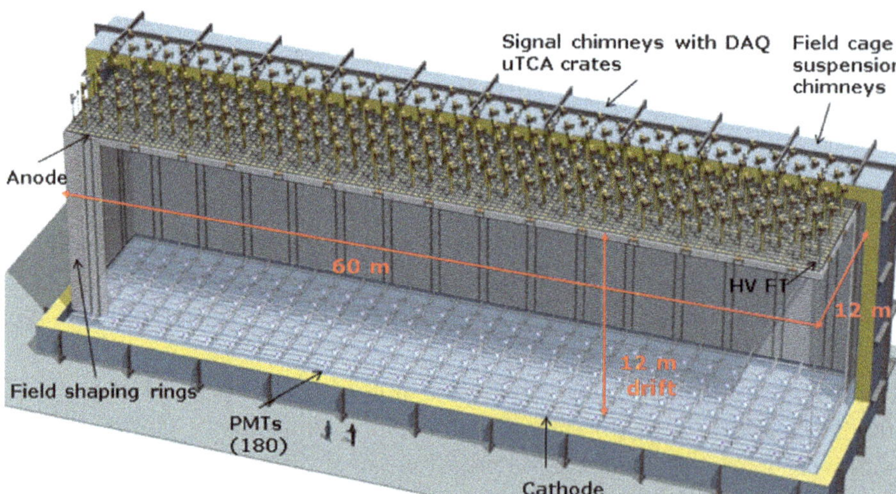

Fig. 15. Open view of the dual-phase 10 kton LAr TPC module design for the DUNE experiment.

printed circuit board placed at ground potential just above the LEM can collect the electrons on copper tracks arranged in two orthogonal collection views. This configuration (Fig. 14) typically improves the S/N ratio by one order of magnitude ($S/N \sim 100$ for minimum ionizing particles) with respect to the charge readout in liquid. It makes accessible the construction of detectors with 10–20 m drift path, which are fully active and have a reduced number of readout channels distributed between two equivalent collection views.

The dual-phase design was first considered in the European neutrino oscillations project LBNO[72] aiming at the determination of the mass hierarchy and search for CP violation. LBNO was including two detectors for a total LAr fiducial mass of 70 kton at 2300 km distance from the neutrino beam source at CERN. The proposed LBNO experiment was designed to make large use of the spectral information[73] by looking at the first and second oscillations' maxima. This was possible given the wide band beam spectrum covering the few GeV region and the excellent neutrino energy resolution of the detector (~10%). The same technology, as well as the one based on a single-phase design extended from ICARUS, is considered in the DUNE experiment[74–77] on which European and USA efforts have been jointly focusing since 2014. DUNE will look for neutrino oscillations searching for CP violation and for the determination of the mass hierarchy with a total target mass of 40 kton, fractioned in four 10 kton LAr TPC detector modules (Fig. 15), to be installed in the SURF underground laboratory at 1300 km from neutrino beam source at Fermilab. A full scale demonstrator of the dual-phase technology, the WA105 (*ProtoDUNE dual-phase*) experiment,[78] is being built at CERN together with a module of similar dimensions based on the wires readout single-phase technology (*ProtoDUNE single-phase*). It will have an active volume of $6 \times 6 \times 6\,\text{m}^3$ and it will be exposed in 2018 to beams of electrons and charged hadrons allowing for its calibration and for a detailed assessment of the performance in showers reconstruction.

References

1. F. J. Hasert *et al.* [Gargamelle Neutrino Collaboration], Observation of neutrino like interactions without muon or electron in the Gargamelle neutrino experiment, *Phys. Lett. B* **46** (1973) 138. doi:10.1016/0370-2693(73)90499-1.
2. D. A. Glaser, Some effects of ionizing radiation on the formation of bubbles in liquids, *Phys. Rev.* **87** (1952) 665. doi:10.1103/PhysRev.87.665.
3. D. Glaser and D. Rahm, Characteristics of Bubble Chambers, *Phys. Rev.* **97** (1955) no.2, 474. doi:10.1103/PhysRev.97.474.

4. H. Bradner, Bubble chambers, *Ann. Rev. Nucl. Part. Sci.* **10** (1960) 109. doi:10.1146/annurev.ns.10.120160.000545.
5. D.A. Glaser, Elementary particles and bubble chambers, http://www.nobelprize.org/nobel_prizes/physics/laureates/1960/glaser-lecture.pdf.
6. N. Armenise *et al.*, Search for neutrino oscillations in Gargamelle at SPS, *Phys. Lett.* **B100**, (1981) 182.
7. J. Boger *et al.* (SNO), The Sudbury Neutrino Observatory, *Nucl. Instrum. Meth.* **A449** (2000) 172, nucl-ex/9910016.
8. Y. Abe *et al.* (Double Chooz), Reactor electron antineutrino disappearance in the Double Chooz experiment, *Phys. Rev.* **D86** (2012) 052008, arXiv:1207.6632 [hep-ex].
9. Fengpeng An *et al.* (JUNO), Neutrino Physics with JUNO, *J. Phys.* **G43** (2016) 030401, arXiv:1507.05613 [physics].
10. J. Timmer [UA1 Collaboration], The UA1 Detector.
11. F. Vannucci, The NOMAD experiment at CERN, advances in high energy physics, **2014**(20), Article ID 129694. doi:10.1155/2014/129694.
12. J. Altegoer *et al.* (NOMAD), A search for $\nu_\mu \to \nu_\tau$ oscillations using the NOMAD detector, *Phys. Lett.* **B431** (1998) 219.
13. D. Macina *et al.* (CHORUS), The CHORUS experiment, *Nucl. Phys. Proc. Suppl.* **48** (1996) 183.
14. A. Ereditato, The Study of Neutrino Oscillations with Emulsion Detectors, *Adv. High Energy Phys.* **2013** (2013) 382172. doi:10.1155/2013/382172.
15. C. H. Albright and R. E. Shrock, Signals for tau-neutrino Interactions in a Beam Dump Experiment, *Phys. Lett. B* **84** (1979) 123. doi:10.1016/0370-2693(79)90665-8
16. J. Altegoer *et al.* (NOMAD), The NOMAD experiment at the CERN SPS, *Nucl. Instrum. Meth.* **A404** (1998) 96.
17. D. Autiero *et al.*, The electromagnetic calorimeter of the NOMAD experiment, *Nucl. Instrum. Meth.* **A373** (1996) 358.
18. G. Bassompierre *et al.*, A large area transition radiation detector for the NOMAD experiment, *Nucl. Instrum. Meth.* **A403** (1998) 363.
19. P. Astier *et al.* (NOMAD), Search for $\nu_\mu \to \nu_e$ oscillations in the NOMAD experiment, *Phys. Lett.* **B570** (2003) 19, hep-ex/0306037.
20. C. Athanassopoulos *et al.* (LSND), Evidence for $\bar{\nu}_\mu \to \bar{\nu}_e$ oscillation from the LSND experiment at the Los Alamos Meson Physics Facility, *Phys. Rev. Lett.* **77** (1996) 3082, nucl-ex/9605003.
21. R. Becker-Szendy *et al.* (IMB), IMB-3: a large water Cerenkov detector for nucleon decay and neutrino interactions, *Nucl. Instrum. Meth.* **A324** (1993) 363.
22. M. Koshiba, Kamiokande experiment, 1988, SLAC Summer Institute, July 1988.
23. M. Koshiba, Observational neutrino astrophysics, *Phys. Rep.* **220** (1992) 229.
24. K. Hirata *et al.* (Kamiokande), Observation of a neutrino burst from the supernova SN1987a, *Phys. Rev. Lett.* **58** (1987) 1490.

25. R. M. Bionta *et al.* (IMB), Observation of a neutrino burst in coincidence with supernova SN1987A in the Large Magellanic Cloud, *Phys. Rev. Lett.* **58** (1987) 1494.
26. K. S. Hirata *et al.* (Kamiokande), Observation in the Kamiokande-II detector of the neutrino burst from supernova SN1987A, *Phys. Rev.* **D38** (1988) 448.
27. K. S. Hirata *et al.* [Kamiokande-II Collaboration], Experimental Study of the Atmospheric Neutrino Flux, *Phys. Lett. B* **205** (1988) 416. doi:10.1016/0370-2693(88)91690-5.
28. M. Koshiba, Birth of Neutrino Astrophysics, Nobel Lecture, December 8, 2002 https://www.nobelprize.org/nobel_prizes/physics/laureates/2002/koshiba-lecture.pdfl.
29. S. Fukuda *et al.* (SuperKamiokade), The Super-Kamiokande detector, *Nucl. Instrum. Meth.* **A501** (2003) 418.
30. T. Kajita, Discovery of atmospheric neutrino oscillations, Nobel Lecture, December 8, 2015 https://www.nobelprize.org/nobel_prizes/physics/laureates/2015/kajita-lecture-slides.pd.
31. S. H. Ahn *et al.* (K2K), Detection of accelerator produced neutrinos at a distance of 250-km, *Phys. Lett.* **B511** (2001) 178, hep-ex/0103001.
32. M. H. Ahn *et al.* (K2K), Measurement of neutrino oscillation by the K2K experiment, *Phys. Rev.* **D74** (2006) 072003, hep-ex/0606032.
33. N. Okamura *et al.*, Prospects of very long base-line neutrino oscillation experiments with the JAERI-KEK high intensity proton accelerator, arXiv preprint hep-ph/0104220.
34. K. Abe *et al.* (T2K), Observation of electron neutrino appearance in a muon neutrino beam, *Phys. Rev. Lett.* **112** (2014) 061802, arXiv:1311.4750.
35. T. Kajita, Y. Totsuka, Observation of atmospheric neutrinos, *Rev. Mod. Phys.* **73** (2001) 85.
36. N. Abgrall *et al.* (T2K), The T2K Experiment, *Nucl. Instrum. Meth.* **A659** (2011) 106, arXiv:1106.1238 [physics].
37. K. Abe *et al.* [Hyper-Kamiokande Working Group Collaboration], A Long Baseline Neutrino Oscillation Experiment Using J-PARC Neutrino Beam and Hyper-Kamiokande, arXiv:1412.4673 [physics.ins-det].
38. K. Kodama *et al.* [DONUT Collaboration], Observation of tau neutrino interactions, *Phys. Lett. B* **504** (2001) 218 doi:10.1016/S0370-2693(01)00307-0.
39. A. Ereditato, P. E. Strolin and K. Niwa, OPERA: An emulsion detector for a long baseline nu/mu nu/tau oscillation search, *Nucl. Phys. Proc. Suppl.* **66** (1998) 423. doi:10.1016/S0920-5632(98)00079-6.
40. M. Guler *et al.* (OPERA), OPERA: An appearance experiment to search for $\nu_\mu \to \nu_\tau$ oscillations in the CNGS beam. Experimental proposal CERN/SPSC (2000).
41. R. Acquafredda, *et al.* The OPERA experiment in the CERN to Gran Sasso neutrino beam, *JINST* **4** (2009) P04018. doi:10.1088/1748-0221/4/04/P04018.
42. N. Agafonova *et al.* (OPERA), Observation of a first ν_τ candidate in the OPERA experiment in the CNGS beam, *Phys. Lett.* **B691** (2012) 138, arXiv:1006.1623 [hep-ex].

43. E. Gschwendtner, K. Cornelis, I. Efthymiopoulos, A. Pardons, H. Vincke, J. Wenninger, and I. Krtschmer, CNGS, CERN neutrinos to Gran Sasso, five years of running a 500 kilowatt neutrino beam facility at CERN, *Conf. Proc.* Vol. 130512, No. CERN-Acc-2013-0266.
44. F. Dydak *et al.* (CDHSW), A search for ν_μ oscillations in the Δm^2 range 0.3–90 eV^2, *Phys. Lett.* **B134** (1984) 281.
45. C. Arroyo *et al.* (CCFR), A Precise measurement of the weak mixing angle in neutrino nucleon scattering, *Phys. Rev. Lett.* **72** (1994) 3452, hep-ex/9405008.
46. A. N. Diddens *et al.* (CERN-Hamburg-Amsterdam-Rome-Moscow), A detector for neutral current interactions of high-energy neutrinos, *Nucl. Instr. Meth.* **178** (1980) 27.
47. K. De Winter *et al.* (CHARM-II), Experimental results obtained from a low Z, fine grained electromagnetic calorimeter, *Nucl. Instrum. Meth.* **A277** (1989) 83.
48. D. G. Michael *et al.* (MINOS), Observation of muon neutrino disappearance with the MINOS detectors and the NuMI neutrino beam, *Phys. Rev. Lett.* **97** (2006) 191801, hep-ex/0607088.
49. P. Adamson *et al.*, MINOS Technical Design Report, NuMI-L-337.
50. J. Evans New results from the MINOS+ experiment, Neutrino 2016 Conference http://neutrino2016.iopconfs.org/IOP/media/uploaded/EVIOP/event_948/14.00___2_.pdf.
51. R. Bayes, A. Laing, F. J. P. Soler, A. Cervera Villanueva, J. J. Gomez Cadenas *et al.*, The Golden Channel at a neutrino factory revisited: improved sensitivities from a magnetised iron neutrino detector, *Phys. Rev.* **D86** (2012) 093015, arXiv:1208.2735 [hep-ex].
52. P. Adamson *et al.* (NOvA), First measurement of electron neutrino appearance in NOvA, *Phys. Rev. Lett.* **116** (2016) 151806, arXiv:1601.05022.
53. D. Drakoulakos *et al.* (Minerva), Proposal to perform a high-statistics neutrino scattering experiment using a fine-grained detector, hep-ex/0405002.
54. D. Autiero and Y. Declais, Accelerator neutrinos, *Comptes Rendus Physique* **6**, 758 (2005). doi:10.1016/j.crhy.2005.07.009
55. N. Agafonova *et al.* [OPERA Collaboration], New results on $\nu_\mu \to \nu_\tau$ appearance with the OPERA experiment in the CNGS beam, *JHEP* **1311** (2013) 036.
56. C. Rubbia, The liquid argon time projection chamber: A new concept for neutrino detectors, 1977, CERN-EP-INT-77-08.
57. B. A. Dolgshein *et al.*, New method of registration of ionizing-particle tracks in condensed matter, PETP LETTERS, 1970.
58. A. Marchionni, Status and new ideas regarding liquid argon detectors, *Ann. Rev. Nucl. Part. Sci.* **63** (2013) 269, doi:10.1146/annurev.nucl.012809.104445 [arXiv:1307.6918 [physics.ins-det]].
59. P. Aprili *et al.* (ICARUS), The ICARUS experiment: A second-generation proton decay experiment and neutrino observatory at the Gran Sasso laboratory. Cloning of T600 modules to reach the design sensitive mass. (Addendum), 2002, CERN-SPSC-2002-027.

60. S. Amerio *et al.* (ICARUS), Design, construction and tests of the ICARUS T600 detector, *Nucl. Instrum. Meth.* **A527** (2004) 329.
61. C. Rubbia *et al.* Underground operation of the ICARUS T600 LAr-TPC: first results, *JINST* **6** (2011) P07011, arXiv:1106.0975 [hep-ex].
62. B. Baibussinov *et al.* Free electron lifetime achievements in liquid argon imaging TPC, *JINST* **5** (2010) P03005, arXiv:0910.5087 [physics].
63. M. Antonello *et al.* Precise 3D track reconstruction algorithm for the ICARUS T600 liquid argon time projection chamber detector, *Adv. High Energy Phys.* **2013** (2013) 260820, arXiv:1210.5089 [physics].
64. A. Ankowski *et al.* (ICARUS), Energy reconstruction of electromagnetic showers from π^0 decays with the ICARUS T600 liquid argon TPC, *Acta Phys. Polon.* **B41**, 103 (2010), arXiv:0812.2373 [hep-ex].
65. D. Autiero *et al.* Large underground, liquid based detectors for astro-particle physics in Europe: Scientific case and prospects, *JCAP* **0711** (2007) 011 doi:10.1088/1475-7516/2007/11/011 [arXiv:0705.0116 [hep-ph]].
66. A. Badertscher *et al.* Operation of a double-phase pure argon large electron multiplier time projection chamber: comparison of single and double phase operation, *Nucl. Instrum. Meth. A* **617** (2010) 188 doi:10.1016/j.nima.2009.10.011 [arXiv:0907.2944 [physics.ins-det]].
67. A. Badertscher *et al.* First operation of a double phase LAr large electron multiplier time projection chamber with a two-dimensional projective readout anode, *Nucl. Instrum. Meth. A* **641** (2011) 48 doi:10.1016/j.nima.2011.02.100 [arXiv:1012.0483 [physics.ins-det]].
68. A. Badertscher *et al.* First operation and drift field performance of a large area double phase LAr Electron Multiplier Time Projection Chamber with an immersed Greinacher high-voltage multiplier, *JINST* **7** (2012) P08026 doi:10.1088/1748-0221/7/08/P08026 [arXiv:1204.3530 [physics.ins-det]].
69. A. Badertscher *et al.* First operation and performance of a 200 lt double phase LAr LEM-TPC with a $40x76cm^2$ readout, *JINST* **8** (2013) P04012 doi:10.1088/1748-0221/8/04/P04012 [arXiv:1301.4817 [physics.ins-det]].
70. C. Cantini *et al.* Long-term operation of a double phase LAr LEM time projection chamber with a simplified anode and extraction-grid design, *JINST* **9** (2014) P03017 doi:10.1088/1748-0221/9/03/P03017 [arXiv:1312.6487 [physics.ins-det]].
71. C. Cantini *et al.* Performance study of the effective gain of the double phase liquid argon LEM time projection chamber, *JINST* **10** (2015) no. 03, P03017 doi:10.1088/1748-0221/10/03/P03017 [arXiv:1412.4402 [physics.ins-det]].
72. A. Stahl *et al.* Expression of Interest for a very long baseline neutrino oscillation experiment (LBNO), CERN-SPSC-2012-021, SPSC-EOI-007.
73. S. K. Agarwalla *et al.* [LAGUNA-LBNO Collaboration], The mass-hierarchy and CP-violation discovery reach of the LBNO long-baseline neutrino experiment, *JHEP* **1405** (2014) 094 doi:10.1007/JHEP05(2014)094 [arXiv:1312.6520 [hep-ph]].
74. R. Acciarri *et al.* [DUNE Collaboration], Long-baseline neutrino facility (LBNF) and deep underground neutrino experiment (DUNE): Volume 1: The LBNF and DUNE Projects, arXiv:1601.05471 [physics.ins-det].

75. R. Acciarri *et al.* [DUNE Collaboration], Long-baseline neutrino facility (LBNF) and deep underground neutrino experiment (DUNE): Volume 2: The physics program for DUNE at LBNF, arXiv:1512.06148 [physics.ins-det].
76. J. Strait *et al.* [DUNE Collaboration], Long-baseline neutrino facility (LBNF) and deep underground neutrino experiment (DUNE): Volume 3: Long-Baseline Neutrino Facility for DUNE June 24, 2015, arXiv:1601.05823 [physics.ins-det].
77. R. Acciarri *et al.* [DUNE Collaboration], Long-baseline neutrino facility (LBNF) and deep underground neutrino experiment (DUNE): Volume 4 The DUNE Detectors at LBNF, arXiv:1601.02984 [physics.ins-det].
78. L. Agostino *et al.* LBNO-DEMO: Large-scale neutrino detector demonstrators for phased performance assessment in view of a long-baseline oscillation experiment, arXiv:1409.4405 [physics.ins-det].

Chapter 4

Solar Neutrinos and Matter Effects

A. Y. Smirnov

Max-Planck-Institute for Nuclear Physics,
Heidelberg D-69117, Germany
smirnov@mpi-hd.mpg.de

Abstract

Studies of the solar neutrinos move to new phase of precision (%–sub% level) measurements. With this perspective, production of the solar neutrinos and their properties (spectra, fluxes) are described. Matter effects relevant for the propagation of solar neutrinos inside the Sun and the Earth are presented in the complete 3ν framework. Those include, in particular, the adiabatic flavor conversion and oscillations in multilayer medium. The LMA MSW solution of the solar neutrino problem is described in details. We discuss results from the solar neutrino experiments and their contributions to establishing the solution of the solar neutrino problem. Status of the LMA MSW solution and existing tensions between the data and predictions are summarized. New matter effects behind these tensions can be due to nonstandard neutrino interactions or presence of light sterile neutrinos. The main issues of future studies of solar neutrinos include precision measurements of all the fluxes, further tests of the LMA MSW solution, exploration of the intermediate energy range, searches for new physics effects, tests of the flux–luminosity relation, solution of the solar metallicity problem. In this connection, physics reach of future experiments is outlined.

1. Introduction

The idea that the Sun produces flux of neutrinos emerged from understanding that nuclear reactions are the source of energy of the Sun and stars. However, in the original papers,[1,2] neutrinos did not even appear in formulas of nuclear reactions! The reasons were that neutrinos had not been experimentally discovered yet being still hypothetical particles and the concept of lepton number conservation was not introduced. Pontecorvo mentioned solar neutrinos in his report devoted to the Cl–Ar method[3]:

"Neutrino flux from the Sun is of the order 10^{10} cm^{-2}s^{-1}. However these neutrinos (*pp*–, NO–) are not very energetic". So, he did not consider them as a realistic case for the neutrino discovery.

The key step in developments of the field was the establishment by Holmgren and Johnston[4] that the cross section of ^{3}He + ^{4}He with production of ^{7}Be is substantially larger than what was sought before. After that, detection of neutrinos from two other *pp*–chains of nuclear reactions in the Sun have been considered by Fowler[5]: (i) electron capture by ^{7}Be which produces the Be neutrinos and (ii) proton capture by ^{7}Be with formation of ^{8}B. The β decay of the latter produces the high-energy boron neutrinos which could be detected.

This had triggered intensive studies of the solar neutrino fluxes and developments of what we call the Standard Solar Model (SSM).[6,7] At the same time, computations of the cross section of Cl–Ar transition have been refined. In particular, it was realized by Bahcall[8] that due to transitions to exited state of Ar, the number of expected events is about 20 times larger than the earlier estimations. At this point, the proposal of Cl–Ar experiment became realistic: for feasible size of the detector and computed neutrino flux, the number of detected neutrino events was large enough. Construction of the Homestake laboratory and Davis detector started. The main goal was to test nuclear origin of the solar energy.

In the middle of 1960s, the idea of complete spectroscopy of solar neutrinos was put forward by Zatsepin and Kusmin and gallium experiment based on the Ga–Ge radiochemical method was proposed.[9] The low threshold of this experiment allows to detect the main component of the spectrum — the *pp* neutrinos.

The solar neutrino problem — the deficit of signal with respect to expectations was anticipated by Pontecorvo. In his 1967 paper,[10] he has marked: "If the oscillation length is large... from the point of view of detection possibilities an ideal object is the Sun." The ν_e signal can be suppressed due to (averaged) oscillations into ν_μ or ν_s. The first Homestake result published after a year, indeed, showed the deficit of signal: about 1/3 of the expected number of events has been detected.[11]

The Homestake result had triggered vast theoretical developments in attempts to explain the deficit. The main directions were (i) astrophysics (e.g., lower temperature in the center of the Sun), (ii) nuclear physics, (iii) new neutrino physics based on nonstandard neutrino properties.

Apart from averaged vacuum oscillations, the solutions based on neutrino spin precession in the magnetic fields of the Sun,[12,13] neutrino decay,[14,15] "just-so" vacuum oscillations with oscillation length comparable with distance between the Sun and the Earth[16] have been proposed. These solutions required neutrino mass and mixing and/or large magnetic moments, *etc.*

In 1978, studying a possibility of oscillations of the massless neutrinos, which require nonstandard interactions, Wolfenstein[17,18] introduced the refraction effect on neutrinos propagating in matter. In the case of massive neutrinos, usual interactions with matter should modify vacuum oscillations as pointed out by Wyler. In the latter case, for solar neutrinos, the conclusion was that matter suppresses oscillations and therefore oscillations should be considered starting from the surface of the Sun. Wolfenstein also thought that due to adiabaticity, it should not be any transition inside the Sun and solar matter just suppresses the oscillations. Barger *et al.*[19] marked that there is always some energy where ν or $\bar{\nu}$ matter mixing is maximal, but the phenomenon of resonance was not recognized.

In 1985, Mikheyev and Smirnov, following Wolfenstein's ideas have realized that the mixing in matter has a resonance character.[20,21] Two effects have been uncovered: the resonance enhancement of oscillations in matter with constant density and the adiabatic (or partially adiabatic) conversion in matter with varying density. The latter allows us to explain the deficit of solar neutrinos in certain range of oscillation parameters. This is what we now call the MSW solutions of the solar neutrino problem.[17,18,20,21] Theory of neutrino flavor transformations in different media has been elaborated[22–26] (see reviews[27–30]).

Matter effects and resonance have been incorporated in other solutions of the problem, e.g., resonance spin-flavor precession.[31,32] Furthermore, the explanations of the ν_e deficit based on the resonance matter effects due to nonstandard neutrino interactions have been proposed.[33,34]

It took more than 17 years to identify the correct solution. With data from Kamiokande-II, SAGE, Gallex, GNO, Super-Kamiokande (SK), a number of solutions have been excluded.[35] Evidences accumulated that new neutrino physics rather than astrophysics is behind the solar ν_e deficit. After publication of the first SK results, the large mixing angle (LMA) MSW solution characterized by mass-squared difference $\Delta m_{21}^2 = (3-10)\cdot 10^{-5}\,\mathrm{eV}^2$ and mixing $\sin^2\theta_{12} = 0.25-0.35$ became the favored one.[36] Decisive results have been obtained by SNO[37] which showed that *neutrino flavor transformations* take place in the Sun, thus completely excluding astrophysical solutions as well as some particle physics solutions.

Ironically, the unique solution of solar neutrino problem had been identified by the nonsolar neutrino experiment KamLAND[38–40] which actually was motivated by a possibility to test the LMA solution. It is the

KamLAND reactor experiment that discovered the vacuum oscillations of $\bar{\nu}_e$ from atomic reactors that put a full stop in the story (in assumption of the CPT invariance): Values of Δm_{21}^2 and θ_{12} measured by KamLAND were in agreement with the LMA MSW solution.

Later, BOREXINO performed direct measurements of the ^{7}Be, *pp* and *pep* neutrinos which improved further an overall agreement of the LMA MSW with data. Establishing the day–night asymmetry of the signal by SK provides an evidence of the matter effects on oscillations and a complete check of the whole picture of the solar neutrino conversion.

In Section 2, we describe production of the solar neutrino fluxes and properties of their energy spectra. We comment on total normalization of the flux and on ratios of fluxes of different components and their dependence on characteristics of the Sun. In Section 3, we consider matter effects relevant for propagation of solar neutrinos (adiabatic flavor conversion, oscillations in multilayer medium). In Section 4, we summarize results from different experiments and their role in establishing the solution of the solar neutrino problem. In Section 5, the present status of the LMA MSW solution is outlined, some existing tensions are discussed and possible effects behind the tensions based on new matter effects are described. In Section 6, we discuss future studies of the solar neutrinos: the main issues, problems, and in this connection, physics potential of forthcoming and planned experiments. We conclude in Section 7.

2. The Sun and solar neutrino fluxes

2.1. *Fundamental reaction; luminosity–flux relation*

Solar neutrinos are by-products of the nuclear reactions that produce energy in the Sun (see Ref. 6 for basics, Ref. 7 for historical perspective, and Refs. 41–44 for recent reviews and computations[45,46]).

In the Sun, the hydrogen burning proceeds via chains of nuclear reactions that eventually convert four protons into helium:

$$4p \rightarrow ^4\mathrm{He} + 2e^+ + 2\nu_e + Q. \tag{1}$$

The energy release, $Q = 26.73\,\mathrm{MeV}$, is in the form of kinetic energy of (light) particles and photons, as well as energy released in annihilation of positrons with electrons of plasma. According to Eq. (1), only electron neutrinos are produced. The transformation (1) proceeds mainly via the *pp* chain of reactions which provides 98.4% of the energy generation in the Sun. About 1.6% of energy is produced in the CNO bi-cycle.

The fundamental reaction which starts the dominant *pp* chain is

$$p + p \to \nu_e + d + e^+. \tag{2}$$

It can be considered as β^+ decay of a proton in the presence of another proton which leads to the formation of deuteron. The reaction proceeds via the weak interaction and has the Coulomb suppression. So, its rate has double suppression: it is one of the slowest processes in the chain with typical time $t_{pp} = 10^{10}$ years which exceeds the age of the Sun. Other reactions of the chain (apart from *pep*) are much faster: typically about 10^6 years which gives the time of termination of the chain in the center of the Sun. It is the reaction (2) that determines the time of hydrogen burning and nuclear evolution of the Sun. And it is the weak interaction that determines evolution of stars. The energy release in Eq. (2) is rather small: $Q_{pp} + 2m_e$, where annihilation of positron is taken into account. The cross section of Eq. (2) can be reliably computed with accuracy better than 2%. Reaction (2) being the first one in the chain produces dominant flux of solar neutrinos.

Neutrino flux luminosity relation. The average energy of solar neutrinos equals

$$\bar{E}_\nu = \sum_i E_i \frac{F_\nu^i}{F_\nu^{\text{tot}}} \approx 0.265\,\text{MeV}, \tag{3}$$

where E_i and F_ν^i are the average energy and flux of neutrinos produced in the ith reaction. $F_\nu^{\text{tot}} = \sum_i F_\nu^i$ is the total neutrino flux. Since the *pp* neutrino flux dominates, $\bar{E}_\nu$ is close to $\bar{E}^{pp} = 0.24\,\text{MeV}$. The average energy (Eq. (3)) and the energy release Q allow to establish relation between the total flux of neutrinos $F_\odot^\nu$ and total luminosity of the Sun $L_\odot$. This relation is based on several assumptions:

1. The time of diffusion of photons from the central parts of the Sun where nuclear reactions occur to its surface is about $t_{\text{diff}} = 10^5$ years. Therefore, the present day luminosity of the Sun is determined by the energy release 10^5 years ago. It is assumed (and SSM confirms this) that changes in nuclear energy release and energy transport during such a period can be neglected.
2. Most of neutrinos are emitted from completed (terminated) chains of reactions (Eq. (1)). This is not correct for nuclear reactions that occur in outer (peripheral) regions of the solar core with lower temperatures. In these regions, reactions with high Coulomb barrier and strong dependence on temperature have low probability and time larger than the age of the Sun. It is assumed that contribution to the total energy release of unterminated

chains is negligible. The thermal energy from each terminated chain equals $Q - 2\bar{E}_\nu$ since two neutrinos are emitted in each termination. Consequently, the total flux of neutrinos is $2L_\odot/(Q - 2\bar{E}_\nu)$. Then the density of neutrino flux at the Earth equals

$$F_\nu^{\rm tot} = \frac{2L_\odot}{4\pi d^2(t)(Q - 2\bar{E}_\nu)}, \tag{4}$$

where $d(t)$ is the distance from the Sun to the Earth which changes by about 7% due to eccentricity of the Earth orbit. The relation in Eq. (4) only very weakly depends on the solar model via $\bar{E}_\nu^i$ which, in turn, depends on branchings of reactions.

Testing experimentally the neutrino flux-luminosity relation, we check the assumptions (1) and (2). Among reasons for possible deviations from Eq. (4) are: (i) fast time dependence of energy release due to some instabilities, (ii) wrong estimation of $\bar{E}_\nu$, (iii) significant amount of unterminated chains, (iv) additional sources of energy which do not accompany neutrino emission.

Neutrinos freely escape from the Sun without affecting dynamics of the Sun's evolution.

2.2. *General features of generation of the neutrino fluxes*

To find neutrino signals at the Earth, we need to compute neutrino fluxes $F^k(r)$ generated in different reactions ($k = pp, pep, \ldots$) from different shells in the Sun characterized by radius r. Then total flux of neutrinos of the type k equals $F^k = \int dr F_\nu^k(r)$. As we will see, the relevant probability of neutrino transformations, namely, the survival probability $\nu_e \to \nu_e$, P_{ee}, depends on density at which neutrino is produced: $P_{ee}(n(r))$. Therefore, the number of ν_e events in a detector is determined by the integral

$$\int dr P_{ee}(n(r)) F_\nu^k(r).$$

The present fluxes of neutrinos are determined by the rates of the corresponding reactions in the present epoch $t_\odot$. The flux $F_\nu^k(r)$ is given by the rate of collisions of the parent particles i and j (determined by k):

$$F_\nu^k(r) = 4\pi r^2 R_{ij}(r, t_\odot).$$

In turn, the rate of collisions of particles i and j equals

$$R_{ij}(r, t_\odot) = \langle v_{ij}(T)\sigma_{ij}\rangle\, n_i(r, t_\odot) n_j(r, t_\odot) \frac{1}{1+\delta_{ij}}, \tag{5}$$

where $T(r,t)$ is the temperature, $v_{ij}(T)$ is the relative velocity, σ_{ij} is the cross section, $n_i(r,t)$ is number density of particles i. The Kronecker index δ_{ij} allows the avoidance of double counting when two identical particles collide ($i = j$). In Eq. (5), $\langle \ldots \rangle$ means averaging over the thermal (Maxwell–Boltzmann) distribution. The cross section is parametrized as

$$\sigma_{ij} = \frac{S_{ij}(E)}{E} e^{-2\pi\eta_{ij}}, \tag{6}$$

where $S_{ij}(E)$ is the astrophysical factor, E is the kinetic energy of colliding nuclei, and in the solar environment, we have $S_{ij}(E) \approx S_{ij}(0)$. The exponent with $\eta_{ij} \approx Z_i Z_j \alpha / v_{ij}$ is the Gamow penetration factor which describes the Coulomb suppression. Here Z_i and Z_j are the electric charges of nuclei, and α is the fine-structure constant. The Coulomb barrier for a single charged particle is about 1 MeV, whereas typical kinetic energy is $E \sim 3kT$ ($\sim$ few keV). So, the interaction occurs due to quantum mechanical tunneling.

Exception is the Be-neutrino flux produced in the electron capture which has no Coulomb barrier. In this case, the rate is determined essentially by the production rate of ^{7}Be nuclei itself which has cross section (Eq. (6)) with the Coulomb suppression.

To compute the local rates R_{ij} and fluxes, one needs to know the temperature $T(r,t)$ and density $n_i(r,t)$ profiles in the present epoch $t_\odot$. These profiles are obtained by solving the system of differential equations which include[6,44]

- equations of hydrostatic equilibrium (balance of the thermal particle and radiation pressure and gravity);
- equations of state, which are close to the equation of ideal gas;
- equations of energy (photons) transport, which depend on opacity, and the latter, in turn depends on number densities of nuclei as well as on diffusion in radiation zone and convection in the outer regions (convective zone).

The initial condition for the evolution of the Sun is a slowly contracting protostar with just started nuclear reactions (i.e., a star appearing in the main sequence of Hertzsprung–Russell diagram). The initial conditions include a mass and certain uniform chemical composition.

To find the number densities of different nuclei, one should solve the system of differential equations that describe the change of $n_i(t,r)$ due to creation of a given nucleus in all possible reactions and its destruction due to possible decay and interaction with other nuclei. The thermal and gravitational diffusions of elements are estimated to be small. These equations, in turn, are coupled with equations which describe temperature and density in the Sun (energy transport, equation of state). Inversely, the equations for energy transport and the equation of state are related to chemical composition via opacity and also directly via the chemical potential. The equations should be integrated starting from the initial moment with the initial abundances of elements.

As we have mentioned, rates of different nuclear reactions and their relative contributions to the neutrino flux depend on conditions (temperature, densities) and therefore change with the radius of the shell. According to the SSM, in the neutrino production region, $R = (0\text{–}0.3)R_\odot$, the temperature decreases by a factor of 3: from $T = 15 \cdot 10^6$ K to $5 \cdot 10^6$ K and the total density decreases from $150\,\text{g}\,\text{cm}^{-3}$ down to $10\,\text{g}\,\text{cm}^{-3}$. Due to strong dependence of reaction rates on T (see later), the rates can be strongly suppressed in the outer shells. As a result, neutrinos from different reactions are produced in regions with different effective radiuses. This should also be taken into account while computing flavor conversion effect of neutrinos, which depends on the density at production. The energy spectra of neutrinos produced in different shells change significantly with the radius of the shell.

Scanning of all possible interactions of nuclei present in the Sun starting from the initial composition, one realizes that, in fact, the number of possibilities which produce substantial effects is not big, and actually the evolution (relevant for neutrino production) is reduced to few chains and cycles of reactions (see Fig. 1). This opens up a possibility to understand neutrino production following transformations (in time) of a given nucleus instead of finding local densities of nuclei. In this case, the key elements of analysis are a total normalization of flux and branchings for different possible reactions.

In what follows, we will establish relations between different neutrino fluxes and find out what these relations depend on without running numerical codes. This will give an idea about main dependences and uncertainties in fluxes and implications of their measurements.

2.3. *The pp chain and CNO bi-cycle: branchings and terminations*

Branchings of chains, spatial distribution of neutrino sources and terminations of chains are the key elements to understand properties of fluxes. Notice that if there is no branching of chains and all the chains are terminated, there would be no uncertainties: fluxes of neutrinos of different types would be equal and determined solely by the luminosity constraint (4). It is the

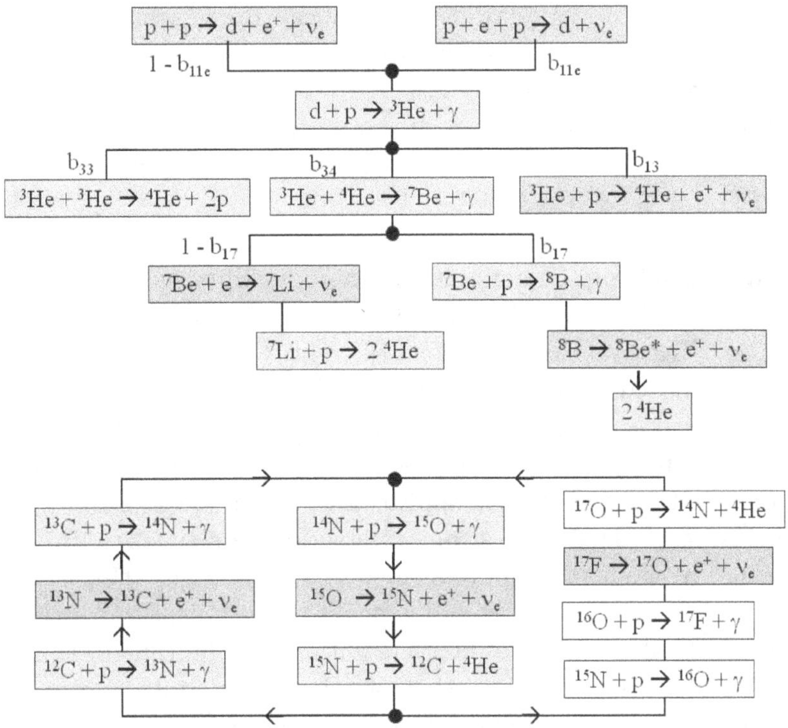

Fig. 1. The *pp* chains of solar nuclear reactions (upper panel). Values of branchings b_{ij} are given in the text. The CNO bi-cycle (lower panel).

branching and existence of unterminated chains that produces complications. The branching fractions $b_{ij} = b_{ij}(T, n)$ and number of unterminated chains strongly depend on nuclear cross sections and physical conditions in the Sun, and therefore on distance from the center. This is how the astrophysical and nuclear uncertainties enter the "game".

In what follows, we will discuss issues relevant for solar neutrino studies. The branching fractions averaged over the nuclear reaction regions are taken from Ref. 46.

The *pp* and *pep* reactions: Apart from reaction in Eq. (2), deuteron is produced in the *pep* reaction (see Fig. 1). It can be considered as branching, characterized by parameter b_{11e}, for a given proton to interact with another proton p or with the pair ep. Although the energy release is larger and Coulomb barrier is partially screened by electron, its rate is two orders of magnitude smaller than the rate of reaction (2) because of three particles' collision in the initial state: $b_{11e} = 0.24\%$. The rate changes as n^3 and therefore the corresponding branching b_{11e} is proportional to n and it does not depend on T immediately. Time of the process is $t_{pep} = 10^{12}$ years.

The dominant processes for deutron is a capture of proton and formation of ^{3}He. The process proceeds via the strong and electromagnetic (EM) interaction and therefore is very fast, $t_{dp} = 10^{-8}$ years, in spite of Coulomb suppression. So, its precise rate is not important.

There are two branchings of the *pp* chain: ^{3}He and ^{7}Be ones (see Fig. 1).

The ^{3}He branching: Scanning of all the possibilities shows that ^{3}He has two main channels with comparable branchings and very rarely a third one (see Fig. 1). With branching fraction $b_{33} = 84.6\%$, ^{3}He interacts with another ^{3}He created by sequence of *pp*, *pep* and *pd* reactions. Initial abundance of ^{3}He disappeared, so the ^{3}He density in a given epoch is determined by the equilibrium of nuclear processes. The reaction provides termination of the chain. (Notice that two protons appear as catalysts of this chain.) Although the process proceeds via the strong interactions, its time, $t_{33} = 10^5$ years, is large due to Coulomb suppression, which is stronger than that for the processes in Eq. (2), as well as due to small number density of ^{3}He.

In $b_{34} = 15.4\%$ cases ^{3}He interacts with ^{4}He: The time of this process, $t_{34} = 10^6$ years, is larger than that of ^{3}He–^{3}He in spite of much smaller density of ^{3}He nuclei and equal Coulomb factors. The reason is that in ^{3}He–^{4}He also EM interaction is present besides the strong interactions, and the phase volume (energy release) is much smaller. The branching does not depend on temperature but does depend on ratio of ^{3}He/^{4}He densities which, in turn, depends on radius of shell.

Interaction of ^{3}He with proton gives ^{4}He, thus terminating the chain and producing the *hep* neutrino. Reaction ^{3}He–p is similar to the *pp* reaction: it proceeds via the weak interactions, and ^{3}He plays the role of one of the protons. Its rate is suppressed in comparison to Eq. (2) by larger Coulomb barrier and smaller number density of ^{3}He: $b_{31} = 2.5 \cdot 10^{-5}\%$.

The ^{7}Be branching: Produced ^{7}Be has two possibilities: In almost 100% of cases, it captures an electron from the continuum of ionized electrons. The transition is due to weak interactions but the Coulomb suppression is absent, so the time of process, $t_{e\text{Be}} = 10^{-1}$ year, is rather short.

In a very small, $b_{17} = 0.11\%$, number of cases ^{7}Be captures a proton forming the ^{8}B nuclei. Correspondingly, the time of the p-Be reaction is three orders of magnitude larger: $t_{p\text{Be}} = 10^2$ years. The cross section of this reaction S_{17} has the largest uncertainty (8%): it was computed but also measured directly in laboratory which allowed to reduce the uncertainty. In turn, the Boron nuclei quickly (with time $t = 10^{-8}$ years) decays with exited (highly unstable) Be nuclei in final state which, in turn, quickly decays $^8Be^* \to 2^4He$ terminating the chain.

The CNO bi-cycle: C, N and O nuclei appear as catalysts of reactions and the chain of reactions form the cycle. The contribution of the cycle to the total energy generation is about 1.6%.[6] The chain is composed of alternate patterns of reactions of the two sequential proton captures and one β decay.

The densities of C and O nuclei are given approximately by initial densities, whereas the abundance of ^{14}N changes substantially in the course of evolution.

Starting from the lightest nuclei, ^{12}C, with number density fraction $2.8 \cdot 10^{-4}\%$, which captures p, we obtain the cycle shown in Fig. 1. In the end, ^{15}N captures a proton giving ^4He and ^{12}C (as in the beginning), thus accomplishing the cycle. The Coulomb suppression of proton captures and small abundance of metals determine the small contribution of the cycle to the energy release.

In about 1% of cases, p–^{15}N interaction leads to formation of ^{16}O with release of large energy. (This reaction gives, however, very small contribution to initial abundance of ^{16}O.) The nuclei ^{16}O can interact with proton and produce ^{17}F which decays emitting the F neutrino: Finally ^{17}O captures p transforming to ^{14}N, thus accomplishing the side branch (part of the second cycle).

2.4. *Solar neutrino spectra*

The neutrino spectra are shown in Fig. 2. The kinetic energies of parents of neutrinos are given by the temperature in the central regions of the Sun (~few keV). Since the energy release in reactions $Q \gg 3kT$ (~4 keV in the center), spectra of individual neutrino components are determined by kinematics of elementary processes — masses of nuclei and electron involved, features of the final state, etc.

For the Be, *pep* neutrinos, the thermal effects lead to widening of the lines, with specific shapes. For example, for the ^7Be neutrino line, the width is 1.6 keV. The influence of thermal motion on continuous spectra is less visible.

The relative values of the neutrino fluxes from different reactions are given by branchings considered in Section 2.3 in assumption that all chains are terminated. In turn, the branching is determined by physical conditions in the Sun (temperature, density, chemical composition). Values of neutrino fluxes can be computed with respect to the *pp* neutrino flux.

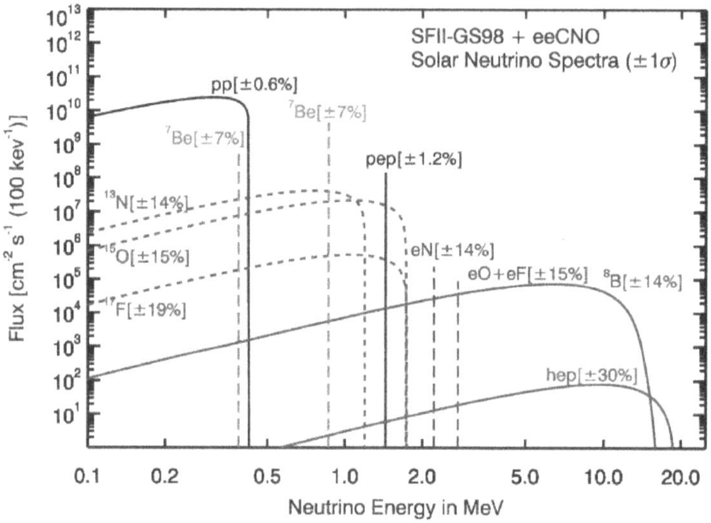

Fig. 2. Energy spectrum of solar neutrinos. The electron capture fluxes are given in units $\text{cm}^{-2}\ \text{s}^{-1}$. Also shown are the subleading neutrino lines from e captures on N, O and F nuclei.[44]

The pp neutrinos (Eq. (2)) have continuous spectrum with maximal energy $Q_{pp} = 0.420\,\text{MeV}$ ($Q = 2M_p - M_d - m_e$). The flux rises slowly with E, peaks at 0.31 MeV and then sharply decreases. The average energy equals $\bar{E}_{pp} = 0.24\,\text{MeV}$. According to SSM, the pp neutrinos compose the majority (92%) of solar flux and therefore the average energy of all neutrinos is close to $\bar{E}_{pp}$. Total number of the pp neutrinos per termination is $1 + b_{33} = 1.85$. The pp neutrino production peaks at $0.09R_\odot$ ($T_c = 14.4 \cdot 10^6\text{K}$) with width of the production region at half height: $(0.04–0.2)R_\odot$. The dependence of flux on the central temperature can be parametrized as $F_{pp} \propto T_c^{-1.2}$. The power is negative because of the luminosity constraint and fast increase of rates of other neutrino fluxes with T, in particular the ^{7}Be neutrino flux. So, F_{pp} decreases because F_{Be} increases.

The pep neutrino line has $\bar{E}_{pep} = 1.441\,\text{MeV}$. The shape of the line is determined by thermal conditions in plasma and the width is of the order of 2 keV. The relative flux of the pep neutrinos is given by branching for p to interact with another proton or with the pair ep:

$$\frac{F_{pep}}{F_{pp}} \approx b_{11e} = 2.4 \cdot 10^{-3}, \tag{7}$$

and the latter is proportional to the density of electrons.

The hep neutrinos have the most extended production region $(0.06–0.21)R_\odot$ with peak at $0.13R_\odot$ and temperature in the peak $12 \cdot 10^6\text{K}$. The flux depends on the central temperature as $F_{hep} \propto T_c^\beta$, where $\beta = 3–6$.

The energy spectrum is nearly symmetric with maximum at 9.5 MeV and the end point at 18.773 MeV. The flux equals $1.33 \cdot 10^{-7}$ of the pp neutrino flux (and three orders of magnitude smaller than the Boron neutrino flux).

The *hep* neutrinos trace distribution of ^{3}He which has the following interesting feature: Production rate of ^{3}He decreases with increase of distance from the center of the Sun. Due strong dependence of ^{3}He–^{3}He and ^{3}He–^{4}He reactions on T, the rate of burning of ^{3}He decreases with increase of R much faster than the production. So, in the center of the Sun, the density of ^{3}He is small because of fast burning, it increases with R because of fast decrease of burning rate with T, it reaches maximum at $R = 0.2R_\odot$ and then decreases because of decrease in production.

Essentially absent branching for ^{7}Be due to quickness of the e capture means that the ^{7}Be-neutrino production rate is given by the rate of production of ^{7}Be itself. The flux can be estimated with respect to the pp neutrino flux as

$$\frac{F_{\rm Be}}{F_{pp}} \sim \frac{b_{34}}{1 + (1 - b_{34})} = 0.081. \tag{8}$$

Recall that $b_{34}/(1-b_{34}) \propto n(^3\text{He})/n(^4\text{He})$. The ^{7}Be neutrino line with energy $E = 0.862$ MeV from decay to the ground level is produced in 89.7% of cases. The low energy line $E = 0.384$ MeV appears in 10.3% of cases. The asymmetric lines have width 1.6 keV due to collisions in plasma and Doppler effect. The production region in the Sun peaks at $0.06R_\odot$ with half-peak width: $(0.03–0.10)R_\odot$. The dependence of flux on the central temperature can be parametrized as $F_{\rm Be} \propto T_c^8$.

Neutrinos produced in the ^{8}B decay have a continuous spectrum with the end point energy about 14 MeV. Maximum is at $E = 6.4$ MeV and the shape is rather symmetric with respect to maximum which is related to big width of the final ^{8}Be* state. The ratio of fluxes of the ^{8}B and ^{7}Be neutrinos is determined by the branching of interaction of ^{7}Be:

$$\frac{F_{\rm B}}{F_{\rm Be}} \approx b_{17}, \tag{9}$$

which strongly depends on temperature. With respect to the pp neutrino flux we have $F_{\rm B}/F_{pp} = b_{17}b_{34}/(2 - b_{34})$. The production region of ^{8}B neutrinos peaks at $0.05R_\odot$ with width at half height $(0.0 - 0.07)R_\odot$. The dependence of flux on the central temperature can be parametrized as $F_{\rm B} \propto T_c^{18}$.

There is a hierarchy of neutrino fluxes. The pp neutrinos and then ^{7}Be neutrinos dominate: they appear from two main branches related to ^{3}He interactions. Their fluxes are determined by the branching in Eq. (8) and by solar luminosity. The *pep*, ^{8}B and *hep* neutrinos have subleading fluxes and their relative changes with parameters of medium (T, ρ) are restricted by luminosity very weakly.

Neutrinos from CNO cycle have the following features: Since ^{13}N quickly decays, the rate of N-neutrino production is determined by the rate of p^{12}C reaction. The N neutrinos have continuous spectrum with maximal energy 1.2 MeV. The spectrum is nonsymmetric with slow increase of flux with energy below maximum at $E = 0.76$ MeV and then sharply decrease. It is similar to the spectrum of the *pp* neutrinos.

Neutrinos generated in ^{15}O decay have a continuous spectrum with end point 1.73 MeV. The shape is similar to that for N neutrinos. The rate is determined by the rates of p^{13}C and p^{14}N reactions.

For $T > 10^{7}$ K, the time of p^{14}N reaction becomes larger than solar age $\tau_{\odot}$ due to higher Coulomb barrier than for processes with carbon. Therefore for outer regions of the solar core the cycle is not completed. This has two consequences: lower flux of O neutrinos: $F_{\mathrm{N}}/F_{\mathrm{O}} = 1.4$ and accumulation of ^{14}N.

The flux of F neutrinos is determined by the rate of p^{16}O process. It is suppressed by high Coulomb barrier and relatively low (in comparison to protons) abundance of ^{16}O (which is mostly given by initial abundance). The total flux is about the ^{8}B neutrino flux. The energy spectrum is similar to the one of ^{16}O neutrinos with very close maximal energy (1.740 MeV) but its flux is two orders of magnitude smaller. So, it will be almost impossible to identify its contribution.

The overall solar neutrino flux strongly decreases with energy which determines features of its detection: low-energy neutrinos have high fluxes, but small cross sections and high background, whereas high-energy neutrinos have low fluxes, larger cross section and lower background.

Table 1 from Ref. 46 presents the neutrino fluxes with their theoretical uncertainties predicted in models with different metallicities (Table 1). Values of fluxes (and their ratios) can be reproduced qualitatively well on the basis of above consideration in terms of branchings and luminosity constraints. These fluxes are compared with neutrino fluxes extracted from experimental measurements.[47]

2.5. *Solar metallicity problem and solar neutrinos*

3D models of the solar atmosphere developed in years 2006–2009 include effects of stratification, inhomogeneities, etc. (see Refs. 42, 44, 45 and recent discussion in Refs. 48, 49). The models improved agreement with absorption line shapes, and their results are more consistent with observations of

Table 1. Solar neutrino fluxes from models with high (GS98) and low (AGSS09met) metallicity and fluxes extracted from experimental data. Units are $10^{10}(pp)$, $10^{9}(^{7}\mathrm{Be})$, 10^{8} (pep, ^{13}N, ^{15}O), 10^{6} (^{8}B, ^{17}F) and 10^{3} (hep) cm^{-2}s^{-1}.[46]

Flux	GS98	AGSS09met	Experiment
pp	$5.98(1 \pm 0.006)$	$6.03(1 \pm 0.005)$	$5.97^{(1+0.006)}_{(1-0.005)}$
pep	$1.44(1 \pm 0.01)$	$1.46(1 \pm 0.009)$	$1.45^{(1+0.009)}_{(1-0.009)}$
hep	$7.98(1 \pm 0.30)$	$8.25(1 \pm 0.30)$	$19^{(1+0.63)}_{(1-0.47)}$
^{7}Be	$4.93(1 \pm 0.06)$	$4.50(1 \pm 0.06)$	$4.80^{(1+0.050)}_{(1-0.046)}$
^{8}B	$5.46(1 \pm 0.12)$	$4.50(1 \pm 0.12)$	$5.16^{(1+0.025)}_{(1-0.017)}$
^{13}N	$2.78(1 \pm 0.15)$	$2.04(1 \pm 0.14)$	≤ 13.7
^{15}O	$2.05(1 \pm 0.17)$	$1.44(1 \pm 0.16)$	≤ 2.8
^{17}F	$5.29(1 \pm 0.20)$	$3.26(1 \pm 0.18)$	≤ 85

neighboring stars. At the same time, the models predict ~40% lower abundance of heavy elements (heavier than ^{4}He) in photosphere. This, in turn, affects temperature and density gradients in the Sun, and consequently, the sound velocity profile (propagation of p modes). It turns out that the profiles are not in agreement with helioseismological measurement.

On the other hand, low metallicity affects neutrino fluxes. According to Table 1, the N and O neutrino fluxes directly and linearly depend on the metallicity and therefore decrease by about 40%. For other neutrinos, the effect of metallicity is indirect and related to change of the density and temperature profiles. Lower value of Z leads to lower opacity and therefore to smaller temperature gradient. Consequently, in models with low Z, the central temperature in the Sun is lower, thus reducing the fluxes of the ^{7}Be and ^{8}B neutrinos. Correspondingly, the flux of pp neutrinos increases to satisfy the luminosity constraint.

Thus, measurements of the CNO neutrinos are crucial to resolve the problem.[46,49] As follows from Table 1 the present theoretical uncertainties on the N and O neutrino fluxes are 14–16%. They can be reduced down to 10% or smaller if correlations between CNO and ^{8}B neutrino fluxes are used (both are subleading and have similar dependence on environmental factors, in particular, strong dependence on temperature). Therefore precise measurements of the ^{8}B neutrino flux will allow to sharpen predictions for the CNO neutrino fluxes.

The ^{7}Be neutrino flux decreases with decrease of metallicity by ~10% which is comparable with theoretical uncertainty. Still combined restrictions on the ^{7}Be- and ^{8}B- neutrino fluxes in future can contribute to resolution of the problem.[46]

3. Propagation of neutrinos in matter

Here we will describe the neutrino flavor transformations which are relevant to solar neutrinos. Keeping in mind the new phase of the field when accuracy of measurements achieved is in percent or even subpercent level, we will present the 3ν formalism in complete and self-explanatory form (see reviews in Refs. 27–30, 50).

3.1. *Mixing in matter*

The vacuum mixing means that the flavor neutrinos $\nu_f \equiv (\nu_e, \nu_\mu, \nu_\tau)^T$ are coherent combinations of the neutrinos $\nu_{\rm mass} \equiv (\nu_1, \nu_2, \nu_3)^T$ with definite masses m_1, m_2, m_3:

$$\nu_f = U_{\rm PMNS}\nu_{\rm mass}. \tag{10}$$

Here $U_{\rm PMNS}$ is the PMNS mixing matrix, and we use the standard parametrization of $U_{\rm PMNS}$

$$U_{\rm PMNS} = U_{23}\Gamma_\delta U_{13}\Gamma_\delta^* U_{12}, \tag{11}$$

where $U_{ij} = U_{ij}(\theta_{ij})$ is the rotation in the space of the ij states on the angle θ_{ij}, $\Gamma_\delta \equiv {\rm diag}(1, 1, e^{i\delta})$ is the diagonal matrix with the CP-violation phase δ. The matrix Γ_δ^* on the RHS of U_{13} in Eq. (11) can be permuted with U_{12} and consequently absorbed in rephasing of the mass states. The matrix $U_{\rm PMNS}$ diagonalizes the mass matrix of neutrinos in the flavor basis M, defined via the term of Lagrangian $\mathcal{L} = -\bar{\nu}_R M \nu_f$ (in the case of Majorana neutrinos, $\bar{\nu}_R = \nu_L^T C^{-1}$) so that

$$U_{PMNS}^\dagger M^\dagger M U_{PMNS} = |M^{diag}|^2 = {\rm diag}(0,\ \Delta m_{21}^2,\ \Delta m_{31}^2). \tag{12}$$

Here $\Delta m_{k1}^2 \equiv |m_k|^2 - |m_1|^2$, $k = 2, 3$, and we have subtracted the matrix proportional to the unit matrix $|m_1|^2 I$ which does not affect the flavor transformation.

In the course of propagation in vacuum, evolution of the flavor states is described by the Schrödinger-like equation,

$$i\frac{d\nu_f}{dx} = H_0\nu_f \tag{13}$$

$(x \approx t)$, where in the ultrarelativistic case the effective Hamiltonian equals

$$H_0 \approx \frac{M^\dagger M}{2p} = \frac{1}{2p} U_{\text{PMNS}} |M^{\text{diag}}|^2 U_{\text{PMNS}}^\dagger. \tag{14}$$

This expression is a generalization to the 3ν case of the expression for energy of a single neutrino, $E = \sqrt{p^2 + m^2} \approx p + m^2/2p$. In Eq. (14), we have omitted the term proportional to the neutrino momentum pI. In the second equality in Eq. (14), the mass matrix squared is expressed in terms of the diagonal mass matrix squared and the mixing matrix using relation (12).

In the case of neutrino propagation in matter, one should take into account neutrino interactions. At low energies, the elastic forward scattering dominates and the inelastic interactions can be neglected. The forward scattering leads to refraction phenomena and can be described by refraction index or the effective potential V. In the case of standard interactions, the matrix of potentials in the flavor basis V is diagonal. The difference of potentials has a physical meaning, and in the standard interaction case, this difference $V_e \equiv V(\nu_e) - V(\nu_a)$ $(a = e,\ \mu)$ is due to the charged current (CC) scattering of ν_e on electrons $(\nu_e e \to \nu_e e)$[17]:

$$V_e = \sqrt{2} G_F n_e. \tag{15}$$

Here G_F is the Fermi constant and n_e is the number density of electrons. In the lowest order, we have $V(\nu_\mu) = V(\nu_\tau)$, and therefore the matrix of the potentials in the flavor space can be written as $V = \text{diag}(V_e, 0, 0)$.

The potential V_e leads to appearance of additional phase difference in neutrino system: $\Delta\phi_{\text{matter}} \equiv V_e t$, $(l \approx t)$. The distance over which this "matter" phase equals 2π determines the *refraction length*:

$$l_0 \equiv \frac{2\pi}{V_e} = \frac{\sqrt{2}\pi}{G_F n_e}. \tag{16}$$

At the distances $l \sim l_0$, the matter effects can be of the order 1.

The Hamiltonian which describes the flavor evolution in matter, H, is given by the sum of the vacuum (Eq. (14)) and matter parts:

$$H = H_0 + V \approx \frac{1}{2p} M^\dagger M + V. \tag{17}$$

We introduce the eigenstates $\nu_m \equiv (\nu_{1m}, \nu_{2m}, \nu_{3m})^T$ and the corresponding eigenvalues (H_{1m}, H_{2m}, H_{3m}) of the Hamiltonian in matter:

$$H \nu_{im} = H_{im} \nu_{im}. \tag{18}$$

In nonuniform medium, V, and therefore H, change with distance (time). In this case, we speak about the eigenstates and eigenvalues of the instantaneous Hamiltonian: $\nu_{im}(t)$ and $H_{im}(t)$. In vacuum, $\nu_{im} \to \nu_i$, $H_{im} \to m_i^2/2p$ and $U^m \to U_{\rm PMNS}$. The evolution of flavor states is described by

$$i\frac{d\nu_f}{dx} = H\nu_f, \tag{19}$$

with total Hamiltonian in Eq. (17).

The key notion is the *flavor mixing in matter.* In general (vacuum, medium), the flavor mixing described by the mixing matrix U_H is determined with respect to ν_H — the eigenstates of the total Hamiltonian of system: $\nu_f = U_H \nu_H$. In vacuum, $\nu_H = \nu_{\rm mass}$ and $U_H = U_{\rm PMNS}$, in matter $\nu_H = \nu_m$ and $U_H = U^m$, so that

$$\nu_f = U^m \nu_m. \tag{20}$$

Using Eqs. (18) and (20), one can show that the matrix U^m is determined via diagonalization of the Hamiltonian:

$$U^{m\dagger} H U^m = H^{\rm diag} = {\rm diag}(H_{1m}, H_{2m}, H_{3m}). \tag{21}$$

Since the Hamiltonian in matter H depends on matter density and neutrino energy $H = H(n, E)$, the mixing in matter U^m (as well as the eigenvalues H_{im}) depends on n and E too. Therefore in matter, the mixing angles, which parametrize U^m, become the *dynamical variables,* whereas in vacuum, θ_{ij} are the fundamental parameters. In matter, the effective masses (eigenvalues of the Hamiltonian) also become dynamical variables.

According to Eq. (20), the mixing matrix gives the eigenstate content of the flavor states. Inverting Eq. (20), we obtain equality

$$\nu_m = U^{m\dagger} \nu_f, \tag{22}$$

which determines the flavor content of the eigenstates.

3.2. *Determination of oscillation parameters: resonance*

In the 3ν case, the explicit dependence of $U^m(n, E)$ is rather complicated, but for many realistic situations, the problem can be reduced to the 2ν problem. This reduction provides very good approximation and allows one to clarify the physics of the conversion.

Keeping in mind applications to the solar neutrinos, we will reduce 3ν to 2ν evolution using the following relations realized at low energies:

$$2EV_e \sim \Delta m_{21}^2 \ll \Delta m_{31}^2.$$

As we will see, under these conditions, the third eigenstate nearly decouples from the rest of the system and evolves independently. We find an approximate expression for U^m in few steps which allow us to make a number of important conclusions:

Step 1. We transform from the flavor basis to the *propagation* basis, $\tilde{\nu}$, defined as $\nu_f \equiv U_{23}\Gamma_\delta\tilde{\nu}$. In this basis, the electron neutrino is not mixed: $\tilde{\nu} = (\nu_e, \tilde{\nu}_\mu, \tilde{\nu}_\tau)$, and the matter potential is not changed. Then according to Eqs. (14) and (17), the Hamiltonian becomes

$$\tilde{H} = U_{13}U_{12}H^dU_{12}^TU_{13}^T + V. \tag{23}$$

This means that transition $\nu_e \to \nu_e$ does not depend on the 2-3 mixing and the CP-phase $\delta_{\rm CP}$ (also see Ref. 51 for general consideration).

Step 2. Performing the 1-3 rotation U_{13} of $\tilde{H}$ on the vacuum angle θ_{13}, we obtain from Eq. (23) the Hamiltonian

$$U_{12}H^dU_{12}^T + U_{13}^TVU_{13}. \tag{24}$$

According to Eq. (24) in the new basis, the mixing of the third state (having dominant term in H) with rest of the system is proportional to the potential and small 1-3 mixing: Vs_{13}. The 1-1 component of the potential relevant for solar neutrinos is reduced by a factor of c_{13}^2: c_{13}^2V.

Step 3. To decouple the heaviest state, we perform an additional 1-3 rotation on the angle $\delta\theta_{13}$ which vanishes the 1-3 element in Eq. (24). The angle $\delta\theta_{13}$ is given by

$$\tan 2(\delta\theta_{13}) = \frac{V_e \sin 2\theta_{13}}{\Delta_{31} - \Delta_{21}\sin^2\theta_{12} - V_e\cos 2\theta_{13}}, \tag{25}$$

where

$$\Delta_{ij} \equiv \frac{\Delta m_{ij}^2}{2E}.$$

It equals approximately $\delta\theta_{13} = \theta_{13}\epsilon_{13}$[52] with

$$\epsilon_{13} \equiv \frac{2V_eE}{\Delta m_{31}^2} = \frac{V_e}{\Delta_{31}}. \tag{26}$$

The small 1-3 rotation $\delta\theta_{13}$ being proportional to the matter potential V_e is nothing but the matter correction to the 1-3 vacuum mixing. The two sequential 1-3 rotations can be combined in a single 1-3 rotation

$$U_{13}^m = U_{13}(\theta_{13})U_{13}(\delta\theta_{13}) = U_{13}(\theta_{13} + \delta\theta_{13}) \tag{27}$$

over the angle $\theta_{13}^m = \theta_{13} + \delta\theta_{13}$. In the case of normal 1-3 mass ordering, $\Delta_{31} > 0$, the matter correction $\delta\theta_{13}$ is positive, thus enhancing mixing. In the case of inverted mass hierarchy, it is negative.

After these two rotations, the new basis $\nu' = (\nu'_e, \tilde{\nu}_\mu, \nu'_\tau)$ is connected to the flavor basis as

$$\nu_f \equiv U_{23}\Gamma_\delta U_{13}^m \nu'. \tag{28}$$

The $U_{13}(\delta\theta_{13})$ rotation, vanishing the H_{13} element, in turn, generates the nonzero 2-3 elements of the size $H_{23} = (1/4)\sin 2\theta_{13}\sin 2\theta_{13} V r$ ($r \equiv \Delta m_{21}^2/\Delta m_{31}^2$), which can be safely neglected. As a result, in the ν' basis, the state ν'_τ decouples and the Hamiltonian becomes

$$H' = \begin{pmatrix} H^{(2)} & 0 \\ 0 & H_{3m} \end{pmatrix},$$

where the 2ν Hamiltonian, $H^{(2)}$, equals

$$H^{(2)} = \frac{\Delta m_{21}^2}{4E} \begin{pmatrix} -\cos 2\theta_{12} + 2\epsilon_{21} c_{13}^{m2} & \sin 2\theta_{12} \\ \sin 2\theta_{12} & \cos 2\theta_{12} \end{pmatrix}. \tag{29}$$

Here

$$\epsilon_{12} \equiv \frac{2V_e E}{\Delta m_{21}^2}, \tag{30}$$

and H_{3m} is the third eigenvalue

$$H_{3m} = \Delta_{31} - 0.5\Delta_{21} + V\cos^2\theta_{13}^m.$$

In Eq. (29) we have subtracted the unit matrix proportional to Δ_{21} to make $H^{(2)}$ traceless in vacuum.

Step 4. The Hamiltonian $H^{(2)}$ is diagonalized by rotation $U_{12}^m = U_{12}(\theta_{12}^m)$, where the angle θ_{12}^m is determined by

$$\sin^2 2\theta_{12}^m = \frac{\sin^2 2\theta_{12}}{(\cos 2\theta_{12} - c_{13}^2\epsilon_{12})^2 + \sin^2 2\theta_{12}}. \tag{31}$$

(Here we take $c_{13}^m \sim c_{13}$.)

The difference of the eigenvalues of $H^{(2)}$,

$$\Delta_{21}^m \equiv H_{2m} - H_{1m} = \Delta_{21}\sqrt{\left[\cos 2\theta_{12} - c_{13}^2\epsilon_{12}\right]^2 + \sin^2 2\theta_{12}}, \tag{32}$$

determines the oscillation length in matter:

$$l_{21}^m = \frac{2\pi}{\Delta_{21}^m}.$$

Combining rotations (Eq. (28)) and $U_{12}(\theta_{12}^m)$ we obtain the total mixing matrix in matter:

$$U^m \approx U_{23}\Gamma_\delta U_{13}^m U_{12}^m, \tag{33}$$

where θ_{12}^m and θ_{13}^m are determined in Eqs. (31) and (27), respectively.

Resonance: Dependence of the mixing parameter $\sin^2 2\theta_{12}^m$ (Eq. (31)) on the neutrino energy and density of matter (on product EV_e) has a resonance character[20,21] (see Fig. 3). According to Eq. (31) at

$$c_{13}^2 V_e(n) \approx \cos 2\theta_{21}\frac{\Delta m_{21}^2}{2E}, \tag{34}$$

the mixing becomes maximal: $\sin^2 2\theta_{21}^m = 1$ or $\theta_{21}^m = \pi/4$. The mixing decreases with increase of density: $\theta_{21}^m \to \pi/2$, so high density suppresses

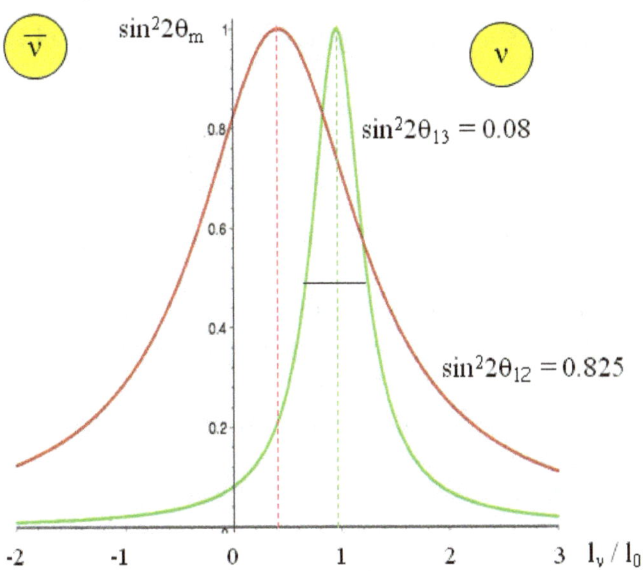

Fig. 3. The resonance. Dependence of the mixing parameter $\sin^2 2\theta_{12}^m$ on the ratio $l_\nu/l_0 \propto EV_e$, i.e., on neutrino energy and/or density of matter for two different mixing angles in vacuum. The left part of the figure ($l_\nu/l_0 < 0$) corresponds to antineutrinos.

the mixing. On the other hand, for low density, $V_e(n) \to 0$, the mixing converges to the vacuum one. The resonance condition can be written as $\epsilon_{21} = \cos 2\theta_{21}/\cos^2\theta_{13}$. The energy (density), which satisfies the condition in Eq. (34), is called the resonance energy (density).

The width of the resonance (at half height) is proportional to the vacuum mixing:

$$\Delta V_e = 2\tan 2\theta_{21} V_e^R, \tag{35}$$

where V_e^R is the value of the potential in resonance. So, the smaller the vacuum mixing, the narrower is the resonance. Similarly, one can consider dependence of the mixing on energy for a fixed density of the medium.

The resonance condition can be rewritten in terms of the oscillation length in vacuum, $l_\nu \equiv 4\pi E/\Delta m_{21}^2$, and the refraction length:

$$l_\nu = l_0 \cos 2\theta_{21} \frac{1}{c_{13}^2}. \tag{36}$$

For small mixings, we have $l_\nu \approx l_0$ which elucidates physical meaning of resonance: the length (inverse frequency) which characterizes system coincides with the length (eigenfrequency) of medium. Due to large 1-2 mixing, the resonance is shifted to smaller energies/densities. For $\Delta m_{21}^2 > 0$, the resonance condition is fulfilled in the neutrino channel with $V_e > 0$ only. For antineutrinos, $V_e < 0$.

In resonance, the splitting of eigenstates (Eq. (32)) is minimal,

$$\Delta_{21}^m = \Delta_{21} \sin 2\theta_{21}, \tag{37}$$

and correspondingly, the oscillation length $l_m^R = l_\nu/\sin 2\theta_{21}$ becomes maximal.

For the 1-3 mixing, the resonance is at much higher energies since $E_R \propto \Delta m_{31}^2$.

Manifestation of resonance depends on profile (dependence on distance or time) of the matter's potential. In matter with constant resonant density, oscillations proceed with maximal depth — *resonance enhancement of oscillations* takes place. In matter with varying density, as we will see, in resonance, the depth of oscillations can be very small, but the flavor transitions in the resonance layer are most strong.

3.3. *Flavor evolution in matter: adiabaticity*

In uniform medium, $n = const$, the mixing $U^m(n, E)$ is constant. Inserting Eq. (20) into Eq. (19) and using Eq. (21), we obtain the evolution equation for the eigenstates:

$$i\frac{d\nu_m}{dx} = H^{\text{diag}}\nu_m. \tag{38}$$

This equation splits into three independent equations for three eigenstates and their integration is trivial giving

$$\nu_{km}(t) = e^{-i\phi_k(t)}\nu_{km}(0), \quad \phi_k(t) = H_{km}t \tag{39}$$

($x \approx t$). Using this result we can immediately write up solution for the flavor states. If ν_α is produced, it can be presented in terms of the eigenstates as $\nu_\alpha = \sum_k U^m_{\alpha k}\nu_{km}(0)$. According to Eq. (39), this state evolves as

$$\nu_\alpha(t) = \sum_k U^m_{\alpha k}\nu_{km}(t) = \sum_k U^m_{\alpha k}e^{-i\phi_k(t)}\nu_{km}(0). \tag{40}$$

Then the probability to find ν_β in the moment t equals

$$P_{\beta\alpha} = |\langle\nu_\beta|\nu_\alpha(t)\rangle|^2 = \left|\sum_k U^m_{\alpha k}U^{m\dagger}_{\beta k}\, e^{-i\phi_k(t)}\right|^2. \tag{41}$$

Since the mixing matrix elements are constants, the only thing that happens is that the phase increases with time. Consequently, dynamics of propagation and expressions for probabilities are similar to that in vacuum, although values of oscillation parameters differ from the vacuum ones. The mixing matrix elements $U^m_{\beta k}$ are determined in Eq. (33).

If density changes on the way of neutrinos, $V = V(x)$, the Hamiltonian depends on $x \approx t$. Correspondingly, the mixing in matter becomes time dependent: $U^m = U^m(n(t))$. The evolution equation for the eigenstates of the instantaneous Hamiltonian can be obtained inserting Eq. (20) into Eq. (19):

$$i\frac{d\nu_m}{dt} = \left(H^{\text{diag}} + iU^{m\dagger}\frac{dU^m}{dt}\right)\nu_m, \tag{42}$$

and H^{diag} is defined in Eq. (21). The new term on the RHS of the equation is off-diagonal (due to unitarity of U^m), thus describing transitions between the eigenstates $\nu_{km} \leftrightarrow \nu_{jm}$.

If density changes slowly so that

$$U^{m\dagger}\frac{dU^m}{dt} \ll \Delta H^{\text{diag}}, \tag{43}$$

the term proportional to derivative can be neglected. The inequality (Eq. (43)) is called the *adiabaticity condition.* The matrix condition

(Eq. (43)) is reduced to three inequalities

$$\left(U^{m\dagger}\frac{dU^m}{dt}\right)_{ij} \ll H_{im} - H_{jm}, \quad i \neq j, \tag{44}$$

and one can consider the adiabaticity in different ij channels, essentially reducing it to 2ν adiabaticity. The adiabaticity can be broken in one channel but fulfilled in others. In the adiabatic case, when Eq. (43) is satisfied, Eq. (42) splits and the evolutions of the instantaneous eigenstates become independent of each other as in the constant density case.[21–23] The solution of the equation (42) is

$$\nu_{km}(t) = e^{-i\phi_k(t)}\nu_{km}(n(t)), \tag{45}$$

where the phase equals

$$\phi_k = \int^t dt' H_{km}(t'). \tag{46}$$

The dependence of $\nu_{km}(t) = \nu_{km}(n(t))$ on time is due to dependence of the mixing on matter density. Indeed, expressing the eigenstates in terms of flavor states, we have

$$\nu_k(n(t)) = \sum_{\beta=e,\mu,\tau} U^{m\dagger}_{\beta k}(n(t))\nu_\beta, \tag{47}$$

i.e., the flavor of a given eigenstate (the flavor composition) changes with time according to density change. Thus, in contrast to the vacuum and constant density cases, now a *new degree of freedom* is involved — the mixing.

With Eq. (45), the solution for flavor transitions can be written immediately. Similar to Eq. (40), the state produced as ν_α at the moment t becomes:

$$\nu_\alpha(t) = \sum_k U^m_{\alpha k}(n_i)e^{-i\phi_k(t)}\nu_k(n(t)). \tag{48}$$

In the course of propagation, two things happen simultaneously:

- The eigenstates acquire the phases $\phi_k(t)$;
- The flavor content of the eigenstates determined by mixing changes according to density change (Eq. (47)).

The amplitude of probability to find ν_β in the state (Eq. (48)) in the layer with density $n(t)$ equals

$$\langle \nu_\beta | \nu(t)_\alpha \rangle = \sum_k U^m_{\alpha k}(n_i) e^{-i\phi_k(t)} U^{m\dagger}_{\beta k}(n_f),$$

where we have taken into account that according to Eq. (20), $\langle \nu_\beta | \nu_k(n_f) \rangle = U^{m\dagger}_{\beta k}(n_f)$. Finally, the probability of $\nu_\alpha \to \nu_\beta$ transition is

$$P_{\beta\alpha} = |\langle \nu_\beta | \nu_\alpha(t) \rangle|^2 = \left| \sum_k U^m_{\alpha k}(n_i) U^{m\dagger}_{\beta k}(n_f) e^{-i\phi_k} \right|^2. \tag{49}$$

The difference from the constant density case (Eq. (41)) is that mixing matrix elements and their conjugates should be taken at different densities. Furthermore, the phase should be computed by integrating along the neutrino trajectory (Eq. (46)). The expression in Eq. (49) gives the adiabatic transition probability (see Fig. 4).

If the phases are averaged out, we obtain

$$P_{\beta\alpha} = \sum_k |U^m_{\alpha k}(n_i)|^2 \left| U^m_{\beta k}(n_f) \right|^2. \tag{50}$$

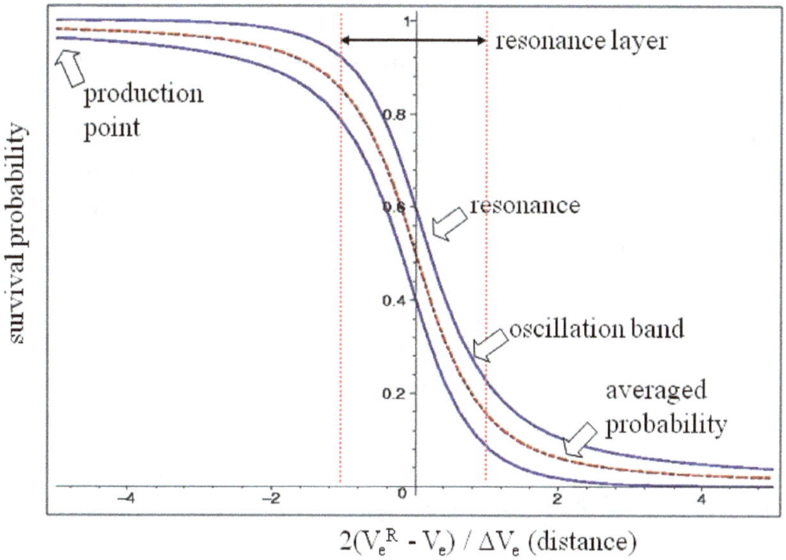

Fig. 4. The adiabatic evolution. Dependence of the survival probability on the matter potential (density) in the units of the width of resonance layer: $2(V_e^R - V_e)/\Delta V_e$. The oscillatory curve is inscribed into the band shown by solid blue lines. Red dashed line shows the averaged probability. For monotonous decrease of the potential with distance, the horizontal axis represents the distance.

This adiabatic probability is determined by mixing in the initial and final points and does not depend on density distribution in between (Fig. 4).

If initial state is ν_e and final state of the adiabatic evolution is a mass state in vacuum ν_k, we have

$$P_{ek} = |U_{ek}^m(n_i)|^2. \tag{51}$$

Nonoscillatory transition:[20,21] For very large initial densities, the mixing is suppressed which means that the initial flavor state consists essentially of a single eigenstate. In this case, the phase disappears in probability and oscillations are absent. There are two cases when it can happen:

(1) The initial value of the potential is in between two resonance values: $V_e^{R,12} \ll V_e \ll V_e^{R,13}$. Then $\nu_e \approx \nu_{2m}$, and the latter transforms into ν_2, when propagating to small densities, so in vacuum, we obtain $P_{ee} = |U_{e2}|^2 = c_{13}^2 \sin^2\theta_{12}$.
(2) The initial value is above the 1-3 resonance value: $V_e \gg V_e^{R,13}$. Now $\nu_e \approx \nu_{3m} \to \nu_3$. In this case, $P_{ee} = |U_{e3}|^2 = \sin^2\theta_{13}$.

In nonoscillatory case, flavor transformations are only due to change of mixing, that is, flavor of eigenstate.

The strongest change of flavor of whole the state is realized when the initial density is much larger and final density is much smaller than the resonance density. Resonance manifests itself via dependence of $U_{ej}^m(n^0)$ on energy (see Section 3.5).

Let us consider the adiabatic condition in detail.[20,21] In the 1-2 channel (relevant for solar neutrinos), we can use the 2ν Hamiltonian (Eq. (29)). The Hamiltonian in the evolution equation for the eigenstates (Eq. (42)) is reduced to

$$H^{(2)} = \begin{pmatrix} H_{1m} & i\dot{\theta}_{12}^m \\ -i\dot{\theta}_{12}^m & H_{2m} \end{pmatrix}, \tag{52}$$

where θ_{12}^m is defined in Eq. (31). The adiabaticity condition, the condition under which the off-diagonal terms can be neglected, $H_{2m} - H_{1m} \gg \dot{\theta}_{12}^m$, becomes

$$\gamma \equiv \frac{2\dot{\theta}_{12}^m}{H_{2m} - H_{1m}} \ll 1, \tag{53}$$

where γ is called *the adiabaticity parameter.*

The adiabatic condition is the most crucial in resonance where $\dot{\theta}^m$ is the biggest one, whereas $H_{2m} - H_{1m}$ takes the smallest value. It can be rewritten as[20]

$$l_R < \Delta r_R, \tag{54}$$

where $l_R = l_\nu / \sin 2\theta_{12}$ is the oscillation length in resonance and $\Delta r_R = h \tan 2\theta_{12}$ is the half width of the resonance layer and $h \equiv n/(\mathrm{d}n/\mathrm{d}x)$ is the scale of density change. So, in the resonance,

$$\gamma^R = \frac{1}{4\pi}\frac{l_R}{\Delta r_R} = \frac{1}{4\pi}\frac{l_\nu}{h(x)}\frac{\cos 2\theta_{12}}{\sin^2 2\theta_{12}}, \tag{55}$$

and inequality (54) corresponds to $\gamma^R < 1/4\pi$. According to Eq. (54), the oscillation length in resonance should be smaller than half of the width of the resonance layer. This ensures that a system characterized by the eigenlength l_ν has time to adjust itself to the change of external conditions determined by the scale of density change, h, which is common notion of adiabaticity.

In the 1-3 channel for densities much below the 1-3 resonance density, the adiabaticity is fulfilled even much better for the same density profile.

According to Eqs. (55), the adiabaticity breaks with increase of neutrino energy and/or decrease of mixing angle and Δm^2_{21}. For the LMA solution, the adiabaticity breaking occurs at 10^3 MeV.

If the density changes on the way of neutrinos fast so that $\gamma > 1$, the transitions between the eigenstates $\nu_{1m} \leftrightarrow \nu_{2m}$ cannot be neglected. For evolution from large density to zero density, we can introduce the amplitudes of $\nu_j^m \to \nu_i$ transitions A_{ji}. In terms of A_{ji}, the $\nu_e \to \nu_1$ probability equals

$$P_{e1} = |U^m_{e1} A_{11} + U^m_{e2} A_{21}|^2, \tag{56}$$

where U^m_{ei} are the mixing elements at the initial point. If the interference between the amplitudes is averaged out, $\langle A^*_{11} A_{21} \rangle = 0$, which indeed takes place when propagation occurs over long distances, we obtain from Eq. (56)

$$P_{e1} = |U^m_{e1}|^2 - \Delta P, \tag{57}$$

where the contribution from adiabaticity violation equals

$$\Delta P = P_{21}(|U^m_{e1}|^2 - |U^m_{e2}|^2), \tag{58}$$

and $P_{21} = |A_{21}|^2$ is the jump probability. Similarly for $\nu_e \to \nu_2$ transition, we have

$$P_{e2} = |U_{e2}^m|^2 + \Delta P. \tag{59}$$

For small mixing, the jump probability P_{12} is given by the Landau-Zenner formula.[25,26] The probability can be obtained considering jump between the eigenstates as the quantum mechanical process of tunneling through the barrier. For small mixing, the adiabaticity is broken, mainly, in resonance and the size of the barrier is given by splitting between the eigenstates in resonance Δ_{21}^R, whereas the energy of system which drives the transition is proportional to $\dot{\theta}_m$. Therefore

$$P_{21} = \exp(-\Delta_{21}^R/\dot{\theta}_m) = \exp(-2\pi/\gamma).$$

The precise formula which is also valid for large mixing angles has been obtain.[53] Adiabaticity violation can be realized in the Sun if, for example, hypothetical very light sterile neutrino exists, which mixes very weakly with the electron neutrino (see Section 4.5).

Adiabaticity can be broken by perturbations of the density profile of the Sun. This can be due to (thermal) density fluctuations, or regular variations induced, e.g., by oscillations of the Sun itself. The typical length of these perturbations, λ_{pert}, should be comparable with the oscillation l_m length, i.e., (10–500) km (depending on neutrino energy). If $\lambda_{\text{pert}} \ll l_m$, perturbations are averaged out; on the contrary, if $\lambda_{\text{pert}} \gg l_m$, they become adiabatic and do not affect result of transformation. Random perturbations can be considered as white noise.[54–57] Usually they lead to weakening of flavor transformations. The solar neutrino data allow to put an upper bound on the amplitude of perturbations as a function of the length at the level of few %. From the solar physics side, no such strong perturbations are expected.

3.4. *Oscillations in multilayer medium*

Another matter effect relevant for solar neutrinos is propagation in the multilayer medium: The Earth density profile can be considered as a sequence of layers with constant or slowly (adiabatically) changing density within layers and sharp density change between the layers where the adiabaticity is broken maximally.

Due to loss of coherence, the fluxes of mass states arrive at the surface of the Earth. On entering the Earth, the mass states split into eigenstates in the matter of the Earth ν_{im} and start to oscillate.[58–65] Therefore the channels of relevance for solar neutrinos are $\nu_k \to \nu_e$ and the important characteristic is the regeneration factor f_{reg} defined as

$$f_{\text{reg}}(\theta_z) = |U_{e1}|^2 - P_{1e}^E(\theta_z) \approx -[|U_{e2}|^2 - P_{2e}^E(\theta_z)], \tag{60}$$

where $P_{ie}^0 \equiv |U_{ei}|^2$ $(i = 1, 2)$ are the probabilities without Earth matter effect, whereas $P_{ie}^E(\theta_z)$ are the probabilities with the matter effect as functions of the zenith angle. In the second equality in Eq. (60), we took into account that matter effect on the 1-3 mixing is negligible, and consequently, $P_{3e}^0 \approx P_{3e}(\theta_z)$. So, the state ν_3 decouples and evolution is reduced to the 2ν evolution. For underground detectors, the matter effect is always present but of course it is large during nights.

The factor $f_{\text{reg}}(\theta_z)$ can be computed taking into account that within layers, an evolution is described by adiabatic formulas. At the borders of layers, a sudden change in eigenstates in matter occurs corresponding to density change from n_i to n_f, while flavor of state does not change. This is described by matrix

$$U^{m\dagger}(n_f)U^m(n_i) = U^m(\theta_{k-1}^{mf} - \theta_k^{mi}), \tag{61}$$

where θ_{k-1}^{mf} is the mixing angle in matter at the end of layer $k-1$ just before density jump, and θ_k^{mi} is the angle in the beginning of the layer k. Here U are 2×2 unitary matrices of rotations. Then for the matrix of oscillation amplitudes, we have

$$S = U^m(\theta_n^{mf})\Pi_{k=n,\ldots,1} D_k U^m(\theta_{k-1}^{mf} - \theta_k^{mi}). \tag{62}$$

Here n is the number of layers, $U^m(\theta_n^{mf})$ is the flavor mixing matrix in the last layer which projects the evolved system onto the flavor states, D_k is the evolution matrix of eigenstates in the layer k:

$$D = \text{diag}(e^{-i\phi_m^k/2}, \ \ e^{i\phi_m^k/2}),$$

where ϕ_m^k is the adiabatic phase acquired in the layer k:

$$\phi_k^m(E) \equiv \int_{x_{k-1}}^{x_k} dx \ \Delta_{21}^m(x). \tag{63}$$

The diagonal character of D_k reflects the adiabaticity of propagation. The probability of transition equals

$$P_{1e}^E = |A(\nu_1 \to \nu_e)|^2 = |S_{e1}|^2. \tag{64}$$

In low-density limit, we can use in Eqs. (62) and (64) the lowest-order terms in $V_e(x)$ or ϵ_{12} which leads to the regeneration factor[61,62]

$$f_{\rm reg} = \frac{1}{2} c_{13}^4 \sin^2 2\theta_{12} \int_0^L dx V_e(x) \sin \phi^m_{x\to L}. \tag{65}$$

Here $\phi^m_{x\to L}$ is the phase acquired from a given point of trajectory x to a detector:

$$\phi^m_{x\to L} = \int_x^L dy \Delta^m_{21}(y)\,,$$

where the difference of eigenvalues is given in Eq. (32). During the day, when the effect of oscillations inside the Earth can be neglected, $P_{1e} = U_{e1}^2 = c_{13}^2 c_{12}^2$. Some insight into the results can be obtained in the constant density approximation. Then Eq. (65) gives

$$f_{\rm reg} = \sin^2 2\theta_{12} \left(\frac{c_{13}^4 V_e}{\Delta^m_{21}} \right) \sin^2 \frac{1}{2} \Delta^m_{21} L\,. \tag{66}$$

The result in Eq. (65) can be obtained using the adiabatic perturbation theory.[66] Second-order correction in ϵ_{21} to $f_{\rm reg}$ has been found in Ref. 64.

Attenuation effect: The formula in Eq. (65) allows us to elucidate the attenuation effect immediately[61] which appears due to finite energy resolution of a detector or finite width of the original energy spectrum. The integration of the regeneration factor (Eq. (65)) with the energy resolution function $R(E, E')$ gives

$$\bar{f}_{\rm reg}(E) = \int dE' R(E, E') f_{\rm reg}(E'). \tag{67}$$

Inserting expression for $f_{\rm reg}$ from Eq. (65) and permuting integrations over E and x, Eq. (67) can be written as

$$\bar{f}_{\rm reg} = \frac{1}{2} c_{13}^4 \sin^2 2\theta \int_{x_0}^{x_f} dx\ V(x) F(x_f - x) \sin \phi_m(x \to x_f), \tag{68}$$

where $F(x_f - x)$ is called the *attenuation* factor.[61] For the Gaussian function $R(E, E')$ with width σ_E, we obtain

$$F(d) = e^{-2\left(\frac{d}{\lambda_{att}}\right)^2}, \tag{69}$$

where

$$\lambda_{\rm att} = l_\nu \frac{E}{\pi \sigma_E}$$

is the *attenuation length*. For $d = 0$, $F(0) = 1$ and it decreases with distance. This means that contributions of structures of the profile at $d > \lambda_{\text{att}}$ to the integral in Eq. (68) are suppressed.

Consider some structure in the point x of the trajectory at the distance $d \equiv x_f - x$ from a detector. According to Eq. (68), the potential $V_e(x)$ is integrated with $\sin \phi_m(d)$. This leads to averaging of oscillations between the structure and a detector. The larger the d, and therefore, the larger the $\phi_m(d)$, the stronger is averaging effect when integration over the energy is performed. The attenuation is, however, not only due to averaging but also due to projection onto certain initial and final states. Thus, for the $\nu_e \to \nu_1$ transition, the attenuation is absent.[61]

According to attenuation effect, a detector with the energy resolution ΔE cannot "see" (be sensitive to) remote structures of the density profile for which distance to the detector is larger than the attenuation length $\lambda_{\text{att}} \propto 1/\Delta E$. The better the energy resolution, the larger is the attenuation length, and consequently, deeper structures can be seen by a neutrino "microscope". This explains, for example, why for the solar neutrinos the zenith angle dependence of the Earth matter effect is flat and there is no enhancement of the regeneration for the core crossing trajectories in spite of 2–3 times larger density. Indeed, for the solar neutrinos with $E \sim 10$ MeV and $\Delta E/E = 0.3$, we obtain $\lambda_{\text{att}} = 1000\,\text{km}$ and therefore the contribution of the core is attenuated. On the contrary, small structures ($\sim$10 km) near the surface can produce strong effect. The attenuation length increases with energy.

3.5. *MSW LMA solution of the solar neutrino problem and its status*

Results obtained in the previous subsections allow one immediately to write up the probability for the MSW LMA solution. Neutrinos ν_e are produced in the Sun and later detected in the Earth detectors; therefore the survival probability $\nu_e \to \nu_e$, P_{ee} should be computed. ν_e produced in the central high density regions of the Sun propagate adiabatically toward the surface with $n \approx 0$.

1. The scale of density change in the Sun decreases from $h \to \infty$ in the center down to $h \approx 0.1 R_\odot$ at $R = 0.1 R_\odot$ and then stays nearly constant till the very low density near the surface where h drops down sharply. With this h and for oscillation parameters that correspond to the LMA MSW, the adiabaticity condition is very well satisfied till densities at which the mixing angles in matter are very close to the vacuum mixing

angles $\theta_{ij}^m \approx \theta_{ij}$. Indeed, from Eq. (31) we find $\theta_{12}^m - \theta_{12} \approx 0.5 \sin 2\theta_{12}\epsilon_{12}$, and for $E = 10\,\text{MeV}$, this gives $\theta_{12}^m/\theta_{12} - 1 = 4 \cdot 10^{-4}$.

2. The adiabatic evolution in the Sun is described according to Eq. (48) as

$$\nu(0) = \nu_e = \sum_j U_{ej}^m(n_e^0)\ \nu_{jm}(n_e^0) \rightarrow \sum_j U_{ej}^m(n_e^0)\ \nu_j\,, \tag{70}$$

where $U_{ej}^m(n_e^0)$ are the elements of mixing matrix in the production point with density n_e^0.

3. Due to loss of coherence, the phases between the mass states in Eq. (70) are irrelevant. Therefore incoherent fluxes of mass states arrive at the surface of the Earth. On crossing the surface of the Earth, the neutrino mass eigenstates "split" into the eigenstates of Hamiltonian in matter of the Earth, ν_{km}: $\nu_j = U(\theta_{12} - \theta_{12}^m)_{jk}\,\nu_{km}$, and oscillate.[58–60] Consequently, the survival probability

$$P_{ee} = \sum_j |U_{ej}^m(n_e^0)|^2 P_{ej}^E\,, \tag{71}$$

where P_{je} are the probabilities of oscillation transitions $\nu_j \rightarrow \nu_e$ in the Earth. During the day, $P_{ej} \approx |U_{ej}|^2$. Figure 5 shows the complete scheme of evolution of the solar neutrinos in the Sun and in the Earth.

In the standard parametrization of the mixing matrix, the relevant elements are

$$U_{e1}^m = \cos\theta_{13}^m \cos\theta_{12}^m, \quad U_{e2}^m = \cos\theta_{13}^m \sin\theta_{12}^m, \quad |U_{e3}^m| = |\sin\theta_{13}^m|. \tag{72}$$

In Eq. (72), the mixing angles in matter should be taken in the production point: $\theta_{12}^m = \theta_{12}^m(n^0)$, $\theta_{13}^m = \theta_{13}^m(n^0)$. The matter correction to 1-3 mixing parameter $s_{13}^2 \equiv \sin^2\theta_{13}$ can be as large as 12% for the highest neutrino energies.

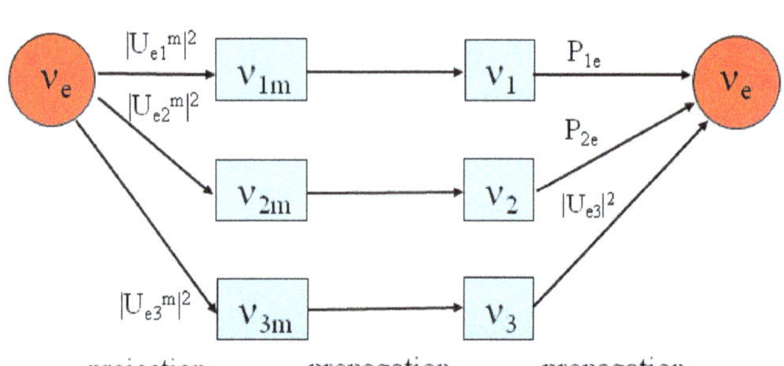

Fig. 5. The scheme of transitions of solar neutrinos.

Effect of matter of the Earth on the 1-3 mixing is very small so that $P_{3e} \approx s_{13}^2$. Therefore the unitarity condition, $\sum_j P_{ej}^E = 1$, becomes $P_{1e}^E + P_{2e}^E = 1 - s_{13}^2$. With this and the parametrization (Eq. (72)), Eq. (71) gives

$$P_{ee} = P_{1e}^E c_{13}^{m2} \cos 2\theta_{12}^m + c_{13}^2 c_{13}^{m2} \sin^2 \theta_{12}^m + s_{13}^2 s_{13}^{m2}. \tag{73}$$

Here for brevity $c_{13}^{m2} \equiv \cos^2 \theta_{13}^m$, *etc.* Using $P_{1e}^E = |U_{e1}|^2 - f_{\rm reg}$, the probability P_{ee} can be written as

$$P_{ee} = c_{13}^2 c_{13}^{m2} P_2^{\rm ad} + s_{13}^2 s_{13}^{m2} - c_{13}^{m2} \cos 2\theta_{12}^m f_{\rm reg}, \tag{74}$$

where

$$P_2^{\rm ad} = s_{12}^2 + \cos 2\theta_{12} \cos^2 \theta_{12}^m = \frac{1}{2}(1 + \cos 2\theta_{12} \cos 2\theta_{12}^m). \tag{75}$$

The first term in Eq. (75) gives the asymptotic ($E \to \infty$) value of probability which corresponds to the nonoscillatory transition. It provides the smallest value of probability so that $P_{ee} \geq c_{13}^2 c_{13}^{m2} s_{12}^2$. The second term describes effect of residual oscillations. The second term in Eq. (74) is the contribution from the decoupled third state ν_3.

Equations (74) and (75) present the main formulas which describe the flavor transformations of the solar neutrinos in the LMA MSW solution.

The probabilities should be averaged over the production region in the Sun. This does not affect the low-energy vacuum oscillation part as well as high-energy asymptotics, but changes P_{ee} in intermediate energy range. In the first approximation, averaging can be accounted by the effective initial densities $n_e^0 \to \bar{n}_e^0$.[66]

According to Eqs. (74) and (75), the difference of probabilities with and without oscillations in the Earth (the day–night asymmetry) is

$$P_{ee} - P_{ee}^0 = -c_{13}^{m2} \cos 2\theta_{12}^m f_{\rm reg}, \tag{76}$$

and $f_{\rm reg}$ is given in Eq. (65). It is proportional to c_{13}^6 (see Ref. 63), so nonzero 1-3 mixing reduces effect by about 7%.

Oscillations in the Earth reduce the probability P_{1e}^E. According to Eq. (76), for high-energy part of the spectrum with $\cos 2\theta_{12}^m < 0$, the Earth matter effect is positive, thus leading to *regeneration* of the ν_e flux, whereas for low energies, one has $\cos 2\theta_{12}^m > 0$, and the oscillations in the Earth further suppress the ν_e flux (although at low energies the effect itself is very small). What follows from Eq. (66) is that the regeneration effect increases linearly with the neutrino energy.

The survival probability $P_{ee}(E, t)$ in Eqs. (74) and (75) depends on neutrino energy and time. The time dependence is due to oscillations in the Earth since the effect depends on the zenith angle of neutrino trajectory.

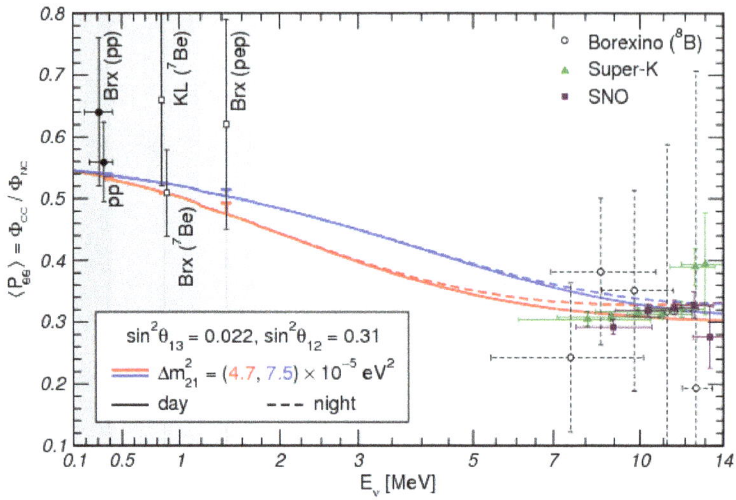

Fig. 6. Dependence of the probability P_{ee} integrated over the day and night time periods for Δm_{21}^2 determined from the global fit of the solar neutrino data only (red) and from the global fit of all oscillation data (blue). Also shown are the results from different experiments. Abbreviations "Brx" for Borexino and "KL" for KamLAND are used.[50]

The main dependence of P_{ee} on the neutrino energy is in $\theta_{12}^m(n_0)$, and much weaker one is in $\theta_{13}^m(n_0)$ and P_{1e}. Figure 6 shows dependence of the probabilities $P_{ee}(E)$ integrated over the day and the night times. Let us consider its properties.

(i) At low energies (vacuum-dominated part), neglecting the ν_e regeneration in the Earth, one has[52]

$$P_{ee} \approx c_{13}^4(1 - 0.5\sin^2 2\theta_{12}) - 0.5c_{13}^6 \cos 2\theta_{12} \sin^2 2\theta_{12}\epsilon_{12}(n_e^0)\,. \tag{77}$$

With decrease of energy: $P_{ee} \to P_{ee}^{\text{vac}}$ (the first term). For the best fit value of the 1-2 mass splitting, deviations of the probability (Eq. (77)) from its vacuum values are 6% for the ^{7}Be neutrinos and 2% for the pp neutrinos with $E = 0.3$ MeV.

(ii) For high energies (matter dominated region), we obtain

$$P_{ee} = c_{13}^4 \sin^2\theta_{12} + \frac{1}{4\epsilon_{12}^2(n_e^0)} \cos 2\theta_{12} \sin^2 2\theta_{12} + c_{12}^2 f_{\text{reg}}\,, \tag{78}$$

where the first term corresponds to the nonoscillatory transition in the Sun.

(iii) The intermediate energy region between the vacuum- and matter-dominated limits is the region where the resonance turns on in the Sun. The middle of this region (before averaging) corresponds to the MSW resonance at maximal density in the Sun (the slope reflects the resonance curve). Value of θ_{12} determines sharpness of the transition, that is, the size of transition region in E. The larger the θ_{12}, the bigger is the size of the region. Integration over the neutrino production region in the Sun smears the transition.

The survival probability in Eqs. (74) and (75) shows certain *scaling* properties.[50] It is a function of three dimensionless parameters: $P_{ee} = P_{ee}(\epsilon_{12}, \epsilon_{13}, \phi_E)$, where $\phi_E \approx \Delta m_{21}^2 L/2E$ is the phase of oscillations in the Earth and ϵ_{12}, ϵ_{13} are defined in Eqs. (30) and (26). Consequently, the probability is invariant with respect to rescaling

$$\Delta m_{21}^2 \to b\Delta m_{21}^2, \quad \Delta m_{31}^2 \to b\Delta m_{31}^2, \quad E \to bE.$$

The adiabatic probability averaged over oscillations does not depend on any spatial scale of the density profile. So, the only dependence on distance is in the phase ϕ_E. If oscillations in the Earth are averaged, then the whole probability, $P_{ee} = P_{ee}(\epsilon_{12}, \epsilon_{13})$, is independent on spatial scale. In this case, P_{ee} is invariant with respect to rescaling:

$$\Delta m_{21}^2 \to a\Delta m_{21}^2, \quad \Delta m_{31}^2 \to a\Delta m_{31}^2, \quad V_e \to aV_e. \tag{79}$$

In particular, $a = -1$ corresponds to change of the signs of mass-squared differences and potentials. Since oscillations in the Earth are invariant with respect to $\phi_E \to -\phi_E$, the rescaling (Eq. (79)) holds also for nonaveraged probability. The dependence of P_{ee} on ϵ_{13} is weak, and if neglected, the probability will depend on only one combination of the parameters: ϵ_{12}.

Let us estimate effect of the adiabaticity breaking. The probability $P_{\rm ee}^{\rm na}$ with nonadiabatic corrections can be obtained by substituting in Eq. (71) $|U_{ei}^m(n_e^0)|^2 \to P_{ei}$ from Eqs. (56) and (59). Then the corrections to the adiabatic case equal

$$\Delta P_{ee} = P_{ee}^{\rm na} - P_{ee} = -\Delta P(|U_{e1}|^2 - |U_{e2}|^2 - 2f_{\rm reg}), \tag{80}$$

where ΔP is defined in Eq. (58). Neglecting regeneration and expressing the mixing matrix elements in terms of mixing angles, we arrive at the known formula for the correction[25,26]:

$$\Delta P_{ee} \approx -P_{12} c_{13}^2 c_{13}^{m2} \cos 2\theta_{12}^m \cos 2\theta_{12}.$$

If the initial density is above the resonance one so that $\cos 2\theta_{12}^m < 0$, the correction is positive, which means that adiabaticity violation weakens suppression of the original ν_e flux. Corrections to the leading order adiabatic approximation (adiabaticity violation effect) can be expressed in terms of the adiabaticity parameter (Eq. (53)) as[66]

$$\frac{\Delta P_{ee}}{P_{ee}} \approx \frac{\gamma^2 \cos 2\theta_{12}}{4 \sin^2 \theta_{12}}.$$

For $E = 10\,\text{MeV}$, the correction is about 10^{-8}, i.e., negligible.

4. Detection of the solar neutrinos: solar neutrino problem and its solution

We summarize here the main experimental results which are used in the present global fits emphasizing their role in establishing solar neutrino problem and its resolution. Technical details and experimental features are described in a number of reviews (see, e.g., Ref. 35).

4.1. *Establishing the solar neutrino problem*

4.1.1. *Homestake experiment*

The radiochemical Cl–Ar method of detection of neutrinos proposed by Pontecorvo[3] (and later elaborated by Alvarez[67]) is based on counting of the Ar atoms produced in the reaction

$$\nu_e +^{37}\text{Cl} \rightarrow^{37}\text{Ar} + e. \tag{81}$$

Its energy threshold is $E^{\text{th}} = 0.814\,\text{MeV}$. The total Ar production rate, Q_{Ar}, is integrated over the energy spectrum above the threshold and averaged over the exposure time of about a month determined by the lifetime of Ar. E^{th} makes the experiment sensitive, mainly, to ^{8}B neutrinos. According to SSM (without flavor conversion), contributions to Q_{Ar} of different components of the spectrum equal 77% (^{8}B), 14% (^{7}Be), 5% (NO), 2.5% (*pep*) 0.4% (*hep*).

The average Ar-production rate for whole period of observations from 1970 to 1995 (108 runs) is[68]

$$Q_{\text{Ar}}^{\text{exp}} = 2.56 \pm 0.16(stat) \pm 0.16(syst)\ \text{SNU}, \tag{82}$$

where 1 SNU (the solar neutrino unit) corresponds to one capture per second in 10^{36} atoms of target material. Total error equals 0.23 SNU.

The measured rate (Eq. (82)) is about three times smaller than the SSM prediction: $Q_{\text{Ar}}^{\text{SSM}} = 8.0 \pm 1.0$ SNU (for latest computations). The error bar is mainly related to uncertainty in prediction of the Boron neutrino flux. This deficit of solar ν_e flux,

$$R_{Ar} \equiv \frac{Q_{\text{Ar}}^{\text{exp}}}{Q_{\text{Ar}}^{\text{SSM}}} = 0.32 \pm 0.05,$$

was the essence of the *solar neutrino problem.*

The rate in Eq. (82) is about 2σ smaller than the rate expected in LMA MSW solution (see below): $Q_{\text{Ar}}^{\text{LMA}} = 3.1 \pm 0.1$ SNU. The discrepancy can be (i) a statistical fluctuation, (ii) some unknown systematics, (iii) new physics in the intermediate part of the spectrum. The latter interpretation is rather restricted after the direct measurements of ^{7}Be- and *pep*- neutrino fluxes by BOREXINO.

Homestake data showed time variations within about a 11-year period which cannot be explained as statistical fluctuations. The variations were in anticorrelation with the solar activity (sunspot number) and could be explained by neutrino spin-flavor precession in solar magnetic fields.[12,13,31,32] This interpretation required very big magnetic moments of neutrinos. Furthermore, time variations of signals have not been observed by Kamiokande and gallium experiments (operating partially during the same period of time as the Homestake experiment) and also later by SK, SNO, BOREXINO.

The small rate $Q_{\text{Ar}}^{\text{exp}}$ and the variations of signal could be related if, for example, in some periods the efficiency of detection was lower than accepted.

4.1.2. *Kamiokande-II, III*[69–72] *and Super-Kamiokande*

These water Cherenkov detectors observed the solar neutrinos via the neutrino–electron scattering:

$$\nu + e \to \nu + e. \tag{83}$$

For each event, the detection time, kinetic energy and direction of the electron are recorded.

The experiment is sensitive to the ^{8}B neutrinos (the *hep* neutrino contribution is negligible). All three neutrino species contribute: for ν_e, the reaction in Eq. (83) proceeds via the charged and neutral currents; whereas for ν_μ and ν_τ, via neutral currents (NCs) with 6.5 times smaller cross section. Kamiokande-II, III has observed the reduced signal, thus confirming the solar neutrino problem. The ratio of the observed to expected signal/flux (in assumption of energy independent suppression) was[72]

$$R_{\nu e} \equiv \frac{\Phi_{\text{B}}^{\text{exp}}}{\Phi_{\text{B}}^{\text{SSM}}} = 0.49 - 0.64. \tag{84}$$

However, the suppression was weaker than that in Homestake $R_{\nu e} > R_{\rm Ar}$, indicating the additional contribution to Kamiokande signal from ν_μ and ν_τ scattering which appear due to $\nu_e \to \nu_\mu, \nu_\tau$ transformation in the Sun.[73]

4.1.3. *Ga–Ge experiment*

The radiochemical Ga–Ge experiment is based on reaction

$$\nu_e + {}^{71}\mathrm{Ga} \to e^- + {}^{71}\mathrm{Ge}, \tag{85}$$

being sensitive to ν_e only. A low threshold, $E^{\rm th} = 0.233\,\mathrm{MeV}$, allows the detection of all the components of the neutrino spectrum including the *pp* neutrinos. According to SSM, contributions of different components equal 53% (*pp*), 26% (Be), 7.5% (NO), 2% (*pep*), 11% (B). The method is based on counting of number of ^{71}Ge atoms. Two experiments have been realized: SAGE (Baksan laboratory) and GALLEX–GNO (Gran-Sasso laboratory). The germanium production rates averaged over the whole period of observations are

$$Q_{\rm Ge} = 65.4^{+3.1}_{-3.0}(stat)^{+2.6}_{-2.8}(syst)\ \mathrm{SNU} \tag{86}$$

from 168 runs of SAGE experiment in the period 1990–2007,[74,75] and

$$Q_{\rm Ge} = 67.13^{+4.64}_{-4.63}\ \mathrm{SNU} \tag{87}$$

from GALLEX (1991–1997)[76] and GNO (1997–2003).[77,78] Results from Eqs. (86) and (87) are in very good agreement, and the combined rate from the two experiments equals $Q_{\rm Ge} = 66.2 \pm 3.1$ SNU.

Comparison of results (Eqs. (86) and (87)) with the SSM prediction, $Q_{\rm Ge}^{\rm SSM} = 128 \pm 5$ SNU, according to GS98-11, shows the deficit

$$R_{\rm Ge} \equiv Q_{\rm Ge}^{\rm exp}/Q_{\rm Ge}^{\rm SSM} = 0.52 \pm 0.03,$$

confirming the solar neutrino problem. The deficit is weaker than that in Kamiokande-II and Homestake, thus indicating that the suppression of the ν_e flux becomes weaker with the decrease of energy.

The results (Eqs. (86) and (87)) are at the level of SSM contribution to $Q_{\rm Ge}$ from the *pp* neutrinos $Q_{\rm Ge}^{pp} = 67.8$ SNU (which is well restricted by luminosity). If the contributions from boron and beryllium neutrino fluxes measured by Kamiokande-II and Homestake are also added, the expected rate would be $Q_{\rm Ge}^{\rm min} = 83$ SNU (minimal astrophysically possible value), which is more than 3σ higher than $Q_{\rm Ge}^{\rm exp}$. This result further indicated that the solution of the problem is in neutrino physics and not in astrophysics.

The calibration experiments with artificial sources have shown some deficit of signal — the gallium anomaly.[79–82] This, however, should not significantly affect conclusions from the solar experiments.

4.1.4. *Super-Kamiokande (SK) experiment*

The presently achieved experimental energy threshold is about 3.5 MeV for kinetic energy of the electron ($E > 4$ MeV for neutrinos). The main results obtained by SK from 1996 to 2015[83,84] can be summarized in the following way:

(i) The ratio of the observed to expected signal (in assumption of energy-independent suppression) is

$$R_{\nu e} = 0.43 \pm 0.06 \tag{88}$$

in agreement with Kamiokande-II. The deficit of signal is weaker than the one of Homestake which testifies (as mentioned before) for additional contribution to the SK signal from ν_μ and ν_τ.

(ii) Measured energy spectrum of recoil electrons is in agreement with flat (energy independent) suppression of the solar neutrino flux. Some increase at high energies is due to the Earth matter effect. The expected (according to the LMA solution) upturn of the spectrum at low energies is not seen and actually disfavored at about 2σ level. Although in SK-IV data, some indication of the upturn exists.

(iii) Dependence of the integrated over the energy rate on the zenith angle of neutrino trajectory (fixed by the time of detection) has been measured. It shows higher rate during the night. The averaged rate of events during the day, N_D, and the night, N_N, determines the day–night asymmetry[84]

$$A_{\rm DN} \equiv 2\frac{N_D - N_N}{N_D + N_N}. \tag{89}$$

The measured value

$$A_{\rm DN} = -[3.3 \pm 1.0(stat) \pm 0.5(syst)]\% \tag{90}$$

is negative and differs from 0 by about 3σ. During the night, the rate of events is consistent with constant. Figure 7 shows the zenith angle dependence of the ratio of measured rate of events and prediction without oscillations. Solar neutrinos in the last night bin pass through the Earth's outer core. However, no enhancement occurs due to the attenuation effect.

Seasonal variations of signal are in agreement with annual (7%) variations of distance between the Sun and the Earth.

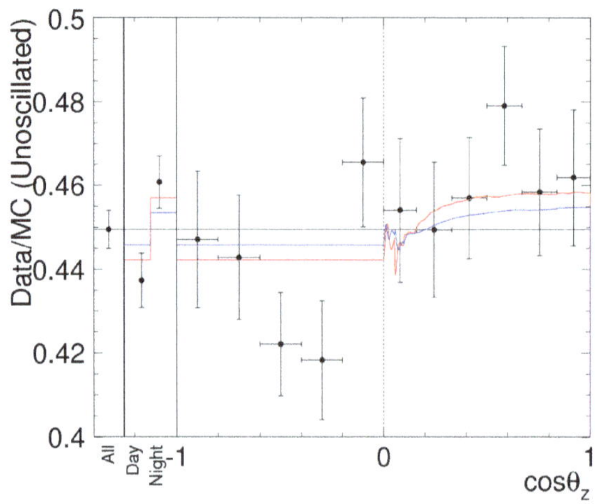

Fig. 7. SK-IV zenith angle dependence of the ratio of rates data/MC (unoscillated) in the energy interval (4.49–19.5) MeV. Red (blue) lines are LMA MSW predictions for small (large) value of Δm^2_{21}. The error bars shown are only statistical uncertainties.[84]

Already after initial SK results (no distortion of the energy spectrum, some indication of higher signal during the night, absence of peak of rate for the core-crossing trajectories expected in the SMA solution), the LMA MSW solution becomes the favored one.[36]

Study of the Earth matter effect provides a complete (integrated) check of the solution of the solar neutrino problem, since all the phases of evolution are involved.

4.2. *Resolving the problem*

Sudbury Neutrino Observatory (SNO) observed neutrino interaction in 1 kton of heavy water via three different processes:

The CC scattering on deuteron[85]

$$\nu_e + d \to e + p + p \tag{91}$$

has been detected by Cherenkov radiation produced by electron. It is sensitive only to ν_e and therefore measures the total flux of ν_e, Φ_e. The neutrino energy can be well determined which allows the measurement of neutrino energy spectrum. With energy threshold $E = 5\,\text{MeV}$, SNO is sensitive to ^{8}B neutrinos only.

NC scattering on deuteron[86,87]:

$$\nu + d \to \nu + p + n, \tag{92}$$

where only neutron has been detected by three different methods: (1) the capture by deuteron: $n + d \to {}^3\text{He} + \gamma$ (2.3 MeV) (phase 1), (2) capture on a salt dissolved in the heavy water, (3) interaction with ^{3}He in the He counters allocated in the detector. Efficiency of the detection increased from (1) to (3). Only total rate of the process (Eq. (92)) has been measured. The reaction in Eq. (92) is equally sensitive to all neutrino species, and therefore is not affected by the flavor transformations. It allows the measurement of the total flux of active solar neutrinos $\Phi_{\rm NC}$.

Neutrino scattering on electrons (Eq. (83)) is in agreement with the SK result.

The main results are:

(i) The deficit of the ν_e flux is established in practically solar-model-independent way: In assumption that the spectra of neutrinos are not distorted by the flavor transformations, the ratio of fluxes measured by CC and NC equals

$$R = \Phi_{\rm CC}/\Phi_{\rm NC} = \Phi_e/\Phi_{\rm tot} = 0.34 \pm 0.04. \tag{93}$$

Since only ν_e flux is produced in the Sun, this result means that the ν_e flux at the Earth is suppressed by a factor of 1/3, and correspondingly, 2/3 of the original ν_e flux is transformed into ν_μ and ν_τ fluxes.

(ii) The $\nu{-}e$ scattering confirms this conclusion. The results can be presented in the plane of $\Phi_e{-}\Phi_{\mu\tau}$ fluxes. Three experimental bands determined by processes in Eqs. (91) and (92) and $\nu{-}e$ scattering intersect in the same region of this plane.

(iii) The energy spectrum of the CC events does not show significant distortion being in agreement with flat suppression. No expected upturn of the spectrum at low energies has been observed. Furthermore, the lowest energy points (although with large errors) show turn down of spectrum (i.e., stronger suppression).

(iv) The energy spectra of the events detected during the days and nights (day and night spectra) have been reconstructed separately.[88,89] They do not show substantial difference and are in agreement with flat suppression.

The extracted day–night asymmetry in Eq. (90) depends on assumption about variations of total flux (i.e., in NC events). If there is no asymmetry in NC, $A_{\rm DN}^{\rm CC} = -(1.5 \pm 5.8 \pm 2.7)\%$. Without fixing asymmetry in NC, $A_{\rm DN}^{\rm CC} = -(5.6 \pm 7.4 \pm 5.3)\%$, but $A_{\rm DN}^{\rm NC} = -(4.2 \pm 8.6 \pm 7.2)\%$, so the asymmetry is mostly in the total flux. The result hints ν_e-flux regeneration although it is not conclusive due to large errors.

SNO provided final confirmation of the *flavor transformations* of solar neutrinos. At the same time, the mechanism of transformations was not identified uniquely. The data further favored LMA solution; however, several other possibilities still existed: LOW solution, solution based on nonstandard neutrino interactions (even for massless neutrinos), spin-flavor precession in the magnetic fields of the Sun, etc.

It is KamLAND reactor experiment[38–40] that uniquely selected the LMA MSW as the mechanism of flavor transformations of solar neutrinos (in assumption of CPT conservation). KamLAND has observed oscillations of electron antineutrinos and values of Δm^2_{21} and θ_{12} determined by KamLAND are in agreement with the LMA MSW solution.

4.3. *Consolidating results*

BOREXINO, a 1 kt scintillation detector, studies the $\nu{-}e$ scattering with fiducial mass about 0.1 kt.[90,91] The energy spectrum of the recoil electrons is measured down to 0.2 MeV. Very low background and high energy resolution allow to disentangle contributions from different components of spectrum as well as contributions from background.

Results from the first phase, and partially the second phase, can be summarized as follows.

(i) The beryllium neutrinos produce[92] $46.0 \pm 1.5^{+1.5}_{-1.6}$ counts/(day 100 ton) in the energy interval (0.2–0.8) MeV, which corresponds to the ratio of the measured to expected flux

$$R = \Phi^{\rm exp}_{\rm Be}/\Phi^{\rm SSM}_{\rm Be} = 0.63 \pm 0.05. \tag{94}$$

This result is in agreement with the LMA MSW expectation, and it shows that there is no drastic modification of the solution in the intermediate energy region.

(ii) For the *pep*-neutrino flux with the rate $(3.1 \pm 0.6 \pm 0.3)$ counts/(day 100 ton),[93] similar suppression is obtained:

$$R = \Phi^{\rm exp}_{pep}/\Phi^{\rm SSM}_{pep} = 0.70 \pm 0.15. \tag{95}$$

(iii) The outstanding BOREXINO result is the first direct measurement of the *pp*-neutrino flux[94]:

$$R = \Phi^{\rm exp}_{pp}/\Phi^{\rm SSM}_{pp} = 0.64 \pm 0.12. \tag{96}$$

This result is in good agreement with the LMA prediction (nearly averaged vacuum oscillations) as well as with suppression 0.56 ± 0.06 found from the global fit of solar neutrino data (without BOREXINO *pp* result). This can be obtained by subtracting from the total Ge production rate the contribution of boron and Be neutrinos using SK, BOREXINO (Be) as well as Homestake results. The latter comparison shows that reduction of the error in Eq. (96) by a factor of 2 is needed to affect substantially the global fit of the solar neutrino data.

(iv) No day–night asymmetry has been found for the Be-neutrino flux[95]

$$A_{\rm DN} = -0.001 \pm 0.012(stat) \pm 0.007(syst) \tag{97}$$

as is expected in the case of LMA solution. This result alone excludes the LOW solution.

(v) Seasonal variations are in agreement with variations related to eccentricity of the Earth orbit.[96]

(vi) The upper bound on rate of events induced by neutrinos from the CNO cycle is established[91]: <7.9 counts/(day · 100 ton) at 95% CL which is still larger than the expected rate by a factor of 2.

The BOREXINO phase II should be considered as independent experiment with lower background and different systematics. The first simultaneous spectroscopy of the low energy components (*pp*, ^{7}Be and *pep*) has been performed[97] with the following results:

$$R_{Be} = 0.62^{+0.015}_{-0.017} \tag{98}$$

for low metallicity and $R_{Be} = 0.53 \pm 0.014$ for high metallicity;

$$R_{pep} = 0.43^{+0.07}_{-0.075}, \tag{99}$$

which is lower than in (95), and the combined result

$$R_{pep} = 0.57^{+0.05}_{-0.06} \tag{100}$$

for both metallicities. New results have 2–3 times smaller error bars and will contribute to the global fit of data substantially. New upper bound on the CNO neutrino rate: <8.1 counts/(day 100 tons).

For ^{8}B neutrinos with 11.5 times higher statistics than in the phase I BOREXINO finds[98] at the effective neutrino energy $E \sim 7.9$ MeV

$$R_B = 0.470^{+0.03}_{-0.04} \tag{101}$$

for low Z, and $R_B = 0.36 \pm 0.03$ for high Z.

All the results are in agreement with the phase I results and with the LMA MSW predictions. They prefer the high metallicity model.

KamLAND scintillator detector also measured the ^{7}Be- and ^{8}B-neutrino fluxes.[99] Although the size of the detector is bigger than BOREXINO, due to a much higher background, the precision of measurements is lower.

All this leads to consolidation of the experimental results and further confirmation of the LMA MSW solution. Agreement with the expected energy profile of the LMA solution has been improved with the inclusion of BOREXINO results substantially.

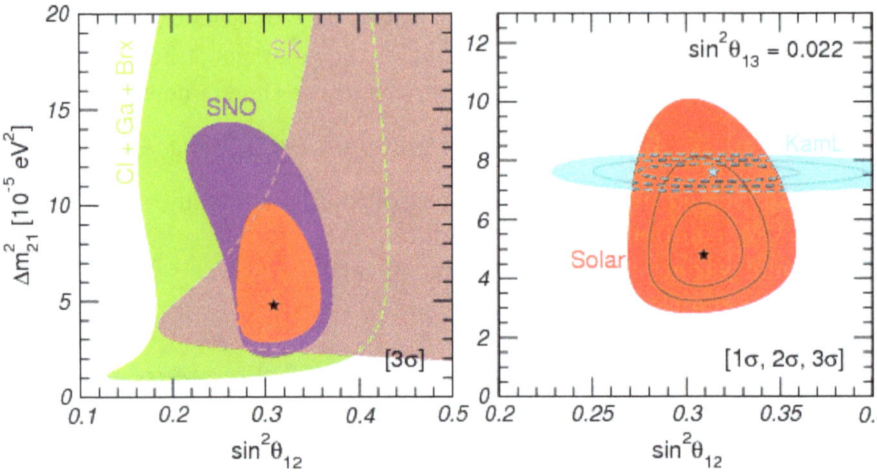

Fig. 8. Allowed regions of θ_{12} and Δm^2_{21} from the global fit of the solar neutrino data (red) as well as KamLAND (blue), for θ_{13} fixed to the best fit of the reactor experiments. In the left panel, also shown are regions restricted by individual experiments.[50]

4.4. *Status of the LMA MSW solution*

The LMA MSW solution provides a good description of all available solar neutrino data with their present accuracy. In Fig. 8, we show the result of the global fit of the data in the $(\theta_{12}, \Delta m^2_{21})$ plane with θ_{13} fixed to the best fit value from the reactor experiments. In the left panel, the regions restricted by individual solar neutrino experiments are shown, whereas in the right panel, we compare the solar and KamLAND allowed regions.

Solar neutrinos allow us to fix the sign of Δm^2_{21} (i.e., the 1-2 mass hierarchy) for the standard value of V_e. The sign determines the resonance channel (neutrino or antineutrino) and the mixing in matter. Due to smallness of the 1-3 mixing, the problem is reduced approximately to the 2ν problem. This, coupled with the fact that suppression of signal averaged over the oscillations at high energies is stronger than 1/2, suggests that $\Delta m^2_{21} > 0$. That corresponds to the normal ordering (hierarchy) when the electron flavor is mostly present in the lightest state.

There are few (probably related) facts which bring a mild ($\sim 2\sigma$) tension:

1. Absence of the expected "upturn" of the spectrum, that is, increase of the ratio of the measured spectrum to the SSM one toward low energies. No one experiment has showed the upturn. The SNO experimental points even turn down at low energies.

2. Large day–night asymmetry (Eq. (90)). For the best fit values from the global solar neutrino fit, one expects the D–N asymmetry $A_{\rm DN} = -2.8\%$. However, for values of Δm^2_{21} from global fit of all oscillation data (dominated by KamLAND), the predicted asymmetry equals $A_{\rm DN} = -1.7\%$. SK gives larger value: 3.3% (Eq. (90)), and even larger asymmetry, -4.2%, has been obtained in assumption of the undistorted energy spectrum.
3. Discrepancy of Δm^2_{21} determined from the solar neutrino data and KamLAND. The 1-2 mass splitting extracted from the global fit of the solar neutrino data $\Delta m^2_{21} = (4.7^{+1.6}_{-1.1}) \times 10^{-5}$ eV^2 is about 2σ smaller than the value measured by KamLAND (antineutrino channel) as well as the global fit value of all oscillation data[100] $\Delta m^2_{21} = (7.50^{+0.19}_{-0.17}) \times 10^{-5}$ eV^2.
   Due to scaling of $EV_e/\Delta m^2_{21}$, with the increase of Δm^2_{21}, the upturn and the ν_e regeneration peak shift to higher energies, which leads to stronger distortion of the spectrum at low energies and smaller D–N asymmetry.
4. Large value of the matter potential. One can consider V_e as a free parameter and extract it by fitting the solar neutrino data.[50,101,102] The result is $V_e^{\rm fit} = 1.6 V_e$, i.e., $V_e^{\rm fit}$ larger than the standard value by a factor of 1.6.[50] This is directly related to the difference of Δm^2_{21} obtained from the solar and KamLAND data due to approximate scaling $EV_e/\Delta m^2_{21}$.

When all the data (SK, SNO, BOREXINO), not just SK ones, are used in the global fit, the inclusion of the Earth matter effect reduces χ^2 by $\Delta\chi^2 \sim 14$, i.e., the effect is seen at about 4σ.[50]

4.5. *Search for effects beyond the 3ν paradigm*

Standard matter effects can be modified due to (i) nonstandard neutrino interaction; (ii) presence of new neutrino species — sterile neutrinos; (iii) interactions which transform neutrinos to antineutrinos. The latter happens, e.g., due to interaction of the magnetic moment of neutrino with magnetic field of the Sun (leading to the spin-flavor precession[31,32,103–106]). Antineutrinos have opposite sign of the potential and therefore matter effect is modified. As a consequence of the spin-flavor precession and usual flavor mixing, ν_e can be converted into $\bar{\nu}_e$. Bound on the $\bar{\nu}_e$, flux from the Sun has been obtained.[107,108]

Here we consider (i) and (ii) which can be motivated by tensions described above. Effects of new physics on propagation can be listed by adding new terms, $H_{\rm NP}$, to the standard Hamiltonian in Eq. (19): $H \to H_{\rm LMA} + H_{\rm NP}$. In addition, new physics can affect production and interactions of neutrinos.

Nonstandard interactions (NSI): They can modify the LMA MSW solution and inversely be restricted by solar neutrinos.[109] In the presence of NSI, the matter potential, V, in the evolution (Eq. (19)) includes an extra term

$$V_{\alpha\beta} = V_e \delta_{\alpha e}\delta_{\beta e} + \sqrt{2}G_F \sum_f n_f \, \epsilon^f_{\alpha\beta},$$

where n_f is the number density of fermions f ($f = u, d, e$), and $\epsilon^f_{\alpha\beta} = \epsilon^{fL}_{\alpha\beta} + \epsilon^{fR}_{\alpha\beta}$. We assume that the second term is smaller than the first one. Due to Hermiticity, $\epsilon^f_{\beta\alpha} = \epsilon^{f*}_{\alpha\beta}$, the diagonal entries $\epsilon^f_{\alpha\alpha}$ must be real. As it was done in Section 3.2, we can perform transformations to the propagation basis but now the 1-3 rotations depend on the NSI parameters and 2-3 mixing angle. After decoupling of the third eigenstate, the matter term for the rest of the system can be written as

$$H^{(2)}_{\rm NP} = \sqrt{2}G_F \sum_f n_f \begin{pmatrix} -\epsilon^f_D & \epsilon^f_N \\ \epsilon^{f*}_N & \epsilon^f_D \end{pmatrix}, \tag{102}$$

where ϵ^f_D and ϵ^f_N are linear combinations of the original parameters, $\epsilon^f_{\alpha\beta}$ and parameters of transformation (see Ref. 102).

In the case of NSI with *electrons* ($f = e$), both the standard and non-standard terms in Eq. (102) scale with the same density profile n_e. This implies that a positive value of ϵ^e_D large enough can "flip the sign" of the matter term so that the resonance will be realized for the inverted 1-2 hierarchy, $\Delta m^2_{21} < 0$, in contrast to the standard case. (The same is possible for scattering on quarks if their couplings are tuned appropriately.) Therefore there is a degeneracy between the sign of Δm^2_{21} and that of the matter potential: only their *relative* sign can be determined by oscillation experiments. For the nondiagonal couplings with quarks, this ambiguity is only approximate: however, the present data are unable to resolve it.

With new interactions, the evolution inside the Sun is still adiabatic, apart from the region where nondiagonal elements of the Hamiltonian vanish due to cancellation of the vacuum mixing and NSI terms. The latter does not produce significant effect and therefore the results are determined by the mixing at the production point. The latter is modified by NSI. The mixing at the exit from the Sun coincides with the vacuum mixing.

If NSIs are diagonal, their effect is reduced to change of V_e. In turn, the latter is equivalent to rescaling of Δm^2_{21}. According to scaling properties, $V = 1.6V_e$ allows to increase Δm^2_{21} by the same factor 1.6 thus achieving agreement with the KamLAND value. Furthermore, in this case the D–N asymmetry will be enhanced by 1.6 in agreement with SK measurements. Since in isotopically neutral medium (like in the Earth) the number of quarks is six times larger than electrons, the values $\epsilon \sim 0.1$ are enough to get such a modification of the potential.

Qualitatively new effects appear in the presence of nonzero off-diagonal potentials. In this case, at large values of the potential, the mixing is not suppressed. In asymptotics, $E \to \infty$, the vacuum contribution $\Delta m^2_{21}/2E$ is small, and one obtains from Eq. (102)

$$\cos 2\theta_m \approx -\frac{c_{13}^2 - 2r_f\epsilon_D^f}{\sqrt{(2r_f\epsilon_D^f - c_{13}^2)^2 + (2r_f\epsilon_N^f)^2}}, \tag{103}$$

where $r_f \equiv n_f/n_e$ and c_{13}^2 terms are left over from the standard potential term. According to Eq. (103), $\cos 2\theta_m \neq -1$, and therefore for the same vacuum mixing, a suppression at high energies is always weaker than that without NSI. In the limit of very low energies, P_{ee} approaches the vacuum value as in the standard case. The strongest modification of P_{ee} appears in the intermediate energy region. In general, the probability is given by Eq. (74) with

$$\cos 2\theta_m \approx -\frac{A}{\sqrt{A^2 + (2r_f\epsilon_N^f + \sin 2\theta_{12}\epsilon_{12}^{-1})^2}}, \tag{104}$$

where $A \equiv -\cos 2\theta_{12}\epsilon_{12}^{-1} + c_{13}^2 - 2r_f\epsilon_D^f$. The absolute minimum of the probability is achieved when the second term in denominator of Eq. (104) vanishes, i.e. when

$$\epsilon_N^f = -\frac{\sin 2\theta_{12}}{2r_f\epsilon_{12}}. \tag{105}$$

This corresponds to the vanishing value of the off-diagonal elements of Hamiltonian (Eq. (102)). For $E = 3\,\text{MeV}$, we obtain $\epsilon_N^f = -0.3$. The region of the strongest transition shifts from $E \to \infty$ (in the standard case) to finite energy, thus forming the dip.

In Fig. 9, we plot the P_{ee} survival probability for the cases of NSIs with up-quarks and down-quarks. The presence of NSIs allows to considerably flatten the spectrum above 3 MeV. Moreover, NSIs can also generate large day–night asymmetries: (4–5)% at 10 MeV for both cases, $f = u$ and $f = d$. This allows to increase Δm^2_{21} and therefore relax the tension with KamLAND data.

Sterile neutrinos: Sterile neutrinos, being singlets of the standard model symmetry group, have zero potential $V_S = 0$. Therefore in the case of one

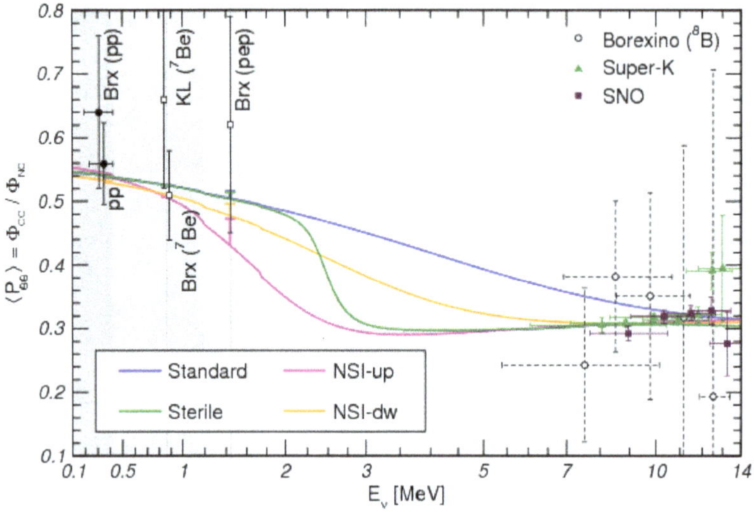

Fig. 9. Survival probability as a function of neutrino energy with and without new physics. Blue line: standard LMA MSW solution with $\sin^2\theta_{12} = 0.31$ and $\Delta m^2_{21} = 7.5 \times 10^{-5}$ eV^2). Green line: effect of inclusion of sterile neutrino with mixing $\sin^2 2\alpha = 0.005$ and $\Delta m^2_{01} = 1.2 \times 10^{-5}$ eV^2. Pink line: effect of inclusion of NSI with up quarks having $\epsilon^u_D = -0.22$, $\epsilon^u_N = -0.30$. Yellow line: effect of inclusion of NSI with down quarks having $\epsilon^d_D = -0.12$, $\epsilon^d_N = -0.16$. In all the cases, 1-3 mixing is fixed to be $\sin^2\theta_{13} = 0.022$. See text for details. Abbreviations "Brx" for Borexino and "KL" for KamLAND are used.[50]

sterile neutrino ν_s in the flavor basis $\nu_f = (\nu_s, \nu_e, \nu_\mu, \nu_\tau)^T$, the matrix of potentials equals $V = \text{diag}(0, V_e + V_n, V_n, V_n)$, where $V_n = -(1/\sqrt{2})G_F n_n$ with n_n being the number density of neutrons (see, e.g., Ref. 30). ν_s can mix with active neutrinos via Yukawa interaction with Higgs boson field. The mass state basis can be written as $\nu_m = (\nu_0, \nu_1, \nu_2, \nu_3)^T$, where ν_0 has mass m_0. There are different phenomenological possibilities depending on the mass-squared differences for ν_0.

The possibility $\Delta m^2_{01} < \Delta m^2_{21}$ was motivated by relatively low Ar-production rate in the Homestake experiment and an absence of the spectral upturn.[112,113] With an extra sterile state, the simplest mixing matrix which connects ν_f and ν_m states is $U = U_{\text{PMNS}} U_{01}(\alpha)$.[112] The value of new mixing angle is assumed to be very small: $\sin^2 2\alpha \sim 10^{-3}$ and the 0-1 mass splitting was taken to be $\Delta m^2_{01} \sim (0.2 - 0.3)\Delta m^2_{21}$. Then neutrinos propagating inside the Sun encounter *two* resonances: one is associated with the 1-2 mass splitting, as in the standard case, and another one related to the 0-1 mass splitting. With values of α and Δm^2_{01} defined above the adiabaticity is broken in the 0-1 resonance. This leads to narrow dip in the survival probability, as in the case of SMA solution, with sharp low-energy edge and nonadiabatic edge at high energies. For the chosen value of Δm^2_{01}, the dip appears at the intermediate energies, $E \sim 3\,\text{MeV}$, thus suppressing the upturn (see Fig. 9). This alleviates the tension between solar and KamLAND data. There is no enhancement of A_{DN}.

In the limit $\Delta m^2_{01} \gg \Delta m^2_{21}$ (see Refs. 114,115 for latest discussions), all Δm^2_{ij} other than Δm^2_{21} can be treated as being infinite, and in the propagation basis, the neutrino evolution is described by the sum of $H^{(2)}_{\text{LMA}}$ in Eq. (29) and

$$H^{(2)}_{\text{NP}} = \sqrt{2}G_F \frac{n_n}{4} \begin{pmatrix} -\xi_D & \xi_N e^{-i\delta_{12}} \\ \xi_N e^{i\delta_{12}} & \xi_D \end{pmatrix}. \tag{106}$$

Here ξ_D, ξ_N are combinations of the mixing matrix elements $U_{\alpha i}$.

The term (106) proportional to n_n is induced by the decoupling of heavy neutrino states. The off-diagonal elements "induced" by decoupling of the ν_0 are proportional to the product of mixing elements between ν_s and the flavor states ν_e: $\xi_N \propto U_{\mu 0} U_{e0}$. The phase δ_{12} appearing in $H^{(2)}_{\text{NP}}$ originates from the phases of the general $(3+1)$ mixing matrix. This phase does not produce CP asymmetry but affects neutrino propagation in matter.

Sterile neutrinos modify the oscillation probabilities in two different ways:

1. mixing of ν_e with the "heavy" states leads to a flattening of the energy dependence of conversion probabilities, in analogy with effect of θ_{13} in the standard case;
2. mixing of the sterile states with $\nu_{1,2}$ leads to overall disappearance of active neutrinos, so that $P_{ee} + P_{\mu e} + P_{\tau e} \neq 1$. Phenomenologically, this effect is restricted by SNO results: the precise NC measurement confirms that the total flux of active neutrinos from the Sun is compatible with the expectations of the SSM.

5. Future studies of solar neutrinos

The next phase in the field comprises

- precision (at sub %–% level) measurements of neutrino fluxes,
- further developments of detection techniques,
- elaboration of phenomenology of subleading effects both from standard and nonstandard physics. Small-size effects will be accessible and cannot be neglected as before.

We will first formulate the main issues/problems of future studies and then discuss physics reach of forthcoming and planned experiments.

5.1. *Main issues in the solar neutrino studies*

1. Test of the "neutrino flux–luminosity" relation. This requires precise measurement of the *pp* neutrino and ^{7}Be-neutrino fluxes. The test allows to check assumptions behind the relation, i.e., to (i) reveal/restrict possible existence of additional energy loss or generation mechanisms, (ii) measure the average energy of neutrinos, (iii) restrict fraction of nonterminated chains, (iv) search for time variations of the energy release and energy transport in the Sun. All this will check the fundamental principles of solar and, in general, star evolution.
2. Determination of the oscillation parameters. Accurate measurements of *pp*-, *pep*- and ^{7}Be-neutrino fluxes will substantially contribute to the global fits of the solar neutrino data and to further checks of the LMA solution.

   Precise reconstruction of the energy profile of the effect can be done: At low energies P_{ee} will be extracted from the rate of the *pp*-neutrino events. To extract P_{ee} at high energies, one should use the data from both νe scattering experiment and the CC interactions on nuclei (like in SNO, DUNE) since the flux of ^{8}B neutrinos is not known with high precision.
3. Exploration of physics in the intermediate-energy region; resolution of the "missing upturn problem". Here the *pep* flux plays a crucial role since its value is well predicted by SSM, being related to the *pp*-neutrino flux. At the energy E_{pep}, the difference of values of survival probability for standard MSW effect with two different Δm^2_{21} is 6%. For the global best fit value of Δm^2_{21}, the probability $\langle P_{ee}\rangle$ is 8% smaller than the probability of the averaged vacuum oscillations. The NSI can reduce $\langle P_{ee}\rangle$ by 15%. Effect of sterile neutrino can range from zero to 30% depending on Δm^2_{01}.
4. Detailed studies of the Earth matter effects. Oscillation tomography of the Earth with solar neutrinos will be possible.
5. Measurements of the CNO neutrino fluxes, resolution of the solar metallicity problem with implications to theory of star evolution. This is correlated to neutrino transformation in the intermediate-energy range.
6. High sensitivity searches for time variations of the fluxes. Searches for $\bar{\nu}_e$ flux from the Sun. Better bounds on the magnetic moment of neutrinos and magnetic fields of the Sun can be obtained.
7. Test of theory of neutrino oscillations and flavor conversion in matter; searches for new neutrino physics effects. They will allow to put more stringent bounds on nonstandard neutrino properties.

Results of analysis of the solar neutrino data can be compared with results from nonsolar experiments which have different environment, type of neutrino, energy, etc. For instance, comparing results from the KamLAND measurement and solar neutrino experiments, one can search for effects of CPT violation, presence of NSIs, etc.

8. The solar neutrino fluxes for the particle physics. Laboratory experiments with known neutrino fluxes from the Sun. The Sun becomes the source of neutrino fluxes with well-known properties: energy spectra, flavor. This has already been used by BOREXINO: Study of neutrino scattering at low energies allowed to establish bound on the neutrino magnetic moment. The presence of NSI can affect the neutrino cross sections relevant for neutrino production and detection. Stringent bounds on the NSI of neutrinos with electrons have been derived from the data on $\nu-e$ scattering in the ^{7}Be-neutrino energy window.[111]

5.2. *Forthcoming and planned experiments*

In what follows, we consider different future experiments and summarize how they can address the issues listed in Section 5.1.

5.2.1. *SNO+*

SNO+ will use the SNO acrylic vessel (12 m diameter) filled in by 780 t of liquid scintillator.[117] Solar neutrinos will be detected via the νe scattering. Background is expected to be as in the BOREXINO phase 1. Cosmogenic background will be substantially lower than that in BOREXINO. For the solar neutrinos, fiducial mass is about 50% of total mass, i.e., 390 t which is about four times larger than in BOREXINO. Using this, one can obtain estimations for SNO+ rescaling the BOREXINO results.

In the tellurium phase, SNO+ will be able to detect the ^{8}B neutrinos above 3 MeV. Measurements of low-energy solar neutrinos will not be possible. These measurements will require further background mitigation in the dedicated scintillator phase. Physics's reach (contribution) of SNO+ in resolving existing problems is the following:

(i) Measurements of the *pep* neutrinos (predicted with 1.2% accuracy). In 1 year, SNO+ will reach accuracy better than 8.9% being restricted mainly by statistics. To shed light onto physics in the transition region, an accuracy better than 5% is required, and therefore at least 4 years of operation of SNO+ in the scintillator mode are needed.

(ii) The ^{7}Be neutrino signal will be measured with 4% accuracy (in 1 year); however possible effects of new physics are substantially smaller. Therefore it will marginally contribute to solution of the intermediate region problem. Combined analysis of the ^{7}Be- and *pep*- neutrino data will reinforce the sensitivity.

(iii) ^{8}B-neutrino flux will be measured with 7% accuracy (in 1 year). The accuracy is dominated by statistics and by correlations between signals from different components of neutrino spectrum and background. The accuracy of reconstruction of the neutrino energy spectrum can be about two times better than that in BOREXINO. This may not be enough to resolve the problem of missing upturn, although decrease of the probability at 5 MeV can be as large as 30–40%.

(iv) The CNO flux together with ^{210}Bi background can be measured with 4.4% accuracy. This will certainly contribute to the solution of the metallicity problem. About 30% lower metallicity would lead to reduction of the CNO flux by 30%. However, the same amount of the effect can be due to new physics.

(v) There is a potential of SNO+ to measure the *pp*-neutrino flux with few percent level accuracy.

5.2.2. *JUNO*

JUNO is a 20 kt liquid scintillator detector at rather small depth (700 m of rock).[119] JUNO experiments will affect the solar neutrino physics in two different ways directly and indirectly via very precise measurement of Δm_{21}^2 with antineutrino flux from reactors. After six years with pulls of all other parameters taken into account, the accuracy of measurements will reach $\pm 0.05 \cdot 10^{-5}$ eV^2 (3σ) which should be compared with difference of the present central values of Δm_{21}^2 (solar–KamLAND) $\pm 2.50 \cdot 10^{-5}$ eV^2. Confirmation of the KamLAND value of Δm_{21}^2 will leave the issue open: still some nonstandard physics may affect the solar neutrinos. The problem will be solved if the value of Δm_{21}^2 shifts toward the solar value.

Concerning direct measurements of the solar neutrino flux via the νe scattering, the problem is that rather shallow JUNO laboratory has large cosmogenic background. Another background is natural intrinsic radioactivity of liquid scintillator as well as vessel and instruments. A high level of purification should be achieved in very large volume. Realistic radiopurity is at the level of the KamLAND solar experiment which provides 1:3 signal to background ratio. For the more demanding 10 times higher radiopurity such as in BOREXINO phase 1, the ratio could be 2:1.

The advantages of JUNO are high rate of events and very good energy resolution. This will allow the measurement of the shape of energy spectrum of events with high accuracy. Detection of the ^{8}B neutrinos (with low threshold), as well as the ^{7}Be and *pep* neutrinos is under consideration.

5.2.3. *Hyper-Kamiokande*

Hyper-Kamiokande[118] will detect the boron neutrinos via the νe scattering with $E_\nu > 6.5$ MeV. According to last version of the proposal, the fiducial volume of HK will be 10 times larger than that of SK, so about 150 events with $E_\nu > 6.5$ MeV will be observed in a day.

(i) Large number of events will allow very precise measurements of the electron energy spectrum and search for the upturn (in spite of high energies and therefore smaller effect). Precise determination of the total boron neutrino flux in combination with low-energy data will allow us to search for new physics effects.
(ii) The day–night asymmetry will be measured with high precision. After 10 years, 0.5% accuracy will be achieved so that the present central value of A_{DN} will be established at a level more than 6σ. The zenith angle distribution of events will be measured with some accuracy which will allow us to start the oscillation tomography of the Earth.

5.2.4. *JinPing*

JinPing proposal is a 4 kt liquid scintillator or water-based liquid scintillator detector in CJPL — the deepest laboratory in the world with an overburden, 2400 m of rock (6720 m.v.e.).[120] For solar neutrinos, fiducial mass is 2 kt, that is, 20 times larger than in BOREXINO. JinPing will have much lower external background. So, it is optimized for detection of low-energy neutrinos having high fluxes: *pp*, Be, *pep*, CNO. The detection is based on νe scattering. Results of simulations strongly depend on the energy resolution. As a benchmark, the following results have been obtained for a 5-year run. For *pp* neutrinos with 0.2–0.3 MeV, the statistical error is 0.5%. Total uncertainty will be dominated by systematic errors and the goal is to achieve total error below 1%. This will provide a crucial test of the "flux–luminosity" relation. For the ^{7}Be neutrinos the statistical error is 0.4%, and with systematics error, the accuracy can be about 1%. With this, JingPing measurement can contribute to resolution of the missing upturn problem.

For the *pep* neutrinos statistical uncertainty equals 3% (which is smaller than the difference of predictions from different nonstandard scenarios). So, it can substantially contribute to solution of the intermediate range problem. JinPing will be able to measure CNO neutrinos with less than in 30% error. Studies of the high-energy ^{8}B neutrinos are restricted by small volume: statistical error can be 2%. Systematics has comparable value. With these errors, the study of the Earth matter effect will not be possible.

5.2.5. *ASDC-THEIA*

As future developments, one can consider the advanced scintillation detector concept (ASDC) which uses water-based liquid scintillator (WbLS) and ultrafast timing photosensors.[121] Such a detector will register both prompt Cherenkov light and the delayed scintillation, thus employing advantages of both techniques: directionality of event and high-energy resolution and low threshold. Ultrafast photosensors will allow us to separate the two signals. The WbLS can be loaded by metallic ions which will allow us to detect the CC processes. For solar neutrinos, 0.5–1% doping by ^{7}Li is considered. The reaction

$$\nu_e + {}^7\mathrm{Li} \to {}^7\mathrm{Be} + e \tag{107}$$

has two main transitions: one to the ground state with threshold of 862 keV and the other to the exited (430 keV) state with further emission of a photon. If e and $e + \gamma$ are resolved, the energy of neutrino can be reconstructed precisely. Recoil energy and direction of electron will be measured.

The detector may have 30 kt fiducial mass (50 kt total mass) and 1% uploading. Apart from the reaction in Eq. (107), the νe scattering will be detected. Due to the relatively small concentration of ^{7}Li, the νe signal will dominate. The latter contribution can, however, be suppressed if one uses the angular cut excluding directions toward the Sun.

The detector will be able to measure via CCs the *pep*, CNO and B neutrinos and in addition, the ^{7}Be neutrinos via νe scattering. Comparison of signals from νe and reaction in Eq. (107) will provide some interesting tests.

Using spectral information, it may be possible to measure independently the N and O neutrino fluxes. The boron neutrino spectrum can be measured down to 2 MeV. In 1 year, the following statistics of events may be collected: $5 \cdot 10^6$ (^{7}Be), $5 \cdot 10^5$ (CNO), 10^5 (*pep*) and $5 \cdot 10^4$ (^{8}B) neutrinos.

5.2.6. *DARWIN*

DARWIN the ultimate dark matter detector, which will be sensitive to the neutrino floor,[122] is a multiton (50 t total, 30 t fiducial mass) liquid xenon time projection chamber (TPC).[123] It will detect solar neutrinos via νe scattering. 2000 *pp*-neutrino events per year are expected. In the absence of a background (which is under investigation), better than 1% precision of measurement of the *pp*-neutrino flux can be achieved after 5 years. The number of events due to ^{7}Be neutrinos is eight times smaller.

5.2.7. *DUNE*

DUNE is a liquid argon (LAr) TPC.[124] It consists of four modules, each with 17.1 kt of LAr and fiducial mass of 11.6 kt for solar neutrinos. The depth is 1478 m. DUNE is the only foreseen detector (apart from Theia) which will use neutrino nuclei scattering

$$\nu_e + {}^{40}\mathrm{Ar} \to e^- + {}^{40}\mathrm{K}^*, \tag{108}$$

where K^* is an exited state. In the energy interval of 5–15 MeV in a 40 kt LAr TPC, a rate of about 122 events/day (with 82 from Eq. (108)) is expected. So, in 5 years, 130000 events will be detected. The background, however, will be rather high. Still one can perform a detailed study of the Earth matter effect and D–N asymmetry of the signal.

6. Concluding remarks

Solar neutrino studies have provided the fundamental contributions to physics and astrophysics:

1. The detection of solar neutrinos had confirmed the nuclear reactions mechanism of the energy generation in the Sun and in the stars. Furthermore, a detailed picture with all the *pp*-chain reactions and whole nuclear evolution in Sun have been confirmed. Neutrinos from all three *pp* branches (including subleading ones) have been discovered to be in agreement with the expectations (once flavor transformations are taken into account).

   The CNO neutrinos with potentially important implications for the metallicity problem and understanding of neutrino conversion in the intermediate energy range are still missing.
2. The LMA MSW solution of the solar neutrino problem has been established as the unique (dominant?) solution of the solar neutrino problem. The LMA MSW is the adiabatic flavor conversion of neutrinos in the resonance channel. In contrast to oscillations which are driven by the phase change, the adiabatic conversion is related to another degree

of freedom — the mixing angle in matter. In matter, the angle becomes a dynamical variable — a function of density and energy.

Nature has selected the simplest scale-invariant solution which depends essentially on density in the neutrino production region. In the Sun, oscillations are irrelevant: due to loss of coherence, the flavor evolution can be treated as an incoherent production and independent propagation of the eigenstates during which they transform into the corresponding mass states.

Solar neutrinos oscillate inside the Earth. The picture of these oscillations determined by density profile is much more complicated than the picture of transformations in the Sun. Results of these oscillations depend strongly on energy and energy resolution of a detector due to the attenuation effect. For different trajectories, parametric enhancement or suppression of oscillations can be realized.

Identification of the LMA MSW solution allowed to determine Δm_{21}^2 (value and sign) and θ_{21} which are in agreement with results of the KamLAND experiment. Historically, identification of the MSW solution regions provided the first correct evaluation of the neutrino oscillation parameters.

3. Existence of matter effects on neutrino flavor evolution (conversion, oscillations) has been established by (i) the energy dependence of suppression of the ν_e signal or by nonzero value of the matter potential extracted from the solar neutrino data, (ii) the day–night asymmetry of signals.
4. It is not excluded that corrections to neutrino transformations from some additional subleading effects exist. The energy profile of the LMA MSW effect is in agreement with all available data with some pulls at about 2σ level. The latter includes (i) low Ar production rate in the Homestake experiment, (ii) an absence of the upturn of the spectrum at low energies. (iii) larger observed D–N asymmetry than the one predicted with the KamLAND value of Δm_{21}^2. The tension disappears for low value of Δm_{21}^2, and so the problem can be formulated as the tension between Δm_{21}^2 extracted from the solar and KamLAND data. It can be fixed by 1.6 times larger matter potential.

   Future precision (percent and subpercent level) measurements of all the components of the solar neutrino spectrum will bring us to a new level of checks of solar models, physics of neutrino propagation and transformations.

References

1. H. A. Bethe and C. L. Critchfield, *Phys. Rev.* **54**, 248 (1938).
2. H. A. Bethe, *Phys. Rev.* **55**, 434 (1939).
3. B. Pontecorvo, Chalk River Laboratory Report PD-205 (1946).
4. H. D. Holmgren and R. L. Johnston, *Phys. Rev.* **113**, 1556 (1959).
5. W. A. Fowler, *Astrophys. J.* **127**, 551 (1958).
6. John N. Bahcall, *Neutrino Astrophysics*, Cambridge University Press (1989).
7. *Solar Neutrinos. The first thirty years*, Westview Press, 1994, Edited by J. N. Bahcall, R. Davis, Jr. P. Parker, A. Smirnov, R. Ulrich.
8. J. N. Bahcall, *Phys. Rev. Lett.* **12**, 300 (1964). doi:10.1103/PhysRevLett.12.300.
9. V. A. Kuzmin, *Sov. Phys. JETP* **22**, 1051 (1966); [*Zh. Eksp. Teor. Fiz.* **49**, 1532 (1965)].
10. B. Pontecorvo, *Sov. Phys. JETP* **26**, 984 (1968); [*Zh. Eksp. Teor. Fiz.* **53**, 1717 (1967)].
11. R. Davis, Jr., D. S. Harmer and K. C. Hoffman, *Phys. Rev. Lett.* **20**, 1205 (1968).
12. A. Cisneros, *Astrophys. Space Sci.* **10**, 87 (1971).
13. L. B. Okun, M. B. Voloshin, and M. I. Vysotsky, *Sov. Phys. JETP* **64**, 446 (1986) [Zh. Eksp. Teor. Fiz. **91**, 754 (1986)].
14. J. N. Bahcall, N. Cabibbo, and A. Yahil, *Phys. Rev. Lett.* **28**, 316 (1972).
15. Z. G. Berezhiani and M. I. Vysotsky, *Phys. Lett.* **B199**, 281 (1987).
16. S. L. Glashow and L. M. Krauss, *Phys. Lett.* **B190**, 199 (1987).
17. L. Wolfenstein, *Phys. Rev.* **D17**, 2369 (1978); *Phys. Rev.* **D20**, 2634 (1979).
18. L. Wolfenstein, in *Neutrino -78*, Purdue Univ. C3 (1978).
19. V. D. Barger, K. Whisnant, S. Pakvasa, and R. J. N. Phillips, *Phys. Rev.* **D22**, 2718 (1980).
20. S. P. Mikheyev and A. Y. Smirnov, *Sov. J. Nucl. Phys.* **42**, 913 (1985); *Nuovo Cim.* **C9**, 17 (1986).
21. S. P. Mikheev and A. Y. Smirnov, *Sov. Phys. JETP* **64**, 4 (1986).
22. H. A. Bethe, *Phys. Rev. Lett.* **56**, 1305 (1986).
23. A. Messiah, In *Tignes 1986, Proceedings, '86 massive neutrinos* 373.
24. S. P. Rosen and J. M. Gelb, *Phys. Rev.* **D34**, 969 (1986).
25. W. C. Haxton, *Phys. Rev. Lett.* **57**, 1271 (1986).
26. S. J. Parke, *Phys. Rev. Lett.* **57**, 1275 (1986).
27. T. K. Kuo and J. T. Pantaleone, *Rev. Mod. Phys.* **61**, 937 (1989).
28. S. M. Bilenky and S. T. Petcov, *Rev. Mod. Phys.* **59**, 671 (1987) Erratum: [*Rev. Mod. Phys.* **61**, 169 (1989)] Erratum: [*Rev. Mod. Phys.* **60**, 575 (1988)].
29. S. P. Mikheev and A. Y. Smirnov, *Sov. Phys. Usp.* **30**, 759 (1987) [*Usp. Fiz. Nauk* **153**, 3 (1987)].
30. S. P. Mikheyev and A. Y. Smirnov, *Prog. Part. Nucl. Phys.* **23**, 41 (1989).
31. C. S. Lim and W. J. Marciano, *Phys. Rev.* **D37**, 1368 (1988).
32. E. K. Akhmedov, *Phys. Lett.* **B213**, 64 (1988).
33. E. Roulet, *Phys. Rev.* **D44**, 935 (1991).
34. M. M. Guzzo, A. Masiero, and S. T. Petcov, *Phys. Lett.* **B260**, 154 (1991).
35. L. Ludhova, *Eur. Phys. J. A* **52** no. 4, 82 (2016).

36. J. N. Bahcall, P. I. Krastev, and A. Y. Smirnov, *Phys. Rev.* **D60**, 093001 (1999). [hep-ph/9905220].
37. A. B. McDonald, *Rev. Mod. Phys.* **88** no. 3, 030502 (2016).
38. K. Eguchi *et al.* [KamLAND Collaboration], *Phys. Rev. Lett.* **90**, 021802 (2003). [hep-ex/0212021].
39. T. Araki *et al.* [KamLAND Collaboration], *Phys. Rev. Lett.* **94**, 081801 (2005). [hep-ex/0406035].
40. S. Abe *et al.* [KamLAND Collaboration], *Phys. Rev. Lett.* **100**, 221803 (2008). [arXiv:0801.4589 [hep-ex]].
41. W. C. Haxton, R. G. H. Robertson, and A. M. Serenelli, *Ann. Rev. Astron. Astrophys.* **51**, 21 (2013). [arXiv:1208.5723 [astro-ph.SR]].
42. A. M. Serenelli, W. C. Haxton, and C. Pena-Garay, *Astrophys. J.* **743**, 24 (2011). [arXiv:1104.1639 [astro-ph.SR]].
43. V. Antonelli, L. Miramonti, C. Pena-Garay, and A. Serenelli, *Adv. High Energy Phys.* **2013**, 351926 (2013). [arXiv:1208.1356 [hep-ex]].
44. A. Serenelli, *Eur. Phys. J.* **A52** no. 4, 78 (2016). [arXiv:1601.07179 [astro-ph.SR]].
45. A. Serenelli, C. Penn-Garay, and W. C. Haxton, *Phys. Rev.* **D87** no. 4, 043001 (2013). [arXiv:1211.6740 [astro-ph.SR]].
46. N. Vinyoles *et al.*, *Astrophys. J.* **835** no. 2, 202 (2017). [arXiv:1611.09867 [astro-ph.SR]].
47. J. Bergstrom, M. C. Gonzalez-Garcia, M. Maltoni, C. Pena-Garay, A. M. Serenelli, and N. Song, *JHEP* **1603**, 132 (2016). [arXiv:1601.00972].
48. S. Vagnozzi, K. Freese, and T. H. Zurbuchen, arXiv:1603.05960 [astro-ph.SR].
49. A. Serenelli, P. Scott, F. L. Villante, A. C. Vincent, M. Asplund, S. Basu, N. Grevesse, and C. Pena-Garay, *Mon. Not. Roy. Astron. Soc.* **463** no. 1, 2 (2016). [arXiv:1604.05318 [astro-ph.SR]].
50. M. Maltoni and A. Y. Smirnov, *Eur. Phys. J.* **A52** no. 4, 87 (2016). [arXiv:1507.05287 [hep-ph]].
51. H. Minakata and S. Watanabe, *Phys. Lett.* **B468**, 256 (1999). [hep-ph/9906530].
52. S. Goswami and A. Y. Smirnov, *Phys. Rev.* **D72**, 053011 (2005). [hep-ph/0411359].
53. S. T. Petcov, *Phys. Lett.* **B200**, 373 (1988).
54. C. P. Burgess, *Prog. Part. Nucl. Phys.* **40**, 137 (1998). [hep-ph/9711425].
55. C. P. Burgess and D. Michaud, *Annals Phys.* **256**, 1 (1997). [hep-ph/9606295].
56. C. P. Burgess, N. S. Dzhalilov, T. I. Rashba, V. B. Semikoz, and J. W. F. Valle, *Mon. Not. Roy. Astron. Soc.* **348**, 609 (2004). [astro-ph/0304462].
57. C. Burgess, N. S. Dzhalilov, M. Maltoni, T. I. Rashba, V. B. Semikoz, M. A. Tortola, and J. W. F. Valle, *Astrophys. J.* **588**, L65 (2003). [hep-ph/0209094].
58. S. P. Mikheyev and A. Y. Smirnov, Proc. of the 6th Moriond Workshop on massive Neutrinos in Astrophysics and Particle Physics, Tignes, Savoie, France Jan. 1986 (eds. O. Fackler and J. Tran Thanh Van) p. 355 (1986).

59. J. Bouchez, M. Cribier, J. Rich, M. Spiro, D. Vignaud, and W. Hampel, *Z. Phys.* **C32**, 499 (1986).
60. M. Cribier, W. Hampel, J. Rich, and D. Vignaud, *Phys. Lett.* **B182**, 89 (1986).
61. A. N. Ioannisian and A. Y. Smirnov, *Phys. Rev. Lett.* **93**, 241801 (2004). [hep-ph/0404060].
62. E. K. Akhmedov, M. A. Tortola, and J. W. F. Valle, *JHEP* **0405** 057 (2004). [hep-ph/0404083].
63. T. Ohlsson and H. Snellman, *Phys. Lett.* **B474**, 153 (2000). [hep-ph/9912295].
64. A. N. Ioannisian, N. A. Kazarian, A. Y. Smirnov, and D. Wyler, *Phys. Rev. D* **71**, 033006 (2005). [hep-ph/0407138].
65. A. Ioannisian, A. Smirnov, and D. Wyler, arXiv:1702.06097 [hep-ph].
66. P. C. de Holanda, W. Liao, and A. Y. Smirnov, *Nucl. Phys.* **B702**, 307 (2004). [hep-ph/0404042].
67. L. W. Alvarez, doi:10.2172/929771.
68. B. T. Cleveland, T. Daily, R. Davis, Jr., J. R. Distel, K. Lande, C. K. Lee, P. S. Wildenhain, and J. Ullman, *Astrophys. J.* **496**, 505 (1998).
69. K. S. Hirata *et al.* [Kamiokande-II Collaboration], *Phys. Rev. Lett.* **63**, 16 (1989).
70. K. S. Hirata *et al.* [Kamiokande-II Collaboration], *Phys. Rev. Lett.* **65**, 1297 (1990).
71. K. S. Hirata *et al.* [Kamiokande-II Collaboration], *Phys. Rev.* **D44**, 2241 (1991). Erratum: [*Phys. Rev.* **D45**, 2170 (1992)].
72. Y. Fukuda *et al.* [Kamiokande Collaboration], *Phys. Rev. Lett.* **77**, 1683 (1996).
73. V. Berezinsky, G. Fiorentini, and M. Lissia, *Phys. Lett.* **B365**, 185 (1996). [astro-ph/9509116].
74. J. N. Abdurashitov *et al.* [SAGE Collaboration], *Phys. Rev.* **C80**, 015807 (2009). [arXiv:0901.2200 [nucl-ex]].
75. J. N. Abdurashitov *et al.* [SAGE Collaboration], *J. Exp. Theor. Phys.* **95**, 181 (2002). [*Zh. Eksp. Teor. Fiz.* **122**, 211 (2002)] [astro-ph/0204245].
76. W. Hampel *et al.* [GALLEX Collaboration], *Phys. Lett.* **B447**, 127 (1999).
77. F. Kaether, W. Hampel, G. Heusser, J. Kiko, and T. Kirsten, *Phys. Lett.* **B685**, 47 (2010). [arXiv:1001.2731 [hep-ex]].
78. M. Altmann *et al.* [GNO Collaboration], *Phys. Lett.* **B616**, 174 (2005). [hep-ex/0504037].
79. P. Anselmann *et al.* [GALLEX Collaboration], *Phys. Lett.* **B342**, 440 (1995).
80. W. Hampel *et al.* [GALLEX Collaboration], *Phys. Lett.* **B420**, 114 (1998). doi:10.1016/S0370-2693(97)01562-1
81. J. N. Abdurashitov *et al.* [SAGE Collaboration], *Phys. Rev.* **C59**, 2246 (1999). [hep-ph/9803418].
82. J. N. Abdurashitov *et al.*, *Phys. Rev.* **C73**, 045805 (2006). doi:10.1103/PhysRevC.73.045805 [nucl-ex/0512041].
83. A. Renshaw *et al.* [Super-Kamiokande Collaboration], *Phys. Rev. Lett.* **112** no. 9, 091805 (2014). [arXiv:1312.5176 [hep-ex]].

84. K. Abe *et al.* [Super-Kamiokande Collaboration], *Phys. Rev.* **D94** no. 5, 052010 (2016). [arXiv:1606.07538 [hep-ex]].
85. Q. R. Ahmad *et al.* [SNO Collaboration], *Phys. Rev. Lett.* **87**, 071301 (2001). [nucl-ex/0106015].
86. Q. R. Ahmad *et al.* [SNO Collaboration], *Phys. Rev. Lett.* **89**, 011301 (2002). [nucl-ex/0204008].
87. S. N. Ahmed *et al.* [SNO Collaboration], *Phys. Rev. Lett.* **92**, 181301 (2004). [nucl-ex/0309004].
88. B. Aharmim *et al.* [SNO Collaboration], *Phys. Rev.* **C72**, 055502 (2005). [nucl-ex/0502021].
89. Q. R. Ahmad *et al.* [SNO Collaboration], *Phys. Rev. Lett.* **89**, 011302 (2002). [nucl-ex/0204009].
90. G. Bellini *et al.* [Borexino Collaboration], *Phys. Rev.* **D82**, 033006 (2010). [arXiv:0808.2868 [astro-ph]].
91. G. Bellini *et al.* [Borexino Collaboration], *Phys. Rev.* **D89** no. 11, 112007 (2014). [arXiv:1308.0443].
92. G. Bellini *et al.*, *Phys. Rev. Lett.* **107**, 141302 (2011). [arXiv:1104.1816 [hep-ex]].
93. G. Bellini *et al.* [Borexino Collaboration], *Phys. Rev. Lett.* **108**, 051302 (2012). [arXiv:1110.3230 [hep-ex]].
94. G. Bellini *et al.* [BOREXINO Collaboration], *Nature* **512**, 7515 (2014)., 383.
95. G. Bellini *et al.* [Borexino Collaboration], *Phys. Lett.* **B707**, 22 (2012). [arXiv:1104.2150 [hep-ex]].
96. M. Agostini *et al.* [BOREXINO Collaboration], arXiv:1701.07970 [hep-ex].
97. M. Agostini *et al.* [BOREXINO Collaboration], arXiv:1707.09279 [hep-ex].
98. M. Agostini *et al.* [BOREXINO Collaboration], arXiv:1709.00756 [hep-ex].
99. A. Gando *et al.* [KamLAND Collaboration], *Phys. Rev.* **C92** no. 5, 055808 (2015). [arXiv:1405.6190 [hep-ex]].
100. I. Esteban, M. C. Gonzalez-Garcia, M. Maltoni, I. Martinez-Soler, and T. Schwetz, *JHEP* **1701**, 087 (2017). [arXiv:1611.01514 [hep-ph]].
101. G. L. Fogli, E. Lisi, A. Marrone, and A. Palazzo, *Phys. Lett.* **B583**, 149 (2004). [hep-ph/0309100].
102. M. C. Gonzalez-Garcia and M. Maltoni, *JHEP* **1309**, 152 (2013). [arXiv:1307.3092].
103. E. K. Akhmedov and J. Pulido, *Phys. Lett.* **B553**, 7 (2003). [hep-ph/0209192].
104. H. Minakata and H. Nunokawa, *Phys. Rev. Lett.* **63**, 121 (1989).
105. O. G. Miranda, T. I. Rashba, A. I. Rez, and J. W. F. Valle, *Phys. Rev.* **D70**, 113002 (2004). [hep-ph/0406066].
106. W. Grimus, M. Maltoni, T. Schwetz, M. A. Tortola, and J. W. F. Valle, *Nucl. Phys.* **B648**, 376 (2003). [hep-ph/0208132].
107. G. Bellini *et al.* [Borexino Collaboration], *Phys. Lett.* **B696**, 191 (2011). [arXiv:1010.0029 [hep-ex]].
108. K. Eguchi *et al.* [KamLAND Collaboration], *Phys. Rev. Lett.* **92**, 071301 (2004). [hep-ex/0310047].
109. A. Friedland, C. Lunardini, and C. Pena-Garay, *Phys. Lett.* **B594**, 347 (2004). [hep-ph/0402266], A. Palazzo and J. W. F. Valle, *Phys. Rev.* **D80**,

091301 (2009). [arXiv:0909.1535 [hep-ph]], A. Bolanos, O. G. Miranda, A. Palazzo, M. A. Tortola, and J. W. F. Valle, *Phys. Rev.* **D79**, 113012 (2009). [arXiv:0812.4417 [hep-ph]], O. G. Miranda, M. A. Tortola, and J. W. F. Valle, *JHEP* **0610**, 008 (2006). [hep-ph/0406280], M. M. Guzzo, P. C. de Holanda, and O. L. G. Peres, *Phys. Lett.* **B591**, 1 (2004). [hep-ph/0403134].

110. Y. Farzan, arXiv:1505.06906 [hep-ph].
111. S. K. Agarwalla, F. Lombardi, and T. Takeuchi, *JHEP* **1212**, 079 (2012). [arXiv:1207.3492 [hep-ph]].
112. P. C. de Holanda and A. Y. Smirnov, *Phys. Rev.* **D69**, 113002 (2004). [hep-ph/0307266].
113. P. C. de Holanda and A. Y. Smirnov, *Phys. Rev.* **D83**, 113011 (2011). [arXiv:1012.5627 [hep-ph]].
114. A. Palazzo, *Phys. Rev.* **D83**, 113013 (2011). [arXiv:1105.1705 [hep-ph]].
115. J. Kopp, P. A. N. Machado, M. Maltoni, and T. Schwetz, *JHEP* **1305**, 050 (2013). [arXiv:1303.3011 [hep-ph]].
116. G. Bellini *et al.* [Borexino Collaboration], *Phys. Rev.* **D88**, 7 (2013)., 072010 [arXiv:1311.5347 [hep-ex]].
117. S. Andringa *et al.* [SNO+ Collaboration], *Adv. High Energy Phys.* **2016**, 6194250 (2016). [arXiv:1508.05759 [physics.ins-det]].
118. E. Kearns *et al.* [Hyper-Kamiokande Working Group], arXiv:1309.0184 [hep-ex].
119. F. An *et al.* [JUNO Collaboration], *J. Phys.* **G43**, no. 3, 030401 (2016). [arXiv:1507.05613 [physics.ins-det]].
120. J. F. Beacom *et al.* [Jinping Collaboration], *Chin. Phys.* **C41** no. 2, 023002 (2017). [arXiv:1602.01733 [physics.ins-det]].
121. J. R. Alonso *et al.*, arXiv:1409.5864 [physics.ins-det].
122. J. Billard, L. E. Strigari, and E. Figueroa-Feliciano, *Phys. Rev.* **D91**, 9 (2015)., 095023. [arXiv:1409.0050 [astro-ph.CO]].
123. J. Aalbers *et al.* [DARWIN Collaboration], *JCAP* **1611** no. 11, 017 (2016). [arXiv:1606.07001 [astro-ph.IM]].
124. R. Acciarri *et al.* [DUNE Collaboration], arXiv:1601.05471 [physics.ins-det].

Chapter 5

Atmospheric Neutrinos

K. Okumura

Research Center for Cosmic Neutrino, ICRR, University of Tokyo
5-1-5 Kashiwa-no-ha Kashiwa, Chiba, Japan
okumura@icrr.u-tokyo.ac.jp

Abstract

Atmospheric neutrinos are the most abundant natural source of $O(1)$ GeV neutrinos and play an important role in the study of neutrino physics. They are distributed across a wide range of energies and pathlengths relative to a terrestrial detector, the parameters controlling neutrino oscillation. This allowed the discovery of neutrino oscillation to be made in the atmospheric sector, which implies that at least some neutrinos have a nonvanishing mass, with a mass splitting of $\Delta m^2 \approx 2.4 \times 10^{-3}\,\text{eV}^2$. This enabled the first detailed investigations of neutrino oscillation and continuing efforts to probe unrevealed features, such as the mass hierarchy. Since the realization of kilometer-scale detectors, the ability to study atmospheric neutrinos has been extended to higher energies, such as prompt neutrinos. In this section, we will review the atmospheric neutrino flux and related physics, mainly neutrino oscillation, which has been explored by the atmospheric neutrino observations.

1. Atmospheric neutrino flux

Cosmic rays, dominated by protons and helium, are continuously pouring onto the Earth. They interact with the molecules of the Earth's atmosphere, producing many secondary particles. Atmospheric neutrinos are produced via the decays of these secondaries, as illustrated in Fig. 1.

For energies below TeV, muon and electron neutrinos, which originate from π and K secondaries, are produced via the following channels:

$$p, A + \text{air} \to \pi^{\pm}, \pi^0, K^{\pm}, K^0_{S,L}$$
$$\pi^{\pm}, K^{\pm} \to \mu^{\pm}\nu_{\mu}(\bar{\nu}_{\mu})$$

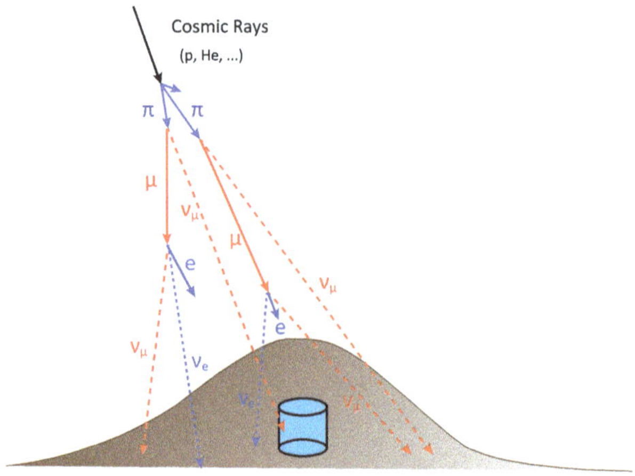

Fig. 1. Diagram of cosmic ray shower development in the atmosphere.

$$\mu^{\pm} \to e^{\pm}\ \nu_e(\bar{\nu}_e)\ \bar{\nu}_\mu(\nu_\mu)$$
$$K^{\pm}, K_L^0 \to [\pi^{\pm}, \pi^0] e^{\pm} \nu_e(\bar{\nu}_e)$$

Such neutrinos are often called "conventional", as opposed to those produced by short-lived mesons such as D at higher energies, which are called "prompt" and will be mentioned later. The contribution of ν_τ flux is negligible in the energy region of the conventional neutrinos because heavier secondaries such as τ leptons and D mesons, which can be the origin of ν_τ, are rarely produced.

To predict the atmospheric neutrino flux on the Earth, many flux calculations have been produced.[1–6] The first calculations were based on one-dimensional tracking, in which all the secondaries are assumed to follow the direction of the parent cosmic ray and the bending effect of the primary/secondary particles by the geomagnetic field was neglected. However, since such bending was known to cause a considerable modification to the horizontal flux, newer flux calculations based on three-dimensional tracking using Monte Carlo methods, where the effect of the transverse momentum of the secondaries and the bending of the particles in the geomagnetic field when they propagate are taken into account, have been essential for precise predictions. Such effects are significant in the low energies below $O(10)$ GeV.

The calculated ν_μ ($\bar{\nu}_\mu$) and ν_e ($\bar{\nu}_e$) fluxes in the energy region below 1 TeV are shown in Fig. 2. The spectrum of the atmospheric neutrino peaks at around $O(1)$ GeV and extends to higher energies with approximately a power-law spectrum, similar to those of the primary cosmic rays and the

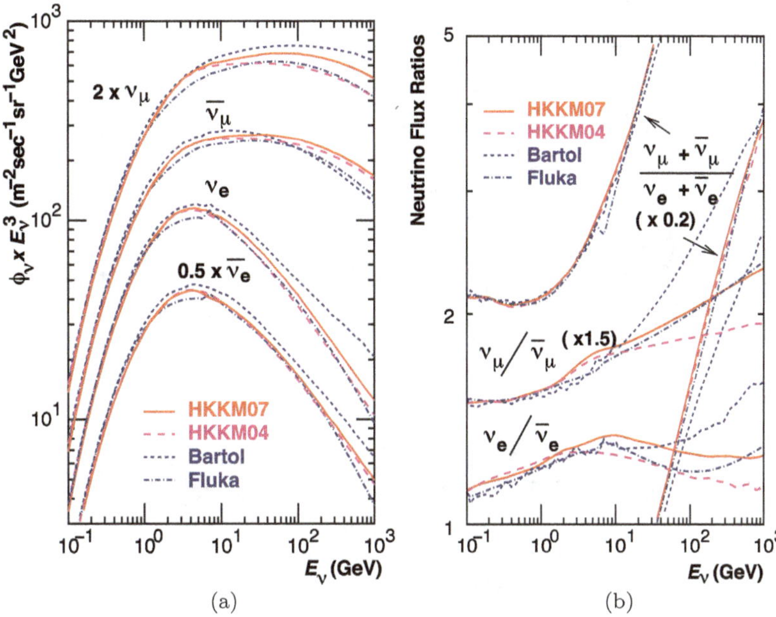

Fig. 2. Calculated atmospheric ν_μ and ν_e fluxes averaged over all directions (a), and the ratios of the same fluxes (b), comparing several flux calculations.[1,2,5,6] Note that the vertical scale in the left figure is $\phi_\nu \times E_\nu^3$ and the flux is enhanced as the energy increased. This figure has been taken from Ref. 2 with minor modifications.

produced secondaries. The typical atmospheric neutrino energy spectrum is $\sim$E$^{-3.7}$, steeper than that of the primary cosmic rays (E$^{-2.7}$) since the energy of a primary cosmic ray particle tends to be divided among many secondaries as the energy becomes higher. The slope is steeper for ν_e compared to that of ν_μ in the high-energy region since the high-energy muons reach the ground before they decay. At such energies, ν_e originating from K, which decays more quickly in the atmosphere, becomes dominant. The flux of all neutrino species sharply decreases below 1 GeV, because of the rigidity cutoff of the primary cosmic rays by the geomagnetic field, which will be described shortly.

The ν_μ/ν_e ratio is approximately 2 for energies up to $O(1)$ GeV, due to the ratio of the number of produced ν_μ and ν_e in the dominant $\pi^\pm$ decay channel. The uncertainty in the ν_μ/ν_e ratio is relatively small (<5%) compared to the absolute flux uncertainty. That flavor ratio was one of the important feature to be investigated, as the ratio was expected to be altered by the oscillation mechanism. The ratio increases as the neutrino energy becomes larger, as the electron neutrino flux becomes smaller.

Due to the transparent nature of matter from the neutrino's perspective, atmospheric neutrinos easily penetrate the Earth, and consequently are

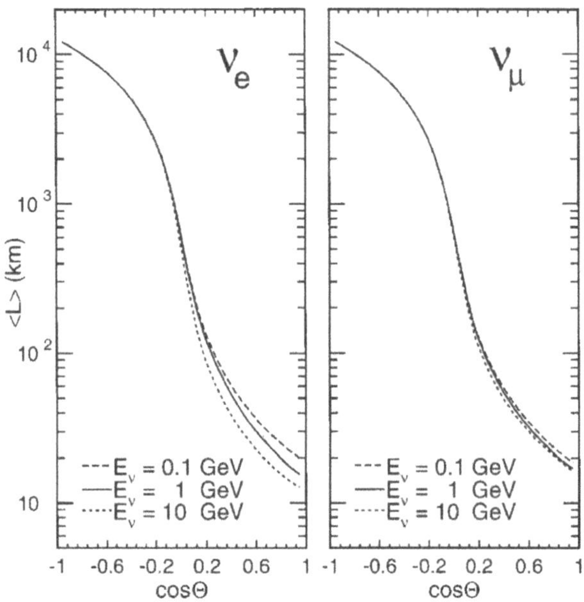

Fig. 3. Neutrino pathlength from the production point in the atmosphere to the detector as a function of cosine of the zenith angle of the arrival direction. This figure has been taken from Ref. 7

observable from all directions at any point on the Earth. The pathlength from production to observation point depends on the zenith angle of the arrival direction, ranging from $O(10)$ km for neutrinos arriving directly downwards from the sky, up to approximately the Earth's diameter for neutrinos arriving from below, as shown in Fig. 3.

Figure 4 shows the zenith angle dependence of the atmospheric neutrino flux. In the horizontal direction ($\cos\theta = 0$) the neutrino flux increases relative to the vertical directions ($\cos\theta = \pm 1$). This is because primary and secondary particles travel further in the atmosphere, thus producing more secondaries, and in turn more neutrinos. This tendency becomes more significant at higher neutrino energies. At low energies a flux asymmetry is seen between upward-going and downward-going directions since the low-energy flux is affected by the geomagnetic field at the neutrino production point. As the energy increases, the geomagnetic effect becomes less significant and the asymmetry disappears.

The geomagnetic field deflects incoming primary cosmic rays, potentially preventing them from reaching the Earth depending on their geomagnetic rigidity, defined as pc/Ze (momentum p, atomic number Z and the elementary charge e). This threshold effect is known as the rigidity cutoff, and depends on the arrival location and direction on the Earth's surface. Only

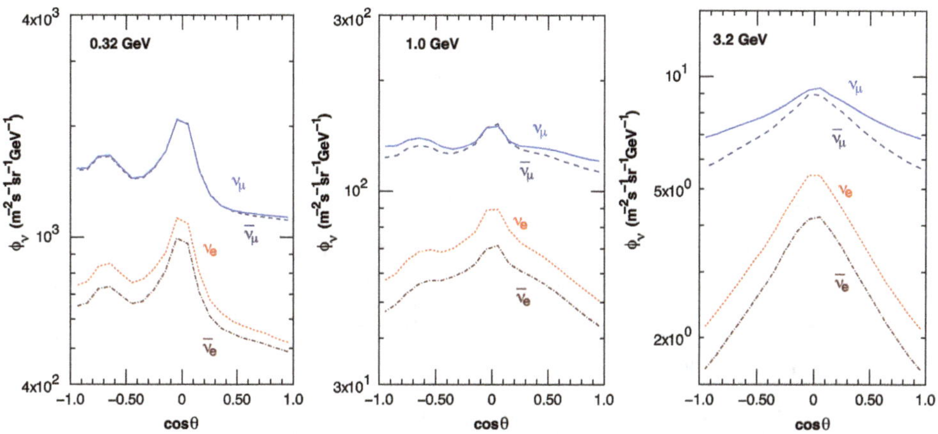

Fig. 4. Zenith angle dependence of atmospheric neutrino flux calculated for Kamioka site. Here θ is the zenith angle of the arrival direction of the neutrino, with $\cos\theta = 1$ for vertically downward-going neutrinos, and $\cos\theta = -1$ for vertically upward-going neutrinos. This figure has been taken from Ref. 3.

cosmic rays with a rigidity above a certain threshold are able to traverse the geomagnetic field to that point, excluding trajectories intersecting the Earth. Generally the cutoff energy becomes lower, and thus the neutrino flux increases, at high latitudes. The structure of the geomagnetic field also causes an East–West oriented anisotropy in cosmic rays, and a similar anisotropy is propagated to the atmospheric neutrino flux, which has been observed experimentally, as shown in Fig. 5.

The low-energy cosmic ray flux is known to be anticorrelated with solar activity[9] because the plasma flux (or "solar wind") from the sun can scatter cosmic rays entering the solar system, and thus the atmospheric neutrino flux is relatively reduced during periods of high solar activity. A reduction at the few-percent level in the sub-GeV flux is expected, though such modulation has not yet been clearly observed.[8]

Seasonal changes in the atmospheric neutrino flux due to temperature variations are also expected, since the atmospheric density increases at higher altitudes in the summer, and thus more neutrinos are created by secondary particles decaying in flight. Such changes are predicted[4] to be strongest at the polar regions, with a normalization change of a few percent around the GeV–TeV scale and becoming negligible at the Equator, where there is less seasonal variation in air density.

As the interaction energy increases above the TeV scale, the π/K secondary hadrons reach the ground before they decay and therefore do not contribute to neutrino production. At this point, the production of neutrinos from shorter-lived hadrons, predominantly D mesons, begins to dominate the

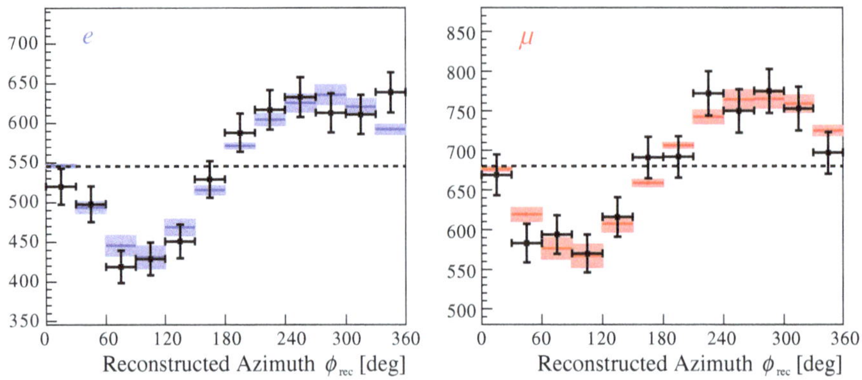

Fig. 5. East–West anisotropy observed in atmospheric neutrinos for e-like (left) and μ-like (right) events, showing the observed data (points with statistical error) and Monte Carlo simulation (boxes with systematic error). Events are selected for $0.4 < E_{rec} < 3.0$ GeV in reconstructed energy and for near-horizontal arrival direction. The azimuthal angle is defined as the horizontal angular distance from the south direction to the reconstructed lepton direction. The angles of 0°, 90°, 180°, 270° correspond to South-going, West-going neutrinas, North-going, and East-going neutrinos, respectively. This figure has been taken from Ref. 8.

spectrum, as shown in Fig. 6. The neutrinos from such hadrons are commonly refered to as the "prompt" flux. However, the prompt neutrino flux has not yet been successfully detected.

At $O(10)$ TeV and higher energies, the Earth becomes opaque even for neutrinos, and neutrino interactions in the Earth will attenuate the upward-going flux. Incorporating these and other effects has become an important part of predicting the expected atmospheric neutrino flux at the various experimental sites.

Figure 7 summarizes the measured ν_μ and ν_e fluxes averaged over all directions in the energy region from sub-GeV to over 100 TeV, by several experiments. The measured fluxes are consistent with the predictions from the flux calculations, once the effect of neutrino oscillation is taken into account.

2. Measurements of atmospheric neutrino oscillation

2.1. *The discovery of muon neutrino oscillation*

Atmospheric neutrinos were first detected in 1965 by two independent groups.[19,20] They observed horizontal muons in detectors deep underground, where detectors are shielded from cosmic muons. In the 1970s, atmospheric neutrinos were observed as a background in searches for proton decay, which is predicted by Grand Unified Theories (GUTs). By comparing neutrino data

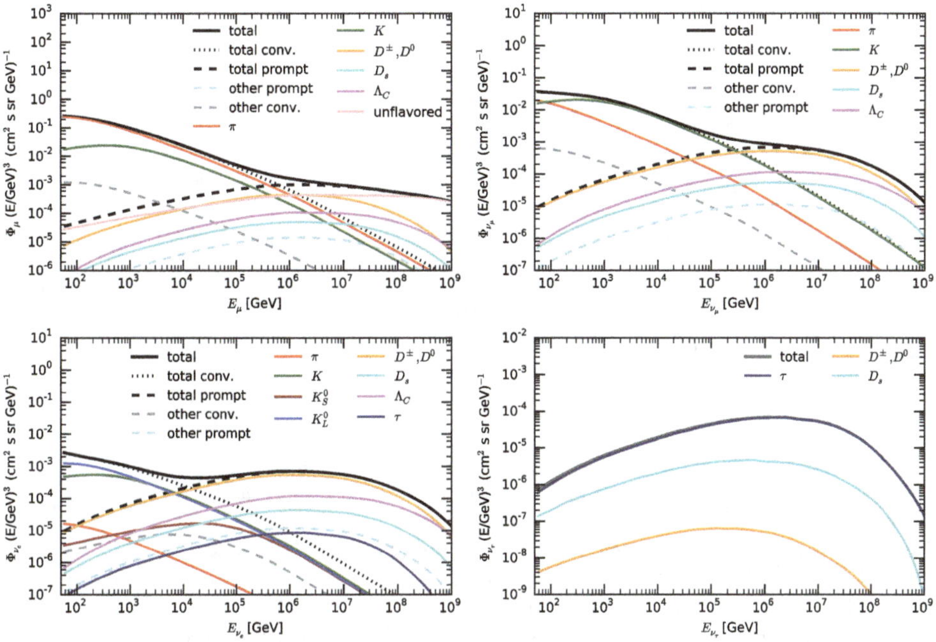

Fig. 6. Calculated contributions of each secondary particle to the flux of atmospheric $\mu^++\mu^-$ (left upper), $\nu_\mu+\bar{\nu}_\mu$ (right upper) $\nu_e+\bar{\nu}_e$ (left lower), and $\nu_\tau+\bar{\nu}_\tau$ (right lower) in the energy above 50 GeV. This figure has been taken from Ref. 10.

with predictions from flux calculations, it was found that the measured flux of each neutrino flavor was not as expected, hinting at the possibility of an additional effect, such as neutrino oscillation.

The first indication of neutrino oscillation in the atmospheric sector was found in 1990. The Kamiokande is a water-Cherenkov detector whose detector size is about 1 k. By using the different pattern of Cherenkov ring images which electrons and muons produce in the water, the observed neutrino events are separated into electron-like (e-like) and muon-like (μ-like), which are induced mainly by ν_e and ν_μ, respectively. They found that the number of observed μ-like events was smaller than expected, while the number of e-like events was in agreement with expectation.[21,22] Due to the relatively small theoretical uncertainty on the ν_μ/ν_e flux ratio, Kamiokande's result indicated the existence of unknown physics, such as neutrino oscillation, neutrino decay, and decoherence. A consistent result was observed by the IMB water-Cherenkov detector.[23,24] On the other hand, NUSEX[25] and Frejus,[26] whose detectors are tracking-type with iron plates, did not observe any significant ν_μ deficit within their statistics. Though the results of later experiments such as Sudan-2[27] and MACRO[28] supported the Kamiokande's findings, the conclusion was unclear and more obvious

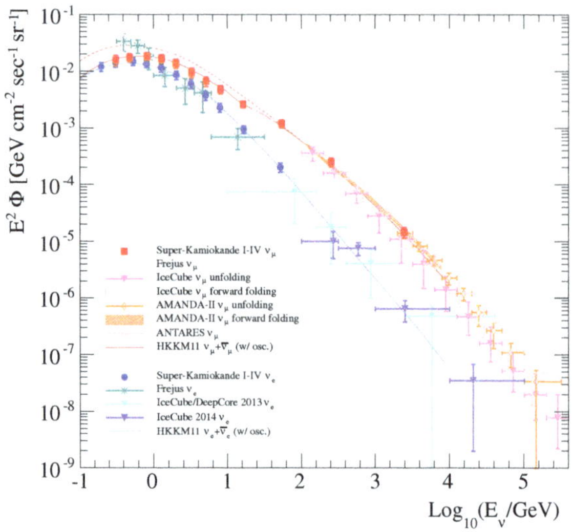

Fig. 7. Measured energy spectra of the atmospheric ν_e and ν_μ fluxes by SK,[8] Frejus,[11] AMANDA-II,[12,13] IceCube,[14–17] and ANTARES.[18] The solid (dashed) lines are the HKKM11 flux[3] predicted for the Kamioka site with (without) neutrino oscillation. Note that the neutrino flux below several GeV is expected to be relatively higher for the Frejus site, due to the lower geomagnetic cutoff of the primary cosmic rays. This figure has been taken from Ref. 8.

evidence was required. Figure 8 shows the summary of the measured ν_μ/ν_e flux ratio compared to the prediction among the experiments described above.

Strong evidence of neutrino oscillation was reported by the Super-Kamiokande (SK) experiment in 1998. SK observed a large deficit of upward-going events induced by muon neutrinos below $O(10)$ GeV. SK, which is a successive detector of Kamiokande, is a water-Cherenkov detector whose fiducial volume is 22.5 k, and ~11 000 20-inch photomultipliers are placed on the inner detector wall.[29] Figure 9 shows the zenith angle distributions of recent SK data.

In the simple case of two-flavor oscillation, the survival probability of a muon neutrino is expressed as

$$P(\nu_\mu \to \nu_\mu) = 1 - \sin^2 2\theta \sin^2\left(\frac{1.267 \ \ \Delta m^2(\mathrm{eV}^2) \ \ \mathrm{L(km)}}{E_\nu(\mathrm{GeV})}\right), \qquad (1)$$

where Δm^2 is the squared difference of the two neutrino masses, and E_ν and L are the neutrino energy and pathlength, respectively. As atmospheric neutrinos typically have energies of O(GeV) and pathlengths of $O(10)$–$O(10000)$ km, according to this formula, neutrino oscillation can

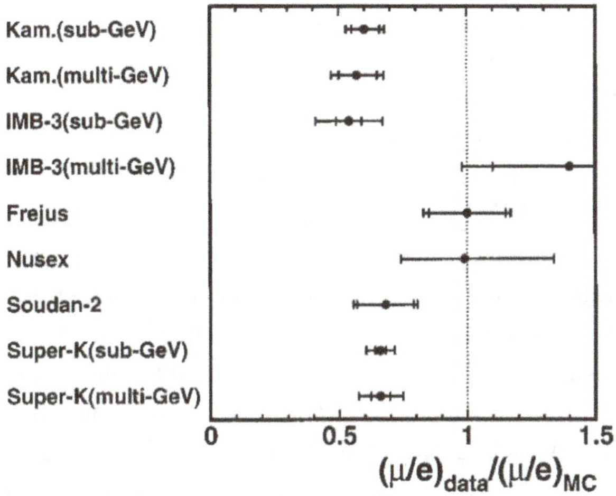

Fig. 8. Summary of the $(\nu/e)_{\rm Data}/(\nu/e)_{\rm MC}$ flux ratio measurements of atmospheric neutrino experiments. Inner (outer) error bars show the statistical (statistical plus systematic) error. This figure has been taken from Ref. 7.

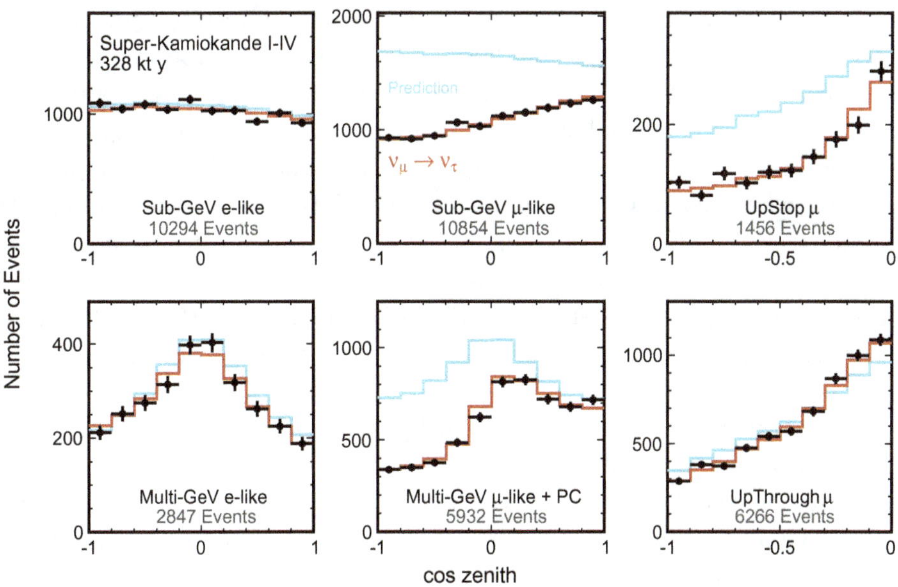

Fig. 9. Zenith angle distributions of SK I–IV data for e-like (left) and μ-like (middle and right) samples. Samples are separated according to the visible energy and the event topology, which can be fully contained sub-GeV and multi-GeV, partially contained (PC), upward-going stopping (UpStop) and through-going (UpThrough) muons. The typical neutrino energy of the sample increases in this order. The events around cos zenith $\sim$ $+1$ (-1) correspond to downward-going (upward-going) events. Figure has been provided by Super-Kamiokande collaboration.

be observed using them if the mass-squared difference is in the region $10^{-4} < \Delta m^2 < 10^{-1}\,\text{eV}^2$. For upward-going events with $E_\nu \approx 1\,\text{GeV}$, the oscillation frequency in the second term of Eq. (1) becomes very rapid, and the oscillation effects are averaged out over experimental scales. This leads to an overall suppression of the flux by a factor of $1 - \frac{1}{2}\sin^2 2\theta$ for $\Delta m^2 \approx 10^{-3}$ eV^2. SK found the mixing angle θ to be relatively large, consistent with the maximal mixing value of $\theta = 45°$. This was surprising because large mixing angles are not observed in the quark sector.

Since large distortions in the e-like sample were not seen, muon disappearance was attributed to $\nu_\mu \to \nu_\tau$ oscillations, where the ν_τ often goes undetected due to the τ high production threshold. In the three-flavor mixing framework which explains neutrino oscillation measured in both solar and atmospheric neutrinos, the mass-squared difference and the mixing angle obtained by the ν_μ deficit correspond to Δm^2_{32} and θ_{23}, respectively.

At the time of SK findings in 1998, possible interpretations of the upward-going muon neutrino deficit in terms of neutrino decay[30,31] or neutrino decoherence[32] were possible. Neutrino oscillation was unambiguously established using both the SK atmospheric results and the later SNO solar neutrino results in 2001, for which the experiments shared the 2015 Nobel Prize. Later, SK was able to exclude these alternative scenarios alone, using a higher statistics dataset, by examining the L/E distribution of the muon neutrino data. For the oscillation hypothesis, a "dip", which corresponds to the first oscillation minimum, is expected around $L/E \approx 230$ km/GeV for $\Delta m^2 = 2.4 \times 10^{-3}\text{eV}^2$, while no dip is expected in the other scenarios. According to the L/E analysis by SK, a "dip" was clearly seen as shown in Fig. 10, and the hypotheses of neutrino decay and neutrino decoherence were disfavored by significance levels greater than 3σ.[33]

After the observation of the upward-going muon neutrino deficit, which indicated $\nu_\mu \to \nu_\tau$ oscillation, the direct detection of oscillation-induced tau neutrino events was a critical component of verifying this standard mixing paradigm. The direct detection of ν_τ appearance in an initially pure muon neutrino beam was reported by the OPERA long-baseline accelerator neutrino experiment,[34] which observed τ decays produced by oscillation-induced ν_τ interactions with 5σ significance.

A complementary measurement was made by SK, which also detected oscillated ν_τ in the upward-going direction of the atmospheric neutrino data. Assuming two-flavor oscillations with $\Delta m^2 = 2.4 \times 10^{-3}\text{eV}^2$, the expected charged-current (CC) ν_τ interaction rate within SK is ~1 per kiloton year, due to the 3.5 GeV production threshold of the tau particle. At SK, the

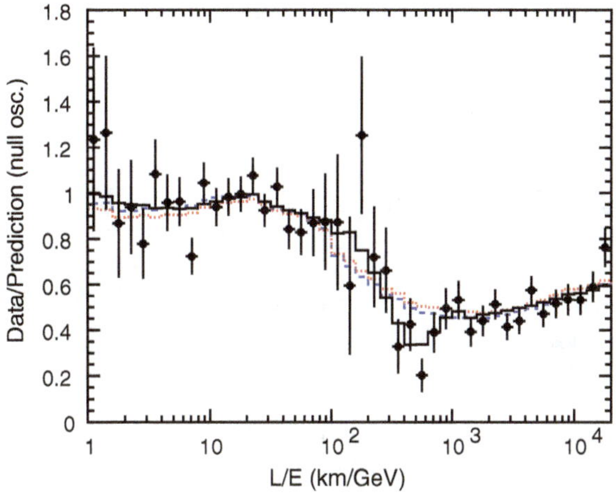

Fig. 10. L/E distribution for the SK I ν_μ-like data sample, where L and E represent the estimated pathlength and the neutrino energy calculated based on the reconstructed observables. The best-fit expectations for two-flavor $\nu_\mu \leftrightarrow \nu_\tau$ oscillations (solid line), neutrino decay (dashed line), and neutrino decoherence (dotted line) are shown. This figure has been taken from Ref. 33.

search for these events focuses on the hadronic decay modes of the τ lepton (with branching ratio 65%). Since the dominant ν_μ disappearance effect is seen in the upward-going data, the signal is expected to appear only in the upward-going direction. By employing a discrimination method using a neural network technique, the presence of the tau signal appears cleanly in the upward-going ν_τ-enhanced sample with 3.8σ significance, as shown in Fig. 11, providing evidence for the appearance of ν_τ from oscillations in the atmospheric data.[35]

The observation of the muon neutrino deficit by SK[36] in conjunction with the SNO solar neutrino results provided the first evidence that at least some neutrinos have a finite mass. While the muon neutrino oscillations of the type seen in the atmospheric neutrino sector were then confirmed by long-baseline experiments, in which accelerator-produced muon neutrinos are utilized, other atmospheric neutrino measurements have provided further supporting results.

MINOS was designed to measure neutrino oscillation using accerator neutrinos over a long baseline. They also measured the ratio of the upward- and downward-going atmospheric ν_μ-induced events. The ratio was found to be significantly smaller than expected, deviating from the unoscillated expectation by more than 6σ.[37]

The Sudbury Neutrino Observatory (SNO) detector contains 1 kt heavy water (D_2O) mainly for detecting solar neutrino oscillation. They measured

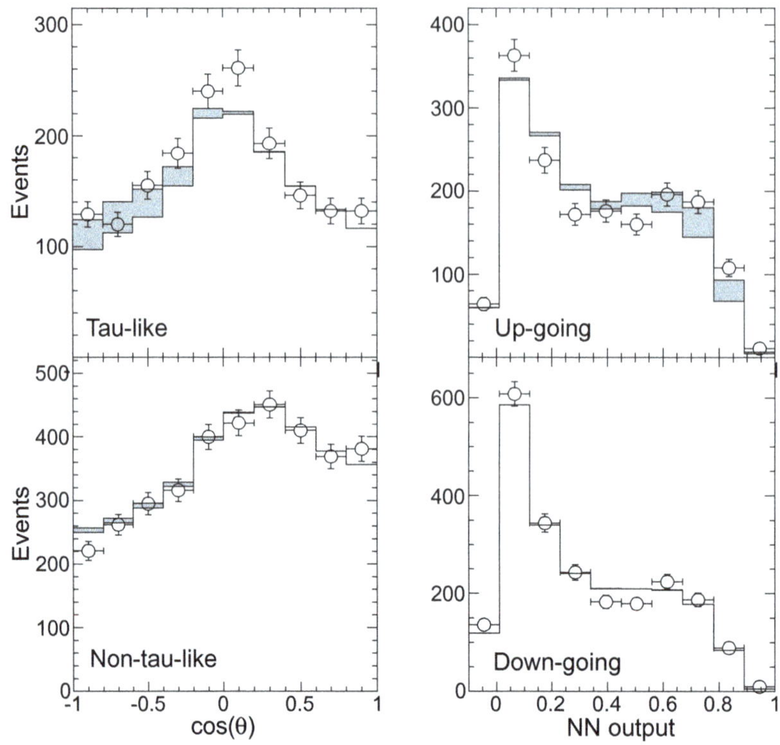

Fig. 11. Zenith angle distributions (left) and neural network classifier output (right) by the SK ν_τ analysis. The left upper (lower) figure shows the zenith angle distribution of tau-like (non-tau-like) events, with the fitted expectation of the tau signal (shaded) overlaid on the background events. The right upper (lower) figure shows the neutral network output of the selected events in the upward-going (downward-going) sample. This figure has been taken from Ref. 35.

a smaller flux of neutrino-induced upward-going muons than the unoscillated prediction for upward-going and horizontal directions.[38]

Recently, kilometer-scale neutrino telescope detectors, which use natural ice and ocean water as the interaction target, have been built with the aim of detecting neutrinos in the PeV energy range originating from astronomical objects. These detectors have also been performing atmospheric neutrino oscillation measurements in the energy region above $O(10)$ GeV. IceCube is a neutrino telescope detector using natural ice at the South pole. They clearly measured the deficit of "track-like" muon events, due to ν_μ oscillation, as shown in Fig. 12. This was realized by lowering the energy threshold with the construction of an inner detector, named "DeepCore", where more photosensors were placed in a dense configuration. ANTARES, a neutrino telescope located deep in the Mediterranean, also observed muon neutrino events and obtained a result consistent with ν_μ oscillation.[39] Figure 13

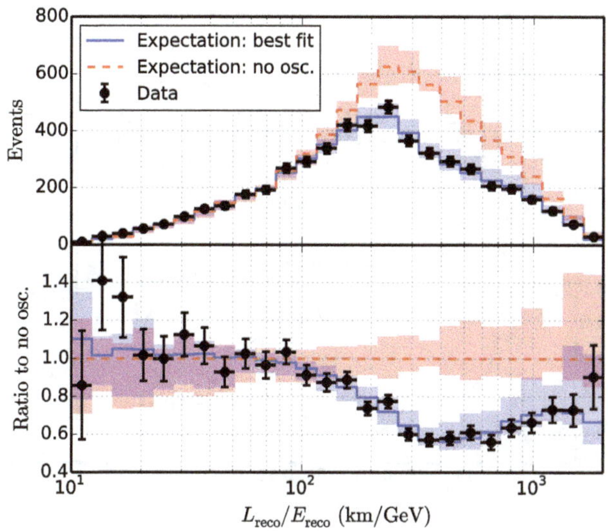

Fig. 12. Number of track-like events and the ratio to the unoscillated expectation plotted as a function of L/E for IceCube–DeepCore data. The solid line shows the best fit to the data while the dashed line shows the no-oscillation scenario. This figure has been taken from Ref. 40.

shows the allowed regions of the atmospheric oscillation parameters, Δm^2_{32} and θ_{23}, from the SK and IceCube/DeepCore experiments and long-baseline accelerator experiments. The measured allowed regions are consistent among experiments.

2.2. *Searches beyond three-flavor mixing*

The three-flavor neutrino oscillation scheme is consistent with a wide range of solar, reactor, accelerator, and atmospheric neutrino experiments. However, some anomalous results, such as those from LSND[44] and MiniBooNE[45] motivate searches beyond three-flavor mixing. The possibility of 3 + 1 neutrino flavors, in which an additional sterile neutrino (ν_s) with $\Delta m^2_{41} \approx$ 1 eV^2 appears, is often discussed in the context of these anomalies. SK gives a constraint on sterile neutrino mixing by looking at both the ν_e and ν_μ data over a wide energy range. In the mass region of $\Delta m^2_{41} > 0.1$ eV^2, where the oscillation is rapidly averaged for atmospheric neutrinos, the limits on the mixing matrix element between the sterile and ν_μ/ν_τ neutrinos are given as $|U_{\mu 4}|^2 < 0.041$ and $|U_{\tau 4}|^2 < 0.18$,[46] where $U_{\mu 4}$ and $U_{\tau 4}$ correspond to the elements of the extended PMNS matrix between ν_μ/ν_τ and ν_s, respectively.

IceCube also searched for sterile neutrinos in the energy range 320 GeV to 20 TeV. In this energy regime, mixing of active and sterile neutrinos

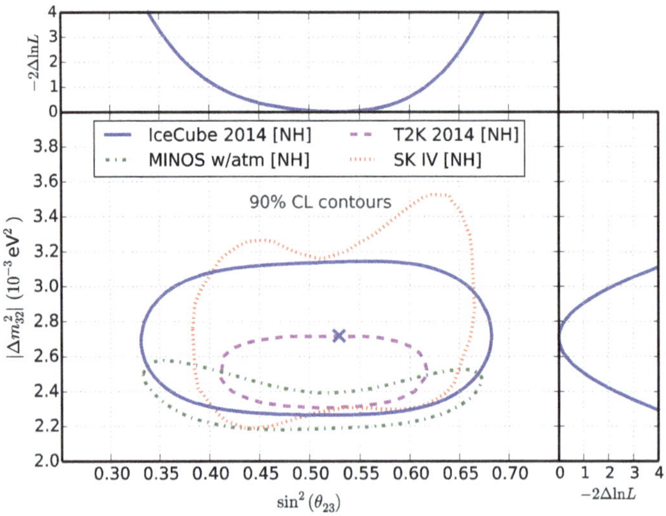

Fig. 13. Allowed regions of the atmospheric neutrino oscillation parameters, Δm^2_{32} and $\sin^2\theta_{23}$, in the 90% confidence level for MINOS,[41] T2K,[42] SK,[43] and IceCube–DeepCore.[40] This figure has been taken from Ref. 40.

could be enhanced by resonance-like oscillations as they propagate through the Earth, which is caused by MSW mechanism,[47,48] producing distinctive, energy-dependent distortions of the measured zenith angle distributions for sterile mass splittings in the range $0.01 < \Delta m^2_{41} < 10$ eV^2. By examining the energy and zenith angle distributions, no such signature of sterile neutrinos was seen in the abovementioned energy range. As a result, a constraint on the parameters of sterile mass and mixing angle between ν_2 and ν_s (Δm^2_{41}, θ_{24}) was produced in the region of $\Delta m^2_{41} \simeq 0.1 \sim 1$ eV^2, as shown in Fig. 14.

The possibility of having different neutrino mass and mixing angle parameters between neutrinos and antineutrinos, which would occur if CPT symmetry is violated, was searched for using atmospheric neutrinos. Though SK cannot generally distinguish neutrinos from antineutrinos on an event-by-event basis, the presence of both in the atmospheric neutrino data allows their separate oscillations to be studied statistically. In the analysis, antineutrino oscillations $\bar{\nu}_\mu \to \bar{\nu}_\mu$ are considered under a two-flavor oscillation scheme, independently of neutrino oscillations using two additional oscillation parameters, $\Delta\bar{m}^2$ and $\bar{\theta}$. The best-fit antineutrino parameters are consistent with the neutrino parameters[51] (Δm^2, θ). The MINOS detector can separate neutrino and antineutrino interactions on an event-by-event basis, using the direction of the muon track which is bent by the magnetized detector. The observed ratio of $\bar{\nu}_\mu$ to ν_μ events is

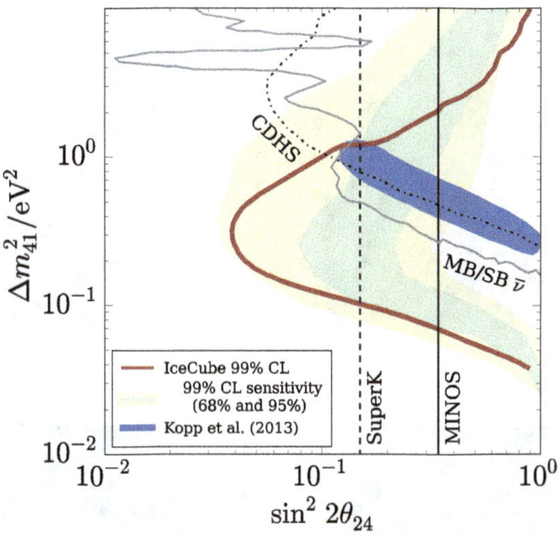

Fig. 14. 99% CL contours for sterile neutrino search in IceCube with bands containing 68% (green) and 95% yellow of 99% contours in sensitivity with the assumption of $3+1$ sterile oscillation. 90% CL allowed region (blue shaded) from [49] and 90% CL exclusion from other experiments are also shown. This figure has been taken from Ref. 50 with modification.

compared with the MC simulations, giving a result consistent with no CPT violation.[37]

3. Future: toward the detection of neutrino mass hierarchy

The nonzero value of the third mixing angle θ_{13} was discovered through $\bar{\nu}_e$ disappearance at reactor neutrino experiments,[52–54] and through $\nu_\mu \to \nu_e$ appearance in accelerator neutrino experiments.[55] The three-flavor oscillation framework, as described by the 3×3 PMNS matrix,[56,57] is consistent with most results and is generally accepted as the standard theory of neutrino oscillation. Nonetheless, some important pieces of that framework are still missing. The octant of θ_{23}, the possibility of a CP-violating phase in the lepton sector $\delta_{\rm CP}$ and the mass ordering between the second and third mass states (the so called mass hierarchy) are still unknown. The mass hierarchy is referred to as either "normal" ($\Delta m^2_{32} > 0$) or "inverted" ($\Delta m^2_{32} < 0$), corresponding to the masses being arranged in the order $m_1 < m_2 \ll m_3$ or $m_3 \ll m_1 < m_2$, respectively. Of these unsolved issues, the determination of the mass hierarchy seems the most likely to be explored by atmospheric neutrino experiments in the near future.

Atmospheric neutrinos, which travel inside the Earth, are affected by a potential as a result of CC coherent forward scattering of ν_e on electrons,

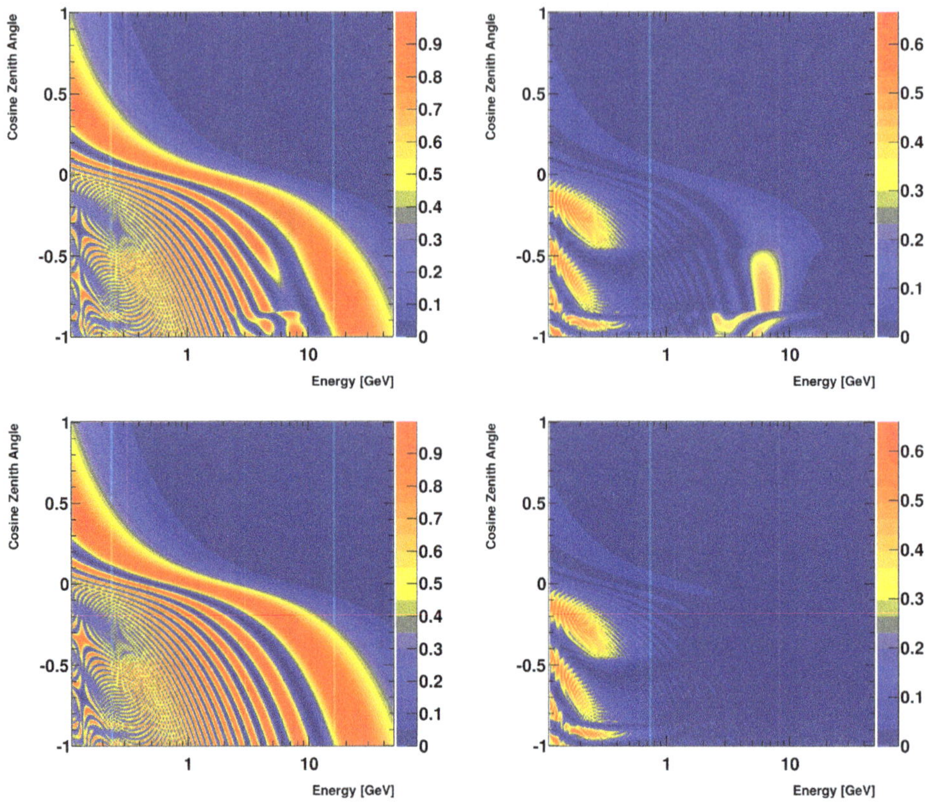

Fig. 15. Oscillation probabilities as a function of the cosine of the neutrino zenith angle, where -1 indicates the upward-going direction, and energy. The left and right figures show $\nu_\mu \to \nu_\tau$ and $\nu_\mu \to \nu_e$ oscillation probabilities, respectively. The top and bottom rows show probabilities for the normal and inverted hierarchies, respectively. For antineutrinos (not shown), the resonance-enhanced oscillations appear in the inverted hierarchy plots instead of the normal hierarchy.

$V_e^{\rm CC} = \pm\sqrt{2}G_F n_e$, where G_F is the Fermi constant and n_e is the local electron number density of matter. It should be noted that the potential changes sign for antineutrinos. Interestingly, through this "matter effect", the oscillation probabilities are changed, in particular, for $\nu_\mu \to \nu_e$ oscillations driven by θ_{13} as shown in Fig. 15 for upward-going neutrinos in the multi-GeV region. Such resonance-like enhancement is caused by MSW mechanism and explained as follows: When neglecting solar term oscillation due to Δm^2_{21} and the effect of CP-voilation phase, the leading term of $\nu_\mu \to \nu_e$ oscillation probability is expressed as

$$P(\nu_\mu \to \nu_e) \approx \sin^2\theta_{23}\sin^2 2\theta_{13}\sin^2\left(\frac{1.267\Delta m^2_{31}L}{E_\nu}\right). \qquad (2)$$

In the propagation of the atmospheric neutrinos inside the Earth, the probability is affected by matter effect and the mixing angle of θ_{13} is changed as follows:

$$\sin^2 2\theta_{13}^m = \frac{\sin^2 2\theta_{13}}{\left(\cos 2\theta_{13} - \frac{2V_e^{CC}E_\nu}{\Delta m_{31}^2}\right)^2 + \sin^2 2\theta_{13}}. \quad (3)$$

The modified mixing angle is enhanced when the resonance condition $E_\nu/\Delta m_{31}^2 \approx \cos 2\theta_{13}/2V_e^{CC}$ is satisfied, which corresponding neutrino energy is between 2 and 10 GeV according to the matter density in the Earth's core and mantle.

Due to the change in sign of the matter potential ($V_e^{CC} < 0$), these resonant features are not present ($\sin^2 2\theta_{13}^m < \sin^2 2\theta_{13}$) for antineutrinos when the mass hierarchy is normal ($\Delta m_{31}^2 > 0$). However, this situation is reversed for an inverted mass hierarchy, where only antineutrino oscillations are enhanced. Note that there is also a distortion of the $\nu_\mu \to \nu_\tau$ oscillation probability in the same energy region. At energies below ~1 GeV, oscillations are driven by the θ_{12} and Δm_{21}^2 mixing parameters. As these structures are sensitive to the size of θ_{23}, δ_{CP}, and the sign of the mass hierarchy, atmospheric neutrinos are in principal sensitive to all of the current open questions in oscillation physics.

So far, SK has been trying to determine the mass hierarchy by looking for an excess in the upward-going electron-like events at multi-GeV energies. The data marginally prefer the normal hierarchy,[58] though the statistical uncertainty is still large due to the low number of multi-GeV events caused by the steep spectrum of the atmospheric neutrino flux. Further improvements to this measurement may be made with larger detectors in the future.

For a final conclusion on the neutrino mass hierarchy in the atmospheric neutrino sector, a detector larger than those currently operating would be necessary. The determination of the mass hierarchy plays an important part of the physics program for future neutrino experiments.

Hyper-Kamiokande (HK) is a proposed water Cherenkov detector designed to study neutrino CP violation using the J-PARC neutrino beam, as well as nucleon decays and atmospheric neutrinos. With a 560 kt fiducial volume, it will accumulate atmospheric neutrinos at 25 times the rate of SK and will therefore offer unprecedented access to atmospheric neutrino physics. The expected sensitivity to the mass hierarchy exceeds 3σ for both hierarchy assumptions, when $\theta_{23} > 0.45$, using 10 years of atmospheric neutrino data. It would also be possible to give a constraint on the octant of θ_{23} at more than 3σ, if $|\theta_{23} - 45| > 4°$.[59,60]

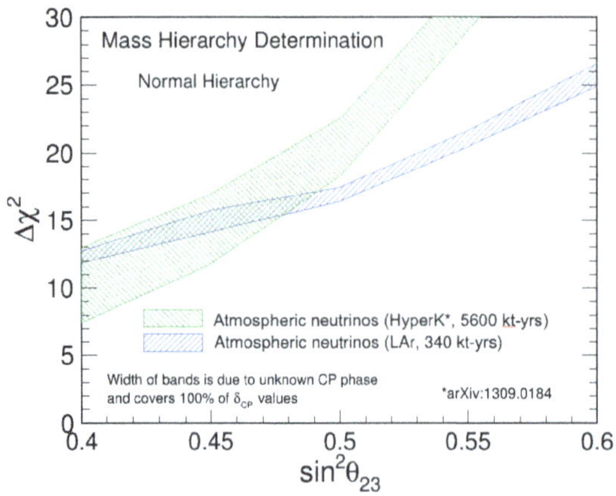

Fig. 16. Sensitivity of the mass hierarchy determination using atmospheric neutrinos as a function of the true value of θ_{23} for HK[61] (green) and DUNE[62] (blue), with the assumption of true normal hierarchy. Approximately the significance of the mass hierarchy sensitivity can be estimated by the square of $\Delta\chi^2$. This figure has been taken from Ref. 62.

Deep Underground Neutrino Experiment (DUNE) is a proposed experiment that will observe both beam and atmospheric neutrinos using a liquid argon time projection chamber (TPC). Though comparatively smaller than HK at 34 kt, its high-resolution imaging offers possibilities for tagging event features that provide statistical discrimination between neutrinos and antineutrinos. According to simulation studies, taking into account the event rate of neutrino energy E_ν and zenith angle $\cos\theta$, depending on the value of θ_{23}, DUNE will have $3.4{-}5\sigma$ sensitivity to the mass hierachy in 10 years of running.[62] Figure 16 shows the expected sensitivities to the mass hierarchy for HK and DUNE.

A 50 kt magnetized iron calorimeter (ICAL) at the India-based Neutrino Observatory (INO) is aiming to use charge discrimination to distinguish between neutrinos and antineutrinos and could provide 2.2σ sensitivity to the normal mass hierarchy, assuming $\sin^2\theta_{23} = 0.5$, with 10 years of data and a 50 kt detector.[63]

The possibility of a mass hierarchy measurement using astrophysical neutrino detectors has also been proposed.[64] Generally the target volumes of these detectors are large enough for atmospheric neutrino measurements; however, they are often required to lower their energy threshold to $O(10)$ GeV in order to be sensitive to the mass hierachy using atmospheric neutrinos. A new inner detector configuration in the IceCube/DeepCore volume, named Precision IceCube Next Generation Upgrade (PINGU), is currently planned in order to make precise measurements of atmospheric neutrino oscillation. PINGU will have a higher photodetector density than

IceCube/DeepCore in order to lower the energy threshold below 10 GeV. A simulation based on the current detector design suggests the ability to identify the mass hierarchy with 3σ median significance within 4 years of operation. In addition, an improvement of the sensitivity of other oscillation-related measurements, such as ν_μ disappearance, the θ_{23} octant and ν_τ appearance are expected.[65]

Oscillations Research with Cosmics in the Abyss (ORCA) is the planned low-energy branch of KM3NeT, an underwater Cherenkov neutrino detector in the Mediterranean Sea. The detector will be a dense array of multi-PMT digital optical modules (DOMs)[66] using technology developed for KM3NeT. According to a Monte Carlo study, the mass hierarchy is expected to be measured at more than 3σ in 3 years, assuming the current world best-fit values of θ_{23}.[67]

Besides neutrino oscillation physics, it is worth mentioning possible applications of the atmospheric neutrino to Earth science, such as neutrino radiography. This has become feasible by the realization of kilometer-scale detectors. Using the absorption characteristics of the high-energy atmospheric neutrino traveling inside the Earth, the density profile of Earth's core and mantle, which is of great interest in Earth science, could be probed by measuring atmospheric neutrino rates above $O(10)$ TeV. In Ref. 68, the sensitivity to distinguish between the preliminary reference Earth model (PREM)[69] and the simple homogeneous model is estimated based on an analytical calculation. In PREM, the Earth consists of a mantle extending to radial distance $r \sim 3000$ km below the Earth's surface and a core under it with a sharp core–mantle transition in density of about a factor 2, while the homogeneous matter density profile is assumed in the latter. The paper concluded that it could be distinguished with more than 3σ significance after 10 years of observation by the IceCube detector.

4. Conclusion

Atmospheric neutrinos are the most abundant source of GeV neutrinos on the Earth. Thanks to the range of energies and pathlengths available, significant signatures of neutrino oscillations emerged, such as ν_μ/ν_e ratio and ν_μ deficit in the upward-going events. The evidence for neutrino oscillations in the atmospheric sector provided evidence of nonvanishing neutrino mass, and showed that the mixing angle between the flavor states is unexpectedly large. This had a great impact on our understanding of particle physics, and the discovery of neutrino oscillations in the atmospheric and solar neutrino sectors was awarded the Nobel Prize in 2015.

The measurement of neutrino oscillations in the atmospheric sector has played a significant role in verifying the three-flavor mixing model and is still having an important role in the determination of the neutrino mass hierarchy. Atmospheric neutrino experiments will continue to contribute toward the elucidation of the whole understanding of neutrino physics in the future.

References

1. M. Honda, T. Kajita, K. Kasahara, and S. Midorikawa, A new calculation of the atmospheric neutrino flux in a 3-dimensional scheme, *Phys. Rev.* **D70**, 043008 (2004). doi: 10.1103/PhysRevD.70.043008.
2. M. Honda, T. Kajita, K. Kasahara, S. Midorikawa, and T. Sanuki, Calculation of atmospheric neutrino flux using the interaction model calibrated with atmospheric muon data, *Phys. Rev.* **D75**, 043006 (2007). doi: 10.1103/PhysRevD.75.043006.
3. M. Honda, T. Kajita, K. Kasahara, and S. Midorikawa, Improvement of low energy atmospheric neutrino flux calculation using the jam nuclear interaction model, *Phys. Rev.* **D83**, 123001 (2011). doi: 10.1103/PhysRevD.83.123001. URL `http://link.aps.org/doi/10.1103/PhysRevD.83.123001`.
4. M. Honda, M. S. Athar, T. Kajita, K. Kasahara, and S. Midorikawa, Atmospheric neutrino flux calculation using the NRLMSISE-00 atmospheric model, *Phys. Rev.* **D92** (2), 023004 (2015). doi: 10.1103/PhysRevD.92.023004.
5. G. Barr, T. Gaisser, P. Lipari, S. Robbins, and T. Stanev, Three-dimensional calculation of atmospheric neutrinos, *Phys. Rev.* **D70**, 023006 (Jul, 2004). doi: 10.1103/PhysRevD.70.023006. URL `http://link.aps.org/doi/10.1103/PhysRevD.70.023006`.
6. G. Battistoni, A. Ferrari, T. Montaruli, and P. Sala, The fluka atmospheric neutrino flux calculation, *Astropart. Phys.* **19**(2), 269–290 (2003). ISSN 0927-6505. doi: http://dx.doi.org/10.1016/S0927-6505(02)00246-3. URL `http://www.sciencedirect.com/science/article/pii/S0927650502002463`.
7. T. Kajita, and Y. Totsuka, Observation of atmospheric neutrinos, *Rev. Mod. Phys.* **73**, 85–118 (2001). doi: 10.1103/RevModPhys.73.85.
8. E. Richard *et al.*, Measurements of the atmospheric neutrino flux by Super-Kamiokande: Energy spectra, geomagnetic effects, and solar modulation, *Phys. Rev.* **D94** (5), 052001 (2016). doi: 10.1103/PhysRevD.94.052001.
9. M. S. Potgieter, Solar modulation of cosmic rays, *Living Rev. Sol. Phys.* **10**(3) (2013). doi: 10.12942/lrsp-2013-3. URL `http://www.livingreviews.org/lrsp-2013-3`.
10. A. Fedynitch, R. Engel, T. K. Gaisser, F. Riehn, and T. Stanev, Calculation of conventional and prompt lepton fluxes at very high energy, *EPJ Web Conf.* **99**, 08001 (2015). doi: 10.1051/epjconf/20159908001.
11. K. Daum, Determination of the atmospheric neutrino spectra with the frejus detector, *Z. Phys.* **C66**, 417–428 (1995).
12. R. Abbasi *et al.*, Determination of the atmospheric neutrino flux and searches for new physics with AMANDA-II, *Phys. Rev.* **D79**, 102005 (2009). doi: 10.1103/PhysRevD.79.102005.

13. R. Abbasi *et al.*, The energy spectrum of atmospheric neutrinos between 2 and 200 TeV with the AMANDA-II detector, *Astropart. Phys.* **34**, 48–58 (2010). doi: 10.1016/j.astropartphys.2010.05.001.
14. R. Abbasi *et al.*, Measurement of the atmospheric neutrino energy spectrum from 100 gev to 400 tev with icecube, *Phys. Rev.* **D83**, 012001 (2011).
15. R. Abbasi *et al.*, A search for a diffuse flux of astrophysical muon neutrinos with the IceCube 40-string detector, *Phys. Rev.* **D84**, 082001 (2011). doi: 10.1103/PhysRevD.84.082001.
16. M. Aartsen *et al.*, Measurement of the atmospheric flux in icecube, *Phys. Rev. Lett.* **110**, 151105 (2013). doi: 10.1103/PhysRevLett.110.151105. URL `http://link.aps.org/doi/10.1103/PhysRevLett.110.151105`.
17. M. G. Aartsen *et al.*, Measurement of the atmospheric ν_e spectrum with IceCube, *Phys. Rev.* **D91** (12), 122004 (2015). doi: 10.1103/PhysRevD.91.122004.
18. S. Adrian-Martinez *et al.*, Measurement of the atmospheric ν_μ energy spectrum from 100 GeV to 200 TeV with the ANTARES telescope, *Eur. Phys. J.* **C73** (10), 2606 (2013). doi: 10.1140/epjc/s10052-013-2606-4.
19. C. V. Achar *et al.*, Detection of muons produced by cosmic ray neutrinos deep underground, *Phys. Lett.* **18**, 196–199 (1965).
20. F. Reines, M. F. Crouch, T. L. Jenkins, W. R. Kropp, H. S. Gurr, G. R. Smith, J. P. F. Sellschop, and B. Meyer, Evidence for high-energy cosmic-ray neutrino interactions, *Phys. Rev. Lett.* **15**, 429–433 (Aug, 1965). doi: 10.1103/PhysRevLett.15.429. URL `http://link.aps.org/doi/10.1103/PhysRevLett.15.429`.
21. K. S. Hirata *et al.*, Experimental Study of the Atmospheric Neutrino Flux, *Phys. Lett.* **B205**, 416 (1988). doi: 10.1016/0370-2693(88)91690-5.
22. K. S. Hirata *et al.*, Observation of a small atmospheric muon-neutrino/electron-neutrino ratio in Kamiokande, *Phys. Lett.* **B280**, 146–152 (1992). doi: 10.1016/0370-2693(92)90788-6.
23. D. Casper *et al.*, Measurement of atmospheric neutrino composition with IMB-3, *Phys. Rev. Lett.* **66**, 2561–2564 (1991). doi: 10.1103/PhysRevLett.66.2561.
24. R. Becker-Szendy *et al.*, The electron-neutrino and muon-neutrino content of the atmospheric flux, *Phys. Rev.* **D46**, 3720–3724 (1992). doi: 10.1103/PhysRevD.46.3720.
25. M. Aglietta *et al.*, Experimental study of atmospheric neutrino flux in the NUSEX experiment, *Europhys. Lett.* **8**, 611–614 (1989). doi: 10.1209/0295-5075/8/7/005.
26. C. Berger *et al.*, Study of atmospheric neutrino interactions with the Frejus detector, *Phys. Lett.* **B227**, 489 (1989). doi: 10.1016/0370-2693(89)90968-4.
27. W. W. M. Allison *et al.*, Measurement of the atmospheric neutrino flavour composition in soudan-2, *Phys. Lett.* **B391**, 491–500 (1997).
28. M. Ambrosio *et al.*, Measurement of the atmospheric neutrino induced upgoing muon flux using MACRO, *Phys. Lett.* **B434**, 451–457 (1998). doi: 10.1016/S0370-2693(98)00885-5.
29. Y. Fukuda *et al.*, The Super-Kamiokande detector, *Nucl. Instrum. Meth.* **A501**, 418–462 (2003). doi: 10.1016/S0168-9002(03)00425-X.

30. V. D. Barger, J. G. Learned, S. Pakvasa, and T. J. Weiler, Neutrino decay as an explanation of atmospheric neutrino observations, *Phys. Rev. Lett.* **82**, 2640–2643 (1999). doi: 10.1103/PhysRevLett.82.2640.
31. V. D. Barger, J. G. Learned, P. Lipari, M. Lusignoli, S. Pakvasa, and T. J. Weiler, Neutrino decay and atmospheric neutrinos, *Phys. Lett.* **B462**, 109–114 (1999). doi: 10.1016/S0370-2693(99)00887-4.
32. E. Lisi, A. Marrone, and D. Montanino, Probing possible decoherence effects in atmospheric neutrino oscillations, *Phys. Rev. Lett.* **85**, 1166–1169 (2000). doi: 10.1103/PhysRevLett.85.1166.
33. Y. Ashie *et al.*, Evidence for an oscillatory signature in atmospheric neutrino oscillation, *Phys. Rev. Lett.* **93**, 101801 (2004). doi: 10.1103/PhysRevLett.93.101801.
34. N. Agafonova *et al.*, Discovery of tau neutrino appearance in the cngs neutrino beam with the opera experiment, *Phys. Rev. Lett.* **115**, 121802 (2015).
35. K. Abe *et al.*, Evidence for the appearance of atmospheric tau neutrinos in Super-Kamiokande, *Phys. Rev. Lett.* **110**(18), 181802 (2013). doi: 10.1103/PhysRevLett.110.181802.
36. Y. Fukuda *et al.*, Evidence for oscillation of atmospheric neutrinos, *Phys. Rev. Lett.* **81**, 1562–1567 (1998). doi: 10.1103/PhysRevLett.81.1562. URL `http://link.aps.org/doi/10.1103/PhysRevLett.81.1562`.
37. P. Adamson, C. Backhouse, G. Barr, M. Bishai, A. S. T. Blake, G. J. Bock, D. J. Boehnlein, D. Bogert, S. V. Cao, J. D. Chapman, S. Childress, J. A. B. Coelho, L. Corwin, D. Cronin-Hennessy, I. Z. Danko, J. K. de Jong, N. E. Devenish, M. V. Diwan, C. O. Escobar, J. J. Evans, E. Falk, G. J. Feldman, M. V. Frohne, H. R. Gallagher, R. A. Gomes, M. C. Goodman, P. Gouffon, N. Graf, R. Gran, K. Grzelak, A. Habig, J. Hartnell, R. Hatcher, A. Himmel, A. Holin, J. Hylen, G. M. Irwin, Z. Isvan, D. E. Jaffe, C. James, D. Jensen, T. Kafka, S. M. S. Kasahara, G. Koizumi, S. Kopp, M. Kordosky, A. Kreymer, K. Lang, J. Ling, P. J. Litchfield, L. Loiacono, P. Lucas, W. A. Mann, M. L. Marshak, M. Mathis, N. Mayer, M. M. Medeiros, R. Mehdiyev, J. R. Meier, M. D. Messier, W. H. Miller, S. R. Mishra, J. Mitchell, C. D. Moore, L. Mualem, S. Mufson, J. Musser, D. Naples, J. K. Nelson, H. B. Newman, R. J. Nichol, J. A. Nowak, W. P. Oliver, M. Orchanian, R. B. Pahlka, J. Paley, R. B. Patterson, G. Pawloski, S. Phan-Budd, R. K. Plunkett, X. Qiu, A. Radovic, J. Ratchford, B. Rebel, C. Rosenfeld, H. A. Rubin, M. C. Sanchez, J. Schneps, A. Schreckenberger, P. Schreiner, R. Sharma, A. Sousa, B. Speakman, M. Strait, N. Tagg, R. L. Talaga, J. Thomas, M. A. Thomson, R. Toner, D. Torretta, G. Tzanakos, J. Urheim, P. Vahle, B. Viren, J. J. Walding, A. Weber, R. C. Webb, C. White, L. Whitehead, S. G. Wojcicki, K. Zhang, and R. Zwaska, Measurements of atmospheric neutrinos and antineutrinos in the minos far detector, *Phys. Rev.* **D86**, 052007 (2012). doi: 10.1103/PhysRevD.86.052007. URL `http://link.aps.org/doi/10.1103/PhysRevD.86.052007`.
38. B. Aharmim *et al.*, Measurement of the cosmic ray and neutrino-induced muon flux at the Sudbury Neutrino Observatory, *Phys. Rev.* **D80**, 012001 (2009). doi: 10.1103/PhysRevD.80.012001.

39. S. Adrian-Martinez *et al.*, Measurement of atmospheric neutrino oscillations with the ANTARES neutrino telescope, *Phys. Lett.* **B714**, 224–230 (2012). doi: 10.1016/j.physletb.2012.07.002.
40. M. G. Aartsen *et al.*, Neutrino oscillation studies with IceCube-DeepCore, *Nucl. Phys.* **B908**, 161–177 (2016). doi: 10.1016/j.nuclphysb.2016.03.028.
41. P. Adamson *et al.*, Measurement of neutrino and antineutrino oscillations using beam and atmospheric data in MINOS, *Phys. Rev. Lett.* **110** (25), 251801 (2013). doi: 10.1103/PhysRevLett.110.251801.
42. K. Abe *et al.*, Precise measurement of the neutrino mixing parameter θ_{23} from muon neutrino disappearance in an off-axis beam, *Phys. Rev. Lett.* **112** (18), 181801 (2014). doi: 10.1103/PhysRevLett.112.181801.
43. A. Himmel, Recent results from Super-Kamiokande, *AIP Conf. Proc.* **1604**, 345–352 (2014). doi: 10.1063/1.4883450.
44. A. Aguilar, L. B. Auerbach, R. L. Burman, D. O. Caldwell, E. D. Church, A. K. Cochran, J. B. Donahue, A. Fazely, G. T. Garvey, R. M. Gunasingha, R. Imlay, W. C. Louis, R. Majkic, A. Malik, W. Metcalf, G. B. Mills, V. Sandberg, D. Smith, I. Stancu, M. Sung, R. Tayloe, G. J. VanDalen, W. Vernon, N. Wadia, D. H. White, and S. Yellin, Evidence for neutrino oscillations from the observation of $\overline{\nu}_e$ appearance in a $\overline{\nu}_\mu$ beam, *Phys. Rev.* **D64**, 112007 (2001). doi:10.1103/PhysRevD.64.112007. URL `http://link.aps.org/doi/10.1103/PhysRevD.64.112007`.
45. A. A. Aguilar-Arevalo, B. C. Brown, L. Bugel, G. Cheng, E. D. Church, J. M. Conrad, R. Dharmapalan, Z. Djurcic, D. A. Finley, R. Ford, F. G. Garcia, G. T. Garvey, J. Grange, W. Huelsnitz, C. Ignarra, R. Imlay, R. A. Johnson, G. Karagiorgi, T. Katori, T. Kobilarcik, W. C. Louis, C. Mariani, W. Marsh, G. B. Mills, J. Mirabal, C. D. Moore, J. Mousseau, P. Nienaber, B. Osmanov, Z. Pavlovic, D. Perevalov, C. C. Polly, H. Ray, B. P. Roe, A. D. Russell, M. H. Shaevitz, J. Spitz, I. Stancu, R. Tayloe, R. G. Van de Water, D. H. White, D. A. Wickremasinghe, G. P. Zeller, and E. D. Zimmerman, Improved search for $\overline{\nu}_\mu \rightarrow \overline{\nu}_e$ oscillations in the miniboone experiment, *Phys. Rev. Lett.* **110**, 161801 (2013). doi: 10.1103/PhysRevLett.110.161801. URL `http://link.aps.org/doi/10.1103/PhysRevLett.110.161801`.
46. K. Abe *et al.*, Limits on sterile neutrino mixing using atmospheric neutrinos in Super-Kamiokande, *Phys. Rev.* **D91**, 052019 (2015). doi:10.1103/PhysRevD.91.052019.
47. S. Mikheev and A. Y. Smirnov, Resonance amplification of oscillations in matter and spectroscopy of solar neutrinos, *Sov. J. Nucl. Phys.* **42**, 913–917 (1985).
48. L. Wolfenstein, Neutrino Oscillations in Matter, *Phys. Rev.* **D17**, 2369–2374 (1978). doi: 10.1103/PhysRevD.17.2369.
49. J. Kopp, P. A. N. Machado, M. Maltoni, and T. Schwetz, Sterile neutrino oscillations: The global picture *JHEP.* **05**, 050 (2013). doi: 10.1007/JHEP05(2013)050.
50. M. G. Aartsen *et al.*, Searches for sterile neutrinos with the IceCube detector, *Phys. Rev. Lett.* **117** (7), 071801 (2016). doi: 10.1103/PhysRevLett.117.071801.

51. K. Abe *et al.*, Search for Differences in oscillation parameters for atmospheric neutrinos and antineutrinos at Super-Kamiokande, *Phys. Rev. Lett.* **107**, 241801 (2011). doi: 10.1103/PhysRevLett.107.241801.
52. F. An *et al.*, Observation of electron-antineutrino disappearance at Daya Bay, *Phys.Rev.Lett.* **108**, 171803 (2012). doi: 10.1103/PhysRevLett.108.171803.
53. J. Ahn *et al.*, Observation of reactor electron antineutrino disappearance in the RENO experiment, *Phys. Rev. Lett.* **108**, 191802 (2012). doi: 10.1103/PhysRevLett.108.191802.
54. Y. Abe *et al.*, Indication for the disappearance of reactor electron antineutrinos in the Double Chooz experiment, *Phys. Rev. Lett.* **108**, 131801 (2012). doi: 10.1103/PhysRevLett.108.131801.
55. K. Abe *et al.*, Observation of electron neutrino appearance in a muon neutrino beam, *Phys. Rev. Lett.* **112**, 061802 (2014). doi: 10.1103/PhysRevLett.112.061802.
56. Z. Maki, M. Nakagawa, and S. Sakata, Remarks on the unified model of elementary particles, *Prog. Theor. Phys.* **28**, 870 (1962).
57. B. Pontecorvo, Neutrino experiments and the question of leptonic-charge conservation, *Sov. Phys. JETP* **26**, 984–988 (1968).
58. S. Moriyama, New atmospheric and solar results from Super-Kamiokande. XXVII International Conference on Neutrino Physics and Astrophysics (NEUTRINO 2016) 4–9 July 2016.
59. K. Abe *et al.*, Letter of Intent: The Hyper-Kamiokande Experiment — Detector Design and Physics Potential (2011).
60. K. Abe *et al.*, Hyper-Kamiokande design report. KEK Preprint 2016-21, ICRR-Report-701-2016-1 .
61. E. Kearns *et al.*, Hyper-Kamiokande physics opportunities. In *Proceedings, Community Summer Study 2013: Snowmass on the Mississippi (CSS2013): Minneapolis, MN, USA, July 29-August 6, 2013* (2013). URL `https://inspirehep.net/record/1252067/files/arXiv:1309.0184.pdf`.
62. R. Acciarri *et al.*, Long-Baseline Neutrino Facility (LBNF) and Deep Underground Neutrino Experiment (DUNE) (2015).
63. S. Ahmed *et al.*, Physics potential of the ICAL detector at the India-based Neutrino Observatory (INO) (2015).
64. M. Ribordy, and A. Y. Smirnov, Improving the neutrino mass hierarchy identification with inelasticity measurement in PINGU and ORCA, *Phys. Rev.* **D87** (11), 113007 (2013). doi: 10.1103/PhysRevD.87.113007.
65. M. G. Aartsen *et al.*, PINGU: A Vision for Neutrino and Particle Physics at the South Pole (2016).
66. S. Adrin-Martnez *et al.*, Deep sea tests of a prototype of the KM3NeT digital optical module, *Eur. Phys. J.* **C74** (9), 3056 (2014). doi: 10.1140/epjc/s10052-014-3056-3.
67. S. Adrian-Martinez *et al.*, Letter of intent for KM3NeT 2.0, *J. Phys.* **G43** (8), 084001 (2016). doi: 10.1088/0954-3899/43/8/084001.

68. M. C. Gonzalez-Garcia, F. Halzen, M. Maltoni, and H. K. M. Tanaka, Radiography of earth's core and mantle with atmospheric neutrinos, *Phys. Rev. Lett.* **100**, 061802 (2008). doi: 10.1103/PhysRevLett.100.061802.
69. A. M. Dziewonski, and D. L. Anderson, *Phys. Earth Planet. Inter.* **25**, 297 (1981).

Chapter 6

Probing the Atmospheric Sector with Accelerator Experiments

C. Pistillo and C. Wilkinson

Laboratory for High Energy Physics, University of Bern
Sidlerstrasse 5, CH-3012 Bern, Switzerland

1. Introduction

By the early 1990s, results from the Kamiokande[1] and IMB[2] experiments had shown a deficit of atmospheric muon neutrinos relative to expectation, but no corresponding deficit of electron neutrinos. With higher statistics, in 1998 Super-Kamiokande (SK)[3] provided compelling evidence in favor of the neutrino oscillation interpretation of ν_μ disappearance, by showing a clear zenith angle (path length) dependence. The deficit was interpreted as $\nu_\mu \rightarrow \nu_\tau$ oscillation as it had been shown by the CHOOZ experiment[4] that it could not be explained through $\nu_\mu \rightarrow \nu_e$ oscillation. The first long-baseline accelerator neutrino experiments were designed to explore the Kamiokande and IMB anomaly, and later to confirm the interpretation of SK results as being due to energy- and distance-dependent neutrino oscillations.

Atmospheric neutrino experiments suffer from relatively large uncertainties on the zenith angle, and therefore on the distance which neutrinos travel between production and detection, and rely on atmospheric neutrino flux predictions to make measurements of oscillation parameters. The atmospheric neutrino flux is continuous, spans a large energy range, and includes contributions from ν_μ, $\bar{\nu}_\mu$, ν_e and $\bar{\nu}_e$. In contrast, accelerator neutrino beams have been used since the 1960s[5] to produce relatively pure beams of muon neutrinos (or antineutrinos). The beam purity is significant because it allows accelerator experiments to search for the appearance of a neutrino flavor which was completely absent in the original, unoscillated beam. They are directional and have a small uncertainty on the neutrino production point and thus on the distance neutrinos travel before reaching a detector. The range of neutrino energies produced can also be tuned, which allows accelerator neutrino experiments to probe oscillation phenomena characterized by different L/E values and to demonstrate the L/E oscillation signature with a single experiment. The pulsed structure of accelerator neutrino beams means that backgrounds can be reduced by many orders of magnitude through GPS synchronization, and they can be further controlled by studying the detector response with the accelerator turned off.

The dominant oscillation observed by SK had a frequency controlled by $\Delta m^2 \approx 2.5 \times 10^{-3}\,\mathrm{eV}^2$, which suggests a characteristic oscillation for an $L/E \sim 500\,\mathrm{km/GeV}$. Typical neutrino beams have peak energies between $0.5 \leq E_\nu \leq 10\,\mathrm{GeV}$, which sets the baseline of the experiments required to probe the atmospheric sector to be on the order of 100–1000 km. Because neutrinos have such a small interaction cross section, the neutrino beam can simply be directed through the Earth towards a detector without significantly reducing the neutrino flux, which also means that the neutrino flux can be sampled by a second detector close to the neutrino production point without affecting the properties of the beam. The event rate, $R(\vec{\mathbf{x}})$, as a function of any reconstructed variable $\vec{\mathbf{x}}$, measured by a neutrino experiment can be generically written as

$$R(\vec{\mathbf{x}}) = \Phi(E_\nu) \times \sigma(E_\nu, \vec{\mathbf{x}}) \times \epsilon(\vec{\mathbf{x}}) \times N_{\mathrm{Targets}} \times P(\nu_A \to \nu_B), \qquad (1)$$

where $\Phi(E_\nu)$ is the neutrino flux as a function of the neutrino energy E_ν, σ is the neutrino cross section for a particular mode which relates E_ν and $\vec{\mathbf{x}}$, ϵ is the detector efficiency and N_{Targets} is the number of target nuclei in the detector. Accelerator neutrino fluxes fall as a function of $1/L^2$, so the intensity of the neutrino flux is a limiting factor for such experiments. To obtain a useable number of events for oscillation physics, long-baseline accelerator neutrino oscillation experiments require a large detector mass (regardless of the detector technology used), at the kiloton scale, and an intense beam of neutrinos, with data collected over a period of years.

In Section 2, accelerator neutrino beams are described. Early long-baseline experiments, such as K2K,[6] MINOS,[7] OPERA[8] which provided the confirmation of atmospheric neutrino oscillation as $\nu_\mu \to \nu_\tau$ are described in Section 3. Section 4 discusses high-precision measurements of atmospheric mixing parameters by current and future experiments. In Section 5, we offer a summary and provide the outlook for the field.

2. Accelerator neutrino beams

Long-baseline neutrino experiments have utilized several different accelerator neutrino beams,[6,9–11] but these have all been of the "conventional" beam design, where pions and kaons produced by colliding protons with a fixed target are focused with magnetic horns, and allowed to decay to produce muon neutrinos. For illustration, this section will make reference to the Neutrinos at the Main Injector (NuMI) beam at Fermilab,[10] but the basic principles and design considerations are largely the same for all current or historical beams. A comprehensive review of alternative neutrino beam designs, which have been used for other purposes and may prove useful to future experiments but will not be discussed here, can be found in Ref. 12.

The basic layout of the NuMI beamline is shown in Fig. 1. Protons from the primary beamline are extracted and guided onto a fixed graphite target. Secondary pions and kaons produced with the desired charge are focused by a series of magnetic horns. The focused secondaries then decay in flight in the decay pipe to produce a predominantly muon neutrino or muon antineutrino beam, depending on the polarity of the magnetic horns. Hadrons and muons produced in these decays, as well as protons which did not interact in the target, are stopped by a large concrete, aluminium and steel beam dump (the absorber). High-energy muons which pass through the beam dump are sampled by the muon monitors in order to measure the direction and relative intensity of the secondary beam. These muons are then absorbed in the rock upstream of any neutrino detector.

The main injector at Fermilab delivers six bunches of protons to the NuMI target over a short period of $\sim 1.6\mu$s, every ~ 2 s. Having a short delivery time with respect to cycle time is a desirable property for neutrino experiments as events outside the beam window can easily be rejected as the background, and thus the background rate (see Eq. 1) can be reduced by

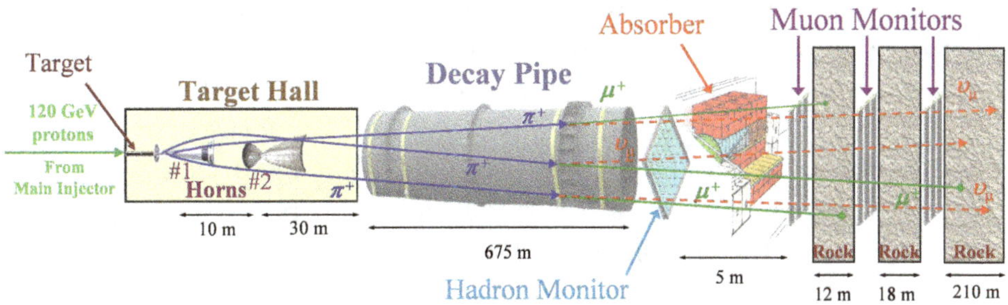

Fig. 1. Schematic of the NuMI beamline. Note that the relevant dimensions are shown, but the components are not to scale. This figure has been reproduced from Ref. 10.

several orders of magnitude. The largest beam-unrelated background sources are cosmic muons, which can be misreconstructed as muons starting within the detector volume and mimic neutrino interactions; low-rate experiments are typically built deep underground because the cosmic muon flux is attenuated as it passes through the Earth. However, in order to achieve a reasonably high event rate, a very intense proton beam is required. Typically, modern experiments operate with beam powers on the order of ~0.1–1 MW at which the intensities of the thermal and mechanical stresses on the target material are large. Ideally, the protons should interact in a small volume of target material. A target which is longer along the beam direction will lead to more primary interactions, but gives more opportunity for secondary particles to reinteract. In particular, the transverse width of the target relative to the beam should be small so that the interaction products, which are produced with some transverse momentum, will leave the target quickly without reinteracting. These are competing requirements; thin targets are less robust, so a compromise is required and some hadronic reinteractions in the thicker target will occur. Simulating the production of secondary hadrons and modeling subsequent interactions within the target is extremely difficult due to limited data, and subsequently a poor understanding of the interaction rates, at the energies typical for neutrino experiments, and is a major source of uncertainty for accelerator produced neutrino beams. For a review of the models used in these simulations, and data taken by dedicated experiments to constrain these models, see Ref. 12.

The secondaries which are focused and decay to produce the neutrino beam are mostly charged pions, with a small proportion of charged and neutral kaons (~10%). Charged pions decay via $\pi \rightarrow \mu\nu_\mu$ 99.9877% of the time,[13] and only a small fraction,[a] $1.23 \times 10^{-4}\%$ decay electronically via $\pi \rightarrow e\nu_e$. Electron neutrino contamination in the beam also comes from charged and neutral kaon decays and from tertiary muons which decay via $\mu \rightarrow e\nu_\mu\nu_e$.[13] Typical beam compositions for neutrino mode running, when positively charged secondaries are focused, is for 90–95% ν_μ, 5–10% $\bar{\nu}_\mu$, ~1–2% $\nu_e + \bar{\nu}_e$. In antineutrino mode, when negatively charged secondaries are focused, there is a significantly larger neutrino contamination due to the larger fraction of positively charged particles in the unfocused beam, giving typical beam compositions of 70–80% $\bar{\nu}_\mu$, 20–30% ν_μ, and ~1–2% $\nu_e + \bar{\nu}_e$.

Electromagnetic horns[15] are used to focus the charged hadrons produced in the target and thereby significantly increase the neutrino flux. Whilst the

[a]The electronic decay mode of the pion is helicity suppressed due to the relatively small electron mass. A full explanation can be found in, for example, Ref. 14.

flux increase depends on the experimental setup and the desired neutrino beam spectrum, the order of magnitude flux increases is typical.[12] They are composed of two axially symmetric conductors. When current passes along the inner conductor and returns along the outer conductor, a toroidal magnetic field is generated between them. The field is usually pulsed to coincide with the proton beam extraction due to the high currents required $O(100\ \text{kA})$ to produce the necessary magnetic field strength. The shape of the magnetic horns means that secondary particles produced in the target will pass through the magnetic field for different periods of time and experience more or less force. In this way, the horns act as a lens, focusing particles of a given charge and defocusing particles with the opposite charge. By modifying the shape of the horn, different particle energies are focused. The sign of the particles focused by the horn can be changed by changing the direction of the current through the conductors. Experiments often have more than one horn, with subsequent horns used to further focus the beam. The NuMI two-horn system[10] is illustrated in Fig. 2 and example particle trajectories illustrate how the two horns work together. Particles which pass through the "neck" of a horn do not pass through the magnetic field and so are unaffected; particles which were not focused by the first horn, or were either over or under focused, are corrected by the second horn. Higher-energy secondary particles are produced in a more forward direction, from which it follows that the average energy of the focused secondary beam, and consequently the average energy

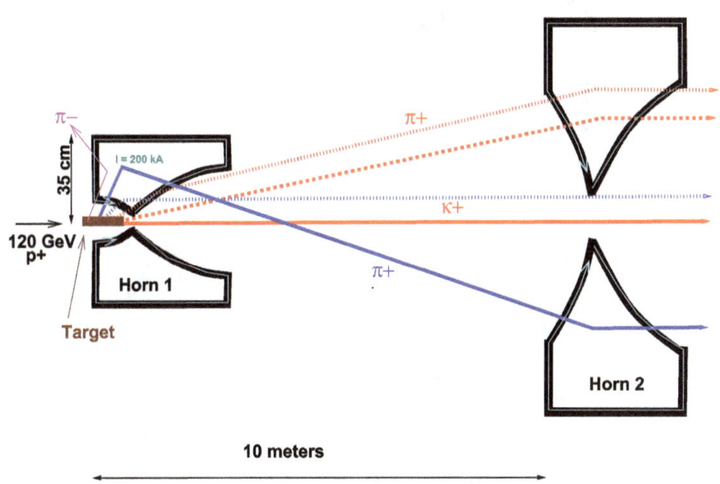

Fig. 2. Diagram of possible hadron trajectories through the NuMI magnetic horns. Solid red: along the central axis of both horns; dashed red: focused by horn 2 only; dotted red: underfocused by horn 1 but corrected by horn 2; dashed blue: focused by horn 1 only; solid blue: overfocused by horn 1 but corrected by horn 2. This figure has been reproduced from Ref. 10.

of the resultant neutrino beam, can be changed by moving the target position relative to the first horn and by changing the distance between subsequent horns. If the target is moved further upstream of the first horn, higher-energy hadrons are focused which would otherwise have passed through its neck. It means that on increasing the distance between subsequent horns, high-energy hadrons which pass through the neck of the first horn are corrected by subsequent horns. Although typically these considerations affect the design of the horn system, the NuMI beam has a moveable target and moveable second horn position to allow different modes of operation, as illustrated in Fig. 3. The reason for this flexibility is historical; when the NuMI beamline was designed and built, the value of Δm^2_{32} was not well known, so some flexibility in the energy spectrum of the beam was desirable. When MINOS, the first experiment to use the NuMI beam started running, the low energy (LE) configuration was used as it was the most sensitive to the Δm^2_{32} values which had been measured by SK at that point. Later, the NOvA experiment exploited this feature of the NuMI beamline and was designed to use the medium energy (ME) setting. The different focusing

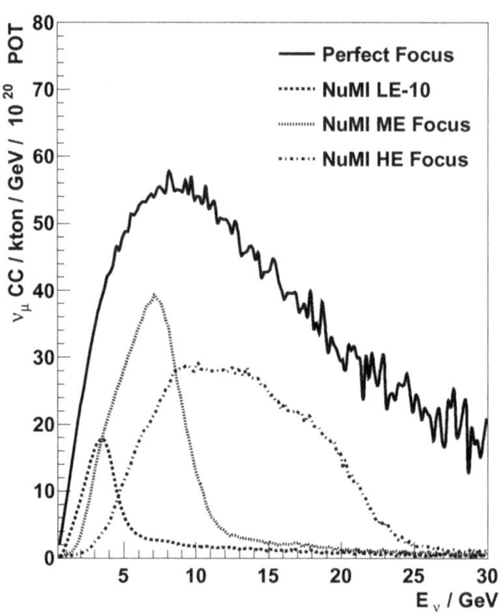

Fig. 3. Predicted neutrino spectra for different NuMI focusing schemes. The difference between "perfect" focusing, with all secondary pions and kaons focused along the beam axis, and the low-, medium- and high-energy NuMI configurations are shown. Note that the predicted event rate is shown for an on-axis detector, so the flux has been convolved with the neutrino charged-current (CC) cross section. This figure has been reproduced from Ref. 10.

schemes highlight the significance of the magnetic horns focusing for the final neutrino beam.

The energy distribution of the neutrino beam can also be tuned by placing the detectors off-axis, away from the central axis of the beam, which results in a lower peak neutrino energy and a narrower beam.[16] The off-axis technique exploits the two-body kinematics of the pion decays which contribute the majority of the neutrino flux. Because there is a maximum transverse momentum which the neutrino can have, very high momentum neutrinos will be strongly boosted along the beam axis, and they do not contribute to the neutrino flux at large angles. By tuning the off-axis angle of a neutrino beam, it is possible to make the neutrino flux peak in a narrow band around the region where the largest oscillation effect is expected. In particular, this technique is useful for experiments which are interesting in observing $\nu_\mu \to \nu_e$ appearance as it shows maximal ν_μ disappearance and ν_e appearance, whilst suppressing higher-energy backgrounds which mimic ν_e events, such as neutral-current (NC) π^0 production. This technique has been used by the T2K experiment,[17] which uses the JPARC neutrino beam,[9] as illustrated in Fig. 4. The baseline for the T2K experiment was set at 295 km by the distance between the existing JPARC beamline and the SK detector (for full details, see Section 4.1). At different off-axis angles, the

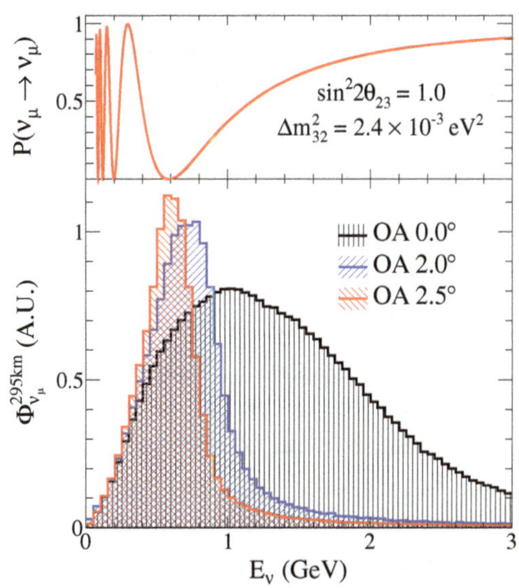

Fig. 4. The tuning of the off-axis angle for the T2K experiment given the expected oscillation probability. The top plot shows the oscillation probability for an approximate set of oscillation parameters. The bottom plot shows the area normalized neutrino fluxes for different values of the off-axis angle. This figure has been reproduced from Ref. 9.

peak neutrino energy takes different values and can be tuned to coincide with the maximum expected oscillation probability. The off-axis spectrum is also significantly narrower, with the majority of the flux in a region where large oscillation effects are expected.

3. Confirmation of atmospheric neutrino oscillation with accelerator experiments

By the mid-1990s, results from Kamiokande[1] and IMB[2] had demonstrated a deficit in the ν_μ/ν_e flux ratio relative to expectation. Early interpretations of these anomalies in terms of neutrino oscillations[18] suggested that the oscillations were driven by mass splittings on the order of $|\Delta m^2| \approx 10^{-2}$ eV^2, somewhat higher than the subsequent results from SK, which meant that the accelerator experiments were typically tuned to have a peak neutrino energy larger than the oscillation maximum. To probe the atmospheric sector with accelerator experiments, three neutrino beamlines were commissioned in Japan (KEK), the United States (NuMI) and Europe (CNGS) in the last decade of the 20th century.

Early long-baseline accelerator experiments can be split into two complementary categories: K2K[6] (in Japan) and MINOS[7] (in the United States) were ν_μ-disappearance experiments which aimed to make measurements of the atmospheric oscillation parameters, Δm^2_{32} and θ_{23}; OPERA[8] (in Europe) was designed to confirm that muon neutrinos oscillate predominantly into τ neutrinos, through the direct observation of ν_τ CC interactions. It would have been difficult to achieve both aims in a single experiment because of several design constraints. To measure ν_μ disappearance, it is important to have an L/E value near the oscillation maximum to observe the characteristic oscillation dip. As the accelerator neutrino flux falls as a function of the distance $\sim 1/L^2$ and the neutrino cross section increases approximately linearly with energy, in order to maximize the event rate at a given L/E value, it is desirable to have as short a baseline as possible, with relatively low energy. In this context, given the small mass splitting, the energy is $O(1\,\mathrm{GeV})$, and the baseline is still $O(100\,\mathrm{km})$. This is well below the τ production threshold ($E_\nu \gtrsim 3.5\,\mathrm{GeV}$), making experiments optimized to measure ν_μ disappearance unsuitable for measuring ν_τ appearance. For OPERA, the L/E value was less crucial. As it was an almost background-free experiment, the detection of even a few ν_τ neutrinos would have been sufficient to demonstrate ν_τ appearance. As a result, it was not necessary to be near the oscillation maximum, and so a higher neutrino beam energy, $O(10\ \mathrm{GeV})$, could be used without having to compensate with a longer baseline. Additionally, the decrease in the appearance probability by being further from the oscillation maximum was compensated for by the increase of the ν_τ interaction cross section.

In this section, oscillation results from the experiments are quoted in terms of $|\Delta m^2|$ and $\sin^2 2\theta$, as in the original publications. These experiments generally used the two-flavor oscillation formula when extracting oscillation parameter values. The survival probability for two-flavor mixing is

$$P(\nu_\mu \to \nu_\mu) = 1 - \sin^2(2\theta)\sin^2\left(\frac{1.27\Delta m^2 L(\mathrm{km})}{E_\nu(\mathrm{GeV})}\right), \tag{2}$$

which is insensitive to the sign of Δm^2, so the absolute value, $|\Delta m^2|$, was given. Modern experiments (as described in Section 4) use the full three-flavor formula when extracting results, and give results separately for the normal and inverted hierarchies. Additionally, $\sin^2\theta$ is a more interesting quantity to measure in the three-neutrino formalism than $\sin^2 2\theta$, so the results of old and new accelerator experiments have to be compared with care. It is worth noting that the degree of approximation when using the two-flavor formula is on the order of a few percent (because of the smallness of θ_{13}), considerably smaller than the sensitivity of these experiments.

3.1. *K2K*

KEK to Kamioka (K2K) was the first long-baseline accelerator neutrino experiment, and it collected data between 1999 and 2004. Muon neutrinos were generated by a 12 GeV proton synchrotron accelerator at the High Energy Accelerator Research Organization (KEK) in Tsukuba and directed toward the far detector, SK, 250 km away from KEK. The peak neutrino energy of the KEK beam was around 1 GeV, and the mean neutrino energy was around 1.4 GeV. This is somewhat higher in energy than the oscillation maximum for the 250 km baseline, but as it has been remarked, this reflects the uncertainty on the oscillation parameters at the time and also practical constraints due to the location of the accelerator and far detector. A near detector was located 300 m downstream from the target inside the KEK laboratory and was used to sample the unoscillated beam. Additionally, a gas-Cherenkov detector was located between the horn system and the decay volume to measure the kinematics of the pions after production and focusing. A muon monitor composed of ionization chambers and silicon pad detectors was located behind the beam dump and designed to measure the beam profile and intensity.

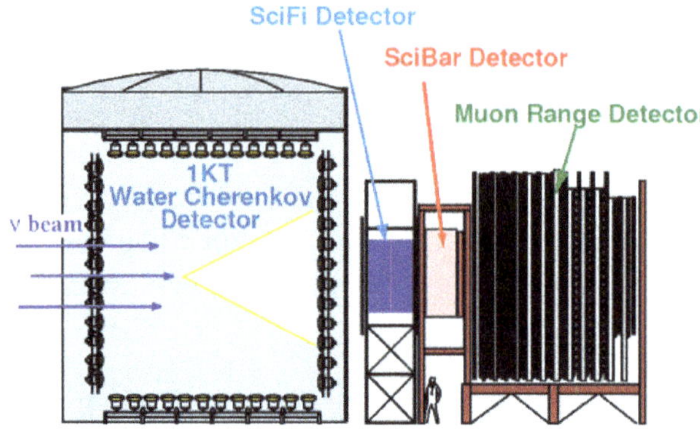

Fig. 5. A representation of the K2K near detectors during the K2K-IIb running period. A fully active tracking plastic scintillator, SciBar, replaced the lead glass calorimeter used in K2K-I. This figure has been reproduced from Ref. 6.

The K2K near detector was composed of a collection of subdetectors with complementary abilities as shown in Fig. 5. It included a 1 kt water-Cherenkov detector which acted as a miniature SK used to constrain the muon neutrino flux and to measure the interaction rate. A series of fine-grained detectors were positioned downstream of the water-Cherenkov detector and were used to resolve particular interaction modes and constrain the cross section model. They also served to monitor the beam direction and to characterize its spectrum and composition. A full description of the near detectors used can be found in Ref. 6.

With the total accumulated statistics of 8.9×10^{19} protons on target, 112 neutrino events were observed in the far detector, compared with a prediction of $158.1^{+9.2}_{-8.6}$ for the no-oscillation case.[6] From the analysis of 58 single-ring μ-like events, a clear energy spectrum deformation was observed in the energy region below 1 GeV, as shown in Fig. 6.

In K2K, as in the other long-baseline experiments, the extraction of oscillation parameters was done by minimizing a likelihood of the form

$$\mathcal{L}(\Delta m^2, \sin^2 2\theta, \vec{\mathbf{f}}) = \mathcal{L}_{\text{norm}}(\Delta m^2, \sin^2 2\theta, \vec{\mathbf{f}}) \times \mathcal{L}_{\text{shape}}(\Delta m^2, \sin^2 2\theta, \vec{\mathbf{f}}) \times \mathcal{L}_{\text{syst}}(\vec{\mathbf{f}}), \quad (3)$$

where the normalization term $\mathcal{L}_{\text{norm}}$ is the Poisson probability to observe N_{obs} events, given the predicted number of events for a set of oscillation and systematic parameters, $N_{\text{exp}}(\Delta m^2, \sin^2 2\theta, \vec{\mathbf{f}})$. The shape term $\mathcal{L}_{\text{shape}}$ is given by the product of the probabilities for each detected single-ring μ-like event, $N_{1\mu}$,

$$\mathcal{L}_{\text{shape}}(\Delta m^2, \sin^2 2\theta, \vec{\mathbf{f}}) = \prod_{i=1}^{N_{1\mu}} p_i(E_i; \Delta m^2, \sin^2 2\theta, \vec{\mathbf{f}}), \quad (4)$$

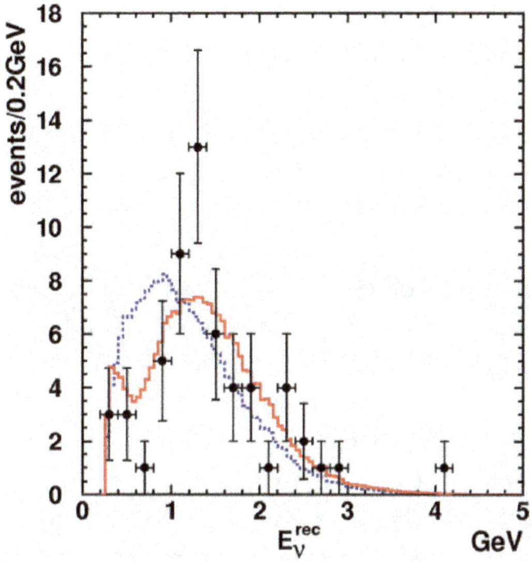

Fig. 6. The neutrino reconstructed energy distribution for the single-ring μ-like sample in K2K. Data are represented by points, the solid line is the best fit spectrum with neutrino oscillation, and the dashed line is the expectation without oscillation. This figure has been reproduced from Ref. 6.

where p_i is the probability that the ith event has reconstructed neutrino energy E_i.[b] The various sources of systematic error, such as the far/near flux ratio, flux and cross section errors, the absolute energy scale and efficiency in SK, including correlations, are denoted $\vec{\mathbf{f}}$ and are included in the $\mathcal{L}_{\text{syst}}$ penalty term.

K2K reported 4.3σ evidence of ν_μ disappearance,[6] with a best-fit value of $(|\Delta m^2|, \sin^2 2\theta) = (2.8 \times 10^{-3}\ \text{eV}^2, 1)$ with a 90% confidence limit of $1.9 \leq |\Delta m^2| \leq 3.5 \times 10^{-3}\,\text{eV}^2$ for the neutrino mass splitting, given $\sin^2 2\theta = 1$.

3.2. *MINOS*

The Main Injector Neutrino Oscillation Search (MINOS) experiment[7] started taking data in 2005, shortly after K2K stopped running, and ran until 2012. MINOS used the NuMI beam at the Fermilab National Accelerator Laboratory in Illinois and sampled the beam with a 0.98 kt near detector and a 5.4 kt far detector, located 735 km from the target in the Soudan mine, Minnesota. The near and far detectors were sampling calorimeters

[b]K2K reconstructed the neutrino energy assuming two-body kinematics of $\nu_\mu + n \to \mu^- + p$ interactions, which will be discussed further in the context of T2K in Section 4.1. Assumptions made in this method were not critical for K2K as it was largely a counting experiment.

designed to be as similar as possible and were composed of magnetized steel plates interleaved with scintillator planes which provided three-dimensional tracking, good energy resolution, and the ability to distinguish events with different final-state particle content. A third detector was also built for test-beam studies at CERN and used for calibration. The NuMI beam was designed to operate at several different energies, with different neutrino energy spectra as described in Section 2. This was motivated by the uncertainty in the value of the neutrino oscillation parameters when the beamline was being designed. As SK indicated a smaller mass splitting than what had been suggested by Kamiokande and IMB, MINOS ran exclusively in the low-energy mode in order to maximize its sensitivity, with a neutrino energy of 3 GeV at the flux peak.

MINOS collected a total of 10.71×10^{20} protons on target (POT) with neutrino mode running and 3.36×10^{20} POT in antineutrino mode. Additionally, they collected 37.88 kt-yr of atmospheric neutrinos using their far detector and included the atmospheric neutrino samples in all results using the full dataset. Intermediate results were given using 7.25×10^{20} POT of neutrino mode running,[19] which had around 4000 CC neutrino interactions in the far detector sample. The reconstructed neutrino energy spectra for this intermediate sample is shown in Fig. 7, and the best-fit oscillation and no oscillation predictions are shown for comparison. The ratio between the observed event rate and the unoscillated expectation is also shown, which demonstrates the energy dependence of the oscillation probability (given the fixed baseline). The disappearance effect is striking, although it is also obvious that the oscillation maximum is around the lower energy limit of the MINOS flux, and a more sensitive measurement could be achieved with a lower energy in the same baseline (which partially motivated the NOvA experiment described in Section 4.2).

The final MINOS results on the atmospheric oscillation parameters with the two-flavor formalism used a combined fit including the neutrino mode, antineutrino mode and atmospheric neutrino samples to get the best constraint possible.[20] Assuming that the same oscillation parameters govern ν_μ disappearance and $\bar{\nu}_\mu$ disappearance gave $|\Delta m^2| = \left(2.41^{+0.09}_{-0.10}\right) \times 10^{-3}\,\mathrm{eV}^2$ and $\sin^2(2\theta) = 0.950^{+0.035}_{-0.036}$. Additionally, a fit to antineutrino data only gave $|\Delta\overline{m}^2| = \left(2.50^{+0.23}_{-0.25}\right) \times 10^{-3}\,\mathrm{eV}^2$ and $\sin^2(2\overline{\theta}) = 0.97^{+0.03}_{-0.08}$, which does not indicate any tension between the neutrino and antineutrino samples, although the statistics are limited in the antineutrino case. The 90% confidence intervals from the final two-flavor results are shown in Fig. 8. The final MINOS dataset was later used in a full three-flavor oscillation fit, including

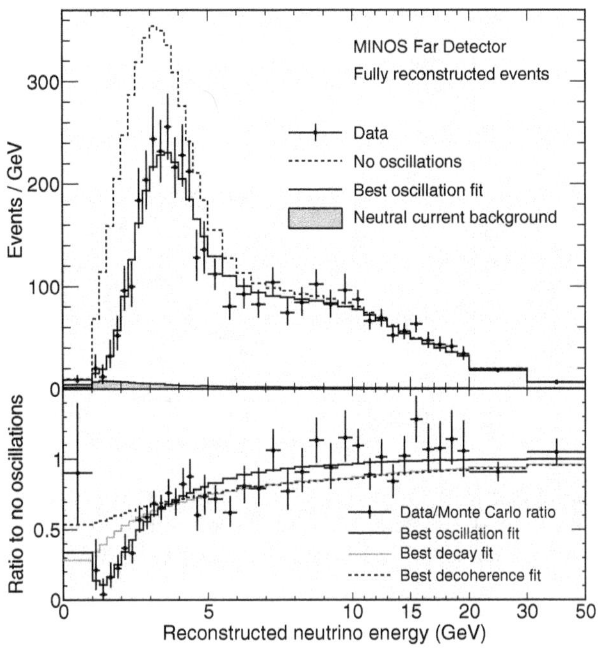

Fig. 7. Top: reconstructed neutrino energy spectrum for CC ν_μ-like events in the MINOS far detector. The solid line represents the best fit of the oscillation hypothesis; the dashed line is extrapolated near detector prediction assuming no oscillations. The simulated background due to NC interactions is shown as the shaded area. Bottom: ratio between data and the no-oscillation hypothesis. The best fits for oscillations, neutrino decay and decoherence effects (introduced as a possible explanation for the atmospheric neutrino anomaly) are also reported. This figure has been reproduced from Ref. 19.

the ν_e-appearance and $\bar{\nu}_e$-appearance samples.[21] The three-flavor results were $2.28 \leq \Delta m^2_{32} \leq 2.46 \times 10^{-3}$ eV^2 and $0.35 \leq \sin^2\theta_{23} \leq 0.65$ ($2.32 \leq \Delta m^2_{32} \leq 2.53 \times 10^{-3}$ eV^2 and $0.34 \leq \sin^2\theta_{23} \leq 0.67$) at 90% confidence for the normal (inverted) hierarchy. The three-flavor results will be used for comparisons with more recent measurements in subsequent sections.

3.3. *OPERA*

To complement ν_μ-disappearance measurements of the atmospheric oscillation parameters from K2K and MINOS, the European program focused on making a ν_τ-appearance measurement with the CERN Neutrino to Gran Sasso (CNGS) beam. The aim was to conclusively prove that the ν_μ disappearance observed by the atmospheric experiments could be explained by $\nu_\mu \to \nu_\tau$, rather than some more exotic scenario. OPERA[8] was proposed in 2000 to sample the CNGS beam 732 km away from CERN, in the Gran Sasso INFN laboratory in Italy, and collected data from 2008 to 2012.

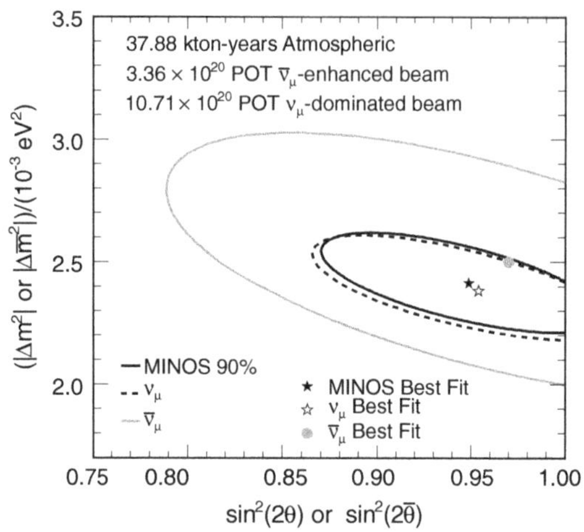

Fig. 8. Best-fit value and likelihood contours for mass splitting and mixing angle for the final MINOS two-flavor analysis. Results are shown for a fit to ν_μ only data, $\bar{\nu}_\mu$ only data, and their combined fit. This figure has been reproduced from Ref. 20.

The primary goal of OPERA was to detect τ leptons produced through charged current ν_τ interactions from the oscillated beam. As previously remarked, the process has a threshold of around 3.5 GeV, which would require a baseline of several thousand kilometers to be at the oscillation maximum. However, the OPERA design had to maximize the number of expected ν_τ interactions, so the cross section, which increases with the energy, as well as the oscillation probability, had to be considered. The CNGS beam energy was tuned to maximize the number of detected ν_τ by matching the neutrino flux to the convolution of the interaction cross section with the oscillation probability, well above the tau production threshold, and far away from the oscillation maximum (for a baseline of 732 km), with a mean neutrino energy of $\sim$17 GeV.

OPERA required a kiloton-scale target in order to have a sufficiently high event rate to make a measurement possible, which would still only be a few to a few tens of events, and had to be relatively background free in order to claim discovery of $\nu_\mu \to \nu_\tau$ oscillations with only a handful of τ leptons observed. OPERA also required a very high position resolution as the mean path length of τ leptons produced by the CNGS beam would only be a few hundred micrometers. The solution was to use emulsion cloud chambers (ECCs), which had already been used successfully in the DONUT experiment to observe ν_τ interactions for the first time.[22] ECC bricks are a sandwich of passive lead layers to provide target mass, and nuclear emulsion layers which provide micrometric tracking capabilities. The final OPERA design[8]

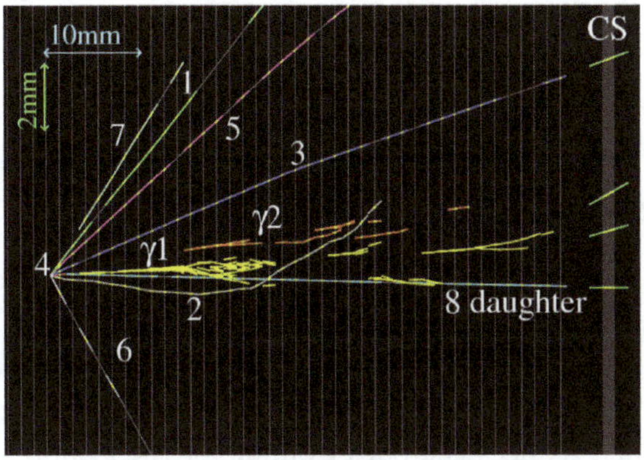

Fig. 9. The first ν_τ candidate in OPERA. The primary neutrino interaction produces seven tracks of which one (labeled 4) exhibits a "kink" topology and is identified as the tau lepton. This figure has been reproduced from Ref. 23.

was for a hybrid nuclear emulsion and electronic detector, where a total of 150,000 ECC bricks were organized in walls perpendicular to the incoming neutrino beam direction to provide the target mass of ~1.35 kt. The walls of ECC bricks were separated by scintillator planes to identify those bricks in which interactions had occurred and a muon spectrometer to measure muon charge and momentum. When a neutrino interaction was triggered by the electronic detectors, the corresponding brick was extracted and its emulsion sheets sent to laboratories equipped with automatic scanning microscopes for full event characterization.

Out of 5408 fully reconstructed neutrino interactions in the total OPERA exposure of 1.8×10^{20} protons on target, five ν_τ candidates were indentified. A picture of the first event is shown in Fig. 9, which was interpreted as a $\tau^- \to \rho^- \nu_\tau$ decay where the ρ^- subsequently decays into a π^0 and a π^-. For $(|\Delta m^2|, \sin^2 2\theta_{23}) = (2.44 \times 10^{-3}\ \text{eV}^2, 1)$, 2.64 ± 0.53 signal interactions were expected, with a background of 0.25 ± 0.05. The observation of 5 ν_τ candidates is therefore interpreted as evidence of ν_τ-appearance in a muon neutrino beam at 5.1σ.[23]

4. Precision measurements of the atmospheric sector

Long-baseline accelerator neutrino experiments built to address other areas of neutrino oscillation physics are also able to make precision measurements of the atmospheric mixing parameters θ_{23} and Δm^2_{32}. Indeed, in a muon neutrino (or antineutrino) beam, θ_{23} and Δm^2_{32} are convolved with other oscillation parameters of interest, and the strongest constraints come from a joint fit to both muon and electron samples. Experiments looking for electron neutrino appearance, CP violation or attempting to determine the mass hierarchy or the θ_{23} octant, must constrain these terms to avoid large uncertainties in other measurements.

Two experiments, T2K and NOvA, which are both taking data at the time of writing are already placing stronger limits on θ_{23} and Δm^2_{32} than atmospheric or previous long-baseline results (described in Section 3), despite having only collected a fraction of their design statistics. Both experiments were designed with the primary goal of measuring ν_e appearance in a muon neutrino beam. In order to maximize sensitivity to the subdominant oscillation channel, both experiments utilize the off-axis technique described in Section 2, which reduces the backgrounds for the ν_e appearance search by suppressing the high-energy tail. As they were planned after the confirmation of neutrino oscillation by SK, and after initial data was taken by previous long-baseline experiments, they were able to tune their L/E by changing the off-axis angle and thereby changing the peak neutrino energy of their narrow band beam.

4.1. *T2K*

The T2K experiment[17] uses the JPARC accelerator neutrino beam,[9] directed 2.5° off-axis from its detectors. This gives a narrow band beam with a flux peak of ~600 MeV, tuned to be close to the predicted oscillation maximum probability (as shown in Fig. 4). SK is used as a far detector,[24] and is located 295 km downstream of the target. It is a 50 kt water-Cherenkov detector located 1 km underground. The off-axis near detector is a magnetized tracking detector, with both carbon and water targets, which allows T2K to tune its flux and cross section model accurately, and can separate the neutrino and antineutrino flux components. Additionally, there is an on-axis near detector which monitors the neutrino beam position, intensity and axial symmetry.

Because the flux peak is at a relatively low energy, the majority of CC interactions at T2K are quasielastic, $\nu_l + n \to l^- + p$ (CCQE). By using the two-body kinematics of the interaction, and assuming that the initial state nucleon is at rest, the neutrino energy can be inferred from the charged-lepton kinematics

$$E_\nu^{\mathrm{QE}} = \frac{2M'_n E_\mu - (M'^2_n + m^2_\mu - M^2_p)}{2(M'_n - E_\mu + \sqrt{E^2_\mu - m^2_\mu}\cos\theta_\mu)}, \tag{5}$$

where E_μ is the muon energy; M_n, M_p and m_μ are the neutron, proton and muon masses, respectively; and $M'_n = M_n - V$, where V is a constant

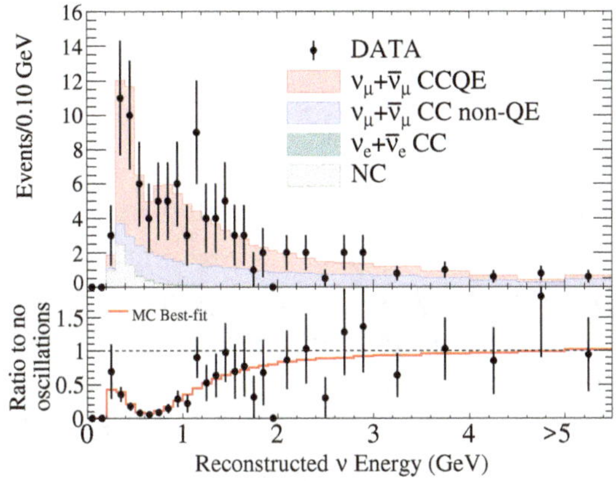

Fig. 10. Reconstructed energy spectrum for ν_μ candidate events at SK. The observed spectrum (black points) and best-fit oscillated predictions (filled histograms) are shown (top). The ratio of observed SK candidate events to the expected unoscillated prediction (black points) along with the best-fit oscillation results is also shown (bottom). This figure has been reproduced from Ref. 25.

binding energy. In reality, the initial state nucleon is not at rest and the binding energy is not a constant, so the equation is approximate, and the reconstructed E_ν^{QE} distribution is smeared with respect to true E_ν. There are also non-CCQE events which are part of the single-ring sample and which also contribute to this smearing, as shown in Fig. 10. Because this smearing "fills in" the characteristic oscillation dip, a good understanding of nuclear effects is essential to produce reliable oscillation results.

The most precise T2K measurement of the oscillation parameters which govern ν_μ disappearance at the time of writing were obtained operating in neutrino mode with 6.6×10^{20} protons on target, which is ~8% of the total planned exposure.[25,26] The reconstructed neutrino energy spectrum for ν_μ candidate events at SK is shown in Fig. 10, along with the ratio of the observed events to the unoscillated prediction. The best-fit oscillation parameters for the normal (inverted) hierarchy are $\Delta m^2_{32} = 2.51 \pm 0.10 \times 10^{-3}\,\text{eV}^2$ ($\Delta m^2_{32} = -2.48 \pm 0.10 \times 10^{-3}$) and $\sin^2\theta_{23} = 0.514^{+0.055}_{-0.056}$ ($\sin^2\theta_{23} = 0.511 \pm 0.055$). The T2K measurement uncertainty on $\sin^2\theta_{23}$ is already smaller than that from atmospheric or previous long-baseline experiments and is expected to reach a precision of ~10%.[27] The measurement of Δm^2_{32} (or Δm^2_{31}) is reaching the same level of precision as MINOS, and with full statistics will reach ~2% precision.[27]

T2K has also released initial results in antineutrino mode for $\bar{\nu}_\mu$ disappearance with 4.01×10^{20} protons on target.[28] The best fit oscillation values of $\sin^2\overline{\theta}_{23} = 0.45$, and $|\Delta \overline{m}^2_{32}| = 2.51 \times 10^{-3}\,\text{eV}^2$ with 68% confidence intervals of $0.38 \leq \sin^2 \overline{\theta}_{23} \leq 0.68$ and $2.26 \leq |\Delta \overline{m}^2_{32}| \leq 2.80 \times 10^{-3}\,\text{eV}^2$ are consistent with the T2K neutrino mode results but have larger uncertainties.

4.2. *NOvA and MINOS+*

The NuMI beamline at Fermilab was reconfigured to operate in medium-energy mode for the NOvA experiment (see Fig. 3) with a peak on-axis neutrino energy of ~7 GeV. The NOvA experiment[29] uses two segmented tracking calorimeter detectors comprised of isolated cells filled with liquid scintillator: a 290-t near detector located 1 km away from the target and a 14-kt far detector 810 km downstream of the target. Both detectors are placed 14.6 mrad off-axis, which results in a narrow band neutrino flux peaked at 2 GeV. As for the MINOS experiment, using the same detector design for the near and far detectors helps to reduce systematics.

Because the flux peak is at a higher neutrino energy than T2K, the event rate is not dominated by CCQE ($\nu_l + n \to l^- + p$, $\bar{\nu}_l + p \to l^+ + n$) interactions, so the neutrino energy reconstruction cannot use Eq. 5. Instead, the signal is defined to be all CC interactions, and the neutrino energy is reconstructed with

$$E_\nu^{\text{rec}} = E_\mu + E_{\text{had}}, \tag{6}$$

where E_μ is the reconstructed energy of the primary muon track, and E_{had} is the total calorimetric energy which is not attributed to the muon track. As for T2K, this method relies on a good understanding of the signal cross section model, which is particularly difficult as the NOvA flux covers the transition region between quasielastic, resonant and deep inelastic scattering events. However, unlike for water-Cherenkov detectors, E_{had} is reasonably well reconstructed for any number of particles produced in the neutrino interaction, which partially mitigates the problem of a badly understood cross section model.

Intermediate neutrino mode results using approximately 20% of the planned NOvA exposure have been reported.[30] The reconstructed neutrino energy distribution for selected far detector events is shown in Fig. 11. As for T2K, the disappearance of muon neutrinos is striking, with 78 ν_μ candidate events selected compared with an unoscillated prediction of 473 ± 30 events. The measured oscillation parameter values assuming the normal (inverted) hierarchy are $\Delta m^2_{32} = 2.67 \pm 0.11 \times 10^{-3}\,\text{eV}^2$ ($\Delta m^2_{32} = -2.72 \pm 0.11 \times 10^3$) and the statistically degenerate values $\sin^2\theta_{23} = 0.404^{+0.030}_{-0.022}$ and $\sin^2\theta_{23} = 0.624^{+0.022}_{-0.030}$ ($\sin^2\theta_{23} = 0.398^{+0.030}_{-0.022}$ and $\sin^2\theta_{23} = 0.618^{+0.022}_{-0.030}$) at 68% confidence. These results are shown in Fig. 12 for the normal hierarchy.

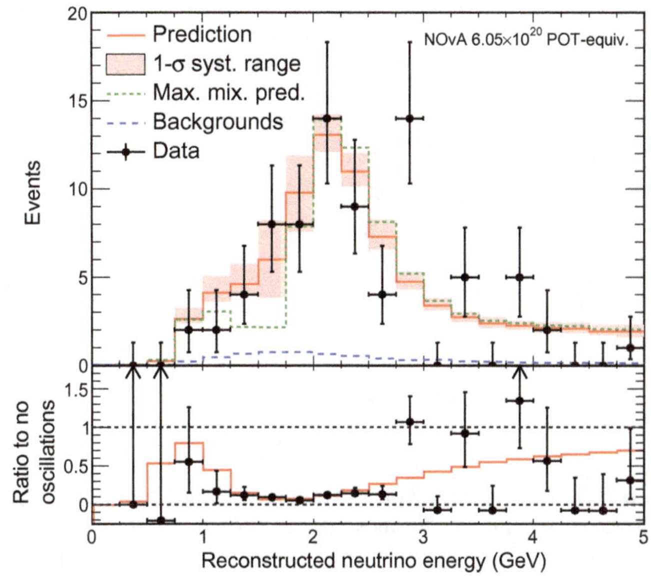

Fig. 11. The reconstructed neutrino energy for NOvA's far detector ν_μ candidate events (black points) and the best fit prediction (red line with shaded uncertainty). The beam and cosmic backgrounds are also shown (blue dashed line), along with the maximal mixing prediction (green short dashed line). The ratio to the unoscillated prediction is also shown. This figure has been reproduced from Ref. 30.

This NOvA result provides the first evidence that ν_μ disappearance is not maximal ($\theta_{23} \neq 45°$) at 2.6σ confidence.

MINOS+ is a rechristening of MINOS, with upgraded detector electronics to deal with the higher-energy events of the NuMI medium-energy beam.[31] Although its broad physics program includes improving the Δm^2_{32}, $\sin^2\theta_{23}$ contours from MINOS, the higher energies of the on-axis beam make it significantly less competitive than NOvA. The majority of the events will be in the high-energy tail for which the three-flavor neutrino oscillation probabilties are small. The main focus of MINOS+ is on sterile neutrinos; however, it will contribute to the understanding of the NuMI beam, which will improve the NOvA results. At the time of writing, MINOS+ has finished collecting data, but it has not yet produced final results.

5. Summary and outlook

Accelerator neutrino beams have proven to be powerful tools for neutrino oscillation physics over the past 20 years. They produce a high-purity, tightly focused, beam of muon neutrinos, with a relatively narrow and well-understood neutrino energy distribution, which can be directed as desired.

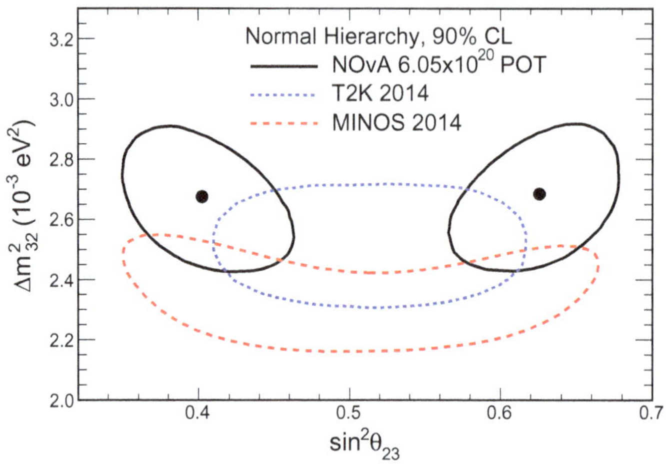

Fig. 12. The best fit NOvA data points are shown (black points) along with the 90% confidence contour (black line). The most recent 90% contours of T2K[25,26] (blue short dashed line) and MINOS[21] (red dashed line) are also shown for comparison. This figure has been reproduced from Ref. 30.

Using accelerator beams, the first generation of long-baseline neutrino oscillation experiments, K2K, MINOS and OPERA, provided comfirmation of atmospheric neutrino mixing as $\nu_\mu \to \nu_\tau$ oscillation and have made high statistics measurements of the oscillation parameters Δm^2_{32} and θ_{23}. The second generation of long-baseline experiments, T2K and NOvA, has benefitted from the improved understanding of oscillation parameters to tune their L/E to be where the oscillation effect is maximal. Although designed primarily to look at subleading order oscillation effects, they are already producing more precise measurements of the atmsopheric parameters with only a fraction of their planned exposure. As we move into the precision era of neutrino oscillation physics, these results will no longer be dominated by low statistics, and a better control of systematic errors will become very important. In particular, the reconstructed neutrino energy spectrum depends heavily on a good understanding of neutrino cross sections, which are likely to become a limiting factor for future experiments, although the method for reconstructing the energy differs depending on the detector design (Eqs. (5) and (6)).

The current limits on the atmospheric mixing parameters Δm^2_{32} and θ_{23} are summarized in Fig. 13, although the most recent NOvA results are not included (shown in Fig. 12). It is clear that accelerator neutrino experiments provide stronger constraints on both mixing parameters than recent atmospheric neutrino analyses (IceCube, SK). In particular, T2K and NOvA have produced the most precise measurements to date, with

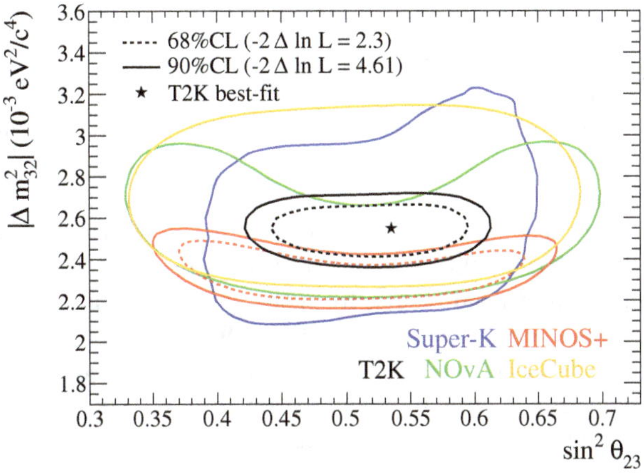

Fig. 13. The most recent results for the atmospheric mixing parameters Δm^2_{32} and θ_{23} are shown for a number of experiments. The best-fit T2K value is shown (black star) along with the 68% and 90% confidence contour (dashed and solid black lines),[32] along with the MINOS+ 68% and 90% contours (dashed and solid red lines),[21] the 90% Super-Kamiokande contour (blue line),[33] the 90% IceCube contour (yellow line),[34] and the 90% NOvA contour (green line).[35] Note that the NOvA results shown used a smaller dataset than those shown in Fig. 12. This figure has been reproduced from Ref. 32.

only a fraction of their expected data collected at the time of writing. The most recent NOvA results[30] show some tension with T2K results and have provided the first evidence that $\theta_{23} \neq 45°$. Further interesting results are expected in the near future.

Future experiments which are currently being designed to test other aspects of neutrino oscillation phenomenology, such as the determination of the mass hierarchy and CP violation in the leptonic sector, will further improve the knowledge of the atmospheric parameters Δm^2_{32} and θ_{23}. These experiments will benefit from more powerful facilities with beam powers at the MW scale, and new detector technologies, but require further work on theoretical and experimental systematic errors. Improved measurements of these parameters is not just a by-product of these experiments but an essential part of their physics program because of the complex interplay between parameters in the three-neutrino oscillation probability. Despite the improved measurements from recent experiments, θ_{23} is currently the mixing angle with the largest uncertainty.[13] The reduction of this uncertainty, together with the determination of its octant (i.e., whether θ_{23} is larger or smaller than $\pi/4$), is essential for understanding the oscillation mechanism and will be a confounding factor for more subtle measurements if not well understood.

References

1. K. S. Hirata *et al.*, Experimental study of the atmospheric neutrino flux, *Phys. Lett.* **B205**, 416 (1988). doi: 10.1016/0370-2693(88)91690-5. [447(1988)].
2. R. Becker-Szendy *et al.*, The electron-neutrino and muon-neutrino content of the atmospheric flux, *Phys. Rev.* **D46**, 3720–3724 (1992). doi: 10.1103/PhysRevD.46.3720.
3. Y. Fukuda *et al.*, Evidence for oscillation of atmospheric neutrinos, *Phys. Rev. Lett.* **81**, 1562–1567 (1998). doi: 10.1103/PhysRevLett.81.1562.
4. M. Apollonio *et al.*, Initial results from the chooz long baseline reactor neutrino oscillation experiment, *Phys. Lett.* **B420**, 397–404 (1998).
5. G. Danby, J. M. Gaillard, K. A. Goulianos, L. M. Lederman, N. B. Mistry, M. Schwartz, and J. Steinberger, Observation of high-energy neutrino reactions and the existence of two kinds of neutrinos, *Phys. Rev. Lett.* **9**, 36–44 (1962). doi: 10.1103/PhysRevLett.9.36.
6. M. H. Ahn *et al.*, Measurement of neutrino oscillation by the K2K experiment, *Phys. Rev.* **D74**, 072003 (2006). doi: 10.1103/PhysRevD.74.072003.
7. I. Ambats *et al.*, The MINOS Detectors Technical Design Report (1998).
8. R. Acquafredda *et al.*, The OPERA experiment in the CERN to Gran Sasso neutrino beam, *JINST.* **4**, P04018 (2009). doi: 10.1088/1748-0221/4/04/P04018.
9. K. Abe *et al.*, The T2K neutrino flux prediction, *Phys. Rev.* **D87**, 012001 (2013). doi: 10.1103/PhysRevD.87.012001, 10.1103/PhysRevD.87.019902.
10. P. Adamson *et al.*, The NuMI neutrino beam, *Nucl. Instrum. Method* **A806**, 279–306 (2016). doi: 10.1016/j.nima.2015.08.063.
11. G. Acquistapace *et al.*, The CERN neutrino beam to Gran Sasso (NGS) (1998).
12. S. E. Kopp, Accelerator-based neutrino beams, *Phys. Rept.* **439**, 101–159 (2007). doi: 10.1016/j.physrep.2006.11.004.
13. K. Olive *et al.*, Review of particle physics, *Chin. Phys.* **C38**, 090001 (2014). doi: 10.1088/1674-1137/38/9/090001.
14. F. Halzen, and A. D. Martin, *Quarks and Leptons: An Introductory Course in Modern Particle Physics.* Wiley, New York, USA (1984).
15. S. van der Meer, A Directive Device for Charged Particles and Its use in an Enhanced Neutrino Beam (1961).
16. A. C. D. Beavis, and I. Chiang, Long baseline neutrino oscillation experiment at the AGS, BNL 52459. (1995).
17. K. Abe *et al.*, The T2K experiment, *Nucl. Instrum. Method Phys. Res. A* **659** (1), 106–135 (2011). ISSN 0168-9002. doi: http://dx.doi.org/10.1016/j.nima.2011.06.067. URL http://www.sciencedirect.com/science/article/pii/S0168900211011910.
18. K. Hidaka, M. Honda, and S. Midorikawa, Neutrino oscillations and the anomalous atmospheric neutrino flux, *Phys. Rev. Lett.* **61**, 1537 (1988). doi: 10.1103/PhysRevLett.61.1537.
19. P. Adamson *et al.*, Measurement of the neutrino mass splitting and flavor mixing by MINOS, *Phys. Rev. Lett.* **106**, 181801 (2011). doi: 10.1103/PhysRevLett.106.181801.

20. P. Adamson *et al.*, Measurement of neutrino and antineutrino oscillations using beam and atmospheric data in MINOS, *Phys. Rev. Lett.* **110**(25), 251801 (2013). doi: 10.1103/PhysRevLett.110.251801.
21. P. Adamson *et al.*, Combined analysis of ν_μ disappearance and $\nu_\mu \to \nu_e$ appearance in MINOS using accelerator and atmospheric neutrinos, *Phys. Rev. Lett.* **112**, 191801 (2014). doi: 10.1103/PhysRevLett.112.191801.
22. K. Kodama *et al.*, Final tau-neutrino results from the DONuT experiment, *Phys. Rev.* **D78**, 052002 (2008). doi: 10.1103/PhysRevD.78.052002.
23. N. Agafonova *et al.*, Observation of a first ν_τ candidate in the OPERA experiment in the CNGS beam, *Phys. Lett.* **B691**, 138–145 (2010). doi: 10.1016/j.physletb.2010.06.022.
24. Y. Fukuda *et al.*, The Super-Kamiokande detector, *Nucl. Instrum. Method* **A501**, 418–462 (2003). doi: 10.1016/S0168-9002(03)00425-X.
25. K. Abe *et al.*, Precise measurement of the neutrino mixing parameter θ_{23} from muon neutrino disappearance in an off-axis beam, *Phys. Rev. Lett.* **112**(18), 181801 (2014). doi: 10.1103/PhysRevLett.112.181801.
26. K. Abe *et al.*, Measurements of neutrino oscillation in appearance and disappearance channels by the T2K experiment with 6.610^{20} protons on target, *Phys. Rev.* **D91**(7), 072010 (2015). doi: 10.1103/PhysRevD.91.072010.
27. K. Abe *et al.*, Neutrino oscillation physics potential of the T2K experiment, *PTEP.* **2015**(4), 043C01 (2015). doi: 10.1093/ptep/ptv031.
28. K. Abe *et al.*, Measurement of muon antineutrino oscillations with an accelerator-produced off-axis beam, *Phys. Rev. Lett.* **116**(18), 181801 (2016). doi: 10.1103/PhysRevLett.116.181801.
29. D. S. Ayres *et al.*, The NOvA Technical Design Report (2007).
30. P. Adamson *et al.*, Measurement of the neutrino mixing angle θ_{23} in NOvA (2017).
31. G. Tzanankos *et al.*, MINOS+: a Proposal to FNAL to run MINOS with the medium energy NuMI beam (2011).
32. K. Abe *et al.*, First combined analysis of neutrino and antineutrino oscillations at T2K (2017).
33. R. Wendell, Atmospheric Neutrino Oscillations at Super-Kamiokande, *PoS.* **ICRC2015 1062** (2015). URL http://pos.sissa.it/cgi-bin/reader/conf.cgi?confid=236.
34. M. G. Aartsen *et al.*, Neutrino oscillation studies with IceCube-DeepCore, *Nucl. Phys.* **B908**, 161–177 (2016). doi: 10.1016/j.nuclphysb.2016.03.028.
35. P. Adamson *et al.*, First measurement of muon-neutrino disappearance in NOvA, *Phys. Rev.* **D93**(5), 051104 (2016). doi: 10.1103/PhysRevD.93.051104.

Chapter 7

The Measurement of θ_{13} with Reactors and Accelerators

F. Di Lodovico

Queen Mary University of London,
Mile End Road, London E1 4NS, UK.
f.di.lodovico@qmul.ac.uk

Abstract

In 2012, the neutrino mixing angle θ_{13} was measured as the culmination of a global effort to measure the last unknown mixing angle. Two main approaches were used: measuring the electron antineutrino disappearance using reactor neutrinos and observing the muon into electron neutrino appearance at long-baseline neutrino experiments. The current result from a global fit with NuFIT that includes the measurements from reactor (Double CHOOZ, Daya Bay and RENO) and long-baseline (MINOS, T2K, NOvA) neutrino experiments is: $\theta_{13}/° = 8.46^{+0.15}_{-0.15}$ $(8.49^{+0.15}_{-0.15})$ for normal (inverted) ordering. θ_{13} is currently the best measured mixing angle and its precision is expected to improve as all the current experiments, apart from MINOS, will continue to take data.

1. Introduction

The observation of neutrino oscillations by Super-Kamiokande (SK)[1] in 1998 and the Sudbury Neutrino Observatory[2,3] in 2002 using atmospheric and solar neutrinos, respectively, was followed by overwhelming evidence from other experiments of flavor neutrino oscillations. However, the third and last mixing angle that governs neutrino oscillations in the framework of three-flavor oscillation model, θ_{13}, was only measured in 2012 by reactor and long-baseline neutrino experiments, as described in this chapter.

Flavor oscillation models rely on the assumption that the neutrino mass and weak interaction eigenstates are not equivalent. As a neutrino propagates, the phases of its components change relative to each other provided the mass eigenstates are non-degenerate. This leads to the possibility that a neutrino interaction may produce a charged lepton of different flavor than the one associated with its production. This process is known as neutrino oscillation. The probability for a neutrino to oscillate into a different flavor depends on the Pontecorvo, Maki, Nakagawa and Sakata (PMNS) matrix elements[4–6] (see Eq. 1), the differences between the squared masses of the eigenstates ($\Delta m^2_{ij} = m^2_i - m^2_j$ where $i, j =$ 1,2,3), the neutrino energy and the distance it has travelled. Since the PMNS matrix is a unitarity matrix, there are only four independent parameters, three real and one imaginary phase called θ_{12}, θ_{13}, θ_{23} and δ_{CP}, respectively.

$$
\begin{aligned}
|\nu_\alpha\rangle &= \sum_{i=1}^{3} U^*_{\alpha i} |\nu_i\rangle\,, \qquad \alpha = e, \mu, \tau \\
U &= \begin{bmatrix} 1 & 0 & 0 \\ 0 & c_{23} & s_{23} \\ 0 & -s_{23} & c_{23} \end{bmatrix} \begin{bmatrix} c_{13} & 0 & s_{13}e^{-i\delta_{CP}} \\ 0 & 1 & 0 \\ -s_{13}e^{+i\delta_{CP}} & 0 & c_{13} \end{bmatrix} \\
&\quad \times \begin{bmatrix} c_{12} & s_{12} & 0 \\ -s_{12} & c_{12} & 0 \\ 0 & 0 & 1 \end{bmatrix}, \qquad \begin{matrix} s_{13} \equiv \sin\theta_{13} \\ c_{13} \equiv \cos\theta_{13} \end{matrix}
\end{aligned} \tag{1}
$$

For historical reasons, the parameters θ_{12} and Δm^2_{12} are called "solar" parameters, whilst θ_{32} and Δm^2_{32} are called "atmospheric" parameters.

Before 2011, measurements of the antineutrinos emitted from nuclear reactors had only shown that $\sin^2 2\theta_{13} < 0.17$ at 90% C.L.[7,8]

A new campaign of experiments was launched to determine θ_{13} by either $\overline{\nu}_e$ disappearance with reactors or $\nu_\mu \to \nu_e$ ($\overline{\nu}_\mu \to \overline{\nu}_e$) appearance with accelerators.

In the case of reactor experiments, carefully designed experiments with near and far detectors were built to minimize the uncertainty on the reactor flux, thus improving the precision with respect to the earlier reactor neutrino experiments.

In the case of accelerators, the ν_e signal has lower statistics than reactor neutrino experiments. Its background comes from both neutral current (NC) events and ν_e contamination in the ν_μ beam. Therefore the new experiments were built with a higher power than the earlier experiments, a large far detector with excellent electron identification and a very precise near detector (ND) to reduce the systematics.

In the following, we will discuss the observation, current status and future prospects for the measurement of θ_{13} in both reactor and long-baseline experiments.

2. Neutrino reactor experiments

Reactors are pure and powerful sources of electron antineutrinos, $\overline{\nu}_e$. They derive their power from fission and radioactive decay of the corresponding fission fragments. In power reactors used in the neutrino reactor experiments, ~99.9% of the thermal power and reactor antineutrinos come from the fission of ^{235}U, ^{239}Pu, ^{241}Pu and ^{238}U.

The reactor antineutrinos are produced at a rate of $\sim 10^{21}$ per second per gigawatt of reactor thermal power (GW_{th}) and provide a distinct signal via the inverse beta decay (IBD) reaction $\overline{\nu}_e + p \to n + e^+$.

An IBD event is detected as a pair of coincident signals consisting of a prompt signal induced by the positron ionisation and annihilation inside the detector, such as a liquid scintillator (LS) detector, and a delay signal produced by the neutron capture on proton or nucleus such as gadolinium (Gd). The scintillator in the target is usually doped with 0.1% of natural Gd by mass. The neutron capture on Gd would release multiple gammas with a total energy of ~8 MeV and a delay of about 30 μs with respect to the prompt signal.

The delay-coincidence nature of the signal helps to significantly reduce the background for detection of $\overline{\nu}_e$ interactions, such as radioactive background, which mostly consist of only a single signal.

The measurement of the energy of the positron in the IBD reaction allows the $\overline{\nu}_e$ energy to be determined via the kinematic relationship $E_\nu \approx E_e + m_e + 0.78$ MeV, which is an attractive feature for measurements of neutrino oscillations that require knowledge of the antineutrino energy.

The observable reactor neutrino spectrum is the multiplication of the antineutrino flux per fission and the cross section, shown in Fig. 1 for the four main thermal fissile isotopes ^{235}U, ^{239}Pu, ^{241}Pu and ^{238}U.[9]

A reactor-based experiment is a disappearance experiment since the antineutrinos produced by the reactor do not have enough energy to produce muons (or taus) and the NC reactions of the "oscillated" $\bar{\nu}_\mu$ or $\bar{\nu}_\tau$ or $\overline{\nu}_e$ NC interactions have very low cross-section values and are indistinguishable from the backgrounds.

In the three-flavor oscillation model, the survival probability of $\overline{\nu}_e$ with energy E at a distance L for reactor experiments with $L/E < 1\,\mathrm{km/MeV}$[10]

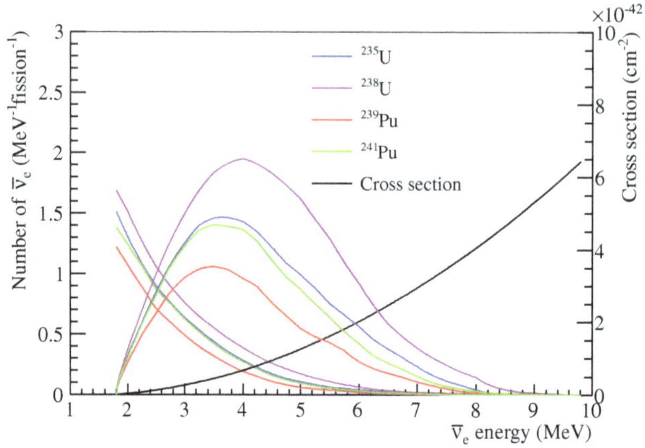

Fig. 1. Neutrino yield per fission (black line), interaction cross section of the inverse beta decay and the observables spectra for the four main thermal fissile isotopes ^{235}U, ^{239}Pu, ^{241}Pu and ^{238}U.[9]

is written as

$$P(\bar{\nu}_e \to \bar{\nu}_e) \approx 1 - \cos^4\theta_{13}\sin^2 2\theta_{12}\sin^2\Delta_{21} - \sin^2 2\theta_{13}\sin^2\Delta_{ee},$$

where

$$\begin{aligned}\Delta m^2_{ee} &\equiv \cos^2\theta_{12}\ \Delta m^2_{31} + \sin^2\theta_{12}\ \Delta m^2_{32}\\ &= \Delta m^2_{31} - \sin^2\theta_{12}\ \Delta m^2_{21} = \Delta m^2_{32} + \cos^2\theta_{12}\ \Delta m^2_{21}\\ &= m^2_3 - (\cos^2\theta_{12} m^2_1 + \sin^2\theta_{12} m^2_2).\end{aligned}$$

Δm^2_{ee} acts as the effective atmospheric Δm^2 associated with θ_{13} oscillations.

This probability, using the current experimental data, allows for an accurate determination of the mixing angle θ_{13} and the atmospheric mass splitting $|\Delta m^2_{ee}|$, independent from the atmospheric mass ordering, and is only very weakly dependent on our current knowledge of the solar parameters, through the solar term. Moreover, this oscillation probability does not depend on the CP phase $\delta_{\rm CP}$.

The interplay between atmospheric and solar oscillation frequency is illustrated in Fig. 2. The black curve corresponds to $\theta_{13} = 0$. In this case, the oscillation dip is at about ~60 km in the gray-shaded solar oscillation window. Small changes in Δm^2_{21} will move this dip. This is illustrated by the red-dashed curve by increasing Δm^2_{21} by 20%, which means that such an experiment is very sensitive to Δm^2_{21}. If $\theta_{13} > 0$ (yellow curve), the faster atmospheric oscillation will be superimposed leading to sensitivity to θ_{13} in the atmospheric oscillation window at about 1–2 km. At longer

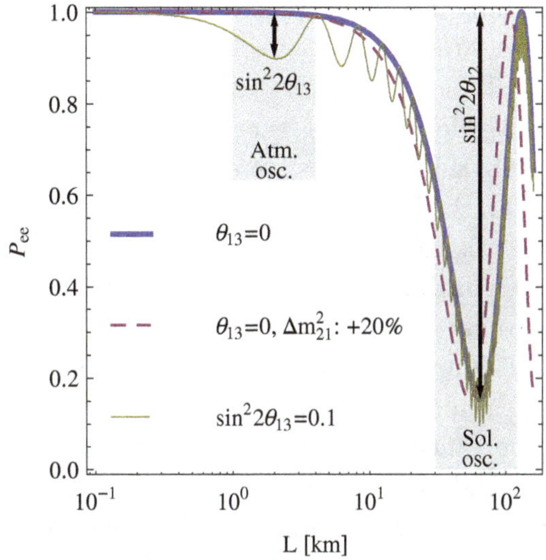

Fig. 2. Reactor electron antineutrino disappearance probability,[11] illustrated for different sets of parameters and perfect resolution.

baselines, the atmospheric oscillations cannot be resolved in practice and are averaged out.

2.1. *Determination of θ_{13} from reactor experiments*

Nuclear reactors have played a prominent role in our understanding of the neutrino. Neutrinos were discovered in an experiment at the Savannah River nuclear reactor in 1956 through the IBD reaction.[12]

Since the Savannah River experiment, several similar experiments have been set up to measure the spectrum of neutrinos involved in IBD reactions at various distances from reactors focussing on the measurement of the third neutrino mixing angle θ_{13}. Whilst the Savannah River experiment used cadmium to capture neutrons, the current experiments use gadolinium in a liquid scintillator detector.

The first experiment of this type was the CHOOZ experiment[13] in the late 1999s and early 2000s. The CHOOZ power station had two pressurised-water reactors (PWRs) with a total thermal power of 8.5 GW_{th}. The first reactor reached full power in May 1997, the second in August 1997. The detector was located in an underground laboratory with 300 m.w.e. (meter water equivalent) rock overburden at about 1 km from the neutrino source. It consisted of a central 5 t target filled with 0.09% Gd-loaded liquid scintillator, surrounded by an intermediate 17 t and outer 90 t regions filled with Gd-free liquid scintillator. The CHOOZ experiment found no evidence for neutrino oscillations in the disappearance mode $\overline{\nu}_e \to \overline{\nu}_x$ setting a limit of $\sin^2 2\theta_{13} < 0.12$ at 90% C.L.[7,8]

Similarly, no oscillations were observed by the Palo Verde experiment[14] that used the Palo Verde Nuclear Generating Station, located about 70 km west of Phoenix, Arizona that consists of three identical PWRs producing a total thermal power of 11.63 GW_{th}. The Gd-doped liquid scintillator segmented detector was located at a distance of 890 m from two of the reactors and 750 m from the third in an underground laboratory at a shallow (32 m.w.e.) underground site.

After the establishment of atmospheric and solar neutrino oscillations, and the $\overline{\nu}_e$ disappearance limits placed by CHOOZ and Palo Verde in the 1990s, the importance of the reactor neutrino oscillation experiment to measure θ_{13} was widely recognised. This led to the realization of three new reactor experiments: the Double CHOOZ experiment[15] in the Ardennes region of France, the Daya Bay experiment[16] near Dapeng City in the Guangdong province of China and the Reactor Experiment for Neutrino Oscillations (RENO)[17] near Yeonggwang city in South Korea.

The main improvement over previous experiments is the use of multiple detectors to make a relative measurement of antineutrino disappearance to obtain higher precision over the previous experiments limited by uncertainties of a few percent in the knowledge of the absolute flux of $\overline{\nu}_e$ emitted from nuclear reactors.

All three experiments use PWRs. The ND is typically at an average distance of ~200/300 m from the nuclear reactor cores and the far detector is at ~1000 m.

The Double CHOOZ experiment has the simplest configuration with two PWRs, achieving 4.25 GW_{th} each, and two detectors of 8.3 t fiducial mass each, situated at ~395 m and ~1050 m from the reactors along a curve of isoflux such that the ratio of neutrino events from each reactor is equal in the two detectors.

Daya Bay consists of three separate sites, each hosting two reactors of 2.95 GW_{th} power, where eight detector modules of 20 fiducial tons each are distributed, with four detectors placed at the far site and two detectors each at the near sites. Daya Bay's far detectors lie between ~1500 and ~1900 m from the sites' reactors whilst the NDs have baselines from ~360 to ~560 m.

The RENO experiment power station hosts six reactors of ~2.8 GW_{th} power each for a total thermal power of 16.8 GW_{th}. The RENO near and far detectors are located along a line running perpendicular to and through the midpoint of the line of reactors, at distances of 294 m and 1383 m. The near

Table 1. Main characteristics of the Double CHOOZ, Daya Bay and RENO neutrino reactor experiments. The names of the experiments are in the first column, the power of the reactors and the number of reactors in the second column, the baselines for the near and far detectors, the overburden and finally the mass of the far detectors and the number of detectors.

Experiment	Power (reactors) (GW_{th})	L near (m)	L far (m)	Depth near/far (m.w.e.)	Target × detectors (t)
Double CHOOZ	4.25 (2)	395	1050	120/300	8.3×1
Daya Bay	2.9 (6)	360–560	1500–1900	280/300/880	20×4
RENO	2.6 (6)	294	1383	120–410	16×1

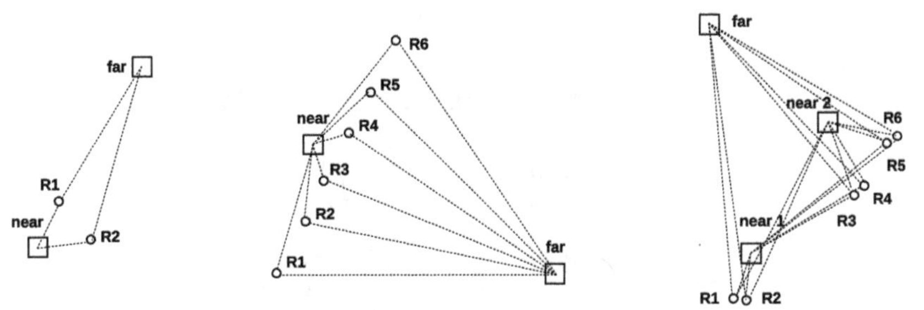

Fig. 3. Configurations of near and far detectors (squares) and nuclear reactors (circles) for the Double CHOOZ (left), RENO (middle) and Daya Bay (right) experiments.[18] The dotted lines represent the baselines between the detectors and reactor as summarised in Table 1.

(far) detector is installed underground with an overburden of 120 (450) m.w.e.

A summary of the main characteristics of the three reactor experiments is presented in Table 1.

Schematic configurations of the near and far detector and the nuclear reactors for the three experiments[18] are shown in Fig. 3.

The experiments can use essentially identical detectors at their near and far sites as they see the same flux differently from the long-baseline experiments where the near and far baselines are seeing a different flux. Thus, the near and far detector in reactor experiments are designed to have exactly the same fiducial mass of 10 or 20 t of liquid scintillator.

Moreover, the detectors of the reactor neutrino experiments are all quite similar in terms of basic principles; see Fig. 4 for a cross-sectional view of the Double CHOOZ, Daya Bay and RENO detectors.

A detector is usually subdivided into three volumes, an inner veto and an outer veto (OV) system. At the centre of each detector, there is a

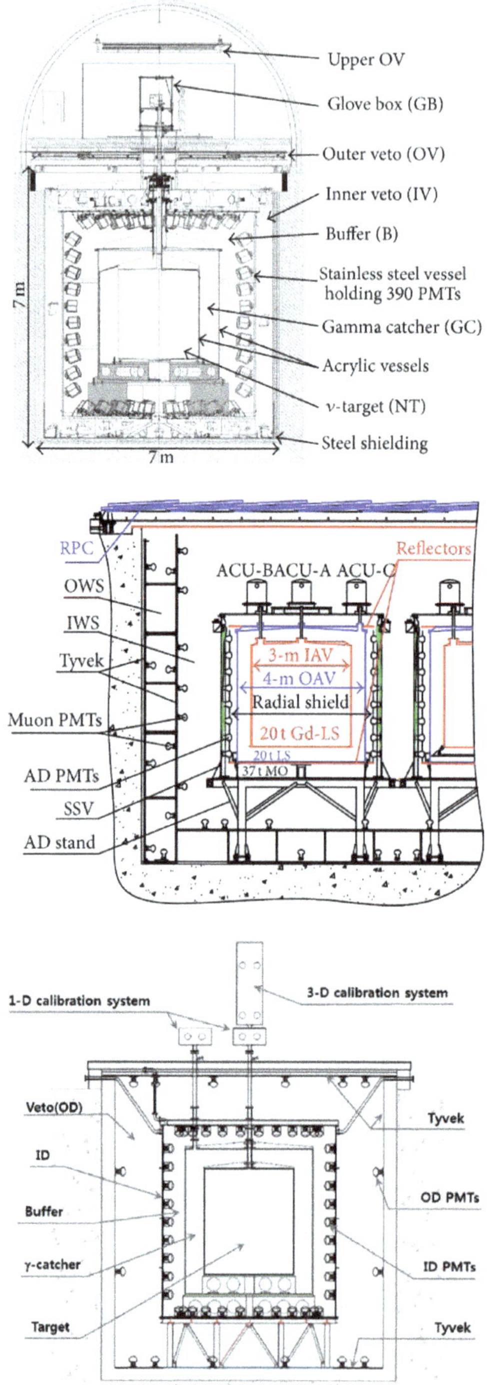

Fig. 4. Cross-sectional view of the Double CHOOZ[19] (left), Daya Bay[20] (middle) and RENO[21] (right) detectors.

multi-ton antineutrino target ("target region") made of liquid scintillator. The scintillator in the target contains 0.1% of natural gadolinium by mass. An ~8 MeV gamma cascade is generated by the neutron capture in Gd.

The target region is nested within an additional multi-ton liquid scintillating region which has no Gd, so it does not serve as the primary antineutrino target. Its purpose is to increase the detector efficiency by absorbing gamma rays which happen to escape the target region ("γ catcher"). This region is further surrounded by an inactive zone of mineral oil which shields the scintillating regions from external radiation. Photomultiplier tubes (PMTs) mounted within this region detect the scintillation light emitted by particle interactions within the detector. These three volumes and the PMTs constitute the inner detector (ID).

Outside the ID, and optically separated from it, there is an "inner veto" liquid scintillator that functions as a cosmic muon veto and as a shield to spallation neutrons produced outside the detector. It is equipped with generally smaller PMTs than the ID. The whole detector is further surrounded by demagnetised steel or a water pool equipped with PMTs to suppress external γ rays.

Finally, the main detector is covered by an OV system made either of plastic scintillator strips or a system of resistive plate chamber modules to veto against muons which do not pass through the main detector, but which could act as precursors to cosmogenic backgrounds. Since the far detector is generally positioned deeper underground than the ND to allow more natural shielding from cosmic muons, the OV detector is bigger for the near than for the far detector.

The experiment is calibrated using light sources, radioactive point sources, cosmic rays and natural radioactivity. Deployments in the target are carried out both using motorised systems and glovebox position on the top of the detector.

Although Double CHOOZ and Daya Bay were still partially constructed at the time of starting the data taking, all the three reactor experiments reported results on $\overline{\nu}_e$ disappearance between the end of 2011 and beginning of 2012.

In November 2011, Double CHOOZ reported an indication of $\overline{\nu}_e$ disappearance[19] from 101 days of data collected with just the far detector. The reactor antineutrino flux prediction used the Bugey4 flux measurement[22] after correcting for differences in core composition. Double CHOOZ observed a total of 4121 events with 4344 ± 165 expected with a signal to background ratio of ~11:1. This deficit led to

$\sin^2 2\theta_{13} = 0.086 \pm 0.041(\text{stat}) \pm 0.030(\text{syst})$ based on an analysis using rate and energy spectrum information and thus disfavoring $\theta_{13} = 0$ at 94.6% C.L.

In March 2012, Daya Bay reported an observation of $\overline{\nu}_e$ disappearance[20] from 55 days' data taking with six detectors deployed in two near and one far underground experimental halls. An earlier study of the rate measured by side-by-side detectors helped to confirm studies of the relative efficiency between detectors.[23] A total of 10416 reactor $\overline{\nu}_e$ were observed at the two far detectors. The ratio of the observed to expected $\overline{\nu}_e$ at the far detector, assuming no oscillations, was measured to be $R = 0.940 \pm 0.011(\text{stat}) \pm 0.004(\text{syst})$, indicating a deficit of 6% with respect to the expectations. A rate-only analysis yielded $\sin^2 2\theta_{13} = 0.092 \pm 0.016(\text{stat}) \pm 0.005(\text{syst})$, which means that the neutrino mixing angle θ_{13} is non-zero with a significance of 5.2σ.

An improved measurement[24] within the same year using 139 days of data was performed. A total of 28909 (205308) $\overline{\nu}_e$ candidates were detected at the far (near) detectors leading to a ratio of the observed to expected number of $\overline{\nu}_e$ at the far hall, assuming no oscillations, of $R = 0.94 \pm 0.007(\text{stat}) \pm 0.003(\text{syst})$. Also, $\sin^2 2\theta_{13} = 0.089 \pm 0.010(\text{stat}) \pm 0.005(\text{syst})$ was measured and the value of $\theta_{13} = 0$ was excluded at 7.7σ.

In April 2012, soon after Daya Bay, the RENO experiment also measured a deficit of the number of observed $\overline{\nu}_e$ at the far detector.[21] A total of 17102 (154088) $\overline{\nu}_e$ events with a background fraction of 5.5% (2.7%) were detected at the far (near) detector after 229 days of data taking. The ratio of measured to expected events at the far detector in absence of oscillations was $R = 0.920 \pm 0.009(\text{stat}) \pm 0.014(\text{syst})$. Using a rate-only analysis, a value of $\sin^2 2\theta_{13} = 0.113 \pm 0.013(\text{stat}) \pm 0.019(\text{syst})$ was obtained. A value of θ_{13} equaling zero was excluded at 4.9σ.

Since the initial observation of $\overline{\nu}_e$ disappearance, a series of improved results have been reported by the three collaborations.

The updated results include simultaneous fits to the rate and shape of the selected signal sample for all the experiments.[25,26] They also include new studies of neutron capture in hydrogen,[27,28] both as a cross check of the Gd-based analysis and to provide a further independent sample, and furthermore, overall improvements to the analyses, in particular the background and systematic error reduction (e.g., reactor-off data, new analyis techniques, new calibration, background reduction, etc.).[25,26,29–33]

Construction of all the experiments is now complete with all the planned near and far detectors.

Since December 2014, the Double CHOOZ ND has been taking data. The latest results from Double CHOOZ[34] include the ND. Double CHOOZ ran for a total of ~488 and ~384 days without and with the ND, respectively. The analysis is a combination of the IBD selection in Gd (and a small component of C) and hydrogen. Hydrogen captures, possible thanks to the low background level, occur in the undoped LS in addition to the Gd-doped region. This increases the fiducial volume for the analyses by three times. Furthermore, the two samples provide two different data sets with partially different systematic uncertainties, thus increasing the total statistics available for the analysis. A combination of the selections in Gd and hydrogen and also including the data taken with the ND is presented in the latest results from Double CHOOZ. Figure 5 shows the energy spectrum of the prompt signal superimposed on the best-fit prediction and background components, where the best fit gives $\sin^2 2\theta_{13} = 0.119 \pm 0.016$ with $\chi^2/ndof = 236.2/114$. The dominant systematic uncertainty associated with the IBD signal detection is due to the number of free protons in the detector, mostly from the "γ catcher", which was originally not considered as a target for IBD interactions, and hence affects the detection of capture in hydrogen.

Figure 6 shows the ratio of the data to the null oscillation prediction after subtraction of the background as a function of the visible energy of the prompt signal. An energy-dependent deficit is clearly seen in the data below 4 MeV, which is consistent with the expectation from reactor neutrino oscillation. Also a spectrum distortion is observed at high energy above 4 MeV, which can be characterised by an excess around 5 MeV and a deficit around 7 MeV. Although the impact of the deviation in the observed energy spectrum on the $\sin^2 2\theta_{13}$ measurement is not significant, it will be discussed further in Section 2.2.

The latest results from Daya Bay[35] are based on the combination of 217 days of operation of six antineutrino detectors (December 2011 to July 2012) with a subsequent 1013 days using the complete configuration of eight detectors (October 2012 to July 2015) for a total of more than 2.5 million $\overline{\nu}_e$ IBD interactions observed. The $\overline{\nu}_e$ rate observed at the far detectors relative to the NDs showed a significant deficit, $R = 0.949 \pm 0.002(\text{stat}) \pm 0.002(\text{syst})$. This can be seen in the reconstructed positron energy spectra for the $\overline{\nu}_e$ candidates shown in Fig. 7, together with the ratio of the background-subtracted spectra to prediction assuming no oscillation.

In Fig. 8, the measured $\overline{\nu}_e$ survival probability sampled over almost one full cycle is shown. It demonstrates distinct evidence in support of neutrino flavor oscillation.

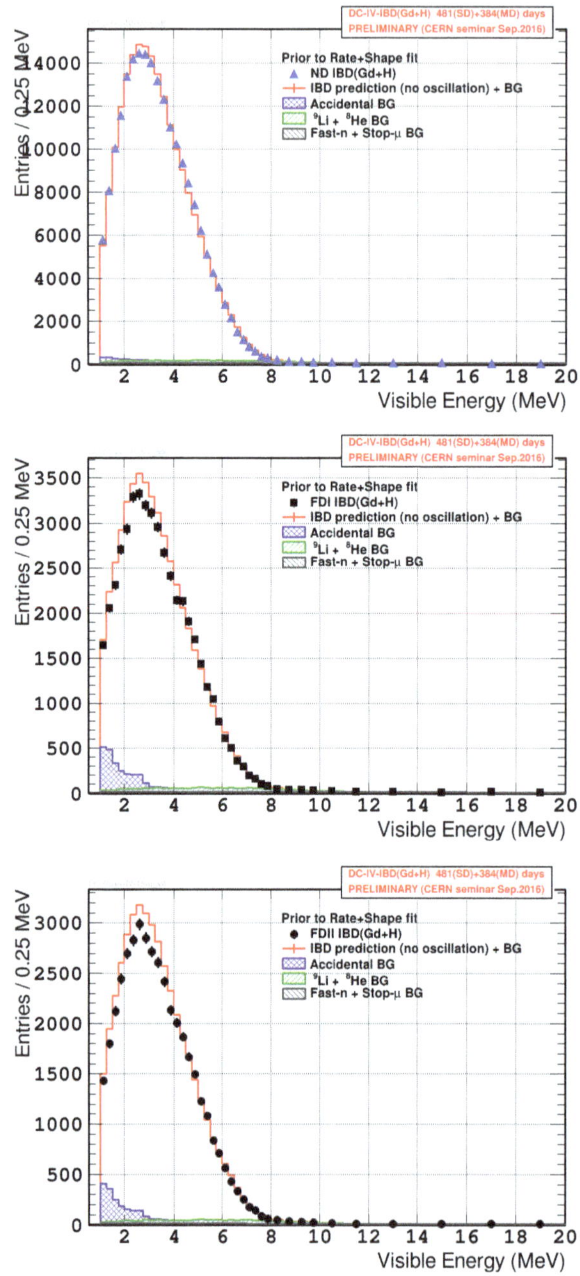

Fig. 5. Measured energy spectrum of the prompt signal (black points) superimposed on the predictions without neutrino oscillations plus background (red line) for the ND (left), far detector before the ND became operative (middle) and far detector after the ND became operative (right) from the Double CHOOZ analysis.[34] Background components after the fit are also shown (blue cross-hatched).

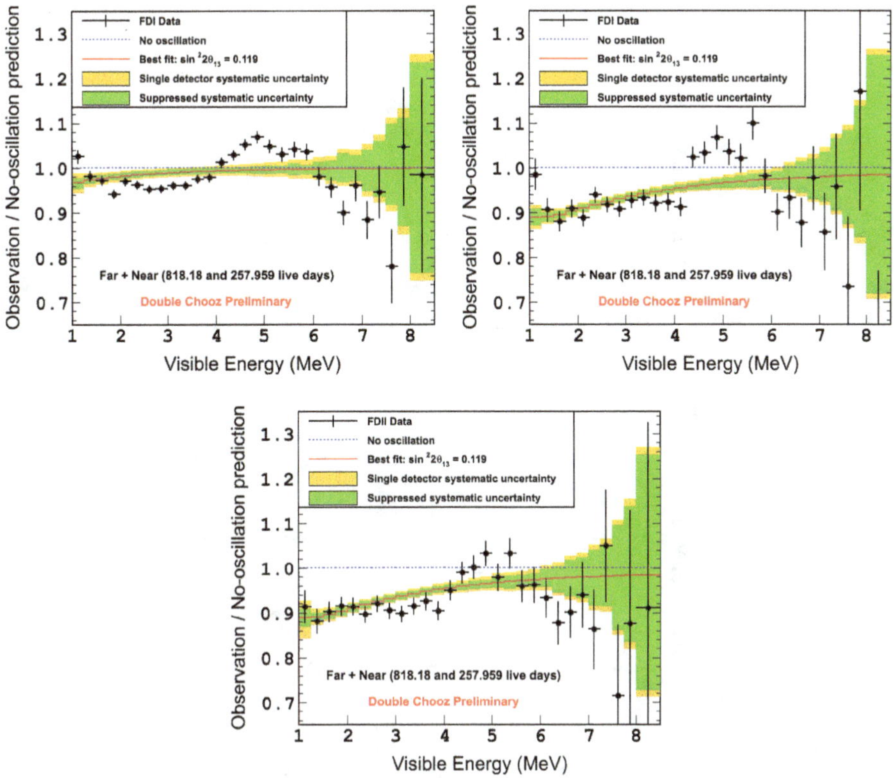

Fig. 6. Ratio of the data, after subtraction of the background, to the non-oscillation predictions, represented by the black points, as a function of the visible energy of the prompt signal for the ND (left), far detector before the ND became operative (middle) and far detector after the ND became operative (right) for Double CHOOZ.[34] The overlaid red line corresponds to the rate of the best fit to the non-oscillation prediction with the reactor flux uncertainty (green) and the total systematic uncertainty (yellow).

Finally, an analysis was performed using an approximation for the three-flavor oscillation probability that yielded the flavor-mixing angle $\sin^2 2\theta_{13} = 0.0841 \pm 0.0027(\text{stat}) \pm 0.0019(\text{syst})$ and the effective neutrino mass-squared difference of $|\Delta m^2_{ee}| = (2.50 \pm 0.06(\text{stat}) \pm 0.06(\text{syst})) \times 10^{-3}\,\text{eV}^2$.

The corresponding confidence regions of $\sin^2 2\theta_{13}$ and $|\Delta m^2_{ee}|$ are shown in Fig. 9.

The analysis by Daya Bay included only neutrons captured in the Gd as the n-H sample suffered from larger uncertainties in the target volume and detector response, as well as a much more significant background contamination, as it is not possible to clearly discriminate whether the $\overline{\nu}_e$ interacted within the target region or in the gamma catcher region.

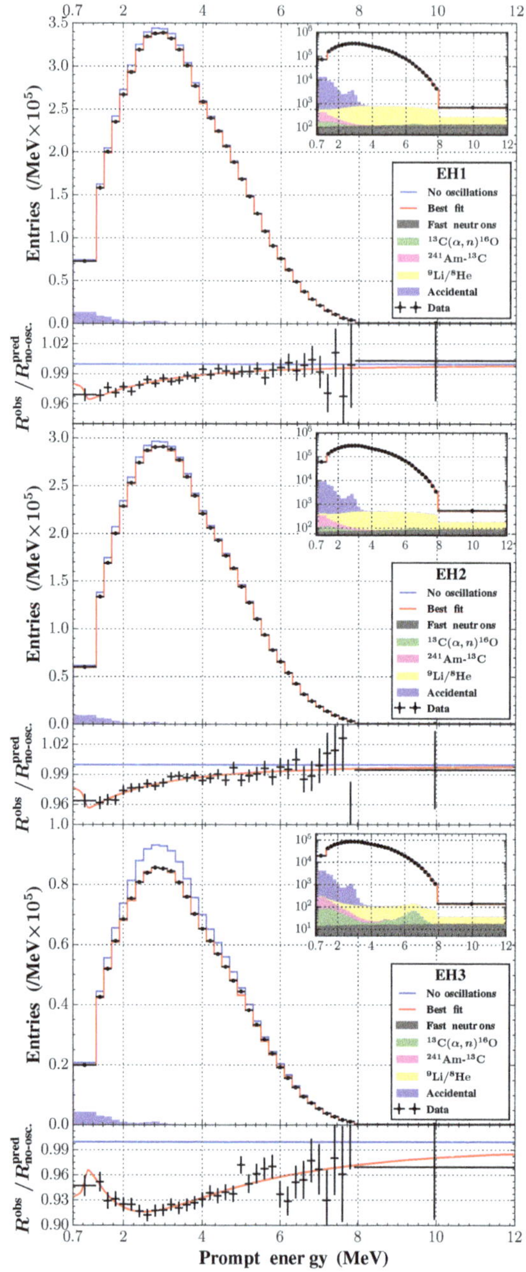

Fig. 7. The reconstructed positron energy spectra for the $\overline{\nu}_e$ candidate interactions (black points) for the three experimental halls (EH): EH1 (top, NDs), EH2 (middle, NDs), and EH3 (bottom, far detectors) in Daya Bay.[35] The ratio of the background-subtracted spectra to prediction assuming no oscillation is shown for each experimental hall data beneath the energy spectrum distribution. The data are represented by the cross; the prediction, assuming no oscillation, is the blue line and the best-fit three-flavor neutrino oscillation model is the red line. The inset in semilogarithmic scale shows the backgrounds.

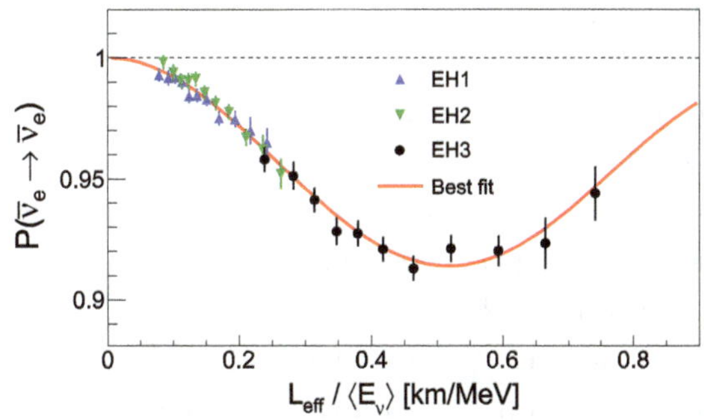

Fig. 8. The oscillation survival probability versus L_{eff}/E_ν at Daya Bay.[35] L_{eff} is the effective propagation distance and it is estimated for each experimental hall from the distribution of reactors contributing to the signal. The average true $\overline{\nu}_e$ energy $\langle E_\nu \rangle$ is determined for each bin in the observed prompt positron spectrum based on the model of the detector response, whilst the corresponding survival probability is given by the observed signal in each bin divided by the prediction assuming no oscillation.

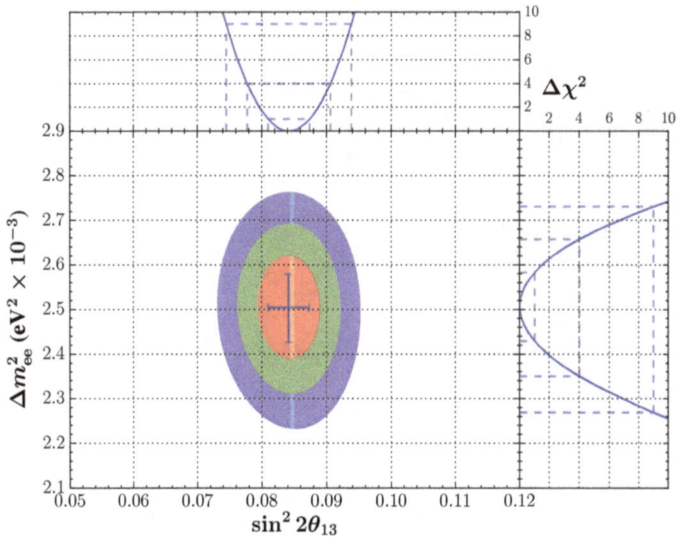

Fig. 9. Confidence regions of $\sin^2 2\theta_{13}$ and $|\Delta m^2_{ee}|$ from a combined analysis of the prompt positron spectra and rates at Daya Bay.[35] The point marks the best estimates, and the error bars display their one dimensional 1σ confidence intervals. The 1σ, 2σ, and 3σ two dimensional confidence regions are estimated using $\Delta\chi^2$ values of 2.30 (red), 6.18 (green) and 11.83 (blue) relative to the best fit. The upper panel provides the one-dimensional $\Delta\chi^2$ for $\sin^2 2\theta_{13}$ obtained by profiling $|\Delta m^2_{ee}|$ (blue line), and the dashed lines mark the corresponding 1σ, 2σ, and 3σ intervals. The right panel is the same, but for $|\Delta m^2_{ee}|$, with $\sin^2 2\theta_{13}$ profiled.

However, independent measurements of neutrino oscillations by the Daya Bay collaboration[28,36] where the neutron is captured on hydrogen showed consistent results with the analysis of signals identified by n-Gd capture, albeit with less precision.

RENO has been using two identical detectors since August 2011. The latest results[31] were published using roughly 500 live days of data through January 2013. A total of 290 775 (31514) reactor $\overline{\nu}_e$ IBD events with 2.8 (4.9)% background in the near (far) detector were observed. Using both the rate and spectrum information, values of $\sin^2 2\theta_{13} = 0.082 \pm 0.009(\text{stat}) \pm 0.006(\text{syst})$ and $|\Delta m^2_{ee}| = (2.62^{+0.21}_{-0.23}(\text{stat})^{+0.12}_{-0.13}(\text{syst})) \times 10^{-3}\,\text{eV}^2$ with $\chi^2/ndof = 58.9/66$ were measured.

Figure 10 shows the background-subtracted observed spectrum at the far detector compared to the one expected with no oscillation and the one expected with the best-fit oscillation at the far detector. Beneath this, the fractional difference between data and prediction is shown.

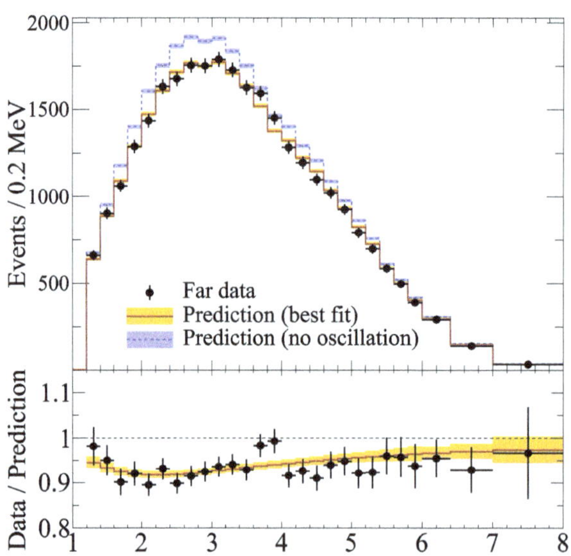

Fig. 10. Observed IBD prompt spectrum by RENO[31] at the far detector (black points) with the three-flavor neutrino oscillation model best-fit (red-line histogram) and with the no-oscillation prediction (blue-shaded histogram) obtained from the measurement in the ND. The plot underneath shows the corresponding ratio of reactor $\overline{\nu}_e$ events measured in the far detector to the no-oscillation prediction (black points) and the ratio from the simulation with best-fit results folded in (yellow band) as a function of the prompt energy spectrum. Errors are statistical uncertainties only although both statistical and systematic uncertainties are included in the χ^2 fitting.

The observed spectrum at the far detector indicates a clear energy-dependent disappearance of reactor $\overline{\nu}_e$ events consistent with neutrino oscillations.

Figure 11 shows the confidence level allowed regions for the neutrino oscillation parameters $\sin^2 2\theta_{13}$ and $|\Delta m^2_{ee}|$.

A clear $L_{\rm eff}/E_\nu$-dependent disappearance of reactor $\overline{\nu}_e$ is observed in Fig. 12. Because of the observed 5 MeV excess, the expected $L_{\rm eff}/E_\nu$ distribution is obtained from the measured spectrum at the ND instead of the IBD simulated spectrum.

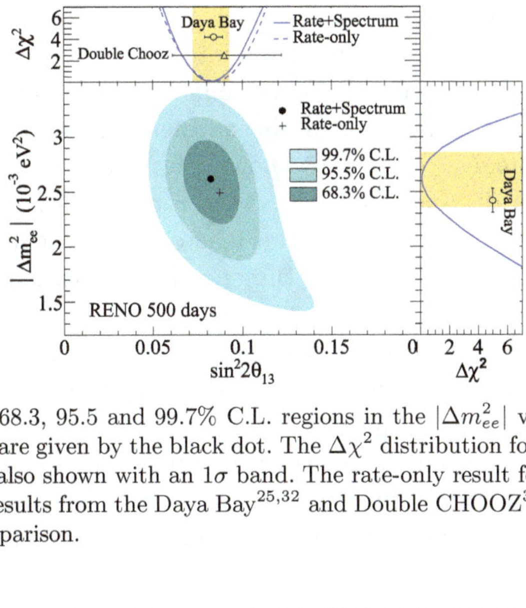

Fig. 11. RENO's 68.3, 95.5 and 99.7% C.L. regions in the $|\Delta m^2_{ee}|$ vs. $\sin^2 2\theta_{13}$ plane.[31] The best-fit values are given by the black dot. The $\Delta\chi^2$ distribution for $\sin^2 2\theta_{13}$ (top) and $|\Delta m^2_{ee}|$ (right) are also shown with an 1σ band. The rate-only result for $\sin^2 2\theta_{13}$ is shown by the cross. The results from the Daya Bay[25,32] and Double CHOOZ[37] collaborations are also shown for comparison.

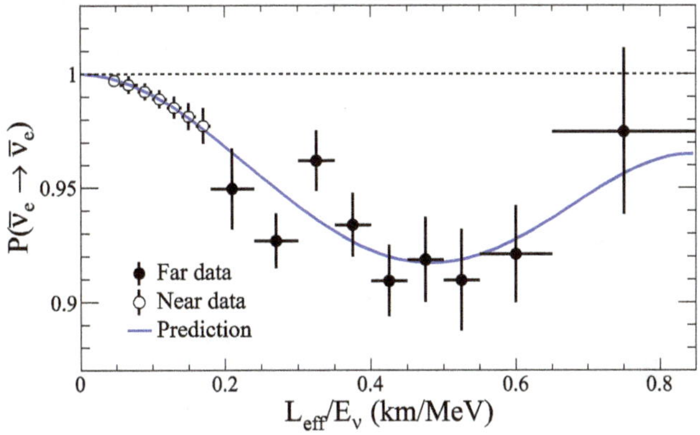

Fig. 12. The measured reactor ν_e survival probability in the far detector as a function of $L_{\rm eff}/E_\nu$ from RENO.[31] The curve is a predicted survival probability, obtained from the observed probability in the ND, for the best-fit values of $|\Delta m^2_{ee}|$ and $\sin^2 2\theta_{13}$. The $L_{\rm eff}/E_\nu$ value of each data point is given by the average of the counts in each bin.

2.2. *Reactor antineutrino fluxes*

The $\overline{\nu}_e$ spectrum from a nuclear reactor is determined either using an *ab initio* summation or the electron spectrum conversion method.[38–40]

Whilst the first method is based on the integral beta spectrum of all the fission fragments from which the neutrino spectrum can be reconstructed, the second method is based on the electron spectrum associated with the fission of each individual actinide, which will then need to be converted into the $\overline{\nu}_e$ spectra. In both cases, determining the uncertainties is quite difficult as there are many issues involved.

In 2011, in preparation for the new reactor experiments starting to take data, the $\overline{\nu}_e$ spectrum based on integral beta spectra measured in the 1980s[43–45] at the Institute Laue-Langevin was re-evaluated.[41,42] This resulted in an upward revision of the antineutrino signal by ∼6%, called the "reactor anomaly." Figure 13 shows the measured reactor $\overline{\nu}_e$ rate as a function of the distance from the reactor normalised to the theoretical predictions.[41,42] Consequently, all the reactor experiments from $L > 10\,\text{m}$ are missing approximately 6% of the signal, independent of the distance from the reactor.

However, although previous searches for the $\overline{\nu}_e$ disappearance due to θ_{13}[7,14] were affected by the uncertainty on the $\overline{\nu}_e$ flux from the reactor, the use of the ND in the current experiments allows a differential comparison between the near and far detectors overcoming the uncertainty due to the knowledge of the neutrino flux.

Furthermore, a "bump" at neutrino energies of 5–7 MeV (see Section 2.1), thereafter called the "5 MeV anomaly," not predicted in the calculation spectrum, and possibly not related to the "reactor anomaly,"[40] was observed by the experiments, see Fig. 14. The largest excess was observed by RENO but all the experiments observed an excess.

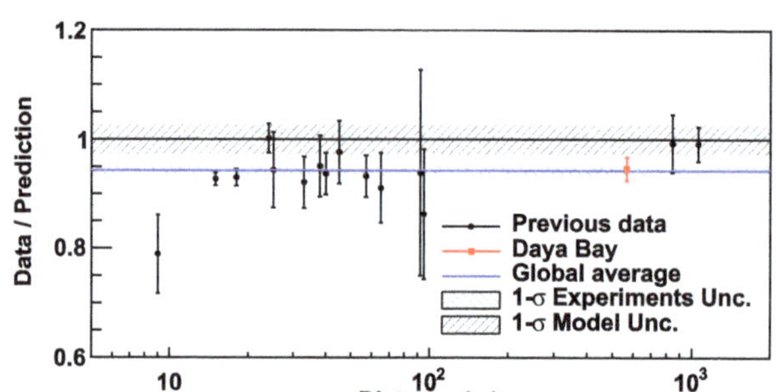

Fig. 13. The measured reactor $\overline{\nu}_e$ rate as a function of the distance from the reactor,[40] normalised to the theoretical prediction.[41,42]

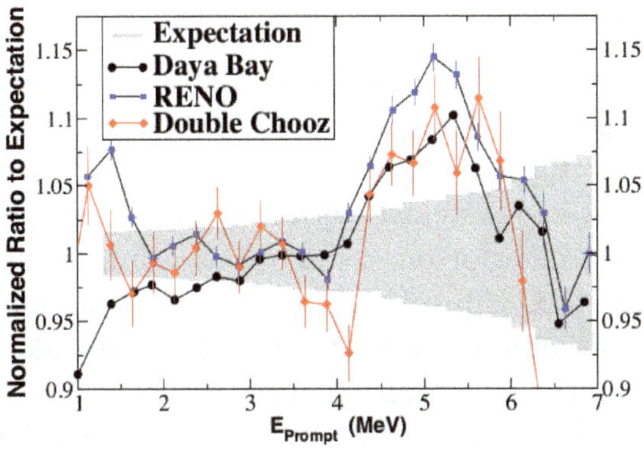

Fig. 14. Neutrino energy distributions,[46] after background subtraction for Daya Bay, Double CHOOZ and RENO compared to the expectations.[41,42]

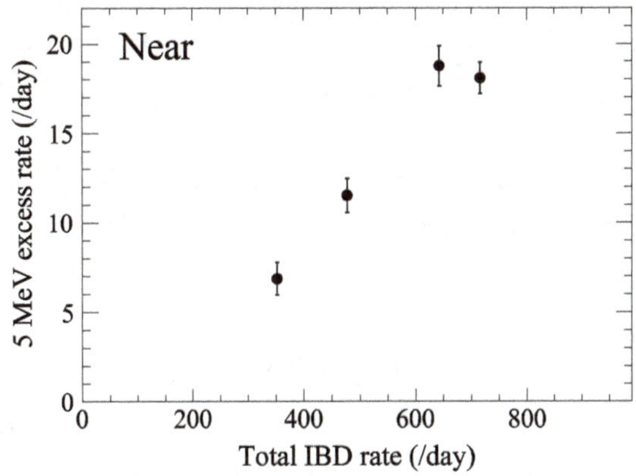

Fig. 15. Correlation between the 5 MeV excess rate and the total IBD rate from RENO.[31] The data are shown with their statistical uncertainties.

This discrepancy was found to be time-independent and correlated with the reactor thermal power, thus disfavoring hypotheses involving detector response and unknown backgrounds.[31,47] Figure 15 shows a clear correlation between the 5 MeV excess rate and the total IBD rate from RENO[31] that corresponds to the reactor thermal power.

The bump has little effect on the extraction of the neutrino oscillation parameters from the reactor experiments as the ND is used to predict the flux at the far detector. The impact on the Double CHOOZ results without an ND was evaluated to be insignificant with respect to the uncertainty.[30,37]

Further theoretical inputs and especially experimental measurements will be needed to improve the $\overline{\nu}_e$ spectrum prediction, although the NDs provide a good way to minimize the effect.

2.3. *Future experiments*

Based on the measured value of θ_{13}, a new round of more precise neutrino experiments are needed to find CP-violation effects.

The current reactor experiments will continue to accumulate data and improve the analysis and systematic errors in the coming years. A prediction of the reduction of the error on $\sin^2(2\theta_{13})$ as a function of the years, and thus the integrated flux, is presented by the three collaborations.

Figure 16 (top) shows the Double CHOOZ actual and predicted 1σ error on $\sin^2(2\theta_{13})$ since the beginning of data taking,[48] whereas the different curves represent the different improved analyses by the collaboration. Double CHOOZ has run with just the far detector for ~3 years and it is expected to run with both the near and far detectors for 3 more years to achieve the nominal exposure and sensitivity.

Daya Bay presents the expected precision on both $\sin^2(2\theta_{13})$ and $|\Delta m^2_{ee}|$ as a function of time[49] in Fig. 16 (middle). Daya Bay is expected to operate until 2020; by then, the precision should be ~3% for both $\sin^2(2\theta_{13})$ and $|\Delta m^2_{ee}|$.

RENO's goal for sensitivity for $\sin^2(2\theta_{13})$ and $|\Delta m^2_{ee}|$ is to achieve a 6% precision in 5 years. The projected sensitivity for $\sin^2(2\theta_{13})$ as a function of time[50] is presented in Fig. 16 (bottom).

3. Reactors vs. accelerator experiments

Reactor neutrino experiments take advantage of the isotropic emission of $\overline{\nu}_e$ from the reactor cores; the large IBD cross section and the particular signature of the IBD events in a Gd-loaded liquid scintillator detector that help substantially improve the signal to background ratio. Neutrinos from reactors have a mean energy of 3 MeV, much lower than the energy of the neutrinos produced by accelerator experiments, which means reactor experiments must reduce low-energy background. Similarly the reactor experiments have a much shorter baseline than accelerator experiments. An important difference between reactor and accelerator experiments is that the oscillation probability for reactor-based neutrino oscillation experiments depends on fewer parameters. The nondependence upon the $\delta_{\rm CP}$ violating phase and mass ordering makes the reactor neutrino experiment the "cleanest" and most direct way to measure the neutrino mixing angle θ_{13}. However, the current reactor experiments cannot measure the $\delta_{\rm CP}$ violating phase and/or mass ordering.[a] Moreover, in the case of neutrinos produced by accelerators,

[a] A new reactor experiment is being constructed, JUNO,[9] that will address the mass ordering measurement but will not be sensitive to the measurement of θ_{13}.

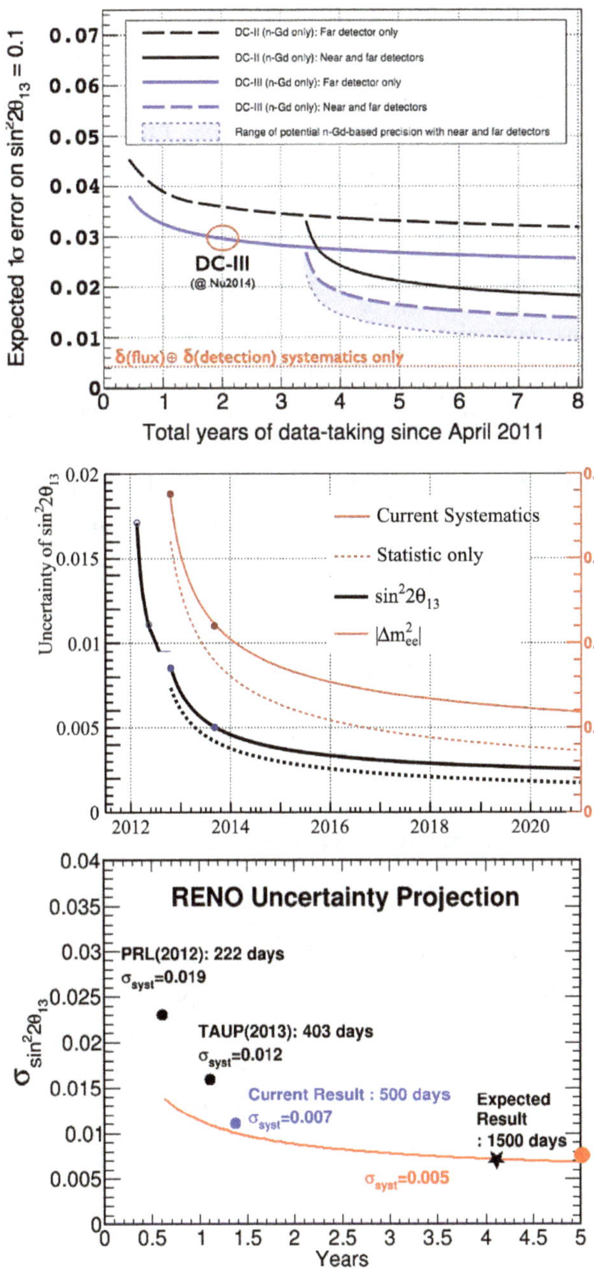

Fig. 16. Projected sensitivities for Double CHOOZ[48] (top), Daya Bay[49] (middle) and RENO[50] (middle).

the experiment has full control over the beam (e.g., the neutrino and antineutrino beam energies, their intensity and when they are produced), whilst the flux of $\overline{\nu}_e$ from a nuclear reactor cannot be changed. However, in practice, a PWR gets refuelled every 12 months. This allows the experiment to have a period of data taking with a different neutrino flux and this is used to validate background estimations.

4. Accelerator experiments

In this section, we consider experiments with accelerator-produced beams of neutrinos, mainly $\overset{(-)}{\nu}_\mu$ (where the parentheses above ν_μ indicate that we refer to both neutrinos and antineutrinos), and with baselines longer than 100 km.

Accelerator experiments use the $\overset{(-)}{\nu}_\mu \to \overset{(-)}{\nu}_e$ oscillation channel to search for appearance of $\overset{(-)}{\nu}_e$ after hundreds or thousand kilometres in a "far" detector. The probability for $\overset{(-)}{\nu}_\mu \to \overset{(-)}{\nu}_e$ oscillation as a function of neutrino propagation distance L and energy E is shown in Eq. 2.

$$\begin{aligned} P(\overset{(-)}{\nu}_\mu \to \overset{(-)}{\nu}_e) \simeq \sin^2\theta_{23}\sin^2 2\theta_{13}\sin^2\frac{\Delta m^2_{31}L}{4E} \\ \overset{(+)}{-}\frac{\sin 2\theta_{12}\sin 2\theta_{23}}{2\sin\theta_{13}}\sin\frac{\Delta m^2_{21}L}{4E} \\ \times \sin^2 2\theta_{13}\sin^2\frac{\Delta m^2_{31}L}{4E}\sin\delta_{\mathrm{CP}} \\ +(\text{CP-even, solar, matter effect terms}) \end{aligned} \tag{2}$$

The comprehensive nature of this probability is both an advantage and disadvantage for long-baseline neutrino experiments. Measurements of accelerator $\overset{(-)}{\nu}_e$ appearance provide sensitivity to all parameters, but it is difficult to determine the value of any one parameter. Measurements of both ν_e and $\overline{\nu}_e$ appearance allow experiments to break some degeneracy in the parameters and will be relevant for the measurement of θ_{13}.

5. Long-baseline neutrino experiments

Long-baseline neutrino experiments follow the same basic principle:

- beam of $\overset{(-)}{\nu}_\mu$ from pion decays;
- ND to measure interaction rates;
- far detector to look for oscillations.

The current experiments, the Main Injector Neutrino Oscillation Search (MINOS),[51] the Tokai-2-Kamioka (T2K)[52] and the NuMI Off-Axis ν_e

Appearance Experiment (NOvA)[53] use similar beamlines to produce muon neutrinos.

Each beamline consists of a graphite target placed in a proton beam in order to produce pions. The target sits inside the first of two (MINOS, NOvA) or three (T2K) magnetic horns, which focus charged secondary particles of a given charge sign. The geometry of the horns and decay volume with respect to the target determines the momentum range for which secondaries will be focussed.

The focussed secondary beam consists mostly of pions, and these are allowed to decay in flight in a helium-filled decay pipe to produce predominantly $\mu^{\pm}$ and $\overset{(-)}{\nu}_{\mu}$. Downstream of the decay volume, instrumentation allows characterisation of the beam by looking at the profile of muons and remaining hadrons. The beams are angled a few degrees downward due to the curvature of the Earth. Depending on whether we focus positive or negative pions, we have either a neutrino or antineutrino beam with a few percent contamination of electron neutrino or antineutrinos.

In T2K and NOvA, an additional trick is used. The detectors are placed slightly off of the beam axis ($\sim$2.5° at the target), and the on-axis neutrino beam energy is tuned a little way above the desired energy. This gives a slightly narrower flux peak compared to an on-axis beam tuned for the same energy, and drastically reduces the tail out at higher energies. Both of these effects are useful for $\overset{(-)}{\nu}_{e}$ appearance experiments; the latter one particularly as feed down from neutral current interactions of high energy neutrinos is a potentially large background to the $\overset{(-)}{\nu}_{e}$ charged-current (CC) interaction signal.

Whilst MINOS is an older experiment, and was optimized for studying $\overset{(-)}{\nu}_{\mu}$ disappearance, both T2K and NOvA were optimized to study $\overset{(-)}{\nu}_{e}$ appearance.

5.1. *MINOS analysis*

The MINOS experiment started detecting neutrinos from the Main Injector (NuMI) neutrino beam at the Fermi National Accelerator Laboratory from 2005 to 2012 at a typical beam power of 350 kW. The experiment had both a near and far detector with a baseline of 1.04 km and 735 km, respectively.

The two detectors were functionally identical. They were both made of alternating layers of 1.00 cm thick plastic-scintillator and 2.54 cm thick steel planes. However, the ND was smaller with a mass of 0.98 kt, whilst the far detector was bigger with a mass of 5.4 kt to compensate the lower flux. As neutrinos travelled through the detector, they interacted with the iron nuclei, and the charged final-state particles travelled through the scintillator depositing energy which was read out as light.

The granularity of the MINOS detectors made resolving any topology from electromagnetic showers almost impossible. Therefore, all shower-like candidate events were further classified using a library-event matching (LEM) algorithm.[54,55] The final discriminating variable between signal and background, α_{LEM}, comes from a neural network that uses the LEM variables and the reconstructed energy of the event.

Furthermore, MINOS couldn't distinguish ν_e and $\overline{\nu}_e$ events directly; however, the relative number of neutrino and antineutrino interactions in the neutrino-dominated and antineutrino-enhanced beams was well known.

In 2011, the MINOS experiment reported a slight excess of ν_e-like interactions,[56] although with a large background. Figure 17 shows the

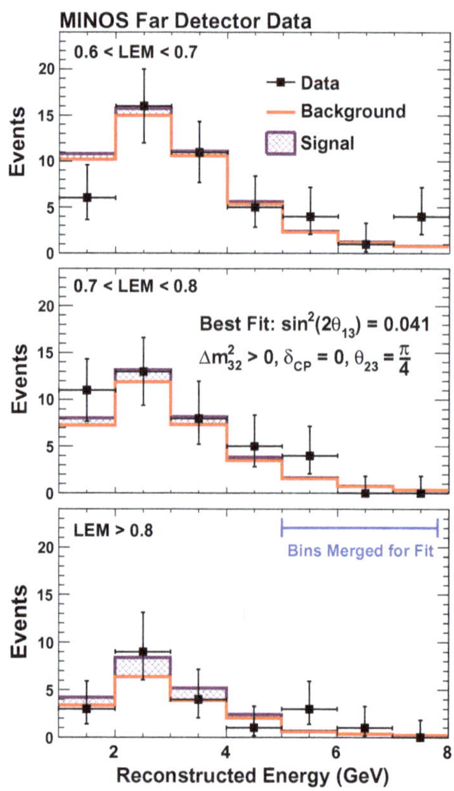

Fig. 17. MINOS reconstructed energy spectra for ν_e CC candidate events in 2011 in the far detector[56] for three bins in the final discriminating variable. The black points indicate the data with their statistical uncertainties. The histogram indicates the expected background (unfilled area) together with the contribution of $\nu_\mu \rightarrow \nu_e$ signal (hatched area) for the best-fit value of $\sin^2 2\theta_{13} = 0.041$.

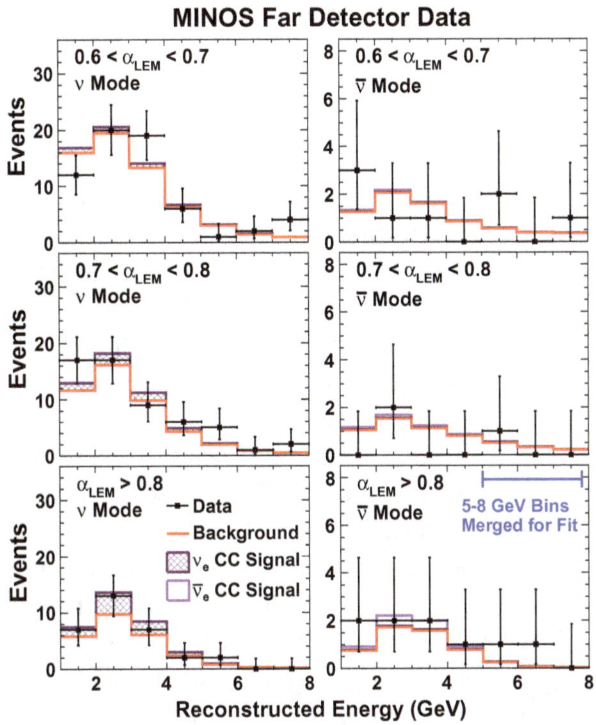

Fig. 18. The MINOS reconstructed energy distributions in 2013[57] for three bins in the LEM discriminating variable for a neutrino (left) and antineutrino (right) enhanced beam. The data are represented by the black dots with their statistical uncertainties. The histogram indicates the expected background (unfilled area) together with the contribution of $\nu_\mu \rightarrow \nu_e$ neutrino and antineutrino signal (hatched area) for the best-fit value of $\sin^2 2\theta_{13} = 0.051$ and $\Delta m^{32}_2 > 0$, $\delta_{\rm CP} = 0$ and $\theta_{23} = \pi/4$.

energy spectra of these events, divided into bins of the LEM discriminant variable. The data were fit to extract a measurement of θ_{13}. In 2013, MINOS updated the results with the full statistics,[57] 3.3×10^{20} protons on target, and continued to see a slight excess (see Fig. 18). A total of 161.4 background events were expected at the far detector in the neutrino-dominated beam, and 21.4 events in the antineutrino-enhanced beam. A total of 152 and 20 events, respectively, were observed in data for a value of $\alpha_{\rm LEM}$ greater than 0.6.

The measured value of θ_{13} depends on the CP-violating phase $\delta_{\rm CP}$, which directly affects the ν_e and $\overline{\nu}_e$ appearance probabilities, and the mass ordering, which affects the appearance probabilities through the interactions of the neutrinos with the matter in the Earth's crust. Assuming $\delta_{\rm CP} = 0$, and $\theta_{23} < \pi/4$, MINOS cites best-fit values for $2\sin^2 2\theta_{13}\sin^2\theta_{23}$ of $0.051^{+0.038}_{-0.030}$ under the normal ordering assumption and $0.093^{+0.054}_{-0.049}$ for the inverted ordering.

The best-fit values show a very weak dependence on the choice of octant for θ_{23}.

5.2. *T2K analysis*

The T2K experiment started data taking in 2010 and has run continuously apart from an ∼1 year interruption due to the Great East Japan Earthquake in 2011. T2K uses an almost pure muon flavor $\overset{(-)}{\nu}_\mu$ off-axis at the target beam produced at J-PARC with a beam power that has been gradually increasing up to almost 500 kW at the beginning of 2017. The T2K experiment ran with a neutrino-dominated beam until 2015; after that, it started to run alternating with an antineutrino-enhanced beam as well.

The near detector, ND280, is off-axis with a baseline of 280 m and is a magnetised tracking detector. An on-axis detector, INGRID, composed of iron-absorber/scintillator-tracker sandwich modules, is used to monitor the beam stability and the off-axis angle as the neutrino beam direction.

T2K makes use of the 22.5 kt (fiducial) SK water-Čherenkov detector as its far detector. The flux peak is in the hundred–MeV region, and at these energies, the ν_e signal is from the CC quasielastic interaction, $\nu_e+n \to p+e^-$ on ^{16}O nuclei, which is detected as a fuzzy Čherenkov ring.

The importance of exclusive interaction channels, both as signal and sources of background, means that T2K needs a good understanding of cross-sections. ND280 uses fine-resolution scintillator detectors as neutrino target regions and time-projection chambers for momentum measurement and particle identification. This can be directly used to provide a much better estimate of interaction rates at SK than from simulation alone.

To improve the flux predictions in T2K, a replica target was provided to and used in the NA61 hadroproduction experiment[59,60] at CERN. The primary beam energy at NA61 is also very close to that used by T2K, meaning there is very little extrapolation to be made from NA61 data to T2K flux inputs.

The first evidence of nonzero θ_{13} was reported[58] in the $\nu_\mu \to \nu_e$ appearance channel in 2011 (see Fig. 19).

In 2013, with more data collected by T2K, the $\nu_\mu \to \nu_e$ transition was observed.[61] With the fixed parameters $|\Delta m^2_{32}| = 2.4 \times 10^{-3}$ eV^2, $\sin^2\theta_{23} = 0.5$, $\delta_{\rm CP} = 0$ and $\Delta m^2_{32} > 0$ ($\Delta m^2_{32} < 0$), a best-fit value of $\sin^2 2\theta_{13} = 0.140^{+0.038}_{-0.032}(0.170 \pm 0.045)$ was obtained with a significance of 7.3σ for a non-nonzero θ_{13}.

More recently, the reconstructed neutrino energy spectra for the ν_e and $\overline{\nu}_e$ samples was measured[62] for 7.482×10^{20} (7.471×10^{20}) protons-on-target for

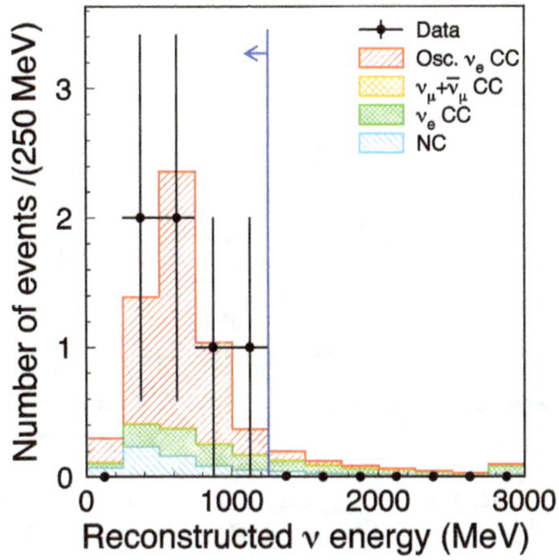

Fig. 19. The 2011 T2K reconstructed neutrino energy at the far detector[58] for the ν_e candidate samples. The data are shown using points with error bars (statistical only) and the predictions from simulation are in shaded histograms, corresponding to oscillated ν_e charged current signal and various background sources for $\sin^2 2\theta_{13} = 0.1$. The vertical line shows the applied selection cut.

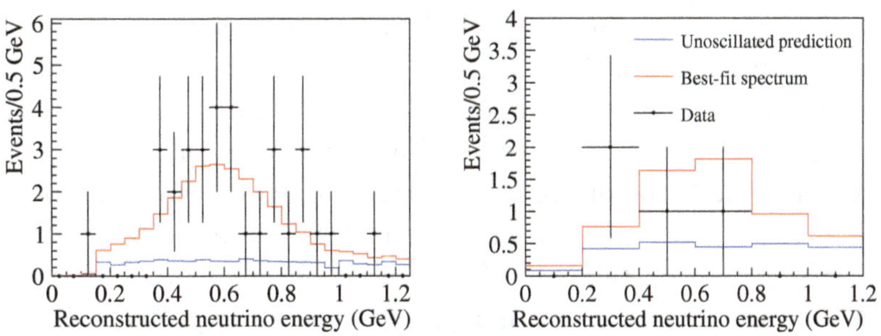

Fig. 20. The T2K reconstructed neutrino energy[62] at the far detector for the ν_e (left) and $\overline{\nu}_e$ (right) candidate samples is shown together with the expected distribution without oscillation (blue histogram) and the best fit (red histogram).

a neutrino (antineutrino) enhanced beam as shown in Fig. 20. These results will contribute to the combined appearance and disappearance neutrino and antineutrino fits described in Section 5.4.

5.3. *NOvA analysis*

NOvA uses the Fermilab NuMI neutrino beam, upgraded to allow 700 kW maximum beam power. The NOvA experiment has two detectors located 1 km and 810 km from the NuMI beam target with total masses of 193 t and 14 kt, respectively. Both are sited 14.6 mrad off the central axis of the beam, as measured from the average neutrino production point, where they observe

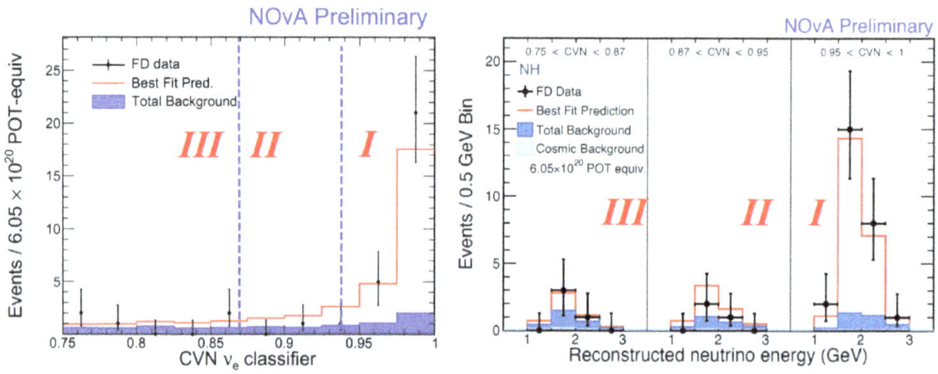

Fig. 21. The NOvA CVN distribution (left) and reconstructed energy spectrum (right) of ν_e CC selected events[64] in the far detector (black points), the best fit-prediction (red histogram) and the background simulation (blue-shaded histograms).

neutrinos mainly in a narrow range of energies between 1 and 3 GeV. The NOvA detectors are functionally equivalent tracking calorimeters.

A first measurement of electron neutrino appearance in NOvA was performed at the beginning of 2016,[63] to be then replaced in the Summer 2016 with updated results.[64]

A convolutional neural network (CVN) based algorithm[65] has been implemented in NOvA to serve as the primary event identifier. It uses pixels as inputs and the output is a variable that describes the probability to be a ν_e CC event with a range 0–1. Using the CVN selection with 6.05×10^{20} protons-on-target NuMI neutrino beam data, a total of 33 ν_e candidates in the far detector are observed, and the expected background is 8.2 ± 0.8 events (see Fig. 21). The significance of ν_e appearance is greater than 8σ.

The data used for this analysis also include a period when the far detector was still being constructed, so the effective fiducial mass varied from 2.3 kt for 4.0 kt of total mass to 10 kt for the full 14 kt during the period considered for this analysis.

5.4. *Combined appearance and disappearance neutrino and antineutrino fits*

To gain maximum sensitivity to the mass ordering and $\delta_{\rm CP}$, one needs to perform a full three-flavor analysis combining ν_μ and $\overline{\nu}_\mu$ disappearance with ν_e and $\overline{\nu}_e$ appearance. Full three-flavor analyses have been presented by MINOS,[67] T2K[62,68] and NOvA.[64]

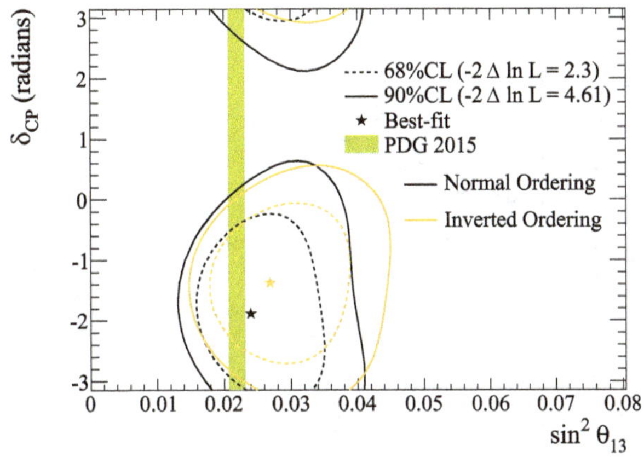

Fig. 22. The T2K 68% (90%) constant $-2\Delta \ln L$ confidence regions in the $\delta_{\text{CP}}-\sin^2\theta_{13}$ plane[62] are shown by the dashed (continuous) lines for the normal (black) and inverted (yellow) mass ordering. The best-fit point is shown by a star. The 68% $\sin^2\theta_{13}$ confidence region from reactor experiments from PDG2015[66] is indicated by the yellow vertical band.

The latest T2K results,[62] which also include antineutrinos, are shown in Fig. 22. Confidence regions in the $\sin^2\theta_{13}-\delta_{\text{CP}}$ plane are computed independently for both mass-ordering hypotheses. The addition of antineutrino samples at SK gives the first sensitivity to δ_{CP} from T2K data alone. There is good agreement between the T2K result and the reactor measurement for $\sin^2\theta_{13}$. For both mass-ordering hypotheses, the best-fit value of δ_{CP} is close to $\pi/2$.

6. Global oscillation fits

A global analysis of neutrino data in the framework of three-neutrino oscillations is needed to fully interpret the results from the experiments and to combine them.

Earlier global analyses of the available oscillation data in a three-neutrino framework indicated that θ_{13} was nonzero,[69,70] which was then experimentally measured. The current analyses mainly focus on the yet unknown parameters like δ_{CP}.

An up-to-date (as of Fall 2016) global analysis from the NuFIT collaboration is performed[71] using the current data from the modern reactor experiments Double CHOOZ,[72] Daya Bay,[35] and RENO[31] and the accelerator long-baseline experiments MINOS,[57] T2K[73,74] and NOvA[64] where earlier results than the ones we presented for Double CHOOZ[72] and T2K[73,74] are used in the global fit. Alternative recent global fits have been presented[75,76] as well. The combined global analysis starts to show some

Table 2. Three-flavor oscillation parameters from the NuFIT collaboration.[71] The numbers in the first (second) column are obtained assuming normal (inverse) ordering, whereas in the third column, the results are for any mass ordering. Both the best-fit point and the 3σ range are shown.

	Normal ordering (best fit)		Inverted ordering ($\Delta\chi^2 = 0.83$)		Any ordering
	best fit $\pm 1\sigma$	3σ range	best fit $\pm 1\sigma$	3σ range	3σ range
$\sin^2\theta_{12}$	$0.306^{+0.012}_{-0.012}$	$0.271 \to 0.345$	$0.306^{+0.012}_{-0.012}$	$0.271 \to 0.345$	$0.271 \to 0.345$
$\theta_{12}/^\circ$	$33.56^{+0.77}_{-0.75}$	$31.38 \to 35.99$	$33.56^{+0.77}_{-0.75}$	$31.38 \to 35.99$	$31.38 \to 35.99$
$\sin^2\theta_{23}$	$0.441^{+0.027}_{-0.021}$	$0.385 \to 0.635$	$0.587^{+0.020}_{-0.024}$	$0.393 \to 0.640$	$0.385 \to 0.638$
$\theta_{23}/^\circ$	$41.6^{+1.5}_{-1.2}$	$38.4 \to 52.8$	$50.0^{+1.1}_{-1.4}$	$38.8 \to 53.1$	$38.4 \to 53.0$
$\sin^2\theta_{13}$	$0.02166^{+0.00075}_{-0.00075}$	$0.01934 \to 0.02392$	$0.02179^{+0.00076}_{-0.00076}$	$0.01953 \to 0.02408$	$0.01934 \to 0.02397$
$\theta_{13}/^\circ$	$8.46^{+0.15}_{-0.15}$	$7.99 \to 8.90$	$8.49^{+0.15}_{-0.15}$	$8.03 \to 8.93$	$7.99 \to 8.91$
$\delta_{\mathrm{CP}}/^\circ$	261^{+51}_{-59}	$0 \to 360$	277^{+40}_{-46}	$145 \to 391$	$0 \to 360$
$\dfrac{\Delta m^2_{21}}{10^{-5}\,\mathrm{eV}^2}$	$7.50^{+0.19}_{-0.17}$	$7.03 \to 8.09$	$7.50^{+0.19}_{-0.17}$	$7.03 \to 8.09$	$7.03 \to 8.09$
$\dfrac{\Delta m^2_{3\ell}}{10^{-3}\,\mathrm{eV}^2}$	$+2.524^{+0.039}_{-0.040}$	$+2.407 \to +2.643$	$-2.514^{+0.038}_{-0.041}$	$-2.635 \to -2.399$	$\begin{bmatrix} +2.407 \to +2.643 \\ -2.629 \to -2.405 \end{bmatrix}$

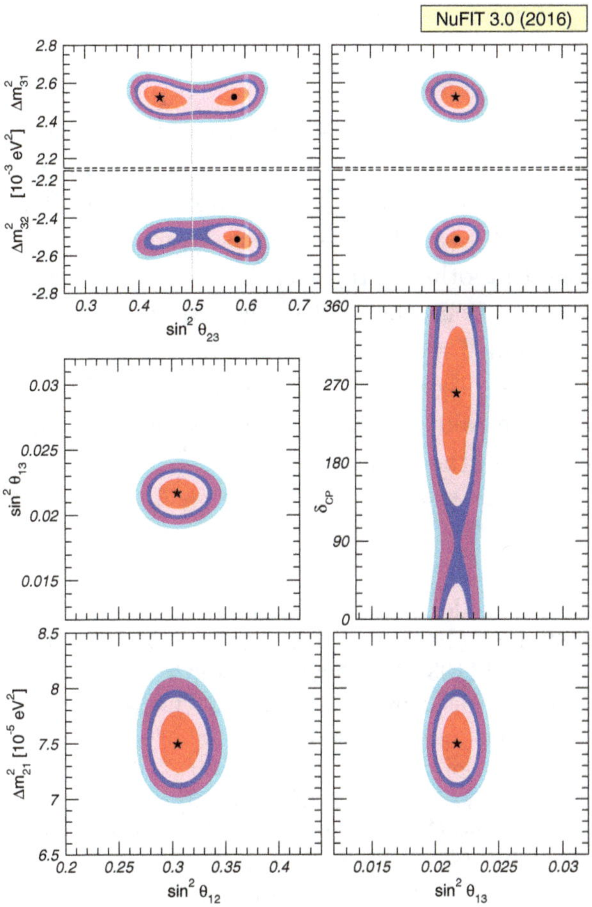

Fig. 23. Global oscillation analysis from NuFIT.[71] Each panel shows the two-dimensional projection of the allowed region after marginalisation with respect to the undisplayed parameters. The different contours correspond to 1σ, 90%, 2σ, 99%, 3σ C.L. The regions in the four lower panels are obtained from $\Delta\chi^2$ minimised with respect to the mass ordering.

sensitivity to subtle effects like the θ_{23} octant or the $\delta_{\rm CP}$ phase (though still at low statistical significance). A total of six parameters are fit and the results are shown in Table 2, where a value of $\sin^2\theta_{13} = 0.02166^{+0.00075}_{-0.00075}$ $(0.02179^{+0.00076}_{-0.00076})$ or $\theta_{13}/^\circ = 8.46^{+0.15}_{-0.15}$ $(8.49^{+0.15}_{-0.15})$ for normal (inverted) ordering is obtained for θ_{13} from the fit. The corresponding confidence intervals of the six parameters are shown in Fig. 23.

7. Summary

In the space of 10 months in 2011–2012, five experiments have produced results consistent with a value of $\sin^2(2\theta_{13}) \sim 0.1$ which means that θ_{13} is nonzero.

Since then, more data and experiments have contributed and now the measurement is considered well established.

Whilst the reactor experiments provide precise measurements that dominate the current value, the results from the long-baseline experiments can be used to study CP violation, θ_{23} and even the mass ordering.

The field as a whole has rapidly transitioned from θ_{13} as an unknown parameter towards θ_{13} as the most precisely measured mixing angle and input for investigating other parameters.

The final result from the NuFIT global fit[71] is: $\theta_{13}/° = 8.46^{+0.15}_{-0.15}$ $(8.49^{+0.15}_{-0.15})$ for normal (inverted) ordering.

The experiments described in this text, apart from MINOS, are all expected to continue taking data, so the precision and information on θ_{13} will further improve in the coming years.

References

1. Y. Fukuda *et al.*, Evidence for oscillation of atmospheric neutrinos, *Phys. Rev. Lett.* **81**, 1562–1567 (1998). doi: 10.1103/PhysRevLett.81.1562. URL http://link.aps.org/doi/10.1103/PhysRevLett.81.1562.
2. Q. R. Ahmad *et al.*, Direct evidence for neutrino flavor transformation from neutral-current interactions in the sudbury neutrino observatory. URL http://link.aps.org/doi/10.1103/PhysRevLett.89.011301 (2002).
3. S. N. Ahmed *et al.*, Measurement of the total active ^{8}B solar neutrino flux at the Sudbury Neutrino Observatory with enhanced neutral current sensitivity, *Phys. Rev. Lett.* **92**, 181301 (2004). doi: 10.1103/PhysRevLett.92.181301. URL http://link.aps.org/doi/10.1103/PhysRevLett.92.181301.
4. B. Pontecorvo, Mesonium and anti-mesonium, *Sov. Phys. JETP.* **6**, 429 (1957). [Zh. Eksp. Teor. Fiz. 33549 (1957)].
5. B. Pontecorvo, Neutrino experiments and the problem of conservation of leptonic charge, *Sov. Phys. JETP.* **26**, 984–988 (1968). [Zh. Eksp. Teor. Fiz. 531717 (1967)].
6. Z. Maki, M. Nakagawa, and S. Sakata, Remarks on the unified model of elementary particles, *Prog. Theor. Phys.* **28**, 870–880 (1962). doi: 10.1143/PTP.28.870.
7. M. Apollonio *et al.*, Limits on neutrino oscillations from the CHOOZ experiment, *Phys. Lett.* **B466**, 415–430 (1999). doi: 10.1016/S0370-2693(99)01072-2.
8. M. Apollonio *et al.*, Search for neutrino oscillations on a long baseline at the CHOOZ nuclear power station, *Eur. Phys. J.* **C27**, 331–374 (2003). doi: 10.1140/epjc/s2002-01127-9.
9. F. An *et al.*, Neutrino Physics with JUNO, *J. Phys.* **G43**(3), 030401 (2016). doi: 10.1088/0954-3899/43/3/030401.
10. S. Parke, What is Δm^2_{ee}? *Phys. Rev.* **D93**(5), 053008 (2016). doi: 10.1103/PhysRevD.93.053008.

11. W. Winter, Lectures on neutrino phenomenology, *Nucl. Phys. Proc. Suppl.* **203–204**, 45–81 (2010). doi: 10.1016/j.nuclphysbps.2010.08.005.
12. F. Reines, C. L. Cowan, F. B. Harrison, A. D. McGuire, and H. W. Kruse, Detection of the free anti-neutrino, *Phys. Rev.* **117**, 159–173 (1960). doi: 10.1103/PhysRev.117.159.
13. M. Apollonio *et al.*, Initial results from the CHOOZ long baseline reactor neutrino oscillation experiment, *Phys. Lett.* **B420**, 397–404 (1998). doi: 10.1016/S0370-2693(97)01476-7.
14. F. Boehm *et al.*, Final results from the Palo Verde neutrino oscillation experiment, *Phys. Rev.* **D64**, 112001 (2001). doi: 10.1103/PhysRevD.64.112001.
15. F. Ardellier *et al.*, Double Chooz: A Search for the neutrino mixing angle θ_{13} (2006). hep-ex/0606025.
16. X. Guo *et al.*, A precision measurement of the neutrino mixing angle θ_{13} using reactor antineutrinos at Daya-Bay (2007). hep-ex/0701029.
17. J. K. Ahn *et al.*, RENO: An experiment for neutrino oscillation parameter θ_{13} using reactor neutrinos at yonggwang (2010). hep-ex/1003.1391.
18. A. Cucoanes, P. Novella, A. Cabrera, M. Fallot, A. Onillon, M. Obolensky, and F. Yermia, Reactor neutrino flux uncertainty suppression on multiple detector experiments (2015). hep-ex/1501.00356.
19. Y. Abe *et al.*, Indication for the disappearance of reactor electron antineutrinos in the Double Chooz experiment, *Phys. Rev. Lett.* **108**, 131801 (2012). doi: 10.1103/PhysRevLett.108.131801.
20. F. P. An *et al.*, Observation of electron-antineutrino disappearance at Daya Bay, *Phys. Rev. Lett.* **108**, 171803 (2012). doi: 10.1103/PhysRevLett.108.171803.
21. J. K. Ahn *et al.*, Observation of reactor electron antineutrino disappearance in the RENO Experiment, *Phys. Rev. Lett.* **108**, 191802 (2012). doi: 10.1103/PhysRevLett.108.191802.
22. Y. Declais *et al.*, Study of reactor antineutrino interaction with proton at bugey nuclear power plant, *Physics Letters B.* **338**(2), 383–389 (1994). ISSN 0370-2693. doi: http://dx.doi.org/10.1016/0370-2693(94)91394-3. URL http://www.sciencedirect.com/science/article/pii/0370269394913943.
23. F. P. An *et al.*, A side-by-side comparison of Daya Bay antineutrino detectors, *Nucl. Instrum. Meth.* **A685**, 78–97 (2012). doi: 10.1016/j.nima.2012.05.030.
24. F. P. An *et al.*, Improved measurement of electron antineutrino disappearance at Daya Bay, *Chin. Phys.* **C37**, 011001 (2013). doi: 10.1088/1674-1137/37/1/011001.
25. F. P. An *et al.*, Spectral measurement of electron antineutrino oscillation amplitude and frequency at Daya Bay, *Phys. Rev. Lett.* **112**, 061801 (2014). doi: 10.1103/PhysRevLett.112.061801.
26. J. H. Choi *et al.*, Observation of energy and baseline dependent reactor antineutrino disappearance in the RENO experiment, *Phys. Rev. Lett.* **116**(21), 211801 (2016). doi: 10.1103/PhysRevLett.116.211801.
27. Y. Abe *et al.*, First measurement of θ_{13} from delayed neutron capture on hydrogen in the Double Chooz experiment, *Phys. Lett.* **B723**, 66–70 (2013). doi: 10.1016/j.physletb.2013.04.050.

28. F. P. An *et al.*, Independent measurement of the neutrino mixing angle θ_{13} via neutron capture on hydrogen at Daya Bay, *Phys. Rev.* **D90**(7), 071101 (2014). doi: 10.1103/PhysRevD.90.071101.
29. Y. Abe *et al.*, Background-independent measurement of θ_{13} in Double Chooz, *Phys. Lett.* **B735**, 51–56 (2014). doi: 10.1016/j.physletb.2014.04.045.
30. Y. Abe *et al.*, Measurement of $_{13}$ in Double Chooz using neutron captures on hydrogen with novel background rejection techniques, *JHEP.* **01**, 163 (2016). doi: 10.1007/JHEP01(2016)163.
31. H. Seo *et al.*, Spectral measurement of the electron antineutrino oscillation amplitude and frequency using 500 live days of RENO data (2016). hep-ex/1610.04326.
32. F. P. An *et al.*, New measurement of antineutrino oscillation with the full detector configuration at Daya Bay, *Phys. Rev. Lett.* **115**(11), 111802 (2015). doi: 10.1103/PhysRevLett.115.111802.
33. F. P. An *et al.*, Improved measurement of the reactor antineutrino flux and spectrum at Daya Bay, *Chin. Phys.* **C41**(1), 013002 (2017). doi: 10.1088/1674-1137/41/1/013002.
34. A. C. Serra. Double Chooz Improved Multi-Detector Measurements. EP Seminar, CERN, 20 Sept. 2016. .
35. F. P. An *et al.*, Measurement of electron antineutrino oscillation based on 1230 days of operation of the Daya Bay experiment (2016). hep-ex/1610.04802.
36. F. P. An *et al.*, New measurement of θ_{13} via neutron capture on hydrogen at Daya Bay, *Phys. Rev.* **D93**(7), 072011 (2016). doi: 10.1103/PhysRevD.93.072011.
37. Y. Abe *et al.*, Improved measurements of the neutrino mixing angle θ_{13} with the Double Chooz detector, *JHEP.* **10**, 086 (2014). doi: 10.1007/JHEP02(2015)074,10.1007/JHEP10(2014)086. [Erratum: JHEP02,074(2015)].
38. P. Vogel. Evaluation of reactor neutrino flux: Issues and uncertainties. In *Prospects in Neutrino Physics (NuPhys2015) London, UK, December 16–18, 2015* (2016). URL http://inspirehep.net/record/1436503/files/ arXiv:1603.08990.pdf.
39. P. Huber, Reactor antineutrino fluxes: Status and challenges, *Nucl. Phys.* **B908**, 268–278 (2016). doi: 10.1016/j.nuclphysb.2016.04.012.
40. A. C. Hayes and P. Vogel, Reactor neutrino spectra, *Ann. Rev. Nucl. Part. Sci.* **66**, 219–244 (2016). doi: 10.1146/annurev-nucl-102115-044826.
41. T. A. Mueller *et al.*, Improved predictions of reactor antineutrino spectra, *Phys. Rev.* **C83**, 054615 (2011). doi: 10.1103/PhysRevC.83.054615.
42. P. Huber, Determination of antineutrino spectra from nuclear reactors, *Phys. Rev. C.* **84**, 024617 (2011). doi: 10.1103/PhysRevC.84.024617. URL http://link.aps.org/doi/10.1103/PhysRevC.84.024617.
43. F. von Feilitzsch, A. Hahn, and K. Schreckenbach, Experimental beta-spectra from 239pu and 235u thermal neutron fission products and their correlated antineutrino spectra, *Phys. Lett. B.* **118**(1), 162–166 (1982). ISSN 0370-2693. doi: http://dx.doi.org/10.1016/0370-2693(82)90622-0. URL http://www.sciencedirect.com/science/article/pii/0370269382906220.

44. K. Schreckenbach, G. Colvin, W. Gelletly, and F. V. Feilitzsch, Determination of the antineutrino spectrum from 235u thermal neutron fission products up to 9.5 mev, *Phys. Lett. B.* **160**(4), 325–330 (1985). ISSN 0370-2693. doi: http://dx.doi.org/10.1016/0370-2693(85)91337-1. URL http://www.sciencedirect.com/science/article/pii/0370269385913371.
45. A. Hahn, K. Schreckenbach, W. Gelletly, F. von Feilitzsch, G. Colvin, and B. Krusche, Antineutrino spectra from 241pu and 239pu thermal neutron fission products, *Phys. Lett. B.* **218**(3), 365–368 (1989). ISSN 0370-2693. doi: http://dx.doi.org/10.1016/0370-2693(89)91598-0. URL http://www.sciencedirect.com/science/article/pii/0370269389915980.
46. A. Hayes. Reactor Neutrino Fluxes: Theory and Uncertainty. NuPhys2016: Prospects in Neutrino Physics, London (UK), 12–14 Dec. 2016 .
47. F. P. An *et al.*, Measurement of the reactor antineutrino flux and spectrum at Daya Bay, *Phys. Rev. Lett.* **116**(6), 061801 (2016). doi: 10.1103/PhysRevLett.116.061801,10.1103/PhysRevLett.118.099902. [Erratum: *Phys. Rev. Lett.* **118**, no. 9, 099902 (2017)].
48. J. I. Crespo-Anadn, Double Chooz: Latest results, *Nucl. Part. Phys. Proc.* **265–266**, 99–104 (2015). doi: 10.1016/j.nuclphysbps.2015.06.025.
49. J. Cao and K.-B. Luk, An overview of the Daya Bay reactor neutrino experiment, *Nucl. Phys.* **B908**, 62–73 (2016). doi: 10.1016/j.nuclphysb.2016.04.034.
50. K. Joo. New results from RENO & prospects with RENO-50. The XXVII International Conference on Neutrino Physics an Astrophysics, London (UK), 5–9 July 2016.
51. D. Michael *et al.*, The magnetized steel and scintillator calorimeters of the {MINOS} experiment, *Nucl. Instrum. Method. Phys. Res.* **596**(2), 190–228 (2008). ISSN 0168-9002. doi: http://dx.doi.org/10.1016/j.nima.2008.08.003. URL http://www.sciencedirect.com/science/article/pii/S016890020 8011613.
52. K. Abe *et al.*, The {T2K} experiment, *Nucl. Instrum. Method. Phys. Res.* **659**(1), 106–135 (2011). ISSN 0168-9002. doi: http://dx.doi.org/10.1016/j.nima.2011.06.067. URL http://www.sciencedirect.com/science/ article/pii/S0168900211011910.
53. D. S. Ayres *et al.*, The NOvA Technical Design Report (2007).
54. J. P. Ochoa. Ph.D. Thesis, California Institute of Technology, FERMILAB-THESIS-2009-44 (2009).
55. R. Toner. R. Toner, Ph.D. Thesis, University of Cambridge, FERMILAB-THESIS-2011-53 (2011).
56. P. Adamson *et al.*, Improved search for muon-neutrino to electron-neutrino oscillations in MINOS, *Phys. Rev. Lett.* **107**, 181802 (2011). doi: 10.1103/PhysRevLett.107.181802.
57. P. Adamson *et al.*, Electron neutrino and antineutrino appearance in the full MINOS data sample, *Phys. Rev. Lett.* **110**(17), 171801 (2013). doi: 10.1103/PhysRevLett.110.171801.
58. K. Abe *et al.*, Indication of electron neutrino appearance from an Accelerator-produced off-axis muon neutrino beam, *Phys. Rev. Lett.* **107**, 041801 (2011). doi: 10.1103/PhysRevLett.107.041801.

59. N. Abgrall *et al.*, Measurements of cross sections and charged pion spectra in proton-carbon interactions at 31 GeV/c, *Phys. Rev. C.* **84**, 034604 (2011). doi: 10.1103/PhysRevC.84.034604. URL http://link.aps.org/doi/10. 1103/ PhysRevC.84.034604.
60. N. Abgrall *et al.*, Measurement of production properties of positively charged kaons in proton-carbon interactions at 31 GeV/c, *Phys. Rev. C.* **85**, 035210 (2012). doi: 10.1103/PhysRevC.85.035210. URL http://link. aps.org/doi/10. 1103/PhysRevC.85.035210.
61. K. Abe *et al.*, Observation of electron neutrino appearance in a muon neutrino beam, *Phys. Rev. Lett.* **112**, 061802 (2014). doi: 10.1103/PhysRevLett.112. 061802.
62. K. Abe *et al.*, First combined analysis of neutrino and antineutrino oscillations at T2K (2017). hep-ex/1701.00432.
63. P. Adamson *et al.*, First measurement of electron neutrino appearance in NOvA, *Phys. Rev. Lett.* **116**(15), 151806 (2016). doi: 10.1103/PhysRevLett. 116.151806.
64. P. Vahle. New Results from NOvA. NEUTRINO2016, XXVII International Conference on Neutrino Physics and Astrophysics, 4–9 July 2016, London, UK.
65. A. Aurisano *et al.*, A convolutional neural network neutrino event classifier, *JINST.* **11**(09), P09001 (2016). doi: 10.1088/1748-0221/11/09/P09001.
66. K. Olive and P. D. Group, Review of particle physics, *Chin. Phys. C.* **38**(9), 090001 (2014). URL http://stacks.iop.org/1674-1137/38/i=9/a=090001.
67. P. Adamson *et al.*, Measurement of neutrino and antineutrino oscillations using beam and atmospheric data in MINOS, *Phys. Rev. Lett.* **110**(25), 251801 (2013). doi: 10.1103/PhysRevLett.110.251801.
68. K. Abe *et al.*, Measurements of neutrino oscillation in appearance and disappearance channels by the T2K experiment with 6.610^{20} protons on target, *Phys. Rev.* **D91**(7), 072010 (2015). doi: 10.1103/PhysRevD.91.072010.
69. G. L. Fogli, E. Lisi, A. Marrone, A. Palazzo, and A. M. Rotunno, Hints of $\theta_{13} > 0$ from global neutrino data analysis, *Phys. Rev. Lett.* **101**, 141801 (2008). doi: 10.1103/PhysRevLett.101.141801.
70. G. L. Fogli, E. Lisi, A. Marrone, A. Palazzo, and A. M. Rotunno, Evidence of $\theta_{13} > 0$ from global neutrino data analysis, *Phys. Rev.* **D84**, 053007 (2011). doi: 10.1103/PhysRevD.84.053007.
71. I. Esteban *et al.*, Updated fit to three neutrino mixing: Exploring the accelerator-reactor complementarity, *JHEP.* **01**, 087 (2017). doi: 10.1007/ JHEP01(2017)087.
72. M. Ishitsuka. New results of Double Chooz. Rencontres de Moriond EW 2016, La Thuile, Italy, March 1219, 2016.
73. K. Iwamoto. Recent Results from T2K and Future Prospects. International Conference on High Energy Physics, Chicago, USA, August 310, 2016 .
74. A. Cervera. Latest Results from Neutrino Oscillation Experiments. SUSY 2016 Conference, Melbourne, Australia, July 38, 2016.

75. F. Capozzi, E. Lisi, A. Marrone, D. Montanino, and A. Palazzo, Neutrino masses and mixings: Status of known and unknown 3ν parameters, *Nucl. Phys.* **B908**, 218–234 (2016). doi: 10.1016/j.nuclphysb.2016.02.016.
76. D. V. Forero, M. Tortola, and J. W. F. Valle, Neutrino oscillations refitted, *Phys. Rev.* **D90**(9), 093006 (2014). doi: 10.1103/PhysRevD.90.093006.

Chapter 8

Neutrinos from Supernovae and Other Astrophysical Sources

K. Scholberg

Department of Physics, Duke University,
Durham 27708, NC, USA
schol@phy.duke.edu

Abstract

Stellar core collapses are prodigious sources of neutrinos with a few to a few tens of MeV energy. These neutrinos are messengers of both physics and astrophysics. Thanks to the weakness of their interactions, neutrinos can bring information out from the depths of astronomical objects, from which light cannot escape, to tell us about hidden processes in dense matter. Neutrino-bright objects are also sources of neutrinos which can reveal information about the nature of the neutrinos themselves. This chapter will focus primarily on core-collapse supernovae, the expected neutrino signal and the harvest of physics and astrophysics from them but will mention other potential low-energy neutrino sources as well.

1. Neutrinos from supernovae

1.1. *Supernovae*

Supernovae are violent explosions which can shine, for a brief time, with the electromagnetic luminosity of an entire galaxy.[a] Supernovae come in a variety of observational types, classified by their electromagnetic spectra and their light curves (luminosity as a function of time).[1] In broad brush, Type I supernovae do not have hydrogen lines in their spectra, whereas Type II supernovae do. However each of these supernova types have subcategories depending on other spectral and observational features, e.g., whether

[a]Well, a smallish galaxy... a typical peak supernova luminosity is $(10^9-10^{10})L_\odot$, whereas a galaxy like Milky Way shines with $\sim 10^{10} L_\odot$.

their light curves drop off linearly or plateau and what the explosion energies are. While many of the feature details are determined by the progenitor's envelope and environment, the different observational categories are thought to arise from two main physical mechanisms. In particular, the Type Ia explosion mechanism is thought to be thermonuclear. Such supernovae should produce relatively small neutrino fluxes (see Section 4). In contrast, the mechanism behind Type II supernovae, and some of the other observational types, is certainly the collapse of a massive star to produce a compact remnant, such as a neutron star (NS) or a black hole (BH). These core collapses are cataclysmic events in which nearly all of the gravitational binding energy of the resulting compact remnant is converted to neutrinos. The neutrinos from core-collapse supernovae are the primary focus of this chapter.

Core collapses are expected to occur about three times per century in the Milky Way.[2] The most likely distance will be at about 10 kpc,[3] a bit farther than the center of the Milky Way. As commonly done in the literature, we will take 10 kpc as the standard distance for considering neutrino fluxes and interaction rates.

1.2. *Core-collapse models*

While the overall physical picture is reasonably well understood, there are still many mysteries surrounding the astrophysical mechanisms of core collapses.

The classic (simplified) narrative about the collapse-induced death of a star is as follows: a star of mass greater than about $8\,M_\odot$, after burning successively heavier elements arranged in concentric onion-skin layers, accumulates iron at its core. Fusion of iron does not generate energy, given that iron has the highest binding energy of all nuclei. The iron ash builds up, supported by electron degeneracy pressure, until the core's dead mass exceeds $1.4\,M_\odot$, the Chandrasekhar mass — this is the mass at which gravitational pull exceeds degeneracy pressure and the core can no longer hold itself up. At this moment, the core collapses. Protons and electrons get squeezed together to make neutrons (and neutrinos) and an NS is born. When the collapse reaches "maximum scrunch" (density of nuclear matter), the star rebounds, makes a shock wave that races outwards and blows the star up.

While this story does represent many core collapses, the truth may be rather more complicated — there are now different models with different versions of this picture differing in some significant details. But in all cases, a collapse is followed by the star's release of a blast of neutrinos. Because weakly interacting neutrinos *can* escape, they *do*, and make off with most of the compact remnant's binding energy.

Supercomputer modeling of core collapse has been increasing in sophistication over the past several decades, although many aspects are not yet fully understood. The interested reader is directed to Refs. 4 and 5.

Because the core is dense enough to trap neutrinos for a while, information about core collapse processes is imprinted on the time-dependent neutrino spectra. Therefore observation of the neutrino flux will tell us about the core collapse mechanism.

1.3. *Core-collapse neutrino production*

During a core collapse, a star-size object suddenly shrinks to a kilometer-scale object. The vast binding energy has to be radiated away. Only about 0.01% of this energy ends up in photons (which, recall, shine as bright as a small galaxy for some period of time). About 1% of the energy ends up as kinetic energy of the expanding remnant. The remaining ~99% gets radiated as neutrinos.

How are the neutrinos produced? Some of them, primarily during infall, are ν_e resulting from production of neutrons as protons and electrons get squeezed together, $n + p \rightarrow e^- + \nu_e$. However the great bulk of the flux is produced by thermal processes, including: pair production ($e^+ + e^- \rightarrow \nu + \bar{\nu}$), photo-annihilation ($e^- + \gamma \rightarrow e^- + \nu + \bar{\nu}$), plasmon decay ($\gamma^* \rightarrow \nu + \bar{\nu}$), and nucleon–nucleon bremsstrahlung ($e^- + (Z, A) \rightarrow (Z, A) + e^- + \nu + \bar{\nu}$).[6] All six neutrino flavors (ν_e, ν_μ, ν_τ, and their antiflavors) are produced, with roughly equal luminosities. We typically consider ν_μ, ν_τ, $\bar{\nu}_\mu$ and $\bar{\nu}_\tau$ together, and designate them collectively as ν_x because their fluxes are nearly degenerate in the supernova — they have nearly equal production and interaction rates in the supernova.[b] Furthermore the different ν_x flavor components are observationally essentially impossible to distinguish (see Section 2.1).

The core collapse and subsequent explosion proceeds in several phases on different timescales corresponding to distinctive neutrino signatures. These are summarized in Table 1 and the neutrino luminosities for an example model are shown in Fig. 3. Neutrinos from ν_e ramp up during infall. Neutrino trapping manifests itself as a small notch before the bright

[b] Note that equality is inexact, due to weak magnetism that affects neutrinos and antineutrinos differently.

Table 1. Phases of a core collapse and features of the neutrino signal.

	Timescale	Flavors (unoscillated)	Physics/Features
Infall	-1 ms to -0	ν_e	Core deleptonizes, ν_e turns on
Neutronization	0–60 ms	Mostly ν_e	Trapping notch, ν_e spike
Accretion	60 ms–1 s	ν_e, other flavors turning on	Structure related to SASI, explosion; difference between ν_e, $\bar{\nu}_e$ and ν_x
Cooling	1 s to tens of seconds	All flavors, spectral differences decreasing	Decreasing luminosity and energy

neutronization burst. This burst, also known as the "breakout burst", lasts a few tens of milliseconds and is composed primarily of ν_e from neutronization ($n + p \rightarrow e^- + \nu_e$) which are able to stream from their neutrinosphere.[c] Around this time, the other flavors start to turn on. The next phase, the "accretion phase", lasts around a second, and has complex flavor, time and energy structure, representing dynamics of the protoneutron-star matter and envelope. Features are apparent especially in ν_e as matter falls back on to the core. Neutrino spectra harden and then soften. The SASI, or "Standing Accetion Shock Instability", may also become apparent in this time frame. SASI refers to a dynamical instability resulting in 100–200 Hz "sloshing" oscillations which may be reflected in the neutrino fluxes. Measurements of neutrino fluxes in this phase are likely to give insight into the explosion mechanism. Explosion happens at the "accretion-to-cooling" transition. In the final stage, cooling of the protoneutron star occurs over a few tens of seconds. During this phase, all flavors are emitted and gradually decrease in luminosity and average energy.

Note that there might not always be a bright supernova subsequent to a core collapse. Some core collapses may be electromagnetic "damp squibs" but are nevertheless still neutrino extravaganzas. However these failed supernovae may end up as the most interesting events: the core collapse may result in a BH rather than a NS. The neutrino signal would exhibit sharp cutoff of the neutrino luminosity after about a second (or perhaps longer).

Interesting physics signatures can happen at all times during the event. And of course, the collapse may not unfold at all according to this standard picture — there may be surprises!

[c]The "neutrinosphere" is the surface of last coupling of an emitted neutrino, i.e., the radius at which the mean free path of the neutrino is infinity.

At any given moment in time, the produced neutrino spectra are thought to be reasonably well described by the quasiempirical empirical parameterization for each flavor as follows[7]:

$$\phi(E_\nu) = N_0 \frac{(\alpha+1)^{\alpha+1}}{\langle E_\nu \rangle \Gamma(\alpha+1)} \left(\frac{E_\nu}{\langle E_\nu \rangle}\right)^\alpha \exp\left[-(\alpha+1)\frac{E_\nu}{\langle E_\nu \rangle}\right], \tag{1}$$

where E_ν is the neutrino energy, $\langle E_\nu \rangle$ is the mean neutrino energy, α is a "pinching parameter", Γ is the gamma function, and N_0 is the total number of neutrinos emitted. This functional form produces a roughly Fermi–Dirac-like spectrum, where α controls how much the tails of the distribution are suppressed: a larger value of α means more suppression, i.e., a more pinched spectrum, and $\alpha < 1$ indicates an "antipinched" spectrum (enhanced tail) with respect to a simply thermal distribution.[d] The spectrum is therefore described by three parameters — L, $\langle E_\nu \rangle$, α — where L is the total luminosity, related to the normalization by $N_0 = L/\langle E_\nu \rangle$.

Figure 1 shows some examples of pinched spectra with different α parameters. Figure 2 shows fluences (fluxes integrated over the time interval of the burst) for a specific model. The overall time-dependent burst spectrum and flux can be described well by specifying the three parameters $(L, \langle E_\nu \rangle, \alpha)$ as a function of time for each flavor (ν_e, $\bar{\nu}_e$, ν_x). Figure 3 shows an example of a model flux specified by these parameters over the timescale of the burst, and Fig. 4 shows the energy- and time-dependent fluxes of the three-flavor components on three different timescales corresponding to neutronization, accretion and cooling.

The luminosities of each flavor are approximately equal but details of the spectra exiting the star depend on the interactions in the protoneutron-star matter, which differ for ν_e, $\bar{\nu}_e$ and ν_x. The ν_e have the most interactions with the protoneutron-star matter; hence they decouple farthest out (i.e., they have the largest neutrinospheres), where the matter is coolest. The $\bar{\nu}_e$ have fewer interactions than ν_e, because neutrons are more abundant than protons, so ν_e charged-current (CC) interactions are suppressed; hence $\bar{\nu}_e$ decouple a bit deeper and tend to be a bit hotter. The ν_x have only neutral-current (NC) interactions and so decouple at the deepest neutrinosphere and tend to be hottest. This leads to the standard flavor–energy hierarchy, $\langle E_{\nu_e} \rangle < \langle E_{\bar{\nu}_e} \rangle < \langle E_{\nu_x} \rangle$, which holds for most models over much of the burst.

There are additional time-dependent modulations of luminosities and spectra due to neutrino oscillation, which will be discussed in Section 1.5.

[d] Pinching can have quite a large effect on event rates, since cross sections often increase quadratically with energy.

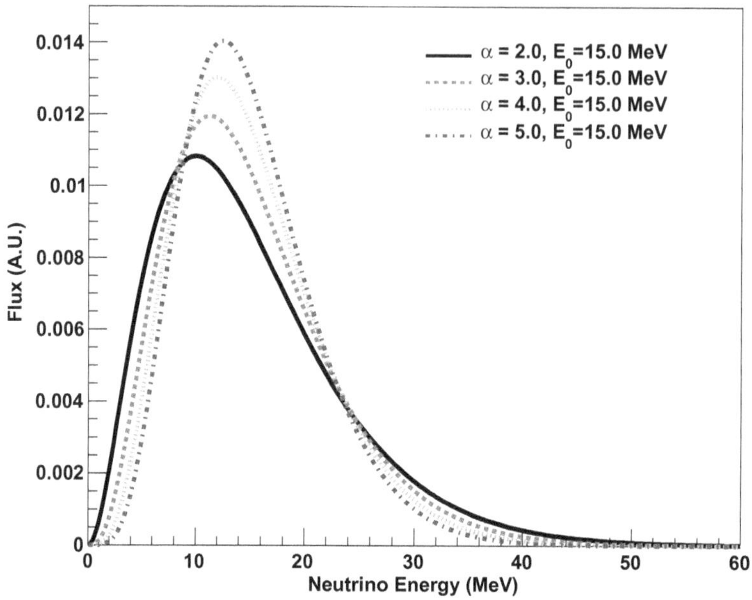

Fig. 1. Examples of spectral shapes (normalized) described using Eq. 1 for constant average energy, showing the effect of the pinching parameter α for values within a plausible range for core-collapse spectra.

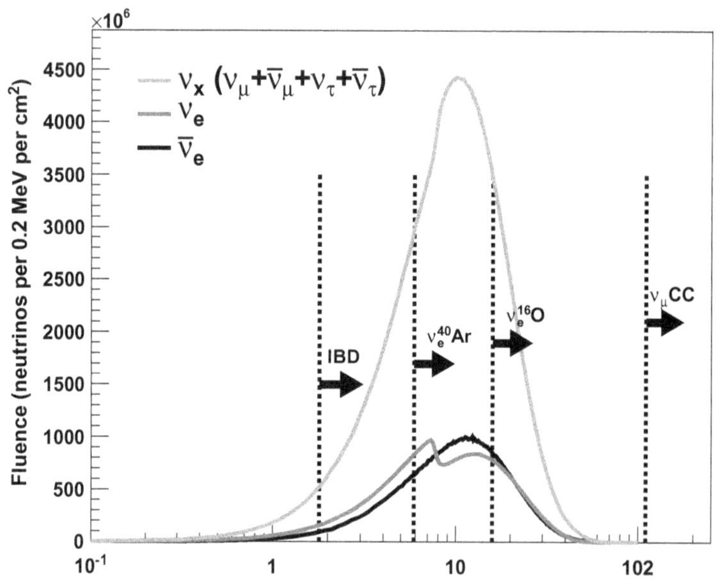

Fig. 2. Neutrino spectra for an example model,[8] integrated over the timescale of the burst. In this particular model, collective effects create spectral swap features. Also shown are interaction thresholds for several neutrino interactions relevant for current detectors (see Section 2.1).

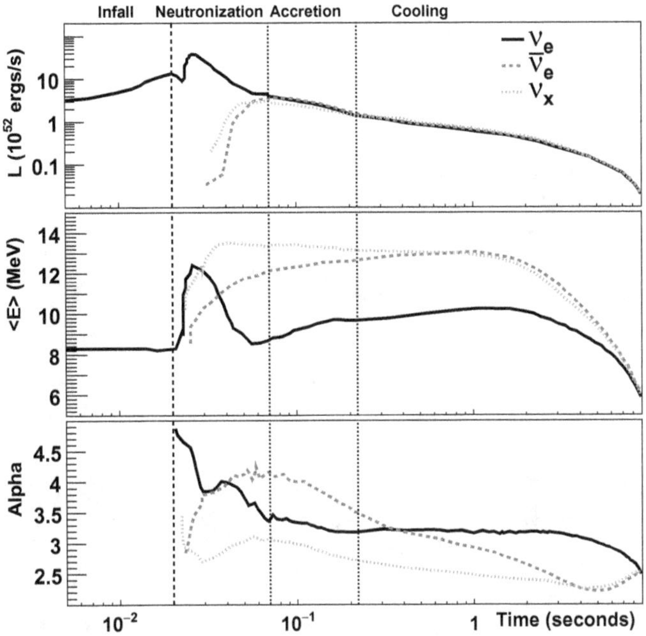

Fig. 3. Expected time-dependent signal for a specific flux model for an electron-capture supernova[10] at 10 kpc. No oscillations are assumed. The top plot shows the luminosity as a function of time, the second plot shows average neutrino energy, and the third plot shows the α (pinching) parameter. The vertical dashed line at 0.02 s indicates the time of core bounce, and the vertical lines indicate different eras in the supernova evolution. The leftmost time interval indicates the infall period. The next interval, from core bounce to 50 ms, is the neutronization burst era, in which the flux is composed primarily of ν_e. The next period, from 50 to 200 ms, is the accretion period. The final era, from 0.2 to 9 s, is the protoneutron-star cooling period. This figure and its caption have been reproduced from Ref. 9.

1.4. *SN1987A*

There has been a single celebrated observation of neutrinos from a core collapse. This was SN1987A in the Large Magellanic Cloud, a satellite galaxy of the Milky Way, about 55 kpc from Earth. A total of about two dozen clean events were recorded from two water Cherenkov detectors, Kamiokande II in Japan and IMB in the US.[11,12] Also reported were observations from Baksan, a scintillator detector in Russia,[13] and the LSD detector in Mont Blanc.[14] The latter events are puzzling and controversial as they occurred out of time with the others. However their association with the supernova cannot be fully ruled out. Vissani[15] gives a comprehensive examination and interpretation of all of the SN1987A neutrino data.

The basic features of the observed neutrino burst — energy range, luminosity, time scale — were roughly consistent with expectation for

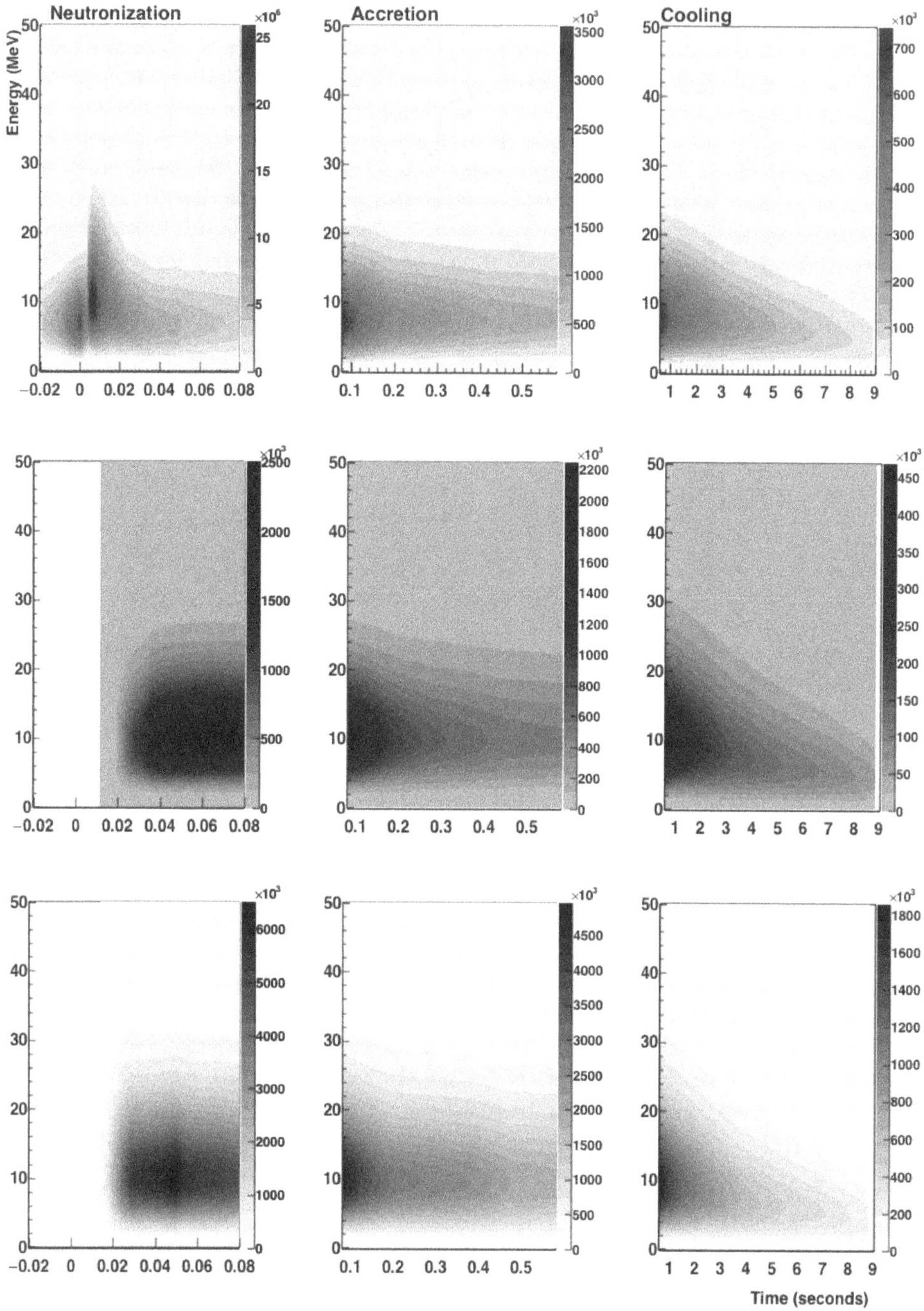

Fig. 4. Example of time-dependent spectra for the electron-capture supernova model,[10] parameterized in Fig. 3, on three different timescales. The z-axis units are neutrinos per cm^2 per millisecond per 0.2 MeV. Top: ν_e. Center: $\bar{\nu}_e$. Bottom: ν_x. Oscillations are not included here; note they can have dramatic effects on the spectra.

a core-collapse supernova. However they were insufficient to discriminate between core-collapse models. They did, however, bring information about neutrinos and about particle physics — see Ref. 16 for a review. For example, the time spread of the events enabled, for the time, some of the best constraints on absolute neutrino mass. The neutrino burst was recorded about 2.5 hours before the first observable light but was not used to alert astronomers. Next time, we will do better (see Section 2.5).

A very large number of papers have been written about the SN1987A neutrino burst. This was a rather atypical supernova with an atypical progenitor (Sanduleak — 69°202, a blue supergiant). No associated compact remnant such as a pulsar has been observed; however that does not rule out one's existence. We await the next nearby supernova to tell us more.

1.5. *Flavor transitions of supernova neutrinos*

The phenomenon of neutrino oscillations has been multiply confirmed (and, in 2015, honored by a Nobel Prize) and explored by many experiments. We now know definitively that neutrinos change flavors as they propagate, and the three-flavor picture describes most experimental results quite well, as described in detail elsewhere in this volume. Neutrino oscillation will certainly happen in a supernova and will affect the flavor–energy–time profile of the burst. This is both a drawback and an opportunity. It is a drawback in that signatures of core-collapse astrophysics may be obscured by oscillation effects, but it is an opportunity in that one may learn about neutrino parameters — in particular, the mass ordering (hierarchy) — from the flavor modulation of the flux.

There are several flavor transition regimes corresponding to different zones of the supernova. Flavor transitions in general depend on density and can be affected by shock wave propagation. Some effects are well understood and others are less well understood. A full review can be found in Ref. 17. In high neutrino-number-density regions of the supernova, *neutrino–neutrino* interactions result in so-called "collective effects", including flavor swaps above or below a particular energy threshold. The nonlinear phenomenology of these self-induced transitions is complex. A review can be found in Ref. 18 although the reader should keep in mind that the phenomena are the subject of ongoing theoretical investigation and are not yet fully understood.

"Ordinary" matter-induced (Mikheyev–Smirnov–Wolfenstein, MSW) flavor transition effects can occur as well in the supernova; matter effects are described in detail elsewhere in this volume. Resonant transitions occur in layers with specific densities according to the mass ordering. For a static progenitor profile, the fluxes F can be approximated by (see Ref. 17)

$$F_{\nu_e} = F^0_{\nu_x} \quad \text{(normal)}, \tag{2}$$

$$F_{\nu_e} = \sin^2\theta_{12} F^0_{\nu_e} + \cos^2\theta_{12} F^0_{\nu_x} \quad \text{(inverted)}, \tag{3}$$

or, for $\bar{\nu}_e$,

$$F_{\bar{\nu}_e} = \cos^2\theta_{12} F^0_{\bar{\nu}_e} + \sin^2\theta_{12} F^0_{\bar{\nu}_x} \quad \text{(normal)}, \tag{4}$$

$$F_{\bar{\nu}_e} = F^0_{\bar{\nu}_x} \quad \text{(inverted)}, \tag{5}$$

where F^0 represents the unoscillated flux for a given flavor, and θ_{12} is the "solar" mixing angle. This approximation should be appropriate for describing flavor transitions of the neutrinos in the neutronization burst, and possibly at other times as well. Clearly, suppression of the different flavors depends on the mass ordering, and strong ν_e suppression in the neutronization burst is one of the most robust signatures of normal mass ordering.[19]

After the neutrinos leave the star, they will typically traverse very long distances before reaching Earth such that they decohere (i.e., no longer oscillate). The decoherence length at supernova energies is much longer than the distance to a typical star. Therefore, neutrinos will be described by mass eigenstates when they reach Earth.

Finally, matter-induced flavor transitions will also occur in the Earth, resulting in mass-ordering-dependent modulation of the spectra, in principle observable in detectors with good energy resolution.[20–22]

1.6. *Physics and astrophysics from core-collapse neutrinos*

A tremendous amount will be learned from the next high-statistics supernova burst if observed in multiple flavors with reasonable energy and time resolution.

The huge burst of neutrinos will give us insight into the astrophysics of core collapse if we can track the fluxes through the phases of the supernova's evolution. The neutronization burst will signal the formation of an NS, and tell-tale signatures of shock waves, accretion, SASI, and NS cooling should all be visible, giving insight into the explosion mechanism as well (or will help to understand nonexplosion for core collapses that do not end up as supernovae). As noted above, we should be able to tell if the collapse resulted in a BH by observing the abrupt cessation of neutrino fluxes. There will be information about supernova nucleosynthesis to be had from observing ratios of neutrinos and antineutrinos.

The burst of neutrinos will furthermore bring knowledge of neutrino physics. While measurements of absolute neutrino mass based on energy-dependent spreading of the signal are not likely to be highly competive with near-future terrestrial kinematic experiments, under some circumstances, some may come close.[23–26] There will be multiple observables related to neutrino mass ordering (e.g., suppression of the neutronization burst, as discussed in Section 1.5). There will be signatures of or limits on other particle physics as well (see, e.g., Ref. 27) based on the principle that neutrino emission is a measure of the total energy of the collapse; therefore properties of unobserved exotic new particles or interactions, which would drain additional energy from the supernova, are constrained given a neutrino burst energy consistent with expectation.

The more we can understand about the astrophysics, the sharper will be limits on particle physics; similarly, the better the understanding of neutrino properties, the cleaner will be interpretation of the signal for understanding of core-collapse processes. The more the detectors are able to record the burst, preferably with multiple flavor sensitivities, the better we will understand the signal.

In addition to observations in all wavelengths, gravitational wave observations will be extremely interesting.[28–31] Gravitational waves will be emitted promptly. Although the specific gravitational wave signature from a core collapse is not yet precisely known, a combined neutrino–gravitational wave observation would be extremely exciting.

2. Detection of supernova neutrinos

The weak interaction strength of neutrinos is both a blessing and a curse. It is a blessing, because it enables neutrinos from deep inside ultradense matter to make off with valuable information about the goings-on there. But it is a curse as well, because neutrinos barely deign to interact with matter and require enormous detectors to intercept only a handful of them.

Here is a back-of-the-envelope estimate of the approximate order-of-magnitude of detector mass needed to see ~100 events. For a typical core-collapse supernova, the neutrino fluence F (flux integrated over the time window of the burst) can be estimated by assuming that the entire binding energy of the resulting NS is released in neutrinos equally distributed among flavors.[32] The binding energy is $E_b \sim 2.5\times10^{53}$ ergs, and if each neutrino has $E_\nu \sim 15\,\text{MeV}$, then the fluence per flavor is $F = \frac{E_b}{4\pi d^2 E_\nu} \sim 1.5\times10^{11}\,\text{cm}^{-2}$ at a typical distance of $d = 10\,\text{kpc}$. For a neutrino cross section of $\sigma \sim 10^{-41}\,\text{cm}^2$, using $N_{\text{int}} = \sigma F N_t$, we then require $N_t \sim 10^{32}$ targets for $N_{\text{int}} = 100$, which for a typical material corresponds to some kilotonnes of material. Therefore, in spite of the huge flux from a core collapse, one needs detectors of at least several kilotonne-scale to observe the order of ~100 events.

2.1. *Interactions of neutrinos with matter in the few to few tens-of-MeV range*

In the tens-of-MeV range, the best understood interactions are elastic scatters on the simplest targets: electrons and protons. These interactions have well-known cross sections and can proceed via CC and NC processes. Neutrinos can also interact via CC and NC processes with nuclei. The rates of these neutrino–nucleus interactions are more complicated to determine, as they depend on the structure of the target nucleus involved, and they may proceed to excited final states. Specific nuclear reactions have thresholds (up to a few tens of MeV) and therefore may not be accessible to the low energy portion of the flux. CC interactions must create a lepton of the flavor corresponding to the neutrino flavor in the final state. Heavy-flavor leptons (μ, τ and their antiparticles) have high neutrino CC interaction thresholds: ν_μ CC interactions require neutrino energies of more than 110 MeV, and ν_τ CC interactions have thresholds of more than 3.5 GeV. Because nearly all of supernova neutrinos have less than 100 MeV, only ν_e and $\bar{\nu}_e$ are able to create leptons. Therefore the only sensitivity to the muon- and tau-flavor components of the flux must be via NC interactions. Table 2 summarizes relevant neutrino interactions.

An important note: very few of these cross sections have been measured in the laboratory.[33] While inverse beta decay (IBD) on protons, $\bar{\nu}_e + p \to n + e^+$, and elastic scattering on electrons have been used by neutrino physicists for decades, and are theoretically well understood, the only other existing measurements at the 10% level are on ^{12}C. Additional measurements are highly desirable. Neutrinos from pion decay at rest, which range up to 50 MeV, are particularly suitable for such measurements.[34]

Table 2. Summary of relevant neutrino interactions.

	Electrons	Protons	Nuclei
CC	Elastic Scattering $\nu_x + e^- \to \nu_x + e^-$	Inverse Beta Decay $\bar{\nu}_e + p \to n + e^+$	Nuclear Interactions $\nu_e + (N, Z) \to (N-1, Z+1) + e^-$ $\bar{\nu}_e + (N, Z) \to (N+1, Z-1) + e^+$
NC	Elastic Scattering $\nu_x + e^- \to \nu_x + e^-$	Elastic Scattering $\nu_x + p \to \nu_x + p$	Nuclear Interactions $\nu_x + A \to \nu_x + A^*$ $\nu_x + A \to \nu_x + A$

2.2. *Detector types*

For the purpose of determining supernova neutrino observability, not only is the interaction rate relevant, but the ability of a given detector to *tag* interactions as ν_e, $\bar{\nu}_e$ or NC via the specific final states of a given channel, as well as energy, time and direction resolution, is also important.

Detectors with an abundance of free protons, such as water- and scintillator-based detectors, will observe the $\bar{\nu}_e$ component of the supernova neutrino flux via IBD's relatively large cross section.[e] The positron is the primary observable; in water detectors, one detects the Cherenkov ring from the e^+, and in scintillator detectors, one detects the scintillation light from the energy loss of the positron. The IBD final-state neutron is also potentially observable, primarily by gamma rays emitted via its capture on a nucleus. The neutron will thermalize and be captured on a timescale of tens to hundreds of microseconds. In pure water, the capture is on hydrogen, which results in a 2.2-MeV gamma ray. Doping with some material with high neutron capture cross section, such as gadolinium, can increase visible energy for the capture. For example, Gd releases a cascade of gammas totalling about 8 MeV worth of energy after swallowing a neutron. Gammas are detectable in this energy range primarily via Compton scatters. The gammas from positron annihilation in coincidence with the lepton are usually observable in scintillator as well, although this will yield gammas below Cherenkov threshold for water. Scintillation detectors have more light per energy loss than Cherenkov detectors (by a factor of several tens) and do not suffer from a Cherenkov threshold. They typically have much higher efficiency for secondary neutron detection. However neutron detection efficiency can still be reasonable with water. Super-Kamiokande (SK), for example, has $\sim$18% efficiency without Gd doping and $\sim$90% is expected with Gd.[35]

All detector materials have electrons and will experience elastic scatters of neutrinos on electrons, although not all detectors can take advantage of this interaction, depending on the observability of the electron's energy loss.

Detectors can also exploit CC interactions of ν_e and $\bar{\nu}_e$ on nuclei (oxygen in water and carbon in scintillator), although for water and scintillator these tend to be subdominant processes. NC interactions are of special interest due to their sensitivity to all flavor components of the flux. The NC nuclear excitation of carbon in scintillator, which de-excites via a

[e] All, or very nearly all, of the SN1987A observed interactions were IBD.

15-MeV gamma, is particularly interesting as a tagged NC channel. Also of special interest is the ν_e interaction on argon, $\nu_e + {}^{40}\text{Ar} \rightarrow e^- + {}^{40}\text{K}^*$, which is one of the most promising ways to learn about the electron flavor component of the burst flux. In principle, the interaction channel in argon can be tagged by the de-excitation pattern of gammas. Other interesting interaction channels for detection of supernova burst neutrinos are CC and NC interactions on lead: $\nu_e + {}^{208}\text{Pb} \rightarrow e^- + {}^{208}\text{Bi}^*$ and $\nu_x + {}^{208}\text{Pb} \rightarrow \nu_x + {}^{208}\text{Pb}^*$, which can eject one or two neutrons. Energy thresholds for double-neutron emission are higher than for single-neutron emission, so ejected neutron counts can in principle provide information about the neutrino spectrum.[36]

All detector materials will also experience elastic scatters on nuclei. While these tend to have high cross sections. Coherent elastic neutrino-nucleus scattering (CEvNS) is enhanced by a factor of N^2, where N is the number of neutrons in the nucleus — the kinematics is such that the scatters produce only very low energy recoils. It's like hitting a bowling ball with a ping pong ball; it's easy to hit the target but hard to get it rolling very fast. Only detectors specialized for detection of very low energy recoils will be able to observe scattering on nuclei. Scintillators, C_nH_{2n}, typically have some sensitivity to proton recoils, and dark-matter direct-detection detectors (with primary aims of weakly interacting massive particle searches) can see the recoils of nuclei[37] from neutrino collisions.

2.3. *Pointing to the supernova with neutrinos*

Anisotropic neutrino interactions can be used to point back to the supernova (of clear value for the early alert described in Section 2.5). Water Cherenkov detectors have the advantage of an intrinsically directional signal — Cherenkov light forms a cone in the direction of the charged particle's trajectory. Since neutrinos elastically scatter electrons forward (about a $\sim 25°$ cone at supernova neutrino energies), the direction back to the supernova can be reconstructed using the light pattern. Resolution is limited by statistics, $\sim 25°/\sqrt{N}$, where N is the number of scatters. In practice, direction determination is worsened by a substantial near-isotropic background of IBD events. SK expects pointing of 3–6° for a 10-kpc supernova.[38] Addition of Gd may reduce this background and improve the pointing[39] by enabling tagging of IBD neutrons. Tracking detectors such as liquid argon time projection chambers can also exploit elastically scattered electrons to point to the supernova.[40] Scintillator detectors will record a signal, but due to isotropy of the emitted scintillation photons, they will have rather poor directionality.[f] Overall, elastic scattering in a directional detector is likely the best prospect for pointing to a supernova. Other methods, such as triangulation, are not likely to do as well.[42]

2.4. *Existing and future detectors*

Existing and future instances of the detector classes described in the previous section are listed in Table 3.

The primary water detector currently in operation is SK,[38] which will observe $\sim 10^3 - 10^4$ events for a 10-kpc supernova. A planned upgrade to dope the water with gadolinium, SK-Gd,[43] will improve the neutron tagging capability and will help to disentangle the flavor composition of the burst. The next planned large water Cherenkov detector is Hyper-Kamiokande (HK),[44] which will have a truly enormous statistics for a Galactic burst. The current planned configuration for HK is two modules with a total fiducial volume of 374 k[45] and higher photomultiplier density than SK in order to achieve a lower threshold.

Table 3. Current and proposed supernova neutrino detectors as of the time of this writing. Neutrino event estimates are approximate for 10 kpc; note that there is significant variation by supernova model. An asterisk indicates a surface detector, which may not be self-triggering due to background. Numbers in parentheses indicate special cases; see text.

Detector	Type	Mass (kt)	Location	Events	Status
Super-Kamiokande	H_2O	32	Japan	7000	Running
LVD	C_nH_{2n}	1	Italy	300	Running
KamLAND	C_nH_{2n}	1	Japan	300	Running
Borexino	C_nH_{2n}	0.3	Italy	100	Running
IceCube	Long string	(600)	South Pole	(10^6)	Running
Baksan	C_nH_{2n}	0.33	Russia	50	Running
HALO	Pb	0.08	Canada	30	Running
Daya Bay	C_nH_{2n}	0.33	China	100	Running
NOνA*	C_nH_{2n}	15	USA	4000	Running
MicroBooNE*	Ar	0.17	USA	17	Running
SNO+	C_nH_{2n}	0.8	Canada	300	Near future
DUNE	Ar	40	USA	3000	Future
Hyper-Kamiokande	H_2O	374	Japan	75 000	Future
JUNO	C_nH_{2n}	20	China	6000	Future
RENO-50	C_nH_{2n}	18	Korea	5400	Future
PINGU	Long string	(600)	South Pole	(10^6)	Future

[f] One can exploit relative positions of positron and neutron signals from IBD to give some directional information.[41] However this is not likely to do as well as elastic scattering.

Another kind of water Cherenkov detector has long strings of photomultipliers embedded in ice or water. The main current instance of a long-string detector with supernova sensitivity is IceCube.[46] This detector cannot observe neutrinos with good event-by-event information, because the photosensors are sparse and rarely record more than one photon per interaction. However the diffuse photons produced by neutrino-induced Cherenkov light in the ice are numerous enough that overall the detectors can track the timing curve of the supernova to good precision.[47]

There are a number of scintillator detectors currently online of 1-k scale, which expect several hundred IBD events for a 10-kpc supernova. These detectors also expect to see NC events from elastic scattering on protons (with some neutrino spectral information from the recoil spectra) and from ^{12}C excitation. A very promising future detector is JUNO,[48] which will have 20 k of scintillator. Even some scintillator detectors on the surface, such as NOvA,[49] may have reasonable sensitivity to a burst, in spite of cosmogenic background. Although not listed in the table, also relevant are smaller surface scintillator detectors such as Daya Bay.[50] Such detectors can suppress background by requiring coincident counts in multiple detectors.

The most promising prospect for large enhancement of global supernova flavor sensitivity is DUNE,[9,51] which is sensitive to the ν_e component of the flux via CC absorption on argon. DUNE will have 40 k of argon in four large time projection chambers, offering precision tracking of charged interaction products.

HALO is an instance of a lead-based detector[52] with neutron-detection capability and sensitivity to ν_e.

WIMP dark-matter-search detectors (e.g., xenon-based[53,54]) will also record NC CEvNS events (especially valuable as a measure of the total flux), and next-generation tonne-scale detectors may provide promising statistics.

Figure 5 shows expected events as a function of energy and time with nominal detector resolution assumptions[55] in three future large detectors, for the supernova model corresponding to Figs. 3 and 4.

2.5. *The early alert*

Because neutrinos decouple from the star almost completely within a few tens of seconds, whereas it takes some time — hours or even some fraction of a day — for the photons to make their way out (shock wave propagation time) — the observation of neutrinos can serve as an early warning that an optical supernova may soon appear. The SuperNova Early Warning

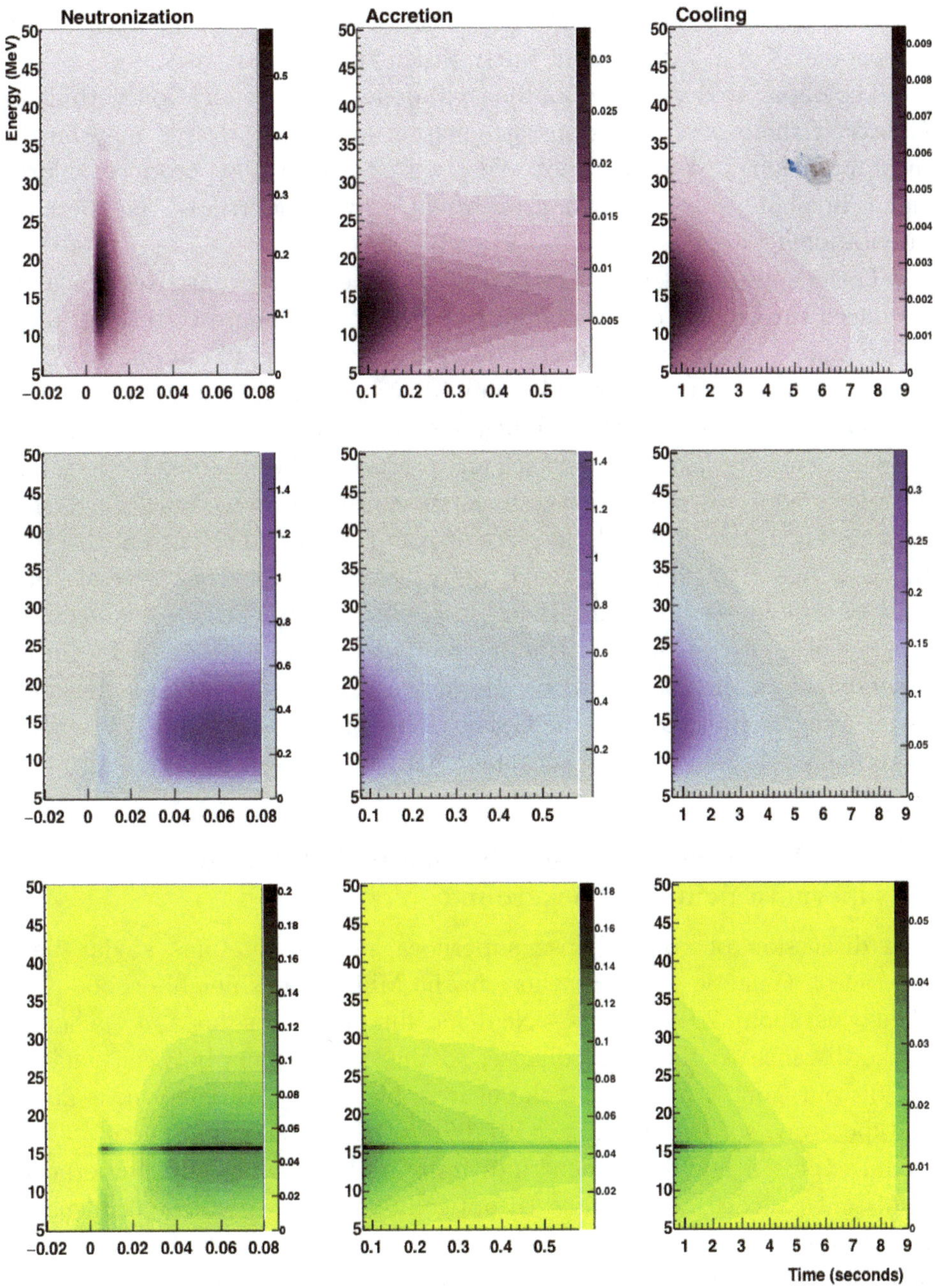

Fig. 5. Example of time-dependent *observed* spectra in detectors for the fluxes in Fig. 4 on three different timescales. All interaction channels are summed. The z-axis units are events per 0.5 MeV of observed energy per millisecond. Top: 40 k of liquid argon (mostly $\bar{\nu}_e$). Center: 374 k of water (mostly $\bar{\nu}_e$). Bottom: 20 k of scintillator (mostly $\bar{\nu}_e$, but note the NC 15-MeV γ signal). The detector assumptions approximate DUNE, HK and JUNO, respectively.[55]

System (SNEWS) is an international network of detectors prepared to alert astronomers[56,57] if a coincident burst of neutrinos is observed.

The observability of the photons will depend on the location within the Galaxy[31]; there may be serious attenuation and near infrared may be the most promising. Note that not every core collapse may produce a large signal in photons — some may be bright only in neutrinos (and perhaps gravitational waves).

There is also a chance for a *true* supernova early warning. Neutrinos are produced throughout a star's life during stellar burning processes. For most of this time, the flux is too small to observe on Earth for all but our own Sun. However, one expects an uptick in luminosity and warming of the spectrum in the final stages of stellar burning before core collapse, over its final few days or hours. For very nearby stars, such an uptick could be observed by detectors with low energy thresholds, such as large scintillator detectors.[58–61] Such a presupernova alert is already in place at KamLAND.[62] The current distance range is less than 1 kpc (a range nevertheless occupied by several core-collapse candidates, including Betelgeuse, about 180 pc away). A detectable number of presupernova neutrino interactions could be observed by future large detectors such as JUNO or HK in the hours before collapse for stars as far away as the center of the Galaxy. Here the experimental challenge is to recognize the signature of an impending collapse over background.

3. Very distant supernovae: mini-bursts and diffuse supernova neutrino background

The discussion of core-collapse supernova signals here has so far been focused on Galactic events or those in the Milky Way's neighborhood. The Galaxy is about 20 kpc across. Neutrino fluxes decrease as $1/d^2$, where d is the distance to the core collapse, so the cosmic gaps between galaxies are discouraging. Andromeda, the nearest big galaxy to our own, is about 700 kpc away, so one expects $\sim(10/700)^2$ of the events, and therefore one would expect a few dozen neutrinos from Andromeda in HK. Supernovae in distant galaxies, while observed optically very often, will be challenging neutrino sources to observe. However it is worth recognizing that with large detectors, a reasonable number of "minibursts" — one or a few neutrinos from extragalactic supernovae — could be acquired over decades.[63,64]

Going even farther out, one can consider the Diffuse Supernova Neutrino Background (DSNB), also known as the Supernova Relic Neutrino (SRN) background. The DSNB, reviewed in Refs. 65, 66, is a diffuse flux of supernova neutrinos which have been emitted from supernovae over the history of the universe. This redshifted flux is small but in principle observable, as it occupies a regime more energetic than solar neutrinos (which range only up to about 15 MeV), and yet with less energy than most atmospheric neutrinos. The issue for observing this flux is background rejection, since one cannot associate the neutrinos with observable astronomical events, and they come one by one rather than in bursts. Still, experiments with good background rejection can hope to observe the signal. The best limits so far are from SK[35,67]; there are good future prospects for SK-Gd[43] and HK.[44]

4. Other astrophysical sources of low-energy neutrinos

Bursts of neutrinos in the supernova energy range are expected from some other potential extrasolar sources as well. Unfortunately none of these quite match the spectacular prospects of a core-collapse supernova — they either have a comparable local occurrence rate but are much dimmer in neutrinos or else have comparable neutrino brightness yet are expected to be extremely rare. Still, the experimentalist should keep these potential neutrino sources in mind:

- *Type Ia supernovae*: Type Ia supernovae, of cosmological-standard-candle fame, are thought to occur when a white dwarf accretes enough matter from a companion to ignite a thermonuclear explosion. However the exact explosion mechanism is quite uncertain. Of comparable electromagnetic energy scale to core-collapse supernovae and occurring at a reasonable human-lifetime-scale rate in our Galaxy (~30% of the core-collapse rate), they are subtle neutrino events — the expected neutrino energies are mostly lower than ~10 MeV and fluxes are down by few orders of magnitude with respect to core-collapse supernovae. Nevertheless, a Type Ia supernova within ~1 kpc of Earth would be detectable in next-generation detectors (e.g., HK), and observation of the neutrino flux could give insight into the explosion mechanism (Refs. 68–71).[g]
- *NS and BH mergers*: One expects also neutrinos to be emitted from the accretion disks that form when BHs and NSs merge.[72] Spectra and neutrino fluxes would not be dramatically different from those of a core collapse; however the time profile of neutrino emission would be different. One expects ~0.1–1 s bursts from BH–NS or NS–NS mergers.[72,73] Unfortunately, however, the rate of these merger events within our Galaxy,

[g]Locations of potential nearby progenitors for Type Ia supernovae are poorly known.

for which the neutrino flux is detectable, is expected to be very low.[h] NS–NS mergers should happen 10^3 times less often that core collapses, and BH–NS mergers should be even rarer, occurring at 10^{-4} the rate of core collapses.

5. Big Bang relic neutrinos

A different kind of "relic" neutrino flux is that left over from the Big Bang, which should have a current temperature of 1.95 K and a number density of 113 per cm^{-3} per flavor.[74] Depending on the absolute mass scale of neutrinos, they may even be nonrelativistic and clustered gravitationally.[75] While we have clear evidence of the existence of these relic neutrinos from cosmological measurements[76,77] (indeed, some of the most stringent limits on the neutrino mass scale come from cosmology), there has never been any direct detection of relic Big Bang neutrinos. Direct detection of the cosmic neutrino background is a fiendishly difficult experimental problem, as neutrino interaction rates are so terribly low at such low neutrino energies. There have been a few ideas for detection strategies, but none are really feasible with the current-generation experiments. Among the ideas are attempts to exploit coherent scattering: here the idea is that the energies are so low that the neutrino de Broglie wavelength is of macroscopic scale so that one could observe coherent scattering off macroscopic objects.[78] However this is probably not feasible with current or near-future technology. Another idea is to exploit "Z-bursts"[79,80]; here the idea is high-energy neutrinos at the Z pole interact with cosmic relic neutrinos, leading to either an observable cascade of particles or an absorption dip in the neutrino spectrum. This requires very large, ultrahigh-energy neutrino detectors to sample the very small flux in this range; again, technology does not currently exist. Perhaps the most promising (although still not quite in reach) idea is the "zero-threshold reaction".[81–84] The idea is that neutrinos interact with nuclei which are unstable against beta decay such that the CC interactions ($\nu_e + A_Z \to e^- + A^*_{Z+1}$ or $\bar{\nu}_e + A_Z \to e^+ + A^*_{Z-1}$) that require no threshold neutrino energy occur. The signature is a slight excess of events near the beta spectrum endpoint. The experimental difficulty is that this requires unprecedented energy resolution to separate capture events from beta decays. Tritium is the most promising candidate for this measurement, but even the state-of-the-art tritium-endpoint KATRIN experiment has poor prospects for measuring the expected background[85] (although it could set limits). The experimental

[h]Note that gravitational wave observatories have ranges well beyond the Milky Way and can probe these merger events in distant galaxies.

collaboration PTOLEMY[86] is actively working towards this challenging goal, which will require significant advances in technology.

6. Summary

In summary, there is tremendous potential for physics and astrophysics from observation of neutrinos from nearby core-collapse supernovae (and possibly even other astrophysical sources). The current worldwide sensitivity is to the $\bar{\nu}_e$ component of the flux; however, there are excellent prospects to expand flavor sensitivity in the future. Multimessenger prospects, such as coincident observation with gravitational waves, are also exciting. We are ready to gather knowledge from the next neutrino burst now, and capabilities will steadily advance over the coming decades.

References

1. M. Turatto, Classification of supernovae, *Lect. Notes Phys.* **598**, 21 (2003). doi: 10.1007/3-540-45863-8_3.
2. S. M. Adams, C. S. Kochanek, J. F. Beacom, M. R. Vagins, and K. Z. Stanek, Observing the next galactic supernova, *Astrophys. J.* **778**, 164 (2013). doi: 10.1088/0004-637X/778/2/164.
3. A. Mirizzi, G. G. Raffelt, and P. D. Serpico, Earth matter effects in supernova neutrinos: Optimal detector locations, *JCAP.* **0605**, 012 (2006).
4. H.-T. Janka, K. Langanke, A. Marek, G. Martinez-Pinedo, and B. Mueller, Theory of core-collapse supernovae, *Phys. Rept.* **442**, 38–74 (2007). doi: 10. 1016/j.physrep.2007.02.002.
5. H.-T. Janka, Explosion mechanisms of core-collapse supernovae, *Ann. Rev. Nucl. Part. Sci.* **62**, 407–451 (2012). doi: 10.1146/annurev-nucl-102711-094901.
6. S. Shapiro and S. Teukolsky, *Black Holes, White Dwarfs, and Neutron Stars.* Wiley, New York (1983). ISBN 047187316.
7. H. Minakata, H. Nunokawa, R. Tomas, and J. W. F. Valle, Parameter degeneracy in flavor-dependent reconstruction of supernova neutrino fluxes, *JCAP.* **0812**, 006 (2008). doi: 10.1088/1475-7516/2008/12/006.
8. J. Gava, J. Kneller, C. Volpe, and G. C. McLaughlin, A dynamical collective calculation of supernova neutrino signals, *Phys. Rev. Lett.* **103**, 071101 (2009). doi: 10.1103/PhysRevLett.103.071101.
9. R. Acciarri *et al.*, Long-Baseline Neutrino Facility (LBNF) and Deep Underground Neutrino Experiment (DUNE) (2015).
10. L. Huedepohl, B. Muller, H. T. Janka, A. Marek, and G. G. Raffelt, Neutrino signal of electron-capture supernovae from core collapse to cooling, *Phys. Rev. Lett.* **104**, 251101 (2010). doi: 10.1103/PhysRevLett.104.251101.
11. K. Hirata *et al.*, Observation of a neutrino burst from the supernova sn1987a, *Phys. Rev. Lett.* **58**, 1490–1493 (1987).
12. R. M. Bionta *et al.*, Observation of a neutrino burst in coincidence with supernova sn1987a in the large magellanic cloud, *Phys. Rev. Lett.* **58**, 1494 (1987).

13. E. N. Alekseev, L. N. Alekseeva, V. I. Volchenko, and I. V. Krivosheina, Possible detection of a neutrino signal on 23 february 1987 at the baksan underground scintillation telescope of the institute of nuclear research, *JETP Lett.* **45**, 589–592 (1987).
14. M. Aglietta *et al.*, On the event observed in the mont blanc underground neutrino observatory during the occurrence of supernova 1987a, *Europhys. Lett.* **3**, 1315–1320 (1987).
15. F. Vissani, Comparative analysis of SN1987A antineutrino fluence, *J. Phys.* **G42**, 013001 (2015). doi: 10.1088/0954-3899/42/1/013001.
16. D. N. Schramm and J. W. Truran, New physics from Supernova SN1987A, *Phys. Rept.* **189**, 89–126 (1990). doi: 10.1016/0370-1573(90)90020-3.
17. A. Mirizzi, I. Tamborra, H.-T. Janka, N. Saviano, K. Scholberg, R. Bollig, L. Hudepohl, and S. Chakraborty, Supernova neutrinos: Production, oscillations and detection, *Riv. Nuovo Cim.* **39**(1–2), 1 (2016). doi: 10.1393/ncr/i2016-10120-8.
18. H. Duan, G. M. Fuller, and Y.-Z. Qian, Collective neutrino oscillations, *Ann. Rev. Nucl. Part. Sci.* **60**, 569–594 (2010). doi: 10.1146/annurev.nucl.012809.104524.
19. M. Kachelriess, R. Tomas, R. Buras, H. T. Janka, A. Marek, and M. Rampp, Exploiting the neutronization burst of a galactic supernova, *Phys. Rev.* **D71**, 063003 (2005). doi: 10.1103/PhysRevD.71.063003.
20. C. Lunardini and A. Y. Smirnov, Supernova neutrinos: Earth matter effects and neutrino mass spectrum, *Nucl. Phys.* **B616**, 307–348 (2001). doi: 10.1016/S0550-3213(01)00468-0.
21. A. S. Dighe, M. T. Keil, and G. G. Raffelt, Identifying earth matter effects on supernova neutrinos at a single detector, *JCAP.* **0306**, 006 (2003). doi: 10.1088/1475-7516/2003/06/006.
22. B. Dasgupta, A. Dighe, and A. Mirizzi, Identifying neutrino mass hierarchy at extremely small theta(13) through Earth matter effects in a supernova signal, *Phys. Rev. Lett.* **101**, 171801 (2008). doi: 10.1103/PhysRevLett.101.171801.
23. J. F. Beacom and P. Vogel, Mass signature of supernova nu/mu and nu/tau neutrinos in SuperKamiokande, *Phys. Rev.* **D58**, 053010 (1998). doi: 10.1103/PhysRevD.58.053010.
24. J. F. Beacom, R. N. Boyd, and A. Mezzacappa, Black hole formation in core collapse supernovae and time-of-flight measurements of the neutrino masses, *Phys. Rev.* **D63**, 073011 (2001). doi: 10.1103/PhysRevD.63.073011.
25. J.-S. Lu, J. Cao, Y.-F. Li, and S. Zhou, Constraining absolute neutrino masses via detection of galactic supernova neutrinos at JUNO, *JCAP.* **1505**(05), 044 (2015). doi: 10.1088/1475-7516/2015/05/044.
26. F. Rossi-Torres, M. M. Guzzo, and E. Kemp, Boundaries on Neutrino Mass from Supernovae Neutronization Burst by Liquid Argon Experiments (2015).
27. G. G. Raffelt, Particle physics from stars, *Ann. Rev. Nucl. Part. Sci.* **49**, 163–216 (1999). doi: 10.1146/annurev.nucl.49.1.163.
28. G. Pagliaroli, F. Vissani, E. Coccia, and W. Fulgione, Neutrinos from Supernovae as a trigger for gravitational wave search, *Phys. Rev. Lett.* **103**, 031102 (2009). doi: 10.1103/PhysRevLett.103.031102.

29. F. Vissani, G. Pagliaroli, and F. Rossi-Torres, The use of supernova neutrinos to monitor the collapse, to search for gravity waves, to probe neutrino masses (2010). doi: 10.1142/S0218271811019888. [*Int. J. Mod. Phys.* **D20**, 1873 (2011)].
30. I. Leonor *et al.*, Searching for prompt signatures of nearby core-collapse supernovae by a joint analysis of neutrino and gravitational-wave data, *Class. Quant. Grav.* **27**, 084019 (2010). doi: 10.1088/0264-9381/27/8/084019.
31. K. Nakamura, S. Horiuchi, M. Tanaka, K. Hayama, T. Takiwaki, and K. Kotake, Multi-messenger signals of long-term core-collapse supernova simulations: synergetic observation strategies (2016). doi: 10.1093/mnras/stw1453.
32. J. Bahcall, *Neutrino Astrophysics*. Cambridge University Press, Cambridge (1989). ISBN 052137975-X.
33. J. A. Formaggio and G. P. Zeller, From eV to EeV: Neutrino cross sections across energy scales, *Rev. Mod. Phys.* **84**, 1307 (2012). doi: 10.1103/RevModPhys.84.1307.
34. A. Bolozdynya *et al.*, Opportunities for Neutrino Physics at the Spallation Neutron Source: A White Paper (2012).
35. H. Zhang *et al.*, Supernova relic neutrino search with neutron tagging at Super-Kamiokande-IV, *Astropart. Phys.* **60**, 41–46 (2015). doi: 10.1016/j.astropartphys.2014.05.004.
36. D. Vaananen and C. Volpe, The neutrino signal at HALO: learning about the primary supernova neutrino fluxes and neutrino properties, *JCAP.* **1110**, 019 (2011). doi: 10.1088/1475-7516/2011/10/019.
37. C. J. Horowitz, K. J. Coakley, and D. N. McKinsey, Supernova observation via neutrino nucleus elastic scattering in the CLEAN detector, *Phys. Rev.* **D68**, 023005 (2003). doi: 10.1103/PhysRevD.68.023005.
38. K. Abe *et al.*, Real-time supernova neutrino burst monitor at Super-Kamiokande, *Astropart. Phys.* **81**, 39–48 (2016). doi: 10.1016/j.astropartphys.2016.04.003.
39. R. Tomas, D. Semikoz, G. G. Raffelt, M. Kachelriess, and A. S. Dighe, Supernova pointing with low- and high-energy neutrino detectors, *Phys. Rev.* **D68**, 093013 (2003). doi: 10.1103/PhysRevD.68.093013.
40. A. Bueno, I. Gil Botella, and A. Rubbia, Supernova neutrino detection in a liquid argon TPC (2003).
41. V. Fischer *et al.*, Prompt directional detection of galactic supernova by combining large liquid scintillator neutrino detectors (2015). doi: 10.1088/1475-7516/2015/08/032. [JCAP1508,032(2015)].
42. J. F. Beacom and P. Vogel, Can a supernova be located by its neutrinos? *Phys. Rev.* **D60**, 033007 (1999). doi: 10.1103/PhysRevD.60.033007.
43. J. F. Beacom and M. R. Vagins, GADZOOKS! Antineutrino spectroscopy with large water Cherenkov detectors, *Phys. Rev. Lett.* **93**, 171101 (2004). doi: 10.1103/PhysRevLett.93.171101.
44. K. Abe *et al.*, Letter of Intent: The Hyper-Kamiokande Experiment Detector Design and Physics Potential (2011).
45. H.-K. Collaboration. Hyper-k design report. KEK Preprint 2016-21, ICRR-Report-701-2016-1 (2016).

46. R. Abbasi *et al.*, IceCube Sensitivity for Low-Energy Neutrinos from Nearby Supernovae (2011).
47. F. Halzen and G. G. Raffelt, Reconstructing the supernova bounce time with neutrinos in IceCube, *Phys. Rev.* **D80**, 087301 (2009). doi: 10.1103/PhysRevD.80.087301.
48. F. An *et al.*, Neutrino physics with JUNO, *J. Phys.* **G43**(3), 030401 (2016). doi: 10.1088/0954-3899/43/3/030401.
49. D. S. Ayres *et al.*, NOvA proposal to build a 30-kiloton off-axis detector to study neutrino oscillations in the Fermilab NuMI beamline (2004).
50. H. Wei, L. Lebanowski, F. Li, Z. Wang, and S. Chen, Design, characterization, and sensitivity of the supernova trigger system at Daya Bay, *Astropart. Phys.* **75**, 38–43 (2016). doi: 10.1016/j.astropartphys.2015.10.011.
51. A. Ankowski *et al.*, Supernova Physics at DUNE. (2016). URL http://inspirehep.net/record/1484266/files/arXiv:1608.07853.pdf.
52. C. A. Duba *et al.*, HALO: The helium and lead observatory for supernova neutrinos, *J. Phys. Conf. Ser.* **136**, 042077 (2008). doi: 10.1088/1742-6596/136/4/042077.
53. K. Abe *et al.*, Detectability of galactic supernova neutrinos coherently scattered on xenon nuclei in XMASS (2016).
54. R. F. Lang, C. McCabe, S. Reichard, M. Selvi, and I. Tamborra, Supernova neutrino physics with xenon dark matter detectors: A timely perspective (2016).
55. http://www.phy.duke.edu/~schol/snowglobes.
56. P. Antonioli *et al.*, SNEWS: The SuperNova Early Warning System, *New J. Phys.* **6**, 114 (2004). doi: 10.1088/1367-2630/6/1/114.
57. K. Scholberg, The SuperNova Early Warning System, *Astron. Nachr.* **329**, 337–339 (2008).
58. A. Odrzywolek and A. Heger, Neutrino signatures of dying massive stars: From main sequence to the neutron star, *Acta Phys. Polon.* **B41**, 1611–1628 (2010).
59. K. M. Patton and C. Lunardini, Presupernova neutrinos: realistic emissivities from stellar evolution (2015).
60. C. Kato, M. D. Azari, S. Yamada, K. Takahashi, H. Umeda, T. Yoshida, and K. Ishidoshiro, Pre-supernova neutrino emissions from ONe cores in the progenitors of core-collapse supernovae: Are they distinguishable from those of Fe cores? *Astrophys. J.* **808**(2), 168 (2015). doi: 10.1088/0004-637X/808/2/168.
61. T. Yoshida, K. Takahashi, H. Umeda, and K. Ishidoshiro, Presupernova neutrino events relating to the final evolution of massive stars, *Phys. Rev.* **D93**(12), 123012 (2016). doi: 10.1103/PhysRevD.93.123012.
62. K. Asakura *et al.*, KamLAND sensitivity to neutrinos from pre-supernova stars, *Astrophys. J.* **818**(1), 91 (2016). doi: 10.3847/0004-637X/818/1/91.
63. S. Ando, J. F. Beacom, and H. Yuksel, Detection of neutrinos from supernovae in nearby galaxies, *Phys. Rev. Lett.* **95**, 171101 (2005). doi: 10.1103/PhysRevLett.95.171101.

64. M. D. Kistler, H. Yuksel, S. Ando, J. F. Beacom, and Y. Suzuki, Core-collapse astrophysics with a five-megaton neutrino detector, *Phys. Rev.* **D83**, 123008 (2011). doi: 10.1103/PhysRevD.83.123008.
65. C. Lunardini, Diffuse supernova neutrinos at underground laboratories, *Astropart. Phys.* **79**, 49–77 (2016). doi: 10.1016/j.astropartphys.2016.02.005.
66. J. F. Beacom, The diffuse supernova neutrino background, *Ann. Rev. Nucl. Part. Sci.* **60**, 439–462 (2010). doi: 10.1146/annurev.nucl.010909.083331.
67. K. Bays *et al.*, Supernova relic neutrino search at Super-Kamiokande, *Phys. Rev.* **D85**, 052007 (2012).
68. A. Odrzywolek and T. Plewa, Probing thermonuclear supernova explosions with neutrinos, *Astron. Astrophys.* **529**, A156 (2011). doi: 10.1051/0004-6361/201015133.
69. I. R. Seitenzahl, M. Herzog, A. J. Ruiter, K. Marquardt, S. T. Ohlmann, and F. K. Rpke, Neutrino and gravitational wave signal of a delayed-detonation model of Type Ia supernovae, *Phys. Rev.* **D92**(12), 124013 (2015). doi: 10.1103/PhysRevD.92.124013.
70. W. P. Wright, G. Nagaraj, J. P. Kneller, K. Scholberg, and I. R. Seitenzahl, Neutrinos from type Ia supernovae: The deflagration-to-detonation transition scenario, *Phys. Rev.* **D94**(2), 025026 (2016). doi: 10.1103/PhysRevD.94.025026.
71. W. P. Wright, J. P. Kneller, S. T. Ohlmann, F. K. Roepke, K. Scholberg, and I. R. Seitenzahl, Neutrinos from Type Ia supernovae: The gravitationally confined detonation scenario (2016).
72. O. L. Caballero, G. C. McLaughlin, R. Surman, and R. Surman, Detecting neutrinos from black hole neutron stars mergers, *Phys. Rev.* **D80**, 123004 (2009). doi: 10.1103/PhysRevD.80.123004.
73. L. Dessart, C. Ott, A. Burrows, S. Rosswog, and E. Livne, Neutrino signatures and the neutrino-driven wind in binary neutron star mergers, *Astrophys. J.* **690**, 1681 (2009). doi: 10.1088/0004-637X/690/2/1681.
74. P. Peebles, *Principles of Physical Cosmology.* Princeton University Press, Princeton (1993). ISBN 0691019339.
75. A. Ringwald and Y. Y. Y. Wong, Gravitational clustering of relic neutrinos and implications for their detection, *JCAP.* **0412**, 005 (2004). doi: 10.1088/1475-7516/2004/12/005.
76. K. N. Abazajian *et al.*, Cosmological and astrophysical neutrino mass measurements, *Astropart. Phys.* **35**, 177–184 (2011). doi: 10.1016/j.astropartphys.2011.07.002.
77. Y. Y. Y. Wong, Neutrino mass in cosmology: status and prospects, *Ann. Rev. Nucl. Part. Sci.* **61**, 69–98 (2011). doi: 10.1146/annurev-nucl-102010-130252.
78. P. F. Smith and J. D. Lewin, Coherent interaction of galactic neutrinos with material targets, *Phys. Lett.* **B127**, 185–190 (1983). doi: 10.1016/0370-2693(83)90873-0.
79. A. Ringwald (2005). arXiv:hep-ph/0505024.
80. G. B. Gelmini, Prospect for relic neutrino searches, *Phys. Scripta.* **T121**, 131–136 (2005). doi: 10.1088/0031-8949/2005/T121/019.

81. A. G. Cocco, G. Mangano, and M. Messina, Capturing relic neutrinos with beta-decaying nuclei, *J. Phys. Conf. Ser.* **120**, 022005 (2008). doi: 10.1088/1742-6596/120/2/022005.
82. R. Lazauskas, P. Vogel, and C. Volpe, Charged current cross section for massive cosmological neutrinos impinging on radioactive nuclei, *J. Phys.* **G35**, 025001 (2008). doi: 10.1088/0954-3899/35/2/025001.
83. M. Blennow, Prospects for cosmic neutrino detection in tritium experiments in the case of hierarchical neutrino masses, *Phys. Rev.* **D77**, 113014 (2008). doi: 10.1103/PhysRevD.77.113014.
84. B. McElrath, Laboratory tests for the cosmic neutrino background using beta-decaying nuclei (2009). arXiv:0901.3491.
85. A. Faessler, R. Hodak, S. Kovalenko, and F. Simkovic, Can one measure the Cosmic Neutrino Background? (2016).
86. S. Betts *et al.*, Development of a Relic Neutrino Detection Experiment at PTOLEMY: Princeton Tritium Observatory for Light, Early-Universe, Massive-Neutrino Yield. In *Proceedings, Community Summer Study 2013: Snowmass on the Mississippi (CSS2013): Minneapolis, MN, USA, July 29-August 6, 2013* (2013). URL http://www.slac.stanford.edu/econf/C1307292/docs/submittedArxivFiles/1307.4738.pdf.

Chapter 9

High-Energy Astrophysical Neutrinos

F. Halzen

Wisconsin IceCube Particle Physics Center,
UW — Madison, Madison, WI, USA
halzen@icecube.wisc.edu

Abstract

Chargeless, weakly interacting neutrinos are ideal astronomical messengers because they travel through space without scattering, absorption, or deflection. They provide the only unobstructed view of cosmic accelerators. But this weak interaction also makes them notoriously difficult to detect, leading to neutrino observatories requiring large-scale detectors. The IceCube experiment discovered PeV-energy neutrinos originating beyond the Sun, with energies bracketed by those of TeV-energy gamma rays and EeV-energy extragalactic cosmic rays. In this chapter, we discuss the IceCube neutrino telescope, the status of the observation of cosmic neutrinos, and what neutrinos can tell us about the nonthermal universe. Besides the search for the sources of galactic and extragalactic cosmic rays, the scientific missions of IceCube and similar instruments under construction in the Mediterranean Sea and Lake Baikal include the observation of galactic supernova explosions, the search for dark matter, and the study of neutrinos themselves.

1. Neutrino astronomy: a brief history

Soon after the 1956 observation of the neutrino,[1] the idea emerged that it represented the ideal astronomical messenger. Neutrinos travel from the edge of the universe without absorption and with no deflection by magnetic fields. Having essentially no mass and no electric charge, the neutrino is similar to the photon, except for one important attribute: its interactions with matter are extremely feeble. So, high-energy neutrinos may reach us unscathed from cosmic distances: from the inner neighborhood of black holes and from the nuclear furnaces where cosmic rays are born. But their weak interactions also make cosmic neutrinos very difficult to detect. Immense particle detectors are required to collect cosmic neutrinos in statistically significant numbers.[2] By the 1970s, it was clear that a cubic-kilometer detector was needed to observe cosmic neutrinos produced in the interactions of cosmic rays with background microwave photons.[3] Subsequent estimates for observing potential cosmic accelerators such as Galactic supernova remnants and gamma-ray bursts (GRBs) unfortunately pointed to the same exigent requirement.[4–6] Building a neutrino telescope has been a daunting technical challenge.

Given the detector's required size, early efforts concentrated on transforming large volumes of natural water into Cherenkov detectors that collect the light produced when neutrinos interact with nuclei in or near the detector.[7] After a two-decade-long effort, building the Deep Underwater Muon and Neutrino Detector (DUMAND) in the sea off the main island of Hawaii unfortunately failed.[8] However, DUMAND paved the way for later efforts by pioneering many of the detector technologies in use today and by inspiring the deployment of a smaller instrument in Lake Baikal[9] as well as efforts to commission neutrino telescopes in the Mediterranean.[10–12] These efforts in turn have led toward the construction of KM3NeT. But the first telescope on the scale envisaged by the DUMAND collaboration was realized instead by transforming a large volume of transparent natural Antarctic ice into a particle detector, the Antarctic Muon and Neutrino Detector Array (AMANDA). In operation beginning in 2000, it represented a proof of concept for the kilometer-scale neutrino observatory, IceCube.[13,14]

Neutrino astronomy has achieved spectacular successes in the past: neutrino detectors have "seen" the Sun and detected a supernova in the Large Magellanic Cloud in 1987. Both observations were of tremendous importance; the former showed that neutrinos have mass, opening the first crack in the Standard Model of particle physics, and the latter confirmed the basic nuclear physics of the death of stars. Figure 1 illustrates the neutrino energy spectrum covering an enormous range, from microwave energies (10^{-12} eV) to 10^{20} eV.[15] The figure is a mixture of observations and theoretical predictions. At low energy, the neutrino sky is dominated by neutrinos produced in the Big Bang. At MeV energy, neutrinos are produced by supernova explosions; the flux from the 1987 event is shown. At yet higher energies, the figure displays the measured atmospheric-neutrino flux up to energies of 100 TeV by the AMANDA experiment.[16] Atmospheric neutrinos are a main player in our story, because they are the dominant background for extraterrestrial searches. The flux of atmospheric neutrinos falls dramatically with increasing energy; events

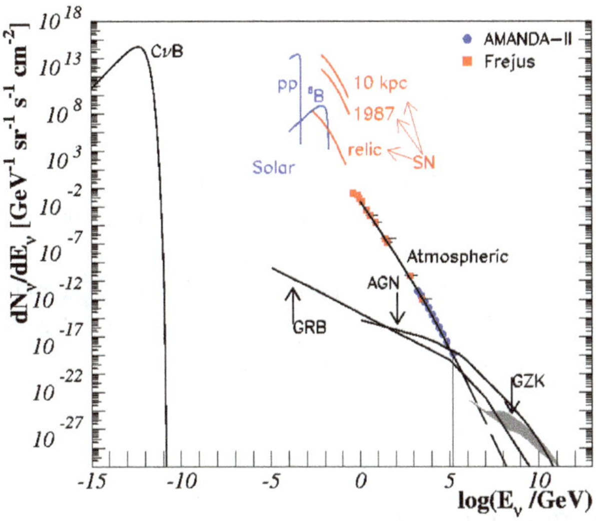

Fig. 1. The cosmic-neutrino spectrum. Sources are the Big Bang (CνB), the Sun, supernovae (SN), atmospheric neutrinos, active galactic nuclei (AGN) galaxies, and GZK neutrinos. The data points are from detectors at the Frejus underground laboratory[17] (red) and from AMANDA[16] (blue).

above 100 TeV are rare, leaving a clear field of view for extraterrestrial sources.

The highest-energy neutrinos in Fig. 1 are the decay products of pions produced by the interactions of cosmic rays with microwave photons.[18] Above a threshold of $\sim 4 \times 10^{19}$ eV, cosmic rays interact with the microwave background introducing an absorption feature in the cosmic-ray flux, the Greisen–Zatsepin–Kuzmin (GZK) cutoff. As a consequence, the mean free path of extragalactic cosmic rays propagating in the microwave background is limited to roughly 75 megaparsecs, and therefore, the secondary neutrinos are the only probe of the still enigmatic sources at longer distances. What they will reveal is a matter of speculation. The calculation of the neutrino flux associated with the observed flux of extragalactic cosmic rays is straightforward and yields one event per year in a kilometer-scale detector. The flux, labeled GZK in Fig. 1, shares the high-energy neutrino sky with neutrinos anticipated from GRBs and active galactic nuclei (AGN).[4–6]

A population of cosmic neutrinos covering the 30 TeV–1 PeV energy region was revealed by the first 2 years of IceCube data. Association of cosmic neutrinos with these, or any other, source candidates is still a work in progress. The goal of this chapter is to discuss these topics in some detail. Subsequently, it will briefly cover other uses of neutrino telescopes.

2. Rationale for the construction of kilometer-scale neutrino detectors

The construction of kilometer-scale neutrino detectors was primarily motivated by the prospect of detecting neutrinos associated with the sources of high-energy cosmic rays. Cosmic accelerators produce particles with energies in excess of 100 EeV; we still do not know where or how;[19] see Fig. 2.[a] The bulk of the cosmic rays are galactic in origin. Any association with our galaxy presumably disappears at EeV energy when the gyroradius of a proton in the galactic magnetic field exceeds its size. The cosmic-ray spectrum exhibits a rich structure above an energy of ~0.1 EeV, but where exactly the transition to extragalactic cosmic rays occurs is a matter of debate.

2.1. *Cosmic-ray accelerators*

The detailed blueprint for a cosmic-ray accelerator must meet two challenges: the highest-energy particles in the beam must reach energies beyond

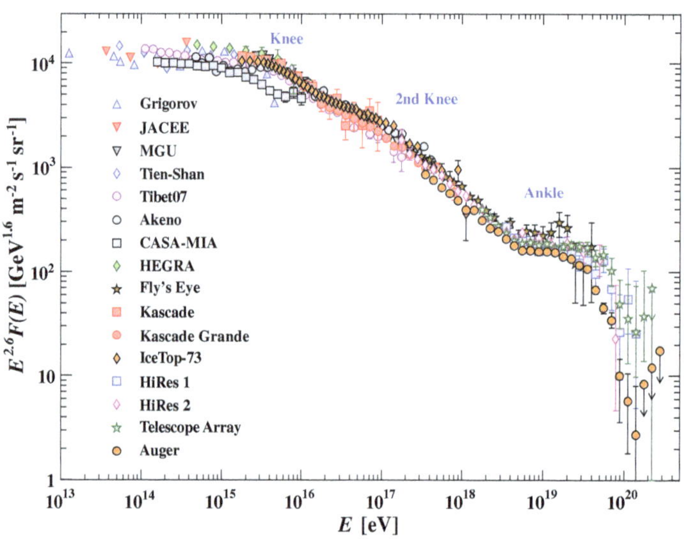

Fig. 2. At the energies of interest here, the cosmic-ray spectrum follows a sequence of three power laws. The first two are separated by the "knee," the second and third by the "ankle." Cosmic rays beyond the ankle are a new population of particles produced in extragalactic sources. Note that the spectrum $F(E)(= dN/dE)$ has been multiplied by a power $E^{2.7}$ in order to visually enhance the structure in the spectrum (data compiled by Particle Data Group[20]).

[a] We will use energy units TeV, PeV and EeV, increasing by factors of 1000 from GeV energy.

Fig. 3. Opportunities exist near intense charged-particle flows, seen as filaments in this X-ray picture of a solar flare, for solar particles to accelerate to GeV energy.

10^3 TeV (10^8 TeV) for galactic (extragalactic) sources and their luminosities must accommodate the observed flux. Both requirements represent severe constraints that have guided theoretical speculations. Acceleration of protons (or nuclei) to TeV energy and above requires massive bulk flows of relativistic charged particles. The blueprint of the accelerator can be copied from solar flares where particles are accelerated to GeV energy by shocks and, possibly, reconnection; see Fig. 3. Requiring that the gyroradius of the accelerated particle be contained within the accelerating B-field region, $E/ZecB \leq R$, leads to an upper limit on the energy of the particle,

$$E \leq ZecBR. \tag{1}$$

Reaching energies much above 10 GeV in solar flares is dimensionally impossible. In a solar flare, the extent R of the accelerating region and the magnitude of the magnetic fields B are not large enough to accelerate particles of charge Ze to energies beyond GeV; their velocity is taken to be the speed of light, c. The central idea for accommodating the higher energies of the galactic and extragalactic cosmic rays observed is that a fraction of the gravitational energy released in a stellar collapse is converted into particle acceleration, presumably by shocks.

Baade and Zwicky[21] suggested as early as 1934 that supernova remnants could be sources of the galactic cosmic rays. It is assumed that, after the collapse, $\sim 10^{51}$ erg of energy is transformed into particle acceleration by diffusive shocks associated with young ($\sim$1000 year old) supernova remnants expanding into the interstellar medium. Like a snowplow, the shock sweeps up the $\sim$1 proton/cm^3 density of hydrogen in the galactic plane. The accumulations of dense filaments of particles in the outer reaches of the shock, clearly visible as sources of intense X-ray emission, are the sites of high magnetic fields; see Fig. 4. It is theorized that particles crossing these structures multiple times can be accelerated to high energies following an approximate power-law spectrum $\mathrm{d}N/\mathrm{d}E \sim E^{-2}$. The mechanism copies solar flares where filaments of high magnetic fields, visible in Fig. 3, are the sites for accelerating nuclear particles to tens of GeV. The higher energies reached in supernova remnants are the consequence of particle flows of much larger intensity powered by the gravitational energy released in the stellar collapse.

This idea has been widely accepted despite the fact that to date no source has been conclusively identified, neither by cosmic rays nor by accompanying gamma rays and neutrinos produced when the cosmic rays interact with galactic hydrogen. Galactic cosmic rays reach energies of at least several PeV, the "knee" in the spectrum; therefore, their interactions should generate gamma rays and neutrinos from the decay of secondary pions reaching hundreds of TeV. Such sources, referred to as PeVatrons, have not been found; see, however, Ref. 22. Nevertheless, Zwicky's suggestion has become the stuff of textbooks, and the reason is energetics: three galactic supernova explosions per century converting a reasonable fraction of a solar mass into particle acceleration can accommodate the steady flux of cosmic rays in the galaxy. It is interesting to note that Zwicky originally assumed that the sources were extragalactic since the most recent supernova in the Milky Way was in 1572. After diffusion in the interstellar medium was understood, supernova explosions in the Milky Way became the source of choice for the origin of galactic cosmic rays,[23] although after more than 50 years the issue is still debated.[24]

Energetics also guides speculations on the origin of extragalactic cosmic rays. By integrating the cosmic-ray spectrum above the ankle at $\sim$4 EeV, it is possible to estimate[25] the energy density in extragalactic cosmic rays as $\sim 3 \times 10^{-19}$ erg cm^{-3}. This value is rather uncertain because of our ignorance of the energy where the transition from galactic to extragalactic sources occurs. The power required for a population of sources to generate this energy density over the Hubble time of 10^{10} years is 2×10^{37} erg s^{-1} per Mpc^3. Long-duration GRBs have been associated with the collapse of massive stars to black holes, and not to neutron stars, as is the case in a

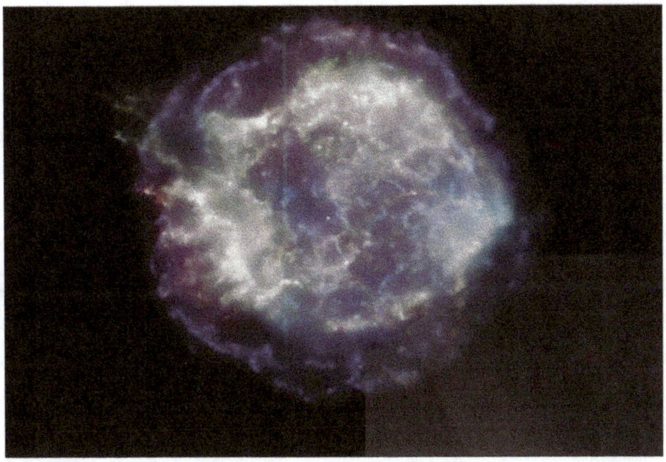

Fig. 4. This X-ray picture of the supernova remnant CasA reveals strong particle flows near its periphery. We believe they are the site for accelerating galactic cosmic rays to energies reaching the "knee" in the spectrum.

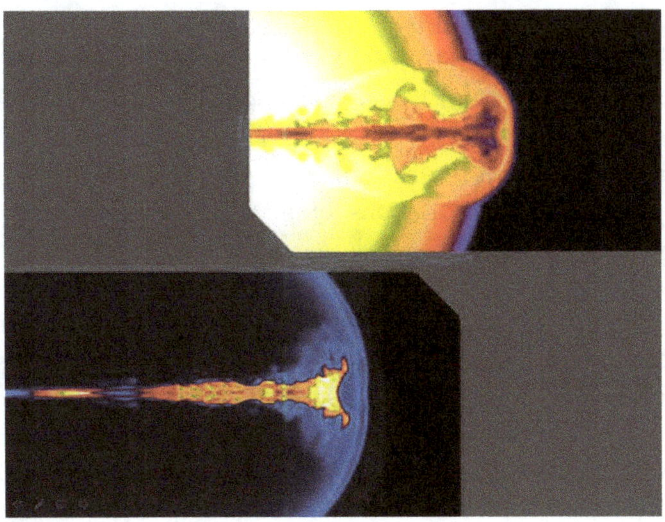

Fig. 5. Colliding shocks in the simulation of a GRB fireball may accelerate cosmic rays to the highest energies observed. The filaments in the particle flow are directed along the rotation axis of the black hole. Animated view at http://www.nasa.gov/centers/goddard/news/topstory/2003/0618rosettaburst.html.

collapse powering a supernova remnant. A GRB fireball converts a fraction of a solar mass into the acceleration of electrons, seen as synchrotron photons. The observed energy in extragalactic cosmic rays can be accommodated with the reasonable assumption that shocks in the expanding GRB fireball convert roughly equal energy into the acceleration of electrons and cosmic rays.[26] It so happens that 2×10^{51} erg per GRB will yield the observed energy density in cosmic rays after 10^{10} years, given that their rate is on the order of 300 per Gpc^3 per year. Hundreds of bursts per year over a Hubble time produce the observed cosmic-ray density, just as three supernovae per century accommodate the steady flux in the Galaxy.

Problem solved? Not really: it turns out that the same result can be achieved assuming that AGN convert, on average, 2×10^{44} erg s^{-1} each into particle acceleration.[15] This is an amount that matches their output in electromagnetic radiation. An AGN is the center of a galaxy that hosts a supermassive black hole. We will return to this point further on.

2.2. *Neutrinos and gamma rays associated with cosmic rays*

Neutrinos will be produced at some level in association with the cosmic-ray beam. Cosmic rays accelerated in regions of high magnetic fields near black holes or neutron stars inevitably interact with radiation surrounding them. Thus, cosmic-ray accelerators are also "beam dumps" producing neutrino beams. The method is what is used for the production of neutrino beams at accelerator laboratories: the beam is dumped in a dense target where it produces pions and kaons that decay into neutrinos. All particles are absorbed in the extended target except the neutrinos. Cosmic rays accelerated in supernova shocks interact with gas in the galactic disk, producing equal numbers of pions of all three charges that decay into pionic photons and neutrinos. A larger source of secondaries is likely to be gas near the sources, for example, cosmic rays interacting with high-density molecular clouds that are ubiquitous in the star-forming regions where supernovae are more likely to explode. For extragalactic sources, the neutrino-producing target may be electromagnetic, for instance, photons radiated by the accretion disk of an AGN or synchrotron photons that coexist with protons in the expanding fireball producing a GRB. In Fig. 6, estimates of astrophysical neutrino fluxes are compared with measurements of atmospheric neutrinos. The shaded band indicates the level of model-dependent expectations for high-energy neutrinos of astrophysical origin. The estimates that we will discuss in more detail further on optimistically predicted a neutrino flux at a level of

$$E_\nu^2 \, \mathrm{d}N_\nu / \mathrm{d}E_\nu \simeq 10^{-8} \, \mathrm{GeV} \, \mathrm{cm}^{-2} \mathrm{s}^{-1} \mathrm{sr}^{-1} \tag{2}$$

per flavor and formed the rationale for building a kilometer-scale detector; this is indeed the magnitude of the cosmic component of the neutrino spectrum above 100 TeV revealed by IceCube's data.

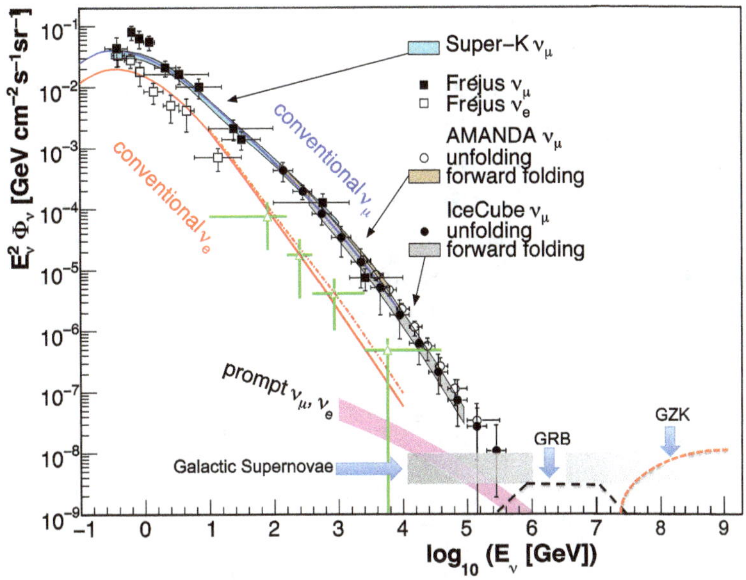

Fig. 6. Anticipated astrophysical neutrino fluxes compared with measured and calculated fluxes of atmospheric neutrinos. Measurements of ν_μ from SK,[27] Frejus,[28] AMANDA,[29,30] and IceCube[31,32] are shown along with the electron-neutrino spectrum at high energy (green open triangles).[33] Calculations of conventional ν_e (red line) and ν_μ (blue line),[34] ν_e (red dotted line)[35] and charm-induced neutrinos (magenta band)[36] are also shown.

How many neutrinos and, inevitably, gamma rays are produced in association with the cosmic-ray beam? Generically, a cosmic-ray source should also be a beam dump. Cosmic rays accelerated in regions of high magnetic fields near black holes may interact with radiation surrounding them, e.g., UV photons in some active galaxies or MeV photons in GRB fireballs. In these interactions, neutral and charged pion secondaries are produced by the processes

$$p+\gamma \to \Delta^+ \to \pi^0 + p \quad \text{and} \quad p+\gamma \to \Delta^+ \to \pi^+ + n. \tag{3}$$

While secondary protons may remain trapped and lose energy in the high magnetic fields, neutrons and the decay products of neutral and charged pions escape with high energy. The energy escaping the source is therefore distributed among cosmic rays, gamma rays and neutrinos, particles produced by the decay of neutrons, neutral pions and charged pions, respectively.

Galactic supernova shocks are in contrast an example of a hadronic beam dump. Cosmic rays mostly interact with the hydrogen in the galactic disk, producing equal numbers of pions of all three charges in hadronic collisions $p+p \to n\,[\pi^0 + \pi^+ + \pi^-] + X$; n is the pion multiplicity.

In a generic cosmic beam dump, accelerated cosmic rays, assumed to be protons for illustration, interact with a photon target. These may be photons radiated by the accretion disk in AGNs and synchrotron photons that co-exist with protons in the exploding fireball producing a GRB. Their interactions produce charged and neutral pions according to Eq. (3), with probabilities of 2/3 and 1/3, respectively. Subsequently, the pions decay into gamma rays and neutrinos that carry, on average, 1/2 and 1/4 of the energy of the parent pion. We further assume that, on average, the four leptons in the decay $\pi^+ \to \nu_\mu + \mu^+ \to \nu_\mu + (e^+ + \nu_e + \bar{\nu}_\mu)$ equally share the charged pion's energy. The energy of the pionic leptons relative to the proton is

$$x_\nu = \frac{E_\nu}{E_p} = \frac{1}{4}\langle x_{p\to\pi}\rangle \simeq \frac{1}{20} \tag{4}$$

and

$$x_\gamma = \frac{E_\gamma}{E_p} = \frac{1}{2}\langle x_{p\to\pi}\rangle \simeq \frac{1}{10}. \tag{5}$$

Here,

$$\langle x_{p\to\pi}\rangle = \left\langle \frac{E_\pi}{E_p} \right\rangle \simeq 0.2 \tag{6}$$

is the average energy transferred from the proton to the pion.

While both gamma-ray and neutrino fluxes can be calculated knowing the density of the accelerated protons and the density of the target material, their relative flux is independent of the details of the production mechanism. The spectral production rates $\mathrm{d}N/\mathrm{d}E\mathrm{d}t$ of neutrinos and gamma rays are related by

$$\frac{1}{3}\sum_{\nu_\alpha} E_\nu \frac{\mathrm{d}N_\nu}{\mathrm{d}E_\nu \mathrm{d}t}(E_\nu) \simeq \frac{K_\pi}{2} E_\gamma \frac{\mathrm{d}N_\gamma}{\mathrm{d}E_\gamma \mathrm{d}t}(E_\gamma). \tag{7}$$

Here, N and E denote the number and energy of neutrinos and gamma rays and ν stands for the neutrino flavor. Note that this relation is solid and depends only on the charged-to-neutral secondary pion ratio, with $K_\pi = 1(2)$ for $\gamma(pp)$ neutrino-producing interactions. In deriving the relative number of neutrinos and gamma rays, one must be aware of the fact that the neutrino flux represents the sum of the neutrinos and antineutrinos, which cannot be separated by current experiments: in short, a π^0 produces two gamma rays for every charged pion producing a $\nu_\mu + \bar{\nu}_\mu$ pair. A more formal derivation of this relation will be given in Section 7.

The production rate of gamma rays described by Eq. (7) is not necessarily the emission rate observed. For instance, in cosmic accelerators that efficiently produce neutrinos via $p\gamma$ interactions, the target photon field can also efficiently reduce the pionic gamma rays via pair production. This is a calorimetric process that will, however, conserve the total energy of hadronic gamma rays. The production of photons in association with cosmic neutrinos is inevitable. The relation is, however, calorimetric; unlike neutrinos, photons reach Earth after propagation in the universal microwave and infrared photon backgrounds to reach our telescopes with TeV energy or less. Also, one must be aware of the fact that inverse-Compton scattering and synchrotron emission by accelerated electrons in magnetic fields in the source have the potential to produce gamma rays; not every high-energy gamma ray is pionic.

The estimates in Fig. 6 of the neutrino flux associated with cosmic rays accelerated in supernova remnants and GRBs are relatively straightforward as both the beam, identified with the observed cosmic-ray flux, and the targets, observed by astronomers, are known. In the case of supernova remnants, the main uncertainty is the availability of nearby target material. In the case of GRBs, the main uncertainty is the fraction of the extragalactic cosmic-ray population that comes from this source. The ongoing search by IceCube for neutrinos in coincidence with and in the direction of GRB alerts issued by astronomical telescopes has limited the GRB neutrino flux to less that 1% of the diffuse cosmic neutrino flux actually observed by the experiment.[37,38] However, this may not conclusively rule out GRBs as a source of cosmic rays; the events that produce the spectacular photon displays catalogued by astronomers as GRBs may not be the stellar collapses that are sources of high-energy neutrinos. We will return to this point further on when we discuss acceleration of cosmic rays in GRB fireballs. Nevertheless, the failure of IceCube to observe neutrinos from GRBs has lately promoted AGNs as the best-bet source of the cosmic neutrinos observed.

Active galaxies are complex systems with many possible sites for accelerating cosmic rays and for targets to produce neutrinos. First, if acceleration occurs mainly at the spectacular termination shocks of the jets in intergalactic space,[39] there would be little target material available and few neutrinos produced. In contrast, production of neutrinos near a black hole,[40] or in collisions with interstellar matter of the accelerated particles diffusing in the magnetic field of the galaxy hosting the black hole,[41] could yield fluxes at the level observed. We will work through these examples further on.

One generic picture in which the neutrino luminosity is directly related to the contribution of the sources to extragalactic cosmic rays arises if acceleration occurs in the jets of AGNs (or GRBs).[42,43] High-energy protons interact in the intense radiation fields inside the jets. In the $p\gamma \to p\pi^0$ channel, the protons remain in the accelerator. In the $p\gamma \to n\pi^+$ channel, however, the neutrons escape and eventually decay to produce cosmic-ray protons, while the pions decay to neutrinos. The luminosity of neutrinos from photopion production is then directly related by kinematics to the cosmic-ray protons that come from decay of the escaping neutrons.

TeV gamma rays are measured from many AGN blazars.[45] Although the observed gamma rays are likely to be from accelerated electrons, which radiate more efficiently than protons, the gamma-ray luminosity may give an indication of the overall cosmic-ray luminosity and hence of the possible level of neutrino production.[46,47] In this context, we introduce Fig. 7[48] showing IceCube upper limits[44] on the neutrino flux from nearby AGNs as a function of their distance. The sources at red shifts between 0.03 and 0.2 are Northern Hemisphere blazars for which distances and intensities are listed in TeVCat[45] and for which IceCube also has upper limits. In several cases, the muon-neutrino limits have reached the level of the TeV photon flux. One can sum the sources shown in the figure into a diffuse flux. The result, after accounting for the distances and luminosities, is 3×10^{-9} GeV cm^{-2} s^{-1} sr^{-1}, or approximately 10^{-8} GeV cm^{-2} s^{-1} sr^{-1} for all neutrino flavors. This is at the level of the generic astrophysical neutrino flux of Eq. (2). At this intensity,

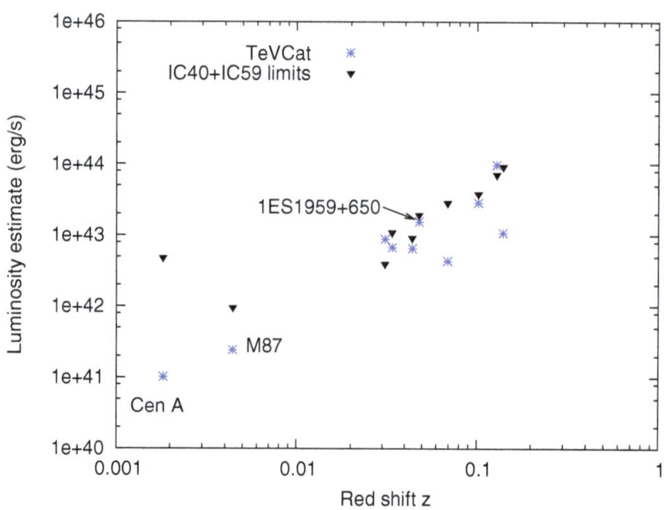

Fig. 7. Limits on the neutrino flux from selected active galaxies derived from IceCube data taken during construction, when the instrument was operating with 40 and 59 strings of the total 86 instrumented strings of DOMs.[44] These are compared with the TeV photon flux for nearby AGNs. Note that energy units are in ergs, not TeV.

neutrinos from theorized cosmic-ray accelerators will cross the steeply falling atmospheric neutrino flux above an energy of ∼300 TeV; see Fig. 6. The level of events observed in a cubic-kilometer neutrino detector is 10–100 ν_μ-induced events per year. Such estimates reinforce the logic for building a cubic-kilometer neutrino detector.[49]

3. Cosmic accelerators and cosmic beam dumps: estimates of the neutrino flux

Next we will discuss two generic blueprints for beam dumps producing neutrinos in association with cosmic-ray accelerators: a beam interacting with nearby gas, radiation fields, or molecular clouds and the relativistically expanding fireball following stellar collapse.

3.1. *Extragalactic neutrino-producing beam dumps*

An active galaxy presents multiple opportunities for particles to accelerate in the inflows and outflows associated with a supermassive black hole. The high-energy particles may subsequently produce neutrinos in interactions with multiple targets such as the dense matter near the black hole, the galactic disk of the galaxy associated with the black hole, or photons produced in the jet or radiated from the accretion disk. It is therefore useful to start by considering a generic beam dump where a beam of protons with an initial flux $j_{\rm p}(E_{\rm p})$ interacts with a target of density n over a distance l. The optical depth of the target after the proton travels a distance l' is given by:

$$\tau' = \frac{l'}{\lambda} = nl'\sigma_{pp}. \tag{8}$$

Each time a proton interacts, it deposits K_pE_p energy into $\langle n_\pi \rangle$ pions of average energy $\langle E_\pi \rangle$; here, K_p is the total proton inelasticity. Energy conservation implies that

$$K_pE_p = E_\pi^{\rm tot} = \langle n_{\pi^\pm} \rangle \langle E_\pi \rangle. \tag{9}$$

The flux of pions produced per energy and time interval, $q_{\pi^\pm}(E_\pi)$, is related to the proton interaction rate in the target, $q_{\rm p}(E_{\rm p}, \tau)$, by[50]

$$q_{\pi^\pm} = \int_{E_{\rm th}}^{\infty} dE_p \int_0^{\tau} d\tau' \, q_p(E_p, \tau') \, \frac{dN_\pi(E_p)}{dE_\pi}. \tag{10}$$

Here, the first factor in the integrand,

$$q_p(E_p, \tau) = j_p(E_p) \exp(-\tau), \tag{11}$$

counts the number of protons interacting in the target for a proton flux $j_p(E_p)$ incident on the target. The second term counts the number of pions produced in the interactions as a function of the proton energy, with

$$\int dE_\pi \frac{dN_\pi(E_p)}{dE_\pi} = \langle n_{\pi^\pm}\rangle. \tag{12}$$

While the physics of this expression is transparent, a more formal discussion can be found in Ref. 51.

Assuming that all pions are produced with the same average energy, the pion distribution in Eq. (10) takes the form

$$\frac{dN_\pi(E_p)}{dE_\pi} = \langle n_{\pi^\pm}\rangle \delta(E_\pi - \langle E_\pi\rangle). \tag{13}$$

Typically, one also makes the approximation that the proton cross section is independent of energy, $\sigma_{pp} \approx 3 \cdot 10^{-26}$ cm^2, and as a result

$$q_{\pi^\pm} = \int_{E_{\rm th}}^{\infty} dE_{\rm p}(1 - \exp(-\tau)) j_p(E_p) \langle n_{\pi^\pm}\rangle \, \delta\left(E_\pi - \langle E_\pi\rangle\right). \tag{14}$$

In typical astrophysical situations, where the optical depth is small, the pion production efficiency $f_\pi = (1 - \exp(-\tau)) \to \tau$, with $\tau = ln\sigma_{pp}$ and

$$q_{\pi^\pm} = nl\sigma_{pp} \int_{E_{\rm th}}^{\infty} dE_p \, j_p\,(E_p)\langle n_{\pi^\pm}\rangle \delta(E_\pi - \langle E_\pi\rangle). \tag{15}$$

The integral can be performed by rewriting the delta function as

$$\delta\left(E_\pi - \frac{K_p}{\langle n_\pi\rangle} E_p\right) = \frac{\langle n_{\pi^\pm}\rangle}{K_p} \delta\left(E_p - \frac{\langle n_{\pi^\pm}\rangle\langle E_\pi\rangle}{K_p}\right) \tag{16}$$

using Eq. (9). We thus obtain the result that

$$q_{\pi^\pm} = nl\sigma_{pp}\langle n_{\pi^\pm}\rangle^2 \frac{1}{K_p} j_p(E_p), \tag{17}$$

and the relation between pion and proton energies is given by

$$\langle E_\pi\rangle^2 q_{\pi^\pm} = f_\pi K_p \frac{K_\pi}{1 + K_\pi} E_p^2 q_p, \tag{18}$$

where the factor of $2/3$, i.e., $K_\pi = 1$ for hadron collisions, selects the fraction of charged pions.

For simple, yet more realistic, calculations we can follow Mannheim[50] who introduces the fact that the multiplicity of pions grows with the proton energy. Starting from a pair of pions $pp \to pp + \pi^+\pi^-$ at threshold $E_{\rm th}$, the multiplicity grows as

$$\langle n_{\pi^\pm}\rangle = 2\left(\frac{E_p - E_{\rm th}}{\rm GeV}\right)^{1/4}, \tag{19}$$

and

$$\langle E_\pi\rangle = \frac{1}{6}(E_{\rm p} - m_p c^2)^{3/4}\,{\rm GeV}. \tag{20}$$

The formalism introduced, combined with these parameterizations, allows us to reproduce the results of detailed simulations. Given the number of protons produced by the accelerator per energy and time interval

$$j_p(E_p) = A_p\left(\frac{E_p - m_p c^2}{\rm GeV}\right)^{-\gamma}, \tag{21}$$

we obtain a pion rate *at the source*:

$$q_{\pi^\pm}(E_\pi) \approx 26 n l A_p \sigma_{pp}\left(\frac{6E_\pi}{\rm GeV}\right)^{-\frac{4}{3}(\gamma-\frac{1}{2})}. \tag{22}$$

The total neutrino rate *at the source* is then given by the sum of the emissivities of the first muon neutrino, directly from the pion, and those of the second muon neutrino and the electron neutrino from the muon decay:

$$q_{\nu,\rm tot} = q^{(1)}_{\nu_\mu} + q^{(2)}_{\nu_\mu} + q_{\nu_e}. \tag{23}$$

As previously discussed, we will make the approximation that the total energy of the pions is distributed equally among the four decay leptons (see e.g., Ref. 52),

$$q_{\nu_i}(E_{\nu_i}) = q_\pi(4E_{\nu_1}) dE_\pi/dE_{\nu_i} = 4q_\pi(4E_{\nu_i}) \tag{24}$$

for each neutrino, $\nu_i = \overline{\nu}_e$ or nu_e, ν_μ, $\overline{\nu}_\mu$. As IceCube does not distinguish between neutrinos and antineutrinos, we will not separate neutrinos from antineutrinos. Another way of stating the approximation introduced is that the number of neutrinos produced per flavor in the energy bin dE_ν arises from the original pion in the energy bin $dE_\pi = 4dE_\nu$; therefore, $q_\nu(E_\nu)\,dE_\nu = q_\pi(4E_\nu)dE_\pi$.

The final result for the total neutrino emission rate at the source is given by

$$q_{\nu,\text{tot}} \approx 3 \times 10^2 N A_\text{p} \sigma_\text{pp} \left(\frac{24 E_\nu}{\text{GeV}}\right)^{-\frac{4}{3}\gamma+\frac{2}{3}}. \tag{25}$$

Equation 25 provides us with an estimate of the total neutrino flux at the source in terms of three key quantities: A_p and γ, the normalization and spectral slope of the flux of the accelerator, and the column density $N \sim ln$.

The spectral index is routinely taken to be $\gamma = 2$, a value obtained from diffusive shock acceleration and, in any case, typical of the spectral flux observed for the spectrum of gamma rays in nonthermal sources. The column density $N = ln$ is usually taken from astronomical information, with l being the distance that the cosmic rays travel through a photon target of density n. For instance, in the case of a galaxy, l could be the diffusion length of the cosmic rays in the magnetic field of the galaxy. This allows the protons to interact with a typical density of dust of $n = 1\,\text{cm}^{-3}$ over an extended pathlength determined by the diffusion time. The typical diffusion distance before escape of the galaxy is given by

$$d_\text{diff} = 2\sqrt{D(E_\text{p})\, t_\text{esc}}, \tag{26}$$

where D is the diffusion coefficient and t_esc is the escape time. In this case, the distance in the calculation is identified with $l = ct_\text{esc}$. Finally, the normalization of the cosmic-ray flux A_p is typically determined from the requirement that the source reproduce the cosmic flux observed at Earth.[40,41]

In the end, the flux of IceCube data can be described by this procedure with the reasonable assumptions that $\gamma = 2$ and $l = 10\,\text{kpc}$, typical of our own galaxy. Additionally, an extragalactic diffuse photon flux is obtained consistent with the Fermi measurements in the GeV–TeV energy range.[41]

Alternatively, one may determine A_p for individual galaxies, determine their neutrino flux, and subsequently do the summation over all sources. An example can be found in Ref. 40 where the cosmic-ray flux is obtained from the following considerations: The radio luminosity of AGNs, L, provides a measure for the AGN luminosity in electrons. The electron luminosity is equal to or larger than the radio luminosity of the source because the latter is produced when electrons are accelerated and emit synchrotron radiation: $L_\text{e} = \chi L$ with $\chi \geq 1$. Hadronic cosmic rays and electrons are connected by a constant fraction f_e: $L_\text{e} = f_e L_\text{p}$, which is on the order of 0.1.[40] With

$$L_\text{p} = \int j_\text{p}(E_\text{p})\, d(E_\text{p}) \approx \frac{\chi L}{f_e}, \tag{27}$$

the normalization of the CR spectrum is

$$A_{\rm p} = A_{\rm p}(L, z) = \frac{\chi}{f_e} \left[\ln(E_{\rm max}/E_{\rm min})\right]^{-1} L\,{\rm GeV}^{-2} \tag{28}$$

for $\gamma = 2$; the generalization for $\gamma \neq 2$ is straightforward.

Finally, the diffuse neutrino flux at Earth is obtained from the single source flux of Eq. (25), the cosmic-ray spectrum normalized by Eq. (28), and

$$\Phi_\nu = \int_L \int_z \frac{q_{\nu,\rm tot}}{4\pi d_L(z)^2} \frac{dn_{\rm AGN}}{dV\,dL} \frac{dV}{dz}\, dz\, dL. \tag{29}$$

Here, d_L is the luminosity distance, $dn_{\rm AGN}/(dV\,dL)$ is the radio luminosity function of the AGN, and dV/dz is the comoving volume at a fixed redshift z. The radio luminosity function is usually represented by the product of a luminosity-dependent and a redshift-dependent function, $dn_{\rm AGN}/(dV\,dL) = g(L)f(z)$. This framework for producing neutrinos in AGNs can accommodate the diffuse flux observed by IceCube; more details can be found in Ref. 40. The neutrino production happens in the relatively dense matter near the black hole, illustrating how two very different mechanisms manage to accommodate the IceCube result. The details may vary, but in general, beam dumps producing similar energies in neutrinos, photons, and cosmic rays match the diffuse neutrino flux observed. We will return to this point when discussing the IceCube results.

In yet another scenario, Wang and Loeb argue that quasar-driven outflows interact with interstellar protons to produce IceCube's neutrinos.[53] While the few examples illustrate that AGNs are plausible sources of the neutrinos observed by IceCube, they do not have the power of proof given the presence of adjustable parameters. This does not mean that a conclusive association is impossible. AGNs are episodic sources that burst, increasing their flux by over one order of magnitude for periods of seconds to days, sometimes even years. A correlation of the arrival of IceCube neutrinos in coincidence with such bursts can provide a smoking gun for their origin; for a recent discussion, see Ref. 54.

3.2. *Galactic neutrino-producing beam dumps*

The rationale for kilometer-scale neutrino detectors is that their sensitivity is sufficient to reveal generic cosmic-ray sources with an energy density in neutrinos comparable to their energy density in cosmic rays[42] and pionic TeV gamma rays.[46] Interestingly, this condition may be satisfied by the sources of Galactic cosmic rays.

The energy density of the cosmic rays in our Galaxy is $\rho_E \sim 10^{-12}\,\mathrm{erg\,cm^{-3}}$. Galactic cosmic rays do not exist forever; they diffuse within microgauss fields and remain trapped for an average containment time of 3×10^6 years. The power needed to maintain a steady energy density requires accelerators delivering 10^{41} erg/s. This happens to be 10% of the power produced by supernovae releasing 10^{51} erg every 30 years (10^{51} erg corresponds to 1% of the binding energy of a neutron star after 99% is initially lost to neutrinos). This coincidence is the basis for the idea that shocks produced by supernovae exploding into the interstellar medium are the accelerators of galactic cosmic rays.

A generic supernova remnant releasing an energy of $W \sim 10^{50}$ erg into the acceleration of cosmic rays will inevitably generate TeV gamma rays by interacting with the hydrogen in the galactic disk. The emissivity in pionic gamma rays is calculated using the formalism previously introduced for the calculation of the neutrino flux with $q_{\pi^0} = \frac{1}{2} q_{\pi^\pm}$; the latter is given by Eq. (15).

Before proceeding, a short discussion of the production of pionic photons follows. For this calculation, we define the production rate Q_{π^0}, per unit volume and time, by converting dl to cdt in the τ' integration in Eq. (15). Additionally one assumes that a single pion is produced carrying a fraction $f_\pi \simeq 0.2$ of the proton momentum. This yields

$$Q_{\pi^0} = cn\sigma_{pp} \int dE_p n_p(E_p) \delta(E_{\pi^0} - f_\pi E_p), \tag{30}$$

which in turn yields the final result

$$Q_{\pi^0}(E_{\pi^0}) = c\sigma_{pp} n \frac{1}{f_\pi} n_p \left(\frac{E_{\pi^0}}{f_\pi} \right). \tag{31}$$

The emissivity Q_{π^0} is simply proportional to the number density of hydrogen targets and the number of cosmic rays is given by $n_p \simeq 4 \times 10^{-14}\,(E_p/\mathrm{TeV})^{-1.7}\,\mathrm{cm^{-3}}$.

The emissivity of photons is obtained from the fact that every pion produces two photons with half its energy; therefore

$$Q_\gamma(E_\gamma) = 2 \times 2Q_{\pi^0}(2E_\gamma) \simeq Q_{\pi^0}, \tag{32}$$

where the latter equality holds for an E^{-2} spectrum.

Our final result is

$$Q_\gamma(E_\gamma) \simeq c \frac{1}{f_\pi} \sigma_{pp} n n_{cr} \left(E_p > \frac{E_\gamma}{f_\pi} \right), \tag{33}$$

or, assuming an E^{-2} spectrum,

$$Q_\gamma(>1\,\text{TeV}) \simeq 10^{-29} \left(\frac{n}{\text{cm}^{-3}}\right) \text{cm}^{-3}\,\text{s}^{-1}. \tag{34}$$

Notice the transparency of this result. The proportionality factor in Eq. (33) is determined by particle physics, where $\lambda_{pp} = (n\sigma_{pp})^{-1}$ is the proton interaction length ($\sigma_{pp} \simeq 40\,\text{mb}$) in a density n of hydrogen atoms. The corresponding luminosity is

$$L_\gamma(>1\,\text{TeV}) \simeq Q_\gamma \frac{W}{\rho_E} \tag{35}$$

where W/ρ_E is the volume occupied by the supernova remnant; given the ambient density $\rho_E \sim 10^{-12}\,\text{erg}\,\text{cm}^{-3}$ of galactic cosmic rays,[4] a supernova with energy $W \sim 10^{50}\,\text{erg}$ in cosmic rays occupies the volume W/ρ_E. We here made the approximation that the density of particles in the remnant is not very different from the ambient energy density.

We thus predict[55,56] a rate of TeV photons from a supernova remnant at a nominal distance d on the order of 1 kpc of

$$\int_{E>1\text{TeV}} \frac{dN_\gamma}{dE_\gamma} dE_\gamma = \frac{L_\gamma(>1\text{TeV})}{4\pi d^2}$$
$$\simeq 10^{-12} \text{ to } 10^{-11} \left(\frac{\text{TeV}}{\text{cm}^2\,\text{s}}\right) \left(\frac{W}{10^{50}\,\text{erg}}\right) \left(\frac{n}{1\,\text{cm}^{-3}}\right) \left(\frac{d}{1\,\text{kpc}}\right)^{-2}. \tag{36}$$

As discussed in the introduction, the position of the knee in the cosmic-ray spectrum indicates that some sources accelerate cosmic rays to energies of several PeV. These PeVatrons therefore produce pionic gamma rays whose spectrum can extend to several hundred TeV without cutting off. For such sources, the gamma-ray flux in the TeV energy range can be parametrized in terms of a spectral slope α_γ, an energy $E_{\text{cut},\gamma}$ where the accelerator cuts off, and a normalization k_γ:

$$\frac{dN_\gamma(E_\gamma)}{dE_\gamma} = k_\gamma \left(\frac{E_\gamma}{\text{TeV}}\right)^{-\alpha_\gamma} \exp\left(-\sqrt{\frac{E_\gamma}{E_{\text{cut},\gamma}}}\right). \tag{37}$$

The estimate in Eq. (36) indicates that fluxes as large as $dN_\gamma/dE_\gamma \sim 10^{-12}$ to 10^{-14} ($\text{TeV}^{-1}\,\text{cm}^{-2}\,\text{s}^{-1}$) can be expected at energies of $\mathcal{O}$ (10 TeV). We will therefore concentrate on the search for PeVatrons, supernova remnants with the required energetics to produce cosmic rays, at least up to the knee in the spectrum.

These sources are well within the sensitivity of existing high-energy gamma-ray detectors such as air Cherenkov telescopes and large acceptance ground-based detectors like Milagro. They may have been revealed by the highest-energy all-sky survey in ∼20 TeV gamma rays using the Milagro detector.[57] A subset of sources, located within nearby star-forming regions in Cygnus and in the vicinity of galactic latitude $l = 40°$, are identified; some cannot be readily associated with known supernova remnants or with nonthermal sources observed at other wavelengths. Subsequently, directional air Cherenkov telescopes were pointed at three of the sources, revealing them as PeVatron candidates with an approximate E^{-2} energy spectrum that extends to tens of TeV without evidence for a cutoff,[58] in contrast with the best studied supernova remnants RX J1713-3946 and RX J0852.0-4622 (Vela Junior).

Some Milagro sources may actually be molecular clouds illuminated by the cosmic-ray beam accelerated in young remnants located within ∼100 pc. Indeed, one expects that multi-PeV cosmic rays are accelerated only over a short time period when the shock velocity is high, i.e., towards the end of the free expansion of the remnant. The high-energy particles can produce photons and neutrinos over much longer periods when they diffuse through the interstellar medium to interact with nearby molecular clouds.[59] An association of molecular clouds and supernova remnants is expected, of course, in star-forming regions. In this case, any confusion with synchrotron photons is unlikely.

Ground-based and satellite-borne instruments with improved sensitivity are able to conclusively pinpoint supernova remnants as the sources of cosmic-ray acceleration by identifying accompanying gamma rays of pion origin. The Fermi Large Area Telescope (LAT) has detected a pion-decay feature in the gamma-ray spectra of two supernova remnants, IC 443 and W44.[60] In contrast, GeV gamma-ray data from Fermi LAT have challenged the hadronic interpretation of the GeV–TeV radiation from one of the best-studied candidates, RX J1713-3946.[61] The most promising PeVatron candidate to date is, instead, the center of the galaxy, as reported by the HESS Collaboration; see Ref. 22. Detecting the accompanying neutrinos from supernova remnants or the galactic center would provide incontrovertible evidence for cosmic-ray acceleration.

Particle physics dictates the relation between pionic gamma rays and neutrinos and basically predicts the production of a $\nu_\mu + \bar{\nu}_\mu$ pair for every two gamma rays seen by Milagro. This calculation can be performed using the formalism discussed with approximately the same outcome. Operating the complete IceCube detector for several years should yield confirmation that some of the Milagro sources produce pionic gamma rays; see Fig. 8.

The quantitative statistics can be summarized in the following. For average values of the parameters describing the flux, we find that

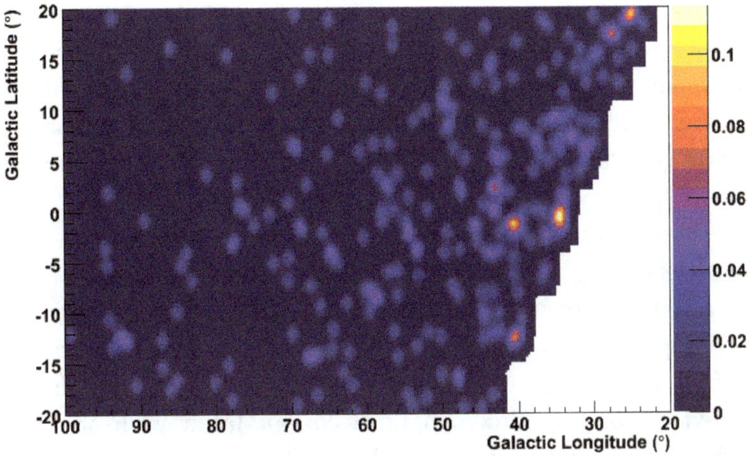

Fig. 8. Simulated sky map of IceCube in galactic coordinates after 5 years of operation of the completed detector. Two Milagro sources are visible with four events for MGRO J1852+01 and three events for MGRO J1908+06 with energy in excess of 40 TeV. These, as well as the background events, have been randomly distributed according to the resolution of the detector and the size of the sources.

the IceCube detector could confirm sources in the Milagro sky map as sites of cosmic-ray acceleration at the 3σ level in less than 1 year and at the 5σ level in 3 years.[55] We here assume that the source extends to 300 TeV or 10% of the energy of the cosmic rays near the knee in the spectrum. These results agree with previous estimates.[62,63] There are intrinsic ambiguities in this estimate of an astrophysical nature that may reduce or extend the time required for a 5σ observation.[55] In particular, the poorly known extended nature of some of the Milagro sources and the value of the cutoff represent a challenge for IceCube analyses that are optimized for point sources. The absence of any observation of an accumulation of high-energy neutrinos in the direction of these sources will seriously challenge the concept that gamma-ray telescopes are seeing actual sources of cosmic rays.

These predictions have been stable over the years and have recently been updated in the context of new gamma-ray information, in particular from the HAWC experiment; see Ref. 58. The predicted fluxes for galactic sources in the Northern Hemisphere are close to IceCube's current upper limits on a point source flux of $10^{-12}\,\mathrm{TeV cm^{-2} s^{-1}}$;[64] see also Ref. 65.

3.3. *Generic fireballs*

Whereas we have confidence that the electromagnetic radiation in some galactic sources is produced by the decay of neutral pions, there is no straightforward gamma-ray path to the neutrino flux expected from extragalactic cosmic-ray accelerators. We presented model calculations to show that AGNs can plausibly accommodate the cosmic neutrino flux observed, assuming that they accelerate protons at the level of the sources of the extragalactic cosmic rays. In fact, we showed how very different blueprints for the beam dump can fit the diffuse neutrino flux observed. As already discussed, there is yet another possibility. Massive stars collapsing to black holes and observed by astronomers as GRBs have the potential to accelerate protons to 100 EeV energy. Neutrinos of 100 TeV–PeV energy should be produced by pion photoproduction when protons and photons coexist in the GRB fireball.[66] As previously discussed, the model is promising because the observed cosmic-ray flux can be accommodated with the assumption that roughly equal energy is shared by electrons, observed as synchrotron photons, and protons. Indeed, the IceCube observations indicate equal extragalactic energy densities of photons and neutrinos, as we will see further on.

The phenomenology that successfully accommodates the astronomical observations is the creation of a hot fireball of electrons, photons, and protons that is initially opaque to radiation. The hot plasma therefore expands by radiation pressure, and particles are accelerated to a Lorentz factor Γ that grows until the plasma becomes optically thin and produces the GRB display. From this point on, the fireball coasts with a Lorentz factor that is constant and depends on its baryonic load. The baryonic component carries the bulk of the fireball's kinetic energy. The energetics and rapid time structure of the burst can be successfully associated with successive shocks (shells), of width ΔR, that develop in the expanding fireball. The rapid temporal variation of the gamma-ray burst, t_v, is on the order of milliseconds and can be interpreted as the collision of internal shocks with a varying baryonic load leading to differences in the bulk Lorentz factor. Electrons, accelerated by first-order Fermi acceleration, radiate synchrotron gamma rays in the strong internal magnetic field and thus produce the spikes observed in the burst spectra.

The usual approach followed in the interpretation of routine IceCube GRB searches[67] has the proton content of the fireball derived from the observed electromagnetic emission. The basic assumption is that a comparable amount of energy is dissipated in fireball protons and electrons, where the latter are observed as synchrotron radiation,

$$E^2\frac{dN_\nu}{dE} = \left(\frac{\epsilon_p}{\epsilon_e}\right)\frac{1}{2}x_\nu\left[E_\gamma^2\,\frac{dN_\gamma}{dE_\gamma}(E_\gamma)\right]_{\rm syn}, \tag{38}$$

where ϵ_p and ϵ_e are the energy fractions in the fireball in protons and electrons,[67] respectively. One can then use the data to determine the baryon loading in the GRB fireball, ϵ_p/ϵ_e. Within this framework, one can estimate the baryon loading or relative abundance of protons and electrons in the fireball from the neutrino observations; no prediction is made. The abundance of protons in the fireball is determined or, at present, limited by the observations.[37]

Although simulations of GRB fireballs have reached a level of sophistication,[68] a simple energy estimate is sufficient to predict the neutrino flux associated with GRB fireballs assuming that they are the sources of the cosmic rays.

Simple relativistic kinematics (see Fig. 9) relates the radius and width R' and $\Delta R'$ to the observed duration of the photon burst $c\Delta t$:

$$R' = \gamma^2 (c\Delta t) \tag{39}$$

$$\Delta R' = \gamma c \Delta t \tag{40}$$

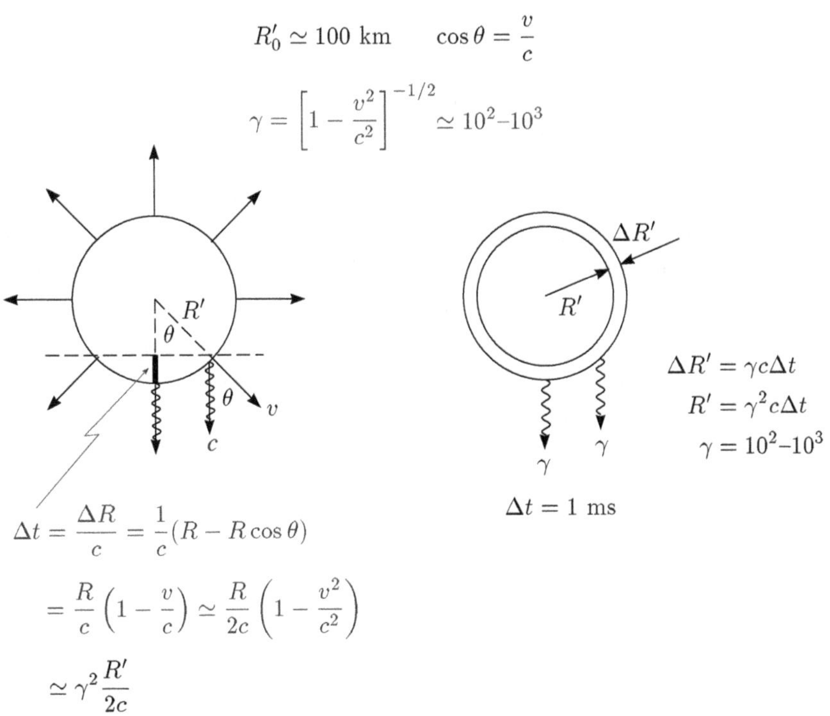

Fig. 9. Kinematics of a relativistically expanding fireball on the left and the resulting shell(s) expanding under radiation pressure on the right.

From the observed GRB luminosity L_γ, we compute the photon energy density in the shell:

$$U'_\gamma = \frac{(L_\gamma \Delta t)/\gamma}{4\pi R'^2 \Delta R'} = \frac{L_\gamma}{4\pi \Delta t^2 c^3 \gamma^6}. \tag{41}$$

The pion production by shocked protons in this photon field is, as before, calculated from the interaction length:

$$\frac{1}{\lambda_{p\gamma}} = N_\gamma \sigma_\Delta \langle x_{p\to\pi}\rangle = \frac{U'_\gamma}{E'_\gamma}\sigma_\Delta \langle x_{p\to\pi}\rangle \quad \left(E'_\gamma = \frac{1}{\gamma}E_\gamma\right). \tag{42}$$

Also as before, σ_Δ is the cross section for $p\gamma \to \Delta \to n\pi^+$ and $\langle x_{p\to\pi}\rangle \simeq 0.2$. The fraction of energy going into π production is

$$f_\pi \cong \frac{\Delta R'}{\lambda_{p\gamma}} \tag{43}$$

$$f_\pi \simeq \frac{1}{4\pi c^2}\frac{L_\gamma}{E_\gamma}\frac{1}{\gamma^4 \Delta t}\sigma_\Delta \langle x_{p\to\pi}\rangle \tag{44}$$

$$f_\pi \simeq 0.14 \left[\frac{L_\gamma}{10^{51}\,\text{ergs}^{-1}}\right]\left[\frac{1\ \text{MeV}}{E_\gamma}\right]\left[\frac{300}{\gamma}\right]^4 \left[\frac{1\ \text{ms}}{\Delta t}\right] \times \left[\frac{\sigma_\Delta}{10^{-28}\,\text{cm}^2}\right]\left[\frac{\langle x_{p\to\pi}\rangle}{0.2}\right]. \tag{45}$$

The relevant photon energy in the problem is 1 MeV, the energy where the typical GRB spectrum exhibits a break. The contribution of higher-energy photons is suppressed by the falling spectrum, and lower-energy photons are less efficient at producing pions. Given the large uncertainties associated with the astrophysics, it is an adequate approximation to neglect the explicit integration over the GRB photon spectrum. The proton energy for production of pions via the Δ resonance is

$$E'_p = \frac{m_\Delta^2 - m_p^2}{4E'_\gamma}. \tag{46}$$

Therefore,

$$E_p = 1.4 \times 10^{16}\,\text{eV}\left(\frac{\gamma}{300}\right)^2 \left(\frac{1\ \text{MeV}}{\text{E}_\gamma}\right) \tag{47}$$

$$E_\nu = \frac{1}{4}\langle x_{p\to\pi}\rangle E_p \simeq 7 \times 10^{14}\,\text{eV}. \tag{48}$$

We are now ready to calculate the neutrino flux:

$$\frac{dN_\nu}{dE_\nu} = \frac{c}{4\pi}\frac{U'_\nu}{E'_\nu} = \frac{c}{4\pi}\frac{U_\nu}{E_\nu} = \frac{c}{4\pi}\frac{1}{E_\nu}\left[\frac{1}{2}f_\pi t_H \dot{E}\right], \tag{49}$$

where the factor 1/2 accounts for the fact that only 1/2 of the energy in charged pions is transferred to $\nu_\mu + \bar{\nu}_\mu$. As before, $\dot{E}$ is the injection rate in cosmic rays beyond the ankle ($\sim 4 \times 10^{44}\,\mathrm{erg\,Mpc^{-3}\,yr^{-1}}$) and t_H is the Hubble time of $\sim 10^{10}$ Gyr. Numerically,

$$\frac{dN_\nu}{dE_\nu} = 2 \times 10^{-14}\,\mathrm{cm^{-2}\,s^{-1}\,sr^{-1}}\left[\frac{7 \times 10^{14}\,\mathrm{eV}}{E_\nu}\right]\left[\frac{f_\pi}{0.14}\right] \times \left[\frac{t_H}{10\ \mathrm{Gyr}}\right]\left[\frac{\dot{E}}{10^{44}\,\mathrm{erg\,Mpc^{-3}\,yr^{-1}}}\right]. \tag{50}$$

The only subtlety here is the γ^2 dependence of the shell radius R'; for a simple derivation, see Ref. 6. The result is insensitive to beaming. Beaming yields more energy per burst, but fewer bursts are actually observed. The predicted rate is also insensitive to the neutrino energy E_ν because higher average energy yields fewer νs, but more are detected. Both effects are approximately linear. Neutrino telescopes are essentially background free for such high-energy events and should be able to identify neutrinos at all zenith angles.

For typical choices of the parameters, $\gamma \sim 300$ and $t_v \sim 10^{-2}$ s, about 100 events per year are predicted in IceCube, a flux that was already challenged[69] by the limit on a diffuse flux of cosmic neutrinos obtained with one-half of IceCube in 1 year.[70] As before, the energy density of extragalactic cosmic rays of $\sim 10^{44}$ TeV Mpc^{-3} yr^{-1} depends on the unknown transition energy between the galactic and extragalactic components of the spectrum. In the end, the predictions can be stretched, but having failed by now to observe high-energy neutrinos in spatial and temporal coincidence with over 1000 GRB observations, IceCube has set a limit that is less than 1% of the PeV cosmic neutrino flux that is actually observed.

We have nevertheless discussed fireballs in some detail to illustrate their generic potential as cosmic accelerators and to point out that they may produce cosmic neutrinos without being the sources of cosmic rays. For instance, they may produce neutrinos of lower energy in events where the boost factor is limited, $\gamma \leq 10$. Candidate events have been observed and are referred to as "low-luminosity GRBs."[71] There is also the possibility that high-energy gamma rays and neutrinos are produced when the shock expands further into the interstellar medium. This mechanism has been invoked as the origin of the delayed high-energy gamma rays. Adapting the previous calculation to the external shock is straightforward.[72] The timescale is seconds rather than milliseconds, and the break in the spectrum shifts from 1 to 0.1 MeV. Although f_π is reduced by two orders of magnitude, in the external shocks, higher energies can be reached, and this increases the neutrino detection efficiency. In the end, the observed rates are an order of magnitude smaller than in internal shocks, but the inherent ambiguities of the estimates are such that it is difficult to establish with confidence their relative neutrino yields.

3.4. *Cosmogenic neutrinos*

The production of neutrinos in the sources that accelerate the high-energy cosmic rays depends on the source environment. In order to efficiently accelerate cosmic rays, any loss mechanism, including pion production in $p\gamma$ and pp interactions, must be suppressed as it reduces the acceleration time. Efficient accelerators are likely to be inefficient beam dumps for producing neutrinos. High-efficiency neutrino production can be achieved by separating the sites of acceleration and neutrino production. For instance, after acceleration, extragalactic cosmic rays propagate over cosmological distances of more than 10 Mpc and can efficiently produce neutrinos on the dilute extragalactic medium.

In this section, we will discuss the production of neutrinos in the interactions of extragalactic cosmic rays with cosmic radiation backgrounds. Soon after the discovery of the cosmic microwave background (CMB), Greisen, Zatsepin and Kuzmin (GZK)[73,74] realized that extragalactic cosmic rays are attenuated by interactions with background photons. Actually, protons interact resonantly via $p\gamma \to \Delta^+ \to \pi^+ n$ with background photons with mean energy $\epsilon \simeq 0.33$ meV at energies $E_p \simeq (m_\Delta^2 - m_p^2)/4/\epsilon \simeq 500$ EeV. The width of the Planck spectrum leads to a significant attenuation of proton fluxes after propagation over distances on the order of 200 Mpc at an energy above $E_{\rm GZK} \simeq 50$ EeV, which is known as the GZK suppression. Also, heavier nuclei are attenuated at a similar energy by photodisintegration of the nucleus by CMB photons via the giant dipole resonance.

The pions produced in GZK interactions decay, resulting in a detectable flux of *cosmogenic* neutrinos first estimated by Berezinsky and Zatsepin[75] in 1969. This *guaranteed* flux of neutrinos became one of the benchmarks for high-energy neutrino astronomy leading early on to the concept of kilometer-scale detectors. The flux of cosmogenic neutrinos peaks at EeV neutrino energy depending on the chemical composition and the evolution with redshift of the unknown sources. The largest neutrino flux results from proton-dominated models.[18,76,77] A particularly strong emission can be expected in such models if the proton spectrum extends below the ankle. Referred to as "dip models," the ankle results from the absorption of protons by Bethe–Heitler pair production on CMB photons. A fit to the observed cosmic-ray spectrum requires relatively strong source evolution with redshift[78–81] that enhances pion production. However, the corresponding electromagnetic emission via neutral pions as well as $e^{\pm}$ pairs is constrained by the isotropic gamma-ray background (IGRB) observed by the Fermi LAT satellite[82,83] and limits the neutrino intensity of these proton-dominated scenarios.[84–89] Recent upper limits on cosmogenic neutrinos resulting from the failure by IceCube to observe EeV neutrinos constrains proton-dominated models.[90]

In contrast, the IceCube constraint can be accommodated by introducing a heavy nuclear composition. Resonant neutrino production still proceeds via the interaction of individual nucleons with background photons, but the threshold of the production is increased to $E_{\rm CR} \gtrsim AE_{\rm GZK}$ for nuclei with mass number A. Therefore, efficient cosmogenic neutrino production would require an injected cosmic-ray flux that extends well above $E_{\rm GZK}$. Especially for heavier nuclear composition of the primary flux, the production of neutrinos on photons of the extragalactic background light (EBL) becomes relatively important.[87,91–99] The interaction with optical photons produces neutrino fluxes in the PeV energy range. However, the overall level is much lower because of the low intensity of the EBL photons. It is unlikely that the PeV neutrino flux observed with IceCube could be related to the neutrino production in the EBL[100] (also see Ref. 101). The observed PeV neutrino flux level is too high to be consistent with associated electromagnetic contributions to the IGRB or upper limits on the EeV neutrino flux.

4. Instrumentation: the first kilometer-scale neutrino detector

As previously discussed, the rationale for kilometer-scale neutrino detectors is that their sensitivity is sufficient to reveal generic cosmic-ray sources with an energy density in neutrinos comparable to their energy density in cosmic and pionic TeV gamma rays. By the turn of the century, a series of first-generation experiments[102,103] demonstrated that high-energy neutrinos with ∼10 GeV energy and above could be detected by observing Cherenkov radiation from secondary particles produced in neutrino interactions inside large volumes of highly transparent ice or water instrumented with a

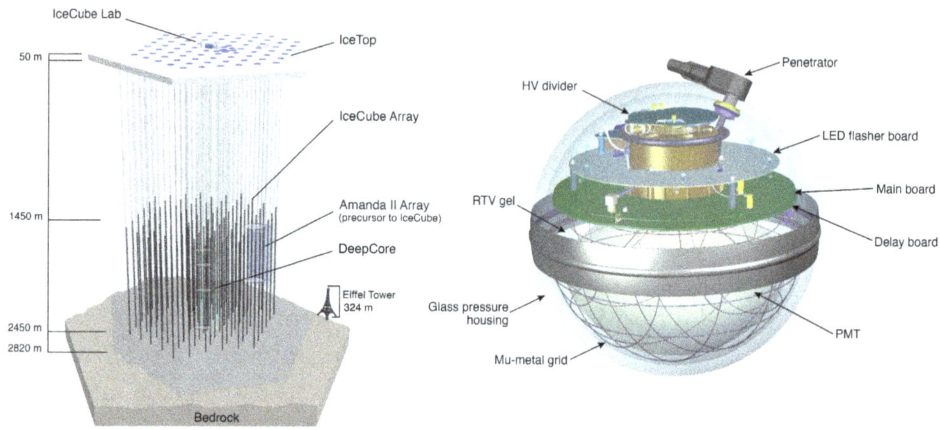

Fig. 10. Sketch of the IceCube observatory (left) and the digital optical module (right).

lattice of photomultiplier tubes. Construction of the first second-generation detector, IceCube, at the geographic South Pole was completed in December 2010[104]; see Fig. 10.

4.1. *IceCube*

IceCube consists of 80 strings, each instrumented with 60 10-inch photomultipliers spaced by 17 m over a total length of 1 k. The deepest module is located at a depth of 2.450 km so that the instrument is shielded from the large background of cosmic rays at the surface by approximately 1.5 km of ice. Strings are arranged at apexes of equilateral triangles that are 125 m on a side. The instrumented detector volume is a cubic kilometer of dark, highly transparent and sterile Antarctic ice. Radioactive background is dominated by the instrumentation deployed into this natural ice.

Each optical sensor consists of a glass sphere containing the photomultiplier and the electronics board that digitizes the signals locally using an on-board computer. The digitized signals are given a global time stamp with residuals accurate to less than 3 ns and are subsequently transmitted to the surface. Processors at the surface continuously collect these time-stamped signals from the optical modules, each of which functions independently. The digital messages are sent to a string processor and a global event trigger. They are subsequently sorted into the Cherenkov patterns emitted by secondary muon tracks, or electron and tau showers, that reveal the direction of the parent neutrino.[105]

As will be discussed further on, similar detectors are planned for deployment in deep transparent Mediterranean water[106] and in Lake Baikal.

4.2. *Detection methods*

Neutrino telescopes exploit the relatively large neutrino cross section and the long muon range above TeV energies to achieve a detection efficiency sufficient to reach the predicted point source and diffuse fluxes previously discussed. At the same time, detecting ν_e and ν_τ neutrinos cannot be ignored; the case has been made in detail in Ref. 13. The detection of neutrinos of all flavors is important in separating diffuse extraterrestrial neutrinos from atmospheric neutrinos. Generic cosmic accelerators produce neutrinos from the decay of pions with admixture ν_e:ν_μ:ν_τ =1:2:0. Over cosmic baselines, neutrino oscillations transform the ratio to 1:1:1, because approximately one-half of the ν_μ convert to ν_τ.

IceCube detects neutrinos by observing the Cherenkov radiation from the charged particles produced by neutrino interactions inside or in the vicinity of the detector. Charged-current (CC) interactions produce a lepton that carries, on average, 50% of the neutrino energy for $E \leq 10\,\mathrm{GeV}$ to 80% at high energies; the remainder of the energy is released in the hadronic shower produced by the target nucleus. Both the secondary lepton and the hadronic shower produce Cherenkov radiation. In neutral-current (NC) interactions, the neutrino transfers a fraction of its energy to a nuclear target, producing a hadronic shower. IceCube can differentiate neutrino flavors on the basis of their topology in the detector, as illustrated in Fig. 11. There are two basic topologies: tracks from ν_μ and cascades from ν_e, ν_τ, and the NC interactions from all flavors. On the scale of IceCube, PeV cascades have a length of less than 10 m and are therefore point sources of Cherenkov light in a detector of kilometer size.

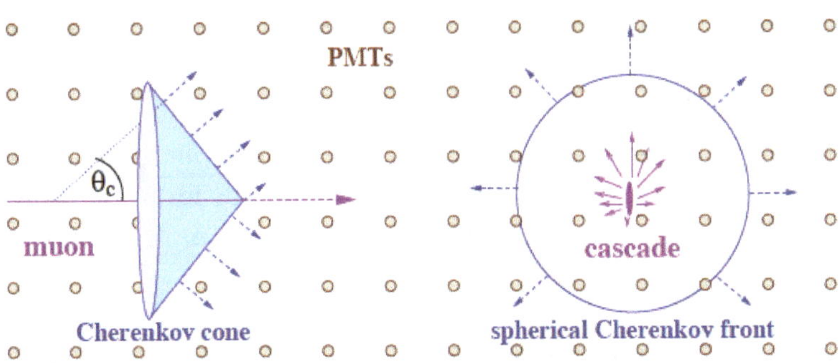

Fig. 11. Contrasting Cherenkov light patterns produced by muons (left) and by showers initiated by electron and tau neutrinos (right) and by NC interactions. The patterns are routinely referred to as tracks and cascades (or showers). Cascades are produced by the radiation of particle showers, whose dimensions are in the tens of meters, i.e., an approximate point source of light with respect to the dimensions of the detector.

Neutrino telescopes also measure neutrino energy. Muons range out, over kilometers at TeV energy to tens of kilometers at EeV energy, generating showers along their track by bremsstrahlung, pair production, and photonuclear interactions. The charged particles produced are the sources of additional Cherenkov radiation. Because the energy of the muon degrades along its track, the energy of the secondary showers decreases, which reduces the distance from the track over which the associated Cherenkov light can trigger a PMT. The geometry of the light pool surrounding the muon track is therefore a kilometer-long cone with a gradually decreasing radius. In its first kilometer, a high-energy muon typically loses energy in a couple of showers having one-tenth of the muon's initial energy. So the initial radius of the cone is the radius of a shower with 10% of the muon energy. At lower energies of hundreds of GeV and less, the muon becomes minimum-ionizing.

Because of the stochastic nature of the muon's energy loss, the relationship between the observed energy loss inside the detector and muon energy varies from muon to muon. Additionally, only the muon energy lost in the detector can be determined; we know neither its energy loss before entering the instrumented volume nor how much energy it carries out upon exiting. An unfolding process is required to determine the neutrino energy based on the observed muon energy; fortunately, it is based on well-understood Standard Model physics. In contrast, for ν_e and ν_τ, the detector is a total energy calorimeter, and the determination of their energy is superior.

The different topologies each have advantages and disadvantages. For ν_μ interactions, the long lever arm of muon tracks, up to tens of kilometers at very high energies, allows the muon direction (and the neutrino direction) to be determined accurately with an angular resolution measured online that is better than $0.4°$. Superior angular resolution can be reached for selected events. Sensitivity to point sources is therefore better as well. The disadvantages are a large background, of atmospheric neutrinos below 100 TeV and cosmic-ray muons at all energies, and the indirect determination of the neutrino energy that must be inferred from sampling the energy loss of the muon when it transits the detector.

Observation of ν_e and ν_τ flavors represents significant advantages. They are detected for both Northern and Southern Hemispheres. (This is also true for ν_μ with energy in excess of 1 PeV, where the background from the steeply falling atmospheric spectrum becomes negligible.) At TeV energies and above, the background of atmospheric ν_e is lower by over an order of magnitude, and there are essentially no atmospheric ν_τ produced. At higher energies, long-lived pions, the source of atmospheric ν_e, no longer decay, and relatively rare K decays become the dominant source of background ν_e. Furthermore, because the neutrino events are totally, or at least mostly, contained inside the instrumented detector volume, the neutrino energy is determined by total-absorption calorimetry. One can establish the cosmic origin of a single event by demonstrating that the energy cannot be reached by muons and neutrinos of atmospheric origin. Finally, ν_τ are not absorbed by the Earth[107]: ν_τ interacting in the Earth produce secondary ν_τ of lower energy, either directly in a NC interaction or via the decay of a secondary tau lepton produced in a CC interaction. High-energy ν_τ will thus cascade down to energies of hundreds of TeV where the Earth becomes transparent. In other words, they are detected with a reduced energy but not absorbed.

Although cascades are nearly pointlike and, in practice, spatially isotropic, the pattern of arrival times of the photons at individual optical modules reveals the direction of the secondary leptons within 3°. While a fraction of cascade events can be reconstructed accurately to within a degree,[108] the precision is inferior to that reached for ν_μ events and typically not better than 10° using the present techniques.

At energies above about 100 PeV, electromagnetic showers begin to elongate because of the Landau-Pomeranchuk-Migdal effect.[109] An extended length scale is associated with the abundant radiation of soft photons that results in interactions of the shower particles on two target atoms. Negative interference in this process results in reduced energy loss.

4.3. *Detector performance*

Cosmic neutrinos must be separated from the large backgrounds of atmospheric neutrinos and atmospheric cosmic-ray muons. This is possible for two classes of events: neutrinos that interact inside the instrumented volume ("starting events") and events where a muon enters the detector from below, created by a neutrino traversing the Earth (throughgoing events), thus pointing back to its origin. In this latter case, the Earth is used as a filter for cosmic-ray muons.

For starting events, the pathlength $l(\theta)$ traversed within the detector volume by a neutrino with zenith angle θ is determined by the detector's geometry. Neutrinos are detected if they interact within the detector volume, i.e., within the instrumented volume of $1\,\text{km}^3$. That probability is

$$P(E_\nu) = 1 - \exp[-l/\lambda_\nu(E_\nu)] \simeq l/\lambda_\nu(E_\nu), \tag{51}$$

where $\lambda_\nu(E_\nu) = [\rho_{\rm ice}\ N_A\ \sigma_{\nu N}(E_\nu)]^{-1}$ is the mean free path in ice for a neutrino of energy E_ν. Here, $\rho_{\rm ice} = 0.9$ g cm^{-3} is the density of the ice, $N_A = 6.022 \times 10^{23}$ is Avogadro's number, and $\sigma_{\nu N}(E_\nu)$ is the neutrino–nucleon cross section. A neutrino flux dN/dE_ν (neutrinos per GeV per cm^2 per second) crossing a detector with energy threshold and cross-sectional area $A(E_\nu)$ facing the incident beam will produce

$$N_{ev} = T \int_{E_\nu^{\rm th}} A(E_\nu)\, P(E_\nu)\, \frac{dN}{dE_\nu}\, dE_\nu \tag{52}$$

events after a time T. The "effective" detector area $A(E_\nu)$ is also a function of the zenith angle θ. It isn't strictly equal to the geometric cross section of the instrumented volume facing the incoming neutrino, because even neutrinos interacting outside the instrumented volume may produce enough light inside the detector to be detected. In practice, $A(E_\nu)$ is determined as a function of the incident neutrino direction and zenith angle by a full-detector simulation, including the trigger.

This formalism applies to contained events. For muon neutrinos, any neutrino producing a secondary muon that reaches the detector (and has sufficient energy to trigger it) will be detected. Because the muon travels kilometers at TeV energy and tens of kilometers at PeV energy, neutrinos can be detected outside the instrumented volume; the probability is obtained by substitution in Eq. (51),

$$l \to \lambda_\mu, \tag{53}$$

thereby giving

$$P = \lambda_\mu / \lambda_\nu. \tag{54}$$

Here, λ_μ is the range of the muon determined by its energy losses. Values for the neutrino nucleon cross section and the range of the muon can be found in Ref. 20

The complete expression for the flux of ν_μ-induced muons at the detector is given by a convolution of the neutrino spectrum ϕ (=dN/dE_ν) with the probability P to produce a muon reaching the detector[4–6]:

$$\phi_\mu(E_\mu^{\rm min}, \theta) = \int_{E_\mu^{\rm min}} P(E_\nu, E_\mu^{\rm min})\, \exp[-\sigma_{\rm tot}(E_\nu)\, N_A\, X(\theta)]\, \phi(E_\nu, \theta) dE_\nu. \tag{55}$$

The additional exponential factor accounts for the absorption of neutrinos along a chord of the Earth of length $X(\theta)$ at zenith angle θ. Absorption becomes important for $\sigma_\nu(E_\nu) \gtrsim 10^{-33}\,\text{cm}^2$ or $E_\nu \gtrsim 100$ TeV. For back-of-the-envelope calculations, the P function can be approximated by

$$P \simeq 1.3 \times 10^{-6} E^{2.2} \quad \text{for } E = 10^{-3} \text{ to } 1\,\text{TeV}, \tag{56}$$

$$\simeq 1.3 \times 10^{-6} E^{0.8} \quad \text{for } E = 1 - 10^{3}\,\text{TeV}. \tag{57}$$

At EeV energy, the increase is reduced to only $E^{0.4}$. Clearly, high-energy neutrinos are more likely to be detected because of the increase with energy of both the cross section and muon range.

Tau neutrinos interacting outside the detector can be observed provided the tau lepton they produce reaches the instrumented volume within its lifetime. In Eq. (51), l is replaced by

$$l \to \gamma c\tau = E/mc\tau, \tag{58}$$

where m, τ, and E are the mass, lifetime, and energy of the tau, respectively. The tau's decay length $\lambda_\tau = \gamma c\tau \approx 50\,\text{m} \times (E_\tau/10^6)$ GeV grows linearly with energy and actually exceeds the range of the muon near 1 EeV. At yet higher energies, the tau eventually ranges out by catastrophic interactions, just like the muon, despite the reduction of the energy-loss cross sections by a factor of $(m_\mu/m_\tau)^2$.

Tracks and showers produced by tau neutrinos are difficult to distinguish from those initiated by muon and electron neutrinos, respectively. To be clearly identified, both the initial neutrino interaction and the subsequent tau decay must be contained within the detector; for a cubic-kilometer detector, this happens for neutrinos with energies from a few PeV to a few tens of PeV.[110]

For an in-depth discussion of neutrino detection, energy measurement, and flavor separation, and for detailed references, see the IceCube Preliminary Design Document[13] and Ref. 2.

4.4. *Atmospheric neutrinos*

Muons and neutrinos from decay of mesons produced by cosmic-ray interactions in the atmosphere are the background in the search for neutrinos of extraterrestrial origin. The 3 kHz trigger rate of IceCube is dominated by atmospheric muons from decay of pions and kaons produced in the atmosphere above the detector. The distribution peaks near the zenith and decreases with increasing angle as the muon energy required to reach the deep detector increases. Most atmospheric muons are easily identified as entering tracks from above and rejected. Because of the large ratio of muons to neutrinos, however, misreconstructed atmospheric muons remain an important source of background for all searches.

Measurement of the spectrum of atmospheric neutrinos is an important benchmark for a neutrino telescope. IceCube detects an atmospheric neutrino every 6 minutes. The spectrum of atmospheric ν_μ has been measured by unfolding the measured rate and energy deposition of neutrino-induced muons entering the detector from below the horizon,[31] as shown in Fig. 6. More challenging is the measurement of the flux of atmospheric electron neutrinos. This has been done by making use of DeepCore, the more densely instrumented subarray in the deep center of IceCube, to identify contained shower events. The known spectrum of ν_μ is used to calculate the contribution of NC interactions to the observed rate of showers. Subtracting the NC contribution leads to the measurement of the spectrum of atmospheric electron neutrinos from 100 GeV to 10 TeV,[33] as shown in Fig. 6.

In general, atmospheric neutrinos are indistinguishable from astrophysical neutrinos. An important exception occurs in the case of muon neutrinos from above when the neutrino energy is sufficiently high and the zenith angle sufficiently small that the muon produced in the same decay as the neutrino is guaranteed to reach the detector.[111] This is used to reject atmospheric neutrinos in analyses where they are a background. Monte Carlo simulation can be used to evaluate the atmospheric neutrino passing rate more generally by also including other high-energy muons produced in the same cosmic-ray shower as the neutrino. In this way, the method can be extended to electron neutrinos. In practice, the passing rate is significantly reduced for zenith angles $\theta < 70°$ and $E_\nu > 100$ TeV.

The spectrum of atmospheric neutrinos becomes one power steeper than the spectrum of primary nucleons at high energy as the competition between interaction and decay of pions and kaons increasingly suppresses their decay. For the dominant kaon channel, the characteristic energy for the steepening is $E_\nu \sim 1$ TeV/cosθ. A further steepening occurs above 100 TeV as a consequence of the knee in the primary spectrum. Astrophysical neutrinos should reflect the cosmic-ray spectrum in the source and are therefore expected to have a significantly harder spectrum than atmospheric neutrinos. Establishing an astrophysical signal above the steep atmospheric background requires an understanding of the atmospheric neutrino spectrum around 100 TeV and above.

Although there is some uncertainty associated with the composition through the knee region,[112] the major uncertainty in the spectrum of atmospheric neutrinos at high energy is the level of charm production.

The short-lived charmed hadrons preferentially decay up to a characteristic energy of 10^7 GeV, producing prompt muons and neutrinos with the same spectrum as their parent cosmic rays. This prompt flux of leptons has not yet been measured. Existing limits[113,114] allow a factor of two or three around the level predicted by a standard calculation[36] (after correction for steepening at the knee). For reasonable assumptions, the charm contribution is expected to dominate the conventional spectrum above ~10 TeV for ν_e, above ~100 TeV for ν_μ, and above ~1 PeV for muons.[115]

The expected hardening in the spectrum of atmospheric neutrinos due to prompt neutrinos is partially degenerate with a hard astrophysical component. However, the spectrum of astrophysical neutrinos should reflect the spectrum of cosmic rays at their sources, which is expected to be harder than the spectrum of cosmic rays at Earth. It should eventually be possible with IceCube to measure the charm contribution by requiring a consistent interpretation of neutrino flavors and cosmic-ray muons for which there is no astrophysical component. An additional signature of atmospheric charm is the absence of seasonal variations for this component.[116]

5. Cosmic neutrinos

There are two primary methods used to identify neutrinos. As previously discussed, neutrino searches have historically focused on the observation of muon neutrinos that interact primarily outside the detector, producing kilometer-long muon tracks that pass through the detection volume. Although this allows observation of neutrinos that interact outside the detector, it is then necessary to use the Earth as a filter in order to remove the huge background of cosmic-ray muons. Even at a depth of 1450 m, IceCube detects atmospheric cosmic-ray muons originating in the Southern Hemisphere at a rate of 3000 per second. This method limits the neutrino view to a single flavor and half the sky. An alternative method exclusively identifies neutrinos interacting inside the detector. It divides the instrumented volume of ice into an outer veto shield and a roughly 500 Mt inner fiducial volume. The advantage of focusing on neutrinos interacting inside the instrumented volume of ice is that the detector then functions as a total absorption calorimeter, measuring energy with a 10–15% resolution. Also, neutrinos from all directions in the sky can be identified, including both muon tracks, produced in muon-neutrino CC interactions, and secondary showers, produced by electron and tau neutrinos as well as in NC interactions of neutrinos of all flavors. The Cherenkov patterns initiated by an electron (or tau) neutrino of 1 PeV, or petaelectronvolt (10^{15} eV), energy and by a

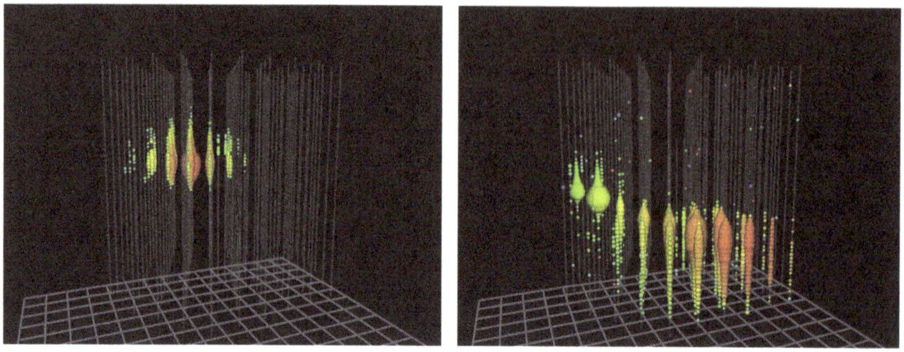

Fig. 12. Light pools produced in IceCube by neutrino interactions. White dots represent sensors with no signal. For the colored dots, the color indicates arrival time, from red (early) to purple (late) following the rainbow, and size reflects the number of photons detected. Left: A shower initiated by an electron or a tau neutrino or by the NC interaction of a neutrino of any of the three flavors. The measured energy of the shower is 1.14 PeV, which represents a lower limit on the energy of the neutrino that initiated the shower. Right: An upgoing muon track traverses the detector at an angle of 11° below the horizon. The deposited energy inside the detector is 2.6 PeV.

muon neutrino depositing 2.6 PeV energy while traversing the detector are contrasted in Fig. 12.

In general, the particle's trajectory is determined from the arrival times of photons at the optical sensors,[117] while the number of photons is a proxy for the amount of energy deposited. The two methods for separating neutrinos from the cosmic-ray muon background have complementary advantages. The long tracks produced by muon neutrinos can point back to their sources with a 0.4° angular resolution. In contrast, the reconstruction of the direction of secondary showers, still in the development stage in IceCube,[118] can be determined to within 10°−15° of the direction of the incident neutrino. Determining the deposited energy from the observed light pool is, however, relatively straightforward, and a resolution of better than 15% is possible.

For neutrino astronomy, the first challenge is to select sufficiently pure samples of neutrinos: roughly 100 000 per year in a background of 10 billion cosmic-ray muons. The second is to identify the small fraction of the sample that are neutrinos of astrophysical origin, expected at the level of tens of events per year. Atmospheric neutrinos are an overwhelming background for cosmic neutrinos, at least at energies below ∼100 TeV. Above this energy, however, the atmospheric neutrino flux is too small to produce events even in a kilometer-scale detector, and every event observed in that range represents the discovery of an astrophysical neutrino.

By now, IceCube has observed cosmic neutrinos using both methods for rejecting background, and each analysis has reached a statistical significance of more than 5σ.[119,120] Based on different methods for reconstruction and energy measurement, the results agree, pointing at extragalactic sources whose flux has equilibrated in the three flavors after propagation over cosmic distances. Its total energy matches that of extragalactic photons and cosmic rays.

5.1. *Evidence for cosmic neutrinos*

IceCube is sensitive to neutrinos with energies above a threshold of approximately 0.1 TeV. Using the Earth as a filter, a flux of neutrinos has been identified that is consistent with atmospheric origin; see Fig. 13. However, in 7 years of data, an excess of events has been observed with 5.6σ significance at energies beyond 100 TeV[120–122] that cannot be accommodated by the atmospheric flux. Whereas IceCube only measures the energy deposited by the secondary muon inside the detector, Standard Model physics allows one to infer the energy spectrum of the parent neutrinos, as illustrated in Fig. 13. For the highest-energy event, shown in Fig. 12, the most likely energy of

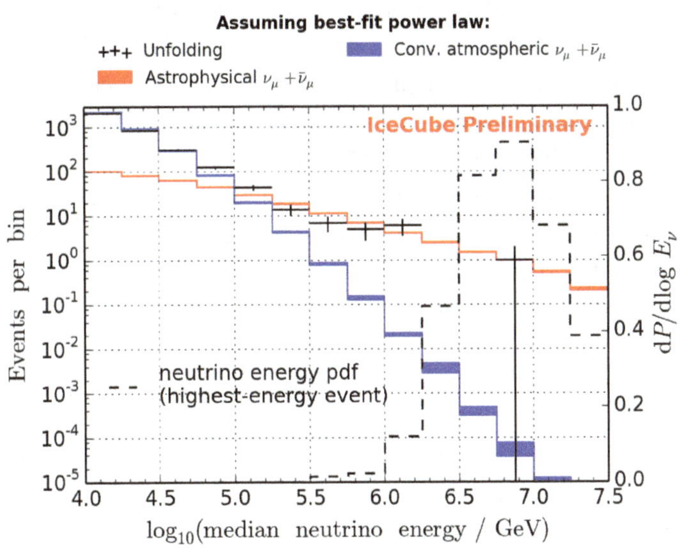

Fig. 13. Spectrum of secondary muons initiated by muon neutrinos that have traversed the Earth, i.e., with zenith angle less than 5° above the horizon, as a function of the energy they deposit inside the detector. For each reconstructed muon energy, the median neutrino energy is calculated assuming the best-fit spectrum. The colored bands (blue/red) show the expectation for the conventional and astrophysical contributions. The black crosses show the measured data. Additionally, the neutrino energy probability density function for the highest energy event assuming the best-fit spectrum is shown (dashed line).

the parent neutrino is almost 10 PeV. The muon energy loss measured in the detector is 2.6 ± 0.3 PeV. The cosmic neutrino flux is well described by a power law with a spectral index $\gamma = 2.13 \pm 0.13$ and a normalization at 100 TeV neutrino energy of $(0.90^{+0.30}_{-0.27}) \times 10^{-18}\,\mathrm{GeV^{-1}cm^{-2}sr^{-1}}$.[123] The error range is estimated from a profile likelihood using Wilks's theorem and includes both statistical and systematic uncertainties. The neutrino energy contributing to this flux covers the range of 200 TeV–9 PeV. As will be discussed further on, we will find that analyses that extend to lower energies reveal an excess of neutrino events over the atmospheric background in the 30−100 TeV energy range, well above the extrapolation of the cosmic muon-neutrino flux to lower energies. The conclusion to be drawn is that the astrophysical neutrino spectrum measured by IceCube is not featureless.

The first hint of cosmic neutrinos emerged when two events were found serendipitously in a search for GZK neutrinos using IceCube data collected between May 2010 and May 2012.[124] The energies of these neutrinos, rather than EeV as expected for GZK neutrinos, were in the PeV range: 1.040–1.140 PeV. The events were particle showers initiated by electron or tau neutrinos interacting inside the instrumented detector volume. Their light pool of roughly one hundred thousand photoelectrons extended over more than 500 m; one of them is shown in Fig. 12. With PeV energy and no trace of accompanying muons from an atmospheric shower, they are unlikely to be atmospheric in origin. It is indeed important to realize that the muon produced in the same pion or kaon decay as an atmospheric neutrino is guaranteed to reach the detector provided that the neutrino energy is sufficiently high and the zenith angle sufficiently small.[111,125] In this case, the atmospheric neutrino provides its own self-veto. This self-veto is routinely applied to IceCube cosmic neutrino candidates that consist exclusively of isolated neutrino events.

The observation of these two neutrinos immediately suggested an analysis that searched for neutrinos originating inside the detector, their well-measured energy allowing a clear separation between neutrinos of atmospheric origin and those of cosmic origin. The geometry of the veto and active signal regions was optimized to reduce the background of atmospheric muons and neutrinos to a handful of events per year while keeping 98% of the cosmic signal. In analyzing the data covering the same time period as the GZK neutrino search, 28 candidate neutrino events were identified with in-detector deposited energies between 30 and 1140 TeV. Of these, 21 were showers. The remaining seven events were muon tracks. The 28 events include the two PeV events previously revealed in the GZK neutrino search.

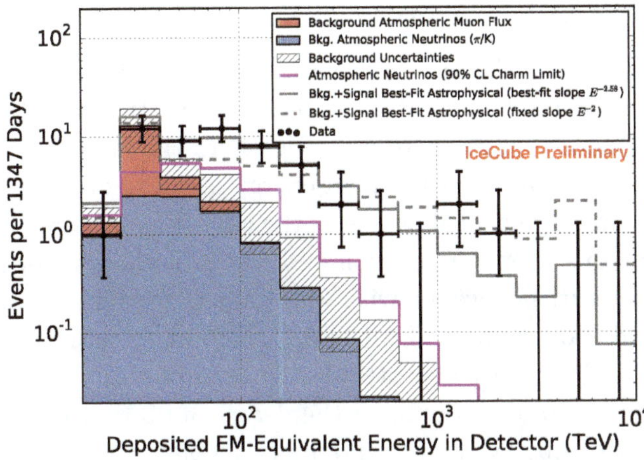

Fig. 14. Deposited energies, by neutrinos interacting inside IceCube, observed in 4 years of data. The hashed region shows uncertainties on the sum of all backgrounds. The atmospheric muon flux (red) and its uncertainty is computed from simulation to overcome statistical limitations in our background measurement and scaled to match the total measured background rate. The atmospheric neutrino flux is derived from previous measurements of both the π, K, and charm components of the atmospheric spectrum.[126] Also shown are two illustrative power-law fits to the spectrum.

The energy and zenith angle dependence observed is consistent with expectations for a flux of neutrinos produced by cosmic accelerators; see Fig. 14. The flavor composition of the flux is, after corrections for the acceptances of the detector to the different flavors, consistent with $\nu_e : \nu_\mu : \nu_\tau \sim 1:1:1$, as anticipated for a flux originating in cosmic sources. It is also consistent with the flux of muon neutrinos penetrating the Earth, as shown in Fig. 13. Subsequently, two additional years of data have been analyzed, doubling the statistics of the published discovery results.[127] A purely atmospheric explanation can be excluded at 7σ. The 4-year data set contains a total of 54 neutrino events with deposited energies ranging from 30 to 2000 TeV. In the fourth year, muon tracks were found that deposited ∼500 TeV energy inside the detector, produced by PeV-energy parent neutrinos. One of them reconstructs through IceTop, IceCube's surface array, with no evidence for an air shower. Combining the absence of an air shower in IceTop with large deposited muon energy results in a high significance for astrophysical origin from a single event.

In summary, both methods for selecting cosmic neutrinos harvest about 10–15 events per year, more if one tolerates a modest background. With 6 years of data from the completed detector, our data sample exceeds one hundred events, approaching the statistics required to identify their origins by matching arrival directions with astronomical maps.[128]

5.2. *Origin of cosmic neutrinos*

Figure 15 shows in galactic coordinates the arrival directions of cosmic neutrinos for 4 years of events with interaction vertices inside the detector and 6 years of muon neutrinos from the Northern Hemisphere. We performed a search for anisotropy using the high-energy starting event (HESE) results and found no significant local excess in the sky; the probability of obtaining the measured anisotropy in randomized pseudoexperiments is 58%. The correlation of neutrino events with the galactic plane is not significant. Letting the width of the plane float freely, the best fit returned a correlation for a value of $\pm 7.5°$ with a posttrial chance probability of 3.3%. Neither probability increased after doubling the data from 2 to 4 years.

In short, the observed neutrino flux is consistent with an isotropic distribution of arrival directions and equal contributions of all neutrino flavors.[130] A subdominant Galactic component of the flux cannot be excluded at this time. Interestingly, a variety of analyses[130,131] suggest that the cosmic neutrino flux dominates the atmospheric background above an energy that may be as low as 30 TeV, with an energy spectrum that cannot be described as a single power as was the case for the muon-neutrino flux through the

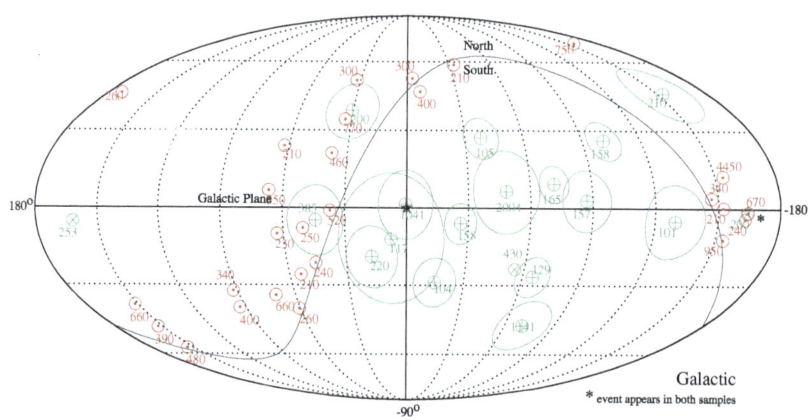

Fig. 15. Mollweide projection in galactic coordinates of the arrival direction of neutrino events. We show the results of the 6-year upgoing track analysis[123] with energy proxy MuEx > 50 (⊙). The red numbers show the most probable neutrino energy (in TeV) assuming the best-fit astrophysical flux of the analysis.[123] The events of the 4-year HESE analysis with deposited energy (green numbers) larger than 60 TeV (tracks ⊗ and cascades ⊕) are also shown.[127,129] Cascade events (⊕) are indicated together with their median angular uncertainty (thin circles). One event (∗) appears in both event samples. The gray-shaded region indicates the zenith angle range where Earth absorption of 100 TeV neutrinos is larger than 90%. The star symbol (★) indicates the galactic center and the thin curved solid black line indicates the horizon. HESE 4 year $E_{\rm dep} > 100$ TeV (green)/classical $v_\mu + \bar{v}_\mu$ 6 year with $E_\mu > 200$ TeV (red).

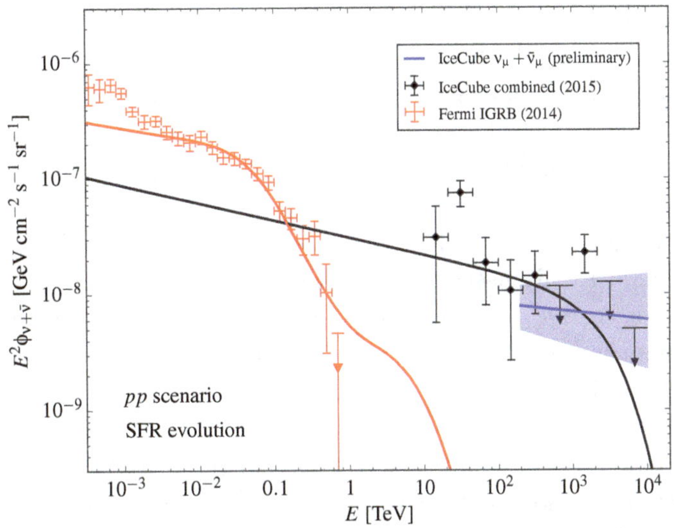

Fig. 16. The figure shows that the astrophysical neutrino flux (black line) observed by IceCube matches the corresponding cascaded gamma-ray flux (red line) observed by Fermi. We here assume that the decay products of neutral and charged pions from *pp* interactions are responsible for the nonthermal emission in the universe.[132] The black data points are combined IceCube results, including the 3-year HESE analysis[127] and a subsequent analysis lowering the energy threshold for events starting in the detector even further.[131] Also shown is the best fit to the flux of high-energy muon neutrinos penetrating the Earth.

Earth for energies exceeding 220 TeV. This observation is reinforced by the fact that fitting the cosmic neutrino flux in different ranges of energy yields values for the power-law exponent that are statistically inconsistent.

Rather than speculate on their origin, we will focus on the multimessenger connection of cosmic neutrinos to cosmic rays and gamma rays; this will turn out to be revealing. Neutrinos are produced in association with the cosmic-ray beam. It is straightforward to apply the multimessenger relation derived in section II to the cosmic neutrino flux observed by IceCube. Figure 16 shows the gamma-ray flux accompanying the observed neutrino flux for two illustrative descriptions of the data that assume $K_\pi = 2$. The black and red lines show the neutrino and gamma-ray spectra after accounting for the cascading of the PeV photons in cosmic radiation backgrounds between source and observation. The black line shows an $E^{-2.15}$ neutrino spectrum with an exponential cutoff around PeV. This scenario actually matches the extragalactic isotropic diffuse gamma-ray background measured by Fermi.[83]

The exercise indicates that the contribution of gamma rays accompanying IceCube neutrinos to Fermi's extragalactic flux is significant, suggesting a common origin of some of the sources at some level. This is intriguing because a recent analysis indicates that blazars, active galaxies with the jet pointing at the observer, dominate the Fermi diffuse flux.[128] Are blazars the final answer? The good news is that IceCube, by accumulating more events, can eventually identify blazars by observing multiple neutrinos from the same sources.[128] On the other hand, a search for neutrinos from Fermi's *identified* blazars[133,134] has come up empty. However, these represent only 50% of the total diffuse background, allowing for the possibility that a partially different source population produces the neutrinos. This could be radio galaxies with misaligned jets, as discussed earlier. Although no definite identification of the sources of cosmic neutrinos has yet emerged, it is rather clear that a multiwavelength path to the neutrino sources looks very promising. Not unexpectedly, evidence for an association has already been claimed by some studies.[135–138]

It is equally remarkable that IceCube's measured energy density in cosmic neutrinos also matches the total energy observed in extragalactic cosmic rays. This is especially interesting because the parents of PeV neutrinos should have energies in the range of 10^{17} eV, well below the ankle in the spectrum at 4×10^{18} eV where traditionally the onset of the extragalactic flux has been theorized.

6. Beyond astronomy

IceCube was designed as a discovery instrument that covers a range of areas in multidisciplinary science. Examples include the search for galactic supernova explosions and the study of neutrinos themselves. With higher energies and high-statistics data samples, opportunities for neutrino physics are varied.

6.1. *Searching for dark matter*

Neutrino telescopes are powerful tools in the search for the particle nature of dark matter. By using the deepest ice and a higher density of optical sensors, IceCube's DeepCore subarray lowers the threshold of the detector to $\sim$10 GeV over a significant fraction of the detector volume; see Fig. 10. It was initially proposed as a way to enhance IceCube's capabilities for detecting lower mass dark-matter particles. It is worth noting that the AMANDA detector, the forerunner and proof of concept for IceCube, received a significant fraction of its initial funding to search for dark matter. Also, in this context, some have considered the isotropic arrival directions of cosmic neutrinos to be a clue that they originate in the galactic halo as a result of the decay of PeV-energy dark-matter particles, a speculation that at this point is perfectly consistent with observations.[139–146]

More traditionally, IceCube and ANTARES search for dark matter by looking for the annihilation of weakly interacting massive particles (WIMPs) wherever they have accumulated to a high density: in the Sun,[147,148] in the Milky Way,[149–151] and in nearby galaxies.[152] For instance, WIMPs are swept up by the Sun as the solar system moves about the galactic halo. Though interacting weakly, they will occasionally scatter elastically with nuclei in the Sun and lose sufficient momentum to become gravitationally bound. Over the lifetime of the Sun, WIMPs may accumulate to a density where equilibrium is established between their capture and annihilation. The annihilation of these WIMPs to final states that can decay to neutrinos represents an indirect signature of halo dark matter. This WIMP annihilation signal is revealed by the neutrinos that escape the Sun with minimal absorption. The neutrinos are, for instance, the decay products of heavy quarks and weak bosons resulting from the annihilation of WIMPs into $\chi\chi \to \tau\bar{\tau}$, $b\bar{b}$, or W^+W^-. Neutrino telescopes are sensitive to such neutrinos because of their relatively high energy, above 20 GeV at this point, reflecting the mass of the decaying WIMP.

The beauty of the indirect detection technique using neutrinos originating from the Sun is that the astrophysics of the problem is understood. The source in the Sun has built up over solar time, sampling the dark matter throughout the galaxy. Therefore, any possible structure in the halo has been averaged out. Given a WIMP mass and properties, one can unambiguously predict the signal in a neutrino telescope; if not observed, the model is ruled out. This is in contrast to other indirect searches whose sensitivity depends critically on the structure of halo dark matter; observation requires cuspy structure near the galactic center or clustering on appropriate scales elsewhere. Observation necessitates not only appropriate WIMP properties but also favorable astrophysical circumstances.

IceCube has established world-leading limits on WIMPs with significant spin-dependent interactions with protons because they result in strong concentrations inside the Sun, a nearby and readily identifiable source.[153,154] An excess of neutrinos, of GeV or higher energy, over the atmospheric neutrino background in the direction of the Sun is the signature of dark matter. There is no alternative astrophysical explanation of such a signal, which represents a smoking gun for dark-matter particles. In most WIMP scenarios, the cross sections for WIMP capture and for WIMP annihilation are large enough so that an equilibrium between capture and annihilation would have been achieved within the age of the solar system.[155] In this case, limits on neutrinos from the Sun can be expressed in terms of the capture cross section. If equilibrium is not reached, weaker limits can still be derived.

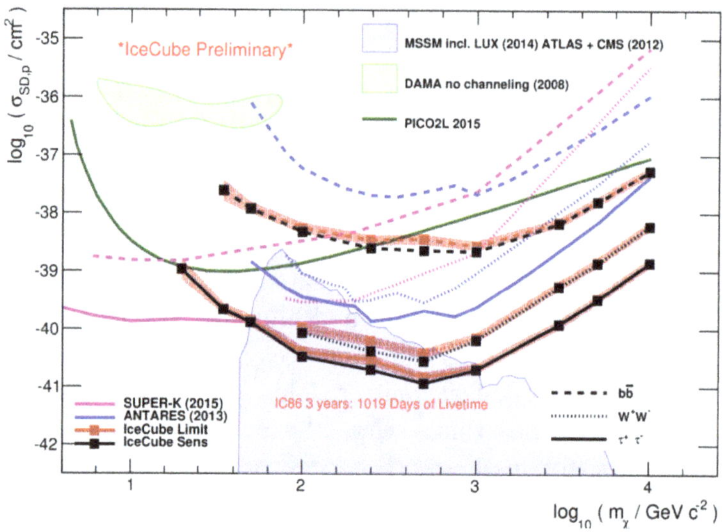

Fig. 17. Upper limits at 90% confidence level on the spin-dependent neutralino-proton cross section assuming that the neutrinos are produced by $b\bar{b}, \tau\bar{\tau}$, and W^+W^- annihilation. Limits from IceCube,[147,153,154] Super-K,[157] and ANTARES[158,159] are shown. Full lines refer to limits on the annihilation channel and dashed lines to the $b\bar{b}$ channel. Direct search results from PICO[160] and tentative signal regions[161–163] (green-shaded area) are included for comparison. The purple-shaded region indicates the allowed parameter space in MSSM supersymmetric dark-matter models that are not ruled out by other experiments.

The current IceCube limits[147,153,154] are shown in Fig. 17. These are derived from 3 years of muon-neutrino observations; no excess of neutrinos over the atmospheric flux has been found in the direction of the Sun. By including events that start inside DeepCore, the mass range for the WIMP search could be extended down to 20 GeV, which overlaps some of the allowed region from the DAMA experiment.[156] Since the exact branching ratios of WIMP annihilation into different channels is model-dependent, experiments usually choose two annihilation channels that give extreme neutrino spectra to show their results. Annihilation into $b\bar{b}$ is chosen as a representative case producing a soft neutrino spectrum; and annihilation into W^+W^- or $\tau\bar{\tau}$, a hard spectrum. Assuming a 100% branching ratio to each of these channels brackets the expected neutrino spectrum for any model with branching to more channels. IceCube and SK reach bounds at the 10^{-40} to 10^{-41}cm^2 level, covering the WIMP mass range, between the two experiments, from a few GeV to 100 TeV. Because of the A^2 coherence factor for scattering on heavy nuclei with atomic number A, the direct detection experiments have an advantage over IceCube for the case of spin-independent interactions. They thus achieve limits superior to those of IceCube.

6.2. *Neutrino oscillations*

The first IceCube neutrino oscillation analysis[164] used data from the 79-string detector from May 2010 to May 2011. The analysis was based entirely on ν_μ-induced muons from below the horizon. By taking advantage of the DeepCore subarray of IceCube, neutrino oscillations were observed over an energy range that includes the oscillation minimum of around 25 GeV for propagation through the diameter of the Earth. Data were divided into two samples: muon tracks reconstructed using the entire IceCube detector ($E_\nu > 100$ GeV) and events starting in DeepCore ($20 < E_\nu < 100$ GeV). The low-energy sample consisted of 719 events, while the high-energy sample included 39 638 events. The high-energy sample, in which standard oscillations do not affect the rates, was used for calibration. A deficit was observed in the low-energy sample, where approximately 25% more events would have been detected in the absence of oscillations. Taking systematic uncertainties into account, including ± 0.05 in the spectral index of the atmospheric neutrino flux at production, the no-oscillation hypothesis was rejected at more than 5σ. The fitted values of the oscillation parameters in a two-flavor fit are $|\Delta m^2_{32}| = 2.3^{+0.5}_{-0.6} \times 10^{-3}$ eV^2 and $\sin^2 2\theta_{23} > 0.93$. For comparison, a recent global three-flavor analysis[165] gives 2.4 eV^2 and 0.95 respectively, with a range of $\pm 5\%$ at 1σ and a slight dependence on the mass hierarchy. While the measured oscillation parameters agree with previous experiments, it is important to realize that they have been measured at a characteristic energy that is higher. The measurement is therefore also sensitive to any new neutrino physics, an important consideration when the precision of the IceCube measurements will be significantly improved. The above results will significantly improve with the development of analysis techniques that are optimized to the selection and analysis of low-energy events. Recently, IceCube has also contributed world-best limits on the existence of eV-mass sterile neutrinos.

6.3. *Supernovae and solar flares*

In addition to the normal acquisition of events that reconstruct as tracks or cascades in the deep array of IceCube and as air showers in IceTop, the rates at which the PMT voltages cross the thresholds of discriminators in the DOMs are continuously monitored. Typical rates for DOMs in the deep ice are 500 Hz (including correlated afterpulses), most of which is noise. Typical rates in the high-gain DOMs of IceTop are 2–5 kHz, most of which is induced by low-energy photons, electrons and muons entering the tanks.

A sudden increase in the total summed counting rate of the deep DOMs would signify a potential supernova explosion in the galaxy. Supernova neutrinos of $\sim$10 MeV interacting within a few meters of a DOM would generate enough hits to cause a sharp increase in the summed counting rate followed by a characteristic decline.[166] IceCube records a DC current that tracks the time evolution of the supernova in microsecond time bins. However, the detector records the time of every photoelectron with nanosecond precision and the binning can therefore be improved offline. This will improve the capability to identify the deleptonization burst. The additional measurement of the rate of neutrino events producing two photons is sensitive to the energy of the supernova neutrinos.

In IceTop, sudden changes in rates occur in response to solar events. Forbush decreases, in which the plasma from a solar flare abruptly reduces the rate of cosmic rays entering the atmosphere, are frequently detected and can be analyzed. More rare are sudden increases caused by solar energetic particles that enter the atmosphere with sufficient energy to generate secondary cosmic rays that reach the IceTop tanks. The event of December 13, 2006, was measured with the 16 tanks (eight stations) then in operation.[167] A flare was subsequently recorded by the completed instrument on May 17, 2012.

6.4. *IceCube, the facility*

During its construction phase, IceCube demonstrated a significant potential for facilitating a range of other research. For example, a dust logger provided measurements with millimeter precision that are not only valuable for event reconstruction in IceCube but also provide a record of surface winds over more than 100 000 years.[168]

Already during construction of AMANDA, antennas forming the RICE detector were deployed in some holes to expand the target volume in the search for cosmogenic neutrinos.[169] An acoustic test setup of receivers in the upper portion of four IceCube holes was deployed in 2007 to explore the acoustic technique for detecting ultrahigh-energy neutrinos. A retrievable transmitter (pinger) was submerged briefly in several newly prepared holes at various depths and distances from the receivers to measure the attenuation of sound in ice. The attenuation length of 300 m is significantly less than had been expected.[170] The Askaryan Radio Array (ARA)[171] plans to take advantage of the kilometer-scale attenuation for radio signals in ice to construct a detector with a 200 km^2 effective area, which should be sufficient to determine the level of production of cosmogenic neutrinos, which at present is highly uncertain. The initially deployed ARA detectors send data to computers housed in the IceCube Lab (ICL) for staging and transmission to the north.

The DM-Ice experiment[172] proposes to repeat the DAMA experiment in the quiet environment of the Antarctic ice. An interesting feature of the observation is the fact that the seasonal modulation of the muon rate has the opposite phase relative to the motion of the Earth through the gas of dark matter as compared to a detector in the Northern Hemisphere. A test detector to explore the noise environment for DM-Ice was deployed at the bottom of an IceCube string in December 2010. Its computers and data transmission are also hosted in the ICL.

The enhancement of the low-energy capabilities of IceCube provided by the DeepCore subarray has led to the PINGU proposal[173] to deploy an additional 40 strings within the existing DeepCore detector. This would lower the threshold to below 5 GeV (<25 m muon track length in ice). In this energy range, matter effects in the Earth lead to resonant oscillations of $\nu_\mu \leftrightarrow \nu_e$ ($\bar{\nu}_\mu \leftrightarrow \bar{\nu}_e$) for normal (inverted) hierarchy[174] that depend on zenith angle. By taking advantage of the fact that the neutrino cross section is larger than that for antineutrinos, coupled with the excess of ν_μ compared to $\bar{\nu}_\mu$, a measurement sensitive to the neutrino mass hierarchy is possible on a relatively short timescale. PINGU would also have sensitivity to ν_μ disappearance, ν_τ appearance, and maximal mixing. The lower energy threshold would also enhance the indirect searches for dark matter with IceCube as well as the sensitivity to neutrinos from supernova explosions. In addition, there is the potential for neutrino tomography of the Earth with PINGU.

The observation of extraterrestrial neutrinos up to the PeV range has stimulated ideas to enhance the discovery potential in that area. One possibility is to improve the atmospheric neutrino veto by expanding the surface array.[175] Expansion of the existing array with strings deployed at a larger separation is also possible in the long term.

7. IceCube-Gen2: from discovery to astronomy

By building IceCube, it was possible to map the optical properties of natural ice over large distances. We made the surprising discovery that the absorption length of the Cherenkov light to which the photomultipliers are sensitive exceeds 100 m. In fact, in the lower half of the detector, it exceeds 200 m. The absorption length dictates the distance by which one can space the strings of sensors without spoiling the uniformity of the detector.

Current modeling indicates that spacings of 250 m, or possibly larger, are acceptable. One can therefore instrument a 10 times larger volume of ice with the same number of strings used to build IceCube. The design of such a next-generation instrument using superior light sensors and data acquisition electronics is in progress.[176] IceCube-Gen2 will turn a fascinating discovery into an instrument for neutrino astronomy.

The larger spacings do of course result in a higher threshold, but this is not necessarily bad. While the 100 000 or so atmospheric neutrinos that IceCube collects above a threshold of 100 GeV every year were useful for calibration, they represent a severe background for isolating the cosmic component of the flux. The peak sensitivity to an E^{-2} spectrum is reached at 40 TeV.[62,63] While the detector has to be efficient below that energy, a threshold much lower than this value introduces a background without a gain in signal.

As discussed in the previous sections, the absence of a strong anisotropy of neutrino arrival directions raises the possibility that the cosmic neutrinos originate from a number of relatively weak extragalactic sources. It is indeed important to keep in mind that the interaction rate of a neutrino is so low that it travels unattenuated over cosmic distances through the tenuous matter and radiation backgrounds of the universe. This makes the identification of individual point sources contributing to the IceCube flux challenging.[128,177–179] Even so, it is also important to realize that IceCube is capable of localizing the sources by observing multiple neutrinos originating in the same location. Not having observed neutrino clusters in the present data raises the question of how many events are required to make such a model-independent identification possible. The answer to this question suggests the construction of a next-generation detector that instruments a 10 times larger volume of ice.[176]

The fact that IceCube does not yet identify sources in the diffuse cosmic neutrino flux[64] leads to constraints on their potential origin. Individual neutrino sources at redshift z contribute a flux (units $\mathrm{GeV}^{-1}\mathrm{cm}^{-2}\mathrm{s}^{-1}$ and summed over flavors)

$$\phi_\nu^{\mathrm{PS}}(E_\nu) = \frac{(1+z)^2}{4\pi d_L^2(z)} \sum_\alpha Q_{\nu_\alpha}((1+z)E_\nu) \tag{59}$$

to the total diffuse flux, where Q_{ν_α} is the neutrino emission rate and d_L is the luminosity distance which, for a flat universe, is given by

$$d_L(z) = (1+z)\int_0^z \frac{\mathrm{d}z'}{H(z')}, \tag{60}$$

with Hubble parameter H. For the standard ΛCDM cosmological model, the Hubble parameter scales as $H^2(z) = H_0^2[(1+z)^3\Omega_{\rm m} + \Omega_\Lambda]$, with $\Omega_{\rm m} \simeq 0.3$, $\Omega_\Lambda \simeq 0.7$, and $c/H_0 \simeq 4.4$ Gpc.[180] The diffuse neutrino flux from extragalactic sources is given by an integral over comoving volume $\mathrm{d}V_c = 4\pi(d_L/(1+z))^2\mathrm{d}z/H(z)$, weighting each neutrino source by its density per comoving volume $\rho(z)$,

$$\phi_\nu(E_\nu) = \frac{c}{4\pi}\int_0^\infty \frac{\mathrm{d}z}{H(z)}\rho(z)\sum_\alpha Q_{\nu_\alpha}((1+z)E_\nu). \tag{61}$$

In the following, we will assume that the neutrino emission rate Q_{ν_α} follows a power law $E^{-\Gamma}$. The flavor-averaged neutrino flux can then be written as

$$\frac{1}{3}\sum_\alpha E_\nu^2\phi_{\nu_\alpha}(E_\nu) = \frac{c}{4\pi}\frac{\xi_z}{H_0}\rho_0\frac{1}{3}\sum_\alpha E_\nu^2 Q_{\nu_\alpha}(E_\nu), \tag{62}$$

where we introduce the redshift factor

$$\xi_z = \int_0^\infty \mathrm{d}z\frac{(1+z)^{-\Gamma}}{\sqrt{\Omega_\Lambda + (1+z)^3\Omega_{\rm m}}}\frac{\rho(z)}{\rho(0)}. \tag{63}$$

A spectral index of $\Gamma \simeq 2.0$ and no source evolution, $\rho(z) = \rho_0$, yields $\xi_z \simeq 0.6$, whereas the same spectral index and source evolution following the star formation rate yields $\rho \simeq 2.4$.

We can now investigate under what circumstances IceCube can detect the neutrino emission from individual, presumably nearby, point sources that contribute to the diffuse emission. Equation (62) relates the average luminosity of individual neutrino sources to the diffuse flux that is measured by the experiment to be $E^2\phi_\nu \simeq 10^{-8}\ \mathrm{GeV cm^{-2}s^{-1}sr^{-1}}$ for energies in excess of ~100 TeV. From the measurement, we can infer the average emission from a single source[128]:

$$\frac{1}{3}\sum_\alpha E_\nu^2 Q_{\nu_\alpha}(E_\nu) \simeq 1.8\times 10^{43}\left(\frac{\xi_z}{2.4}\right)^{-1}\left(\frac{\rho_0}{10^{-8}\,\mathrm{Mpc}^{-3}}\right)^{-1}\mathrm{erg\,s^{-1}}. \tag{64}$$

For a homogenous distribution of sources, we expect, within the partial field of view $f_{\rm sky}$ of the full sky, one source within a distance d_1,

$$4\pi/3d_1^3\,\rho_0\,f_{\rm sky} = 1. \tag{65}$$

Defining F_1 as the point source flux of a source with emission rate (Eq. (64)) at distance r_1, we can write down the probability distribution of the sources:

$$F\frac{\mathrm{d}n}{\mathrm{d}F} = \frac{3}{2}\left(\frac{F_1}{F}\right)^{\frac{3}{2}} e^{-\left(\frac{F_1}{F}\right)^{\frac{3}{2}}}. \tag{66}$$

The average flux from the closest source is then $\langle F\rangle \simeq 2.7F_1$ with median of $F_{\rm med} \simeq 1.3F_1$. Assuming the latter as the emission of the closest source we arrive at

$$\frac{1}{3}\sum_\alpha E_\nu^2\phi_{\nu_\alpha} \simeq 9\times 10^{-13}\left(\frac{f_{\rm sky}}{0.5}\right)^{\frac{2}{3}}\left(\frac{\xi_z}{2.4}\right)^{-1} \times\left(\frac{\rho_0}{10^{-8}\,{\rm Mpc}^{-3}}\right)^{-\frac{1}{3}}{\rm TeV\,cm^{-2}\,s^{-1}}. \tag{67}$$

Interestingly, this value is comparable to IceCube's point-source sensitivity at the level of $5\times 10^{-13}\,{\rm TeV\,cm^{-2}\,s^{-1}}$ in the Northern Hemisphere.[64,181]

In Figure 18, we show the inferred local density and luminosity of neutrino source candidates from Ref. 182. The gray-shaded region is excluded by nonobservation of sources as individual point sources, assuming the aforementioned sensitivity of IceCube in the Northern Hemisphere. The green lines show the combination of density and luminosity for sources at the level of the observed IceCube flux, assuming a source density evolution following the star formation rate (solid line) or no evolution (dotted line). We conclude that IceCube is presently sensitive to source populations with local source densities lower than about $6\times 10^{-8}\,{\rm Mpc}^{-3}$, and much lower local densities, like FSRQs, are excluded by the nonobservation of individual sources. Some source classes, like Fanaroff–Riley (FR) radio galaxies, have an estimated neutrino luminosity that is too low for the observed flux. Note that these estimates depend on the evolution parameter ξ_z, and the exact sensitivity estimate depends on the particular redshift evolution of the source luminosity density. In addition, this simple estimate can be refined by considering not only the closest source of the population but also the combined emission of *known* local sources; see, e.g., Ref. 128.

Is it possible that the sources of the extragalactic cosmic rays are themselves neutrino sources? From the measured cosmic-ray spectrum, one can derive that the emission rate density of nucleons is at the level of [99,183]

$$\mathcal{L}_p = \rho_0 E_p^2 Q_p(E_p) \simeq (1-2)\times 10^{44}\,{\rm erg\,Mpc^{-3}\,yr^{-1}}. \tag{68}$$

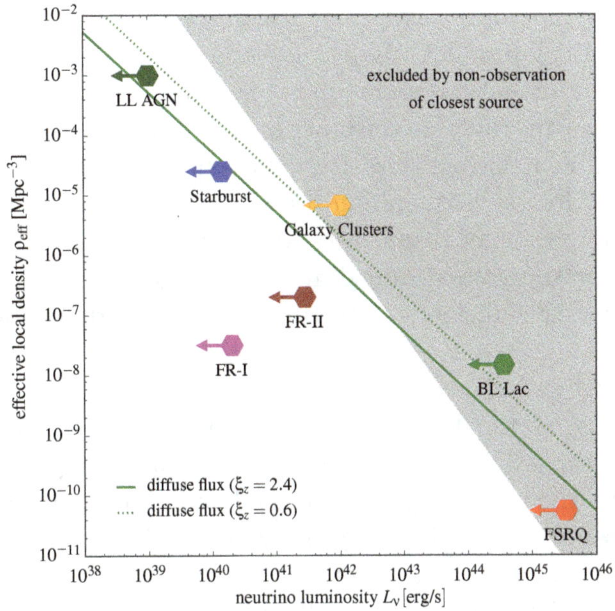

Fig. 18. The effective local density and (maximal) neutrino luminosity of various neutrino source candidates from Ref. 182. The green solid (green dotted) line shows the local density and luminosity of the population of sources responsible for the diffuse neutrino flux of $E^2\phi \simeq 10^{-8}$ GeV cm^{-2} s^{-1} sr^{-1} observed with IceCube, assuming source evolution following the star-formation rate ($\xi_z \simeq 2.4$) or no source evolution ($\xi_z \simeq 0.6$), respectively. The gray-shaded area indicates source populations that are excluded by the nonobservation of point sources in the Northern Hemisphere ($f_{\rm sky} \simeq 0.5$) with sensitivity $E^2\phi^{\rm PS} \simeq 5 \times 10^{-13}$ TeV cm^{-2} s^{-1}.

Combining this with Eq. (18) we can derive the diffuse neutrino flux:

$$\frac{1}{3}\sum_{\alpha} E_\nu^2 \phi_{\nu_\alpha}(E_\nu) \simeq f_\pi \frac{\xi_z K_\pi}{1+K_\pi}(2-4)\times 10^{-8}\,\mathrm{GeV\,cm^{-2}\,s^{-1}\,sr}. \tag{69}$$

Here, ξ_z is the evolution factor previously introduced. The equation has been rewritten and some notation adjusted in order to accommodate both pp and γp interactions. Counting particles, we derive that

$$\frac{1}{3}\sum_{\alpha} E_\nu Q_{\nu_\alpha}(E_\nu) = E_\pi Q_\pi, \tag{70}$$

from which the relation for energy follows by multiplying both sides with E_ν; an additional factor of $1/4$ multiplies the right-hand side by the ratio of neutrino to pion energy within the approximations routinely used throughout.

The requirement $f_\pi \leq 1$ limits the neutrino production by the actual sources of the cosmic rays as pointed out by the seminal work by Waxman and Bahcall.[184] For optically thin sources, $f_\pi \ll 1$, neutrino production is only a small by-product of the acceleration process. The energy loss associated with pion production must not limit the sources' ability to accelerate the cosmic rays. On the other hand, optically thick sources, $f_\pi \simeq 1$, may be efficient neutrino emitters. Realistic sources of this type need different zones, one zone for the acceleration process ($f_\pi \ll 1$) and a second zone for the efficient conversion of cosmic rays to neutrinos ($f_\pi \simeq 1$). An example for this scenario are sources embedded in starburst galaxies, where cosmic rays can be stored over sufficiently long timescales to yield significant neutrino production.

For $\xi_z \simeq 2.4$ and $K_\pi \simeq 1-2$, the upper bound resulting from Eq. (69) and $f_\pi = 1$ is at the level of the neutrino flux observed by IceCube. Therefore, it is possible that the observed extragalactic cosmic rays and neutrinos have the same origin. A plausible scenario is a "calorimeter" in which only cosmic rays with energy below a few 10 PeV interact efficiently. An energy dependence of the calorimetric environment can be introduced by energy-dependent diffusion. If $D(E)$ is the diffusion coefficient, then the timescale of escape from the calorimeter is given by the solution to $6D(E)t = d^2$, where d is the effective size of the region. Typically, we have $D(E) \propto E^\delta$ with $\delta \simeq 0.3$–0.6. In the following, we again consider the case of protons. Taking $\sigma_{pp} \simeq 8 \times 10^{-26}\,\mathrm{cm}^2$ at $E_p = 100$ PeV and the diffusion coefficient of $D(E_p) \simeq D_{\mathrm{GeV}}(E_p/1\mathrm{GeV})^{1/3}$, the pp thickness can be expressed as $\tau_{pp} \simeq ctn_{\mathrm{gas}}\sigma_{pp}$ or

$$\tau_{pp} \simeq 0.18 \left(\frac{d}{100\,\mathrm{pc}}\right)^2 \left(\frac{D_{\mathrm{GeV}}}{10^{26}\,\mathrm{cm}^2/\mathrm{s}}\right)^{-1} \left(\frac{E_p}{10\ \mathrm{PeV}}\right)^{-1/3} \left(\frac{n}{100\,\mathrm{cm}^{-3}}\right). \tag{71}$$

Here, we have used feasible parameters of starburst galaxies.[132,185] Therefore, depending on the calorimetric environment, it is possible that the flux below a few PeV is efficiently converted to neutrinos and contributes to the TeV–PeV diffuse emission observed by IceCube.

Despite the degraded resolution and the reduced potential for astronomy, the observation of electron and tau neutrinos should still be a priority. They complement the sky coverage of muon neutrinos that, at PeV energy, are mostly detected near the horizon because they are absorbed by the Earth. At high energies, neutrino production can happen in the production and decay of unstable nuclei, e.g., neutrons with $n \to pe^-\bar{\nu}_e$ or mesons, e.g., $\pi^+ \to \mu^+\nu_\mu$. Note that the neutrino production from the decay of muons $\mu^+ \to e^+\nu_e\bar{\nu}_\mu$ can be suppressed relative to the pion decay channel if synchrotron losses are important. Hence, the flavor composition is likely energy dependent and provides insight into the relative energy loss of high-energy pions and muons in the magnetic field of the cosmic accelerator.[186]

Various authors have studied the implications of IceCube's HESE topologies with astrophysical and/or exotic production mechanisms.[188–195] Figure 19 shows the general neutrino flavor phase space $\nu_e{:}\nu_\mu{:}\nu_\tau$ and the expected intrinsic flavor ratio in astrophysical sources from neutron decay (triangle), pion + muon decay (circle), and muon-damped pion decay (square). The observable neutrino flavor ratio is expected to be averaged over many oscillations. This leaves only a very narrow range for the possible flavor composition, which is shown as the line in the center of Fig. 19. The corresponding observable flavor ratios of the three astrophysical production mechanisms are also indicated. The final parameter space is very close to the "tri-bi-maximal" approximation of mixing angles, which predicts that the final flavor ratio depends only on the initial electron neutrino ratio $x = N_{\nu_e}/N_{\nu_{\rm tot}}$ and $(2/3 + x){:}(7/6 - x/2){:}(7/6 - x/2)$.

The precise relation between HESE topologies and flavor composition is nontrivial, since atmospheric backgrounds and detector effects have to be taken into account properly. In a recent IceCube analysis,[130] it was shown that the observation of tracks and cascades is consistent with most

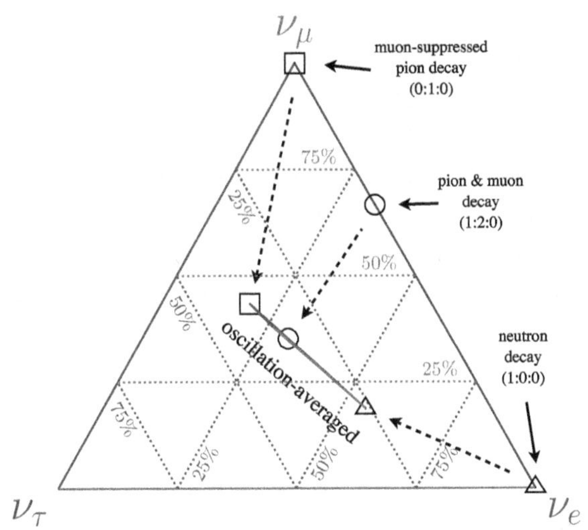

Fig. 19. Neutrino flavor phase space after oscillation. We use the best-fit oscillation parameters $\sin^2\theta_{12} = 0.304$, $\sin^2\theta_{23} = 0.577$, $\sin^2\theta_{13} = 0.0219$, and $\delta = 251°$ following Ref. 187 updated after *Neutrino 2014*. Each position in the triangle parametrizes a general initial flavor ratio $(\nu_e{:}\nu_\mu{:}\nu_\tau)$. We also indicate specific ratios for neutron decay and pion production. The inner triangle is the corresponding observable phase space after decoherence of the neutrino flavor state over large times or distances.

astrophysical scenarios within uncertainties. At sub-PeV energies (before ν_τ events can be distinguished from single cascades), the observation is mostly degenerate in terms of the total $\nu_e + \nu_\tau$ ratio, except for the contribution of prompt tau decays into muons. The expected fraction for tracks out of the total number events is about $7/24 - x/8$, where we take into account that CC interactions are about three times larger than NC interactions at these energies. The uncertainty of the inferred intrinsic electron-neutrino fraction x is hence about eight times higher than the uncertainty of the track fraction. The situation becomes even more challenging if we include backgrounds and systematic uncertainties.

The situation of flavor identification improves at super-PeV neutrino energies. On one hand, the decay length of the τ produced in CC ν_τ interactions becomes resolvable by the detector and can in principle be distinguished from tracks and cascade events as argued before. On the other hand, electron antineutrinos, $\bar{\nu}_e$, can resonantly interact with in-ice electrons via the Glashow resonance, $\bar{\nu}_e e^- \to W^-$, at neutrino energies of about 6.3 PeV. This could be observable as a peak in the cascade spectrum, depending on the relative contribution of $\bar{\nu}_e$ after oscillation. In principle, this will allow us to answer the basic question of whether the cosmic neutrinos are photo- or hadro-produced in the source with different neutrino-to-antineutrino ratios.[196,197]

Construction of a next-generation instrument with at least five times higher sensitivity would likely result in the observation of cosmogenic neutrinos.[176] The rate expected with IceCube currently is only one event per year, assuming that all cosmic rays are protons (and it is difficult to imagine that a significant component of the highest energy neutrinos would not be protons). Obviously, higher sensitivity would also benefit the wide range of measurements performed with the present detector, from the search for dark matter to the precision limits on any violation of Lorentz invariance.

Acknowledgments

I thank my IceCube collaborators for discussions and M. Ahlers, A. Kheirandish and L. Wille for a critical reading of the manuscript. This research was supported in part by the U.S. National Science Foundation under Grants No. OPP-0236449 and PHY-0354776, U.S. Department of Energy under Grant No. DE-FG02-95ER40896 and University of Wisconsin Research Committee with funds granted by the Wisconsin Alumni Research Foundation.

References

1. F. Reines and C. L. Cowan, The neutrino, *Nature* **178**, 446–449 (1956). doi: 10.1038/178446a0.
2. F. Halzen and S. R. Klein, Astronomy and astrophysics with neutrinos, *Phys. Today* **61N5**, 29–35 (2008). doi: 10.1063/1.2930733.
3. A. Roberts, The birth of high-energy neutrino astronomy: A personal history of the DUMAND project, *Rev. Mod. Phys.* **64**, 259–312 (1992). doi: 10.1103/RevModPhys.64.259.
4. T. K. Gaisser, F. Halzen, and T. Stanev, Particle astrophysics with high-energy neutrinos, *Phys. Rept.* **258**, 173–236 (1995). doi: 10.1016/0370-1573(95)00003-Y.
5. J. Learned and K. Mannheim, High-energy neutrino astrophysics, *Ann. Rev. Nucl. Part. Sci.* **50**, 679–749 (2000). doi: 10.1146/annurev.nucl.50.1.679.
6. F. Halzen and D. Hooper, High-energy neutrino astronomy: The cosmic ray connection, *Rept. Prog. Phys.* **65**, 1025–1078 (2002). doi: 10.1088/0034-4885/65/7/201.
7. M. Markov, On high energy neutrino physics, *Proceedings of the 10th International Conference on High-Energy Physics*, pp. 578–581 (1960).
8. J. Babson *et al.*, Cosmic ray muons in the deep ocean, *Phys. Rev.* **D42**, 3613–3620 (1990). doi: 10.1103/PhysRevD.42.3613.
9. V. Balkanov *et al.*, The BAIKAL neutrino project: Status report, *Nucl. Phys. Proc. Suppl.* **118**, 363–370 (2003). doi: 10.1016/S0920-5632(03)01334-3.
10. G. Aggouras *et al.*, A measurement of the cosmic-ray muon flux with a module of the NESTOR neutrino telescope, *Astropart. Phys.* **23**, 377–392 (2005). doi: 10.1016/j.astropartphys.2005.02.001.
11. J. Aguilar *et al.*, First results of the instrumentation Line for the deep-sea ANTARES neutrino telescope, *Astropart. Phys.* **26**, 314–324 (2006). doi: 10.1016/j.astropartphys.2006.07.004.
12. E. Migneco, Progress and latest results from Baikal, Nestor, NEMO and KM3NeT, *J. Phys. Conf. Ser.* **136**, 022048 (2008). doi: 10.1088/1742-6596/136/2/022048.
13. IceCube Collaboration. Icecube preliminary design document (2001). http://www.icecube.wisc.edu/science/publications/pdd/pdd.pdf.
14. J. Ahrens *et al.*, Sensitivity of the IceCube detector to astrophysical sources of high energy muon neutrinos, *Astropart. Phys.* **20**, 507–532 (2004). doi: 10.1016/j.astropartphys.2003.09.003.
15. J. K. Becker, High-energy neutrinos in the context of multimessenger physics, *Phys. Rept.* **458**, 173–246 (2008). doi: 10.1016/j.physrep.2007.10.006.
16. A. Achterberg *et al.*, Multi-year search for a diffuse flux of muon neutrinos with AMANDA-II, *Phys. Rev.* **D76**, 042008 (2007). doi: 10.1103/PhysRevD.76.042008,10.1103/PhysRevD.77.089904. [Erratum: *Phys. Rev.* **D77**, 089904 (2008)].
17. W. Rhode *et al.*, Limits on the flux of very high-energetic neutrinos with the Frejus detector, *Astropart. Phys.* **4**, 217–225 (1996). doi: 10.1016/0927-6505(95)00038-0.

18. R. Engel, D. Seckel, and T. Stanev, Neutrinos from propagation of ultrahigh-energy protons, *Phys. Rev.* **D64**, 093010 (2001). doi: 10.1103/PhysRevD.64.093010.
19. P. Sommers, Ultra-high energy cosmic rays: Observational results, *Astropart. Phys.* **39–40**, 88–94 (2012). doi: 10.1016/j.astropartphys.2012.04.011.
20. Particle Data Group. URL: pdg.lbl.gov .
21. W. Baade and F. Zwicky, Cosmic Rays from Super-novae, *Proc. Nat. Acad. Sci.* **20**, 259–263 (1934). doi: 10.1073/pnas.20.5.259.
22. H.E.S.S Collaboration, *Nature* **531**, 476–479 (2016).
23. V. L. Ginzburg and S. I. Syrovatskii, *The Origin of Cosmic Rays* (1969). New York: Macmillian.
24. Y. Butt, Beyond the myth of the supernova-remnant origin of cosmic rays, *Nature* **460**, 659–772 (2009). doi: 10.1038/nature08127.
25. T. K. Gaisser, Origin of cosmic radiation, *AIP Conf. Proc.* **558**, 27–42 (2001). doi: 10.1063/1.1370778.
26. E. Waxman, *Phys. Rev. Lett.* **75**, 386 (1995).
27. M. Gonzalez-Garcia, M. Maltoni, and J. Rojo, Determination of the atmospheric neutrino fluxes from atmospheric neutrino data, *JHEP* **0610**, 075 (2006). doi: 10.1088/1126-6708/2006/10/075.
28. K. Daum *et al.*, Determination of the atmospheric neutrino spectra with the Frejus detector, *Z. Phys.* **C66**, 417–428 (1995). doi: 10.1007/BF01556368.
29. R. Abbasi *et al.*, Determination of the atmospheric neutrino flux and searches for new physics with AMANDA-II, *Phys. Rev.* **D79**, 102005 (2009). doi: 10.1103/PhysRevD.79.102005.
30. R. Abbasi *et al.*, The energy spectrum of atmospheric neutrinos between 2 and 200 TeV with the AMANDA-II Detector, *Astropart. Phys.* **34**, 48–58 (2010). doi: 10.1016/j.astropartphys.2010.05.001.
31. R. Abbasi *et al.*, Measurement of the atmospheric neutrino energy spectrum from 100 GeV to 400 TeV with IceCube, *Phys. Rev.* **D83**, 012001 (2011). doi: 10.1103/PhysRevD.83.012001.
32. R. Abbasi *et al.*, A search for a diffuse flux of astrophysical muon neutrinos with the IceCube 40-string detector, *Phys. Rev.* **D84**, 082001 (2011). doi: 10.1103/PhysRevD.84.082001.
33. M. Aartsen *et al.*, Measurement of the atmospheric ν_e flux in IceCube, *Phys. Rev. Lett.* **110**(15), 151105 (2013). doi: 10.1103/PhysRevLett.110.151105.
34. M. Honda, T. Kajita, K. Kasahara, S. Midorikawa, and T. Sanuki, Calculation of atmospheric neutrino flux using the interaction model calibrated with atmospheric muon data, *Phys. Rev.* **D75**, 043006 (2007). doi: 10.1103/PhysRevD.75.043006.
35. G. Barr, T. Gaisser, P. Lipari, S. Robbins, and T. Stanev, A three-dimensional calculation of atmospheric neutrinos, *Phys. Rev.* **D70**, 023006 (2004). doi: 10.1103/PhysRevD.70.023006.
36. R. Enberg, M. H. Reno, and I. Sarcevic, Prompt neutrino fluxes from atmospheric charm, *Phys. Rev.* **D78**, 043005 (2008). doi: 10.1103/PhysRevD.78.043005.

37. M. G. Aartsen *et al.*, An all-sky search for three flavors of neutrinos from gamma-ray bursts with the IceCube neutrino observatory, *Astrophys. J.* **824**(2), 115 (2016). doi: 10.3847/0004-637X/824/2/115.
38. S. Adrian-Martinez *et al.*, First search for neutrinos in correlation with gamma-ray bursts with the ANTARES neutrino telescope, *JCAP* **1303**, 006 (2013). doi: 10.1088/1475-7516/2013/03/006.
39. E. Berezhko, Cosmic rays from active galactic nuclei, *Astrophys. J.* **684**, L69–L71 (2008). doi: 10.1086/592233.
40. J. B. Tjus, B. Eichmann, F. Halzen, A. Kheirandish, and S. M. Saba, High-energy neutrinos from radio galaxies, *Phys. Rev.* **D89** (12), 123005 (2014). doi: 10.1103/PhysRevD.89.123005.
41. D. Hooper, A case for radio galaxies as the sources of IceCube's astrophysical neutrino flux, *JCAP* **1609**(09), 002 (2016). doi: 10.1088/1475-7516/2016/09/002.
42. T. Gaisser, Neutrino astronomy: Physics goals, detector parameters (1997).
43. M. Ahlers *et al.*, Neutrinos as a diagnostic of cosmic ray galactic/extra-galactic transition, *Phys. Rev.* **D72**, 023001 (2005). doi: 10.1103/PhysRevD.72.023001.
44. R. Abbasi *et al.*, Time-integrated searches for point-like sources of neutrinos with the 40-string IceCube detector, *Astrophys. J.* **732**, 18 (2011). doi: 10.1088/0004-637X/732/1/18.
45. S. Wakely and D. Horan. Tevcat, an online gamma-ray catalog. URL: http://tevcat.uchicago.edu/.
46. J. Alvarez-Muniz and F. Halzen, Possible high-energy neutrinos from the cosmic accelerator RX J1713.7-3946, *Astrophys. J.* **576**, L33–L36 (2002). doi: 10.1086/342978.
47. J. K. Becker, F. Halzen, A. O'Murchadha, and M. Olivo, Neutrino emission from high-energy component gamma-ray bursts (2010).
48. T. K. Gaisser, Very high energy photons and neutrinos: Implications for UHECR, *EPJ Web Conf.* **53**, 01012 (2013). doi: 10.1051/epjconf/20135301012.
49. F. Halzen, Neutrino astronomy: An update, *Riv. Nuovo Cim.* **36**(03), 81–104 (2013). doi: 10.1393/ncr/i2013-10086-y.
50. K. Mannheim and R. Schlickeiser, Interactions of cosmic ray nuclei, *Astron. Astrophys.* **286**, 983–996 (1994).
51. F. W. Stecker, *Cosmic Gamma Rays* (1971). Baltimore: Monobook Company.
52. S. R. Kelner, F. A. Aharonian, and V. V. Bugayov, Energy spectra of gamma-rays, electrons and neutrinos produced at proton-proton interactions in the very high energy regime, *Phys. Rev.* **D74**, 034018 (2006). doi: 10.1103/PhysRevD.74.034018,10.1103/PhysRevD.79.039901. [Erratum: *Phys. Rev.* **D79**, 039901 (2009)].
53. X. Wang and A. Loeb, Cumulative neutrino background from quasar-driven outflows (2016).
54. F. Halzen and A. Kheirandish, High Energy Neutrinos from Recent Blazar Flares (2016).

55. M. Gonzalez-Garcia, F. Halzen, and S. Mohapatra, Identifying galactic PeVatrons with neutrinos, *Astropart. Phys.* **31**, 437–444 (2009). doi: 10.1016/j.astropartphys.2009.05.002.
56. M. Ahlers, P. Mertsch, and S. Sarkar, On cosmic ray acceleration in supernova remnants and the FERMI/PAMELA data, *Phys. Rev.* **D80**, 123017 (2009). doi: 10.1103/PhysRevD.80.123017.
57. A. A. Abdo *et al.*, Discovery of TeV gamma-ray emission from the Cygnus Region of the galaxy, *Astrophys. J.* **658**, L33–L36 (2007). doi:10.1086/513696.
58. F. Halzen, A. Kheirandish, and V. Niro, Prospects for Detecting Galactic Sources of Cosmic Neutrinos with IceCube: An Update (2016).
59. S. Gabici and F. A. Aharonian, Searching for galactic cosmic ray pevatrons with multi-TeV gamma rays and neutrinos, *Astrophys. J.* **665**, L131 (2007). doi: 10.1086/521047.
60. M. Ackermann *et al.*, Detection of the characteristic pion-decay signature in supernova remnants, *Science* **339**, 807 (2013). doi: 10.1126/science.1231160.
61. A. A. Abdo, Observations of the young supernova remnant RX J1713.7-3946 with the Fermi Large Area Telescope, *Astrophys. J.* **734**, 28 (2011). doi: 10.1088/0004-637X/734/1/28.
62. F. Halzen, A. Kappes, and A. O'Murchadha, Prospects for identifying the sources of the galactic cosmic rays with IceCube, *Phys. Rev.* **D78**, 063004 (2008). doi: 10.1103/PhysRevD.78.063004.
63. A. Kappes, F. Halzen, and A. O. Murchadha, Prospects of identifying the sources of the galactic cosmic rays with IceCube, *Nucl. Instrum. Method* **A602**, 117–119 (2009). doi: 10.1016/j.nima.2008.12.049.
64. M. G. Aartsen *et al.*, All-sky search for time-integrated neutrino emission from astrophysical sources with 7 years of IceCube data (2016).
65. S. Adrian-Martinez *et al.*, Constraints on the neutrino emission from the galactic ridge with the ANTARES telescope, *Phys. Lett.* **B760**, 143–148 (2016). doi: 10.1016/j.physletb.2016.06.051.
66. E. Waxman and J. N. Bahcall, High-energy neutrinos from cosmological gamma-ray burst fireballs, *Phys. Rev. Lett.* **78**, 2292–2295 (1997). doi: 10.1103/PhysRevLett.78.2292.
67. D. Guetta, D. Hooper, J. Alvarez-Muniz, F. Halzen, and E. Reuveni, Neutrinos from individual gamma-ray bursts in the BATSE catalog, *Astropart. Phys.* **20**, 429–455 (2004). doi: 10.1016/S0927-6505(03)00211-1.
68. S. Hummer, P. Baerwald, and W. Winter, Neutrino emission from gamma-ray burst fireballs, revised, *Phys. Rev. Lett.* **108**, 231101 (2012). doi: 10.1103/PhysRevLett.108.231101.
69. M. Ahlers, M. C. Gonzalez-Garcia, and F. Halzen, GRBs on probation: Testing the UHE CR paradigm with IceCube, *Astropart. Phys.* **35**, 87–94 (2011). doi: 10.1016/j.astropartphys.2011.05.008.
70. R. Abbasi *et al.*, Constraints on the extremely-high energy cosmic neutrino flux with the IceCube 2008–2009 data, *Phys. Rev.* **D83**, 092003 (2011). doi: 10.1103/PhysRevD.84.079902,10.1103/PhysRevD.83.092003.

71. N. Senno, K. Murase, and P. Meszaros, Choked jets and low-luminosity gamma-ray bursts as hidden neutrino sources, *Phys. Rev.* **D93**(8), 083003 (2016). doi: 10.1103/PhysRevD.93.083003.
72. M. Bottcher and C. D. Dermer, High-energy gamma-rays from ultrahigh-energy cosmic ray protons in gamma-ray bursts, *Astrophys. J.* **499**, L131–L134 (1998). doi: 10.1086/311366.
73. K. Greisen, End to the cosmic ray spectrum? *Phys. Rev. Lett.* **16**, 748–750 (1966). doi: 10.1103/PhysRevLett.16.748.
74. G. Zatsepin and V. Kuzmin, Upper limit of the spectrum of cosmic rays, *JETP Lett.* **4**, 78–80 (1966).
75. V. Berezinsky and G. Zatsepin, Cosmic rays at ultrahigh-energies (neutrino?), *Phys. Lett.* **B28**, 423–424 (1969). doi: 10.1016/0370-2693(69)90341-4.
76. S. Yoshida and M. Teshima, Energy spectrum of ultrahigh-energy cosmic rays with extragalactic origin, *Prog. Theor. Phys.* **89**, 833–845 (1993). doi: 10.1143/PTP.89.833.
77. R. Protheroe and P. Johnson, Propagation of ultrahigh-energy protons over cosmological distances and implications for topological defect models, *Astropart. Phys.* **4**, 253 (1996). doi: 10.1016/0927-6505(95)00039-9.
78. V. Berezinsky, A. Z. Gazizov, and S. I. Grigorieva, On astrophysical solution to ultrahigh-energy cosmic rays, *Phys. Rev.* **D74**, 043005 (2006). doi: 10.1103/PhysRevD.74.043005.
79. Z. Fodor, S. D. Katz, A. Ringwald, and H. Tu, Bounds on the cosmogenic neutrino flux, *JCAP* **0311**, 015 (2003). doi: 10.1088/1475-7516/2003/11/015.
80. H. Yuksel and M. D. Kistler, Enhanced cosmological GRB rates and implications for cosmogenic neutrinos, *Phys. Rev.* **D75**, 083004 (2007). doi: 10.1103/PhysRevD.75.083004.
81. H. Takami, K. Murase, S. Nagataki, and K. Sato, Cosmogenic neutrinos as a probe of the transition from Galactic to extragalactic cosmic rays, *Astropart. Phys.* **31**, 201–211 (2009). doi: 10.1016/j.astropartphys.2009.01.006.
82. A. A. Abdo *et al.*, The spectrum of the isotropic diffuse gamma-ray emission derived from first-year fermi large area telescope data, *Phys. Rev. Lett.* **104**, 101101 (2010). doi: 10.1103/PhysRevLett.104.101101.
83. M. Ackermann *et al.*, The spectrum of isotropic diffuse gamma-ray emission between 100 MeV and 820 GeV, *Astrophys. J.* **799**, 86 (2015). doi: 10.1088/0004-637X/799/1/86.
84. V. Berezinsky, A. Gazizov, M. Kachelriess, and S. Ostapchenko, Restricting UHECRs and cosmogenic neutrinos with Fermi-LAT, *Phys. Lett.* **B695**, 13–18 (2011). doi: 10.1016/j.physletb.2010.11.019.
85. M. Ahlers, L. Anchordoqui, M. Gonzalez-Garcia, F. Halzen, and S. Sarkar, GZK neutrinos after the Fermi-LAT diffuse photon flux measurement, *Astropart. Phys.* **34**, 106–115 (2010). doi: 10.1016/j.astropartphys.2010.06.003.
86. G. B. Gelmini, O. Kalashev, and D. V. Semikoz, Gamma-ray constraints on maximum cosmogenic neutrino fluxes and UHECR source evolution models, *JCAP* **1201**, 044 (2012). doi: 10.1088/1475-7516/2012/01/044.

87. G. Decerprit and D. Allard, Constraints on the origin of ultrahigh energy cosmic rays from cosmogenic neutrinos and photons, *Astron. Astrophys.* **535**, A66 (2011). doi: 10.1051/0004-6361/201117673.
88. J. Heinze, D. Boncioli, M. Bustamante, and W. Winter, Cosmogenic neutrinos challenge the cosmic ray proton dip model, *Astrophys. J.* **825**(2), 122 (2016). doi: 10.3847/0004-637X/825/2/122,10.3204/PUBDB-2016-01612.
89. A. D. Supanitsky, Implications of gamma-ray observations on proton models of ultrahigh energy cosmic rays, *Phys. Rev.* **D94**(6), 063002 (2016). doi: 10.1103/PhysRevD.94.063002.
90. M. G. Aartsen *et al.*, Constraints on ultra-high-energy cosmic ray sources from a search for neutrinos above 10 PeV with IceCube (2016).
91. D. Hooper, A. Taylor, and S. Sarkar, The impact of heavy nuclei on the cosmogenic neutrino flux, *Astropart. Phys.* **23**, 11–17 (2005). doi: 10.1016/j.astropartphys.2004.11.002.
92. M. Ave, N. Busca, A. V. Olinto, A. A. Watson, and T. Yamamoto, Cosmogenic neutrinos from ultra-high energy nuclei, *Astropart. Phys.* **23**, 19–29 (2005). doi: 10.1016/j.astropartphys.2004.11.001.
93. D. Hooper, S. Sarkar, and A. M. Taylor, The intergalactic propagation of ultrahigh energy cosmic ray nuclei, *Astropart. Phys.* **27**, 199–212 (2007). doi: 10.1016/j.astropartphys.2006.10.008.
94. D. Allard *et al.*, Cosmogenic neutrinos from the propagation of ultrahigh energy nuclei, *JCAP* **0609**, 005 (2006). doi: 10.1088/1475-7516/2006/09/005.
95. L. A. Anchordoqui, H. Goldberg, D. Hooper, S. Sarkar, and A. M. Taylor, Predictions for the cosmogenic neutrino flux in light of new data from the pierre auger observatory, *Phys. Rev.* **D76**, 123008 (2007). doi: 10.1103/PhysRevD.76.123008.
96. R. Aloisio, V. Berezinsky, and A. Gazizov, Ultra high energy cosmic rays: The disappointing model, *Astropart. Phys.* **34**, 620–626 (2011). doi: 10.1016/j.astropartphys.2010.12.008.
97. K. Kotera, D. Allard, and A. Olinto, Cosmogenic neutrinos: Parameter space and detectabilty from PeV to ZeV, *JCAP* **1010**, 013 (2010). doi: 10.1088/1475-7516/2010/10/013.
98. M. Ahlers and J. Salvado, Cosmogenic gamma-rays and the composition of cosmic rays, *Phys. Rev.* **D84**, 085019 (2011). doi: 10.1103/PhysRevD.84.085019.
99. M. Ahlers and F. Halzen, Minimal cosmogenic neutrinos, *Phys. Rev.* **D86**, 083010 (2012). doi: 10.1103/PhysRevD.86.083010.
100. E. Roulet, G. Sigl, A. van Vliet, and S. Mollerach, PeV neutrinos from the propagation of ultra-high energy cosmic rays, *JCAP* **1301**, 028 (2013). doi: 10.1088/1475-7516/2013/01/028.
101. L. Yacobi, D. Guetta, and E. Behar, Implication of the non-detection of gzk neutrinos, *Astrophys. J.* **823**(2), 89 (2016). doi: 10.3847/0004-637X/823/2/89.
102. C. Spiering. High energy neutrino astronomy: Status and perspectives, *AIP Conf. Proc.* **1085**, 18–29 (2008). doi: http://dx.doi.org/10.1063/1.3076635. URL: http://scitation.aip.org/content/aip/proceeding/aipcp/10.1063/1.3076635.

103. U. Katz and C. Spiering, High-energy neutrino astrophysics: Status and perspectives, *Prog. Part. Nucl. Phys.* **67**, 651–704 (2012). doi: 10.1016/j.ppnp.2011.12.001.
104. F. Halzen and S. R. Klein, IceCube: An instrument for neutrino astronomy, *Rev. Sci. Instrum.* **81**, 081101 (2010). doi: 10.1063/1.3480478.
105. F. Halzen, Astroparticle physics with high energy neutrinos: From AMANDA to IceCube, *Eur. Phys. J.* **C46**, 669–687 (2006). doi: 10.1140/epjc/s2006-02536-4.
106. KM3NeT Coll. URL: http://www.km3net.org/.
107. F. Halzen and D. Saltzberg, Tau-neutrino appearance with a 1000 megaparsec baseline, *Phys. Rev. Lett.* **81**, 4305–4308 (1998). doi: 10.1103/PhysRevLett.81.4305.
108. E. Middell. Search for atmospheric neutrino induced particle showers with IceCube 40. In *Proceedings, 32nd International Cosmic Ray Conference (ICRC 2011): Beijing, China, August 11-18, 2011*, vol. 4, pp. 246–249 (2011). doi: 10.7529/ICRC2011/V04/1097. URL http://www.ihep.ac.cn/english/conference/icrc2011/paper/proc/v4/v4_1097.pdf.
109. S. R. Klein. Cascades from nu(E) above 10**20 eV. In *INFN Eloisatron Project 44th Workshop on QCD at Cosmic Energies: The Highest Energy Cosmic Rays and QCD Erice, Italy, August 29-September 5, 2004* (2004).
110. J. G. Learned and S. Pakvasa, Detecting tau-neutrino oscillations at PeV energies, *Astropart. Phys.* **3**, 267–274 (1995). doi: 10.1016/0927-6505(94)00043-3.
111. S. Schönert, T. K. Gaisser, E. Resconi, and O. Schulz, Vetoing atmospheric neutrinos in a high energy neutrino telescope, *Phys. Rev.* **D79**, 043009 (2009). doi: 10.1103/PhysRevD.79.043009.
112. T. K. Gaisser, T. Stanev, and S. Tilav, Cosmic ray energy spectrum from measurements of air showers, *Front. Phys.* (*Beijing*). **8**, 748–758 (2013). doi: 10.1007/s11467-013-0319-7.
113. M. Aglietta *et al.*, Upper limit on the prompt muon flux derived from the LVD underground experiment, *Phys. Rev.* **D60**, 112001 (1999). doi: 10.1103/PhysRevD.60.112001.
114. A. Schukraft, A view of prompt atmospheric neutrinos with IceCube, *Nucl. Phys. Proc. Suppl.* **237–238**, 266–268 (2013). doi: 10.1016/j.nuclphysbps.2013.04.105.
115. T. K. Gaisser, Atmospheric leptons, *EPJ Web Conf.* **52**, 09004 (2013). doi: 10.1051/epjconf/20125209004.
116. P. Desiati and T. K. Gaisser, Seasonal variation of atmospheric leptons as a probe of charm, *Phys. Rev. Lett.* **105**, 121102 (2010). doi: 10.1103/PhysRevLett.105.121102.
117. J. Ahrens *et al.*, Muon track reconstruction and data selection techniques in AMANDA, *Nucl. Instrum. Method.* **A524**, 169–194 (2004). doi: 10.1016/j.nima.2004.01.065.
118. M. Aartsen *et al.*, Energy reconstruction methods in the IceCube neutrino telescope, *JINST* **9**, P03009 (2014). doi: 10.1088/1748-0221/9/03/P03009.
119. C. Kopper, W. Giang, and N. Kurahashi. Observation of astrophysical neutrinos in four years of IceCube data. *Proc. ICRC2015* **1081** (2015).

120. S. Schoenen and L. Rädel, *Proc. ICRC2015* **1079**, 642 (2015).
121. M. G. Aartsen *et al.*, Evidence for astrophysical muon neutrinos from the northern sky with IceCube, *Phys. Rev. Lett.* **115**(8), 081102 (2015). doi: 10.1103/PhysRevLett.115.081102.
122. C. Weaver. Icecube results for diffuse muon neutrinos. APS April Meeting 2014, Savannah, Georgia, USA (2014).
123. M. G. Aartsen *et al.*, Observation and characterization of a cosmic muon neutrino flux from the northern hemisphere using six years of IceCube data, *Astrophys. J.* **833**(1), 3 (2016). doi: 10.3847/0004-637X/833/1/3.
124. M. Aartsen *et al.*, First observation of PeV-energy neutrinos with IceCube, *Phys. Rev. Lett.* **111**, 021103 (2013). doi: 10.1103/PhysRevLett.111.021103.
125. T. K. Gaisser, K. Jero, A. Karle, and J. van Santen, A generalized self-veto probability for atmospheric neutrinos, *Phys. Rev.* **D90**, 023009 (2014). doi: 10.1103/PhysRevD.90.023009.
126. M. Aartsen *et al.*, Search for neutrino-induced particle showers with IceCube-40, *Phys. Rev.* **D89**, 102001 (2014). doi: 10.1103/PhysRevD.89.102001.
127. M. Aartsen *et al.*, Observation of high-energy astrophysical neutrinos in three years of IceCube data, *Phys. Rev. Lett.* **113**, 101101 (2014). doi: 10.1103/PhysRevLett.113.101101.
128. M. Ahlers and F. Halzen, Pinpointing extragalactic neutrino sources in light of recent IceCube observations, *Phys. Rev.* **D90**(4), 043005 (2014). doi: 10.1103/PhysRevD.90.043005.
129. M. G. Aartsen *et al.* The IceCube Neutrino Observatory — Contributions to ICRC 2015 Part II: Atmospheric and Astrophysical Diffuse Neutrino Searches of All Flavors. In *Proceedings, 34th International Cosmic Ray Conference (ICRC 2015) : The Hague, The Netherlands, July 30-August 6, 2015* (2015). URL http://inspirehep.net/record/1398539/files/arXiv:1510.05223.pdf.
130. M. G. Aartsen *et al.*, Flavor ratio of astrophysical neutrinos above 35 TeV in IceCube, *Phys. Rev. Lett.* **114**(17), 171102 (2015). doi: 10.1103/PhysRevLett.114.171102.
131. M. G. Aartsen *et al.*, Atmospheric and astrophysical neutrinos above 1 TeV interacting in IceCube, *Phys. Rev.* **D91**(2), 022001 (2015). doi: 10.1103/PhysRevD.91.022001.
132. K. Murase, M. Ahlers, and B. C. Lacki, Testing the hadronuclear origin of PeV neutrinos observed with IceCube, *Phys. Rev.* **D88**(12), 121301 (2013). doi: 10.1103/PhysRevD.88.121301.
133. T. Glüsenkamp, Analysis of the cumulative neutrino flux from Fermi-LAT blazar populations using 3 years of IceCube data (2015).
134. S. Adrian-Martinez *et al.*, Search for muon-neutrino emission from GeV and TeV gamma-ray flaring blazars using five years of data of the ANTARES telescope, *JCAP* **1512** (12), 014 (2015). doi: 10.1088/1475-7516/2015/12/014.
135. P. Padovani and E. Resconi, Are both BL Lacs and pulsar wind nebulae the astrophysical counterparts of IceCube neutrino events? *Mon. Not. Roy. Astron. Soc.* **443**, 474–484 (2014). doi: 10.1093/mnras/stu1166.
136. P. Padovani, E. Resconi, P. Giommi, B. Arsioli, and Y. L. Chang, Extreme blazars as counterparts of IceCube astrophysical neutrinos, *Mon. Not. Roy. Astron. Soc.* **457**(4), 3582–3592 (2016). doi: 10.1093/mnras/stw228.

137. F. Krauß *et al.*, TANAMI blazars in the IceCube PeV neutrino fields, *Astron. Astrophys.* **566**, L7 (2014). doi: 10.1051/0004-6361/201424219.
138. M. Kadler *et al.*, Coincidence of a high-fluence blazar outburst with a PeV-energy neutrino event (2016).
139. B. Feldstein, A. Kusenko, S. Matsumoto, and T. T. Yanagida, Neutrinos at IceCube from heavy decaying dark matter, *Phys. Rev.* **D88**(1), 015004 (2013). doi: 10.1103/PhysRevD.88.015004.
140. A. Esmaili and P. D. Serpico, Are IceCube neutrinos unveiling PeV-scale decaying dark matter? *JCAP* **1311**, 054 (2013). doi: 10.1088/1475-7516/2013/11/054.
141. Y. Bai, R. Lu, and J. Salvado, Geometric compatibility of IceCube TeV-PeV neutrino excess and its galactic dark matter origin, *JHEP* **01**, 161 (2016). doi: 10.1007/JHEP01(2016)161.
142. A. Bhattacharya, M. H. Reno, and I. Sarcevic, Reconciling neutrino flux from heavy dark matter decay and recent events at IceCube, *JHEP* **1406**, 110 (2014). doi: 10.1007/JHEP06(2014)110.
143. A. Esmaili, S. K. Kang, and P. D. Serpico, IceCube events and decaying dark matter: Hints and constraints, *JCAP* **1412**(12), 054 (2014). doi: 10.1088/1475-7516/2014/12/054.
144. J. F. Cherry, A. Friedland, and I. M. Shoemaker, Neutrino portal dark matter: From dwarf galaxies to IceCube, *JCAP.* **12**, 054 (2014).
145. K. Murase, R. Laha, S. Ando, and M. Ahlers, Testing the dark matter scenario for PeV neutrinos observed in IceCube, *Phys. Rev. Lett.* **115**(7), 071301 (2015). doi: 10.1103/PhysRevLett.115.071301.
146. A. Esmaili and P. D. Serpico, Gamma-ray bounds from EAS detectors and heavy decaying dark matter constraints, *JCAP* **1510**(10), 014 (2015). doi: 10.1088/1475-7516/2015/10/014.
147. M. Aartsen *et al.*, Search for dark matter annihilations in the Sun with the 79-string IceCube detector, *Phys. Rev. Lett.* **110**(13), 131302 (2013). doi: 10.1103/PhysRevLett.110.131302.
148. S. Adrian-Martinez *et al.*, Limits on dark matter annihilation in the Sun using the ANTARES neutrino telescope, *Phys. Lett.* **B759**, 69–74 (2016). doi: 10.1016/j.physletb.2016.05.019.
149. R. Abbasi *et al.*, Search for dark matter from the galactic halo with the IceCube neutrino observatory, *Phys. Rev.* **D84**, 022004 (2011). doi: 10.1103/PhysRevD.84.022004.
150. R. Abbasi *et al.*, Search for Neutrinos from Annihilating Dark Matter in the Direction of the Galactic Center with the 40-String IceCube Neutrino Observatory (2012).
151. S. Adrian-Martinez *et al.*, Search of dark matter annihilation in the galactic centre using the ANTARES neutrino telescope, *JCAP* **1510**(10), 068 (2015). doi: 10.1088/1475-7516/2015/10/068.
152. M. Aartsen *et al.*, IceCube search for dark matter annihilation in nearby galaxies and galaxy clusters, *Phys. Rev.* **D88**(12), 122001 (2013). doi: 10.1103/PhysRevD.88.122001.

153. M. Rameez and T. Montaruli (for the IceCube Collaboration), Search for dark matter annihilations in the Sun using the completed IceCube neutrino telescope, *PoS.* **ICRC2015**, 1209 (2015).
154. M. Zoll (for the IceCube Collaboration), Improved methods for solar dark matter searches with the IceCube neutrino telescope, *PoS.* **ICRC2015**, 1099 (2015).
155. P. Gondolo, J. Edsjo, P. Ullio, L. Bergstrom, M. Schelke, and E. A. Baltz, DarkSUSY: Computing supersymmetric dark matter properties numerically, *JCAP* **0407**, 008 (2004). doi: 10.1088/1475-7516/2004/07/008.
156. P. Ullio, M. Kamionkowski, and P. Vogel, Spin dependent WIMPs in DAMA? *JHEP* **07**, 044 (2001). doi: 10.1088/1126-6708/2001/07/044.
157. K. Choi *et al.*, Search for neutrinos from annihilation of captured low-mass dark matter particles in the Sun by Super-Kamiokande, *Phys. Rev. Lett.* **114**(14), 141301 (2015). doi: 10.1103/PhysRevLett.114.141301.
158. M. Ageron *et al.*, ANTARES: the first undersea neutrino telescope, *Nucl. Instrum. Method.* **A656**, 11–38 (2011). doi: 10.1016/j.nima.2011.06.103.
159. C. Tönnis *et al.*, Overview of dark matter searches with ANTARES, *PoS.* **ICRC2015**, 1207 (2015).
160. C. Amole *et al.*, Dark matter search results from the PICO-2L C_3F_8 bubble chamber, *Phys. Rev. Lett.* **114**(23), 231302 (2015). doi: 10.1103/PhysRevLett.114.231302.
161. C. Savage, G. Gelmini, P. Gondolo, and K. Freese, Compatibility of DAMA/LIBRA dark matter detection with other searches, *JCAP* **0904**, 010 (2009). doi: 10.1088/1475-7516/2009/04/010.
162. C. E. Aalseth *et al.*, Search for an annual modulation in a P-type point contact germanium dark matter detector, *Phys. Rev. Lett.* **107**, 141301 (2011). doi: 10.1103/PhysRevLett.107.141301.
163. R. Agnese *et al.*, Silicon detector dark matter results from the final exposure of CDMS II, *Phys. Rev. Lett.* **111**(25), 251301 (2013). doi: 10.1103/PhysRevLett.111.251301.
164. M. G. Aartsen *et al.*, Measurement of atmospheric neutrino oscillations with IceCube, *Phys. Rev. Lett.* **111**(8), 081801 (2013). doi: 10.1103/PhysRevLett.111.081801.
165. G. L. Fogli, E. Lisi, A. Marrone, D. Montanino, A. Palazzo, and A. M. Rotunno, Global analysis of neutrino masses, mixings and phases: Entering the era of leptonic CP violation searches, *Phys. Rev.* **D86**, 013012 (2012). doi: 10.1103/PhysRevD.86.013012.
166. R. Abbasi *et al.*, IceCube sensitivity for low-energy neutrinos from nearby supernovae, *Astron. Astrophys.* **535**, A109 (2011). doi: 10.1051/0004-6361/201117810.
167. R. Abbasi *et al.*, Solar energetic particle spectrum on 13 December 2006 determined by IceTop, *Astrophys. J.* **689**, L65–L68 (2008). doi: 10.1086/595679.
168. R. C. Bay, R. A. Rohde, P. B. Price, and N. E. Bramall, South pole paleowind from automated synthesis of ice core records, *J. Geophys. Res.: Atmos.* **115**(D14), (2010). doi: 10.1029/2009JD013741. URL: http://dx.doi.org/10.1029/2009JD013741. D14126.

169. I. Kravchenko, S. Hussain, D. Seckel, D. Besson, E. Fensholt, J. Ralston, J. Taylor, K. Ratzlaff, and R. Young, Updated results from the RICE experiment and future prospects for ultra-high energy neutrino detection at the South Pole, *Phys. Rev.* **D85**, 062004 (2012). doi: 10.1103/PhysRevD.85.062004.
170. R. Abbasi *et al.*, Measurement of acoustic attenuation in South Pole ice, *Astropart. Phys.* **34**, 382–393 (2011). doi: 10.1016/j.astropartphys.2010.10.003.
171. P. Allison *et al.*, Design and initial performance of the Askaryan Radio Array prototype EeV neutrino Detector at the South Pole, *Astropart. Phys.* **35**, 457–477 (2012). doi: 10.1016/j.astropartphys.2011.11.010.
172. J. Cherwinka *et al.*, A search for the dark matter annual modulation in South Pole ice, *Astropart. Phys.* **35**, 749–754 (2012). doi: 10.1016/j.astropartphys.2012.03.003.
173. The IceCube-Gen2 Collaboration. PINGU LOI.
174. E. K. Akhmedov, S. Razzaque, and A. Yu. Smirnov, Mass hierarchy, 2-3 mixing and CP-phase with huge atmospheric neutrino detectors, *JHEP* **02**, 082 (2013). doi: 10.1007/JHEP02(2013)082,10.1007/JHEP07(2013)026. [Erratum: JHEP07,026(2013)].
175. M. G. Aartsen *et al.* The IceCube Neutrino Observatory Part VI: Ice Properties, Reconstruction and Future Developments. In *Proceedings, 33rd International Cosmic Ray Conference (ICRC2013) : Rio de Janeiro, Brazil, July 2-9, 2013* (2013). URL http://inspirehep.net/record/1255632/files/arXiv:1309.7010.pdf.
176. M. Aartsen *et al.*, IceCube-Gen2: A Vision for the Future of Neutrino Astronomy in Antarctica (2014).
177. P. Lipari, Perspectives of high energy neutrino astronomy, *Nucl. Instrum. Method.* **A567**, 405–417 (2006). doi: 10.1016/j.nima.2006.05.249.
178. J. K. Becker, A. Groß, K. Münich, J. Dreyer, W. Rhode, and P. L. Biermann, Astrophysical implications of high energy neutrino limits, *Astropart. Phys.* **28**, 98–118 (2007). doi: 10.1016/j.astropartphys.2007.04.007.
179. A. Silvestri and S. W. Barwick, Constraints on extragalactic point source flux from diffuse neutrino limits, *Phys. Rev.* **D81**, 023001 (2010). doi: 10.1103/PhysRevD.81.023001.
180. C. Patrignani *et al.*, Review of particle physics, *Chin. Phys.* **C40** (10), 100001 (2016). doi: 10.1088/1674-1137/40/10/100001.
181. S. Adrian-Martinez *et al.*, The first combined search for neutrino point-sources in the Southern Hemisphere with the Antares and Icecube neutrino telescopes, *Astrophys. J.* **823**(1), 65 (2016). doi: 10.3847/0004-637X/823/1/65.
182. P. Mertsch, M. Rameez, and I. Tamborra, Detection prospects for high energy neutrino sources from the anisotropic matter distribution in the local universe (2016).
183. B. Katz, E. Waxman, T. Thompson, and A. Loeb, The energy production rate density of cosmic rays in the local universe is $\sim 10^{44-45}$erg Mpc^{-3} yr^{-1} at all particle energies (2013).

184. E. Waxman and J. N. Bahcall, High-energy neutrinos from astrophysical sources: An Upper bound, *Phys. Rev.* **D59**, 023002 (1999). doi: 10.1103/PhysRevD.59.023002.
185. A. Loeb and E. Waxman, The cumulative background of high energy neutrinos from starburst galaxies, *JCAP* **0605**, 003 (2006). doi: 10.1088/1475-7516/2006/05/003.
186. T. Kashti and E. Waxman, Flavoring astrophysical neutrinos: Flavor ratios depend on energy, *Phys. Rev. Lett.* **95**, 181101 (2005). doi: 10.1103/PhysRevLett.95.181101.
187. M. Gonzalez-Garcia, M. Maltoni, J. Salvado, and T. Schwetz, Global fit to three neutrino mixing: Critical look at present precision, *JHEP* **1212**, 123 (2012). doi: 10.1007/JHEP12(2012)123.
188. O. Mena, S. Palomares-Ruiz, and A. C. Vincent, Flavor composition of the high-energy neutrino events in IceCube, *Phys. Rev. Lett.* **113**(9), 091103 (2014). doi: 10.1103/PhysRevLett.113.091103.
189. L. Fu and C. M. Ho, Flavor Ratios and Mass Hierarchy at Neutrino Telescopes (2014).
190. Flavor ratios of extragalactical neutrinos and neutrino shortcuts in extra dimensions (2014).
191. L. Fu, C. M. Ho, and T. J. Weiler, The Propagation Matrix and Flavor Triangle for Cosmic Neutrinos (2014).
192. X.-J. Xu, H.-J. He, and W. Rodejohann, Constraining astrophysical neutrino flavor composition from leptonic unitarity, *JCAP* **1412**, 039 (2014). doi: 10.1088/1475-7516/2014/12/039.
193. S. Palomares-Ruiz, O. Mena, and A. C. Vincent, On the flavor composition of the high-energy neutrinos in IceCube (2014).
194. A. Palladino, G. Pagliaroli, F. Villante, and F. Vissani, What is the flavor of the cosmic neutrinos seen by IceCube? *Phys. Rev. Lett.* **114**(17), 171101 (2015). doi: 10.1103/PhysRevLett.114.171101.
195. A. Palladino and F. Vissani, New parametrization of cosmic neutrino oscillations (2015).
196. L. A. Anchordoqui, H. Goldberg, F. Halzen, and T. J. Weiler, Neutrinos as a diagnostic of high energy astrophysical processes, *Phys. Lett.* **B621**, 18–21 (2005). doi: 10.1016/j.physletb.2005.06.056.
197. W. Winter, Photohadronic origin of the TeV-PeV neutrinos observed in IceCube, *Phys. Rev.* **D88**, 083007 (2013). doi: 10.1103/PhysRevD.88.083007.

Chapter 10

Sterile Neutrinos: An Introduction to Experiments

J. Conrad* and M. Shaevitz†

*Massachusetts Institute of Technology
Cambridge, MA 02139, USA
†Columbia University
New York, NY 10027, USA

1. Theoretical models with sterile neutrinos

Neutrinos are unique in the standard model (SM) in that they interact solely through the weak interaction. As a consequence, in the SM, only the left-handed neutrino is active and part of a weak isospin doublet with its partner charged lepton. A question then arises as to whether right-handed neutrinos exist and if they do, how do they fit into the SM? In the SM, right-handed neutrinos, if they exist, would be weak isospin singlets with no weak interactions except through mixing with the left-handed neutrinos. For this reason, the right-handed neutrinos are referred to as "sterile" neutrinos.

In fact, the sterile neutrino is more broadly defined as a neutral lepton with no ordinary weak interactions except those induced by mixing. So, the right-handed neutrino is a prime example, although not the only case. Note, however, that sterile neutrinos can possibly interact through Yukawa couplings with the Higgs boson or through other new physics interactions beyond the SM. So it is not necessarily true that sterile neutrinos have no interactions at all.

Explaining in detail how sterile neutrinos arise in theories is beyond the scope of this chapter. Here, we briefly review the basic theoretical ideas before embarking on the experimental questions related to sterile-neutrino searches, which is the primary goal of this text. Thus, this section is a bit technical, and students who are not already familiar with the theory of neutrino mass may wish to skip onward to the experimental discussion that follows. A useful tutorial on introducing neutrino mass appears in Ref. 1. For those who want to learn more about the theory than is discussed here, a good detailed review appears in Ref. 2.

In the SM, neutrinos are massless, since a Dirac mass term would require the existence of right-handed neutrinos. On the other hand, neutrino oscillations, which have been unambiguously observed,[3] require that the active neutrinos must have mass. Generating a nonzero neutrino mass can be accomplished by adding a mass term to the SM Lagrangian. A Dirac mass term similar to the charged fermion terms requires the existence of right-handed neutrinos. Then, a term with a right-handed neutrino combined with a left-handed neutrino and a SM Higgs field would produce a Dirac mass. To produce such a Dirac mass for the neutrino below the current upper limit from tritium decay studies of $2\,\mathrm{eV}^3$ would require a Yukawa coupling of order 10^{-12}, which would be very much smaller than that of the other couplings. Explanation of these small couplings has led to theoretical models where the neutrino mass terms involve higher dimensional operators, nonperturbative effects or warped extra dimensions.[2]

If neutrinos are Majorana particles, where the neutrino and antineutrino are the same particle species, then a Majorana mass term is also possible. Such a Majorana term would violate lepton number by two units bringing in the possibility of neutrino to antineutrino transitions. A Majorana term for a left-handed neutrino couples the left-handed neutrino with its conjugate and a Higgs triplet state, since such a transition would violate weak isospin by one unit. Instead of coupling to a Higgs triplet state, a Majorana term could also be constructed of higher-dimensioned terms involving a coupling of the left-handed neutrino to two Higgs doublets. Finally, the right-handed neutrino can also have a Majorana mass term where the neutrino couples with its conjugate since, as a singlet, this term does not violate weak isospin.

The combination of Dirac and Majorana mass terms for neutrinos can lead to an interesting scenario where the diagonalization of the mass matrix leads to active neutrinos with small masses of $(m_{\mathrm{Dirac}})^2/M_{\mathrm{heavy}}$ and also heavier sterile neutrinos that mix with the active neutrinos. This is referred to as the seesaw mechanism and may be the explanation for why neutrino masses are so small relative to the other fermions that have masses on the order of m_{Dirac}. On the other hand, the sterile neutrinos that are most naturally suggested by the seesaw mechanism are very heavy compared to all known particles. Nevertheless, light sterile neutrinos are a possibility within many models using the seesaw framework.[2]

These theoretical speculations about the existence of sterile neutrinos and how they might impact neutrino masses and mixing are key areas currently being investigated in particle physics. The experimental discovery of sterile neutrinos and their properties would have a major impact on these speculations and would likely lead to insights as to how to add them to the SM. In addition, sterile neutrinos can have a major impact on interpreting other neutrino measurements such as the search for CP violation in long-baseline experiments[4] and modeling of neutrino production in core-collapse supernovae.[5]

2. A brief tutorial on signals

Before discussing sterile-neutrino oscillation signals, it is useful to review how oscillation results are presented. We will consider this within a generic oscillation model between two flavors of neutrinos α and β. Then we will expand on this to discuss signals with three active flavors and one or more sterile neutrinos.

2.1. *Reviewing the basics of the two-neutrino oscillation example*

In a two-neutrino oscillation model, the flavor eigenstates (α/β subscripts) can be written as a function of the mass eigenstates (1/2 subscripts) as

$$\nu_\alpha = \cos\theta\ \nu_1 + \sin\theta\ \nu_2,$$
$$\nu_\beta = -\sin\theta\ \nu_1 + \cos\theta\ \nu_2,$$

where θ is the "mixing angle." The implication of mixing is that a pure flavor (weak) eigenstate born through a weak decay can oscillate into the other flavor as the state propagates in space. This oscillation is due to the fact that each of the mass eigenstate components propagates with a different frequency, assuming the masses are different, $\Delta m^2 = \left|m_2^2 - m_1^2\right| > 0$. Using quantum mechanics, one can calculate the two-neutrino oscillation probability for $\nu_\alpha \to \nu_\beta$ as

$$\text{Prob}\,(\nu_\mu \to \nu_e) = \sin^2 2\theta\ \sin^2\left(\frac{1.27\ \Delta m^2\ (\text{eV}^2)\ L\,(\text{km})}{E\,(\text{GeV})}\right). \tag{1}$$

In this equation, L is the distance from the source, and E is the neutrino energy. Examining this equation, one can see that the oscillation wavelength will depend upon L, E, and Δm^2. The amplitude will depend upon $\sin^2 2\theta$.

Looking at Eq. 1, one can see that the designer of an experiment would like to arrange for $1.27\Delta m^2 L/E = \pi/2$ because this maximizes the oscillation probability. The L/E value that maximizes this probability, $L/E = \pi/(2.54\Delta m^2)$, is called the "oscillation maximum" value. Of course, the probability function has a number of maxima but the first one usually dominates and others can be partially washed out due to energy and position smearing.

However, when one is searching for new physics, one does not necessarily know the exact Δm^2 of interest. Thus there is some art associated with choosing the detector location and the beam energy. The experiment must be designed with a relatively large value of the ratio L/E in order to enhance the $\sin^2(1.27\Delta m^2 L/E)$ term. However if L/E is too large in comparison to Δm^2, then oscillations occur rapidly. Because experiments have finite resolution on L and E, and a spread in beam energies, the $\sin^2(1.27\Delta m^2 L/E)$ term averages to 1/2 when $\Delta m^2 \gg L/E$ and one loses sensitivity to Δm^2. Lastly, since the oscillation probability is directly proportional to $\sin^2 2\theta$, if the mixing angle is likely to be small, then a design that results in high statistics is required to observe the tiny oscillation signal.

There are two types of oscillation searches: "disappearance" and "appearance." Consider a pure source of neutrinos of type α. In a disappearance experiment, one looks for a deficit in the expected flux of ν_α at a detector located downstream from the source, hence later in time. Another way to say this is, "If it started as a ν_α, did the neutrino stay a ν_α?" and thus, it is denoted as $\nu_\alpha \to \nu_\alpha$. On the other hand, appearance experiments search for one flavor turning into the other flavor, $\nu_\alpha \to \nu_\beta$, by directly observing interactions of neutrinos of type β. In either case, the signal for oscillations will be most persuasive if the deficit or excess has the (L/E) dependence predicted by the neutrino oscillation formula (Eq. 1).

Now consider how the results of neutrino oscillations can be presented. In this two-neutrino model, which was the first model for oscillations that was developed, there are two theoretical parameters (Δm^2 and the mixing angle) and so the results are traditionally shown on the plane of Δm^2 vs. $\sin^2 2\theta$. Let's say that a hypothetical, perfect (no systematic error) two-neutrino oscillation experiment sees no oscillation signal, based on N expected events without oscillations. The experimenters can rule out the probability for oscillations (either appearance or disappearance) within some error or confidence level (CL). A typical choice of confidence level is 90% CL, so in this case of statistical-uncertainty only, the approximate limiting probability is $P < 1.28\sqrt{N}/N$. The 1.28 factor corresponds to a one-sided Gaussian probability for the expected N value to have an upward fluctuation at the 10% probability level. More rigorous methods for setting CL regions will be described later with respect to global fitting techniques. There is only one measurement and there are two unknowns, so this translates to an excluded region within Δm^2–$\sin^2 2\theta$ space. This is indicated by a solid line, with the excluded region on the right. At high Δm^2, the limit on $\sin^2 2\theta$ is given by twice the above P value, since $\langle \sin^2(1.27\Delta m^2 L/E)\rangle$ averages to 1/2. The L and E of the experiment drive the low Δm^2 limit, and this only improves by the fourth root of the statistics associated with N. Thus to improve reach in the mixing angle, the designer must increase the total number of events, while to improve the reach in low Δm^2, the designer should focus on adjusting the L/E ratio.

If an appearance experiment measures an excess of ν_β events, N_{excess}, with some uncertainty, δN_{excess}, in a $\nu_\alpha \rightarrow \nu_\beta$ search, then the experiment can claim a signal if the excess is greater than its uncertainty. (This is also true for a disappearance signal where a significant deficit in events, N_{deficit}, is observed.) This observed signal can be converted into an average oscillation probability using the number of fully transmuted ν_α events, $N_{\text{fulltrans}}$. $N_{\text{fulltrans}}$ is the number of ν_β events that would be observed in the experiment if all of the ν_α neutrinos changed into ν_β neutrinos. To calculate $N_{\text{fulltrans}}$, one has to take the flux of ν_α neutrinos into the experiment and have them interact with the ν_β cross section and detection efficiency. The average oscillation probability is then given by $P \pm \delta P = N_{\text{excess}}/N_{\text{fulltrans}} \pm \delta N_{\text{excess}}/N_{\text{fulltrans}}$. This type of average oscillation probability is often given as a measure of an experiment's signal with a significance of $N\sigma$ where $N = P/\delta P$.

In reality, imperfections of an experiment affect the limits which can be set. Background sources may introduce ν_β into the ν_α beam. Misidentification of the interacting neutrino flavor in the detector may mimic oscillation signatures. In addition, systematic uncertainties in the relative acceptance versus distance and energy must be understood and included in the analysis of the data. These systematics are included in the 90% CL excluded regions presented by the experiments.

Excluded and allowed regions are determined using a test statistic that depends on the likelihood for obtaining a given data set depending on the assumed oscillation parameters. One then determines the allowed oscillation parameter space where the likelihood is greater than the desired CL. Since the oscillation probability in Eq. 1 is nonlinear in the oscillation parameters, determining the critical value for the test statistic at a given CL usually requires assumptions and studies of simulated data samples.

A common example of a test statistic for a Δm^2 vs. $\sin^2 2\theta$ point is the $\Delta\chi^2 = \chi^2_{\Delta m^2/\sin^2 2\theta} - \chi^2_{\text{BestFit}}$. This statistic can be shown in a simple Gaussian error case to follow a χ^2 distribution with the degrees of freedom (dof) equal to the number of fit parameters, which for this case would be 2 corresponding to Δm^2 and $\sin^2 2\theta$. In reality due to the nonlinear nature of the oscillation probability, the limiting value of $\Delta\chi^2$ for a given CL needs to be determined using simulated data sets. For example, at high Δm^2, the value of Δm^2 does not affect the oscillation probability and, in this region, the $\Delta\chi^2$ test statistic will follow a χ^2 distribution with 1 dof.

The "sensitivity" of an experiment is defined as the average expected limit if the experiment were performed many times with no true signal (only background). This leads to an important complication. When the actual background is low compared to prediction there can be a significant difference between the limit and the experimental sensitivity. In some cases, this is due to statistical fluctuation of the background. In this case, the experiment just "got lucky" and the limit represents the true confidence level. On the other hand, if the true background is due to a systematic effect that leads to an overprediction, then the extracted limit is not representative of the true space explored. To address this concern, in an influential paper on statistics for oscillation experiments, Feldman and Cousins[6] suggested that when an experiment sets a significantly better limit than the sensitivity, the experiment should also indicate the sensitivity on the plot. This allows readers to draw conclusions based upon their own opinion of what is acceptable. An improved approach was recently introduced by the LHC experiments for their electroweak and Higgs searches. A band is defined around the line indicating the sensitivity that encompasses a given percentage, for example 90%, of the simulations with no true signal. One would then expect the result of an experiment to largely lie within this band, at 90% CL. The result may cross from one side to the other side of the band, and may even depart from the band for some best-fit parameters. But overall, the result should be consistent with the band. A result that lies significantly outside of the band, beyond the expectation for the stated CL of the band, indicates either a problematic null result or a signal.

We emphasize that limits are not "hard cutoffs." There is some probability that there is a solution outside of the line defined by the limit. However, that signal will have lower confidence level than the confidence level of the cutoff. This will be an important point below, when we compare results from multiple experiments.

2.2. *Two examples of results interpreted within the two neutrino model*

The above discussion was fairly abstract, so let's now consider two real cases of results from oscillation experiments. The first is an experiment searching

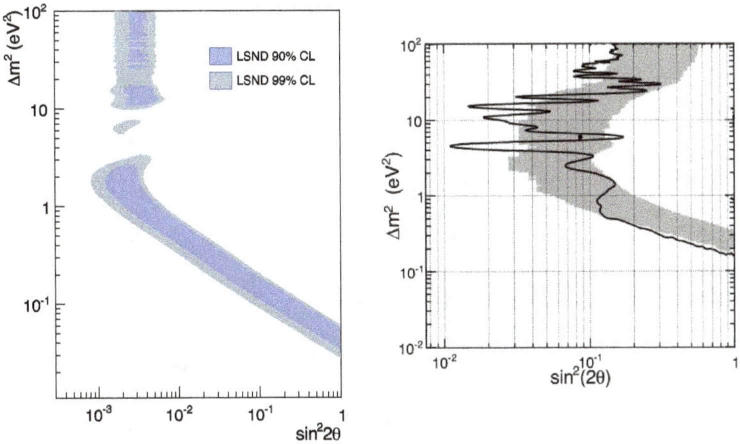

Fig. 1. Two examples of results from oscillation experiments. Left: The LSND allowed region for $\bar{\nu}_\mu \to \bar{\nu}_e$ appearance. The shaded regions indicate 90% (blue) and 99% (gray) CL allowed regions.[7] Right: The SciBooNE/MiniBooNE 90% CL limit for $\bar{\nu}_\mu \to \bar{\nu}_\mu$ disappearance.[8] The region to the right of the solid line is excluded. The shaded area indicates the region where 68% of experiments with no observed signal are predicted to lie, and the sensitivity of the experiment is defined to be the center of the shaded band.

for a $\bar{\nu}_\mu \to \bar{\nu}_e$ appearance signal, the Liquid Scintillator Neutrino Detector (LSND).[7] The second is a pair of detectors, SciBooNE and MiniBooNE, used to search for a disappearance signal of muon antineutrinos, $\bar{\nu}_\mu \to \bar{\nu}_\mu$.[8] The two results are shown in Fig. 1, and are among a large set of experiments we will discuss.

The LSND $\bar{\nu}_\mu \to \bar{\nu}_e$ appearance result was the first of a set of unexpected signals observed at $\Delta m^2 \sim 1\,\text{eV}^2$, and has a significance of 3.8σ. This experiment ran at the Los Alamos LAMPF accelerator National Laboratory from 1993 and 1998. The decay-at-rest (DAR) beam was produced by impinging 800 MeV protons on a beam dump, resulting in π^+s which stop and decay to produce μ^+s, which, in turn, stop and decay to produce $\bar{\nu}_\mu$ and ν_e. Because the π^-s produced in the dump capture, the beam has a $<8\times10^{-4}$ contamination of $\bar{\nu}_e$. The resulting neutrinos were observed in a detector located 30 m downstream of the beam dump. Events with neutrino energies between 20 and 52 MeV were used in the analysis. In the detector, 1220 phototubes surrounded a cylindrical detector that was filled with 167 tons of mineral oil, lightly doped with scintillator. The signature of $\bar{\nu}_e$ appearance was $\bar{\nu}_e + p \to e^+ + n$, a process called "inverse beta decay." This resulted in a two-component signature in the detector: (1) the initial Cherenkov and scintillation light associated with the e^+ and (2) the later scintillation light from the n capture on hydrogen which produces a 2.2 MeV γ. The experiment observed $87.9 \pm 22.4 \pm 6.0$ events[7] above expectation. A plot of

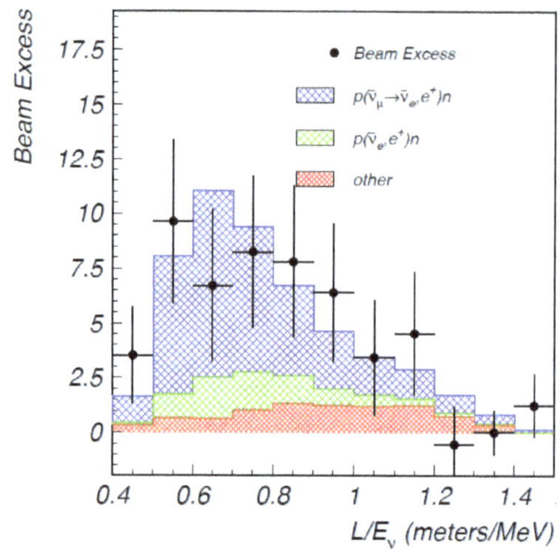

Fig. 2. The LSND excess,[7] indicating possible $\bar{\nu}_\mu \to \bar{\nu}_e$ appearance. The data points have statistical and systematic error added in quadrature. The signal is not background subtracted. The red and green histograms indicate the background from intrinsic $\bar{\nu}_e$ in the beam and from other sources, respectively. The blue histogram is an example of a potential $\bar{\nu}_\mu \to \bar{\nu}_e$ signal that, when added to the predicted backgrounds, is representative of the data. This potential signal corresponds to one point in the allowed region in Δm^2 and $\sin^2 2\theta$.

the data, shown with a stacked plot of background (green and red) and fitted signal (blue), is shown in Fig. 2.

For LSND, the ratio of the distance to the detector to the beam energy, L/E, is approximately 0.8 m/MeV. Thus, from the two-generation oscillation formula, Eq. 1, one can see that this experiment is approximately sensitive to $\Delta m^2 \geq 0.1\,\text{eV}^2$ (this is also seen in Fig. 1). The oscillation probability measured by LSND for the excess was $(2.64 \pm 0.67 \pm 0.45) \times 10^{-3}$. The resulting allowed regions are shown in Fig. 1, left. This plot shows "allowed regions," which are fully enclosed contours, at 90% and 99% CL. Below about $1\,\text{eV}^2$, one can see an extended contour that has the typical slope of -0.5 on the log–log plot corresponding to the region where the oscillatory sine factor that depends on Δm^2 in Eq. 1 has a small argument. This is the range where the experiment is sensitive to the small oscillation probabilities before the first oscillation maximum. The small Δm^2 limit at $\sin^2 2\theta = 1.0$ is given by

$$\Delta m^2 = \frac{\sqrt{\text{Prob}}}{1.27L/E}, \tag{2}$$

which for the above LSND probability gives $0.05\,\text{eV}^2$. Above $1\,\text{eV}^2$, one sees that the LSND result forms islands and exclusion regions because of the oscillatory behavior associated with Eq. 1. At high Δm^2, one reaches the region of rapid oscillations where the oscillatory factor averages to 0.5. The pattern of this LSND allowed region is typical for a single detector experiment combined with a neutrino source that has a small energy spread.

While LSND was a single detector experiment with an allowed region, the SciBooNE/MiniBooNE $\bar{\nu}_\mu \to \bar{\nu}_\mu$ shown in Fig. 1, right, shows typical features of a two-detector experiment where the result is an exclusion. The two-detector design is often used in cases where the flux is rather poorly predicted, which is often the case for decay-in-flight (DIF) beams. To produce DIF beams, accelerated protons are directed onto a target where pions and kaons are produced. The charged pions and kaons are focused magnetically into a gas- or vacuum-filled decay pipe where the mesons decay into muons and muon neutrinos. Generally, this type of beam is mainly composed of muon neutrinos with a small contamination (0.5% to a few percent) of electron neutrinos from muon and kaon decay.

The SciBooNE experiment has a 10.6 t, scintillator-strip neutrino detector located 100 m from the primary proton target that produced the beam. The fine segmentation allowed the position of the neutrino interaction vertex to be well identified. Located at 540 m from the proton target was a second detector, the 800 t MiniBooNE mineral-oil based neutrino detector. In the case of MiniBooNE, events were detected using mainly Cherenkov light associated with the outgoing particles, since the oil was undoped with scintillator. The Cherenkov light formed rings on the 1280 photomultiplier tubes around the periphery of the spherical detector. The amount of light and Cherenkov ring geometry were used to measure the particle energies and direction as well as the particle type: muon, pion, or electron. The combination of MiniBooNE with SciBooNE allows a much more precise search for neutrino disappearance than MiniBooNE could perform independently, since the neutrino flux and energy distribution can be measured in the near SciBooNE detector before oscillations occur. Many systematic uncertainties associated with the neutrino flux and interaction cross sections are reduced or eliminated by comparing a near and far detector. Detector effects are most reduced if the two detectors use the identical detection technology; however, this was not the case for the SciBooNE/MiniBooNE experiment. For practical reasons, it is often true that the near and far detectors are constructed with different technologies.

Fig. 1, right, shows the results for the SciBooNE/MiniBooNE $\bar{\nu}_\mu$ disappearance search. The near and far detectors saw consistent results, so no sign of oscillations was detected. The experimental result is, therefore, a limit at 90% CL as shown in the figure. Oscillations are excluded in the region to the right of the black line. This differs from the LSND example in that there is no fully enclosed space within the parameters, which is to say that the null result, which is off the plot to the left, is allowed. The band in this figure is indicating the expectation if one runs many experiments, as discussed above. This band encompasses 90% of the limits obtained from a collection of simulated data runs. The average of these experimental results, along the center of the band, is the sensitivity (not explicitly shown in this plot). The obtained limit is consistent with the estimated sensitivity and shows classic expected behavior: most of the limit is contained within the band, with several crossings of the sensitivity, depending on the parameter values. Within the statistics, one expects the limit to exit the band for a few parameter values, and this is observed.

For $\Delta m^2 < 1\,\text{eV}^2$, the MiniBooNE/SciBooNE limit curve shows the same characteristics as seen in the LSND result. For this measurement, the disappearance probability limit is at about the 5% level, which would predict a $\Delta m^2 = 0.22\,\text{eV}^2$ at $\sin^2 2\theta = 1.0$. Above $\Delta m^2 = 1\,\text{eV}^2$, the limit shows a complicated curve associated with the oscillatory behavior of Eq. 1 for detectors at two different L values and with uncorrelated statistical fluctuations. At high Δm^2, in this case above $20\,\text{eV}^2$, there are rapid oscillations in both the near and far detectors and the sensitivity grows worse with higher Δm^2 since the two-detector comparison no longer helps. On the other hand, limits can still be set in this region since the absolute neutrino flux is known to some accuracy from the beam and cross section modeling, it is just that the limit is less stringent due to the larger systematic error.

2.3. *Why go beyond a three neutrino model?*

The LSND result shown in Fig. 1, left, is consistent with oscillations with $\Delta m^2 \gtrsim 0.1$ eV^2. This is substantially larger than the Δm^2 values that have been measured for the "solar" and "atmospheric" oscillations: $\Delta m^2_{\text{solar}} = 7.5 \times 10^{-5}\,\text{eV}^2$ and $\Delta m^2_{\text{atmospheric}} = 2.3 \times 10^{-3}\,\text{eV}^2$.[9] The latter are two well-established signals, with significance much higher than 5σ, observed in many experiments and celebrated with the 2015 Nobel Prize in physics. As we discuss in this section, the solar and atmospheric oscillations are consistent with a model of three-neutrinos. We show here that a three-neutrino model cannot explain a third Δm^2 value. As a result, LSND, and other experiments that show oscillations with $\Delta m^2 \sim 1\,\text{eV}^2$ are said to have "anomalous results."

Let's begin by considering only the solar and atmospheric oscillation results. A model that incorporates two distinct solar and atmospheric Δm^2

is consistent with three neutrinos, with mixing between the mass and flavor eigenstates given by

$$\begin{pmatrix} \nu_e \\ \nu_\mu \\ \nu_\tau \end{pmatrix} = \begin{pmatrix} U_{e1} & U_{e2} & U_{e3} \\ U_{\mu 1} & U_{\mu 2} & U_{\mu 3} \\ U_{\tau 1} & U_{\tau 2} & U_{\tau 3} \end{pmatrix} \begin{pmatrix} \nu_1 \\ \nu_2 \\ \nu_3 \end{pmatrix}.$$

The oscillation probability is then

$$\mathrm{Prob}\,(\nu_\alpha \to \nu_\beta) = \delta_{\alpha\beta} - 4 \sum_{j>i} U_{\alpha i} U_{\beta i} U^*_{\alpha\, j} U^*_{\beta\, j} \sin^2 \left(\frac{1.27\ \Delta m^2_{i\,j}\, L}{E} \right) \tag{3}$$

where $\Delta m^2_{i\,j} = \left| m_i^2 - m_j^2 \right|$. Note that there are three different Δm^2 values, but only two are independent. In this model, $\Delta m^2_{12} + \Delta m^2_{23} = \Delta m^2_{13}$. Thus the LSND result cannot be introduced into this model.

This was a slightly simplistic argument. The more general case allows the atmospheric result to be due to a mixture of high (LSND-range) and low (solar-range) Δm^2 values. But global fits have shown that this model disagrees with several data sets.[10] So this is ruled out.

2.4. *Introducing a sterile neutrino: the 3 + 1 models*

As a result, one is forced to introduce new physics if one accepts that LSND and the other consistent but less significant anomalies, which we will discuss in Section 3, are due to oscillations. The simplest next step is to introduce one additional neutrino mass state, connected to one additional flavor state through the unitary mixing matrix:

$$U_{3+1} = \begin{bmatrix} U_{e1} & U_{e2} & U_{e3} & U_{e4} \\ \vdots & & \vdots & U_{\mu 4} \\ \vdots & & \vdots & U_{\tau 4} \\ U_{s1} & U_{s2} & U_{s3} & U_{s4} \end{bmatrix}, \tag{4}$$

where the new flavor is denoted with "s" in the matrix. This explicitly allows the new flavor to mix with the three known neutrino flavors and thereby be produced in neutrino oscillations. Figure 3, left, shows a common illustration of the four mass states (bars) divided up into their four flavor components (colors). Because there are three states that are relatively similar in mass in this model and one mass state that is larger, this is called a $3+1$ model.

Introducing a new neutrino flavor into the theory has important consequences outside of oscillation studies. First, the precise measurements of

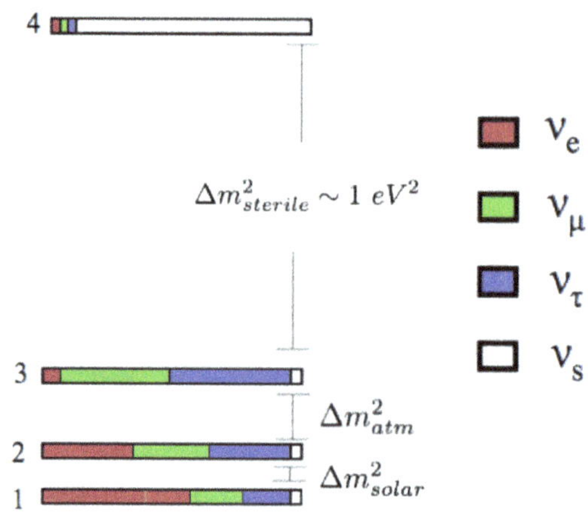

Fig. 3. Commonly used illustration of four mass states and their relationship to four flavor components.

the Z^0 width[3] made at LEP and SLD determined that there are only three families of light-mass, weakly interacting neutrinos. Hence, the additional neutrino, if it is lighter than half the mass of the Z, must not interact via the exchange of W or Z bosons and is, therefore, sterile. Second, we must address the fact that sterile neutrinos would be produced in the early universe through oscillations, thus affecting cosmological fits. Cosmological results are still not precise enough to make strong statements, although recent cosmological fits allow for a single noninteracting sterile neutrino.[11] However, it is also possible that, while sterile neutrinos may not interact weakly, they could have interactions through some new weaker-than-weak interactions that are yet to be observed. In such models, the sterile-neutrinos may not thermalize in the early universe, and so cosmological measurements will be insensitive to a sterile-neutrino signal.[2] Third, the addition of the fourth mass state will affect β decay. A few-eV sterile neutrino will produce a kink in the β decay electron spectrum, since the electron flavor state will have a small admixture associated with the high mass sterile state. The upcoming KATRIN experiment could see a kink in the spectrum for masses greater than a few eV and mixings at the 0.1 level.[12] Higher mass sterile neutrinos in the keV mass range can also affect the β decay spectrum away from the endpoint and sensitivity to small mixings in the keV spectral range may also be possible.[13] Recent astrophysical measurements have shown hints of keV mass objects leading to renewed interest in these higher mass sterile neutrinos.[14]

In this discussion, we will focus on neutrino oscillations. This is the most immediately promising approach to observing evidence for sterile neutrinos, which show up through oscillations with frequency associated with the fourth mass state. Following the arguments of Section 2.1, the optimal L/E to study the additional fourth mass state indicated by LSND is of order $1\,\text{m/MeV}$, while for atmospheric oscillations, it is $1000\,\text{km/GeV}$ and for solar $10\,000\,\text{km/GeV}$. Since most neutrino beams are primarily $E < 5\,\text{GeV}$, this implies an $L < 5\,\text{km}$ is required for studies of this fourth mass state, which is a much shorter baseline than what is called for to study atmospheric and solar oscillations. If one designs the experiment with $\mathcal{O}(1\ \text{m/MeV})$, then the sensitivities to $\Delta m^2_{\text{solar}}$ and $\Delta m^2_{\text{atmospheric}}$ are small. One can make the approximation that $\Delta m^2_{\text{solar}} = \Delta m^2_{\text{atmospheric}} = 0$. This is called the "short baseline (SBL) approximation". Invoking this approximation, we usually call the LSND-related mass splitting Δm^2_{41}.

Disappearance of an active flavor to a sterile flavor at large Δm^2 is direct evidence for these neutrinos. Active-to-active appearance oscillations with a frequency associated with the third Δm^2_{41} splitting is also possible. There are nine possible oscillations that can be observed with this oscillation frequency: $P_{\nu_e\to\nu_e}$, $P_{\nu_\mu\to\nu_\mu}$, $P_{\nu_\tau\to\nu_\tau}$, $P_{\nu_e\to\nu_\mu}$, $P_{\nu_\mu\to\nu_e}$, $P_{\nu_e\to\nu_\tau}$, $P_{\nu_\tau\to\nu_e}$, $P_{\nu_\mu\to\nu_\tau}$ and $P_{\nu_\tau\to\nu_\mu}$. All of these processes must occur with the same oscillation frequency for the model to be consistent.

The probabilities for disappearance and appearance oscillations are given by

$$P(\nu_\alpha \to \nu_\beta) \simeq 4|U_{\alpha 4}|^2|U_{\beta 4}|^2 \sin^2(1.27\Delta m^2_{41}L/E) \tag{5}$$

and

$$P(\nu_\alpha \to \nu_\alpha) \simeq 1 - 4(1-|U_{\alpha 4}|^2)|U_{\alpha 4}|^2 \sin^2(1.27\Delta m^2_{41}L/E). \tag{6}$$

Here we have used $\simeq$ for the relationships to explicitly note that we are employing the SBL approximation and so have dropped the small terms depending on the atmospheric and solar Δm^2 parameters. Throughout the remaining text, we will just call this an equality.

From this, one can see that there are triplets of experiment types, happening with the same oscillation frequency, that depend on pairs of the matrix elements, for example, the set:

$$P_{\nu_e\to\nu_e} = 1 - 4(1-|U_{e4}|^2)|U_{e4}|^2 \sin^2(1.27\Delta m^2_{41}L/E), \tag{7}$$

$$P_{\nu_\mu\to\nu_\mu} = 1 - 4(1-|U_{\mu 4}|^2)|U_{\mu 4}|^2 \sin^2(1.27\Delta m^2_{41}L/E), \tag{8}$$

$$P_{\nu_\mu\to\nu_e} = 4|U_{e4}|^2|U_{\mu 4}|^2 \sin^2(1.27\Delta m^2_{41}L/E). \tag{9}$$

A consistent theory model requires signals in all three of these oscillation modes. As we will see in Section 3, this is a serious problem for 3 + 1 models at present.

The above equations are often written to mimic the form of Eq. 1, for example,

$$P_{\nu_e \to \nu_e} = 1 - \sin^2 2\theta_{ee} \sin^2(1.27\Delta m_{41}^2 L/E), \tag{10}$$
$$P_{\nu_\mu \to \nu_\mu} = 1 - \sin^2 2\theta_{\mu\mu} \sin^2(1.27\Delta m_{42}^2 L/E), \tag{11}$$
$$P_{\nu_\mu \to \nu_e} = \sin^2 2\theta_{e\mu} \sin^2(1.27\Delta m_{41}^2 L/E), \tag{12}$$

where comparison to Eqs. 7, 8 and 9 results in the definitions:

$$\sin^2 2\theta_{ee} = 4(1 - |U_{e4}|^2)|U_{e4}|^2, \tag{13}$$
$$\sin^2 2\theta_{\mu\mu} = 4(1 - |U_{\mu4}|^2)|U_{\mu4}|^2, \tag{14}$$
$$\sin^2 2\theta_{e\mu} = 4|U_{e4}|^2|U_{\mu4}|^2. \tag{15}$$

A similar triplet of equations exists for the other flavor pairings.

The U matrix can be thought of as a rotation matrix. This leads to yet another notation that one finds in the literature. The following definitions of oscillation probabilities can arise from using the rotation angles:

$$P_{\nu_e \to \nu_e} \simeq 1 - \sin^2(2\theta_{14}) \sin^2(1.27\Delta m_{41}^2 L/E), \tag{16}$$
$$P_{\nu_\mu \to \nu_\mu} \simeq 1 - \sin^2(2\theta_{24}) \sin^2(1.27\Delta m_{41}^2 L/E), \tag{17}$$
$$P_{\nu_\mu \to \nu_e} \simeq \frac{1}{4} \sin^2(2\theta_{14}) \sin^2(2\theta_{24}) \sin^2(1.27\Delta m_{41}^2 L/E), \tag{18}$$

and similar definitions follow for the other flavor pairs.

The above presents many definitions of the same probabilities. We discuss these because all of these definitions are used in the literature. One must take care not to confuse axes presenting mixing in one parameter space with another when comparing published results. Another common source of confusion is between plots presented in $\sin^2 2\theta$ and $\sin^2\theta$. The reader must always look carefully at the choice of axis representing the mixing.

Invoking CPT as a good symmetry places an important requirement on any given oscillation triplet. CPT states that $P(\nu_\alpha \to \nu_\beta) = P(\bar{\nu}_\beta \to \bar{\nu}_\alpha)$. Consider the case where α and β are both of electron flavor. Then $P(\nu_e \to \nu_e) = P(\bar{\nu}_e \to \bar{\nu}_e)$. Similarly for all flavors, the oscillation probability for neutrinos and antineutrinos must be the same. In this case,

(1) CPT invariance requires that $\bar{\nu}_\mu \to \bar{\nu}_e$ oscillations and $\nu_e \to \nu_\mu$ oscillations must be identical.

(2) The probability for ν_e disappearance must be as large or larger than the probability for $\nu_e \to \nu_\mu$ oscillations, since this represents only one oscillation channel.
(3) CPT invariance requires that the probability of $\bar{\nu}_e$ disappearance be the same as the probability for ν_e disappearance as argued above.

This then leads to constraints on the relation between appearance and disappearance measurements. For example, if the sensitive region of a $\bar{\nu}_e$ disappearance experiment entirely covers the allowed parameter space associated with another experiment's $\bar{\nu}_\mu \to \bar{\nu}_e$ appearance signal region, then either the disappearance experiment must see a signal or the two experiments are incompatible. This method of testing appearance signals will represent an interesting line of attack for studying sterile-neutrino models as discussed later.

Lastly, it is important to point out that if one invokes the SBL approximation, then there will be no CP violation in the $3+1$ model. CP violation would introduce a difference in the neutrino and antineutrino appearance probabilities in a given channel. CP-violating terms come about as an interference between two or more oscillation frequencies; hence two or more Δm^2 values that are relatively close in magnitude are required for an effect to be observed. In a $3+1$ model that makes use of the SBL approximation, there is only one nonzero Δm^2 and, thus, no CP violation in the model.

2.5. *3 + 2 and 3 + 3*

While the minimal extension to a three neutrino model is to add a single sterile neutrino, in principle, one might expect three sterile neutrinos, reflecting the three-generation structure of the SM. This is called a $3+3$ model. In practice, however, experiments may only have sensitivity to two of the three new neutrinos as we discuss in Section 4. In this case, we must fit to a $3+2$ model.

A $3+2$ model, which is a five-mass-state model, assumes the SBL approximation for the three lowest mass states, and two distinct mass splittings, Δm^2_{41} and Δm^2_{51}. Note that $\Delta m^2_{54} = \Delta m^2_{51} - \Delta m^2_{41}$. The corresponding $3+2$ mixing matrix is an extension of Eq. 4 to a 5×5 matrix. To simplify the appearance equation below, we will use notation where

$$|U_{\alpha i\beta j}| = |U_{\alpha i}||U_{\beta j}|, \tag{19}$$

where i and j refer to the mass states, and

$$\Delta_{ij} = \Delta m^2_{ij}. \tag{20}$$

For a $3+2$ oscillation model, the appearance probability is then given by

$$\begin{aligned} P^{3+2}_{\nu_\alpha \to \nu_\beta} \simeq & -4|U_{\alpha 5 \beta 5}||U_{\alpha 4 \beta 4}| \cos\phi_{54} \sin^2(1.27\Delta_{54} L/E) \\ & + 4(|U_{\alpha 4 \beta 4}| + |U_{\alpha 5 \beta 5}| \cos\phi_{54})|U_{\alpha 4 \beta 4}| \sin^2(1.27\Delta_{41} L/E) \\ & + 4(|U_{\alpha 4 \beta 4}| \cos\phi_{54} + |U_{\alpha 5 \beta 5}|)|U_{\alpha 5 \beta 5}| \sin^2(1.27\Delta_{51} L/E) \\ & + 2|U_{\beta 5 \alpha 5}||U_{\beta 4 \alpha 4}| \sin\phi_{54} \sin(2.53\Delta_{54} L/E) \\ & + 2(|U_{\alpha 5 \beta 5}| \sin\phi_{54})|U_{\alpha 4 \beta 4}| \sin(2.53\Delta_{41} L/E) \\ & + 2(-|U_{\alpha 4 \beta 4}| \sin\phi_{54})|U_{\alpha 5 \beta 5}| \sin(2.53\Delta_{51} L/E). \end{aligned} \tag{21}$$

In this equation, ϕ is a CP phase is given by

$$\phi_{54} = \arg(U_{e5} U^*_{\mu 5} U^*_{e4} U_{\mu 4}). \tag{22}$$

The terms which depend on $\cos\phi_{54}$ are CP conserving since they do not change sign in going from neutrino to antineutrino oscillations, while those that depend on $\sin\phi_{54}$ are CP violating. CP effects only arise in appearance experiments.

The $3+2$ disappearance probability, which is simpler and so we will not use abbreviations, has no dependence on ϕ_{54}:

$$\begin{aligned} P^{3+2}_{\nu_\alpha \to \nu_\alpha} \simeq & 1 - 4|U_{\alpha 4}|^2 |U_{\alpha 5}|^2 \sin^2(1.27\Delta m^2_{54} L/E) \\ & - 4(1 - |U_{\alpha 4}|^2 - |U_{\alpha 5}|^2)(|U_{\alpha 4}|^2 \sin^2(1.27\Delta m^2_{41} L/E) \\ & + |U_{\alpha 5}|^2 \sin^2(1.27\Delta m^2_{51} L/E)). \end{aligned} \tag{23}$$

In other words, disappearance experiments have no CP violating effects. Only appearance experiments can potentially demonstrate CP violation.

In the $\sim$1 eV^2 region, SBL experiments do not constrain $|U_{\tau 4}|$ because the neutrino energy required for the optimal L/E is well below the τ production threshold of 3.48 GeV for the incident neutrino energy needed to produce a τ lepton. Therefore, the global fits to the SBL data sets are constraining Δm^2_{41}, $|U_{e4}|$, $|U_{\mu 4}|$, Δm^2_{51}, $|U_{e5}|$, $|U_{\mu 5}|$, and the CP parameter ϕ_{54}.

The $3+2$ model has introduced more flexibility into the global fits in two ways. The first is to introduce more parameters. The second is through allowing for CP violation, which makes the appearance oscillation probabilities for neutrinos and antineutrinos have different dependencies on the mixing matrix elements. In other words, $P_{\nu_\mu \to \nu_e}$ can be different from $P_{\bar{\nu}_\mu \to \bar{\nu}_e}$ even with the same mixing matrix.

One can take the next step, expanding to a $3+3$ model, with a 6×6 mixing matrix. The appearance probability, expressed using the notation in Eqs. 19 and 20, is then given by:

$$
\begin{aligned}
P^{3+3}_{\nu_\alpha \to \nu_\beta} \simeq & -4|U_{\alpha 5\beta 5}||U_{\alpha 4\beta 4}|\cos\phi_{54}\sin^2(1.27\Delta_{54}L/E) \\
& -4|U_{\alpha 6\beta 6}||U_{\alpha 4\beta 4}|\cos\phi_{64}\sin^2(1.27\Delta_{64}L/E) \\
& -4|U_{\alpha 5\beta 5}||U_{\alpha 6\beta 6}|\cos\phi_{65}\sin^2(1.27\Delta_{65}L/E) \\
& +4(|U_{\alpha 4\beta 4}| + |U_{\alpha 5\beta 5}|\cos\phi_{54} + |U_{\alpha 6\beta 6} \\
& \times|\cos\phi_{64})|U_{\alpha 4\beta 4}|\sin^2(1.27\Delta_{41}L/E) \\
& +4(|U_{\alpha 4\beta 4}|\cos\phi_{54} + |U_{\alpha 5\beta 5}| + |U_{\alpha 6\beta 6}|\cos\phi_{65}) \\
& \times|U_{\alpha 5\beta 5}|\sin^2(1.27\Delta_{51}L/E) \\
& +4(|U_{\alpha 4\beta 4}|\cos\phi_{64} + |U_{\alpha 5\beta 5}|\cos\phi_{65} + |U_{\alpha 6\beta 6}|) \\
& \times|U_{\alpha 6\beta 6}|\sin^2(1.27\Delta_{61}L/E) \\
& +2|U_{\beta 5\alpha 5}||U_{\beta 4\alpha 4}|\sin\phi_{54}\sin(2.53\Delta_{54}L/E) \\
& +2|U_{\beta 6\alpha 6}||U_{\beta 4\alpha 4}|\sin\phi_{64}\sin(2.53\Delta_{64}L/E) \\
& +2|U_{\beta 6\alpha 6}||U_{\beta 5\alpha 5}|\sin\phi_{65}\sin(2.53\Delta_{65}L/E) \\
& +2(|U_{\alpha 5\beta 5}|\sin\phi_{54} + |U_{\alpha 6\beta 6}|\sin\phi_{64})|U_{\alpha 4\beta 4}|\sin(2.53\Delta_{41}L/E) \\
& +2(-|U_{\alpha 4\beta 4}|\sin\phi_{54} + |U_{\alpha 6\beta 6}|\sin\phi_{65})|U_{\alpha 5\beta 5}|\sin(2.53\Delta_{51}L/E) \\
& +2(-|U_{\alpha 4\beta 4}|\sin\phi_{64} - |U_{\alpha 5\beta 5}|\sin\phi_{65})|U_{\alpha 6\beta 6}|\sin(2.53\Delta_{61}L/E).
\end{aligned} \tag{24}
$$

The disappearance equation is much simpler, and so, without abbreviations is:

$$
\begin{aligned}
P^{3+3}_{\nu_\alpha \to \nu_\alpha} \simeq & \, 1 - 4|U_{\alpha 4}|^2|U_{\alpha 5}|^2\sin^2(1.27\Delta m^2_{54}L/E) \\
& -4|U_{\alpha 4}|^2|U_{\alpha 6}|^2\sin^2(1.27\Delta m^2_{64}L/E) - 4|U_{\alpha 5}|^2|U_{\alpha 6}|^2 \\
& \times\sin^2(1.27\Delta m^2_{65}L/E) \\
& -4(1 - |U_{\alpha 4}|^2 - |U_{\alpha 5}|^2 - |U_{\alpha 6}|^2)(|U_{\alpha 4}|^2\sin^2(1.27\Delta m^2_{41}L/E) \\
& +|U_{\alpha 5}|^2\sin^2(1.27\Delta m^2_{51}) + |U_{\alpha 6}|^2\sin^2(1.27\Delta m^2_{61}L/E)).
\end{aligned} \tag{25}
$$

The important point here is that, again, CP phases are only in the appearance formula. With the introduction of another mass splitting, three

CP-violating phases arise:

$$\phi_{54} = \arg(U_{e5}U^*_{\mu5}U^*_{e4}U_{\mu4}), \tag{26}$$

$$\phi_{64} = \arg(U_{e6}U^*_{\mu6}U^*_{e4}U_{\mu4}), \tag{27}$$

and

$$\phi_{65} = \arg(U_{e6}U^*_{\mu6}U^*_{e5}U_{\mu5}). \tag{28}$$

From these equations, one sees that the numbers of parameters in the appearance fits are three for $3+1$ ($|U_{\alpha4}|, |U_{\beta4}|$ and Δm^2_{41}), while the disappearance fits each have one less mixing matrix element to fit. For $3+2$, the number of parameters goes up to seven for appearance, having added in $|U_{\alpha5}|$, $|U_{\beta5}|$, Δm^2_{51} and the CP-violating parameter ϕ_{54}. Again, for disappearance fits, there is only one mixing matrix connecting to the fifth mass state and no CP violation. Lastly, for $3+3$ appearance, the number of parameters rises to 12 for appearance, now including $|U_{\alpha6}|$, $|U_{\beta6}|$, Δm^2_{61}, and two additional CP-violating parameters. Thus, as expected, adding sterile-neutrino states does add parameters, potentially improving fits, but the number of additional parameters turns out to be relatively small compared to the number of data sets and far less than the number of bins in the full fit.

The $3+3$ model is arguably the most natural. However, it has a very large number of parameters, thus requiring more data sets to perform a useful fit. It should be noted that, in the case where Δm^2_{61} is very large, the typical L/E values for SBL experiments are comparatively low so that the terms depending on this splitting will have rapid oscillations, which leads to a flat contribution. If the mixing angles are also small, then this offset will not have a strong effect on the fits and the system will reduce to an effective $3+2$ model.

2.6. *Comparing experimental results in 3 + 1 models*

Experimental searches for sterile neutrinos in either appearance or disappearance mode rely on comparing the measured event rate in some channel to a prediction including backgrounds and a possible sterile-neutrino signal. It is difficult to compare these event rates directly since each experiment has differences in setups, efficiencies, resolutions, and backgrounds. Typically, experiments rely on simulations to do this comparison and then do fits to extract allowed regions or limits in terms of oscillation parameters. For a $3+1$ model, these regions can be displayed for a given experiment as allowed or excluded regions in the Δm^2 vs. $\sin^2 2\theta$ plane. The LSND results, discussed in Section 2.2, are an example of this procedure.

Experimental results presented in this Δm^2 vs. $\sin^2 2\theta$ parameter space can be easily compared via the overlap of regions from the various experiments. This method has the advantage of incorporating all the information from the given experiments and putting the results on a common footing that can be rigorously compared. In particular, the distribution of "true" neutrino energies for any given "reconstructed" neutrino energy can be used to estimate the oscillation regions, and the systematic uncertainties and correlations associated with neutrino flux, backgrounds, and reconstruction at different energies can be correctly applied. Comparing experiments through this type of oscillation phenomenology is the only rigorous method to compare and combine experiments and is the basis of the global fits to multiexperiment results discussed in Section 4.

To provide an example, consider a comparison of the results of the MiniBooNE experiment to LSND. MiniBooNE is a muon-to-electron flavor appearance experiment that we will describe in Section 3. It was designed to follow up on the LSND anomaly, running in both neutrino and antineutrino modes. The L/E of MiniBooNE was selected to allow coverage of the LSND allowed region. MiniBooNE observed an anomalous excess consistent with oscillations in both running modes. Results from the MiniBooNE experiment[15] are shown in Fig. 4. The proper way to compare MiniBooNE to LSND is to overlay the two-neutrino oscillation allowed regions as shown in this figure.

An alternative method to compare experiments[18] that has been used in the literature is to calculate the observed oscillation probability in L/E bins to exploit the expected dependence of Eq. 1. This is a poor choice, because, unfortunately, the true L and true E of an event cannot be determined due to experimental energy resolution and the smearing associated with a finite length neutrino source and event position detection. So, a given L/E bin has contributions from a range of true L/E values. The measured oscillation probability also has uncertainties associated with the modeling of the backgrounds and predicted signal rate. These uncertainties will typically introduce correlations between the various measured data points.

All of the above effects are likely to be experiment-dependent and, thus, render the L/E method not a very robust comparison technique. In order to apply the L/E method to an actual experiment, one must, therefore, use a simulation as a tool to make corrections and assess measured values. Simulated data is binned in measured L/E and the predicted oscillation

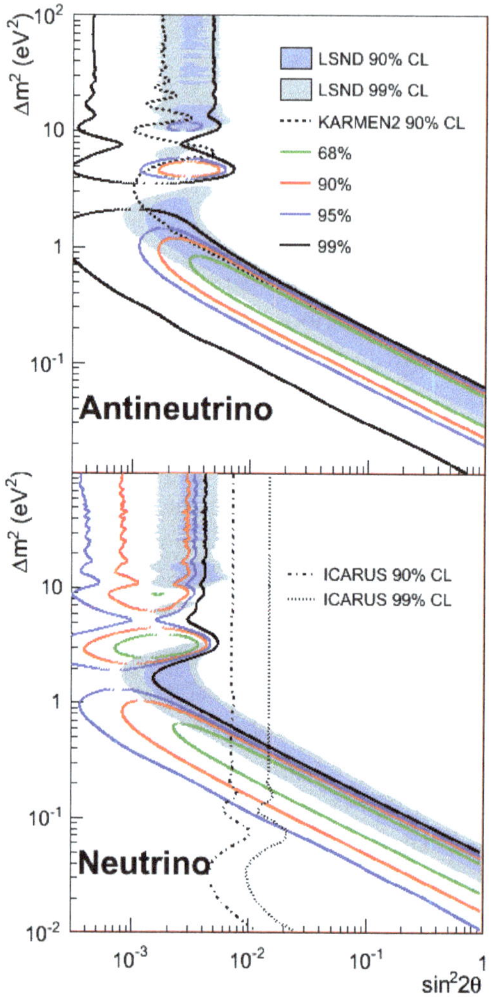

Fig. 4. MiniBooNE[15] allowed regions in antineutrino mode (top) and neutrino mode (bottom) for events with $E_\nu^{QE} > 200\,\text{MeV}$ within a two-neutrino oscillation model. Also shown are the ICARUS[16] and KARMEN[17] appearance limits for neutrinos and antineutrinos, respectively. The shaded areas show the 90% and 99% CL LSND $\bar{\nu}_\mu \to \bar{\nu}_e$ allowed regions.

probability for a given set of oscillation model parameters is calculated. This procedure will be experiment-dependent and so an L/E plot can only be made for a single experiment and used to compare the measured data to various oscillation models. Two example L/E plots (from Ref. 23) are given in Fig. 5 for LSND and in Fig. 6 for MiniBooNE ν and $\bar{\nu}$ results. Because of the bin size compared to the accuracy of the experiment, the LSND experimental measurement of L/E does not need smearing corrections. However, it should be noted that the data points do have correlated

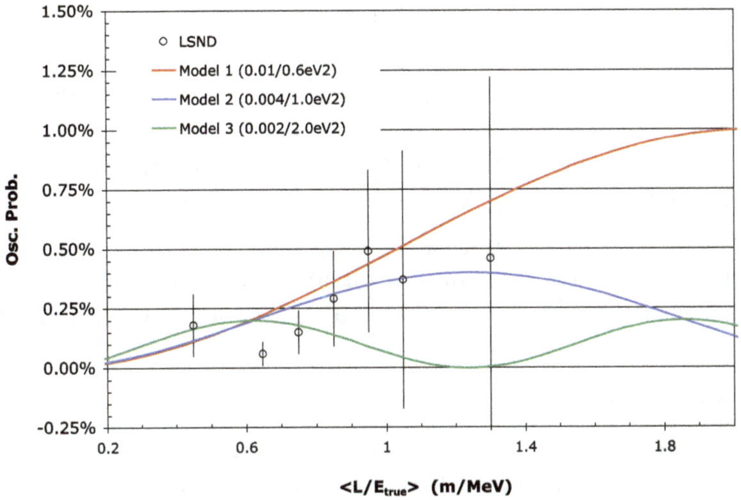

Fig. 5. The LSND-measured oscillation probability as a function of the reconstructed L/E. Three theoretical curves without any energy or flight path smearing are also shown for models with $\sin^2 2\theta/\Delta m^2(eV^2) = 0.01/0.6, 0.004/1.0$, and $0.002/2.0$.[23]

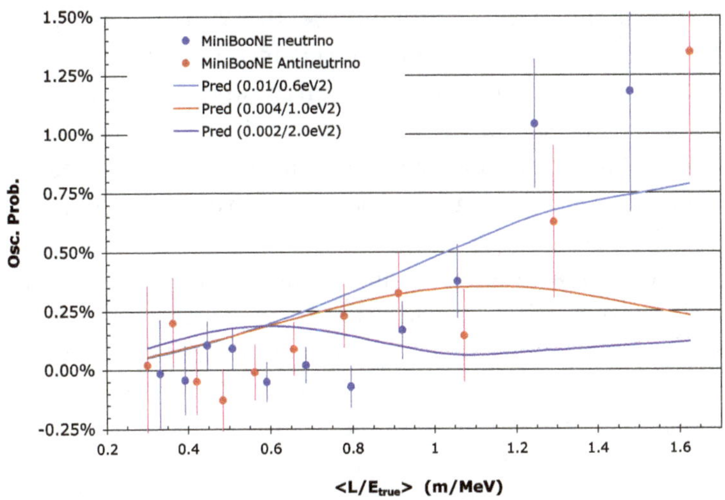

Fig. 6. The MiniBooNE $P_{\rm osc}$ measurements as a function of L/E for neutrino and antineutrino modes of running. The curves are the predicted $P_{\rm osc}$ versus L/E with energy and flight path smearing.[23]

uncertainties associated with the $P_{\rm osc}$ measurement. For MiniBooNE, the smearing effects are significant and the procedure outlined above needs to be used.

Lastly, care must be taken when comparing experiments that are measuring different but related parameters. Consider, for example, the triplet of measurements shown in Eqs. 7 to 9, which are the set of $\nu_e \to \nu_e$, $\nu_\mu \to \nu_\mu$ and $\nu_\mu \to \nu_e$. In order to properly put all results from all three types of experiments onto the same plot, one has to globally fit all of the data to extract $|U_{e4}|$ and $|U_{\mu 4}|$ at every Δm^2 and use the correct values at each point to project into a given plane, for example, the $\sin^2 2\theta_{e\mu}$ plane. This is rarely done in the literature. Most of the time, the best-fit values from some global fit are used to project allowed regions and limits, leading to some distortion of the curves. It is important for readers to be aware of this and to consider such projected curves as qualitative rather than quantitative.

2.7. *Matter effects*

Up to this point, we have only considered vacuum oscillations. This is appropriate for experiments where the neutrinos traverse short distances through matter of low density. This is true of nearly all experiments that are sensitive to oscillations in the $\sim 1\,\text{eV}^2$ region. However there is an important recent exception: the ν_μ disappearance result from IceCube.[19]

This analysis made use of a high-statistics sample of ultrahigh-energy atmospheric neutrino interactions in the 0.4–20 TeV energy range. The pathlength is related to the angle of the incoming neutrinos, and those which are upward-going through the earth were used. The energy and pathlength, given the size of the earth, results in $L/E \sim 1\,\text{m/MeV}$, which is the vacuum-oscillation parameter range of interest for observed SBL sterile-neutrino anomalies.[43] However the resolution of the IceCube detector causes a vacuum-oscillation analysis to be insensitive. The strength of the IceCube result arises from a matter-effect signature in IceCube[21] that predicts a large deficit in the antineutrino flux for the up-going neutrinos that cross the Earth. The matter-magnified signal greatly enhances the IceCube sensitivity to a $3+1$ model consistent with the observed SBL anomalies.

To understand the source of the matter effects in sterile-neutrino searches, let's first consider a known matter effect in the three-neutrino model. If active neutrinos traverse an environment with a high density of electrons, then this environment introduces an additional potential to the Hamiltonian that modifies the ν_e flavor component. The potential V_{CC} is proportional to the Fermi coupling constant G_f and the density of electrons n_e:

$$V_{\text{CC}} = \sqrt{2} G_f n_e. \tag{29}$$

This affects vacuum oscillations involving the ν_e content of the propagating neutrino. Transitions of neutrino flavors within the sun, which has a very high density of electrons, are known to be affected by this potential.[20]

Analogously, in a four-neutrino model, as neutrinos pass through the denser regions of the earth, the propagation of the active flavor components that interact with matter will be affected while the sterile component will not. This introduces an additional potential into the Hamiltonian, leading to a modification of the oscillation probability.[21] The result, for a $3+1$ model, is a substantial predicted deficit for $\nu_\mu \to \nu_\mu$ in specific regions of energy and pathlength (angle) that is in the range of IceCube sensitivity.

3. Existing sterile-neutrino signals, hints, and limits

There have been many searches for oscillations to sterile neutrinos. Table 1 lists a number of the experiments that currently have results based on searches for vacuum oscillations (i.e., experiments without matter effects). This table indicates the name of the experiment, search mode and whether the experiment has a closed contour at 95% CL, which we will call a

Table 1. Oscillation experiments with sensitivity to sterile-neutrino oscillations in the Δm^2 region from 0.01 to $10\,\mathrm{eV}^2$. The experiments are identified as appearance (App) or disappearance (Disapp). They are also identified as "Vac," for vacuum oscillations, as occuring in SBL experiments, and "Mat," for matter dependence, which is at present unique to IceCube. The results of each experiment are categorized either as an observation of a "signal" if there is a closed contour at 95% CL, or a "limit" otherwise. See text for discussion.

	Process	$\nu/\bar{\nu}$	App/Disapp	Vac/Mat	Result
Electron Neutrino Appearance					
LSND[7]	$\bar{\nu}_\mu \to \bar{\nu}_e$	$\bar{\nu}$	App	Vac	Signal
MiniBooNE - ν[24,25]	$\nu_\mu \to \nu_e$	ν	App	Vac	Signal
MiniBooNE - $\bar{\nu}$[15,26]	$\bar{\nu}_\mu \to \bar{\nu}_e$	$\bar{\nu}$	App	Vac	Signal
KARMEN[17]	$\bar{\nu}_\mu \to \bar{\nu}_e$	$\bar{\nu}$	App	Vac	Limit
ICARUS[16]	$\nu_\mu \to \nu_e$	ν	App	Vac	Limit
NOMAD[32]	$\nu_\mu \to \nu_e$	ν	App	Vac	Limit
Electron Neutrino Disappearance					
Bugey and other reactors[27,28]	$\bar{\nu}_e \to \bar{\nu}_e$	$\bar{\nu}$	Disapp	Vac	Signal
Gallium Exps. (SAGE,[29] GALLEX[30])	$\nu_e \to \nu_e$	ν	Disapp	Vac	Signal
KARMEN/LSND XSEC[31]	$\nu_e \to \nu_e$	ν	Disapp	Vac	Limit
Muon Neutrino Disappearance					
MiniBooNE/SciBooNE - ν[35]	$\nu_\mu \to \nu_\mu$	ν	Disapp	Vac	Limit
MiniBooNE/SciBooNE - $\bar{\nu}$[8]	$\bar{\nu}_\mu \to \bar{\nu}_\mu$	$\bar{\nu}$	Disapp	Vac	Limit
IceCube[38]	$\bar{\nu}_\mu \to \bar{\nu}_\mu$	$\bar{\nu}$	Disapp	Mat	Limit
CCFR[33]	$\nu_\mu \to \nu_\mu$	ν	Disapp	Vac	Limit
CDHS[34]	$\nu_\mu \to \nu_\mu$	ν	Disapp	Vac	Limit
MINOS[36,37]	$\nu_\mu \to \nu_\mu$	$\bar{\nu}$	Disapp	Vac	Limit

"signal," or an open contour that includes the null, which we will call a "limit." Note that this is an arbitrary choice, and that KARMEN/LSND XSEC,[31] MiniBooNE/SciBooNE ν,[35] and CDHS[33] have enclosed contours at 90% CL but not at 95% CL. None of the individual experimental results rise to the level of an "observation," which is generally defined as a 5σ signal.

The earliest experiments were looking for oscillations among the three standard neutrinos in the Δm^2 region above $1\,\text{eV}^2$. This was motivated by three-neutrino dark matter models that were viable at the time but are now disfavored by cosmology.[22] Subsequently, the results of these experiments, such as CDHS and CCFR,[33] were used to set limits on ν_μ disappearance associated with possible oscillations to sterile neutrinos.

One of the first dedicated experiments searching for sterile-neutrino oscillations was the LSND experiment at Los Alamos, which has been described previously in Section 2.2. Figure 2 shows the 3.8σ event excess of the data over the background, which fits well with the energy distribution expected for oscillations. Fitting this data to a $3+1$ oscillation model yields the allowed region shown in Fig. 1. The LSND ν_e appearance signal has prompted many follow-up experiments in all the possible channels including ν_e appearance, ν_e disappearance and ν_μ disappearance in the Δm^2 region around $1\,\text{eV}^2$.

As discussed previously, the MiniBooNE experiment was designed to test the LSND oscillation signal by searching for oscillations with a much different experimental setup but holding the L/E value close to the LSND value. For MiniBooNE, L/E averaged around $540\,\text{m}/600\,\text{MeV}$ whereas for LSND the value averaged $35\,\text{m}/40\,\text{MeV}$. MiniBooNE could run in both neutrino and antineutrino modes, so it could search for both $\nu_\mu \to \nu_e$ and $\bar{\nu}_\mu \to \bar{\nu}_e$ oscillations. The systematic uncertainties and backgrounds were also much different than LSND. MiniBooNE had higher relative backgrounds but developed direct data techniques to constrain the impact on the oscillation analysis. Figure 7 shows that the data compared to the expected background indicates a clear excess in both the ν and $\bar{\nu}$ channels. This excess is then displayed in Fig. 8 where the data minus the background is compared to several $3+1$ oscillation models. Notice that when one background subtracts, one can end up with negative values for the oscillation excess — a point that often confuses even seasoned physicists.

The antineutrino excess agrees well with several of the oscillation models but the neutrino data has an excess at low energy above any of the models. This is often referred to as the MiniBooNE low-energy excess and is one of the problems with interpreting the measurements as indications of

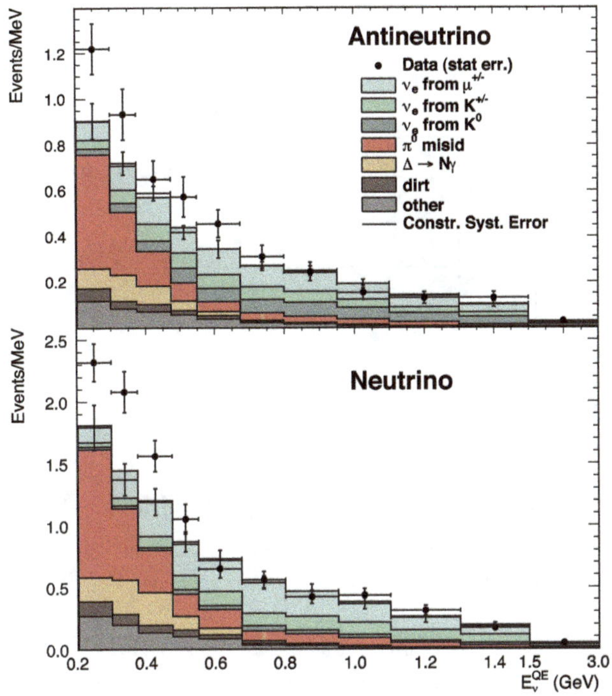

Fig. 7. The antineutrino mode (top) and neutrino mode (bottom) $E_\nu^{\rm QE}$ distributions for ν_e CCQE data (points with statistical errors) and background (histogram with systematic errors).[15]

sterile-neutrino oscillations with a simple $3+1$ model. The low-energy region is also where the γ-ray backgrounds are largest from, for example, "π^0 misid" since the detector did not have the capability to separate electron from γ events. MiniBooNE used direct measurements from the neutrino data to constrain these backgrounds to much below the excess level as displayed by the error bars in Fig. 7, bottom. Nevertheless, the low-energy excess could be from some new type of neutrino process that produces single γ rays and future experiments should have the capability to separate γs from electrons.

The KARMEN,[17] ICARUS,[16] and NOMAD[32] experiments also searched for $\nu_\mu \to \nu_e$ appearance oscillations setting limits but not with the sensitivity to fully exclude the MiniBooNE and LSND signals. As a summary of the current situation, the allowed regions and limits for $\nu_\mu \to \nu_e$ appearance from a $3+1$ model fits to the various data sets are shown in Fig. 9. The red region in the figure shows the allowed region for a combined fit indicating that there are $3+1$ oscillation parameter sets that are compatible with all the appearance data.

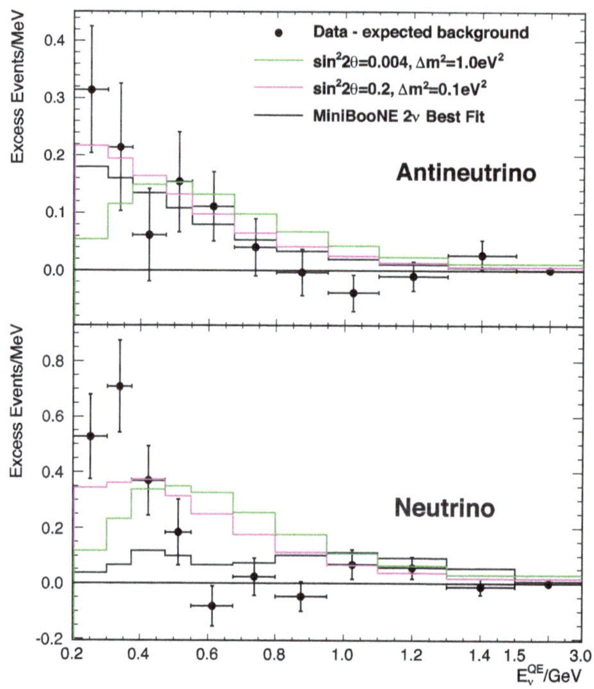

Fig. 8. The antineutrino mode (top) and neutrino mode (bottom) event excesses as a function of E_ν^{QE}. (Error bars include both the statistical and systematic uncertainties.) Also shown are the expectations from the best two-neutrino fit for each mode and for two example sets of oscillation parameters.[15]

In a 3+1 model, ν_e appearance can only come about if both $|U_{e4}|$ and $|U_{\mu 4}|$ are nonzero as shown in Eq. 9. Nonzero values for these elements implies that there must be both ν_e disappearance and ν_μ disappearance, which leads to also consider the current results in those channels.

Two of the ν_e disappearance searches have seen signals. One is a set of results from beams produced by reactors producing antineutrinos. The other comes from studies of very intense radioactive sources producing neutrinos.

The first result, referred to as the "Reactor Neutrino Anomaly",[27,28] uses the measured rate of $\bar{\nu}_e$ events detected from reactors as compared to prediction. When one convolutes the reactor spectrum with the inverse beta decay ($\bar{\nu}_e + p \to e^+ + n$) cross section, one obtains events with neutrino energies between 1.8 MeV and about 8 MeV, with a peak at about 3 MeV. The absolute prediction of this flux is difficult to accomplish because of the complicated production mechanism of $\bar{\nu}_e$'s from the beta decay of reactor fission fragments that must be modeled. Recently, several experiments have observed an unmodeled bump in the higher-energy part of the spectrum indicating some problem with the reactor model.[40] For $|\Delta m^2|$ values near

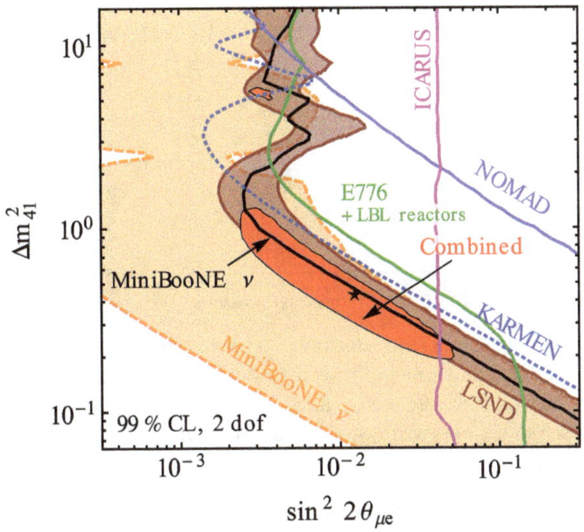

Fig. 9. Allowed regions and upper bounds at 99% CL (2 dof) for muon to electron flavor appearance experiments in the 3 + 1 scheme. The red region corresponds to a combined fit of all $\nu_\mu \to \nu_e$ appearance data sets, with the star indicating the best-fit point.[39]

1 eV^2, given the 3 MeV peak neutrino energy, the first oscillation maximum is near 3 m from the reactor. Experiments near to the reactors are called "SBL reactor experiments." For L values $> \mathcal{O}(10)$ m, the probability should reach the fast-oscillation limiting value of Prob $= \frac{1}{2} \times \sin^2 2\theta$. Experiments in this distance range from the reactor, and extending up to about a kilometer or more, are called "long-baseline experiments." Previous reactor measurements with L values in the 10–100 m range have measured ratios to prediction less than one that average to 0.933 ± 0.021 as shown in Fig. 10. This difference from 1.0, the reactor anomaly, is a possible 3.2σ indication of $\bar{\nu}_e$ disappearance.

A second electron-neutrino disappearance signal is associated with the measured rate of ν_e's produced by sources in the PBq (10^{15} decays/s) range. These are extremely hot sources — for comparison, the potassium decays that naturally occur in the human body are at the kBq level. These sources were used in the "gallium experiments," GALLEX[30] and SAGE,[29] which were solar neutrino detectors. Again the measured rate is compared to a prediction and a deficit is found at the 2.9σ level with $R_{\text{obs}}/R_{\text{pred}} = 0.84 \pm 0.05$. This is referred to as the "gallium anomaly" since both detectors employ gallium in the detection medium. The combination of the reactor and gallium anomalies along with other ν_e disappearance measurements leads to the allowed regions given in Fig. 11. This collection

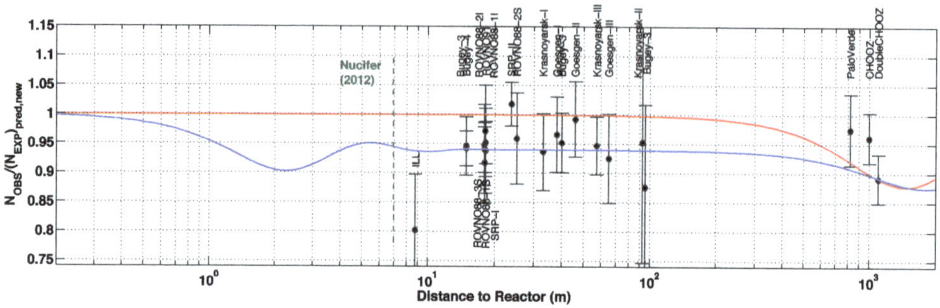

Fig. 10. SBL reactor antineutrino anomaly. The experimental results are compared to the prediction without oscillations, taking into account the new antineutrino spectra, the corrections of the neutron mean lifetime, and the off-equilibrium effects. As an illustration, the red line shows a 3 active neutrino mixing solution and the blue line displays a solution including a new neutrino mass state, such as $|\Delta m_4^2| \approx 1$ eV^2 and $\sin^2 2\theta = 0.12$.[2]

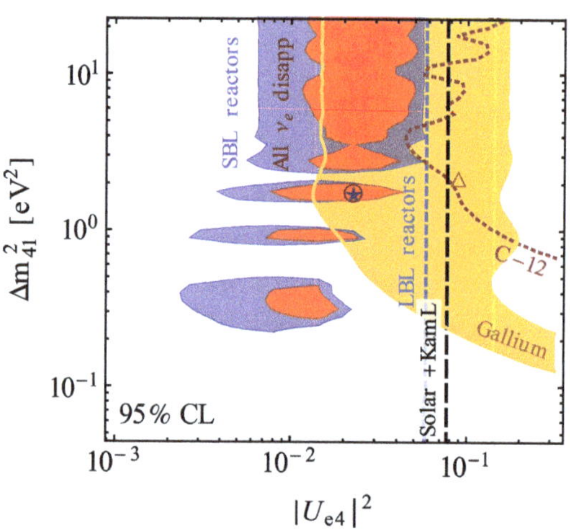

Fig. 11. Allowed regions at 95% CL (2 dof) for 3 + 1 oscillations. Regions shown are SBL reactor data (blue shaded), gallium radioactive source data (orange shaded), ν_e disappearance constraints from comparing ν_e−C-12 scattering data from LSND and KARMEN, called LSND-KARMEN XSEC in Table 1 (dark red dashed), long-baseline reactor data (blue short dashed), and data from solar neutrino experiments, including the KamLAND result which is sensitive to solar oscillations (black long dashed). The red shaded region is the combined region from all of these ν_e and $\bar{\nu}_e$ disappearance limits. Note: $|U_{e4}|^2 \approx \sin^2 2\theta_{ee}/4$.[39]

of ν_e data is well fit by the 3 + 1 neutrino hypothesis, while the no-oscillation hypothesis is disfavored at 99.97% CL (3.6 σ).

For ν_μ disappearance, the present situation is quite different. Currently there are no signals of sterile oscillations in this channel, using our 95% CL definition. Recall that in a 3 + 1 model, the disappearance and appearance oscillation channels are not independent and are coupled through the $|U_{e4}|$

and $|U_{\mu 4}|$ as shown in Eq. 9. These equations lead to the approximate relation between the effective mixing angles given by

$$\sin^2 2\theta_{\mu e} \approx \frac{1}{4} \sin^2 2\theta_{ee} \sin^2 2\theta_{\mu\mu} . \tag{30}$$

Thus the ν_e disappearance and $\nu_\mu \to \nu_e$ appearance signals described above lead to a prediction for ν_μ disappearance. If no disappearance is seen in that range, with high confidence level, then the model is excluded.

The ν_μ disappearance experiments are looking for a deficit in detected muon neutrino events as compared to the predicted number of events. The prediction is best done by using an experimental setup with two detectors, a near detector at short distance to measure the flux and a far detector to search for the deficit. Extrapolating from the near to far detector can bring in systematic uncertainties that depend on modeling the beam flux and divergence as well as detector differences but these uncertainties are typically at the few percent level. Figure 12 shows the current limits for the various measurements. CCFR, CDHS, MINOS, and SciBooNE/MiniBooNE are all two-detector measurements and the MiniBooNE-only limit comes from a shape analysis of the observed events compared to expectation.

The tension between the appearance and disappearance data is displayed in Fig. 13 which shows the oscillation parameters associated with different data sets at the 3σ CL. As given in Eq. 30 ($\sin^2 2\theta_{\mu e} \approx \frac{1}{4} \sin^2 2\theta_{ee} \sin^2 2\theta_{\mu\mu}$), the independent ν_μ and ν_e disappearance limits can be used to give a combined DIS limit on $\sin^2 2\theta_{\mu e}$ for $3+1$ models. One can see that the combined ν_e allowed region (APP) is highly restricted by the DIS limits.

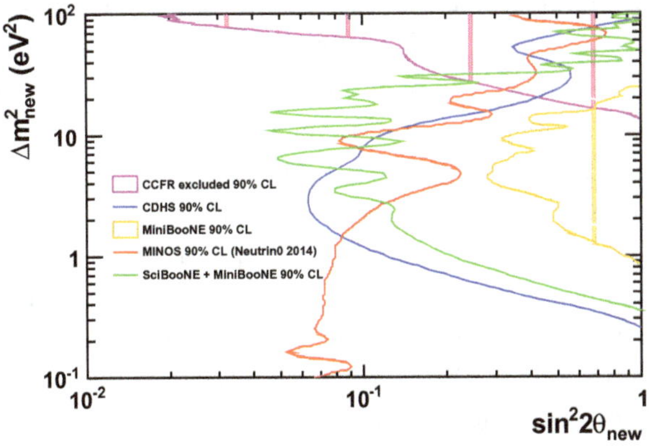

Fig. 12. The current exclusion limits from ν_μ disappearance searches in the 0.1–100 eV^2 Δm^2 region with the mixing angle given by $\sin^2 2\theta_{\text{new}}$. All the exclusion limits are at 90% CL.[41]

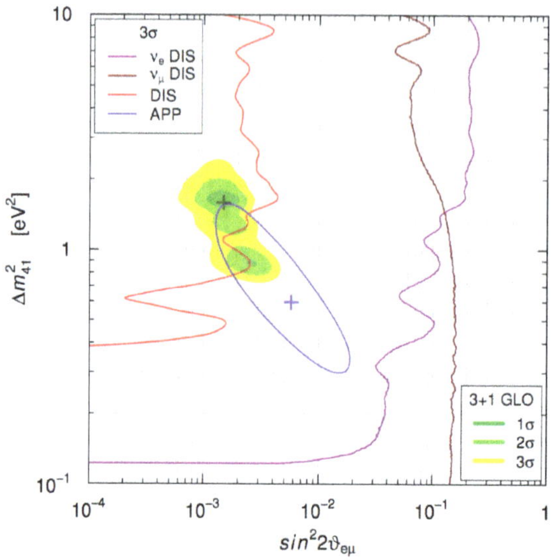

Fig. 13. Allowed regions (at 3σ) in the Δm^2_{41} versus $\sin^2 2\theta_{e\mu}$ plane from global fits of SBL appearance and disappearance oscillation data ($3+1$ GLO) compared with the allowed regions obtained from $\nu_\mu \to \nu_e$ appearance-only data (APP) and the constraints obtained from ν_e disappearance-only data (ν_e DIS), ν_μ disappearance-only data (ν_μ DIS) and the combined disappearance data (DIS). The best-fit points of the GLO and APP fits are indicated by crosses. (Updated plot presented by C. Giunti at the Neutrino 2016 conference associated with Ref. 42.)

One should keep in mind that limits are not hard cutoffs, and there is a lower probability signal extending outside of the defined 3σ CL, where APP and DIS do overlap. A global fit of all data is also shown and indicates the small allowed region left from the APP allowed region.

The IceCube experiment has also done a search for ν_μ and $\bar{\nu}_\mu$ disappearance using the atmospheric muon neutrino spectrum as a function of zenith angle as described in Section 2.7. As a reminder, this experiment depends on the modification of the vacuum oscillations by matter effects. As shown in Fig. 14, these results greatly improve the limits from previous experiments for $\Delta m^2 < 2\,\text{eV}^2$. Note that this figure shows the LSND and MiniBooNE allowed regions, which are for appearance, on the disappearance plot. The appearance allowed regions were transferred to this plot using the best-fit value for $|U_{e4}|$ rather than a point-by-point calculation of $|U_{e4}|$ in a global fit involving IceCube. As was pointed out in Section 2.6, this will lead to some distortion of the appearance allowed region. So the overlay should be regarded as qualitative, not quantitative. Nevertheless, this result only adds the the apparent tension between appearance and disappearance data sets.

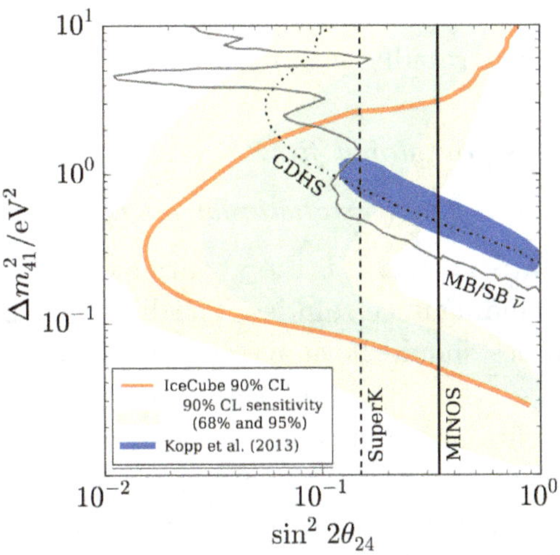

Fig. 14. Results from the IceCube $\nu_\mu/\bar{\nu}_\mu$ disappearance search. The 90% (orange solid line) CL contour is shown with bands containing 68% (green) and 95% (yellow) of the 90% contours in simulated pseudoexperiments. Also shown are the CDHS and SciBooNE/MiniBooNE limits along with the MiniBooNE/LSND 90% CL blue allowed region[39] assuming $|U_{e4}|^2 = 0.023$.[38]

In summary, the current sterile-neutrino oscillation status has several signals in the ν_e appearance channel, and also signals in the ν_e disappearance channel, but no oscillation signal for ν_μ disappearance. A smoking gun for oscillations would be the observation of detected event rates that vary with the expected L/E behavior; currently, this behavior has not been convincingly observed in any of the experiments with $\Delta m^2 \sim 1\,\mathrm{eV}^2$ signals.

4. Global data fits

In order to determine the viability of sterile-neutrino models, it is necessary to use a global fit. This needs to include as many of the relevant data sets as possible. It is best if these data sets are from considerably different experimental designs, as this reduces sensitivity to backgrounds that are under or oversubtracted and to other systematic uncertainties. A few experiments have recently adopted the practice of presenting their limit combined with one other experiment that will give the best reach. This turns out to be misleading and the results overpredict the final result of the global fit. This should be avoided.

In this section, we discuss the techniques of global fits. An important aspect of the fit is the choice of test statistic. This turns out to be a complicated issue, which we must consider first. We then discuss implementation. Throughout, we then show results of global fits to the present data, however our emphasis is on the approach to the problem rather than these specific results, as these change rapidly with time.

4.1. *Test statistics for global fits*

4.1.1. *The problem with the χ^2 statistic for fitting*

The simplest test statistic for a fit is the χ^2 variable, which compares the data to the prediction including uncertainties. The distribution of this variable for Gaussian uncertainties should have a mean equal to the dof and standard deviation equal to $\sqrt{2\ \text{dof}}$. However, this statistic can be very misleading because the global fits are keying in on deviations of the data from the prediction in rather localized data regions that are sensitive to oscillations. For the data space outside of these sensitive region, one can get a good fit to almost any oscillation model. If you add these insensitive bins into the χ^2, then the power to discriminate models will be diluted and $\chi^2/\text{dof} \to 1$.

This problem is enhanced when one is fitting many data sets, which all may have insensitive bins in certain areas of parameter space that dilute the χ^2. As a result, even though data sets may disagree, the χ^2/dof from the global fit may be acceptable.

This is the case with the present global fits. In a $3+1$ model, the $\chi^2/\text{dof} = 359/315$ dof $= 1.14$ for all of the data. However, separate fits to the appearance and disappearance data sets listed on Table 1 sets show poor overlap as seen in Fig. 15. One ends up with a global fit result with allowed regions because the strength of the combined ν_e disappearance and $\nu_\mu \to \nu_e$ appearance results overcomes the lack of signal in ν_μ disappearance. But the χ^2/dof does not reflect the stress due to the large number of bins in all three cases where the best-fit model predicts no signal.

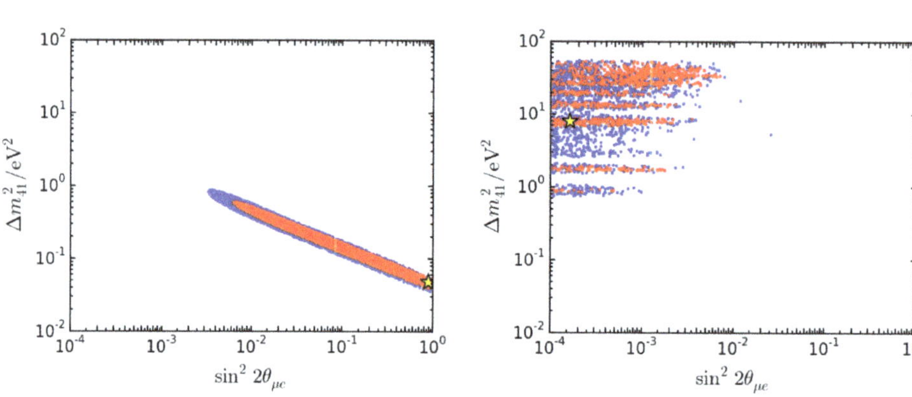

Fig. 15. Left: global fit to the appearance data only; Right: global fit to the disappearance data only. Plot is from Ref. 43.

One does not want to change the selected binning to focus on only the identified signal regions as one performs a fit. This would make a test statistic that could only be understood with extensive frequentist tests using a large set of experiment simulations that, at present, are computationally too expensive to perform. Also, the regions with no signal can add very important information to the fit and be used, for example, to constrain normalization uncertainties.

4.1.2. *The $\Delta\chi^2$ statistic for global fitting*

A powerful method for doing global fitting to determine oscillation parameters and confidence regions is to use the $\Delta\chi^2$ statistic, where

$$\Delta\chi^2 = \chi^2_{\text{TestPoint}} - \chi^2_{\text{BestFit}}. \tag{31}$$

The $\Delta\chi^2$ statistic is related to the likelihood (L) ratio of the best fit with respect to some test point for Gaussian uncertainties with

$$\Delta\chi^2 = -2\ln\left(L_{\text{TestPoint}}/L_{\text{BestFit}}\right). \tag{32}$$

The best fit for a given data set is found by minimizing the χ^2 for the data vs. prediction. For the simplest problems, $\Delta\chi^2$ will follow a χ^2 distribution with the dof value equal to the number of parameters being determined in the fit. To determine a CL region around the best fit, one uses the critical value for the χ^2 distribution with this given dof. For example, with two fit parameters (say Δm^2 and $\sin^2 2\theta$), the critical value for 90% CL would be 4.61 and for a 2σ CL it would be 6.18. One can then find the Δm^2 vs. $\sin^2 2\theta$ region where the $\Delta\chi^2$ value for those points are below this critical value. This region then corresponds to the allowed region at the given CL. An example of CL regions for a toy model with two-dimensional, normally distributed data is shown in Fig. 16.

Limits are a special case of confidence intervals where, for example, the no-oscillation (or "null") model is allowed by the data. On a Δm^2 vs. $\sin^2 2\theta$ plot, a limit curve, at a given CL, separates the allowed region to left of the curve from the excluded region to the right of the curve. The curve corresponds to the points where $\Delta\chi^2$ is equal to the critical value for the given CL. Thus, the points to the right of the curve have $\Delta\chi^2$ values greater than the critical value and are, therefore, excluded at the given CL.

Other parameters that are not directly involved in the oscillation model (*a.k.a.* "nuisance" parameters) such as those that describe the normalization, backgrounds, etc. and their uncertainties, can be included in the fit using extra terms in the χ^2, called "pull" terms. These nuisance parameters can be

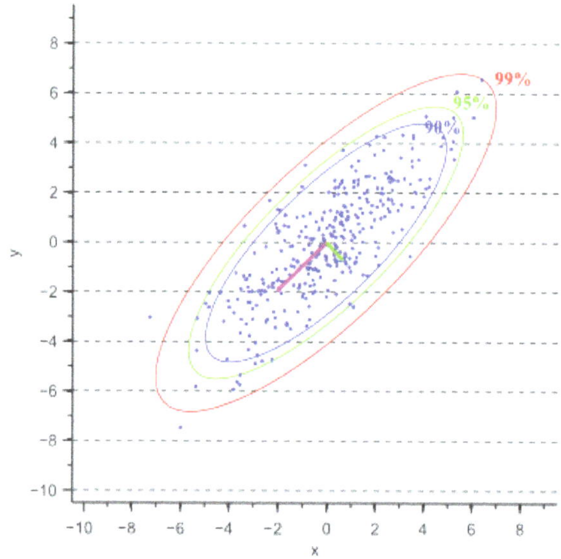

Fig. 16. Example confidence level (CL) regions for a two-dimensional, normally distributed data sample in the variables (x, y). The arrows show the variance of the data along the two covariance (or eigenvector) directions. The ellipses show the CL regions at 90%, 95%, and 99% that contain that fraction of the data points.[44]

included (usually called being "profiled") in the fit by minimizing them at each test point. The fit value for the nuisance parameters can be of interest and comparisons of the values of these parameters with their expectations can give information on how much the fit determines they have been pulled from what is expected. This is commonly referred to as the "pull" for a parameter.

The $\Delta\chi^2$ also solves the problem described in Sec. 4.1.1 where data bins outside of the sensitive region erroneously reduce the χ^2/dof. For the $\Delta\chi^2$ calculation, bins with no or very small sensitivity in the fit will contribute zero to the $\Delta\chi^2$ value since the contribution from the test point and from the best fit will be identical and cancel.

A problem with the $\Delta\chi^2$ statistic is that the probability distribution may not follow a χ^2 distribution with a dof equal to the number of fit parameters. This can come about because the predicted event function is nonlinear in the fit parameters or because the uncertainty in certain bins is hard to determine from first principles or is non-Gaussian. An example of this behavior is given by fits in the high Δm^2 region, much above the value for the first oscillation maximum. In that region, the L/E behavior in the oscillation probability has rapid oscillation and effectively gives an average of 0.5 for the $\sin^2(1.27\Delta m^2 L/E)$ factor. Therefore, in this region the oscillation probability has no dependence on Δm^2 and the effective dof number will be 1.0 instead of 2.0. To quantitatively calculate the dof value as a function of test point parameters and, thus, the $\Delta\chi^2$ critical values for a given CL, one needs to do simulation studies. At every test point, one uses a large set of fake, simulated experiments to determine the $\Delta\chi^2$ distribution numerically from the distribution of simulated values. This is commonly referred to as the "Feldman-Cousins" frequentist method[6] for determining confidence regions.

4.1.3. *Anscombe's quartet as a cautionary tale*

In Section 3, we emphasized that there is tension between the appearance and disappearance data sets. As we will present below, the $\Delta\chi^2$ test statistic will lead to apparently high-quality fits despite this tension. It is reasonable to ask how this happens. The example of Anscombe's Quartet[45] provides an explanation of how this can come about.

Figure 17, reproduced from Ref. 46, shows four example distributions. All four distributions are fit to lines, yielding slopes and intercepts that agree to three and two digits past the decimal, respectively. They also have the same mean, variance, total χ^2, and correlation between x and y.

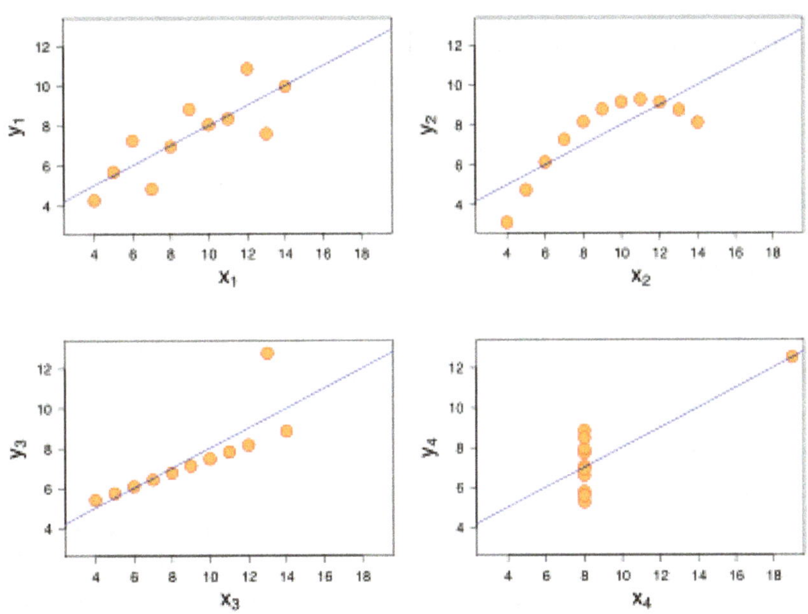

Fig. 17. Anscombe's quartet of data sets (from Ref. 46 and first published in Ref. 45). This shows four data sets that all fit to the same straight line. The data sets have the same mean and variance.

The information from a fit to all data points does not indicate that these are substantially different distributions. However, the differences between these distributions becomes clear if you divide the data sets in certain ways and perform separate fits on each subset. In that case, the fits to the two separate subsets will not agree with each other, nor with the fit to all of the data points.

Having discovered a problem like this, the question then becomes: how do you quantify the probability of the split between two data sets resulting in the inconsistent? The problem is not simple — by selecting how you will cut the data set, you no longer have random samples. You have biased the information by choosing how to cut the data set to maximize the tension. A frequentist fake-data study will tell you how often you will see such a disagreement or worse disagreement. But such a study is prohibitive in the case of the SBL global fits, given today's computing power.

This problem led to the introduction of the Parameter Goodness of Fit, which we describe below. It provides a useful approach to comparing data sets that you have split. However, we will show that this approach fails in certain circumstances. At this point, the meaning of the probability returned by the Parameter Goodness of Fit remains unclear.

4.1.4. *The parameter goodness of fit*

A possible problem with the $\Delta\chi^2$ global fit method to determine confidence regions is that there can be large discrepancies between classes of data sets that go into the fit. This may lead to allowed regions deriving from the $\Delta\chi^2$ prescription, even when no part of the parameter space can well-reproduce the data. An example of this behavior is shown in Fig. 15, where the fits to appearance and disappearance give much different allowed regions but taken together do have allowed regions at reasonable confidence levels. Of course, the global fits should include all the uncertainties in determining the allowed region. Nevertheless, one might also want to quantify how compatible the different data sets that go into this determination are.

In response to this question, the Parameter Goodness-of-fit (PG) test was established.[47] This test was meant to provide a compatibility test for two subsets of data in a given global fit, without bias that may come from irrelevant bins or data sets with many bins. Reference 47 provides a formal, analytic derivation of this new test statistic, the χ^2_{PG}, its degrees of freedom, N_{PG} and the cumulative probability function for this PG test statistic.

As an example of its application, consider the PG test for appearance ($\nu_\mu \to \nu_e$) versus disappearance ($\nu_\mu \to \nu_\mu$ and $\nu_e \to \nu_e$) data sets that underlie a $3+1$ global fit. One performs separate fits on each of the two underlying subsets, as well as a combined fit to the full data set, to obtain three χ^2 values, $\chi^2_{\rm app}$, $\chi^2_{\rm disapp}$ and $\chi^2_{\rm glob}$. One then forms an effective χ^2:

$$\chi^2_{PG} = \chi^2_{\rm glob} - (\chi^2_{\rm app} + \chi^2_{\rm disapp}). \tag{33}$$

In the case of good agreement between underlying data sets, the χ^2 contribution from nonsignal bins is highly reduced because of the subtraction in a similar way as the $\Delta\chi^2$ statistic. The number of degrees of freedom is defined as

$$N_{\rm PG} = (N_{\rm app} + N_{\rm disapp}) - N_{\rm glob}, \tag{34}$$

where each N is the number of independent parameters involved in the given fit. In our $3+1$ example, the appearance fit has two parameters ($|U_{e4}||U_{\mu4}|$ and Δm^2), the disappearance fit has three parameters ($|U_{e4}|$, $|U_{\mu4}|$, and Δm^2), and the global fit has three parameters ($|U_{e4}|$, $|U_{\mu4}|$, and Δm^2). This then leads to the dof calculation for the PG test of $N^{3+1}_{PG} = (2+3)-3 = 2$ dof. The probability that the two data sets are in agreement is that associated with $\chi^2_{\rm PG}$ for $N_{\rm PG}$, or $\chi^2_{\rm PG}$ for 2 dof in the case of a $3+1$ model.

The PG test has been shown to successfully assess the agreement between data sets with systematic uncertainties that are Gaussian-distributed across multiple runs of the experiment. However, in the case of an underlying systematic which is single valued rather than Gaussian-distributed between runs of an experiment, the number of degrees of freedom is not necessarily $N_{\rm PG}$. The deviation occurs when the shape of the systematic effect is correlated to the shape of the oscillation signal. In this case, $\chi^2_{\rm PG}$ for $N_{\rm PG}$ is not a valid estimate of the probability.

As a tangible example, imagine an appearance experiment which has both a $3+1$ oscillation signal and a known background. To fit the data, the background must be subtracted. For simplicity, let's assume the background shape is perfectly known from the SM, but the normalization has an error that comes from past measurements of the cross section. Our imaginary experimentalists will look up the central value for the background assigned by the Particle Data Group,[3] and use this to subtract a background function with this normalization. The assumption is the data set now has zero background, after subtraction, but there is an associated error that comes from the past measurements of the cross section. This error represents the experimentalist's best knowledge of the cross section, and it is presented and treated as a Gaussian-distributed systematic error. In reality, however, the true, natural value of the background is single valued. No matter how many times the experiment is run, this true value is always the same from run to run. Therefore, the experiment has an underlying residual background subtraction which is the same for every run and sis not Gaussian-distributed.

To make our imaginary experiment more concrete, consider the following legitimate set of oscillation parameters for a $3+1$ model:

- $\Delta m^2 = 0.75\,\text{eV}^2$,
- $|U_{e4}|^2 = 0.1$, and
- $|U_{\mu 4}|^2 = 0.1$.

Thus $|U_{e4}||U_{\mu 4}| = 0.1$. These are used to generate the "true" oscillation signals that our imaginary experiment will see. We generate data for $\nu_\mu \to \nu_e$ appearance, ν_μ disappearance and ν_e disappearance in 16 energy bins in the range from 200 to 1000 MeV, assuming $L = 500\,\text{m}$.

The problem occurs when the residual background for our imaginary experiment has the same shape as a legitimate oscillation signal but one with different parameters than the model we describe above. As an example, we consider a background to the appearance signal that is an exponential residual background of the form $N^{\text{backgnd}} = A_{\text{true}} \exp(-E/200)$, where E is the neutrino energy. We will study values of the normalization of the residual background, A_{true}, between 0.0 and 0.4.

(We leave it as an exercise for the student to explore this model with zero background and with a flat residual background as a function of neutrino energy of the form $N_i^{\text{backgnd}} = A_{\text{true}} = \text{constant}$. Note that the latter case is important because a $3+3$ model with large Δm^2_{61} can lead to a flat overall offset. In these two cases, the student will find the PG test succeeds.)

Our imaginary experimenters have assumed that this background has been correctly subtracted, thus $A_{\text{exp}} = 0$. The shape is known and so they fit for the normalization, A_{fit}. The experimenters places a systematic error on this assumption, which, for the purposes of this discussion, will be $\sigma_{A_{\text{exp}}} = 0.15$.

In order to generate the appearance data, the function describing the residual background is added to the function describing the oscillation signal, resulting in a function that describes the total ν_e-like events. The same oscillation parameters and residual background function are used to generate each data set in a study.

The experimental data is then generated using statistical uncertainties about the functions for the appearance (with background) and disappearance predictions. This simulates statistical fluctuations but no other smearing effects are included. We will study the effect of the underlying background by running 1000 fake studies, each of which has an appearance, electron disappearance and muon disappearance experimental data set.

In each of the 1000 fake studies, A_{true} is held to the same value. Each data set is fit for the oscillation parameters with the appearance and global data set also simultaneously fit for the normalization of the residual background, A_{fit}, using a pull term. We define the χ^2 for each fit in the following way:

$$\chi^2_{\nu_e \text{app}} = \sum_{i=1}^{16} \frac{(d_i^{\nu_e \text{app}} - (\text{osc}_i^{\nu_e \text{app}} + b_i^{\nu_e \text{app}}(A_{\text{fit}})))^2}{(\sigma_i^{\nu_e \text{app}})^2} + \frac{(A_{\text{fit}} - A_{\text{exp}})^2}{\sigma^2_{A_{\text{exp}}}} \tag{35}$$

$$\chi^2_{\text{disapp}} = \sum_{i=1}^{16} \frac{\left(d_i^{\nu_\mu \text{disapp}} - \text{osc}_i^{\nu_\mu \text{disapp}}\right)^2}{\left(\sigma_i^{\nu_\mu \text{disapp}}\right)^2} + \sum_{i=1}^{16} \frac{\left(d_i^{\nu_e \text{disapp}} - \text{osc}_i^{\nu_e \text{disapp}}\right)^2}{\left(\sigma_i^{\nu_e \text{disapp}}\right)^2} \tag{36}$$

$$\chi^2_{\text{glob}} = \sum_{i=1}^{16} \frac{(d_i^{\nu_e \text{app}} - (\text{osc}_i^{\nu_e \text{app}} + b_i^{\nu_e \text{app}}(A_{\text{fit}})))^2}{(\sigma_i^{\nu_e \text{app}})^2} + \frac{(A_{\text{fit}} - A_{\text{exp}})^2}{\sigma^2_{A_{\text{exp}}}} + \sum_{i=1}^{16} \frac{\left(d_i^{\nu_\mu \text{disapp}} - \text{osc}_i^{\nu_\mu \text{disapp}}\right)^2}{\left(\sigma_i^{\nu_\mu \text{disapp}}\right)^2} + \sum_{i=1}^{16} \frac{\left(d_i^{\nu_e \text{disapp}} - \text{osc}_i^{\nu_e \text{disapp}}\right)^2}{\left(\sigma_i^{\nu_e \text{disapp}}\right)^2}. \tag{37}$$

Figure 18 provides an example run for the case of a residual exponential background where $A_{\text{true}} = 0.4$. The top, middle and bottom frames, are, respectively, appearance, electron-flavor disappearance, and muon-flavor disappearance signals shown as a function of energy. The points show the generated data with statistical fluctuations. The true oscillation signal is shown in each frame in magenta. The green curve in the appearance data set (top frame) shows the underlying residual background for this run. The

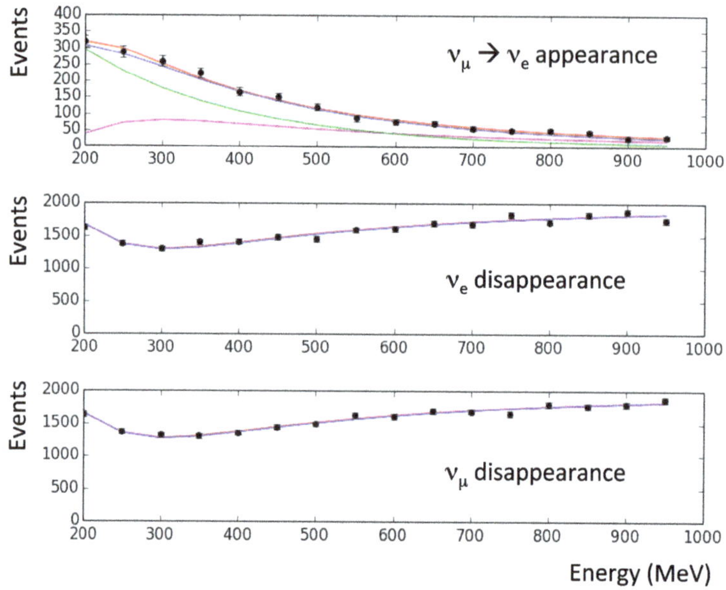

Fig. 18. Example of signals versus energy for appearance (top), electron-flavor disappearance (middle) and muon-flavor disappearance (bottom) from one experiment with an exponential residual background of $A_{\mathrm{true}} = 0.4$. In all plots, the true signal is shown in magenta. In the top plot, the underlying background is shown in green. In all plots, the global fit result is shown in blue. In the top plot, the appearance-only fit, with the residual amplitude included as a pull term, is shown in red.

red curve on the appearance data set, shows the fit obtained from minimizing $\chi^2_{\nu_e \mathrm{app}}$ (Eq. 35). The blue lines on all three data sets show the fits obtained by minimizing χ^2_{glob} (Eq. 37).

In this case of an exponential background, the shape of the oscillation signal and the background are correlated with a correlation parameter of $\rho = 0.6$. This correlation is reflected in the range of values that the parameters can take in any given run, as is illustrated in Fig. 19. These plots are for the case of $A = 0.4$, and the plots shown are for the appearance fit. One can see that the parameters are widely varying from the true value and they are highly correlated.

On the other hand, the global fit performs very well. The oscillation parameters that are returned are consistently in agreement with the input values. We find $A_{\mathrm{fit}} = 0.100 \pm 0.011, 0.0198 \pm 0.013, 0.297 \pm 0.016, 0.395 \pm 0.019$ for A_{true}=0.1, 0.2, 0.3, and 0.4, respectively.

In order to find the degrees of freedom, the χ^2 distributions of the data sets are fit to a standard χ^2 form

$$P_0(x^{(P_1/2)-1})e^{-x/2}, \tag{38}$$

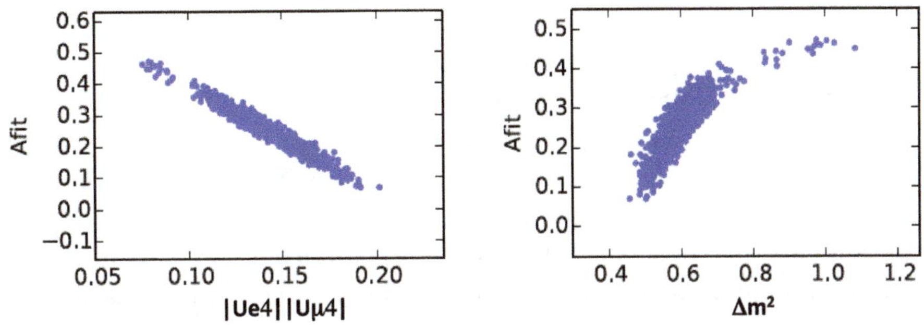

Fig. 19. Parameter values for the appearance fit for an exponential background with $A_{\text{true}} = 0.4$. One can see strong correlations between the parameters, and the fit values deviate far from the true values.

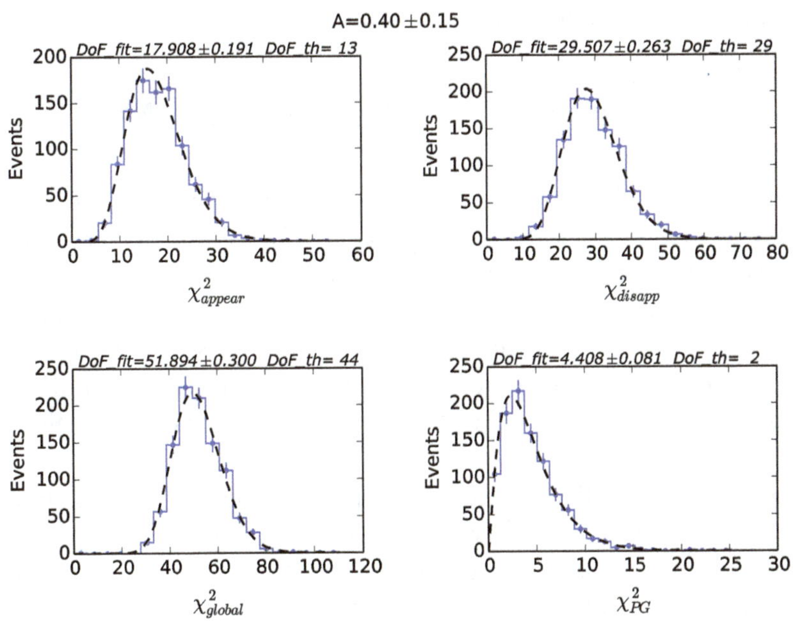

Fig. 20. The χ^2 distributions for an exponential background with $A_{\text{true}} = 0.4$. For each of the χ^2 types, the results of the dof fit are given with uncertainties along with the theoretical expectation. The histogram with error bars are the distribution of values and the dashed curve is the result of the dof fit. The number of dof from the fit for the χ^2_{PG} deviates significantly from the expected value of 2.

where $x = \chi^2$, P_0 is a normalization and P_1 is the number of degrees of freedom indicated by the χ^2 distribution.

As an example of the dof fits, the χ^2 distributions for 1000 experiments for the case of $A_{\text{true}} = 0.4$ are shown in Fig. 20. One can see that in this case, the χ^2_{PG} deviates far from the expectation of two even though there is not an incompatibility of data with a true oscillation model encompassing all three data sets. For $A_{\text{true}} = 0.1$, 0.2, 0.3, and 0.4, respectively, the degrees of freedom of the χ^2_{PG} are found to be 1.92 ± 0.062, 2.484 ± 0.065, 3.361 ± 0.074 and 4.408 ± 0.081. The problem is arising because the high correlation between the background and the signal allows the fit to find a minimum $\chi^2_{\nu_e \text{app}}$ whose distribution is close to the theoretical expectation, even though the fit parameters are far from the inputs. On the other hand, the global fit parameters are constrained by the disappearance data contributions and so the minimum value of the χ^2_{glob} is relatively large. This leads to a χ^2_{PG} distribution that is larger than the expectation for 2 dof. Since the χ^2_{PG} values are larger than what is theoretically expected, the PG test will on average erroneously indicate that the appearance and disappearance data are incompatible. But, in reality, the only issue is that there is a background that mimics a possible oscillation signal.

4.2. *Using Markov chain methods for finding global fit parameters*

The process of finding the best fit in a global analysis comes under the umbrella of algorithms called "optimizers." The most efficient optimizers use a gradient to quickly find the minimum (or maximum) of a function. However, gradient descent is susceptible to being trapped in local minima, and "homes in" on the location of the minimum, without exploring the space around it. We are interested in not only the best fit but also this surrounding region, which characterizes the level of confidence we have in our result.

Ideally, we would like a method of optimization that moves toward the minimum on average but otherwise randomly explores the parameter space. The random walk would allow the method to "tunnel through" local minima to find other regions of the parameter space. Such a process can be modeled by a Markov chain[50] that provides an efficient way to sample the likely regions of the parameter space.

A simple form of a Markov-chain-based fit proceeds as follows. At each step, a proposed set of parameters is selected that is a function of the previous parameters in the Markov chain, and the parameters are accepted based only on the set preceding it. The code makes use of a vector of independent parameters x. Each independent parameter is generated and varied according to a function that depends upon the previous step, x_{old}, in the minimization chain. As a simple example, one might use

$$x = x_{\text{old}} + s(R - 0.5)(x_{\text{min}} - x_{\text{max}}), \tag{39}$$

where $x_{\rm min}$ and $x_{\rm max}$ represent the boundaries on the parameter x; R is a random number between 0 and 1, which is varied as one steps from $x_{\rm old}$ to x; and s is the "stepsize." The acceptance function for x can, for example, be formed as a Boltzmann distribution:

$$P = \min(1, e^{-(\chi^2-\chi^2_{\rm old})/T}) \,, \tag{40}$$

where T is the Markov Chain parameter "temperature." The user-defined step-size and temperature control how quickly the Markov Chain diffuses toward the minimum χ^2 value. At every step along the chain, each of which corresponds to a point in the oscillation space described by the vector of independent parameters, a χ^2 is calculated appropriately for normally or Poisson distributed data. Following a specified number of steps, the minimum χ^2 is found from the list. Then,

$$\Delta\chi^2(\vec{x}) = \chi^2(\vec{x}) - \chi^2_{\rm min}, \tag{41}$$

is calculated for each saved χ^2. These $\Delta\chi^2$ values are used to draw the confidence intervals in plots. All points that satisfy

$$\Delta\chi^2 < {\rm CDF}^{-1}_{\chi^2}(k, p) \tag{42}$$

are drawn inside the interval with probability p. The ${\rm CDF}^{-1}_{\chi^2}$ is the inverse χ^2 cumulative distribution function (CDF) and k is the number of degrees of freedom with $k = 2$ for a two-dimensional plot.

The simple Markov-chain algorithm described above has been used in many of the early fits to SBL data.[10,48,49] However, in the intervening years, much progress has been made in the field of Markov-chain-based optimizers. In particular, algorithms that make efficient use of parallel computing resources have been developed.

Often called Markov chain Monte Carlo (MCMC) methods, these algorithms make probabilistic samples of the parameter space, allowing for Bayesian analysis. An example is to use fitting software with an affine invariant parallel tempering MCMC based on the Emcee fitting package.[51] Details of this approach are described in Ref. 43. This approach runs many Markov chains in parallel, and each chain shares information about the parameters' space with the other chains. Thus, the algorithms can make more efficient proposals, as it adapts to the shape of the χ^2 surface.

Drawing the confidence intervals from the $\Delta\chi^2$ statistic assumes that the statistic is correctly χ^2 distributed. As discussed above, this may not be true, and so the Feldman–Cousins technique[6] was introduced to allow for meaningful confidence intervals in these conditions. However, the method, which involves throwing many fake experiments for each parameter set in the fit, is far too computationally expensive to be practical to use in many global fits. Another way to avoid the issue is to use the log likelihood instead of the χ^2 statistic to determine the probability. This can be done using Bayesian credible intervals, which can be found using the MCMC, as is discussed in Ref. 43.

4.3. *Global fit results*

This section summarizes the status of global fits that make use of the experiments discussed in Section 3. The discussion of specific solutions risks becoming dated quickly, and so here we will be brief and take a "big-picture" approach to the results.

When the data of Table 1 is combined into a global fit, one obtains the result shown in Fig. 21 (left). In this case, three allowed regions are present. These allowed regions are driving the present interest in sterile-neutrino searches. The allowed regions appear at the overlap of the ν_e disappearance and $\nu_\mu \to \nu_e$ appearance signals. This contour is modified by those experiments with limits, especially the ν_μ disappearance limit from SciBooNE/MiniBooNE, which is what is giving the allowed regions the distinct "three island" shape. If one compares the allowed regions of in Fig. 21 (left) to the strongly varying oscillations of the SciBooNE/MiniBooNE result shown in Fig. 1 (right), one sees that the solutions lie between the strongly varying SciBooNE/MiniBooNE contours.

This analysis draws confidence intervals from the $\Delta\chi^2$ statistic, assuming that this statistic is correctly described by a χ^2 distribution with dof equal to the number of fit parameters. In the previous section, we pointed out that

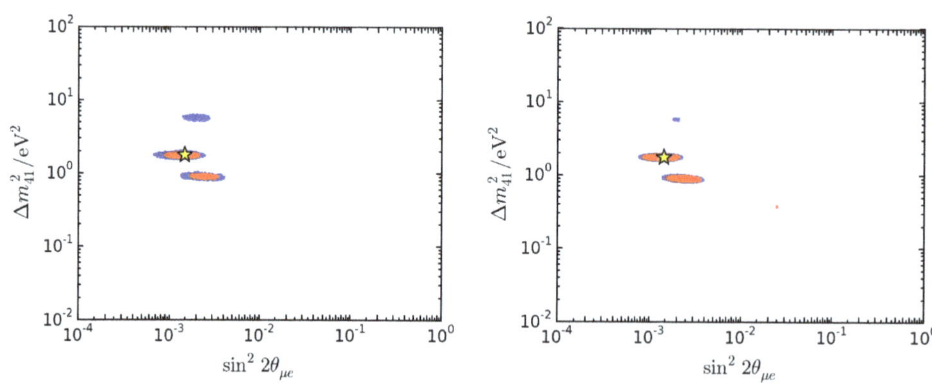

Fig. 21. Left: Global fit to SBL data, showing the frequentist confidence regions. Right: Global fit, showing Bayesian credible intervals.[43]

this assumption may be incorrect for a number of reasons. A comparison to the result from Bayesian credible intervals, Fig. 21 (right), found using the MCMC method previously discussed and described in Ref. 43, shows that the results for the two global fits are very similar. Thus, this check indicates that the $\Delta\chi^2$ distribution assumption must be relatively good.

Including the IceCube result is quite time consuming because of the need to repeatedly propagate the atmospheric flux through the matter profile of the earth, varying the fit parameters with each iteration. Thus, these data were included in the fit by first fitting the SBL data sets, determining the parameter range of interest, and then addressing IceCube for only the viable models. Details of the method are described in Ref. 52. The result of this approach leads to Fig. 22. The IceCube result substantially weakens the $\sim$1 eV^2 allowed region. With this said, it should be noted that including IceCube data in the fits assumes the simplest $3+1$ oscillation model with no additional beyond SM physics. Introducing new physics that would obscure or inhibit the sterile-neutrino related matter effects would weaken the IceCube result. Thus, the IceCube data requires some assumptions beyond the simplest $3+1$ model.

The results of the global fits, with and without IceCube, are summarized in Table 2 results. As per the discussion above, the interesting parameters to compare are the χ^2 for the global best fit point compared to the no oscillations ("null") value. One can see that the difference is on the order of 50 compared to the value 3 or 4 for the dofs. Thus the $\Delta\chi^2$ method is indicating a strong preference for a fit that includes a sterile neutrino. Note that the difference in the null and best-fit χ^2 for the IceCube only fit is small. This is the hallmark of the case where the best-fit point is well outside of

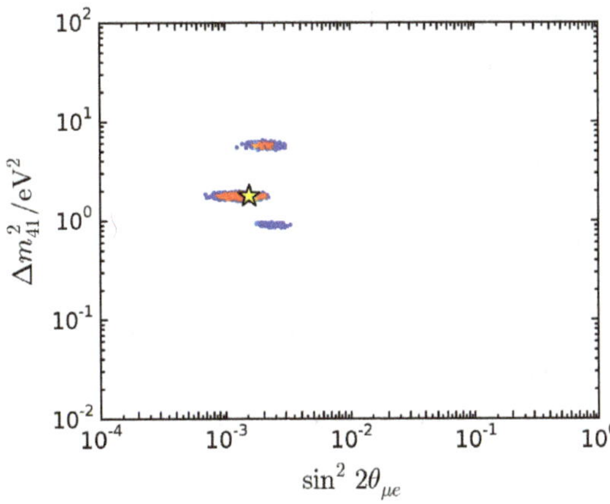

Fig. 22. Frequentist global fit including the IceCube data.[52]

Table 2. The oscillation parameter best-fit points for $3+1$ for the combined SBL and IceCube (IC) data sets compared to SBL alone, as reported in Ref. 52. Units of Δm^2 are eV^2.

$3+1$	Δm^2_{41}	$\|U_{e4}\|$	$\|U_{\mu 4}\|$	$\|U_{\tau 4}\|$	N_{bins}	χ^2_{min}	χ^2_{null}	$\Delta\chi^2$ (dof)
SBL	1.75	0.163	0.117	—	315	306.81	359.15	52.34 (3)
SBL + IC	1.75	0.164	0.119	0.00	524	518.23	568.84	50.61 (4)
IC	5.62	—	0.314	—	209	207.11	209.69	2.58 (2)

the region addressed by the data set. Although the IceCube result has not changed the best-fit point, it does guide future experiments to concentrate on regions that generally have $\Delta m^2 > 1$ eV^2.

At the time of writing, $3+2$ and $3+3$ fits have been performed for the short-baseline data sets only. Including IceCube in the fits is even more computationally time consuming than for $3+1$ due to the additional parameters. However, since the IceCube result did not strongly affect the best fit for $3+1$, it is likely that it will also not strongly affect the $3+2$ and $3+3$ results. It is found that the SBL-fit result for $3+2$ is nearly the same as $3+1$. The $3+2$ fit has a $\Delta\chi^2$[null-min] (dof) of 56.99 (7). Thus adding additional parameters does not affect the conclusions. It is for this reason, and because the fits are quicker and simpler, that, at present, usually only $3+1$ fits are discussed in conferences.

5. Future experiments and requirements

The field of sterile-neutrino experiments is rapidly developing. Follow-up experiments that build on the techniques of the experiments listed in Table 1 are proliferating rapidly. Even more exciting, new and innovative technologies and techniques are being developed to explore electron flavor disappearance (see, for example, IsoDAR[53]), muon flavor disappearance (see, for example, KPIPE[54]), and muon-to-electron appearance experiments (see, for example, MicroBooNE[55]) are underway. A review of all of the planned experiments is beyond the scope of this chapter. Instead, for a review, we refer the reader to articles that are routinely appearing on the arXiv. Instead, here, we approach this discussion of future experiments from an educational point of view. From our summary of the existing results, above, one can identify requirements for future experiments to address the current anomalies and unexplored oscillation regions. We look at these requirements in detail below.

The current muon to electron flavor ($\nu_\mu \to \nu_e$) appearance signals are the strongest indications for oscillations in the Δm^2 around $1\,\text{eV}^2$

region. However, as discussed above, there are tensions between MiniBooNE neutrino data and antineutrino results from MiniBooNE and LSND. Thus, it is essential that the MiniBooNE neutrino result be checked with improved and higher statistics measurements. Here, a new technology, liquid argon time projection chambers (LArTPCs)[56] have the promise of large data samples with improved background rejection and measurement resolution. These systems are large volumes of liquid argon with two subdetectors: a time projection chamber (TPC) for tracking and a light collection system. The MicroBooNE experiment, described in Ref. 55, is an example. LArTPCs have the capability to discriminate electron showers from background γ showers using the dE/dx at the start of the shower. This type of experiment can, therefore, address the potential of a background contamination in the MiniBooNE signal that might arise from electron/γ confusion in a Cherenkov detector. MicroBooNE has just begun an experimental run at Fermilab, on the same beamline and near in location to MiniBooNE. In the future, MicroBooNE is to be combined with a near LArTPC detector (SBND) and a large far LArTPC detector (ICARUS) to form the Short Baseline Neutrino (SBN)[57] program at Fermilab. This program is being designed to have good sensitivity in the regions associated with the current signals by comparing the rates of appearance in the three detectors. This comparison will help constrain systematic uncertainties. The comparison will also have some ability to map out the expected oscillatory behavior of sterile oscillations in that it will measure the observed neutrino rate versus energy at three points.

Improved searches for electron neutrino disappearance are also a prime area for improved experiments. Several radioactive source experiments are being considered using existing large detectors.[58] But the problem with these experiments is that the very hot radioactive sources are short lived, and so the experiment can only run for a modest period of time. As a result, the sensitivity is typically at the level to only just cover the "reactor anomaly" region. Many new very SBL reactor oscillation experiments[59] are also being planned with some prototypes underway. These experiments need to have setups within 10 m from the reactor core and so have to contend with high background rates from neutrons and gammas, which can have confusing position dependence, as well as the large source size of the reactor core, which smears the signal. The reactor flux is also a combination of many decays. It is possible that some decay sources are isolated to specific regions of the reactor, because of the arrangement of fuel. The IsoDAR experiment[53] avoids many of these issues by creating higher energy (8 MeV) electron antineutrinos from ^{8}Li beta decay. This has the advantages of a very intense radioactive source

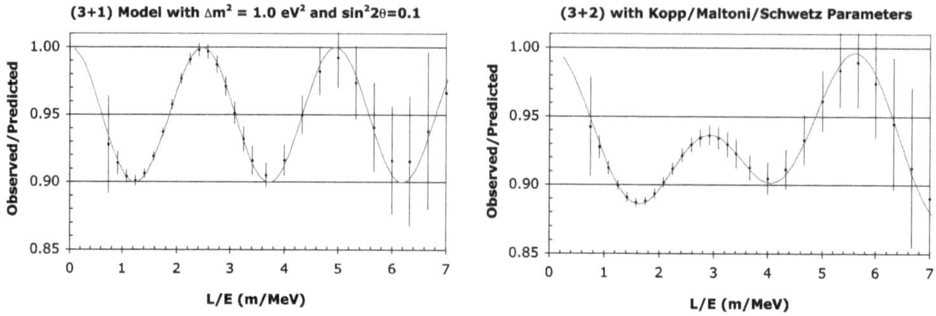

Fig. 23. The L/E dependence of sample data sets for 5 years of running for $3+1$ (left) and $3+2$ (right) oscillation scenarios. The solid curve is the oscillation probability with no smearing in the reconstructed position and energy and the data points with error bars are from simulated events including smearing.

experiments but can run for long periods because this is a "driven source." By this we mean that the ^{8}Li is constantly produced from ^{7}Li using a 60 MeV, 10 mA proton beam from a cyclotron that the collaboration is developing. This ^{8}Li source provides an isotropic source of antineutrinos that would be placed near a large scintillator neutrino detector such as KamLAND or JUNO.[60] For these setups, oscillation signals would be detected by observing the oscillatory behavior within the detector, thus, minimizing systematic uncertainties and providing a definitive signal of oscillations. The oscillatory pattern even has the capability to separate $3+1$ from $3+2$ oscillation models as shown in Fig. 23.

In general, ability to reconstruct the oscillation wave in detail, as shown in Fig. 23, should be a goal for future neutrino experiments. As discussed above, the actual sterile-neutrino model that describes neutrino oscillations may be much more complicated than the $3+1$ world view. However, all of these models will predict oscillation waves. Thus, patterns like these are the "smoking guns" for neutrino oscillations involving sterile neutrinos.

For muon neutrino disappearance, new experiments must improve on the past experiments with better statistics and reduced systematic uncertainties. Improved results from IceCube are expected with somewhat better sensitivity to high Δm^2 and mixing angle. The multidetector SBN program also has good sensitivity to muon neutrino disappearance by comparing the detected rate versus energy in the three detectors and can make measurements with both ν_μ and $\bar{\nu}_\mu$ beam settings. However, the KPIPE experiment[54] is, at present, the one example of a ν_μ disappearance experiment that can observe and use the oscillatory behavior of oscillations to detect a disappearance signal. The neutrino source is from the decay of stopped kaons produced in the J-PARC MLF spallation target, which produces monoenergetic

236 MeV muon neutrinos. Note that a monoenergetic neutrino source that has energy high enough to produce charged-current ν_μ interactions is unique among experiments. In this case, the L/E dependence is reduced to just L dependence, since E is a constant. A 120 m long by 3 m diameter cylindrical liquid-scintillator detector is proposed to be used to detect the neutrino rate as a function of distance. Like IsoDAR, this allows for a search for the oscillatory behavior of the signal, with coverage beyond the other proposed experiments.

These examples give you a taste of the interesting new ideas that are now under discussion. Exploring the possibility that light sterile neutrinos exist is one of the current major goals of particle physics. If sterile neutrinos are established, this will be revolutionary for the field. Previous measurements are a possible guide, particularly when considered within the context of global fits, and have led to an extensive list of new experiments and programs. These experiments are in the small to midscale range and, thus, are real possibilities to be supported and initiated over the next decade. That makes this a great time to be interested in the physics of this thriving field of sterile-neutrino searches.

References

1. https://en.wikipedia.org/wiki/Standard_Model_(mathematical_formulation), See Section 4.3 on "Neutrino Masses".
2. K. N. Abazajian *et al.*, arXiv:1204.5379 [hep-ph].
3. K. A. Olive *et al.* (Particle Data Group), *Chin. Phys. C* **38**, 090001 (2014), and 2015 update.
4. D. Dutta, R. Gandhi, B. Kayser, M. Masud, and S. Prakash, Capabilities of long-baseline experiments in the presence of a sterile neutrino, arXiv:1607.02152 [hep-ph].
5. M. L. Warren, M. Meixner, G. Mathews, J. Hidaka, and T. Kajino, Sterile neutrino oscillations in core-collapse supernovae, *Phys. Rev. D* **90**(10), 103007 (2014) doi:10.1103/PhysRevD.90.103007 [arXiv:1405.6101 [astro-ph.HE]].
6. G. J. Feldman and R. D. Cousins, *Phys. Rev. D* **57**, 3873 (1998) [physics/9711021 [physics.data-an]].
7. A. Aguilar-Arevalo *et al.* [LSND Collaboration], Evidence for neutrino oscillations from the observation of anti-neutrino(electron) appearance in a anti-neutrino(muon) beam, *Phys. Rev. D* **64**, 112007 (2001) [hep-ex/0104049].
8. G. Cheng *et al.* [MiniBooNE and SciBooNE Collaborations], Dual baseline search for muon antineutrino disappearance at $0.1\text{eV}^2 < \Delta m^2 < 100\text{eV}^2$, *Phys. Rev. D* **86**, 052009 (2012) [arXiv:1208.0322].
9. M. C. Gonzalez-Garcia, M. Maltoni, and T. Schwetz, Updated fit to three neutrino mixing: status of leptonic CP violation, *JHEP* **1411**, 052 (2014) doi:10.1007/JHEP11(2014)052 [arXiv:1409.5439 [hep-ph]].

10. M. Sorel, J. M. Conrad, and M. Shaevitz, A Combined analysis of short baseline neutrino experiments in the $(3+1)$ and $(3+2)$ sterile neutrino oscillation hypotheses, *Phys. Rev. D* **70**, 073004 (2004) doi:10.1103/PhysRevD.70.073004 [hep-ph/0305255].
11. A. G. Riess *et al.*, *Astrophys. J.* **826**(1), 56 (2016) [arXiv:1604.01424 [astro-ph.CO]].
12. J. A. Formaggio and J. Barrett, *Phys. Lett. B* **706**, 68 (2011) [arXiv:1105.1326 [nucl-ex]].
13. S. Mertens *et al.*, *JCAP* **1502**(02), 020 (2015) [arXiv:1409.0920 [physics.ins-det]].
14. M. Drewes *et al.*, A White Paper on keV Sterile Neutrino Dark Matter, [arXiv:1602.04816 [hep-ph]].
15. A. Aguilar-Arevalo *et al.*, *Phys. Rev. Lett.* **110**, 161801 (2013).
16. M. Antonello *et al.*, *Eur. Phys. J. C* **73**, 2599 (2013).
17. B. Armbruster *et al.*, *Phys. Rev. D* **65**, 112001 (2002).
18. A. A. Aguilar-Arevalo *et al.*, [MiniBooNE Collaboration], Using L/E Oscillation Probability Distributions, arXiv:1407.3304 [hep-ex].
19. M. G. Aartsen *et al.* [IceCube Collaboration], Searches for sterile neutrinos with the IceCube detector, *Phys. Rev. Lett.* **117**(7), 071801 (2016) doi:10.1103/PhysRevLett.117.071801 [arXiv:1605.01990 [hep-ex]].
20. A. Y. Smirnov, The MSW effect and solar neutrinos, hep-ph/0305106.
21. A. Esmaili and A. Y. Smirnov, Restricting the LSND and MiniBooNE sterile neutrinos with the IceCube atmospheric neutrino data, *JHEP* **1312**, 014 (2013) doi:10.1007/JHEP12(2013)014 [arXiv:1307.6824 [hep-ph]].
22. C. Y. Cardall and G. M. Fuller, Three generation neutrino mixing and LSND dark matter neutrinos, *Nucl. Phys. Proc. Suppl.* **51B**, 259 (1996) doi:10.1016/S0920-5632(96)00510-5 [astro-ph/9606024].
23. A. A. Aguilar-Arevalo *et al.* [MiniBooNE Collaboration], arXiv:1407.3304 [hep-ex].
24. A. A. Aguilar-Arevalo *et al.* [MiniBooNE Collaboration], *Phys. Rev. Lett.* **102**, 101802 (2009) [arXiv:0812.2243 [hep-ex]].
25. A. A. Aguilar-Arevalo *et al.* [MiniBooNE Collaboration], *Phys. Rev. Lett.* **98**, 231801 (2007) [arXiv:0704.1500 [hep-ex]].
26. A. A. Aguilar-Arevalo *et al.* [MiniBooNE Collaboration], *Phys. Rev. Lett.* **105**, 181801 (2010) doi:10.1103/PhysRevLett.105.181801 [arXiv:1007.1150 [hep-ex]].
27. G. Mention, M. Fechner, T. Lasserre, T. A. Mueller, D. Lhuillier, M. Cribier, and A. Letourneau, *Phys. Rev. D* **83**, 073006 (2011) [arXiv:1101.2755 [hep-ex]].
28. Y. Declais, J. Favier, A. Metref, H. Pessard, B. Achkar *et al. Nucl. Phys.* **B434**, 503 (1995).
29. J. N. Abdurashitov *et al.* [SAGE Collaboration], *Phys. Rev. C* **80**, 015807 (2009) [arXiv:0901.2200 [nucl-ex]].
30. F. Kaether, W. Hampel, G. Heusser, J. Kiko, and T. Kirsten, *Phys. Lett. B* **685**, 47 (2010) [arXiv:1001.2731 [hep-ex]].
31. J. M. Conrad and M. H. Shaevitz, *Phys. Rev. D* **85**, 013017 (2012) [arXiv:1106.5552 [hep-ex]].

32. P. Astier *et al.* [NOMAD Collaboration], *Phys. Lett. B* **570**, 19 (2003) [hep-ex/0306037].
33. I. E. Stockdale *et al.*, *Z. Phys. C* **27**, 53 (1985).
34. F. Dydak *et al.*, *Phys. Lett. B* **134**, 281 (1984).
35. K. B. M. Mahn *et al.* [SciBooNE and MiniBooNE Collaborations], *Phys. Rev. D* **85**, 032007 (2012) [arXiv:1106.5685 [hep-ex]].
36. P. Adamson *et al.* [MINOS Collaboration], Submitted to: *Phys. Rev. Lett.* [arXiv:1607.01176 [hep-ex]].
37. P. Adamson *et al.* [MINOS Collaboration], *Phys. Rev. D* **77**, 072002 (2008) [arXiv:0711.0769 [hep-ex]].
38. M. G. Aartsen *et al.* [IceCube Collaboration], *Phys. Rev. Lett.* **117**(7), 071801 (2016) [arXiv:1605.01990 [hep-ex]].
39. J. Kopp, P. A. N. Machado, M. Maltoni, and T. Schwetz, *JHEP* **1305**, 050 (2013) [arXiv:1303.3011 [hep-ph]].
40. P. Huber, Reactor antineutrino fluxes —Status and challenges, *Nucl. Phys. B* **908**, 268 (2016) doi:10.1016/j.nuclphysb.2016.04.012 [arXiv:1602.01499 [hep-ph]].
41. A. Anokhina *et al.*, arXiv:1503.07471 [hep-ph].
42. S. Gariazzo, C. Giunti, M. Laveder, Y. F. Li, and E. M. Zavanin, *J. Phys. G* **43**, 033001 (2016) [arXiv:1507.08204 [hep-ph]].
43. G. H. Collin, C. A. Argüelles, J. M. Conrad, and M. H. Shaevitz, *Nucl. Phys. B* **908**, 354 (2016) [arXiv:1602.00671 [hep-ph]].
44. Vincent Spruyt (Sentiance UK LTD), http://www.visiondummy.com/2014/04/.
45. F. J. Anscombe, "Graphs in statistical analysis." *Am. Stat.* **27**(1), 17 (1973).
46. https://en.wikipedia.org/wiki/Anscombe%27s_quartet.
47. M. Maltoni and T. Schwetz, *Phys. Rev. D* **68**, 033020 (2003) [hep-ph/0304176].
48. G. Karagiorgi, A. Aguilar-Arevalo, J. M. Conrad, M. H. Shaevitz, K. Whisnant, M. Sorel, and V. Barger, Leptonic CP violation studies at MiniBooNE in the $(3+2)$ sterile neutrino oscillation hypothesis, *Phys. Rev. D* **75**, 013011 (2007) [Erratum: [*Phys. Rev. D* **80**, 099902 (2009)]].
49. J. M. Conrad, C. M. Ignarra, G. Karagiorgi, M. H. Shaevitz, and J. Spitz, Sterile neutrino fits to short baseline neutrino oscillation measurements, *Adv. High Energy Phys.* **2013**, 163897 (2013) doi:10.1155/2013/163897 [arXiv:1207.4765 [hep-ex]].
50. J. R. Norris, *Markov Chains*, Cambridge: Cambridge University Press (1998). See sections downloadable at: http://www.statslab.cam.ac.uk/~james/Markov/.
51. D. Foreman-Mackey, D. W. Hogg, D. Lang, and J. Goodman, *Publ. Astron. Soc. Pacific* **125**, 306 (2013), arXiv: 1202.3665.
52. G. H. Collin, C. A. Argelles, J. M. Conrad, and M. H. Shaevitz, First constraints on the complete neutrino mixing matrix with a sterile neutrino, arXiv:1607.00011 [hep-ph].
53. M. Abs *et al.*, IsoDAR@KamLAND: A conceptual design report for the technical facility, arXiv:1511.05130 [physics.acc-ph]; A. Bungau *et al.*, Proposal for

an electron antineutrino disappearance search using high-rate ^{8}Li production and decay, *Phys. Rev. Lett.* **109**, 141802 (2012) [arXiv:1205.4419 [hep-ex]].
54. S. Axani, G. Collin, J. Conrad, M. Shaevitz, J. Spitz, and T. Wongjirad, Decisive disappearance search at high Δm^2 with monoenergetic muon neutrinos, *Phys. Rev. D* **92**(9), 092010 (2015). doi:10.1103/PhysRevD.92.092010 [arXiv:1506.05811 [physics.ins-det]].
55. T. Miceli (for the MicroBooNE collaboration), Exploring Neutrino Interactions with MicroBooNE, arXiv:1411.4572 [physics.ins-det].
56. A. Marchionni, Status and new ideas regarding liquid argon detectors, *Ann. Rev. Nucl. Part. Sci.* **63**, 269 (2013). doi:10.1146/annurev.nucl.012809.104445 [arXiv:1307.6918 [physics.ins-det]].
57. M. Antonello *et al.* [MicroBooNE and LAr1-ND and ICARUS-WA104 Collaborations], A proposal for a three detector short-baseline neutrino oscillation program in the fermilab booster neutrino beam, arXiv:1503.01520 [physics.ins-det].
58. T. Lasserre, Testing the reactor and gallium anomalies with intense (anti)neutrino emitters, *Nucl. Phys. Proc. Suppl.* **235–236**, 214 (2013). doi: 10.1016/ j.nuclphysbps.2013.04.013 [arXiv:1209.5090 [physics.ins-det]].
59. D. Lhuillier, Future short-baseline sterile neutrino searches with reactors, *AIP Conference Proc.* **1666**, 180003 (2015); http://dx.doi.org/10.1063/1.4915600
60. J. M. Conrad and M. H. Shaevitz, Electron antineutrino disappearance at KamLAND and JUNO as decisive tests of the short baseline $\overline{\nu}_\mu \to \overline{\nu}_e$ appearance anomaly, *Phys. Rev. D* **89**(5), 057301 (2014). doi:10.1103/PhysRevD.89.057301 [arXiv:1310.3857 [hep-ex]].

Chapter 11

Dirac and Majorana Neutrinos, Double Beta Decay

J.-L. Vuilleumier

Albert Einstein Center for Fundamental Physics, Sidlerstrasse 5, Bern CH-3012, Switzerland

1. Introduction

As a consequence of the Dirac equation, fermions are known to exist in two forms, particle and antiparticle. As pointed out by Ettore Majorana, however, neutrinos may be an exception. They might be truly neutral, particle and antiparticle being identical. If this is the case, lepton number violating processes are expected to happen. They are entirely suppressed by helicity if the mass is exactly zero. But we now know that the neutrino masses, although tiny, are finite. They break the helicity, with an amplitude proportional to mass divided by energy. Lepton number violation is thus best probed at low energy. The search of neutrinoless double beta decay in nuclei appears to be the most sensitive approach. Many textbooks and review articles[1–6] discuss neutrino properties and double beta decay, and this chapter borrows much from them. They provide additional information. More fundamental concepts can be found in Refs. 7–10.

2. The Dirac equation

Neutrinos only interact through the weak interaction, which conserves the lepton number L ($L = +1$ for the neutrinos and $L = -1$ for the antineutrinos). The wavefunction ν obeys the Dirac equation

$$(i\gamma^{\mu}\partial_{\mu} - m)\nu = 0 \tag{1}$$

with

$$\{\gamma^{\mu}, \gamma^{\nu}\} = 2g^{\mu\nu}. \tag{2}$$

We work in the Weyl or chiral representation, where the γ^μ matrices are

$$\gamma^0 = \begin{pmatrix} 0 & -1 \\ -1 & 0 \end{pmatrix} \quad \gamma^i = \begin{pmatrix} 0 & \sigma^i \\ -\sigma^i & 0 \end{pmatrix}, \; i = 1,\, 2,\, 3.^{\text{a}} \tag{3}$$

We are looking for plane wave solutions $\nu(x)$, describing free neutrinos with 4-momentum $p^\mu = (E, \vec{p})$, in the form

$$\nu(x) = \omega e^{-ipx}, \tag{4}$$

and ω is a spinor of rank 4. There are four independent solutions, two with positive energy ($E > 0$) and two with negative energy ($E < 0$). We remind that in the Stückelberg–Feynman formalism, the interpretation of the positive energy solutions is straightforward, describing physical neutrinos ν. The negative energy solutions propagate backward in time in the equations and describe in the real world antineutrinos $\overline{\nu}$ propagating forward in time with energy $-E > 0$, momentum $-\vec{p}$, spin $-\vec{s}$, and thus unchanged helicity

$$h = \frac{\vec{s}\cdot\vec{p}}{|\vec{s}||\vec{p}|}, \quad \vec{s} = \frac{1}{2}\begin{pmatrix} \vec{\sigma} & 0 \\ 0 & \vec{\sigma} \end{pmatrix}.$$

These four independent solutions describe the so-called Dirac neutrinos and antineutrinos in two different spin or helicity states.

We define

$$\gamma_5 = i\gamma^0\gamma^1\gamma^2\gamma^3 = \begin{pmatrix} 1 & 0 \\ 0 & -1 \end{pmatrix}, \quad \text{with } \{\gamma^\mu, \gamma_5\} = 0,\; \gamma_5^2 = 1 \tag{5}$$

and introduce the operators

$$P_L = \frac{1}{2}(1-\gamma_5) = \begin{pmatrix} 0 & 0 \\ 0 & 1 \end{pmatrix}, \quad P_R = \frac{1}{2}(1+\gamma_5) = \begin{pmatrix} 1 & 0 \\ 0 & 0 \end{pmatrix}. \tag{6}$$

One sees that they satisfy

$$P_L P_R = 0, \quad P_R P_L = 0, \quad P_L^2 = P_L, \quad P_R^2 = P_R, \quad P_L + P_R = 1, \tag{7}$$

and they are projectors. The spinors

$$\nu_L = P_L\nu = \frac{1}{2}(1-\gamma_5)\nu \text{ and } \nu_R = P_R\nu = \frac{1}{2}(1+\gamma_5)\nu$$

are the left-handed and right-handed chiral projections of ν. They verify

$$\nu = \nu_L + \nu_R.$$

[a] The Pauli matrices $\sigma^1 = \begin{pmatrix} 0 & 1 \\ 1 & 0 \end{pmatrix}$, $\sigma^2 = \begin{pmatrix} 0 & -i \\ i & 0 \end{pmatrix}$, $\sigma^3 = \begin{pmatrix} 1 & 0 \\ 0 & -1 \end{pmatrix}$ satisfy $\{\sigma^i, \sigma^j\} = 2\delta_{ij}$.

The left-handed projection ν_L is the one that is active in the weak interaction, whether mediated by charged or neutral currents. The right-handed projection ν_R is inactive.

For future use, we remind that one can build scalars $\overline{\nu}\nu$ starting from $\bar{\nu} = \nu^{*T}\gamma^0$. Considering that γ_5 anticommutes with all the γ^μ, in particular γ^0, we see that

$$\overline{\nu}_L\nu_L = \overline{\nu}_R\nu_R = 0. \tag{8}$$

In contrast $\overline{\nu}_L\nu_R$, $\overline{\nu}_R\nu_L$ are finite so that

$$\overline{\nu}\nu = \overline{\nu}_L\nu_R + \overline{\nu}_R\nu_L. \tag{9}$$

2.1. *Dirac neutrinos*

We now formally solve the Dirac equation with the Ansatz (Eq. (4)) and

$$\omega = \begin{pmatrix} \Phi \\ \chi \end{pmatrix},$$

Φ and χ being spinors of rank 2. The equation becomes

$$\begin{aligned} (-E - \sigma^i p^i)\chi &= m\Phi, \\ (-E + \sigma^i p^i)\Phi &= m\chi. \end{aligned} \tag{10}$$

Combining the two implies $(-E \mp \sigma^i p^i)(-E \pm \sigma^j p^j)\Phi(\chi) = m^2\Phi(\chi)$, or $E^2 = p^2 + m^2$.

With $m = 0$, the two equations decouple. The solutions correspond to states with definite helicity:

$$\begin{array}{ll} 1.\ \chi = 0, \Phi \text{ arbitrary, } E = \sigma^i p^i, & 2.\ \Phi = 0, \chi \text{ arbitrary, } E = -\sigma^i p^i, \\ E > 0\colon\ h = \frac{\sigma^i p^i}{|p|} = +1, & E > 0\colon\ h = \frac{\sigma^i p^i}{|p|} = -1, \\ E < 0\colon\ h = \frac{\sigma^i p^i}{|p|} = -1, & E < 0\colon\ h = \frac{\sigma^i p^i}{|p|} = +1. \end{array} \tag{11}$$

States 2 satisfy $\nu = P_L\nu = \nu_L$ and are eigenstates of P_L. They are fully active in the weak interaction; they describe neutrinos ν with helicity $h = -1$ and $L = +1$ and antineutrinos $\overline{\nu}$ with $h = +1$ and $L = -1$. States 1 satisfy $\nu = P_R\nu = \nu_R$ and are eigenstates of P_R. They are totally inactive in the weak interaction. They correspond to neutrinos ν with $h = +1$ and $L = +1$ and antineutrinos $\overline{\nu}$ with $h = -1$ and $L = -1$.

By a suitable rotation of the reference frame, the spin can be brought to the state $\pm 1/2$. The momentum is then aligned along the

z-axis ($\vec{p} = (0, 0, p)$). The four eigenstates are given by

$$\overbrace{\begin{pmatrix}1\\0\\0\\0\end{pmatrix}}^{E>0,p>0,h=+1} \quad \overbrace{\begin{pmatrix}0\\1\\0\\0\end{pmatrix}}^{E<0,p>0,h=-1} \quad \underbrace{\overbrace{\begin{pmatrix}0\\0\\1\\0\end{pmatrix}}^{E>0,p<0,h=-1} \quad \overbrace{\begin{pmatrix}0\\0\\0\\1\end{pmatrix}}^{E<0,p<0,h=+1}}_{\text{weak interaction}}.$$

In the limit $m \to 0$, the states ν_R are irrelevant and are ignored. The distinction between particle and antiparticle, or $L = +1$ and $L = -1$ states, is redundant with that between $h = -1$ and $h = +1$ and is unnecessary. Helicity implies lepton number conservation.

We know now that neutrinos have masses and that the above picture is only an approximation. We can still classify the states in function of the helicity. We keep $\vec{p} = (0, 0, p)$, pay no attention to the absolute normalization, and solving Eq. (10) find

$$\overbrace{\begin{pmatrix}1\\0\\-\frac{E-p}{m}\\0\end{pmatrix}}^{E>0,\ p>0,h=+1} \quad \overbrace{\begin{pmatrix}0\\1\\0\\-\frac{E+p}{m}\end{pmatrix}}^{E<0,\ p>0,h=-1} \quad \overbrace{\begin{pmatrix}-\frac{E+p}{m}\\0\\1\\0\end{pmatrix}}^{E>0,p<0,h=-1} \quad \overbrace{\begin{pmatrix}0\\-\frac{E-p}{m}\\0\\1\end{pmatrix}}^{E<0,p<0,h=+1}. \tag{12}$$

The chiral projections $\nu_L = P_L\nu$ and $\nu_R = P_R\nu$ are no longer solutions of the Dirac equations. With nonvanishing masses m, the weak interaction no longer strictly selects given helicities.

The neutrino masses are tiny and in most situations we can use the approximation $p = \pm\sqrt{E^2 - m^2} \cong \pm E(1 - \frac{1}{2}\frac{m^2}{E^2})$. With that, Eq. (12) becomes

$$\overbrace{\begin{pmatrix}1\\0\\-\frac{m}{2E}\\0\end{pmatrix}}^{E>0,\ p>0,h=+1} \quad \overbrace{\begin{pmatrix}0\\1\\0\\-\frac{m}{2E}\end{pmatrix}}^{E<0,\ p>0,h=-1} \quad \overbrace{\begin{pmatrix}-\frac{m}{2E}\\0\\1\\0\end{pmatrix}}^{E>0,p<0,h=-1} \quad \overbrace{\begin{pmatrix}0\\-\frac{m}{2E}\\0\\1\end{pmatrix}}^{E<0,p<0,h=+1}. \tag{13}$$

P_L lets through to a large extent states 3 ($E > 0$, $h = -1$) and 4 ($E < 0$, $h = +1$). With $m \ll |E|$, the weak interaction still favors the helicities allowed with $m = 0$. But P_L also lets through the opposite helicity states 1 ($E > 0$, $h = +1$) and 2 ($E < 0$, $h = -1$) with an amplitude of order m/E. They participate in the weak interaction with a probability of order $(m/E)^2$. Nevertheless, in this Dirac neutrino scenario, lepton number violating processes are still forbidden since neutrinos and antineutrinos are distinct.

There is one class of neutrinos which may be at least partially nonrelativistic, the relic neutrinos from the Big Bang today. It is thus interesting to see what happens in the limit $E \to \pm m$, $p \to 0$. One sees from Eq. (12) that all coefficients converge towards ± 1 so that, applying P_L, the neutrinos with $h = +1$ and $h = -1$ are equivalent with regard to the weak interaction. So are the antineutrinos with $h = +1$ and $h = -1$.

2.2. *Charge conjugation*

Before going further, we note that the wave function of a positive energy state with 4-momentum p^μ can be transformed into that of a negative energy state with 4-momentum $-p^\mu$ by the charge conjugation operation and vice versa. In the Weyl representation, it is written as

$$\nu^c = i\gamma^2\nu^* = C\gamma^0\nu^* = C\bar{\nu}^T \tag{14}$$

with

$$i\gamma^2 = \begin{pmatrix} 0 & 0 & 0 & 1 \\ 0 & 0 & -1 & 0 \\ 0 & -1 & 0 & 0 \\ 1 & 0 & 0 & 0 \end{pmatrix} \quad C = i\gamma^2\gamma^0 = \begin{pmatrix} 0 & -1 & 0 & 0 \\ 1 & 0 & 0 & 0 \\ 0 & 0 & 0 & 1 \\ 0 & 0 & -1 & 0 \end{pmatrix}.$$

The spin is inverted. All the additive quantum numbers, for neutrinos the lepton number L, change sign. We note for future use that

$$C = -C^{-1} = -C^T. \tag{15}$$

It implies $C^2 = -1$ and $(\nu^c)^c = \nu$. Also, using the anticommutation rules (Eqs. (2) and (5)) we see that

$$(\nu_L)^c = \frac{1}{2}(1+\gamma_5)\nu^c = (\nu^c)_R, \qquad (\nu_R)^c = \frac{1}{2}(1-\gamma_5)\nu^c = (\nu^c)_L. \tag{16}$$

Thus, the charge conjugate of a left-handed projection is the right-handed projection of the conjugate state and vice versa.

2.3. *Majorana neutrinos*

Neutrinos are electrically neutral. In addition, they may be truly neutral, neutrinos being identical to antineutrinos with a lepton number equal to zero, since the apparent lepton number conservation, taking the Dirac assignment, may be due to helicity. Neutrinos could thus be described by the linear combinations

$$\nu_1^M = \frac{1}{\sqrt{2}}(\nu + \nu^c), \;\; \nu_2^M = \frac{1}{\sqrt{2}i}(\nu - \nu^c), \tag{17}$$

which satisfy the Dirac equation. These forms describe Majorana neutrinos and fulfill the condition

$$(\nu_i^M)^c = \nu_i^M,$$

expressing that neutrinos and antineutrinos are indeed identical. Incidentally, inverting Eq. (17), one sees that Dirac neutrinos can be thought of as superpositions of Majorana neutrinos with degenerate masses.

For illustration, we construct ν_1^M. We consider the case $m \ll E$ and use Eq. (13). We start from the spinors with $E > 0$ and $h = +1$, $h = -1$, use Eq. (14) to evaluate the charge conjugate, and add the two. The coefficient $e^{\pm ipx}$, with $-ipx = -i(Et - pz)$, $E > 0$, is explicitely written since it differs for the various components of the spinor:

$$\overbrace{\begin{pmatrix} e^{-ipx} \\ \frac{m}{2E}e^{ipx} \\ -\frac{m}{2E}e^{-ipx} \\ e^{ipx} \end{pmatrix}}^{h=+1,\,p>0} \; \overbrace{\begin{pmatrix} -\frac{m}{2E}e^{-ipx} \\ -e^{ipx} \\ e^{-ipx} \\ -\frac{m}{2E}e^{ipx} \end{pmatrix}}^{h=-1,\,p<0}. \tag{18}$$

Doing the same operation starting with the spinors with $E < 0$ of Eq. (13) leads to the same result. The components with e^{-ipx} propagate forward in time, while the ones with e^{ipx} propagate backward in time. The first ones act like neutrinos; the latter ones like antineutrinos. Applying P_L projects the weak interaction component:

$$\overbrace{\begin{pmatrix} 0 \\ 0 \\ -\frac{m}{2E}e^{-ipx} \\ e^{ipx} \end{pmatrix}}^{h=+1,\,p>0} \; \overbrace{\begin{pmatrix} 0 \\ 0 \\ e^{-ipx} \\ -\frac{m}{2E}e^{ipx} \end{pmatrix}}^{h=-1,\,p<0}. \tag{19}$$

With $m = 0$, the spinors have only one nonvanishing component. The left one describes Majorana neutrinos with $h = +1$, behaving like Dirac antineutrinos with $h = +1$. The right one corresponds to Majorana neutrinos with $h = -1$, behaving like Dirac neutrinos with $h = -1$. The phenomenology is the same as with massless Dirac neutrinos, and lepton number violating processes, using the usual assignment, are entirely suppressed by helicity.

With m nonvanishing, a second component develops in each spinor, identical to the main one in the other spinor, propagating in opposite direction in time with respect to the main component in its own spinor. It has an amplitude of order m/E. This means that a Majorana neutrino with $h = -1$ behaves essentially like a Dirac neutrino; with an amplitude of order m/E, like a Dirac antineutrino; and conversely for the neutrino with $h = +1$. Lepton number violating processes, assuming the traditional assignment, are expected with a probability of order $(m/E)^2$. The phenomenology is thus different in the Dirac and Majorana cases.

Nuclear double beta decay is a situation where indeed $m \ll E$ is satisfied, but not excessively, and the effect hopefully measurable. It is presently the main line of research to probe lepton number conservation and the nature of neutrinos. It will be discussed in Section 3.10.

Still with the relic neutrinos in mind, we look at what happens in the nonrelativistic limit $p \to 0$, $E \to \pm m$. Starting from Eq. (12), we obtain the same result as Eqs. (18) and (19), with the factor $m/2E$ replaced by 1. The weak interaction equally treats the Majorana neutrinos with $h = -1$ and $h = +1$.

2.4. *Neutrino masses*

To go further and take into account that neutrinos exist in several subspecies, we must be a little more formal and turn to field theory. The basic ingredient is the Lagrange density. For a single neutrino field ν of mass m_D, it is given by

$$\mathcal{L} = \bar{\nu} i\gamma^\mu \partial_\mu \nu - m_D \bar{\nu}\nu.$$

With this, using the Euler–Lagrange equation, one sees that the field obeys the Dirac equation. The form above is not the most general however. As described in Refs. 1 and 4, the Lagrange density may contain one or more mass terms, which must be hermitian and Lorentz invariant. This leaves two possibilities, $m_D\bar{\nu}\nu \equiv m_D\bar{\nu}^c\nu^c$ for Dirac masses, as above, and $m_M\bar{\nu}^c\nu$ plus hermitian conjugate for Majorana masses. For hermiticity, the Dirac mass m_D must be real. The Dirac mass term is invariant under the global gauge transformation; $\nu \to e^{i\alpha}\nu$. The associated conserved quantum number is the lepton number L. Majorana mass terms are not invariant under the gauge transformation; they break the lepton number by two units. They also break the electric charge and are obviously allowed only for neutral particles.

It is interesting to write the Lagrange density in function of the projections ν_L and ν_R. Because of Eq. (9), we see that

$$\bar{\nu}\nu = \bar{\nu}_L\nu_R + \bar{\nu}_R\nu_L.$$

Likewise, using Eq. (16)

$$\bar{\nu}\nu^c = \bar{\nu}_L(\nu_L)^c + \bar{\nu}_R(\nu_R)^c.$$

The Lagrange density may thus contain Dirac mass terms in $\bar{\nu}_L\nu_R$ and $\bar{\nu}_R\nu_L$, as well as Majorana mass terms in $\bar{\nu}_L(\nu_L)^c$ and $\bar{\nu}_R(\nu_R)^c$, not necessarily with the same coefficients since nature treats left-handed and right-handed neutrinos differently, plus hermitian conjugate.

The standard model assumes no neutrino mass and has three left-handed fields $\nu_{\ell L}$, $\ell = e,\, \mu,\, \tau$, corresponding to the flavor states active in the weak interaction. We must extend it to incorporate mass terms by introducing a number n of right-handed fields ν_{jR}. Dropping the kinetic term, the Lagrange density then reads

$$-\mathcal{L}_M = \sum_{i,\ell} M^D_{i\ell}\overline{\nu}_{iR}\nu_{\ell L} + \sum_{i,j}\frac{1}{2}M^M_{ij}\,(\overline{\nu}_{iR})(\nu_{jR})^c + h.c., \quad i,j = 1,\ldots,n. \tag{20}$$

Here, M^D is a $n \times 3$ matrix characterizing a Dirac mass term; and M^M, a $n \times n$ matrix for a Majorana mass term. We introduce

$$\nu_L = \begin{pmatrix} \nu_{\ell L} \\ (\nu_{jR})^c \end{pmatrix}, \tag{21}$$

in which all fields are left-handed since $(\nu_{jR})^c = (\nu_j^c)_L$. The Lagrange density can be written in a more compact form:

$$-\mathcal{L}_M = \frac{1}{2}\,\overline{(\nu_L)^c}M_\nu\nu_L \; + h.c. \text{ with } M_\nu = \begin{pmatrix} 0 & (M^D)^T \\ M^D & M^M \end{pmatrix}.$$

In the basis of the eigenvectors ν'_{iL} of M_ν, with $i = 1\ldots N$, $N = 3+n$, the Lagrange density is

$$-\mathcal{L}_M = \frac{1}{2}\left(\overline{(\nu'_{1L})^c}\,...\,\overline{(\nu'_{NL})^c}\right)\begin{pmatrix} m_1 & & 0 \\ & \ddots & \\ 0 & & m_N \end{pmatrix}\begin{pmatrix} \nu'_{1L} \\ \vdots \\ \nu'_{NL} \end{pmatrix} + h.c. \tag{22}$$

With

$$\nu_i = \nu'_{iL} + (\nu'_{iL})^c \text{ satisfying } \nu_i = \nu_i^c,$$

considering that $\overline{\nu'}_{iL}\nu'_{iL} = \overline{(\nu'_{iL})^c}(\nu'_{iL})^c = 0$, the Lagrange density becomes

$$-\mathcal{L}_M = \frac{1}{2}\left(\bar{\nu}_1 \ldots \bar{\nu}_N\right)\begin{pmatrix} m_1 & & 0 \\ & \ddots & \\ 0 & & m_N \end{pmatrix}\begin{pmatrix} \nu_1 \\ \vdots \\ \nu_N \end{pmatrix}. \tag{23}$$

The fields ν_i correspond to the eigenstates with mass m_i. They describe Majorana neutrinos. The fields ν_L and the left-handed projection $\nu_{iL} = \nu'_{iL}$ are connected by a unitary matrix U ($UU^\dagger = 1$)

$$\begin{pmatrix} \nu_{\ell L} \\ (\nu_{jR})^c \end{pmatrix} = U\begin{pmatrix} \nu_{1L} \\ \vdots \\ \nu_{NL} \end{pmatrix} \text{ such that } \begin{pmatrix} m_1 & & 0 \\ & \ddots & \\ 0 & & m_N \end{pmatrix} = U^T M_\nu U. \tag{24}$$

To give an idea, we restrict ourselves first to the simple case of one flavor ν_{eL} and one right-handed neutrino ν_{1R} ($n = 1$). The matrices in M_ν are then scalars, which we assume real, and the diagonalization gives

$$m_1 = \frac{1}{2}\left(M^M - \sqrt{4(M^D)^2 + (M^M)^2}\right),$$
$$m_2 = \frac{1}{2}\left(M^M + \sqrt{4(M^D)^2 + (M^M)^2}\right).$$

The mixing matrix depends on one mixing angle θ,

$$U = \begin{pmatrix} \cos\theta & \sin\theta \\ -\sin\theta & \cos\theta \end{pmatrix} \text{ with } \tan 2\theta = \frac{2M_D}{M_M}.$$

Let us consider two situations:

(1) $M^M = 0$: $m_1 = -M^D$, $m_2 = M^D$. The mixing is maximal with $\theta = \pi/4$. The negative mass m_1 can be made positive in Eq. (24) by adding a phase to U:

$$U = \begin{pmatrix} \cos\theta & \sin\theta \\ -\sin\theta & \cos\theta \end{pmatrix}\begin{pmatrix} i & 0 \\ 0 & 1 \end{pmatrix}.$$

This rule can be generalized to avoid negative masses. Anyway, ν_{eL} and $\nu_{1R} \equiv \nu_{eR}$ are the two components of a Dirac neutrino ν_e with Dirac mass M^D.

(2) $M^M \gg M^D$: $m_1 \cong -\frac{(M^D)^2}{M^M}$, $m_2 \cong M^M$. Again, the negative mass m_1 can be made positive. In any event, we have a very light neutrino, whose mass is pushed down by M^M, and a very heavy one. The mixing

angle $\theta \cong \frac{M^D}{M^M}$ is very small, and the left-handed flavor projection ν_{eL} is essentially the left-handed projection of a light Majorana neutrino with mass m_1. The right-handed field ν_{1R} is in good approximation the right-handed projection of the field with the mass $\cong M^M$. This is a simple realization of the seesaw mechanism,[11–13] which provides an explanation to the smallness of the neutrino mass.

With three flavors $\ell = e, \mu, \tau$ and n arbitrary, the scale of M^M much larger than that of M^D, one has a more realistic realization of the seesaw mechanism. There are three light Majorana neutrinos with masses of order $\frac{(M^D)^2}{M^M}$. The three components ν_{eL}, $\nu_{\mu L}$ and $\nu_{\tau L}$ are only very weakly admixed to the n heavy neutrinos, the mixing being of order $\frac{M^D}{M^M}$. They are essentially linear combinations of the left-handed projection of three light Majorana neutrinos:

$$\nu_{\ell L} = \sum_{i=1}^{3} U_{\ell i} \nu_{iL}. \tag{25}$$

The matrix U can be described by the usual PMNS matrix, depending on three mixing angles θ_{12}, θ_{13} and θ_{23} and a CP-violating phase δ, multiplied by a matrix containing two additional CP-violating phase differences α_{21} and α_{31} specific of Majorana neutrinos, and manifesting themselves in lepton number violating processes only. With the notation $s_{ij} = \sin\theta_{ij}$, $c_{ij} = \cos\theta_{ij}$, it reads

$$\begin{pmatrix} c_{12}c_{13} & s_{12}c_{13} & s_{13}e^{-i\delta} \\ c_{12}s_{13}s_{23}e^{i\delta} - s_{12}c_{23} & c_{12}c_{23} - s_{12}s_{13}s_{23}e^{i\delta} & c_{13}s_{23} \\ -s_{12}s_{23} - c_{12}s_{13}c_{23}e^{i\delta} & c_{12}s_{23} - s_{12}s_{13}c_{23}e^{i\delta} & c_{13}c_{23} \end{pmatrix} \begin{pmatrix} 1 & 0 & 0 \\ 0 & e^{i\frac{\alpha_{21}}{2}} & 0 \\ 0 & 0 & e^{i(\delta + \frac{\alpha_{31}}{2})} \end{pmatrix}. \tag{26}$$

With three left-handed flavor fields, three right-handed fields ($n = 3$), $M^M = 0$, the left-handed $\nu_{\ell L}$ and right-handed $\nu_{jR} \equiv \nu_{\ell R}$ fields become the left-handed and right-handed components of three Dirac neutrinos ν_ℓ. The mixing between flavor eigenstates ν_ℓ and mass eigenstates ν_i is again the PMNS matrix (Eq. (26)) without the Majorana CP phases α_{21} and α_{31}.

Many more elaborate scenarios have been discussed in the literature (see, for instance, Ref. 4). It is in particular possible to introduce additional sterile neutrinos, should experiments demonstrate their existence, altering the mixing pattern. We do not elaborate any further here but conclude that Majorana neutrinos are an interesting hypothesis, since they are implied by the seesaw mechanism, which provides a nice explanation to the smallness of the masses.

3. Double beta decay as a probe of the nature of neutrinos

To get around the helicity suppression when probing lepton number conservation, it is necessary to look at low-energy processes. Nuclear double beta decay is most appropriate. It can occur in nuclei (A, Z) which are stable with respect to beta decay, because the nucleus $(A, Z+1)$ is heavier, or because of a large spin difference as in ^{48}Ca, but where the nucleus $(A, Z+2)$ is lighter. This is possible for even–even nuclei. An inspection of the table of isotopes shows that the transitions are always from a 0+ ground state to a 0+ ground state, and sometimes to a first excited 2+ state, as shown in Fig. 1.

Two modes of double beta decay are distinguished. First is the two-neutrino double beta decay ($2\nu\beta\beta$),

$$(A, Z) \longrightarrow (A, Z+2) + e^- + e^- + \bar{\nu}_e + \bar{\nu}_e,$$

in which two neutrons (d quarks) undergo simultaneously a beta decay to a proton (u quark), with emission of two e^- and two $\bar{\nu}_e$. This process conserves the lepton number and is allowed by the standard model. The Feynman diagram is shown in Fig. 2. It is a second-order weak interaction process and thus proceeds very slowly. It has nevertheless been observed in several nuclei. Second is the long sought but so far never observed neutrinoless double beta

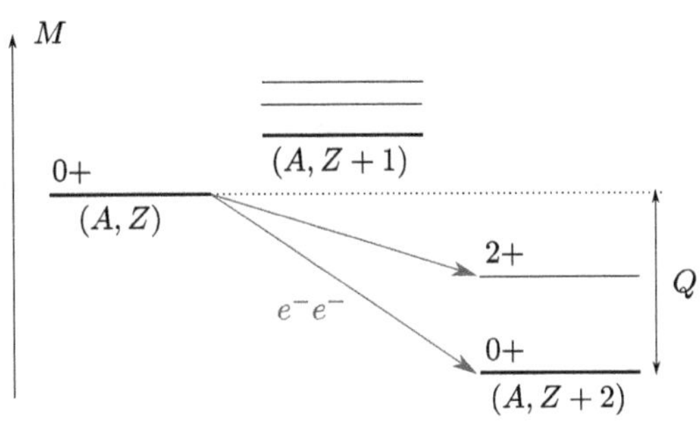

Fig. 1. Decay scheme in double beta decay from the (A, Z) to the $(A, Z+2)$ nucleus.

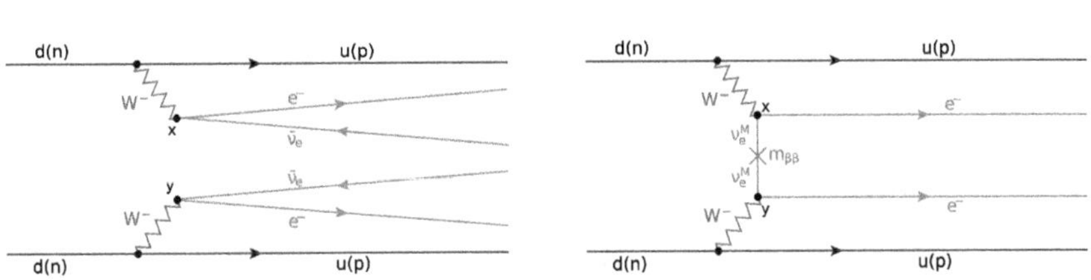

Fig. 2. The Feynman diagrams for $2\nu\beta\beta$ (left) and $0\nu\beta\beta$ (right) decay.

decay ($0\nu\beta\beta$)

$$(A, Z) \longrightarrow (A, Z+2) + e^- + e^-.$$

It violates the lepton number L by two units and can only proceed with Majorana neutrinos. The Feynman diagram is given in Fig. 2. The neutrino emitted (absorbed) at one vertex is absorbed (emitted) at the other one. The neutrino needs to have mass so that at emission it is not purely $h = \pm 1$, depending on how it propagates in time, but has a small $h = \mp 1$ amplitude of order m/E, which allows it to be absorbed at the other vertex or vice versa. The calculation of the half-life for the two double beta decay modes is fairly complicated, and we only present here the basic underlying physics. More information can be found in Refs. 1, 3, 4 and 6. First, the physics of double beta decay is strongly related to that of ordinary beta decay. The leptons are light; the nuclear recoil is negligible. The leptons carry away the entire kinetic energy Q released in the decay, which turns out to be, neglecting the binding energy of the valence electrons, the difference between the atomic mass in the initial and final states:

$$Q = M(A, Z) - M(A, Z+2).$$

The neutrino mass is small enough to be neglected in that context. The phase space of the leptons and thus the decay probability increase rapidly with Q. For this reason, experiments chose nuclei with high Q values. In addition, a high natural isotopic abundance is an advantage. A list of popular candidates is given in Table 1. In principle, double beta plus decay probes the same physics and could also be of interest. But it turns out that the Q values are significantly lower (here, four electron masses must be subtracted from the difference of the atomic masses). This more than offsets the advantage of the

Table 1. List of double beta candidates with natural isotopic abundance.

Transition	Q [MeV]	Abundance[%]	Experiments
$^{48}Ca \rightarrow {}^{48}Ti$	4.271	0.187	dir.
$^{76}Ge \rightarrow {}^{76}Se$	2.040	7.8	dir., geo.
$^{82}Se \rightarrow {}^{82}Kr$	2.995	9.2	dir.
$^{100}Mo \rightarrow {}^{100}Ru$	3.034	9.6	dir., geo.
$^{128}Te \rightarrow {}^{128}Xe$	0.868	31.4	dir., geo.
$^{130}Te \rightarrow {}^{130}Xe$	2.533	34.5	dir.
$^{136}Xe \rightarrow {}^{136}Ba$	2.479	8.9	dir.
$^{150}Nd \rightarrow {}^{150}Sm$	3.367	5.6	dir.
$^{232}Th \rightarrow {}^{232}U$	0.858	100	milking
$^{238}U \rightarrow {}^{238}Pu$	1.145	99.3	milking

better experimental signature for the positrons and makes this option less attractive.

The wavelength of the leptons is large compared to the nuclear radius, so that the long wavelength approximation can be used. For the $0+ \to 0+$ transition, which anyway is expected to dominate because of phase space, the leptonic and the hadronic part separate in first order. One major problem in evaluating the graphs in Fig. 2 is that the quarks are bound in nucleons, themselves bound in nuclei. The nucleus undergoes a transition from the initial 0+ state to a virtual state in the $(A, Z+1)$ nucleus and a second one from there to the final 0+ state. The excitation energy in the intermediate nucleus is expected to be much less in $0\nu\beta\beta$ decay, since the virtual neutrino carries a large part of it. The nuclear transition can be of the Fermi type $\langle 1 \rangle$ or Gamow–Teller type $\langle \vec{\sigma} \rangle$, characterized by corresponding matrix elements $M_F^{0\nu,2\nu}$ and $M_{GT}^{0\nu,2\nu}$. These elements are evaluated in the frame of nuclear models. Conceptually, the shell model is most realistic. It is however numerically very challenging. But elaborate calculations have been made.[14] Another approach is the quasirandom phase approximation (QRPA),[15] which is more phenomenological and describes the nucleon–nucleon interaction in terms of parameters which can be determined from experiments. Other approaches are the Interacting Boson Model[16] and the projected Hartree–Fock–Bogoljubov framework.[17]

3.1. *The half-life for $2\nu\beta\beta$ decay*

The evaluation of the lepton part of the amplitude of $2\nu\beta\beta$ decay does not present conceptual difficulties and can be performed using standard perturbation theory methods. A phase space factor $G^{2\nu}(E_0, Z)$ for the four leptons must be calculated, which includes the Z-dependent Coulomb correction and further depends on the total energy released $E_0 = Q$. With that and the matrix elements $M_{GT}^{2\nu}$ and $M_F^{2\nu}$, the half-life becomes

$$\left[T_{1/2}^{2\nu}(0+ \to 0+)\right]^{-1} = G^{2\nu}(E_0, Z)\left|M_{GT}^{2\nu} - \frac{g_V^2}{g_A^2} M_F^{2\nu}\right|^2, \qquad (27)$$

where g_V and g_A are the vector and axial vector coupling constants of the weak interaction. The electrons only take away part of the kinetic energy. The spectrum of the sum of the kinetic energy of the two electrons E_{ee} can be calculated accurately, its shape being completely determined by the lepton phase space. An example (^{136}Xe) is shown in Fig. 3. The half-life $T_{1/2}^{2\nu}$ was measured for several of the nuclei in Table 1, and one observes a fair agreement with predictions based on QRPA matrix elements.

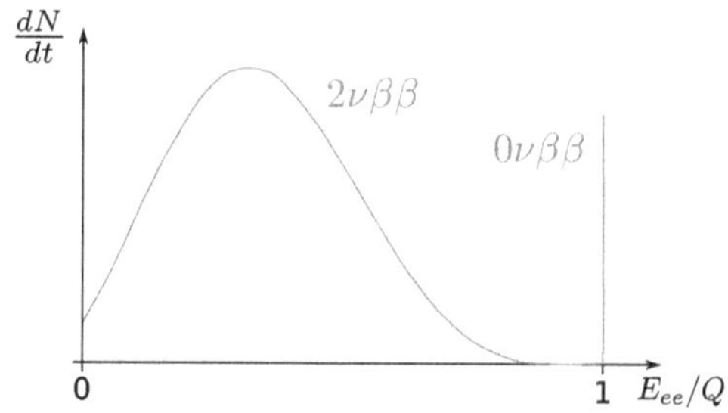

Fig. 3. The spectrum of the sum energy of the two electrons in double beta decay (^{136}Xe, $Q = 2.479$ MeV).

3.2. *The half-life for $0\nu\beta\beta$ and the effective mass*

In the case of $0\nu\beta\beta$ decay, we must consider the leptonic part of the decay amplitude a little more carefully. It is given by the product of two weak charged currents interacting at the vertices x and y:

$$\overline{e}(x)\gamma^{\mu}\nu_{eL}(x)\nu_{eL}^{T}(y)\gamma_{\mu}^{T}\overline{e}^{T}(y).$$

The neutrino is emitted at the vertex x and absorbed at that in y or vice versa. The neutrino line is internal, the neutrino virtual, so that $\nu_{eL}(x)\nu_{eL}^{T}(y)$ are contracted.

The flavor state projection ν_{eL} is a superposition of N left-handed projections of the mass eigenstates ν_i. The line describes the propagation of these from x to y or vice versa. The mixing, given by Eq. (24), is $\nu_{eL} = \sum_{i=1}^{N} U_{ei}\nu_{iL}$, and we get

$$\nu_{eL}(x)\nu_{eL}^{T}(y) = \sum_{i=1}^{N} U_{ei}^{2}\frac{1}{2}(1-\gamma_5)\nu_i(x)\nu_i^{T}(y)\frac{1}{2}(1-\gamma_5).$$

Using Eqs. (14) and (15) along with the Majorana condition $\nu^{c} = \nu$, we find

$$\nu_i^{T} = \overline{\nu}_i C^{T} = -\overline{\nu}_i C,$$

and the expression above becomes

$$-\sum_{i=1}^{N} U_{ei}^{2}\frac{1}{2}(1-\gamma_5)\nu_i(x)\overline{\nu}_i(y)\frac{1}{2}(1-\gamma_5)C.$$

The expression $\nu_i(x)\overline{\nu}_i(y)$, properly time ordered, is the propagator[b] for a fermion of mass m_i:

$$\frac{1}{2\pi^4}\int d^4p\; e^{-ip(x-y)}\frac{\not{p}+m_i}{p^2-m_i^2}$$

with the 4-momentum p^μ carried by the fermion and the notation $\not{p}=\gamma^\mu p_\mu$. Substituting this, dropping the integral, we get

$$-\sum_1^N U_{ei}^2\frac{1}{2}(1-\gamma_5)\frac{\not{p}+m_i}{p^2-m_i^2}\frac{1}{2}(1-\gamma_5)C.$$

We see that

$$\frac{1}{2}(1-\gamma_5)\not{p}\frac{1}{2}(1-\gamma_5)=\frac{1}{2}(1-\gamma_5)\frac{1}{2}(1+\gamma_5)\not{p}=0.$$

This reflects the fact that with massless neutrinos, $0\nu\beta\beta$ decay is entirely suppressed by helicity. The decay amplitude only contains terms proportional to the mass, which break the helicity:

$$-\sum_1^N U_{ei}^2 m_i\frac{1}{p^2-m_i^2}\frac{1}{2}(1-\gamma_5)C.$$

The integration must then be performed over the 4-momentum p^μ, compared to which m_i is always small. In the end, the decay amplitude is proportional to the effective mass

$$m_{\beta\beta}=\sum_{i=1}^N U_{ei}^2 m_i. \tag{28}$$

The effective mass vanishes for pure Dirac neutrinos because of cancellations. This is obvious in example 1 on p. 451, since the two masses have opposite signs, or because of the phase i which appears after giving them the same sign.

In addition, the half-life depends on a phase space factor $G^{0\nu}(E_0,Z)$ including the Coulomb correction, and the matrix elements $M_{GT}^{0\nu}$ and $M_F^{0\nu}$:

$$\left[T_{1/2}^{0\nu}(0+\to 0+)\right]^{-1}=G^{0\nu}(E_0,Z)\left|M_{GT}^{0\nu}-\frac{g_V^2}{g_A^2}M_F^{0\nu}\right|^2|m_{\beta\beta}|^2. \tag{29}$$

In $0\nu\beta\beta$ decay, the entire energy is taken away by the two electrons. Their sum energy spectrum is thus a narrow peak at Q. This is essential, in experiments, to distinguish the two modes.

[b]See Refs. 8 and 10. For a more formal treatment, see Refs. 3 and 18.

Observing $0\nu\beta\beta$ would thus demonstrate lepton number violation, indicate the Majorana nature of neutrinos, and determine $m_{\beta\beta}$, a quantity related to the neutrino masses, a measure of the absolute mass scale. The uncertainty on the matrix elements induces, it is believed, an uncertainty of the order of a factor 2 on $|m_{\beta\beta}|$.

We consider the scenario (Eq. (25)) with effectively $N = 3$, no sterile neutrinos, and the mixing (Eq. (26)):

$$m_{\beta\beta} = c_{12}^2 c_{13}^2 m_1 + s_{12}^2 c_{13}^2 e^{i\alpha_{21}} m_2 + s_{13}^2 e^{i\alpha_{31}} m_3. \tag{30}$$

The Majorana phases α_{21} and α_{31}, about which nothing is known so far, can lead to cancellations. Existing measurements of the mixing angles and differences of squared masses by neutrino oscillation experiments put constraints on $|m_{\beta\beta}|$.[19–21] It is customary to plot the allowed areas in the plane $|m_{\beta\beta}|$ vs. the smallest m_i as shown in Fig. 4. The two possible mass hierarchies, normal (NH) and inverted (IH), are considered. The width of the allowed bands is due to the lack of knowledge about α_{21} and α_{31} and the experimental uncertainties. One sees that, for the normal hierarchy, cancellations can indeed occur in the region $10^{-3} < m_i < 10^{-2}$ eV.

Experiments studying the kinematics in weak decays give information on the masses m_i, independent of their Dirac or Majorana nature. The best results come from the study of ^{3}H decay, which provides an upper limit of order 2 eV.[22,23] This is large in comparison with the differences of squared masses measured in oscillation experiments. The ^{3}H experiments thus probe a mass range where all three masses are nearly degenerate and measure a

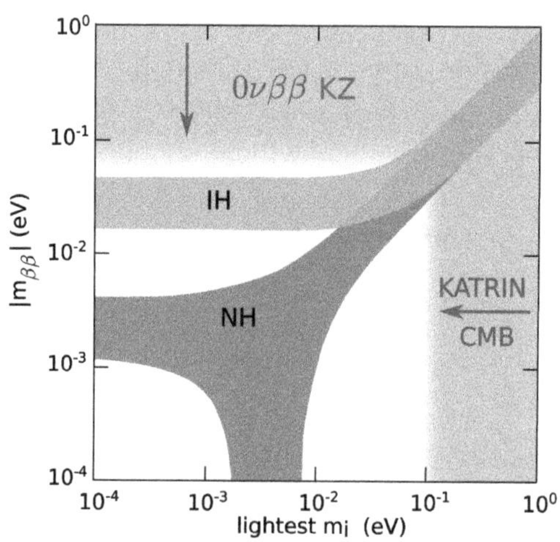

Fig. 4. Allowed areas of the effective mass $|m_{\beta\beta}|$ in function of the smallest mass m_i.

mass that is close to the lightest one. The KATRIN experiment[24] is starting to take data, probing the mass down to 0.1 eV, and its sensitivity is shown in Fig. 4. Studies of the cosmic microwave background (CMB) provide an upper bound on the sum of masses $\sum_{i=1}^{3} m_i < 0.05–0.1$ eV, which can also be considered a bound on the lightest mass.[25] The area excluded by present double beta experiments is also shown. This will be discussed in more detail in the next section.

3.3. *Experimental techniques*

For many decades now, double beta decay has been the subject of intense scrutiny. Investigated nuclei are listed in Table 1, with the experimental technique used.

Geochemical experiments were the first ones in which $2\nu\beta\beta$ decay was observed. In these, one investigates an ore rich in a double beta decay candidate. One attempts to extract the daughter products which may have accumulated since the formation of the rock and identify them. The ore age can be estimated by geological techniques, but is in any case very long, which is an advantage. A disadvantage is that it is not possible to determine in a straightforward way to what extent $0\nu\beta\beta$ and $2\nu\beta\beta$ contributed. The best results were obtained by Kirsten and collaborators.[26] They studied a 4.5 g sample containing 94% of tellurium. It was estimated to be $1.3 \cdot 10^9$ yr old. There are two candidates, ^{128}Te and ^{130}Te decaying to ^{128}Xe and ^{130}Xe. The Q value and phase space factor are higher for ^{130}Te. Xenon as a noble gas is relatively easy to extract. A mass spectrograph analysis revealed a strong excess of ^{130}Xe, compared to atmospheric xenon, and a smaller one, not completely significant, of ^{128}Xe. In any event, these excesses were attributed to $2\nu\beta\beta$ decay and yielded the half-lives $T_{1/2}^{2\nu}(^{130}\text{Te}) = (2.6 \pm 0.3) \cdot 10^{21}$ yr and $T_{1/2}^{2\nu}(^{128}\text{Te}) > 8 \cdot 10^{24}$ yr (95% CL). Later, direct experiments confirmed the assignment and the results.

Milking experiments are somewhat similar. A source is produced by chemical purification, which eliminates daughter products. The source is left alone for a long period after which the daughter product is chemically extracted. The transition $^{238}\text{U} \rightarrow {}^{238}\text{Pu}$ was investigated this way.[27] Uranyle nitrate salt was purified. Over 30 years later, a sample containing the plutonium was chemically extracted. The α activity was measured, and a weak peak from ^{238}Pu was identified. The half-life $T_{1/2} = (2.0 \pm 0.6) \cdot 10^{21}$ yr was deduced. Here again, it is impossible to figure out the relative contributions of $2\nu\beta\beta$ and $0\nu\beta\beta$.

3.4. *Direct counting experiments*

In that sense, direct counting experiments are more powerful. In these, the decay is identified by the detection of the electrons in the final state. Their energy is measured, which allows to distinguish $2\nu\beta\beta$ and $0\nu\beta\beta$ decay. Many experiments have been carried out in the last decades, sensitive to longer and longer half-lives. This has primarily been achieved by increasing the source mass, which has passed from a few kilos to several hundreds of kilos in present experiments, to push up the count rate. Source masses in the range of several tons are considered for next generation experiments. Most setups use a source which is at the same time the active part of the detector. This makes it easier to achieve high masses. In general, source materials enriched in the isotope of interest are used.

Several background sources can mimic a double beta decay event. The increase in mass must be accompanied by a corresponding background suppression.

The most dangerous background sources are nuclear α, β and γ activities. The ubiquitous natural activities, mainly the ^{232}Th and ^{238}U decay chains, and ^{40}K, are important contributors. The radon isotope ^{222}Rn in the ^{238}U chain, with its 3.8 d half-life, is particularly annoying, since it can leak out of materials into the sensitive volume of the detector. Man-made activities such as ^{137}Cs and cosmogenic activations such as ^{60}Co are also a problem. Choosing a $\beta\beta$ candidate with high Q value is a first help since it pushes the $0\nu\beta\beta$ signal to higher energies, where the background is less.

Cosmogenic activations and cosmic muon-induced backgrounds are suppressed to a large extent by conducting experiments in underground laboratories. A small rock overburden is sufficient to absorb secondary cosmic neutrons, responsible for nuclear activations. It is necessary to go much deeper to suppress cosmic muons, a problem because of direct hits, and also because of the production of spallation neutrons and bremsstrahlung γ's. SNOLAB, for instance, is at a depth of over 6000 m water equivalent, and the muon flux is 10^2 m^{-2} yr^{-1}, reduced by a factor 10^7 compared to above-ground value.

The background from the ^{238}U, ^{232}Th chains and ^{40}K can be largely reduced by choosing radio pure materials for the construction of the detector. Impressive progress has been achieved in recent years. Various material screening methods exist, with their advantages and disadvantages:

(1) Low-background gamma ray spectroscopy using Ge detectors
(2) Glow-discharge mass spectroscopy (GDMS)

(3) Inductively Coupled Plasma Mass Spectroscopy (ICPMS)
(4) Neutron Activation Analysis (NAA)

Sensitivities of the order of 10^{-12} g/g impurities for ^{238}U, ^{232}Th and ^{40}K are achieved.[28]

Finally, the background can be fought by having a selective signature for $\beta\beta$ decay events. Here, a variety of techniques have been developed. Energy resolution is an important parameter, since it helps in making the $0\nu\beta\beta$ signal stick out of the underlying background. Ultimately, it is essential to separate $0\nu\beta\beta$ from events in the high-energy tail of $2\nu\beta\beta$, a concern in next generation experiments. For illustration, some of the most competitive experiments in the field will be described in the following.

3.5. *EXO-200*

EXO-200[29] studies ^{136}Xe ($Q = 2457.83 \pm 0.37$ keV). The central detector is a time projection chamber (TPC) filled with liquid xenon enriched to 80.6% in ^{136}Xe. The central vessel is made from highly purified copper, the inner volume being 40 cm in diameter and 44 cm in length. The xenon is constantly circulated and purified. The active volume contains 110 kg of xenon. Two drift regions are separated by a central cathode (Fig. 5) which defines the drift field along with readout planes at ground potential at both ends. The readout planes consist of an induction wire plane (V) followed by a charge collection plane (U). The U and V wires are at an angle of 60°. They measure the ionization deposited by charged particles in the drift volume, as well as the position in the plane XY perpendicular to the vessel axis. Xenon scintillates in the UV, with a broad spectrum peaking at 178 nm. This light is measured in LAAPD planes located behind the wire planes. The LAAPDs have a diameter of 16 mm, and each plane contains 234 of them. The difference in time between the prompt light pulse and the delayed ionization pulse allows one to determine, knowing the drift velocity, the third coordinate Z along the vessel axis. The deposited energy is reconstructed from the total collected light and ionization. The energy resolution is $\sigma/E = 1.5\%$ at the Q value.

The TPC vessel is immersed in a cooling liquid which maintains it at liquid xenon temperature. The cooling liquid also acts as shielding against local activities. It is contained in a cryostat, surrounded by a layer of lead shielding giving additional protection against local activities. A set of scintillators serves to veto the cosmics. The setup is located in the WIPP (NM) underground facility.

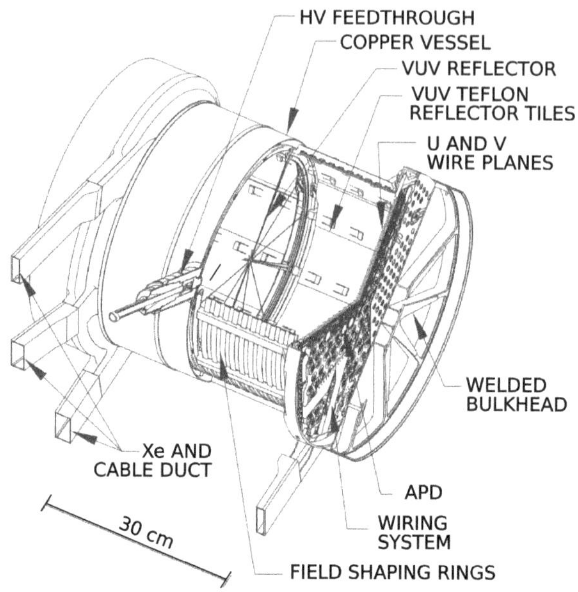

Fig. 5. The central TPC of EXO-200.

Electron tracks at 2.5 MeV are short compared to the spatial resolution, and look like point depositions. $\beta\beta$ events appear as single site (SS) depositions. So do α's from the background, but they have much higher energies and can be identified easily. The remaining but penetrating γ background from detector components in general leads to sequences of Compton scatters in the active volume, followed by photoabsorption. These show mostly as multisite (MS) events in the TPC. The response of the detector to $0\nu\beta\beta$ and $2\nu\beta\beta$ decay is calculated by simulations. So is that of the major background sources. A joint fit of the SS and MS events is then performed, the normalization of each source of events being allowed to vary. The result, using the total data set corresponding to an exposure of over 100 kg·yr, is shown in Fig. 6. The $2\nu\beta\beta$ contribution is prominent and gives a half-life of $T^{2\nu}_{1/2} = 2.165 \pm 0.016(\text{stat}) \pm 0.059(\text{sys}) \cdot 10^{\ 21}$ yr. There is also an indication of a peak centered at the Q value. It is however statistically consistent with a null effect at the 1.2σ level. The limit $T^{0\nu}_{1/2} > 1.1 \cdot 10^{25}$ yr at 90% CL was derived. It leads to the limit $|m_{\beta\beta}| < 190$–450 meV, depending on the matrix elements.

3.6. *KamLAND-Zen*

KamLAND-Zen[30] also investigates ^{136}Xe. It uses the infrastructure and the radiopurification system built for the KamLAND reactor neutrino

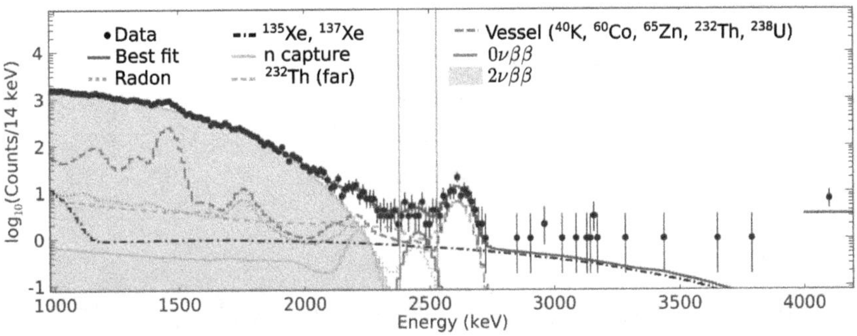

Fig. 6. EXO-200, Energy spectrum of single site events with the best fit and the contributions of various sources. The vertical lines indicate the $\pm 2\sigma$ region around the Q value.

experiment and is operated in the Kamioka underground laboratory. The central component is a spherical inner balloon (IB), optically transparent, 3.08 m in diameter, filled with 13 t of Xe-loaded liquid scintillator. The xenon concentration is $(2.91 \pm 0.04)\%$ by weight. The xenon is enriched to 90.8% in ^{136}Xe. The IB is in the middle of a spherical outer balloon (OB), 13 m in diameter, filled with 1 kt of liquid scintillator. The OB itself is inside a spherical stainless steel containment tank (SCT), 18 m in diameter, filled with transparent buffer oil, instrumented with 1879 inward looking photomultipliers on its inner surface. The photocathode coverage is 34%. Finally, the SCT is immersed in a cylindrical tank containing 3.2 kt of water, instrumented with photomultipliers, used as Čerenkov detector, to veto cosmic muons. Events in the IB and OB are reconstructed from the timing and charge collection on all photomultipliers of the SCT. The vertex spatial reconstruction resolution is better than 2 cm. The energy resolution is $\sigma = (6.6 \pm 0.3\,\%)/\sqrt{E(\mathrm{MeV})}$, deteriorating to 7.3 in later runs because of the failure of some photomultipliers. The outer scintillator volume of the OB is used as veto.

Candidate $\beta\beta$ events are contained in a spherical central fiducial volume, initially 1.35 m in diameter, in the IB. This was raised to 2 m in later runs. Candidate events must be isolated in space and time by 1.7 m and 1.9 ms, respectively, to suppress ^{214}Bi–^{214}Po $\beta-\alpha$ sequences (^{238}U chain). Reactor $\overline{\nu}_e$ interactions producing a prompt e^+ signal and a delayed γ cascade signal following n capture are identified and eliminated.

A data set corresponding to an exposure of 89.5 kg·yr of ^{136}Xe was acquired first. Following this, the liquid scintillator was purified further to extremely high levels, and a second data set was taken, corresponding to 504 kg·yr. A fit was performed to the energy and spatial distribution of events to both data sets, including $0\nu\beta\beta$ and $2\nu\beta\beta$ decay, as well as all known background sources. A $2\nu\beta\beta$ signal compatible with that of EXO-200 was found. No $0\nu\beta\beta$ signal was seen, and the limit $T^{0\nu}_{1/2} > 1.07 \cdot 10^{26}$ yr (90% CL) was derived. It corresponds to an upper bound in the range $|m_{\beta\beta}| <$ 60–165 meV depending on the matrix elements.

3.7. *CUORE-0*

CUORE-0 investigates ^{130}Te ($Q = 2527.518$ keV)[31] in the Gran Sasso underground laboratory, like its predecessor Cuoricino. It uses an original detection technique: bolometry. The basic components of the detector are $^{nat}TeO_2$ crystals, $5 \times 5 \times 5$ cm^3 in size, weighing 750 g each. They are placed in a cryostat and cooled down to a temperature of 12 mK via a weak thermal link. When an energy E is deposited in a crystal, its temperature increases suddenly by $\delta T = E/C(T)$. $C(T)$ is the specific heat, very low at the operating temperature, of order 10–20 μK/MeV. A small neutron transmutation doped (NTD) Ge thermistor is glued to each crystal. The rise in temperature is obtained from the change in resistance. In all, 13 groups of four crystals are arranged in a tower-like structure and held together by a copper frame. The total mass of ^{130}Te is 10.9 kg. The energy resolution, 4.8 keV FWHM at 2614 keV, is a strong point. Good events have an energy deposition in one crystal only. This criterion efficiently suppresses the remaining α background, mostly on the crystal surfaces. Data corresponding to 9.8 kg·yr of ^{130}Te were accumulated. The spectrum around the Q value is shown in Fig. 7. A small peak is seen at 2507 keV due to the simultaneous detection of the 1170 and 1332 keV peaks in ^{60}Co. There is no evidence of a $0\nu\beta\beta$ signal above the flat background, and the limit $T^{0\nu}_{1/2} > 2.7 \cdot 10^{24}$ yr

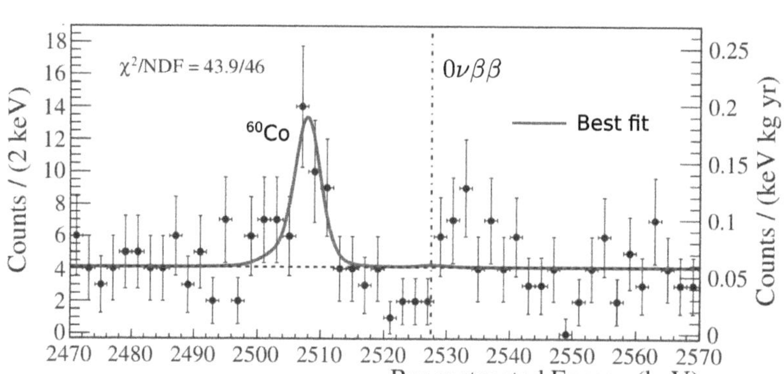

Fig. 7. CUORE-0, energy spectrum around the Q value (vertical dot-dashed line).

(90% CL) was derived. Combined with the Cuoricino bound, it becomes $T^{0\nu}_{1/2} > 4.1 \cdot 10^{24}$ yr. It translates into the limit $|m_{\beta\beta}| < 300-760$ meV.

3.8. *GERDA, phases I and II*

GERDA studies ^{76}Ge ($Q = 2039.061$ keV).[32] The experiment is conducted in the Gran Sasso underground laboratory. An array of germanium crystals is suspended near the center of a 64 m^3 liquid argon (LAr) tank. The crystals are directly immersed in LAr, at the temperature of which they can safely be operated as detectors, instead of being in vacuum as usual. This simplifies the design, which helps in fighting radioactive contaminations. Additionally, LAr serves as shielding. The LAr tank is itself inside a 590 m^3 water tank, providing additional shielding. It is instrumented with photomultipliers to collect the Čerenkov light of cosmic μ hits which can thus be vetoed.

In phase I, eight crystals enriched to 85% in ^{76}Ge in a coaxial geometry were used, with a total mass of 17.7 kg. The energy resolution varied from 4.2 to 5.7 keV FWHM near the Q value. Signals are read out with a flash ADC, and the shape information is used to distinguish SS events, such as $\beta\beta$ decay, from MS events, dominated by background. In the course of the experiment, five crystals, in a new Broad Energy Germanium (BEGe) geometry, with a total mass of 3.63 kg, were added. They provide a better SS–MS discrimination and exhibit a better energy resolution (2.6–4.0 keV).

In phase II, the number of BEGe crystals was raised to 30 (20 kg).[33] Moreover, the volume around the crystals was instrumented with photosensors to detect the scintillation light created by background events, which can be vetoed. First data are available. An analysis combining phase I and II data amounting to a total exposure of 34.3 kg·yr was performed. No excess of events above the flat background in the region of interest around the Q value was seen, and the limit $T^{0\nu}_{1/2} > 5.3 \cdot 10^{25}$ yr was derived. This leads to the limit $|m_{\beta\beta}| < 150-330$ meV, depending on the matrix elements.

3.9. *What next in $0\nu\beta\beta$ decay?*

So far, $0\nu\beta\beta$ decay has not been observed, and the most constraining bound $|m_{\beta\beta}| < 60-165$ meV was obtained by KamLAND-Zen. It is indicated in Fig. 4. One sees that, also taking into account the bound from the CMB, the degenerate mass region, where the mass differences are small compared to the absolute masses, and where the NH and IH allowed bands coincide, is disfavored. Furthermore, the KamLAND-Zen bound is getting close to the IH band. The quest is continuing. Some of the abovementioned experiments will be upgraded, and new projects are starting.

GERDA is continuing to accumulate data in the improved phase II configuration. Also dedicated to ^{76}Ge, the MAJORANA collaboration is building a demonstrator, with 40 kg of enriched germanium crystals to be operated more conventionally in vacuum.[34] The setup will be installed in the SURF underground laboratory. It will serve as a test facility in view of a ton-scale ^{76}Ge experiment.

CUORE-0 was in fact a prototype for CUORE, which will have 19 CUORE-0-like towers in a new cryostat, with 988 TeO_2 detectors for a total mass of 741 kg. The new assembly is being commissioned in the Gran Sasso laboratory. SNO+ will also investigate ^{130}Tl.[35] It uses the SNO solar neutrino detector, installed in the SNOlab underground laboratory. The central acrylic sphere, 12 m in diameter, will be filled with liquid scintillator, loaded with nearly 800 kg of ^{130}Te initially, possibly much more later.

SuperNEMO[36] is an upgraded expansion of NEMO-III,[37] which observed $2\nu\beta\beta$ decay in several nuclei and obtained competitive limits on $0\nu\beta\beta$ decay. Here, the source is not active. It consists in a thin foil (40 mg/cm^2, 4.5$\times$3 m^2), sandwiched between two tracking devices filled with He gas plus a quencher, with a magnetic field, in which the electron tracks can be measured. The advantage is a good particle identification (e^-, e^+, α) and the possibility to measure the angular correlation between the two electrons. The distribution favors a back to back emission in case of $0\nu\beta\beta$ decay. The energy is obtained from scintillators behind the tracking volume, the resolution being of order 7 % FWHM. 20 modules with 5–7 kg of source material each are planned; ^{82}Se, ^{150}Nd and ^{48}Ca are considered. A first prototype module is being commissioned in the Fréjus underground laboratory.

A bold step forward is the proposed nEXO. It is a larger version of EXO-200, with 5 t of xenon enriched in ^{136}Xe. The TPC will have only one drift volume. The energy resolution should be improved,[38] and ways to reduce the background are being explored. Its sensitivity should extend down to $|m_{\beta\beta}| \sim 8-20$ meV. R&D is pursued in view of a vast improvement of the event signature: in addition to the two electrons, it will be attempted to detect the Ba^{++} ion produced in the decay. The ion should change rapidly to Ba^+ in liquid xenon. It is foreseen to extract it electrostatically with a tip and release it in an ion trap, in which it can be identified by resonant laser spectroscopy.[39] This so-called Ba tagging, if it can be made to work, would make the experiment background free, except for the $2\nu\beta\beta$ tail when looking for $0\nu\beta\beta$. The expected sensitivity would extend to $|m_{\beta\beta}| \sim 4.5 - 9$ meV. With this the entire IH band in Fig. 4, and even part of the NH band could be explored.

3.10. *An alternative: CνB?*

Another way to probe the Majorana vs. Dirac nature of neutrinos may be the study of the cosmic neutrino background (CνB). It is made of neutrinos which were in thermal equilibrium with matter in the hot early universe, neutrinos with $h = -1$ and (anti)neutrinos with $h = +1$ of all flavors e, μ, τ. They were produced as wave packets containing superpositions of the mass eigenstates ν_i. They decoupled before the photons, cooled off with the expansion of the universe, maintaining their helicity. With time, the wave packets ran out of coherence. The background presently is made of mass eigenstates ν_i, which can be seen as superpositions of flavor eigenstates. It has a temperature of $kT = 0.168$ meV, and a number density of 56 cm^{-3} per degree of freedom.[40] At least one neutrino type has a mass larger than $\sqrt{\Delta m^2_{31}} = 50$ meV. These neutrinos are nonrelativistic now. We shall assume that they all are, so that the weak interaction pays no attention to helicity. The PTOLEMY experiment[41] is being set up to detect these neutrinos via the charged-current reaction $\nu_e + {}^3\mathrm{H}\,(T_{1/2} = 12.3\ \mathrm{yr}) \rightarrow {}^3\mathrm{He} + e^-$. In the simplest Dirac case, there are three families of neutrinos with $h = -1$, which can trigger the reaction, and three families of antineutrinos with $h = +1$ which cannot. In the Majorana case, there are three families of neutrinos with $h = -1$, and three families of neutrinos with $h = +1$, all of which can trigger the reaction. The count rate is two times higher with Majorana neutrinos (8 yr^{-1} in PTOLEMY) as it is for Dirac neutrinos (4 yr^{-1}). If successfull, PTOLEMY may provide decisive information.

References

1. F. Boehm and P. Vogel, *Physics of Massive Neutrinos* , 2nd edition, Cambridge University Press, 1992.
2. K. Zuber, *Neutrino Physics*, Series in High Energy Physics, Cosmology and Gravitation, Taylor and Francis, 2004.
3. S. M. Bilenky and S. T. Petcov, *Rev. Mod. Phys.* **59** (1987) 671.
4. S. M. Bilenky *et al.*, *Eur. Phys. J. C* **71** (2011) 1754,
5. S. Dell'Oro *et al.*, *Adv. High Energy Phys.* **2016** (2016).
6. R. D. Mc Keown and P. Vogel, *Phys. Rept.* **394** (2004) 315.
7. I. J. R. Aitchison and A. J. G. Hey, *Gauges Theories in Particle Physics*, 2nd edition, Institute of Physics Publishing, London and Philadelphia, 1989.
8. J. D. Bjorken and S. D. Drell, *Relativistic Quantum Mechanics*, McGraw and Hill, 1964.
9. F. Mandl and G. Shaw, *Quantum Field Theory*, John Wiley and Sons, New York, 1991.
10. J.-P. Derendinger, *Théorie des Champs*, Presses polytechniques romandes, 2001.

11. P. Minkowski, *Phys. Lett. B* **67** (1977) 421.
12. M. Gell-Mann, P. Ramond, and R. Slansky, in *Supergravity*, p. 315, edited by F. van Nieuwenhuizen and D. Freedman, North Holland, Amsterdam, 1979.
13. R. N. Mohapatra and G. Senjanović, *Phys. Rev. Lett.* **44** (1980) 912.
14. J. Menendez, A. Poves, E. Caurier, and F. Nowacki, *Nucl. Phys. A* **818** (2009) 139.
15. F. Simkovic *et al.*, *Phys. Rev. C* **87** (2013) 045501.
16. J. Barea, J. Kotila, and F. Iachello, *Phys. Rev. C* **87** (2013) 014315.
17. T. R. Rodriguez and G. Martinez-Pinedo, *Phys. Rev. Lett.* **105** (2010) 252503.
18. J. Gluza and M. Zrafek, *Phys. Rev. D* **45** (1992) 1693.
19. S. M. Bilenky and C. Giunti, *Int. J. Mod. Phys. A* **30** (2015) 1530001.
20. C. Giunti, E. M. Zavanin, *JHEP* **07** (2015) 171.
21. X. Qian and P. Vogel, *Prog. Part. Nucl. Phys.* **83** (2015) 1.
22. C. Kraus *et al.*, *Eur. Phys. J. C* **40** (2005) 447.
23. V. Aseev *et al.*, *Phys. Rev. D* **84** (2011) 112003.
24. M. Arenz *et al.*, The KATRIN collaboration, *JINST* **11** (2016) P04011.
25. K. A. Olive *et al.*, Review of particle physics, *Chin. Phys. C* **38** (2014) 090001.
26. T. Kirsten *et al.*, *Phys. Rev. Lett.* **50** (1983) 474.
27. A. L. Turkevich, T. E. Economou and G. A. Cowan, *Phys. Rev. Lett.* **67** (1991) 3211.
28. D. S. Leonard *et al.*, *Nucl. Inst. Method* **591** (2008) 490.
29. J. B. Albert *et al.*, The EXO collaboration, *Nature* **510** (2014) 229.
30. A. Gando *et al.* (The Kamland collaboration), *Phys. Rev. Lett.* **110** (2013) 062502; arXiv:1605.02889 (2016).
31. C. Alduino *et al.*, CUORE collaboration, *Phys. Rev. C* **93** (2016) 045503.
32. M. Agostini *et al.*, GERDA collaboration, *Phys. Rev. Lett.* **111** (2013) 122503.
33. M. Agostini *et al.*, GERDA collaboration, *Nature* **544** (2017) 5.
34. W. Xu *et al.*, The MAJORANA collaboration, *J. Phys. Conf. Ser.* **606** (2015) 012004.
35. S. Andringa *et al.*, SNO+ collaboration, *Adv. High Energy Phys.* **2016** (2016) 6194250.
36. R. Arnold *et al.*, *Eur. Phys. J. C* **70** (2010) 927.
37. R. Arnold *et al.*, *Phys. Rev. D* 89 (2014) 111101.
38. I. Ostrovskiy *et al.*, *IEEE Trans.* **62**(4) (2015) 1825.
39. B. Flatt *et al.*, *Nucl. Inst. Method A* **578** (2007) 399.
40. A. J. Long, C. Lunardini and E. Sabancilar, *JCAP* **08** (2014) 038.
41. S. Betts *et al.*, arXiv:1307.4738.

Chapter 12

Low-Energy Neutrino Interactions

A. M. Szelc

School of Physics and Astronomy
University of Manchester, Manchester, UK
andrzej.szelc@manchester.ac.uk

Abstract

This chapter will present the models we use to describe neutrino interactions with other particles in the framework of the standard model of particle physics. We will focus on the lower range of neutrino energies ranging from thresholdless interactions up to the quasielastic and resonant interactions with nuclei up to a scale of few GeV.

1. Introduction

Neutrinos are the only particles in the standard model that interact weakly. They neither carry electric charge nor interact via the strong force. This makes studying neutrino properties challenging as only a small fraction of passing neutrinos will interact in any detector we may put in their path. In addition, because they are electrically neutral, we cannot see neutrinos themselves but can see only the results of their interactions with other particles, namely the nuclei and electrons that form the atoms from which our detector is constructed. We use these to infer and understand the properties of the original neutrino. For this reason, a thorough understanding of how neutrinos interact is crucial to measuring any property of these ghost-like particles.

In this chapter, we will outline the formalism used to describe neutrino interactions in the standard model starting with the simplest interactions with leptons and then ramp up with energy and move to neutrino–nucleus interactions. We assume the reader is familiar with basic QED formalism and the Feynman diagram notation. We direct readers wanting to brush up on those subjects to excellent textbooks[1-4] that delve in more depth into electroweak interactions as well as the technical aspects of the calculations. This will be particularly useful in Section 2, which focuses on interactions with leptons, while the later parts of this chapter can provide an understanding of the qualitative state of the field even without a good understanding of the theory.

As it is the only mode of interaction available to the neutrino, the weak force defines the nomenclature of the neutrino interactions we observe. The interaction itself is mediated by the so-called gauge bosons, W^+, W^- and Z^0. These bosons are massive particles, which determine the range of the weak interaction to be extremely short, especially compared to electromagnetic or gravitational interactions.

The two types of gauge bosons split the possible interactions into the charged current (CC) interaction, where the neutrino or antineutrino exchanges a charged W^+ or W^- boson with another particle, see Figs. 1(a) and 1(b), and a neutral current (NC) interaction where the neutral Z boson is exchanged, see Fig. 1(c). In the case of a neutrino interacting via the CC interaction, the transfer of the weak current results in the creation of a charged lepton which, given that the weak interactions conserve lepton number, will be of the same flavour as the incoming neutrino, i.e., $\nu_e \to e$, $\nu_\mu \to \mu$, and $\nu_\tau \to \tau$. Observing the CC interactions is in fact the only

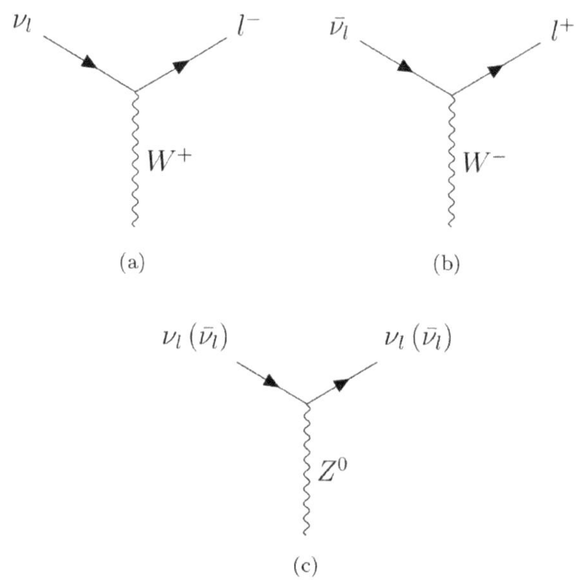

Fig. 1. The CC neutrino (a) and antineutrino interactions (b) and NC neutrino (antineutrino) interaction (c).

way to determine the flavour of the interacting neutrino although one must remember that the neutrino must carry enough energy to create its corresponding charged lepton. This means that we cannot unambiguously identify ν_μ's and ν_τ's with energies below the μ and τ production thresholds, respectively. In the case of NC interactions, the neutrino only transfers a fraction of its energy to the other particle and is therefore available to all neutrino flavors at all energies. It does not provide us with a method of determining the flavor of the incoming neutrino. For this reason, NC interactions can often be difficult backgrounds in experiments aiming to detect a specific neutrino flavour.

The exchange of the $W^\pm$ and Z^0 bosons is a process that is precisely described in the framework of the standard model of particle physics. At lower neutrino energies, the effect of the large intermediate boson masses becomes so big that we can treat these interactions as effectively point-like. This approximation, called the Fermi theory, was historically the first attempt to describe weak interactions. It is best to test this description, and through it the standard model, in a simple system. Neutrino interactions with leptons are a good example and are excellent test-beds to experimentally verify and confirm the weak-interaction model and that is what we will focus on in the first part of the chapter. The theoretical description provided by the standard model becomes significantly more complicated when applying it to neutrino-nucleus interactions due to the increasing role played by nuclear effects. This has led in recent years to a great deal of work toward developing the models of neutrino-nucleus interactions by neutrino theorists, which will be mentioned in the final part of the chapter.

In general, we will focus on how the effects of neutrino interactions could affect what is observable in the detector. We note therefore that in terms of experimental observables, most detectors will be designed to reconstruct some or all of the following properties of the incoming neutrinos: their flavour, their direction and their energy and from these observables reconstruct quantities such as their flux or cross section. We direct readers interested in investigating the subject in more depth to excellent review articles[5] and textbooks on neutrinos.[1]

2. Formalism of weak interactions

The mathematical model we use to describe weak interactions is quite similar to that of quantum electrodynamics (QED) which is not surprising given that the two interactions can be unified into the electroweak force which is one of the cornerstones of the Weinberg–Glashow–Salam standard model of particle physics.[6–8] To use this model, we need to define what the elements in the Feynman diagrams like those in Fig. 1 mean, namely understand what the currents corresponding to the weak boson exchange are and how we define the propagators of these bosons. We can then treat the Feynman diagrams as mathematical recipes that allow the calculation of the so-called matrix elements which gauge the strength of a given interaction. Integrating these matrix elements over the phase space and spin states allowed for an interaction leads to a physical observable, the cross section.

Weak interactions behave differently than QED in a few notable ways. The biggest difference stems from the large mass of the bosons that mediate the interaction leading to a very short effective range of the interaction. The second crucial difference, first predicted by Lee and Yang,[9] and subsequently proven experimentally by Wu,[10] is that they do not conserve parity. This has immediate consequences in the way we can describe the weak interactions. A fully generic Lagrangian describing an interaction would contain scalar (S), pseudoscalar (P), vector (V), axial (A) and tensor (T) currents, which are all bilinear combinations of Dirac γ matrices; see, e.g., Ref. 3 for a more thorough description. Because the vector bosons that mediate the interaction have spin ± 1, the current for weak interactions must be composed from vector γ^μ and axial-vector $\gamma^\mu\gamma^5$ components:

$$j^\mu_W \simeq \bar{\nu}_l(g_V\gamma^\mu + g_A\gamma^\mu\gamma^5)l = g_V j^\mu_V + g_A j^\mu_A, \tag{1}$$

where l and $\bar{\nu}_l$ are Dirac spinors corresponding to the neutrino and lepton, and g_A and g_V are the axial and vector coupling strengths and we have separated out the vector and axial components of the currents: $j^\mu_V = \bar{\nu}_l\gamma^\mu l$ and $j^\mu_A = \bar{\nu}_l\gamma^\mu\gamma^5 l$. Scalar, pseudoscalar and tensor currents require force carriers with spins different than 1 and so are excluded.

In order to generate the parity violation in the weak interactions, it can be shown[2] that both the vector and axial couplings must be nonzero. In order to reproduce the maximal parity violation observed in experiments a stronger constraint is needed: $|g_V| = |g_A|$. This is equivalent to a pure V–A interaction, which corresponds to a $W^\pm$ exchange. Inserting these values for g_V and g_A into Eq. 1 leads to the following form for CC interactions:

$$j^\mu_W = \frac{g}{\sqrt{2}}\bar{\nu}_l\frac{1}{2}\gamma^\mu(1-\gamma^5)l. \tag{2}$$

This formula is useful to demonstrate how the V–A nature of the weak interactions affects the chirality of the particles allowed to couple to them. Noting that the expression $\frac{1}{2}\gamma^\mu(1-\gamma^5)$ is the left-handed chiral projection operator and given that a right-handed field is defined as $u_R = \frac{1}{2}(1+\gamma^5)u$

and the fact that $(1+\gamma^5)(1-\gamma^5) \equiv 0$, we can immediately see that the CC is only nonzero for left-handed particles. This is the mathematical reflection of the surprising property of weak interactions of coupling only to left-handed neutrinos. For antiparticles, this argument is reversed, given that $\bar{u}(1-\gamma^5) = \bar{u}_R$, so only right-handed antineutrinos can be observed through CC weak interactions.

The V–A nature of weak interactions has been confirmed in numerous experiments starting with the classic Goldhaber experiment,[15] which showed that neutrinos are found with only left-handed helicities. Another very important test of the theory comes from studying muon decays. This is done by measuring the energy spectrum and angular distribution of the resulting positrons which, as shown by Michel,[16] can be described by four parameters usually referred to as ρ, δ, η and ξ. These parameters, which are combinations of the coupling constants and therefore have well defined values in the standard model, are today called Michel parameters. The most recent and precise measurements of these parameters come from the TWIST experiment.[17] For example, the Michel parameter ρ, which governs the shape of the energy spectrum, is found to be $\rho = 0.74997 \pm 0.00026$, in agreement with the value 3/4 expected for a pure V–A interaction.

2.1. *The W-boson propagator*

Since the weak interactions are transmitted via massive bosons, this needs to be accounted for in the propagator used to calculate any scattering amplitude. The Feynman propagator corresponding to the exchange of a charged W boson is

$$\frac{-i}{q^2 - m_W^2}\left(g_{\mu\nu} - \frac{q_\mu q_\nu}{m_W^2}\right), \tag{3}$$

where $g_{\mu\nu}$, the four-dimensional metric tensor, and $q_\mu q_\nu$, the components of the vector boson four-momenta, result from the summation of the polarization states of the exchanged boson. In most cases in experimental neutrino physics, we will be in a regime where $q^2 \ll m_W^2$ (and therefore, $q_\mu q_\nu \ll m_W^2$) and the propagator will simplify to

$$\frac{-ig_{\mu\nu}}{q^2 - m_W^2}. \tag{4}$$

2.2. *Neutral current*

The CC model described weak interactions where the leptonic current is exchanged very well; however, it was quickly noticed[6,11] that the calculations using only the CC transfers were divergent. This led to the prediction of the existence of NC interactions where instead of a charged $W^{\pm}$ boson, the neutral Z^0 boson, is exchanged. The NC interactions were observed in a beautiful measurement by the GARGAMELLE bubble chamber experiment at CERN.[12] The three weak intermediate bosons, $W^{\pm}$ and Z^0, combined with the electromagnetic photon form the set of intermediate bosons that transmit the unified electroweak interactions proposed by the Weinberg–Salam theory.

The NC interaction does not select only left-handed particles and so its current is somewhat more complicated than in the case of the CC:

$$j_Z^{\mu} = g_L^{\nu}\bar{\nu}\frac{1}{2}\gamma^{\mu}(1-\gamma^5)\nu + g_L^f\bar{l}\frac{1}{2}\gamma^{\mu}(1-\gamma^5)l + g_R^f\bar{l}\frac{1}{2}\gamma^{\mu}(1+\gamma^5)l, \qquad (5)$$

where due to using the $(1+\gamma^5)$ operator in the last element, we now have right-handed fermions in the interactions. The coupling constants $g_L^{\nu} = 1/2$, $g_L^f = -1/2 + \sin^2\theta_W$ and $g_R^f = \sin^2\theta_W$ are parameters of the theory and the g^f would have different values in the case of quarks.

Having defined the NC and CC parts of the interaction, we can now write down the Lagrangian for the weak interaction as

$$L_{\text{weak}} = L_{\text{CC}} + L_{\text{NC}} = -\frac{g}{2\sqrt{2}}(j_W^{\mu}W_{\mu} + j_W^{\mu,\dagger}W_{\mu}^{\dagger}) - \frac{g}{2\cos\theta_W}j_Z^{\mu}Z_{\mu}, \qquad (6)$$

where W_{μ} and Z_{μ} are the weak-interaction gauge boson fields, g is the weak-coupling constant and the θ_W is the Weinberg or weak mixing angle, which is a parameter of the standard model and $\sin^2\theta_W = 1 - (m_W/m_Z)^2 \simeq 0.239$ at the energies relevant to this chapter.

2.3. *The Fermi theory*

If we take a look back at Eq.(4) in the regime of small momentum transfer q^2, such that $q^2 \ll m_W^2$ we can simplify it even further, which leads to the approximate expression:

$$\frac{ig_{\mu\nu}}{m_W^2}. \qquad (7)$$

This is called the Fermi theory and is historically the first attempt at a model of weak interactions, introduced by Enrico Fermi.[13] In this case, the dependence of the propagator on q^2 disappears and the interaction becomes essentially point-like as shown in Fig. 2. The matrix element which is a mathematical object that corresponds to the strength of this interaction can

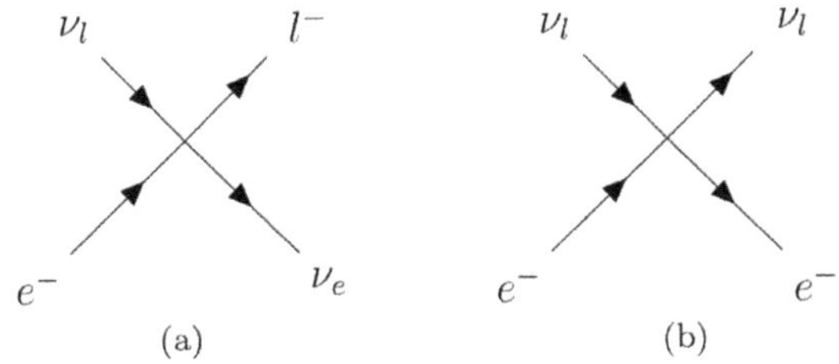

Fig. 2. Fermi point-like interactions for CC (a) and NC (b) exchanges. Note that the NC interaction was introduced into the theory much later.

be written as

$$M_f = \frac{1}{\sqrt{2}} G_F \left([\bar{l}\gamma^\mu(1-\gamma^5)\nu_l][\bar{\nu_e}\gamma_\mu(1-\gamma^5)e] \right), \tag{8}$$

where as in Fig. 2, l denotes the charged lepton field; ν_l, the l-neutrino field; and e and ν_e, the electron and electron-neutrino fields, respectively. The propagators are folded into G_F, the Fermi constant. Its value is related to the weak-coupling constant as

$$G_F = \frac{g^2}{4\sqrt{2}M_W^2}. \tag{9}$$

Treating G_F as a constant of nature, and the weak interaction as point-like, is an excellent approximation for transfer of momenta smaller than the mass of the gauge bosons $W^\pm$ and Z, i.e., practically all of the cases of interactions discussed in this chapter. For such applications, the value of G_F can be determined experimentally by measuring the decay of the muon. The calculation of the matrix element square[2] results in a formula for the decay width of the muon equal to

$$\Gamma(\mu^- \to e^-\nu_\mu\bar{\nu_e}) = \frac{1}{\tau_\mu} = \frac{G_F^2 m_\mu^5}{192\pi^3}. \tag{10}$$

This formula depends on only the muon mass and decay time. These are values that are well known and lead to the value[14] of $G_F = 1.16638 \times 10^{-5}\,\mathrm{GeV}^{-2}$, which defines the coupling strength of weak interactions. The Fermi theory is extremely useful to examine the simplest interactions of the neutrino at lowest energies, for example, scattering on a charged lepton.

2.4. *Neutrino-electron scattering*

The interactions of neutrinos with leptons, specifically electrons, are the simplest interactions to describe using the weak interaction formalism, as

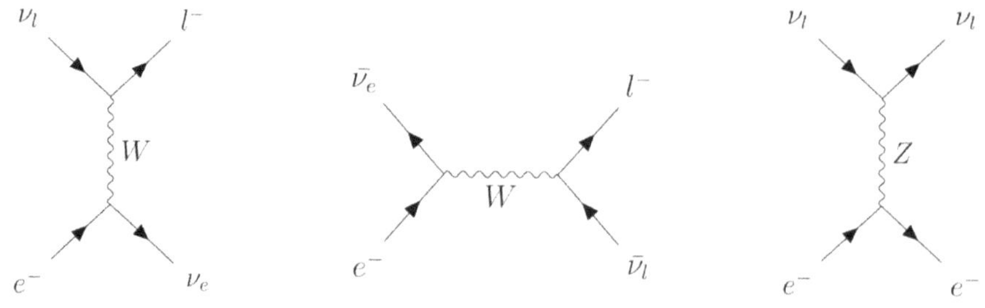

Fig. 3. Neutrino electron (a), antineutrino electron (b) and NC neutrino electron scattering (c).

we do not need to worry about sizes of the particles which will be the case with neutrino–nucleon scattering. In the most general case, a neutrino of any flavor can interact with an electron either via the CC interaction, exchanging a W boson or via the NC interaction exchanging a Z boson, see Fig. 3. The description of this scattering will differ somewhat depending on whether we are calculating neutrino or antineutrino interactions and whether the neutrinos themselves are electron neutrinos or muon or tau neutrinos. This is because of helicity considerations in the former case and the additional diagram available to the electron neutrino, Fig. 4(b) in the latter case. The results of these differences have profound effects on experiments measuring neutrino properties.

2.4.1. *Pure charged-current interaction*

Let us first consider a pure neutrino CC interaction. The corresponding Feynman diagram is shown in Fig. 3(a). This reaction can be written as (assuming $l = \mu, \tau$)

$$\nu_l + e^- \to l^- + \nu_e. \tag{11}$$

This process is sometimes known as inverse muon or tau decay, as we end up with one of these leptons in the final state. We have for now excluded electron neutrinos because of their specific relationship with the electron, we will come back to them shortly. The matrix element of this interaction can be calculated from the Feynman diagram shown in Fig. 3(a) or using Fermi theory as in Fig. 2(a). The result is equal to Eq. (8), so $M_{\text{CC}} = M_f$. Integrating this expression over the allowed spin states and assuming a high neutrino energy leads to the differential cross section as a function of angle. A straightforward derivation of its form is presented, e.g., in Ref. 2 and in the centre-of-mass frame gives the following value:

$$\frac{d\sigma}{d\Omega} = \frac{G_F^2}{4\pi^2}s, \tag{12}$$

where s is one of the Mandelstam variables and represents the $\nu_l - e$ centre-of-mass energy. We can take a step further and integrate this over all of the possible angles, which gives the value of $\sigma_{\nu e} = \frac{G_F^2}{\pi}s$, which is sometimes called the asymptotic cross section. At lower energies, the muon (or tau mass) would need to be taken into account and will result in a threshold for the interaction — this would result in multiplying the above formulas by the factor: $1 - \frac{m_l^2}{s}$, with m_l the mass of the corresponding lepton.

A similar calculation can be performed for antineutrinos, although due to charge and helicity considerations, the diagram is different, see Fig. 3(b). In the high-energy limit, only right-handed antiparticles and left-handed particles participate in the interaction. This leads to a constraint on the angles that the outgoing new lepton can have, and the angular cross section becomes

$$\frac{d\sigma}{d\Omega} = \frac{G_F^2}{4\pi^2}(1 + \cos\theta)^2 s, \tag{13}$$

where θ is the angle of the outgoing lepton. Integrating over $d\Omega$ yields an additional factor of 1/3 compared to the neutrino electron scattering: $\sigma_{\bar{\nu}e} = \frac{G_F^2}{3\pi}s$ — this is a result of helicity suppression only. It is worth noting that the situation would be reversed when comparing antineutrino and neutrino scattering on positrons. This result applies in the same way to interactions with quarks and is an important factor to take into account when designing an experiment observing both neutrinos and antineutrinos.

2.4.2. *Elastic scattering*

A second specific case is when we focus on pure NC interactions and look at the following interaction:

$$\nu_l + e^- \rightarrow \nu_l + e^-. \tag{14}$$

This is called elastic scattering and is shown in Fig. 3(c), and Fig. 4(a) shows the special case for electron neutrino. The only effect of the interaction is a redistribution of the momenta of the interacting particles. An important feature of this interaction from the point of view of experimental physics is that the outgoing electron closely follows the direction of the incoming neutrino. This has been exploited by Cherenkov detectors looking at

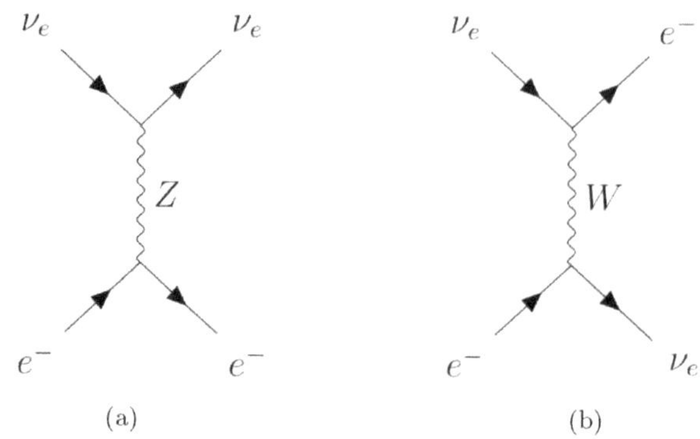

Fig. 4. Elastic scattering diagrams for the electron neutrino (a) and the CC interaction (b) which is indistinguishable.

neutrinos from the sun and atmosphere[18] to determine the direction from which they arrive. The second important thing to note is that we now must take into account the couplings of the right-handed fields. We can write down the matrix element for this interaction using a reformulated version of Eq. (5) where we use axial and vector coupling constants g_A and g_V, respectively:

$$M_{\text{NC}} = -\frac{1}{\sqrt{2}}G_F\left([\bar{\nu}_l\gamma^\mu(1-\gamma^5)\nu_l][\bar{e}\gamma_\mu(g_V^l - g_A^l\gamma^5)e]\right). \tag{15}$$

Applying a similar method as for the CC interaction above allows the calculation of a generic cross section for the pure NC interaction of any neutrino with an electron; see Ref. 5 for a demonstration. Nature, however, is a bit more complicated because the CC interaction of an electron neutrino, see Fig. 4(b), is effectively indistinguishable from a pure NC elastic scattering, shown in Fig. 4(a). This means that when calculating the cross section the ν_e will have an additional diagram through which it can interact; see e.g., Ref. 3 for the full derivation of this cross section. The result of this calculation is that the cross section for electron neutrinos to interact on electrons is higher than for muon neutrinos and tau neutrinos, with the ratio being 6:1. This effect was used by the SNO detector[19] as an independent method of differentiating electron neutrinos from other flavour neutrinos leading to the final evidence of solar neutrino oscillations.

3. Low-energy neutrino–nucleus interactions

Neutrino interactions with nuclei are more complicated than neutrino electron scattering. The principle of the interaction remains the same, namely W and Z bosons are exchanged. The formalism of the exchange of gauge bosons in a neutrino–quark interaction can be calculated in a mode analogous to the neutrino–lepton interaction found in Section 2. However, quarks that could participate in an interaction with a neutrino are not free and are instead bound in nucleons, which in turn are bound in nuclei. As a result, the description of the interaction will depend strongly on the energy transfer from the neutrino to the nucleus. At highest energies, the neutrinos can probe individual quarks ignoring the presence of the nucleons — this is called deep inelastic scattering, or DIS, and the cross section depends linearly on energy in a manner analogous to Eq. (12). At very low energies, the nucleus appears to have no structure and the neutrino interacts with it as whole. The intermediate region, where neutrinos interact with nucleons, is very difficult to model because of the complicated nuclear physics involved. Nuclear effects affect the mechanism of the interaction itself, the intranuclear motion of the resulting nucleons and their binding energies all of which influence nucleons exiting the nucleus after the interaction. This intermediate region is split into several regimes where different mechanisms dominate leading to transition regions which are even harder to model properly. In this section, we will discuss the lower-energy regimes where nuclear effects begin turning on, and the next section will focus on the medium-energy region associated with accelerator neutrinos.

The existence of protons and neutrons as bound states of quarks forces us to replace the simple currents we used up to now with a so-called hadronic transition matrix element, which takes into account the effect of strong interactions. We direct the interested reader to thorough discussions found elsewhere,[1,3,5] limiting the discussion here to a few important key points that will be relevant in the subsequent sections.

The neutrino–nucleon interaction can be split into vector and axial current components. For small momentum transfers, like in (inverse) beta-decay and capture on radioactive nuclei, the vector and axial components correspond to the so-called Fermi and Gamow–Teller transitions, respectively. A Fermi transition leaves the nucleon spin unchanged whereas the Gamow–Teller transitions do allow a change in the nucleon spin as a result of the interaction. To model the finite size of protons and neutrons, we introduce quantities called form factors $F_x(Q^2)$ to account for their structure. The form factors used in the vector component of the current can be measured through electron–nucleon scattering, as they reduce to electric and magnetic form factors of the nucleons. The form factors in the axial component of the current, namely $F_A(Q^2)$ and $F_P(Q^2)$, need to be determined by neutrino scattering experiments. These form-factors have been measured by experiments scattering neutrinos on hydrogen and deuterium.[21,22] The axial-vector form factor, in regimes where it is possible to treat nucleons as free (impulse approximation) has traditionally been parametrized as a dipole:

$$F_A(Q^2) = \frac{g_A}{(1 + Q^2/m_A^2)^2}. \tag{16}$$

This introduces the parameter m_A, the "axial mass". A parameter, which has been a focus of attention in recent years, because of neutrino experiments reporting a higher value than the $m_A \simeq 1$ measured by the hydrogen and deuterium experiments. This will be discussed in Section 5.1.

3.1. *Coherent scattering*

The most basic type of interaction a neutrino could have with a nucleus is an interaction where the neutrino does not probe the structure of the nucleus at all, instead interacting with it as a whole. Such an interaction has been predicted theoretically[23] but, until recently, never observed. As the neutrino scatters off of the nucleus without probing its structure, the only discernible effect becomes the recoil energy of the nucleus. The cross section for this interaction can be approximated as

$$\frac{d\sigma}{dT} = \frac{G_F^2}{2\pi} M \left[2 - \frac{MT}{E^2}\right] \frac{Q_W^2}{4} F^2(Q^2), \tag{17}$$

where E is the incident neutrino energy, T is the energy of the recoiling nucleus and M is the mass of the nucleus. $Q_W = N - (1 - 4\sin^2\theta_W)Z$ is the weak nuclear charge and F is the nuclear form factor for elastic interactions. Coherent scattering should only happen for neutrinos that satisfy the condition for coherence, i.e., the momentum transfer is $Q \lesssim \frac{1}{R}$ which for medium-sized nuclei happens for neutrino energies of up to 50 MeV.

The value of this cross section is actually quite high compared to other standard neutrino–nucleus interactions. The reason this reaction has never been observed is that for neutrinos with energies of the order of 50 MeV, the recoil energy of the nucleus would be very small. So small, as to be comparable to that expected from potential dark matter recoils making it very difficult to detect, except for dark matter detectors where coherent neutrino scattering may become an irreducible background. There are several experiments trying to experimentally measure the existence of coherent neutrino scattering. These experiments are extremely challenging as they must be able to detect these very-low-energy neutrino interactions and filter them out from backgrounds. For this reason, most often the experimental concept consists of placing a detector very similar to one one would use to search for dark matter close to an intense, pulsed source of neutrinos. Examples of such experiments are CEvNS, COGENT, CENNS,[24] and a few others. It is quite likely that coherent scattering of neutrinos may be observed in the next few years by one or all of these experiments.[a]

3.2. *Capture on radioactive nuclei*

A neutrino interacting with a nucleus can stimulate a beta decay, provided the nucleus is a beta-emitting nucleon. This reaction is called either neutrino capture or stimulated beta decay. The reaction can be written as

$$\nu_e + A_N^Z \to e^- + A_{N-1}^{Z+1}. \tag{18}$$

The cross section for this process is closely related to that of beta decay which is quite well understood. From the point of view of experimental neutrino physics, it is worth noting that this reaction may be the best avenue towards direct detection of cosmic background neutrinos.

Capture on nuclei happens for neutrinos of even the lowest energies and produces an electron with an energy high enough to be detected in the final state, making the detection in principle feasible. Tritium has been proposed as the possible element;[27] however, it requires overcoming several technical challenges before the experiment can be built. Any other proposed method must face the challenge of detecting energy transfers lower than the energy of the cosmic microwave background, which seems to be beyond the current and predicted capabilities of experimental apparata for those techniques.[28]

3.3. *Inverse beta decay*

Inverse beta decay or antineutrino scattering on protons is one of the most important and studied channels in neutrino interactions. It is the method by which neutrinos were first discovered. The interaction can be written as

$$\bar{\nu}_e + p \to e^+ + n. \tag{19}$$

The free neutron and positron that remain in the final state can be used to devise a unique experimental signature that has been used in the Reines–Cowan experiment[29] and in many subsequent experiments searching for antineutrinos most often from nuclear reactors. This signature involves detecting the prompt annihilation of the positron into two 512 keV gammas and then registering the neutron capture on a material with a high

[a]Since the writing of this chapter, coherent neutrino scattering has been observed by the COHERENT collaboration.[25]

cross section to capture neutrons like cadmium or gadolinium. This capture results in a second peak with an energy of a few MeV and delayed by several microseconds. This unique two-peak signature with well-defined energies enables suppressing backgrounds very substantially and is being used even today.[30–32] Inverse beta decay does have an energy threshold, which is due to mass difference between the proton and the neutron–positron pair, and will not occur unless $E_\nu \gtrsim 1.806\,\mathrm{MeV}$ in the centre-of-mass frame.

4. Medium-energy neutrino interactions

Several important neutrino measurements currently being devised will use accelerator-created neutrinos. This allows the tuning of the energy of the neutrinos as well as the shape of the spectrum, opening possibilities for much more precise measurements, especially in case of neutrino oscillations where controlling the energy of the neutrino is extremely important. Accelerator neutrinos have energies significantly higher than those discussed until now. Their energies fall much closer to that of atmospheric neutrinos rather than those coming from beta decays and therefore the Sun and nuclear reactors.

Accelerator neutrino energies are comparable, and often higher, than the energy scale of the Fermi motion of nucleons. This means that to understand and describe the effect of neutrino interactions at these energies, we must take into account the effect the neutrinos will have on nucleons or the nucleus as a whole, which we have been ignoring until now. The required complex nuclear calculations are difficult and time consuming. In experimental physics, this problem is most often resolved by using simplified theoretical or phenomenological models. The most basic, yet widely used model to describe the nucleons is the Relativistic Fermi Gas (RFG)[36] model, which treats the nucleus as a set of noninteracting nucleons with a global Fermi momentum and a constant binding energy. An improvement on this is the Relativistic local Fermi Gas (RlFG) model where the Fermi momentum depends on the local density of nucleons. Dedicated software packages, called Monte Carlo generators, use such models to predict the results of a neutrino interaction with a nucleus and the subsequent travel of nucleons or mesons through the nucleus. Their predictions are then compared to the results of the experiment. We will list some of these generators in Section 5.2.

In this section, we will describe the simplified classes of neutrino–nucleus interactions up to an energy of a few GeV. Before that, we will begin with inclusive scattering measurements which are a way to compare and align the measurements from experiments that use different target nuclei which could affect the results of the detailed model scatters. In the next section, we will give a few examples where the outcomes of experiments measuring neutrino scatters have been influenced by nuclear effects making it difficult to point back to these theoretical models.

4.1. *Inclusive scattering*

Before we split neutrino interactions into subcomponents and realize that the size of the nuclei that comprise the detector can affect our interpretation of the neutrino interaction in question, we should note that the general inclusive process,

$$\nu_l + N \to l^- + X, \tag{20}$$

$$\bar{\nu}_l + N \to l^+ + X, \tag{21}$$

can be used as a benchmark and for comparison between the different detector technologies and experimental techniques. This is easiest in the case of muon neutrino interactions, given that an abundant sample of measurements already exists. At higher energies, where neutrino interactions can be treated as with point-like quarks, the per nucleon cross sections divided by energy, as shown in Fig. 5, are in good agreement with the flat dashed line representing the fit of the theoretical predictions. This is the case for different experiments and nuclei. The recent measurements of these inclusive cross sections have been performed by, e.g., ArgoNeuT,[38] T2K[39] and MINERvA.[40] Note that at lower energies, where the interaction can no longer be described as occuring with single quarks and the complicated

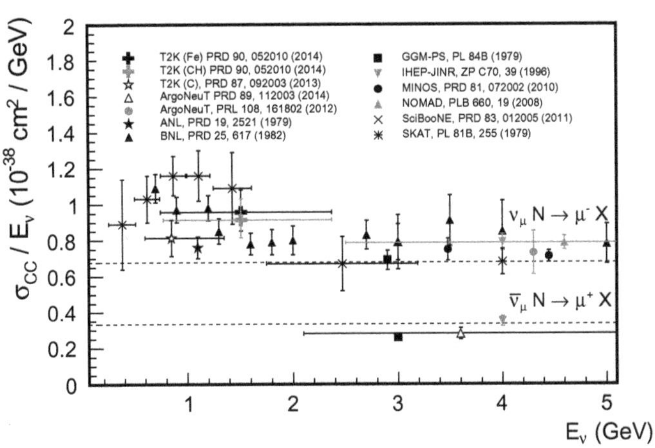

Fig. 5. Measurements of the inclusive per nucleon cross section divided by the neutrino energy for neutrinos and antineutrinos. The dashed lines represent a fit to the inclusive data extending up to much higher energies.[37]

nuclear dynamics start to play a significant role, the linear scaling with energy no longer holds. We will now explore some of the interaction mechanisms that add up to the inclusive scattering cross section in this interesting range.

4.2. *Quasielastic scattering*

Quasielastic scattering is the simplest possible CC interaction that could happen in a nucleus; see Figs. 6(a) and 6(b) for the diagram. It dominates at lower energies of accelerator neutrino interactions (100–600 MeV). In this interaction, the incoming neutrino interacts with only one nucleon, changing its quark content. In the basic model, the resulting nucleon is free to exit the nucleus unhindered. This assumption is easy to fulfill for very light nuclei such as hydrogen or deuterium but becomes somewhat more complicated for heavier nuclei. Through charge and isospin constraints, the neutrino and antineutrino are limited in the nucleons they can interact with:

$$\nu_l + n \to l^- + p, \tag{22}$$

$$\bar{\nu}_l + p \to l^+ + n. \tag{23}$$

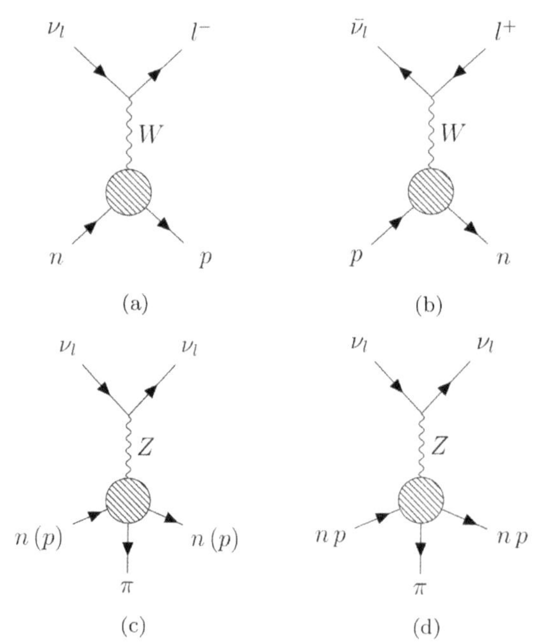

Fig. 6. Diagrams explaining neutrino interactions with a nucleus: CCQE interaction for a neutrino (a) and antineutrino (b). Resonant pion production for CC (c) and NC interactions (d) on neutrons (protons).

Situations where the quasielastic model is applicable have immediate benefits from the experimental point of view. Detecting the outgoing nucleon and identifying it as a proton or neutron determines whether the interacting particle is a neutrino or antineutrino. The relatively simple kinematics of this process allow a calculation of the energy of the original neutrino based only on the angle (θ_μ) and energy of the outgoing lepton (E_μ) as in the following equation:

$$E_\nu = \frac{E_\mu - m_\mu^2/(2M_N)}{1 - (E_\mu - \sqrt{E_\mu^2 - m_\mu^2}\cos\theta_\mu)}, \tag{24}$$

where we have ignored the binding energy and neutrino masses. The equation does not take into account the Fermi motion of the interacting nucleon which will smear the result. It is often used in Cherenkov neutrino detectors, as the majority of protons produced in CCQE neutrino interactions do not satisfy the light production threshold and are therefore invisible to the detector. Neutrino beams aimed at Cherenkov detectors tend to be optimized in terms of energy as to maximize the expected fraction of QE interactions. The CCQE interaction cross section has been measured recently by several detectors,[21,45–47] and the results of these measurements as well as older ones are shown compared to theoretical predictions in Fig. 7.

As we will discuss in the next section, nuclear effects can wash out what should be a clean signature of a QE interaction, as well as the other specific interactions described here. This is especially true for heavier nuclei. The realization of this fact has led to a wide discussion on how to compare CCQE measurements from different detectors and efforts toward a

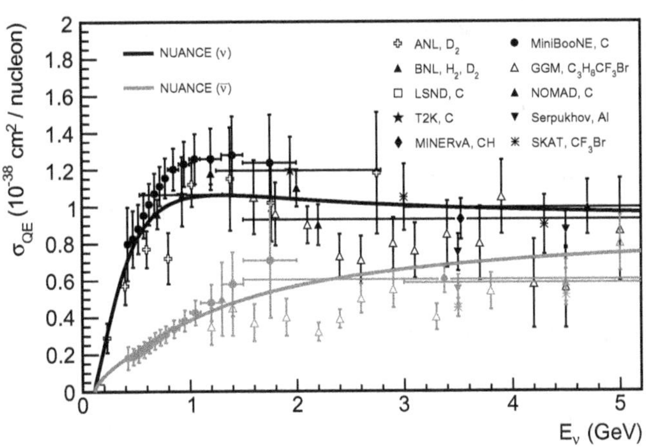

Fig. 7. Comparison of CCQE cross-section measurements overlaid with theoretical predictions for neutrinos and antineutrinos.[37]

model-independent way to present measurements of neutrino interactions. Recent measurements[49] tend to classify the interactions based on the topologies of the final state, with CC 0π $1p$, sometimes called CCQE-like, representing a similar class of events as CCQE having no pions and one proton in the final state.

4.3. *Resonant pion production*

At higher energies, the neutrino can transfer enough energy to excite the nucleon it interacts with into a resonant baryonic state. This resonant state would decay into a similar nucleon as in the QE case with the addition of a charged or neutral pion; see Fig. 6(c) for a diagram.

$$\nu_l + N \to l^- + N^* \to l^- + N + \pi, \tag{25}$$

where N is any nucleon, N^* is a resonance and π can mean either a charged or neutral pion. Neutrino interactions with charged pions in the final state have been observed in many experiments.[53–55] It is worth noting that production of kaons through a similar process is also possible and has been recently observed.[56]

Resonant pion production can also happen in case of NC interactions:

$$\nu_l + N \to \nu_l + N^* \to \nu_l + N + \pi, \tag{26}$$

where π again denotes either a charged or a neutral pion. The diagram for this interaction is shown in Fig. 6(d). These interactions are important from an experimental point of view, because the outgoing pions can be an extremely challenging background for CC interactions of muon neutrinos ($\pi^{\pm}$) or electron neutrinos (π^0). Several recent efforts to measure the production of these mesons experimentally have been made; see, e.g., Refs. 42 and 50–52. Such measurements are in general quite difficult due to the lack of the leading lepton to help select these events from backgrounds.

The energy and rate of the outgoing pions can and will be affected by nuclear interactions, and this is an object of extensive experimental and theoretical studies.

4.4. *Coherent pion production*

A specific case of interactions producing pions is the coherent pion production. If the momentum transfered to the nucleus by the neutrino is small, it is possible to treat the nucleus as a whole. In this case, we would call this a coherent neutrino interaction, similar to that from Section 3.1. At accelerator energies, it is possible that the momentum transfer would be high enough to produce a pion which could be detected in a neutrino detector. In the case of CC interactions, the pion would be charged, but production of neutral pions has been observed in NC interactions.[57] Coherent charged pion production has only recently been observed at lower neutrino energies by the MINERvA, ArgoNeuT and T2K collaborations.[58–60] This has been followed by coherent kaon production observation in MINERvA.[61] Coherently produced pions are discernible from those from resonant processes through the lack of any nuclear interactions in the vertex and a very narrow angle between the resulting charged lepton, e.g., muon and the pion. Because coherent pion production produces pions through a different mechanism than the resonant mechanism, it could affect the agreement of pion production measurements with predictions if it is not identified properly. It is one of the considered reasons for the pion production anomalies described in the next section.

4.5. *Putting the models together*

The CCQE interaction dominates at the lower end of the energy scale for accelerator neutrinos. Moving up in energy, the resonant pion production turns on when the neutrinos have enough energy to excite a nuclear resonance and produce pions. At even higher energies, the Deep Inelastic Scattering mechanism begins to dominate, which leads to the linear dependency seen at the high end of Fig. 5. The picture described above is unfortunately oversimplified because of intranuclear interaction and propagation effects, which lead to a washing out of the signatures of these theoretical models from the point of view of what experiments can reconstruct. This has led to the idea of presenting neutrino cross-section measurements as a function of the properties of the final states observed in the detector, instead of assigning them to a reaction type-like CCQE or resonant pion. This approach is designed to minimize the interpretation of experimental results using models of interaction or the nucleus and instead report only what the detector measures. We now move to discuss a few examples where nuclear effects have recently impacted neutrino scattering experiments.

5. Nuclear effects in cross-section measurements and how to model them

As mentioned earlier, at energies of a few hundred MeV, what we observe as the outcome of neutrino interactions becomes affected by nuclear effects. These can modify the initial state, i.e., what the neutrinos interact with or what happens to the results of the interaction as they travel through the nucleus — in this case, dubbed final-state interactions or FSI. In the former case, this can affect the probabilities of interaction or the number of particles that result from it. In the latter case, the hadrons that are produced in the quasielastic, NC or resonant interactions reinteract with other nuclei on their way out. This can lead to other new particles being observed in the final state or, contrarily, the absorption and therefore absence of the expected particles. Because the outgoing momentum of the final-state particles, or even their nature itself, is changed due to these interactions, the reconstructed properties of the interaction, such as the energy of the neutrino, are immediately affected. Several unexpected phenomena observed in neutrino experiments are now being attributed to nuclear effects.

5.1. *Examples of nuclear effects in neutrino scattering experiments*

Nuclear effects have been observed by electron scattering experiments.[62] However, until a few years ago, similar effects in neutrino interactions were not verified experimentally. They have since been definitely observed and the quest to understand them is one of the most vibrant fields of development in neutrino physics — one that includes theorists and experimentalists alike. Nuclear effects have been suggested as the main cause of a few of the outstanding experimental discrepancies and riddles.

One of the most well known is the MiniBooNE CCQE measurement,[48] (black circles in Fig. 7) which lie significantly higher than most other measurements. At the time, the proposed method to align these results with theoretical predictions was to change the axial mass that resides in the axial form factor defined in Eq. (16) to $M_A \simeq 1.3\,\text{GeV}$, whereas previous measurements on, e.g., deuterium, put it much closer to one. This measurement and few others that required a value of $M_A > 1$ have now been understood to result from simplified treatment of nuclear interactions. The introduction of so-called 2p–2h interactions[63,64] with pairs of correlated nucleons leads to a much better agreement with the new data. The implication of this is that since the nuclear effects affect the outcome of an interaction, the energy calculation that we proposed in Eq. (24) is most likely no longer correct. A recent proposal to circumvent this problem was to select true QE events based on topology and only use this subset to calculate the energy.[65] For more complicated topologies, it will be necessary to reconstruct all of the particles in the final state.

Another puzzle in cross-section measurements is the production of charged pions. A recent measurement by MiniBooNE,[54] was hard to reproduce using the Rein–Sehgal model used in most generators to predict pion production. It is expected that FSI effects felt by pions leaving the nucleus may be the culprit. Implementing such models improves agreement at lowest pion momenta; however, in the higher region, the predictions with FSI are worse than those without it. Recent results from the MINERvA scattering experiment[53] are not entirely compatible with the MiniBooNE measurements when compared in terms of flux-integrated cross section. As such, it is difficult to find a model that would correctly reproduce both data sets. See, e.g., Refs. 34 and 35 for a discussion of possible causes.

Precision detectors such as MINERvA[66] and liquid argon time projection chambers such as ArgoNeuT,[67] the currently running MicroBooNE[68] and in the near future SBND,[79] will be an excellent tool to shed more light on the outcomes of neutrino interactions with nuclei. ArgoNeuT has already observed neutrino events with multiple protons or no protons at all, and antineutrino events with a single proton.[69] This is clearly different from the expectations from quasielastic-like interactions, where one would expect a single proton or none, respectively. MINERvA has been able to show that a significant fraction of their muon+proton events are not clean CCQE events.[49] This ability to precisely reconstruct the energy and angles of the secondary protons and pions is greatly increasing the understanding of how the nucleus affects neutrino interactions. One striking signature of nuclear effects was the observation of back-to-back protons by ArgoNeuT;[70] see Fig. 8, which was interpreted as a possible observation of short-range correlations in neutrino events. This would mean that a neutrino would interact with a pair of nucleons rather than only one leading to the expulsion

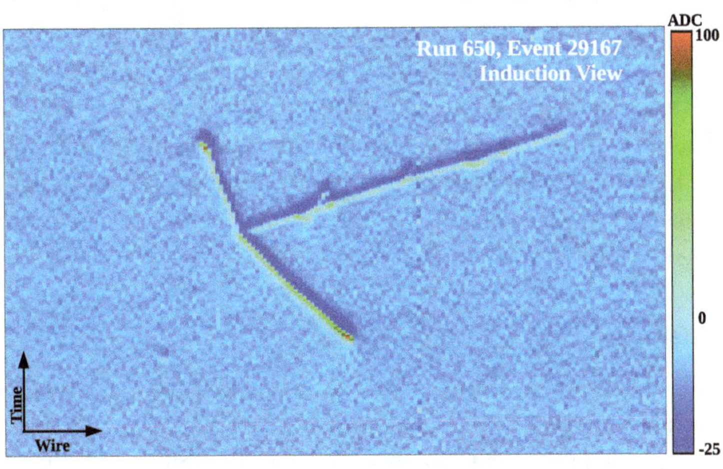

Fig. 8. Back-to-back proton event observed by the ArgoNeuT detector.

of two nucleons in this correlated manner. Given the small sample of events, this observation is far from conclusive but gives hope that the understanding of nuclear processes will be greatly improved by the use of high-resolution detectors which can provide more information about the effects of nuclear physics on neutrino scattering experiments.

5.2. *Modeling neutrino interactions*

As we have seen in the preceding sections, modeling the outcome of neutrino interactions with a nucleus is very challenging. These outcomes will depend on the neutrino energy which determines the processes that dominate, the nucleus itself and the transport effects inside it. From the point of view of an experiment, what we are interested in are the angles and energies of the final-state particles for every event we simulate. The nature and complexity of this problem has led to the development of several different Monte Carlo generators for neutrino interactions, often dedicated to an experiment. Some of the most commonly used generators are the open source GENIE[71] and NEUT.[72] The NUANCE[73] generator was widely used but is not being currently developed. These generators use models to predict the results of different types of interactions. As the theoretical models are being continuously updated to accommodate the newest experimental results, there is an effort to try to implement them in generators as soon as possible. An example is the recent addition of 2p–2h models in GENIE and NEUT. Another recent development was the introduction of the spectral function[75,76] to parametrize the response of the nucleus, which improves on the RFG and RlFG models mentioned earlier. This approach has been first used in the NuWro[74] generator, which is another generator that is becoming widely used. These generators model nuclear transport effects or FSI, through cascade models that have often been homegrown. Another notable approach is the GiBUU[77] project, which numerically solves the semiclassical Boltzmann–Uehling–Uhlenbeck equations[78] that describe the dynamics of the hadronic system as a whole leading to a better treatment of FSI. This approach has been considerably successful in modeling the experimental results described in this section.

6. Conclusions

Understanding the way neutrinos interact is crucial to any physics measurement involving neutrinos. A misunderstanding of the underlying interactions and their effects could affect the results of important measurements, such as neutrino oscillations. The models we use to model neutrino interactions are developing quickly and are benefiting from the new experimental measurements probing the interesting region. New results from high-resolution experiments, such as MINERvA, MicroBooNE and SBND, together with new ideas for water-based measurements such as WAGASCI,[80] nuPRISM[81] and ANNIE,[82] have the potential to drive the theory of neutrino interactions and lead to better models needed by future experiments.

Acknowledgments

The author would like to thank Sam Zeller, Anne Schukraft and Corey Adams for their help in preparing some of the plots. A. M. Szelc is supported by a grant from the Royal Society, UK.

References

1. C. Giunti and C. W. Kim, *Fundamentals of Neutrino Physics and Astrophysics* Oxford University Press, Oxford (2007). ISBN-9780198508717.
2. M. Thomson, *Modern Particle Physics* Cambridge University Press, Cambridge (2013).
3. P. Renton, *Electroweak Interactions: An Introduction to the Physics of Quarks and Leptons* Oxford University Press, Oxford (1990).
4. E. D. Commins and P. H. Bucksbaum, *Weak Interactions of Leptons and Quarks*, Cambridge University Press, Cambridge (1983).
5. J. A. Formaggio and G. P. Zeller, From eV to EeV: neutrino cross sections across energy scales, *Rev. Mod. Phys.* **84**, 1307–1341 (2012).
6. S. L. Glashow, Partial symmetries of weak interactions, *Nucl. Phys.* **22**, 579 (1961).
7. S. Weinberg, A model of leptons, *Phys. Rev. Lett.* **19**, 12641266 (1967).
8. A. Salam, Weak and electromagnetic interactions, *Conf. Proc.* **C680519**, 367–377 (1968).
9. T. D. Lee and C. N. Yang, Question of parity conservation in weak interactions, *Phys. Rev.* **104**, 254 (1956).
10. C. S. Wu *et al.*, Experimental test of parity conservation in beta decay, *Phys. Rev.* **105**, 1413 (1957).
11. A. Salam and J. C. Ward, Electromagnetic and weak interactions, *Phys. Lett.* **13**, 168 (1964).
12. F. J. Hasert *et al.*, Observation of neutrino like interactions without muon or electron in the Gargamelle neutrino experiment, *Phys. Lett.* **B46**, 138140 (1973).
13. E. Fermi, An attempt of a theory of beta radiation, *Z. Phys.* **88**, 161 (1934).
14. C. Patrignani *et al.* (Particle Data Group), Review of particle physics, *Chin. Phys.* **C40**(10), 100001 (2016).
15. M. Goldhaber, L. Grodzins, and A. W. Sunyar, Helicity of neutrinos, *Phys. Rev.* **109**, 1015 (1958).

16. L. Michel, Interaction between four half spin particles and the decay of the μ meson, *Proc. Phys. Soc.* **A63**, 514–531 (1950).
17. R. Bayes *et al.*, Experimental constraints on left-right symmetric models from muon decay, *Phys. Rev. Lett.* **106**, 041804 (2011).
18. Y. Fukuda *et al.*, Evidence for oscillation of atmospheric neutrinos, *Phys. Rev. Lett.* **81**, 1562–1567, (1998).
19. Q. R. Ahmad *et al.*, Direct evidence for neutrino flavor transformation from neutral current interactions in the Sudbury Neutrino Observatory, *Phys. Rev. Lett.* **89**, 011301 (2002).
20. C. J. Horowitz, K. J. Coakley, and D. N. McKinsey, Supernova observation via neutrino nucleus elastic scattering in the CLEAN detector, *Phys. Rev.* **D68**, 023005 (2003).
21. V. Lyubushkin *et al.*, A study of quasi-elastic muon neutrino and antineutrino scattering in the NOMAD experiment, *Eur. Phys. J.* **C63**, 355–381 (2009).
22. S. K. Singh, E. Oset, Quasielastic neutrino (antineutrino) reactions in nuclei and the axial-vector form factor of the nucleon, *Nucl. Phys.* **A542**, 4 (1992).
23. D. Z. Freedman, D. N. Schramm, and D. L. Tubbs, the weak neutral current and its effects in stellar collapse. *Ann. Rev. Nucl. Part. Sci.* **27**, 167–207 (1977).
24. D. Akimov *et al.*, Coherent Scattering Investigations at the Spallation Neutron Source: a Snowmass White Paper, arXiv:1310.0125 [hep-ex].
25. D. Akimov *et al.*, Observation of coherent elastic neutrino–nucleus scattering, *Science* **357**, 6356 (2017).
26. J. Beacom and P. Vogel, Angular distribution of neutron inverse beta decay, $\bar{\nu}_e + p \rightarrow e^+ + n$, *Phys. Rev.* **D60**, 053003 (1999).
27. A. G. Cocco *et al.*, Probing low energy neutrino backgrounds with neutrino capture on beta decaying nuclei, *JCAP* **0706**, 015 (2007); S. Betts *et al.*, Development of a relic neutrino detection experiment at PTOLEMY: Princeton Tritium Observatory for Light, Early-Universe, Massive-Neutrino Yield, arXiv:1307.4738 [astro-ph.IM].
28. G. Gelmini, *Phys. Scripta* **T121**, 131–136 (2005); A. Ringwald, How to detect big bang relic neutrinos? hep-ph/0505024 DESY-05-045, C. Yanagisawa, Looking for cosmic neutrino background, *Front. Phys.* **10** (2014).
29. C. L. Cowan, F. Reines *et al.*, Detection of the free neutrino: a confirmation, *Science* **124**, 103–104 (1956).
30. F. P. An *et al.*, The detector system of the Daya Bay reactor neutrino experiment, *Nucl. Instr. Method* **A811** 133–161 (2016).
31. J. K. Ahn *et al.*, RENO: An Experiment for Neutrino Oscillation Parameter θ_{13} Using Reactor Neutrinos at Yonggwang, arXiv:1003.1391 [hep-ex].
32. Y. Abe *et al.*, Measurement of θ_{13} in double Chooz using neutron captures on hydrogen with novel background rejection techniques, *JHEP* **1601**, 163 (2016).
33. C. H. Llewellyn Smith, Neutrino reactions at accelerator energies, *Phys. Rept.* **3C**, 261 (1972).
34. T. Katori, M. Martini, Neutrino-Nucleus Cross Sections for Oscillation Experiments, arXiv:1611.07770 [hep-ph].

35. L. Alvarez-Ruso, Y. Hayato, J. Nieves, Progress and open questions in the physics of neutrino cross sections, *New J. Phys.* **16**, 075015 (2014).
36. R. A. Smith and E. J. Moniz, *Nucl. Phys.* **B43**, 605 (1972).
37. Anne Schukraft, private communication.
38. C. Anderson *et al.*, First measurements of inclusive muon neutrino charged current differential cross sections on argon, *Phys. Rev. Lett.* **108**, 161802 (2012); R. Acciarri *et al.*, Measurements of inclusive muon neutrino and antineutrino charged current differential cross sections on argon in the NuMI antineutrino beam, *Phys. Rev.* **D89**(11), 112003 (2014).
39. K. Abe *et al.*, Measurement of the inclusive ν_μ charged current cross section on carbon in the near detector of the T2K experiment, *Phys. Rev.* **D87**, 092003 (2013); K. Abe *et al.*, Measurement of the inclusive ν_μ charged current cross section on iron and hydrocarbon in the T2K on-axis neutrino beam, *Phys. Rev.* **D90**, 052010 (2014); K. Abe *et al.*, Measurement of the muon neutrino inclusive charged-current cross section in the energy range of 13 GeV with the T2K INGRID detector, *Phys. Rev.* **D93**, 072002 (2016).
40. J. Devan *et al.*, Measurements of the inclusive neutrino and antineutrino charged current cross sections in MINERvA using the low-ν flux method, *Phys. Rev.* **D94**, 112007 (2016).
41. P. Adamson *et al.*, Neutrino and antineutrino inclusive charged-current cross section measurements with the MINOS near detector, *Phys. Rev.* **D81**, 072002 (2010).
42. R. Acciarri *et al.*, Measurement of muon neutrino and anti-muon neutrino neutral current $\pi^0 \to \gamma\gamma$ production in the ArgoNeuT detector, arXiv:1511.00941 [hep-ex].
43. R. Gran *et al.*, Measurement of the quasielastic axial vector mass in neutrino interactions on oxygen, *Phys. Rev.* **D74**, 052002 (2006).
44. G. A. Fiorentini *et al.*, Measurement of muon neutrino quasielastic scattering on a hydrocarbon target at $E_\nu \sim 3.5$ GeV, *Phys. Rev. Lett.* **111**, 022502 (2013).
45. L. Fields *et al.*, Measurement of muon antineutrino quasielastic scattering on a hydrocarbon target at $E_\nu \sim 3.5$ GeV, *Phys. Rev. Lett.* **111**, 022501 (2013).
46. K. Abe *et al.*, Measurement of the ν_μ charged current quasielastic cross section on carbon with the T2K on-axis neutrino beam, *Phys. Rev.* **D91**, 112002 (2015).
47. A. A. Aguilar-Arevalo *et al.*, First measurement of the muon antineutrino double-differential charged-current quasielastic cross section, *Phys. Rev.* **D88**, 032001 (2013).
48. A. A. Aguilar-Arevalo *et al.*, First measurement of the muon neutrino charged current quasielastic double differential cross section, *Phys. Rev.* **D81**, 092005 (2010); A. A. Aguilar-Arevalo *et al.*, Measurement of muon neutrino quasielastic scattering on carbon, *Phys. Rev. Lett.* **100**, 032301 (2008).
49. T. Walton *et al.*, Measurement of muon plus proton final states in ν_μ interactions on hydrocarbon at $E_\nu = 4.2$ GeV, *Phys. Rev.* **D91**, 071301(R) (2015).
50. K. Abe *et al.*, Measurement of the neutrino-oxygen neutral-current interaction cross section by observing nuclear de-excitation γ rays, *Phys. Rev.* **D90**, 072012 (2014).

51. J. Wolcott *et al.*, Evidence for neutral-current diffractive π^0 production, *Phys. Rev. Lett.* **117**, 111801 (2016).
52. A. A. Aguilar-Arevalo *et al.*, Measurement of the antineutrino neutral-current elastic differential cross section, *Phys. Rev.* **D91**, 012004 (2015); A. A. Aguilar-Arevalo *et al.*, Measurement of the neutrino neutral-current elastic differential cross section on mineral oil at $E_\nu \sim 1\,\text{GeV}$, *Phys. Rev.* **D82**, 092005 (2010); A. A. Aguilar-Arevalo *et al.*, Measurement of ν_μ and $\bar{\nu_\mu}$ induced neutral current single π^0 production cross sections on mineral oil at $E_\nu \sim O\,(1\,\text{GeV})$, *Phys. Rev.* **D81**, 013005 (2010).
53. C. L. McGivern *et al.*, Cross sections for ν_μ and $\bar{\nu_\mu}$ induced pion production on hydrocarbon in the few-GeV region using MINERvA, *Phys. Rev.* **D94**, 052005 (2016).
54. A. A. Aguilar-Arevalo *et al.*, Measurement of neutrino-induced charged-current charged pion production cross sections on mineral oil at $E_\nu \sim 1\,\text{GeV}$, *Phys. Rev.* **D83**, 052007 (2011).
55. K. Abe *et al.*, First measurement of the muon neutrino charged current single pion production cross section on water with the T2K near detector, *Phys. Rev.* **D95**, 012010 (2017).
56. C. M. Marshall *et al.*, Measurement of K^+ production in charged-current ν_μ interactions, *Phys. Rev.* **D94**, 012002 (2016).
57. C. T. Kullenberg, S. R. Mishra *et al.*, A measurement of coherent neutral pion production in neutrino neutral current interactions in NOMAD, *Phys. Lett.* **B682**, 177–184 (2009).
58. A. Higuera *et al.*, Measurement of coherent production of $\pi^\pm$ in neutrino and antineutrino beams on carbon from E_ν of 1.5 to 20 GeV, *Phys. Rev. Lett.* **113**, 261802 (2014).
59. R. Acciarri *et al.*, First measurement of neutrino and antineutrino coherent charged pion production on argon, *Phys. Rev. Lett.* **113**, 261801 (2014).
60. K. Abe *et al.*, Measurement of coherent π^+ production in low energy neutrino-carbon scattering, *Phys. Rev. Lett.* **117**, 192501 (2016).
61. Z. Wang *et al.*, Evidence of coherent K^+ meson production in neutrino-nucleus scattering, *Phys. Rev. Lett.* **117**, 061802 (2016).
62. J. W. Van Orden and T. W. Donnelly, Mesonic processes in deep-inelastic electron scattering from nuclei, *Ann. Phys.* **131**, 451 (1981); W. M. Alberico, M. Ericson, and A. Molinari, The role of two particle-two hole excitations in the spin-isospin nuclear response, *Ann. Phys.* **154**, 356 (1984).
63. M. Martini, M. Ericson, G. Chanfray, and J. Marteau, Unified approach for nucleon knock-out and coherent and incoherent pion production in neutrino interactions with nuclei, *Phys. Rev.* **C80**, 065501 (2009).
64. J. Nieves, I. Ruiz Simo, and M. Vicente Vacas, Inclusive chargedcurrent neutrinonucleus reactions, *Phys. Rev.* **C83**, 045501 (2011); R. Gran, J. Nieves, F. Sanchez, M. Vicente Vacas, Neutrino-nucleus quasi-elastic and 2p2h interactions up to 10 GeV, *Phys. Rev.* **D88**, 113007 (2013).
65. A. P. Furmanski and J. Sobczyk, Neutrino energy reconstruction from one muon and one proton events, arXiv:1609.03530 [hep-ex].

66. L. Aliaga *et al.*, Design, calibration, and performance of the MINERvA detector, *Nucl. Inst. Method* **A743**, 130 (2014).
67. C. Anderson *et al.*, The ArgoNeuT detector in the NuMI low-energy beam line at fermilab, *JINST* **7**, P10019 (2012).
68. R. Acciarri *et al.*, Design and construction of the MicroBooNE detector, *JINST* **12**, P02017 (2017).
69. O. Palamara *et al.*, Exclusive topologies reconstruction in LAr-TPC experiments: a novel approach for precise neutrino-nucleus cross-sections measurements, arXiv:1309.7480 [physics.ins-det].
70. R. Acciarri *et al.*, Detection of back-to-back proton pairs in charged-current neutrino interactions with the ArgoNeuT detector in the NuMI low energy beam line, *Phys. Rev.* **D90**, 012008 (2014).
71. C. Andreopoulos *et al.*, The GENIE neutrino monte carlo generator, *Nucl. Instrum. Method* **A614**, 87–104 (2010).
72. Y. Hayato, A neutrino interaction simulation program library NEUT, *Acta Phys. Polon.* **B40**, 2477 (2009).
73. D. Casper, The nuance neutrino simulation, and the future, *Nucl. Phys. Proc. Suppl.* **112**, 161 (2002).
74. C. Juszczak, J. A. Nowak, and J. T. Sobczyk, Simulations from a new neutrino event generator, *Nucl. Phys. Proc. Suppl.* **159**, 211 (2006).
75. A. M. Ankowski and J. T. Sobczyk, Argon spectral function and neutrino interactions, *Phys. Rev.* **C74**, 054316 (2006).
76. O. Benhar, N. Farina, H. Nakamura, M. Sakuda, and R. Seki, Electron- and neutrino-nucleus scattering in the impulse approximation regime, *Phys. Rev.* **D72**, 053005 (2005).
77. Gallmeister *et al.*, Neutrino-induced reactions on nuclei, *Phys. Rev.* **C94**, 035502 (2016).
78. O. Buss *et al.*, Transport-theoretical description of nuclear reactions, *Phys. Rept.* **512**, 1 (2012).
79. M. Antonello *et al.*, A proposal for a three detector short-baseline neutrino oscillation program in the fermilab booster neutrino beam, arXiv:1503.01520 [physics.ins-det].
80. A. Minamino, WAGASCI and ND280 Upgrades, *JPS Conf. Proc.* **12**, 010038 (2016).
81. S. Bhadra *et al.*, Letter of Intent to Construct a nuPRISM Detector in the J-PARC Neutrino Beamline, arXiv:1412.3086.
82. I. Anghel *et al.*, Letter of Intent: The Accelerator Neutrino Neutron Interaction Experiment (ANNIE), arXiv:1504.01480.

Chapter 13

Theory and Phenomenology of Mass Ordering and CP Violation

P. Coloma* and S. Pascoli†

*Fermi National Accelerator Laboratory
P.O. Box 500, Batavia, IL 60510 USA
pcoloma@fnal.gov
†Institute for Particle Physics Phenomenology,
Department of Physics, Durham University,
South Road, Durham DH1 3LE, United Kingdom
silvia.pascoli@durham.ac.uk

Abstract

Despite the remarkable progress in understanding neutrino properties in recent years, key open questions remain to be answered. Here, we focus on the neutrino mass ordering (MO) and leptonic CP violation (CPV) from theoretical and phenomenological points of view, discussing how they can be searched for in present and future neutrino experiments.

1. Introduction

Since the discovery of neutrino oscillations almost two decades ago, a remarkable progress has been made in neutrino physics. Two mass-squared differences and three mixing angles have been measured with precision, and the first hints of CP violation (CPV) are emerging. Despite these staggering results, compelling questions remain open regarding neutrino properties: are neutrinos Majorana or Dirac? Is the CP symmetry violated in the leptonic sector (as it is for quarks)? What are the absolute values of neutrino masses, and what is mass ordering (MO), normal (NO) or inverted (IO)? What are the precise values of the mixing angles, and which flavor dominates the third mass eigenstate, ν_μ or ν_τ? In addition, we need to thoroughly test the standard three-neutrino scenario and search for possible new physics effects such as sterile neutrinos or the presence of new interactions beyond the known ones. These questions are crucial in order to paint a complete picture of neutrino properties and are essential to hunt for the new physics beyond the standard model (SM) responsible for neutrino masses and mixing.

Here, we focus on two of the above open questions, namely, the neutrino MO and leptonic CPV. In Section 2 we review our current knowledge of the neutrino parameters. In Section 3 we discuss several theoretical issues in the neutrino sector to illustrate how the determination of the MO and leptonic CPV can shed light on the origin of neutrino masses and the flavor problem. Section 4 is devoted to the phenomenological issues concerning their experimental determination including a discussion of long-baseline, atmospheric, reactor neutrino experiments, neutrinoless double-beta decay, cosmological and astrophysical probes. Finally, we draw our summary and conclusions in Section 5.

2. Neutrino mass ordering and CP violation

Neutrino oscillation experiments have so far measured with precision two[a] mass-squared differences $\Delta m^2_{21} \equiv m^2_2 - m^2_1$ and $\Delta m^2_{31} \equiv m^2_3 - m^2_1$. This implies that there are at least three massive neutrinos, ν_i $(i = 1, 2, 3)$, with masses m_i. They are related to the three flavour states ν_α $(\alpha = e, \mu, \tau)$ via a 3×3 unitary matrix U, called the Pontecorvo–Maki–Nakagawa–Sakata[1] (PMNS) matrix, as

$$\nu_{\alpha \mathrm{L}} = \sum_{i=1}^{3} U_{\alpha i}\, \nu_{i\mathrm{L}}, \tag{1}$$

where we have defined $\nu_{\mathrm{L}} \equiv P_{\mathrm{L}}\nu$, with P_{L} being the left-handed projection operator. In general, a 3×3 unitary matrix can be parametrized by three angles and six phases. However, three of these phases are unphysical and can be eliminated from the charged-current (CC) SM Lagrangian via a phase rotation of the charged lepton fields, $\ell_\alpha \to e^{i\phi_\ell}\ell_\alpha$. If neutrinos are Dirac particles, two additional phases can be rotated away via a rephasing of the neutrino fields, analogously to what happens for the Cabibbo–Kobayashi–Maskawa mixing matrix in the quark sector. The remaining phase plays an important role in neutrino oscillation experiments. Conversely, if neutrinos are Majorana particles, the last rephasing is not possible: these

[a]We limit the discussion to "standard" neutrino oscillation results, disregarding the still-controversial results from short-baseline neutrino oscillation experiments pointing toward sterile neutrinos or other nonstandard effects.

two CP-violating phases are physical and enter in processes which rely upon the Majorana nature of neutrinos, such as neutrinoless double beta decay. Three CPV phases are physical in the Majorana case, as opposed to one in the Dirac case.

The PMNS matrix is usually parameterized as[3,4]

$$U_{\text{PMNS}} = \begin{pmatrix} c_{12}c_{13} & s_{12}c_{13} & s_{13}e^{-i\delta} \\ -s_{12}c_{23} - c_{12}s_{23}s_{13}e^{i\delta} & c_{12}c_{23} - s_{12}s_{23}s_{13}e^{i\delta} & s_{23}c_{13} \\ s_{12}s_{23} - c_{12}c_{23}s_{13}e^{i\delta} & -c_{12}s_{23} - s_{12}c_{23}s_{13}e^{i\delta} & c_{23}c_{13} \end{pmatrix} \cdot K,$$

where $c_{ij} \equiv \cos\theta_{ij}$ and $s_{ij} \equiv \sin\theta_{ij}$, with $\theta_{ij} \in [0, 90°]$. In this notation, δ is the Dirac CPV phase $\delta \in [0, 360°]$ and K is a diagonal matrix $K \equiv \text{diag}(1, e^{i\frac{\alpha_{21}}{2}}, e^{i\frac{\alpha_{31}}{2}})$ which contains the two Majorana CPV phases α_{21}, α_{31}.

CP-violating effects due to the Dirac phase can be expressed in a rephasing-invariant manner in terms of the Jarlskog invariant[2]

$$J \equiv \Im[U_{\mu 3}U_{e2}U^*_{\mu 2}U^*_{e3}] = \frac{1}{8}\sin 2\theta_{12}\sin 2\theta_{23}\sin 2\theta_{13}\cos\theta_{13}\sin\delta. \tag{2}$$

It is interesting to notice that J depends on all the three mixing angles, highlighting that Dirac CPV is a genuine three-neutrino mixing effect, and its phenomenological consequences are suppressed by the small angle θ_{13}, as we will discuss in detail in Section 4.

The mixing angles are known with quite good accuracy, thanks to oscillation experiments:[5] $\theta_{12} = 33.56^{+0.77}_{-0.75}$, $\theta_{23} = 41.6^{+1.5}_{-1.2}$ (NO) and $\theta_{23} = 50.0^{+1.1}_{-1.4}$ (IO), although at 3σ $\theta_{23} = 38.4$–52.8 (38.8–53.1) for NO (IO), and $\theta_{13} = 8.46\ (8.49)^{+0.15}_{-0.15}$ for NO (IO), in degrees. The first hints of leptonic CPV have been reported, thanks to the results from long-baseline experiments in combination with the precise determination of θ_{13} by reactor neutrino experiments, as we will discuss in more detail in Section 2. Currently, the fitted values for δ show a preference for large CPV with $\delta = 261^{+51}_{-59}$ (NO) and $\delta = 277^{+40}_{-46}$ (IO) although at 3σ for NO, all values, including the CP-conserving ones $\delta = 0, 180°$, are allowed, and for IO, δ is constrained to be within $145°$ and $391°$. More data is required to test these hints and confirm whether CP is violated in the lepton sector.

The so-called solar mass splitting is determined to be positive from solar neutrino oscillation data,[5] $\Delta m^2_{21} = 7.50^{+0.19}_{-0.17} \times 10^{-5}\ \text{eV}^2$, whereas the sign of the atmospheric splitting Δm^2_{31} is not yet known.[b] In the three-neutrino

[b]The convention of ordering the masses is linked to the definition of the mixing angles, e.g., the correspondence between the solar mixing angle and θ_{12}. The convention considered

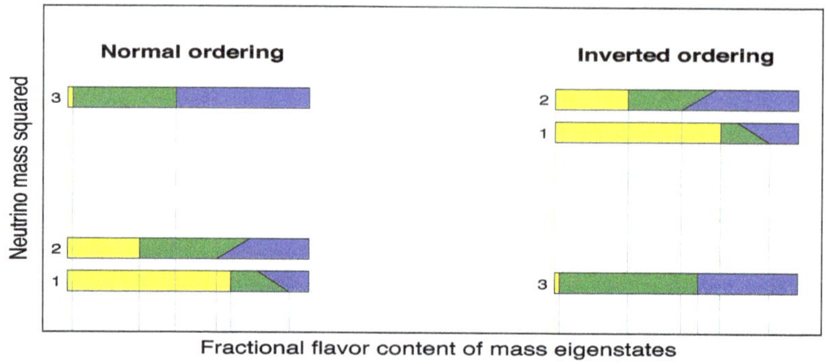

Fig. 1. Fractional flavor content, $|U_{\alpha i}|^2$ ($\alpha = e, \mu, \tau$) of the three mass eigenstates ν_i, as determined by the current best-fit values of the mixing angles[5] and varying δ from 0 (bottom of each colored band) to 180° (top of colored band), for normal and inverted MO schemes on the left and right, respectively. The different colors correspond to the ν_e fraction (yellow), ν_μ (green) and ν_τ (blue).

case, and for positive Δm^2_{21}, two MO schemes are possible:

- normal ordering (NO): $m_1 < m_2 < m_3$, i.e., $\Delta m^2_{31} > 0$,
- inverted ordering (IO): $m_3 < m_1 < m_2$, i.e., $\Delta m^2_{32} < 0$.

The best-fit values for the atmospheric mass-squared splitting differ slightly for the two orderings due to subleading effects in the oscillation probabilities. From a recent global fit to neutrino oscillation data, we have $\Delta m^2_{31} = 2.524^{+0.039}_{-0.040} \times 10^{-3}$ eV^2 (for NO), $\Delta m^2_{32} = -2.514^{+0.038}_{-0.041} \times 10^{-3}$ eV^2 (for IO). The two possible MO schemes are illustrated in Fig. 1. The values of the mixing angles determine the flavor content of the three mass eigenstates, given by $|U_{\alpha i}|^2$, which in the figure are represented by the relative size of the colored bands. As the phase δ is poorly constrained, it has been varied between 0 and 180° in each colored band.

Given the constraints imposed by oscillation data, for each ordering, the three neutrino masses can be expressed in terms of just one unknown parameter, the lightest neutrino mass $m_{\rm MIN}$. The values of the three light neutrino masses are shown in Fig. 2 as a function of $m_{\rm MIN}$. As can be seen in the figure, three different limiting cases can be easily distinguished:

- *Normal Hierarchical (NH)*. For $m_{\rm MIN} \to 0$, for NO we have $m_1 \ll m_2 \ll m_3$ with $m_1 \equiv m_{\rm MIN}$, $m_2 \cong \sqrt{\Delta m^2_{21}}$ and $m_3 \cong \sqrt{|\Delta m^2_{31}|}$.
- *Inverted Hierarchical (IH)*. In the limit $m_{\rm MIN} \to 0$, for IO we have instead $m_3 \ll m_1 < m_2$ with $m_{1,2} \cong \sqrt{|\Delta m^2_{32}|}$ and $m_3 \equiv m_{\rm MIN}$.

here, which is by far the most used in the literature, allows to keep the meaning of the mixing angles to be the same between the NO and IO.

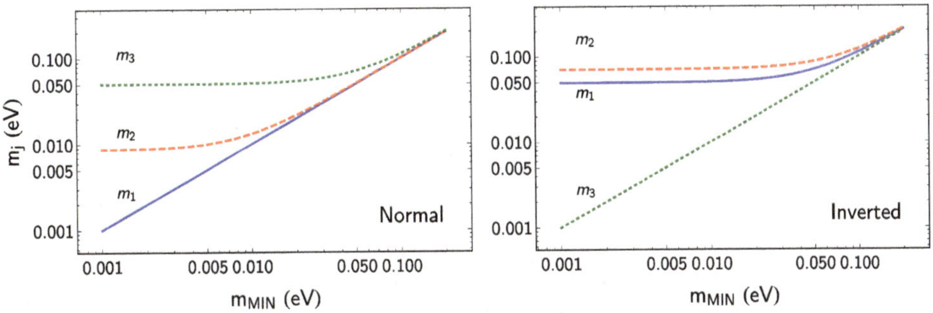

Fig. 2. Values of neutrino masses for NO (left plot) and IO (right plot) as a function of m_{MIN}. The constraints from oscillation data[5] on the values of the solar (Δm^2_{21}) and atmospheric (Δm^2_{31}, or Δm^2_{32}) mass-squared splittings have been imposed; see text for details.

- *Quasidegenerate* (*QD*). For large values of m_{MIN} ($m_{\text{MIN}} \gg \sqrt{|\Delta m^2_{31}|}$) the three mass eigenstates are almost degenerate, $m_j^2 \simeq m_{\text{MIN}}$, $j = 1, 2, 3$.

As these possibilities are still allowed by data, we use the notion of "ordering" instead of "hierarchy" to describe the neutrino mass spectrum since it is not established that neutrino masses are hierarchical.

3. Theoretical aspects of neutrino MO and CPV

One of the outstanding problems of the SM is to explain its flavor structure, the so-called *flavor problem* of the SM. The SM fermion masses span over five orders of magnitude for charged fermions and many more if neutrinos are considered. In addition, neutrino masses may not be very hierarchical, two or even three of them could be quite degenerate, and all mixing angles are rather large (contrary to what happens in the quark sector).

In order to account for neutrino masses, a Dirac and/or a Majorana mass term should be added to the SM Lagrangian. In the following discussion, we focus on Majorana masses, which are by far the most studied case in flavor models. A Majorana mass term can be expressed as

$$\mathcal{L}^M_{\nu\ \text{mass}} = \frac{1}{2}\nu^T_{\beta L} C^\dagger \mathcal{M}_{\nu,\beta\alpha} \nu_{\alpha L} + \text{h.c.}\ ,$$

where C is the charge conjugation matrix and acts on a given field as $\psi^C = C\bar{\psi}^T$. The neutrino mass matrix and its eigenvalues (i.e., the neutrino masses) can be related by a unitary matrix U_ν as

$$\text{diag}(m_1, m_2, m_3) = U^T_\nu\, \mathcal{M}_\nu\, U_\nu\ . \tag{3}$$

The charged lepton mass matrix can be diagonalised using two unitary matrices U_ℓ and V_ℓ, where U_ℓ (V_ℓ) acts on the chiral fields ℓ_L (ℓ_R). The corresponding eigenstates, with definite nonnegative masses, are what we refer to as "electron", "muon" and "tau" leptons, which are differentiated only by their masses. The observable PMNS mixing matrix, which enters the CC interactions, results from the product of the two transformations:

$$U_{\rm PMNS} = U_\ell^\dagger U_\nu. \tag{4}$$

The ordering and absolute values of neutrino masses as well as the mixing angles and CPV phases are determined by the neutrino and, for the latter, also the charged lepton mass matrices and consequently by their structure.

Different theoretical approaches have been invoked to explain specific structures of the PMNS matrix. As the neutrino mixing angles do not exhibit a very strong hierarchy, the simplest assumption would be that of anarchy,[6] in which the values of all the entries of $\mathcal{M}_\nu$ are taken to be of the same order. Anarchy models can reproduce the observed values of the parameters but offer no further insight into the flavor problem.

An alternative approach is to study models in which a guiding principle is applied to explain the observed flavor structure. Most of the work along these lines has concentrated on employing flavor symmetries motivated by the fact that θ_{23} is close to maximal and θ_{13} is considerably smaller than the other two mixing angles. The basic idea is that the leptonic doublets are charged under a flavor symmetry G_f, fixing the allowed entries in the mass terms in the Lagrangian. Typical groups which have been considered are, e.g., the non-abelian discrete groups S_4, A_4, A_5, D_N, $\Sigma(2N^2)$, $\Sigma(3N^3)$, $\Delta(3N^2)$, $\Delta(6N^2)$ (for recent reviews, see Refs. 7–9, for example) and the continuous groups $SU(3)$[10] and $SO(3)$.[11] As charged lepton masses are different, the symmetry must be broken at low energy,[12] typically into Abelian residual symmetries G_e and G_ν in the charged lepton and the neutrino sectors, respectively. These groups constrain the form of the corresponding mass matrices, leading to U_e and U_ν and ultimately to $U_{\rm PMNS}$. It should be noted that these models focus on the lepton sector and the implementation of discrete flavor symmetries in the quark sector is challenging, especially in models which try to have a unified picture for the two sectors.

We refrain from a detailed discussion of flavor models as this is well beyond the scope of this chapter. Instead, we focus on how these models can predict the MO and/or the CP-violating phases, in order to illustrate how a measurement of these quantities can give us some insight into the leptonic flavor problem. We would like to stress that due to the proximity of the value of θ_{23} to maximal, its precise determination is also of key importance from a theoretical point of view and one of the main goals of future neutrino oscillation experiments.

3.1. *The neutrino mass matrix and the MO*

The neutrino MO is determined by the values of neutrino masses and is therefore ultimately related to the structure of $\mathcal{M}_\nu$. In anarchical models, an NH spectrum is often predicted, but an IH case can also be obtained without significant fine-tuning. In these models, the smallness of certain parameters, such as θ_{13} and the ratio $\Delta m^2_{21}/\Delta m^2_{31}$, is down to chance. If instead we assume that a guiding principle is at work to explain the observed flavor structure, then it may be possible to predict the neutrino MO. This is the case, for example, in models with some flavor symmetries.[c] Let us first consider a 2×2 case as an illustration. If the Lagrangian exhibits a L_μ–L_τ symmetry, the neutrino mass matrix will have the following structure:

$$\mathcal{M}_\nu = \begin{pmatrix} a & b \\ b & a \end{pmatrix}, \tag{5}$$

where a and b are two arbitrary parameters. It is immediate to show that, for $a \ll b$, this matrix leads to two quasidegenerate neutrino masses $a \pm b$[d] and maximal mixing $\theta = 45^o$. This symmetry could be used to explain a maximal atmospheric mixing angle and would predict an IH spectrum. If we extend this reasoning to three-neutrino mixing, the most general mass matrix which gives $\theta_{23} = 45^o$ and $\theta_{13} = 0$ is

$$\mathcal{M}_\nu = \begin{pmatrix} x & y & y \\ y & z & w \\ y & w & z \end{pmatrix}. \tag{6}$$

In this case, one will also have $\sin^2 2\theta_{12} = 8y^2/((x-w-z)^2+8y^2)$, taking real parameters for simplicity. Since θ_{13} is exactly zero in this model, corrections to $\mathcal{M}_\nu$ will be needed to account for its observed value. These may also induce a value of δ different from 0 and 180°. The mentioned structure can give different neutrino mass spectra. For instance, $x, y \ll z \simeq w$ leads to NH; $x = z = w \ll y$ corresponds to a L_e–L_μ–L_τ[13] symmetry and implies IH and bimaximal mixing; $y \ll x, w, z$; $x \simeq 2z = -2w$ predicts IH.

[c]It should be noted that many models based on discrete flavor symmetries do not predict the values of the masses and consequently the MO.

[d]A negative sign for the mass can be reabsorbed into a redefinition of the corresponding neutrino field (via its Majorana phase).

The main conclusion derived from this section is that a determination of the neutrino MO will generally help to shed light on the flavor puzzle. It should be kept in mind, however, that in some theoretical scenarios (as in the anarchy one or in models with discrete flavor symmetries), both orderings can be successfully accommodated.

3.2. *CPV and flavor symmetries*

The neutrino mass matrix is also responsible for the mixing structure as discussed above. Motivated by the observed values of the mixing angles, in particular θ_{23} being (nearly) maximal and θ_{13} being much smaller than the other two angles, specific patterns have been considered for the mixing matrix. Among these, the most studied patterns include the tri-bimaximal (TBM) one,[14] with $\sin^2 2\theta_{23} = 1, \sin^2\theta_{12} = \frac{1}{3}, \sin^2\theta_{13} = 0$ and the bimaximal (BM) one,[15] with $\sin^2 2\theta_{23} = 1$, $\sin^2 2\theta_{12} = 1$, $\sin^2\theta_{13} = 0$. Other well-studied examples are the golden ratio, trimaximal or hexagonal mixing; see Ref. 16, for example. The basic idea is to employ a flavour symmetry which can lead to the above structures at leading order, introducing afterward a set of corrections which shift the angles, in particular θ_{13}, bringing them in agreement with their experimentally observed values.[17–19] These corrections typically depend on very few parameters, leading to correlations between the mixing parameters, specifically the mixing angles and the CPV phases. Such correlations are commonly known as *sum rules.* They are typically divided into two main categories:

- atmospheric sum rules[18] of the type $\sin^2\theta_{23} = 1/2+\zeta\sin\theta_{13}\cos\delta$, with ζ a real parameter specified by the model. As a relevant example, we mention trimaximal mixing, which can be obtained from TBM mixing using a rotation in the 2-3 or 1-3 plane leading to the atmospheric sum rules: $\sin^2\theta_{23} = 1/2 - \sqrt{2}\sin\theta_{13}\cos\delta$ and $\sin^2\theta_{23} = 1/2 + 1/\sqrt{2}\sin\theta_{13}\cos\delta$, respectively.
- solar sum rules[19–21] for which $\sin^2\theta_{12} = 1/3 + \zeta'\sin^2\theta_{13}\cos\delta$, with ζ' a real parameter predicted by the model. They typically apply to models in which the leading-order mixing matrix receives corrections from the charged lepton sector. In the first studies, motivated by Grand Unified Theories (GUTs) in which the charged lepton mass matrix is related to the down-type quark one, U_ℓ receives corrections of the order of the Cabibbo angle.

Given the known values of the mixing angles, these sum rules can be rephrased as predictions for the values of the δ phase which are determined by the flavor symmetry. In the case of solar sum rules, in which we require only that there are no 1-3 rotations at leading order in either U_ν or U_ℓ, $\cos\delta$ is predicted to be[22]

$$\cos\delta = \frac{t_{23}\sin^2\theta_{12} + \sin^2\theta_{13}\cos^2\theta_{12}/t_{23} - \sin^2\theta^\nu_{12}(t_{23} + \sin^2\theta_{13}/t_{23})}{\sin 2\theta_{12}\sin\theta_{13}}, \quad (7)$$

where $t_{23} \equiv \tan\theta_{23}$ and θ^ν_{12} is the predicted value in U_ν at leading order, e.g., $\sin^2\theta^\nu_{12} = 1/3$ for TBM and $\sin^2\theta^\nu_{12} = 1/2$ for BM. These relations can be tested at present and/or in future experiments, providing key information on the underlying flavor symmetry or other mechanism responsible for the leptonic flavor structure.

More recently, the possibility to impose a symmetry associated with CP at the fundamental level in the Lagrangian has received great attention, leading to more predictive models for the CPV phases.[23,24,e] A generalized CP symmetry[25] is a combination of charge conjugation and parity which acts in a nontrivial manner on the flavor indices of a field Ψ:

$$\Psi = X\Psi^c, \quad (8)$$

where Ψ^c is the CP conjugate of Ψ. Here, X is a symmetric unitary matrix ($XX^* = X^*X = 1$) so that the original state is recovered after applying the transformation twice. Combining this symmetry, with a discrete flavor symmetry, it is possible to predict the values of the CPV phases. In particular, if one imposes the CP symmetry in the neutrino sector so that the residual symmetry will be given by $G_\nu \times$ CP, the following condition on the neutrino mass matrix needs to be satisfied:

$$X^T \mathcal{M}_\nu X = \mathcal{M}^*_\nu, \quad (9)$$

constraining the Majorana phases relevant for neutrinoless double beta decay. If one chooses $G_\nu = \mathcal{Z}_2$, it is possible to show that, generically, the matrix U_ν, and ultimately $U_{\rm PMNS}$, has the form

$$U_\nu = \Omega_\nu R(\theta) K_\nu,$$

where Ω_ν is a unitary matrix determined by $\mathcal{Z}_2 \times$ CP, $R(\theta)$ is a rotation in a plane (given in terms of one parameter only, θ) and K_ν is a diagonal matrix

---

[e]Generically discrete groups can be classified in groups which admit a physical CP transformation, e.g., S_3, S_4, A_4, A_5, and groups which do not, such as T_7, $\Delta(27)$. It is interesting to notice that, in the latter case, CPV arises directly from group theory.[24] In the former case, in many models, CP is preserved at the fundamental level and is subsequently spontaneously broken, for instance, by the vacuum expectation values of the scalar fields, called flavons, which break the flavor symmetry.

with diagonal elements $\pm 1, i$. Consequently, up to permutations of its rows and columns, $U_{\rm PMNS}$ depends on a single real parameter θ, which can be fixed by matching the mixing angles to the observed values. Thus, the CPV phases are not free parameters but take specific values which can be tested in future experiments. Many models implementing flavor symmetries and a CP symmetry have been studied in recent years, leading to a broad range of predictions for the values of δ and the Majorana phases.

3.3. *CPV and leptogenesis*

One of the most interesting open questions to be addressed in particle physics and cosmology is the origin of the baryon asymmetry of the universe. Assuming that the universe had no initial significant baryon asymmetry (as implied by inflationary models), this means that a process called "baryogenesis" must have taken place at some point in the early universe, generating the excess of matter with respect to antimatter that we observe today. The baryon asymmetry could have been dynamically generated if the conditions suggested by Sakharov in 1967 are satisfied[26]:

- Baryon number violation.
- C and CP violation. If the CP symmetry is conserved, every process which produces a particle will be accompanied by its CP conjugate, with no net baryon number asymmetry generated.
- Departure from thermal equilibrium. This condition is readily satisfied by the expansion of the universe.

Several mechanisms to generate the asymmetry have been proposed. Among these, a very popular and successful alternative is the "leptogenesis"[27] mechanism. It relies on the fact that $B-L$ is conserved both at the perturbative and nonperturbative levels in the SM. Thus, a produced lepton asymmetry would eventually generate a net baryon asymmetry via sphaleron effects,[28] as these conserve $B-L$. The conditions needed to generate a lepton number asymmetry are the same as for baryon number, replacing B by L.

Leptogenesis is particularly appealing because seesaw models,[29] in addition to being able to generate neutrino masses, also satisfy the three Sakharov conditions. Here we focus on the simplest example, namely, leptogenesis in seesaw type I models, where at least two heavy right-handed (RH) neutrinos N_i are added to the SM particle content. These are singlets under the SM gauge group and therefore can have large Majorana masses, which violate lepton number explicitly. The extended Lagrangian contains a Yukawa term which couples the heavy singlets with the Higgs ϕ and the leptonic doublet ℓ in the form $Y_\nu \bar{N} \tilde{\phi} \ell_L$. Once the Higgs gets a vacuum expectation value v, a Dirac mass term is generated, and the full mass Lagrangian in the neutrino sector reads

$$-\mathcal{L}_{\text{mass}} = \overline{N_i}\,(m_D)_{ij}\,\nu_{Lj} + \frac{1}{2}\,\overline{(N_i)^C}\,(M_R)_{ij}\,N_j + \text{h.c.}\,, \tag{10}$$

where $m_D \equiv Y_\nu v$ and M_R are the Dirac and Majorana mass matrices, respectively. The neutrino masses are obtained after diagonalizing the full mass matrix. In the limit $M_R \gg m_D$, this model automatically explains the smallness of neutrino masses, which are suppressed as[29]

$$\mathcal{M}_\nu \simeq -m_D^T\, M_R^{-1}\, m_D\,. \tag{11}$$

In addition, a lepton number asymmetry can be generated in the heavy RH neutrino decays, due to the Yukawa couplings, since the three Sakharov conditions are typically met[27,30–33]: (i) lepton number is violated by the Majorana masses M_R; (ii) CPV takes place through the complex phases in the Y_ν matrix; (iii) the expanding Universe provides the departure of equilibrium condition once the temperature drops below the mass of the heavy neutrinos. The exact value of the generated lepton asymmetry is controlled by the CP asymmetry $\epsilon \equiv (\Gamma(N \to \Phi^+\ell^-) - \Gamma(N \to \Phi^-\ell^+))/(\Gamma(N \to \Phi^+\ell^-) + \Gamma(N \to \Phi^-\ell^+))$ which depends on the Yukawa couplings and ultimately on the high-energy CPV phases. As the mechanism is not instantaneous, part of the CP asymmetry is washed out by processes which partially erase the lepton asymmetry produced in the N decays. The final baryon asymmetry is given by the subsequent conversion of the lepton asymmetry through sphaleron effects and the dilution due to entropy release between production and the present.

A relevant question to ask at this point is whether the CP-violating phases observable at low energies (in the PMNS mixing matrix) and the phases necessary for leptogenesis (in the Yukawa couplings, in the example of the type I seesaw) are related.[34,35] In general, these models contain way more parameters at high energy compared to those which can be accessed at low energy, including several CP-violating phases. Consequently, in a completely general and model-independent way, it is not possible to draw a direct connection between the two. However, in models in which the origin of neutrino masses and flavor structure is explained (for instance, if a specific flavor symmetry is invoked), the same parameters relevant at high energy also control the mixing angles and phases at low energy; see Section 3.2. This provides a direct link and in some cases allows to predict the baryon asymmetry in terms of the value of the low-energy phases. Generically, in many models of neutrino mass generation, the observation of lepton number violation (e.g., neutrinoless double beta decay) and CPV in long-baseline neutrino oscillation experiments and/or possibly neutrinoless double beta decay would constitute a strong indication (even if not a proof) in favor of the leptogenesis mechanism as the origin of the baryon asymmetry of the universe.

4. Determining the MO and CPV

Various experimental strategies can be used to determine the neutrino MO and whether there is CPV in the leptonic sector. These can be divided into several categories:

(1) *Neutrino oscillation experiments.* They are sensitive to the neutrino MO through the observation of matter effects at long-baseline and atmospheric neutrino experiments. Medium-baseline reactor neutrino experiments can also be sensitive to the MO through the interference of the solar and atmospheric contributions to the amplitude in vacuum. Long-baseline neutrino oscillation experiments are sensitive to CPV by looking at subleading terms in the appearance oscillation probability $\nu_\mu \to \nu_e$.

(2) *Neutrinoless double-beta-decay experiments.* If neutrinos are Majorana particles, the rate of this process depends on the values of neutrino masses, mixing angles and CPV phases in a combination called the effective Majorana mass parameter, $|\langle m \rangle|$. In case of a positive signal, a measurement of $|\langle m \rangle|$ would allow to get information on the neutrino mass spectrum and, at least in principle, on CPV.

(3) *Direct mass searches.* The KATRIN experiment can test the quasidegenerate neutrino mass spectrum by looking at the impact of the effective mass of the electron neutrino on the end point of the electron spectrum observed in the beta decay of tritium.

(4) *Supernova neutrinos.* Supernova neutrinos can also reveal the type of neutrino ordering through the impact of matter effects occurring in neutrino oscillations during a supernova explosion.

(5) *Cosmological observations.* After neutrinos decouple from the rest of the primordial plasma they stream freely and contribute to the hot dark matter in the Universe. If their contribution to the energy density of the universe is too large, the formation of the structures in the universe (such as galaxies and clusters of galaxies) would be suppressed at small scales in contradiction with cosmological observations. This allows to

impose very stringent constraints on the sum of neutrino masses $\sum_i m_i$, a quantity which depends on the neutrino MO for sufficiently small masses.

In the rest of this section, we will discuss in more in detail some of the searches outlined above. We will present first the phenomenological aspects, and then we will briefly review the current and future experiments.

4.1. *Neutrino oscillations and matter effects*

Neutrino oscillations may be modified when neutrinos travel through matter, due to their interactions with nuclei and electrons. Dimensional arguments allow to estimate the cross section of a neutrino with energy E to interact with a nucleon at rest with mass m_n as $\sigma_n \sim G_F^2 E^2$ for a neutrino energy $E \ll m_n$ and $\sigma \sim G_F^2 E m_n$ for $E \gg m_n$. For a neutrino interacting with the particles in the background as it propagates, the resulting mean free path can be estimated as

$$l \sim \frac{1}{\mathcal{N}\sigma} \propto \frac{1}{\mathcal{N} G_F^2},$$

where $\mathcal{N}$ is the particle number density. Thus, it will typically be very large due to the very strong suppression of the cross section with G_F^2. For instance, the Earth is opaque to neutrinos (i.e., it absorbes them) only if their energy is above 100 TeV.

Alternatively, if the wavelength associated to the neutrino, λ, is large compared to the space between interactions, the effect of the medium will be that of a refraction index. If λ is also large compared to the size of the scattering particle so that diffraction can be ignored, one can describe the propagation of a neutrino through matter using geometrical optics.[f] The transmitted wave is then formed by a superposition of amplitudes for the interactions with atoms lying within a distance of order λ, with a negligible momentum transfer. This is the so-called *forward coherent scattering*. The resulting effect is that, since the SM interactions affect differently the three neutrino flavors (as we will see below), they will affect oscillations as neutrinos propagate through matter.

Charged-current (CC) forward coherent scattering of ν_μ or ν_τ in matter is negligible since there are no muons or taus present. Thus, they will

[f] The wavelength for a neutrino of momentum p is $\lambda = h/p \sim 200(1\,\mathrm{MeV}/p)(10^{-13}\,\mathrm{cm})$. Therefore, both conditions are usually satisfied.

only feel the matter potential due to neutral-current (NC) interactions, while ν_e will feel both CC and NC contributions. The potential will be proportional to the forward coherent scattering amplitude as well as the density of fermions N_f that the neutrino encounters as it propagates in the medium[36]:

$$V = \sum_f 2\sqrt{2} G_F N_f g^f E,$$

where g^f is the charge of the fermion f under the SM weak interaction. For CC interactions, we have $g^f = 1$, whereas for NC interactions, $g^f = I_3^f - Q^f \sin^2\theta_W$ with I_3^f and Q^f being the isospin and electric charge of the fermion f. Summing over the contributions of electrons, protons and neutrons, the potentials for each type of interaction are obtained as

$$V_{\rm CC} = \pm 2\sqrt{2} G_F N_e E, \quad V_{\rm NC} = -\sqrt{2} G_F N_n E,$$

where the $\pm$ sign corresponds to neutrinos/antineutrinos. It should be noted that the electron and proton contributions in the NC matter potential cancel out in an electrically neutral medium, and as a result, the NC potential only depends on the density of neutrons. Moreover, the NC contribution to the potential can be subtracted as it does affect the oscillations between different active flavors. Therefore, the only contribution that will have a physical effect for oscillations between active neutrinos[g] will be that associated to CC interactions with electron neutrinos, $V_{\rm CC}$.

On the flavor basis, the full Hamiltonian then reads:

$$H = \frac{1}{2E} U_{\rm PMNS} \begin{pmatrix} m_1^2 & 0 & 0 \\ 0 & m_2^2 & 0 \\ 0 & 0 & m_3^2 \end{pmatrix} U_{\rm PMNS}^\dagger \pm \sqrt{2} G_F N_e \begin{pmatrix} 1 & 0 & 0 \\ 0 & 0 & 0 \\ 0 & 0 & 0 \end{pmatrix}, \quad (12)$$

where $U_{\rm PMNS}$ has been used to bring the vacuum Hamiltonian from the mass basis to the flavor one in which the matter potential term is diagonal.

The Hamiltonian in Eq. (12) has to be diagonalized in order to compute the transition probabilities. This is rather complicated if we consider three families, but it can be easily done in the two-family case. In this regime, the effective mass-squared difference, i.e., the difference of the energy

[g]For oscillations between active and sterile neutrinos, as the latter have no interactions and therefore no matter potential, the NC potential will also be relevant.

eigenvalues divided by $2E$ and the effective mixing angle in matter are

$$\begin{aligned}\Delta m_M^2 &= \Delta m^2\sqrt{\sin^2 2\theta + (\cos 2\theta \mp A)^2},\\ \sin^2 2\theta_M &= \frac{\sin^2 2\theta}{\sin^2 2\theta + (\cos 2\theta \mp A)^2},\end{aligned} \tag{13}$$

where the $\mp$ corresponds to neutrinos and antineutrinos, respectively, Δm^2 and θ are the corresponding values in vacuum. We have also defined

$$A \equiv \frac{V_{\rm CC}}{\Delta m^2}.$$

For constant matter, the following oscillation probabilities are readily obtained by solving the Schrödinger equation:

$$\begin{aligned}P(\nu_\alpha \to \nu_\beta) &= \sin^2 2\theta_M \sin^2\left(\frac{\Delta m_M^2 L}{4E}\right),\\ P(\nu_\alpha \to \nu_\alpha) &= 1 - P(\nu_\alpha \to \nu_\beta),\end{aligned} \tag{14}$$

where α, β are neutrino flavor indices, L is the distance traveled by the neutrino and E is the neutrino energy.

In the case with constant matter density (as is the case for neutrinos traveling through the Earth), an interesting phenomenon arises when

$$A = \pm\cos 2\theta \Leftrightarrow E = \pm\frac{\Delta m^2}{2V_{CC}}\cos 2\theta. \tag{15}$$

For this value of A, the mixing angle in matter suffers a resonant enhancement and reaches maximal mixing, regardless of the actual mixing angle in vacuum. This dramatic amplification is the so-called Mikheyev–Smirnov–Wolfenstein (MSW) effect.[37] Notice that this resonance takes place only for neutrinos or only for antineutrinos, depending on the sign of Δm^2. Even for values of the energy for which there is only a partial cancellation in the denominator of Eq. (13), the mixing angle θ_M (and, consequently, the probability in matter) is enhanced for neutrinos for positive Δm^2 and suppressed for negative Δm^2. As we have explained in Section 2, the MO is related to the sign of Δm^2. This offers one of the main avenues to distinguish the MO at neutrino oscillation experiments: by observing the resonant or enhanced behaviour either in the neutrino or the antineutrino oscillation channels due to matter potential.

The main proposals to determine the neutrino MO through the observation of matter effects on the oscillation probabilities are divided into two main categories: long-baseline neutrino oscillation experiments and atmospheric neutrino oscillation experiments. We will discuss them in Sections 4.2 and 4.3 separately.

4.2. Long-baseline neutrino oscillations

4.2.1. General phenomenology

Long-baseline experiments aim to measure the $\nu_\mu \to \nu_e$ and $\bar{\nu}_\mu \to \bar{\nu}_e$ oscillation probabilities precisely. These can be conveniently expressed after performing a perturbative expansion on the following small parameters: $\Delta m^2_{21}/\Delta m^2_{31}$ and $\sin\theta_{13}$. Keeping terms up to second order,[h] the oscillation probability for neutrinos in the presence of matter with a constant density can be written as[39]

$$P_{\mu e} = 4s^2_{23}s^2_{13}c^2_{13}\left(\frac{\Delta_{13}}{B_\mp}\right)^2 \sin^2\left(\frac{B_\mp L}{2}\right) + 4c^2_{23}s^2_{12}c^2_{12}\left(\frac{\Delta_{12}}{A}\right)^2 \sin^2\left(\frac{AL}{2}\right)$$
$$+ \tilde{J}\,\frac{\Delta_{12}}{A}\frac{\Delta_{13}}{B_\mp}\sin\left(\frac{AL}{2}\right)\sin\left(\frac{B_\mp L}{2}\right)\cos\left(\pm\delta + \frac{\Delta_{13}L}{2}\right), \tag{16}$$

where $B_\mp \equiv |A \mp \Delta_{13}|$, $\tilde{J} \equiv c_{13}\sin 2\theta_{13}\sin 2\theta_{12}\sin 2\theta_{23}$ and $\Delta_{ij} \equiv \Delta m^2_{ij}/(2E)$. The upper/lower signs correspond to the neutrino/antineutrino channels.

The first term in Eq. (16) is usually referred to as the "atmospheric term", as its oscillation frequency is dictated by the atmospheric mass splitting, Δm^2_{31}. It is the leading order term in the probability. Neutrino oscillation experiments are usually designed so that $L/E \sim 500$ km/GeV (also known as the *first oscillation maximum*), which is the smallest value of L/E that maximizes this term. It provides the strongest sensitivity to matter effects and, thus, to the neutrino MO. The second term in Eq. (16) is the so-called solar term and its frequency depends on the solar mass splitting instead, Δm^2_{21}. At the first oscillation maximum, the solar part of the oscillation is not yet fully developed and therefore this term is quadratically suppressed with respect to the atmospheric term. Finally, the third term arises as the interference between the solar and atmospheric contributions to the amplitude of the oscillation. It contains the dependence on the CP-violating phase δ. CPV effects are encoded in the Jarlskog invariant and arise from genuine three-neutrino mixing as can be seen from Eq. (2). If we

[h]Given the large value of θ_{13}, higher-order terms for s_{13} should be kept in the expansion. However, this would lead to a more complicated expression than the one shown here. For more accurate expressions of the oscillation probabilities, we refer the reader to Ref. 38.

neglect one mass-squared difference, there are no Dirac CPV effects as a 2×2 mixing matrix contains at most one Majorana CPV phase (which does not enter oscillations). Therefore, the CPV term needs to depend on both mass-squared differences and is suppressed with respect to the atmospheric term as $\Delta m^2_{21}/\Delta m^2_{31}$. Moreover, the Jarlskog invariant depends on all three mixing angles and therefore CPV effects are suppressed by the relative smallness of θ_{13}. This implies that the determination of CPV at long-baseline experiments is challenging as it requires extremely good precision in the measurement of the oscillation probability in this channel.

As can be seen from Eq. (16), the appearance oscillation probabilities depend simultaneously on several unknown parameters, namely, the sign of Δm^2_{31}, the value of δ, and the octant of θ_{23}. This leads to the so-called *degeneracy problem* in neutrino oscillations[40–42]: for a given measured value of the oscillation probability, multiple choices of the unknown parameters are generally able to fit the data. A very relevant degeneracy concerns the simultaneous determination of the CP-violating phase δ and the neutrino MO[41] using long-baseline neutrino oscillation facilities. It can be easily shown that, in vacuum, the set of transformations[43]

$$\begin{aligned} &\Delta m^2_{31} \to -\Delta m^2_{31} + \Delta m^2_{21} = -\Delta m^2_{32}, \\ &\sin\theta_{12} \leftrightarrow \cos\theta_{12}\,, \quad \delta \to \pi - \delta \end{aligned} \tag{17}$$

brings the Hamiltonian $H_{\text{vac}} \to -H^*_{\text{vac}}$, where H_{vac} is the Hamiltonian in vacuum. This renders the evolution of the system invariant,[44] and the two sets of solutions in Eq. (17) will lead to the same values for all oscillation probabilities.

In the presence of matter effects, however, the degeneracy is broken since the Hamiltonian also contains the matter potential; see Eq. (12). For instance, from solar neutrino data, for which matter effects are very important, we know that $\theta_{12} < 45°$, which does not allow for the full transformation in Eq. (17), partially breaking the degeneracy. However, long-baseline experiments are largely insensitive to the solar mixing parameters and, thus, the degeneracy remains even in this case[41] (unless the experiment is also affected by sizable matter effects). This is illustrated in Fig. 3, where we show the neutrino oscillation probabilities in the $\nu_\mu \to \nu_e$ channel for $\delta = 90°$ (solid lines) and $\delta = -90°$ (dotted lines). The blue (red) lines correspond to NO (IO), and the two panels have been obtained for different baselines, as indicated by the labels. The right panel corresponds to a baseline short enough so that matter effects are practically negligible and, consequently, the degeneracy is almost perfect. As seen in the figure, the

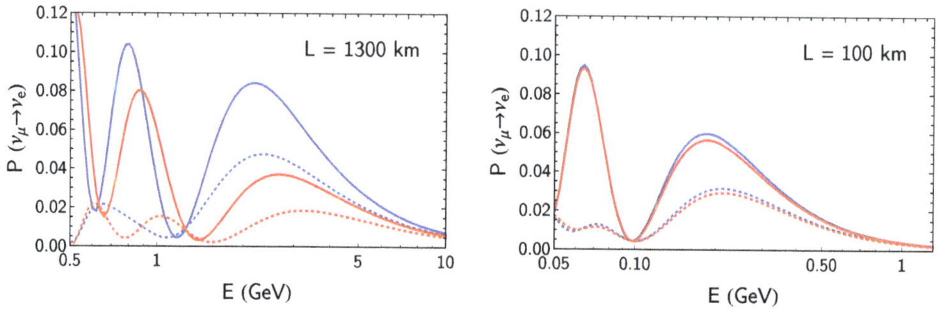

Fig. 3. Probabilities in the $\nu_\mu \to \nu_e$ channel as a function of the neutrino energy (in GeV), for two different baselines as indicated in the panels. Red (blue) lines correspond to NO (IO). Solid lines correspond to $\delta = -90°$, while dotted lines have been obtained for $\delta = 90°$.

probability for NO and $\delta_1 = 90°$ is very similar to the probability obtained for IO and $\delta_2 = 180° - \delta_1 = 90°$ (the two dotted lines). The overlap is also significant for other values of δ (e.g., the two solid lines). In contrast, in the left panel, we can see that matter effects break this degeneracy: the probabilities for NO get enhanced, while the lines corresponding to IO get suppressed. This breaks the two sets of lines apart, lifting the degeneracy. It should be mentioned that for intermediate baselines, $L \sim \mathcal{O}(500)$ km, the matter effects will not be strong enough to break the degeneracy completely and it is sometimes possible to find a degenerate solution. The value of δ at which the degeneracy takes place will no longer coincide with the vacuum case in Eq. (17), and will generally be a function of the matter density at which the experiment is performed.

In the context of CPV searches, the so-called generalized intrinsic degeneracy[40,45] (highlighted in Ref. 46) is also important, as it simultaneously involves the θ_{23}, δ and θ_{13} parameters. Generically, the combination of information obtained at different neutrino energies and/or different oscillation channels eventually lifts some of these degeneracies. The combination with other measurements, such as those performed at reactor or atmospheric neutrino experiments, is another avenue to reduce the impact of this issue, for instance, providing an independent determination of the MO if the long-baseline experiment is not able to do so.

Experimentally, neutrino oscillation experiments using very long baselines have to overcome several practical difficulties. The first one is the small interaction cross sections of the neutrinos at the detector. At low energies, the weak interaction cross section increases quadratically with the neutrino energy and, thus, higher energies will generally be preferred. These can be obtained from the decays of highly boosted particles in a decay pipe or a storage ring: (1) meson decays, typically pions and kaons, for example, in K2K, MINOS, T2K, NOvA; (2) beta decay of radioactive ions for beta-beams; and (3) muon decays for neutrino factories. Finally, very intense neutrino sources will be needed in order to achieve a large enough number of events at a detector located at a distance of $\mathcal{O}(500)$ km/GeV. This poses a formidable task for neutrino factories and beta beams, since the acceleration and storage of muons and unstable ions using conventional accelerators is technically very challenging. Conversely, neutrino beams obtained from meson decays are typically much easier to produce. The disadvantage is, however, that the neutrino fluxes produced in this case cannot be computed analytically (unlike in the case of muon decay or beta decay of radioactive ions). Instead they are predicted simulating the meson production in the target material and the subsequent focusing in the beamline and are thus subject to larger systematic uncertainties.[i] Near detectors are crucial to reduce their impact on the analysis, as they can be used to determine the flux before any oscillations have taken place.[j]

Typically, all neutrino oscillation experiments are performed at distances matching the first oscillation peak, as mentioned above. However, the relative size of the CP-violating interference increases for values of L/E beyond this region,[48] since the solar part of the amplitude develops further. The enhancement can be understood from the analytical dependence of the oscillation probability in Eq. (16), and is clearly seen from the comparison between the solid and dotted lines in Fig. 3: the difference between the two increases at lower energies (corresponding to larger values of L/E). However, practical issues make experiments observing the second oscillation maximum technically more challenging, as these imply either much longer baselines (with a consequent reduction in the number of events as $1/L^2$ due to the beam divergence) or much lower energies (with a consequent reduction in the number of events as E^2 due to the lower cross section).

The relevant backgrounds to the appearance channels $\nu_\mu \to \nu_e$ depend on the detector technology employed. As we will see below, most long-baseline oscillation experimental designs use either water Cherenkov, liquid scintillator or liquid argon detectors. The main backgrounds for water Cherenkov and scintillator detectors are misidentified π^0 produced in NC

[i]These can be partially reduced using hadron production data from external experiments such as NA61/SHINE.[47]

[j]This is technically not correct though. Near detectors can determine only a convolution of the flux *times* the neutrino detection cross section. Disentangling the two effects is a very challenging task.

interactions. These events can easily be misidentified as charged-current quasielastic (CCQE) neutrino events if one of the photons escapes detection or if the Cherenkov rings of the two photons overlap and cannot be resolved. At liquid argon detectors, these events are also problematic, although in this case the rejection efficiency is an order of magnitude better due to: (i) the better resolution for the two photons and (ii) the different way in which a photon and an electron deposit energy in the detector, allowing for a good discrimination even if one of the two photons is undetected. Other relevant sources of background, depending on the beam and type of detector, include ν_e and $\bar{\nu}_e$ intrinsic contamination of the beam if it is produced from meson decays (coming mainly from the decays of kaons produced in the target), CC ν_μ events in which the muon is misidentified as an electron in water Cherenkov detectors and, finally, CC ν_τ events in which the τ decays leptonically if the energy is sufficiently high.

Finally, neutrino oscillation experiments using conventional beams from meson decays can be divided into two main categories: on-axis (or wide-band) beams and off-axis (or narrow-band) beams. In the former, the neutrino beam is aimed directly at the detector, while in the latter kind the detector is intentionally placed off-axis with respect to the beam direction. By using an off-axis design, the peak energy is shifted at lower values and the high-energy tail of the flux is highly reduced due to the kinematics of pion decay. This will suppress the number of NC background events in the region of the oscillation, enhancing the signal significance. The resulting flux at off-axis locations will therefore have a much narrower bandwidth with a peak that matches the first oscillation maximum very well. The disadvantage of using this approach is that the number of events at the detector is severely reduced with respect to wide-band designs and they sample a narrow range of energies. On the other hand, wide-band beams can study not only the first oscillation maximum region of the probability but also a wider range, even extending to the second oscillation maximum at experiments with energy thresholds low enough and are therefore typically more powerful in disentangling parameter degeneracies.

4.2.2. *Current experiments*

Two long-baseline experiments are currently taking data: T2K and NOvA. Their main physics goals are the determination of the MO and the search for CPV in the lepton sector.

T2K[49]**:** The T2K experiment takes place in Japan. A conventional beam produced from meson decays is sourced at the J-PARC accelerator complex.

The neutrino beam is pointed at the Super-Kamiokande (SK) detector located at a distance of $L = 295$ km from the source and at an off-axis angle of 2.5°. At this off-axis angle, the flux peaks at around 600 MeV, which matches the first oscillation maximum for this distance very well. The SK detector, with a fiducial mass of 22.5 kt, uses the water-Cherenkov technology. Two near detectors are used to measure the beam composition and spectrum before oscillations: INGRID (on-axis) and ND280 (off-axis). A combination of neutrino and antineutrino runs (combined with the precise measurements of θ_{13} coming from reactor experiments) will allow T2K to reach a 2σ hint disfavoring the CP-conservation hypothesis if the true value of δ is in the region $\delta \sim -90°$; see Ref. 50, for example. Its short baseline makes it largely insensitive to the MO.

NOvA[51]**:** The NOvA experiment uses a conventional neutrino beam, the NuMI beam, with 700 kW beam power, and a 14 kt detector located at a distance of 810 km and 0.8° off-axis. The detector is filled with mineral oil and uses scintillation light to reconstruct the particles produced in a given event. NOvA also uses a near detector to determine the flux before oscillations, located at the same off-axis angle but at a distance of 1 km from the source. At this off-axis angle, the flux peak is at roughly 2 GeV. Its longer baseline and higher energies make NOvA more sensitive to matter effects than T2K and, therefore, to the MO. Its final sensitivity, however, will depend on the true value of θ_{23}, being worse for higher values of θ_{23}, as well as on the true value of δ due to the presence of sign degeneracies. The combination with T2K partially resolves some of these, leading to a CPV sensitivity at 1.5–2σ for a broader range of values of δ.[52] The sensitivity to the MO would reach the 3σ level for values of $\delta \sim -90°$ and is mainly determined by NOvA.

Both T2K and NOvA have already started taking data. Currently, the T2K experiment has collected data in both neutrino and antineutrino modes with a similar number of protons on target in each mode, $\sim 7.5 \times 10^{20}$ PoT.[49] The NOvA experiment, on the other hand, has accumulated $\sim 6 \times 10^{20}$ PoT in neutrino mode and will start taking data in antineutrino mode in 2017.[51] The combination of both data sets, together with the precise determination of θ_{13} from reactor experiments, currently shows a preference for values of $\delta \sim -90°$.[5] Although the significance of this hint rejecting the CP-conservation hypothesis is still low (1.05σ for normal ordering), values of $\delta \sim 90°$ are already disfavored at around 99% CL for normal ordering and at higher CL ($\Delta\chi^2 \simeq 14$) for inverted ordering.[5]

More data will be needed to confirm this hint at higher statistical significance. However, even at their nominal exposure, the results of both

T2K and NOvA will still be limited by statistics. Thus, they would considerably benefit from an extension in their planned running times, especially if the true value of δ is in the region favored by the present hints.

4.2.3. *Future proposals*

The future proposals for neutrino long-baseline oscillation experiments will aim at the 5σ determination of the MO and, if δ is close to being maximally CP violating, a 5σ discovery of CPV in the lepton sector. These experiments employ more powerful beams and larger detectors and therefore will generally be limited not by a small number of events but by systematic errors instead. A strong effort is currently being carried out to reduce systematic uncertainties down to the few-percent level in order to guarantee that their physics goals are met.

DUNE[53]**:** The DUNE experiment will use the high intensity beam of 1.07 MW power provided by the LBNF facility, with a planned doubling of the power after 6 years. The neutrinos produced at Fermilab will then be observed at the underground Sanford Underground Research Facility (SURF) in South Dakota, 1300 km away. Unlike T2K and NOvA, the DUNE detector will be on-axis. The neutrino flux will peak around 3 GeV, corresponding to the energy of the first oscillation maximum for this baseline, but it will extend down to 1 GeV in order to cover the second oscillation maximum as well. The liquid argon TPC technology has been identified as the most suited choice at these energies, given its excellent track reconstruction and particle identification for both quasielastic and nonquasielastic events. The far detector will consist of four modules of 10 kt mass each. A near detector will be located on the Fermilab site, although its design is still undecided. The very long baseline and high energies imply that the appearance oscillation probabilities are affected by stronger matter effects compared to the NOvA experiment. This results in an MO sensitivity at the 3σ level for all values of δ and θ_{23} and in a 5σ determination in a large fraction of the parameter space. This strong sensitivity to the MO lifts the degeneracies affecting CPV searches. The much larger number of events will also increase its sensitivity to CPV with respect to that of current experiments: DUNE will reach 3σ for 75% of the values of δ after an exposure of 1320 (850) kt MW years using the CDR (optimised) beam and 5σ for 50% of the values of δ after 810 (550) kt MW years.[53]

J-PARC to Hyper-Kamiokande long-baseline experiment (T2HK)[54]**:** The design of this experiment is very similar to that of the T2K experiment, with the same off-axis angle and distance to the neutrino source, albeit with a much larger detector: Hyper-Kamiokande (HK) will be a water Cherenkov detector consisting of two modules, each with a total (fiducial) mass of 258 kt (187 kt). The choice of detector technology is optimal for T2HK: first, water Cherenkov detectors offer an excellent energy resolution for CC quasielastic interactions, which will be the major contribution to the event sample at T2HK; second, the choice of water as target allows to scale the detector mass by a large factor while keeping the cost at a reasonable level. The beam power will also be much higher than for T2K, 1.3 MW. For 10 years of data taking, the T2HK sensitivity to CPV reaches the 3σ (5σ) level for 78% (62%) of the true values of δ. The sensitivity to the MO is however quite limited (as in the T2K case) due to the choice of baseline and neutrino energies. However, the atmospheric data sample collected at HK will provide some sensitivity to this observable as explained in the next section. Recently, the option to add a water-Cherenkov module in South Korea at a distance of 1000–1300 km (at an off-axis angle between 1.0° and 2.5°) has been proposed (T2HKK[55]). The much longer baseline for this second detector would translate into a better sensitivity to the neutrino MO.

ESSνSB[56]: This experiment differs radically from other options as it is designed to exploit the information from events at the second oscillation maximum. Given the much larger values of L/E, the number of events will be suppressed due to the low detection cross section at low energies. Thus, the design of this facility requires a very intense neutrino beam, which would be provided by the European Spallation Source (ESS) in Sweden. The 10 MW beam would produce neutrinos with energies around 150–200 MeV. A very massive detector would be needed to compensate for the low cross section. Several mines have been identified as possible underground locations for a Mt (500 kt fiducial) mass water Cherenkov detector at distances ranging between 300 km and 500 km. The optimal setup for this experiment would be able to discover CPV at 5σ for more than 50% of the true values of δ. As for T2HK, the sensitivity to the MO would be limited due to its short baseline; however, this could be eventually determined using the atmospheric events collected.

Neutrino factory[57]: At neutrino factories, muons would be accelerated at high energies and circulated in a storage ring until they decay, producing a collimated beam of ν_μ and $\bar{\nu}_e$ ($\bar{\nu}_\mu$ and ν_e) in the case of μ^- (μ^+) decay. A magnetized detector would be needed in order to distinguish the oscillated neutrinos from $\bar{\nu}_e \to \bar{\nu}_\mu$ from the ν_μ present in the beam. The neutrino energy is directly proportional to the energy of the stored muons. Several configurations have been considered in the literature, with muon energies ranging between 5 and 50 GeV and several detector configurations. The optimal design was identified in the IDS-NF study[58] given the value of $\theta_{13} \sim 9^\circ$: a setup with 10 GeV muons and $L = 2000$ km using a 100 kt magnetized iron neutrino detector. With 10^{21} muon decays per year, this setup would achieve a superior performance for CPV searches reaching 5σ for more than 70% of the values of δ.[58] However, it presents significant experimental challenges mainly from the point of view of flux production as already mentioned.

4.3. *Atmospheric neutrinos*

Atmospheric neutrinos are produced when cosmic rays hit the upper layers of the atmosphere. As a result of such collisions, pions (and also other mesons) are produced which then decay producing muon neutrinos or antineutrinos depending on the pion charge. The atmospheric neutrino flux will also contain a large ν_e and $\bar{\nu}_e$ component coming from muon decays. For instance, in the case of π^+, we get the following decay chain:

$$\pi^+ \to \mu^+ + \nu_\mu \to e^+ + \bar{\nu}_\mu + \nu_e + \nu_\mu.$$

After being produced, atmospheric neutrinos can travel through the Earth before arriving at a neutrino detector. This allows for the determination of the neutrino MO, using the enhancement or suppression of the oscillation probability for neutrinos and antineutrinos due to the matter potential, as we describe in some detail below.

For baselines in the range of thousands of kilometers (such as those crossing the Earth), the matter effects are maximized for neutrinos with energies of a few GeV. At these energies, the oscillation probabilities can be safely approximated neglecting Δm^2_{21} and CPV effects[k,61]:

$$\begin{aligned}
P_{ee} &\simeq 1 - P_{ex}, \\
P_{\mu e} &\simeq s^2_{23} P_{ex}, \\
P_{e\tau} &\simeq c^2_{23} P_{ex}, \\
P_{\mu\mu} &\simeq 1 - \frac{1}{2}\sin^2 2\theta_{23} - s^4_{23} P_{ex} + \frac{1}{2}\sin^2 2\theta_{23} \mathrm{Re}\left\{e^{-i\phi} \mathcal{A}^{2\nu}_{\tau\tau}\right\}, \\
P_{\mu\tau} &\simeq 1 - P_{\mu\mu} - P_{\mu e},
\end{aligned} \tag{18}$$

where P_{ex} is the two-neutrino probability for $\nu_e \to \nu_x$, with $\nu_x = s_{23}\nu_\mu + c_{23}\nu_\tau$. It is controlled by Δm^2_{31} and θ_{13} and is affected by matter effects. Here,

[k] The possibility of searching for CPV with atmospheric neutrinos would require very large Cherenkov detectors.[59,60]

ϕ is a complex phase associated to the $\nu_\mu \to \nu_\tau$ transition and $\mathcal{A}^{2\nu}_{\tau\tau}$ is the amplitude of the oscillation probability $\nu_\tau \to \nu_\tau$ in two families. In the limit $\theta_{13} \to 0$, one has $P_{ex} \to 0$ and $\mathrm{Re}\left\{e^{-i\phi}\mathcal{A}^{2\nu}_{\tau\tau}\right\} = \cos(\Delta m^2_{31} L/2E)$: oscillations only take place in the $\mu{-}\tau$ sector, and they are identical for both neutrino mass orderings. As long as $\theta_{13} \neq 0$, P_{ex}, ϕ and $\mathcal{A}^{2\nu}_{\tau\tau}$ are affected by the matter potential and depend on the specific profile viewed by the neutrinos through the Earth. Consequently, atmospheric neutrino experiments are sensitive to the neutrino MO through the impact of matter effects on the probabilities, as in the case of long-baseline neutrino experiments. However, several important issues should be noted. First, unlike in the case of long-baseline facilities, in atmospheric neutrino experiments, the distance that the neutrino has traveled through the Earth is not fixed; however, it can be easily inferred from the angle at which the neutrino enters the detector. Thus, angular reconstruction is an important requirement for this type of experiments. Second, atmospheric neutrinos can pass not only through the Earth's mantle but also through its core where the matter density (and, consequently, the matter potential) is much larger. In this case, in addition to the MSW effects discussed earlier, new enhancements of the oscillation probability can arise due to the presence of different densities and step functions in the density profile.[62] An example showing the impact of the neutrino MO on the $\nu_\mu \to \nu_\mu$ oscillation probability for atmospheric neutrinos is shown in Fig. 4 as a function of the neutrino energy and zenith angle. The left (right) panel corresponds to NO (IO). The colors indicate the value of the $P_{\mu\mu}$ probability, going from 1 (red regions) to 0 (purple regions). The effect of the matter

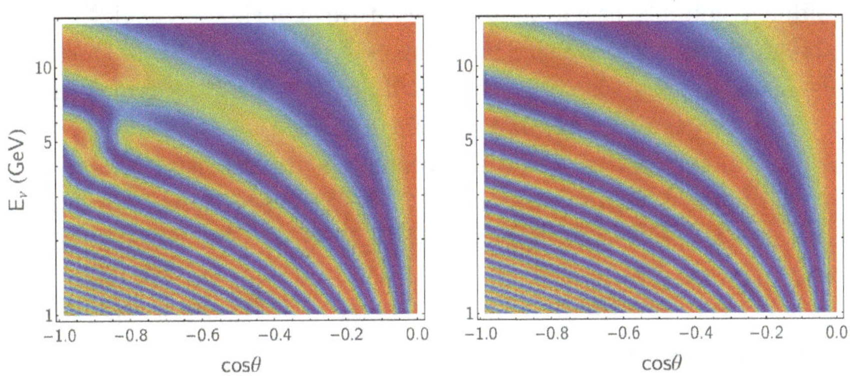

Fig. 4. Oscillograms of neutrinos crossing the Earth, in the $\nu_\mu \to \nu_\mu$ oscillation channel. The x axis shows the cosine of the zenith angle (neutrinos that cross the Earth have $\cos\theta = -1$), while the y axis shows the neutrino energy in log scale (in GeV). Red regions correspond to values of θ and E which give $P_{\mu\mu} = 1$; as the colors change, this probability is reduced reaching 0 for the purple regions. The left (right) panel corresponds to NO (IO).

resonance is largest for neutrinos with energies between 5 and 10 GeV and trajectories with $\cos\theta \lesssim -0.6$, as can be seen from the comparison between the two panels.

Atmospheric neutrino experiments can be divided into two main categories, according to whether their detector is magnetized or not. At a nonmagnetized detector, the number of, e.g., muon neutrino events receive contributions from four different oscillation channels[l]:

$$N_\mu \propto \sigma^{\rm CC}_{\nu_\mu}(\phi^0_{\nu_\mu} P_{\mu\mu} + \phi^0_{\nu_e} P_{\mu e}) + \sigma^{\rm CC}_{\bar\nu_\mu}(\phi^0_{\bar\nu_\mu} \bar P_{\mu\mu} + \phi^0_{\bar\nu_e} \bar P_{\mu e}) ,$$

where $\phi^0_{\nu_\alpha}$ is the initial flux of ν_α and $\sigma^{\rm CC}$ is the CC detection cross section. The event sample includes contributions from both neutrinos and antineutrinos, which are affected in an opposite manner by the matter potential as mentioned in Section 4.1. Thus, the impact of matter effects is partially washed out at nonmagnetized detectors, although not completely since the cross section for antineutrinos is roughly a factor of two smaller than that of neutrinos and, therefore, the contribution of the latter to the number of events is more significant. On the other hand, nonmagnetized detectors can be easily built on the several-hundred-kt scale, providing a large number of events, while magnetized detector are usually smaller.

The atmospheric neutrino oscillation experiments currently taking data are SK[63] and Icecube/Deepcore.[64] Both are not magnetized and use Cherenkov radiation to detect the muons produced in CC neutrino interactions. This allows for an excellent particle identification and angular reconstruction, and the water or ice Cherenkov technology permits for very large fiducial volumes at a low cost. The future experiment ICAL@INO[65] will use a magnetized iron detector. In addition to being sensitive to the charge of muons, iron detectors also offer the possibility of including information from the hadron showers in the analysis, which leads to a better determination of the neutrino energy and an enhancement in the sensitivity to the MO.[66] If only the muon information is used in the analysis, ICAL will reach a 3σ sensitivity to the MO after approximately 15 years of data taking (for true NO and maximal θ_{23}), while the sensitivity goes up to 3.7σ for the same exposure.[65] Other future experiments include the HK water Cherenkov and the PINGU proposals,[67] an extension of the Icecube experiment at the South Pole with a detection threshold of a few GeV. In both cases, the detectors would not be magnetized. At HK, the atmospheric data sample would allow a determination of the MO at 3σ in 5 years of data taking.[68] At PINGU,

[l]We neglect the contribution from CC ν_τ events where the τ decays producing a muon.

instead, the expected sensitivity would reach 3.7σ in 5 years of data taking.[67] It should be noted that these sensitivities are strongly affected by systematic uncertainties due to energy and angular resolution.

4.4. *Reactor neutrinos*

A different way in which neutrino oscillation experiments can be sensitive to the neutrino MO is through the observation of the interference pattern in the oscillation probabilities between the solar and the atmospheric contributions to the oscillation amplitude. For example, the oscillation probability in the $\bar{\nu}_e \to \bar{\nu}_e$ channel can be written as[69]

$$P_{ee} = 1 - c_{13}^4 \sin^2 2\theta_{12} \sin^2\left(\frac{\Delta m_{21}^2 L}{4E}\right) - \sin^2 2\theta_{13}\left[c_{12}^2 \sin^2\left(\frac{\Delta m_{31}^2 L}{4E}\right) + s_{12}^2 \sin^2\left(\frac{\Delta m_{32}^2 L}{4E}\right)\right]. \tag{19}$$

While the first and second terms correspond to the solar and atmospheric contributions to the probability, the interference among the two is given by the last term. Thus, if the three contributions are distinguishable, the MO can be determined simply by comparing the size of the Δm_{31}^2 and $\Delta m_{32}^2 \equiv \Delta m_{31}^2 - \Delta m_{21}^2$ oscillation frequencies. The two last terms in Eq. (19) can be in principle distinguished by studying the evolution of the oscillation probability as a function of L/E. However, this requires a detector with an energy resolution roughly at the $\Delta m_{21}^2/\Delta m_{31}^2 \sim 3\%$ level. It should also be noted that the impact of the interference term on the observable event rates will be suppressed by $\sin^2 2\theta_{13} \sim 0.1$. The effect of the MO on the expected number of events at a medium-baseline reactor experiment is illustrated in Fig. 5. Here, we show the product of the flux, inverse beta decay cross section and oscillation probability in the $\bar{\nu}_e \to \bar{\nu}_e$ channel. This gives the expected number of events (up to a normalization constant determined by the total exposure) for a detector with perfect efficiency and energy resolution. Results are shown for both MOs, as indicated in the legend, for a detector distance $L = 55$ km. The dip in the number of events observed at around 3 MeV is due to the solar term in the oscillation probability in Eq. (19), which appears at leading order. The "wiggles", on the other hand, take place due to the atmospheric contribution and their amplitude is proportional to $\sin^2 2\theta_{13}$.

This approach to determine the neutrino MO does not rely on the observation of matter effects. Thus, the measurement can be performed in vacuum, without the need of very long baselines or high neutrino energies.

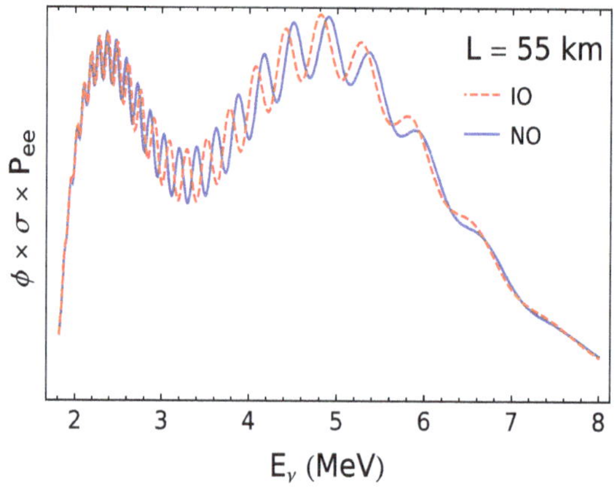

Fig. 5. Normalized distribution of the number of events as a function of the neutrino energy (in MeV), for a reactor experiment with a baseline $L = 55$ km.

The requirement of an interference large enough between the solar and atmospheric amplitude contributions implies that $L/E \sim 10^4$ km/GeV. A very intense neutrino source will also be required in order to achieve an event sample large enough at the detector. Moreover, excellent energy resolution capabilities will be needed as outlined above. All these features make reactor experiments the best place to perform this measurement. Optimization studies have been performed to determine the optimal distance to the detector which maximizes the signal significance. They concluded that it should be in the range $L \sim 40-50$ km depending on the achievable energy resolution. In all cases, the energy resolution should be at least or better than $\delta E_{\rm vis}/E_{\rm vis} \sim 0.03\sqrt{1\ {\rm MeV}/E_{\rm vis}}$ in order to reach a 3σ signal significance using a 20 kt detector, a 36 GW power reactor and 5 years of data taking ($\sim$3600 kt $\times$ GW $\times$ year). This is a very challenging requirement, a factor of two better than the energy resolution achieved in KamLAND. Energy nonlinearities need to be kept at the 1% level as well.[70]

Two experiments have been proposed to perform this measurement: the JUNO[70] and the RENO-50[71] experiments in China and South Korea, respectively. The JUNO setup will use a 20 kt detector placed at a distance of 53 km from the reactors. The RENO-50 will consist of a 18 kt detector placed at a similar distance from the neutrino source. In both cases, the detectors will be filled with liquid scintillator. In addition to the neutrino MO determination, the two detectors aim to bring the current uncertainties on the solar mixing parameters, θ_{12} and Δm^2_{21}, below the percent level. The precise determination of θ_{12} has great importance in distinguishing flavor models. For instance, we recall from Section 3.2 that its leading-order value for TBM is $\sin^2\theta_{12} = 1/3$ as opposed to $\sin^2 2\theta_{12} = 1$ for BM. The value of θ_{12} also enters in flavor sum rules; see Eq. (7), for example.

4.5. *Neutrinoless double beta decay*

Neutrinoless double beta decay takes place in nuclei when two neutrons simultaneously decay into two protons and two electrons with no emission of neutrinos. Contrary to two-neutrino double beta decay (2DBD), in which two antineutrinos are also present in the final state, this process violates lepton number by two units and is not allowed by SM interactions. Its discovery would be of paramount importance as it would shed light on the fundamental symmetries of Nature.

4.5.1. *Theoretical aspects of neutrinoless double beta decay*

In the simplest model, namely the SM augmented by massive neutrinos, this process is mediated by light Majorana neutrinos. However, there exist more complex scenarios in which this process is due to, e.g., R-parity violating supersymmetry, lepto-quarks, new gauge bosons, light and heavy sterile neutrinos. Here, we restrict our discussion to the simplest case of only light Majorana neutrinos. In this case, the decay rate reads

$$T^{-1}_{\beta\beta_{0\nu}} \simeq \frac{G_{0\nu}}{m_e} \left|\langle m\rangle\right|^2 M^2_{\rm NUCL}, \tag{20}$$

where $G_{0\nu}$ is a well-known phase-space factor, m_e is the electron mass, $M_{\rm NUCL}$ is the nuclear matrix element (NME) and $|\langle m\rangle|$ is the *effective Majorana mass parameter*. The latter parameter embeds all the dependence on neutrino physics. It depends on the neutrino parameters as

$$|\langle m\rangle| \equiv \left| m_1|U_{\rm e1}|^2 + m_2|U_{\rm e2}|^2\, e^{i\alpha_{21}} + m_3|U_{\rm e3}|^2\, e^{i(\alpha_{31}-2\delta)}\right|. \tag{21}$$

Here, m_i, $i = 1, 2, 3$, are the three light neutrino masses, and U_{ei} are the elements of the first row of the PMNS matrix.

From Eq. (21), we see that the predicted value of $|\langle m\rangle|$ depends critically on the neutrino mass spectrum and on the values of the two Majorana phases in the PMNS matrix, α_{21} and α_{31} (see, Refs. 72–74, for example). For the three limiting cases of the neutrino mass spectrum discussed in Section 2,

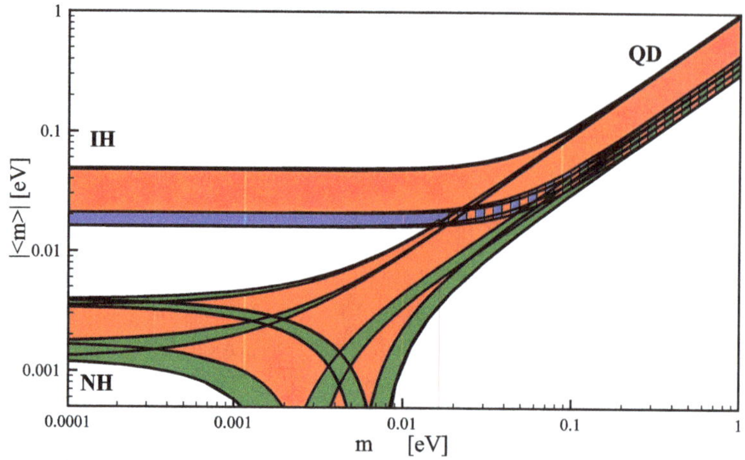

Fig. 6. The effective Majorana mass $|\langle m\rangle|$ at 2σ as a function of the smallest neutrino mass $m_{\rm MIN}$. We have used the current best-fit values and the 1σ errors of the oscillation parameters from Ref. 5. The Majorana phases α_{21} and α_{31}, as well as δ, are varied within their allowed intervals $[0, 180°]$. The blue (for IO) and green (for NO) regions are obtained for CP-conserving values of the CPV phases, while in the red regions the latter take necessarily CP-violating values.

we have that

$$|\langle m\rangle|^{\rm NH} \simeq \left|\sqrt{\Delta m^2_{21}}\sin^2\theta_{12}\cos^2\theta_{13} + \sqrt{\Delta m^2_{31}}\sin^2\theta_{13}e^{i(\alpha_{32}-2\delta)}\right|, \quad (22)$$

$$|\langle m\rangle|^{\rm IH} \simeq \left|\sqrt{|\Delta m^2_{32}|}\cos^2\theta_{13}\sqrt{1-\sin^2 2\theta_{12}\sin^2\left(\frac{\alpha_{21}}{2}\right)}\right|, \quad (23)$$

$$|\langle m\rangle|^{\rm QD} \simeq m_0\left|(\cos^2\theta_{12}+\sin^2\theta_{12}e^{i\alpha_{21}})\cos^2\theta_{13}+e^{i(\alpha_{31}-2\delta)}\sin^2\theta_{13}\right|. \quad (24)$$

It is possible to evaluate the predicted values of $|\langle m\rangle|$ by substituting in Eq. (21) the measured values for Δm^2_{21}, Δm^2_{31}, θ_{12} and varying the CPV phases in their allowed ranges. In Fig. 6, we show the current predictions for $|\langle m\rangle|$ for all values of the lightest neutrino mass and the two MOs.

As it was noticed in Ref. 72 (also see Refs. 73 and 75), in the case of *large but nonmaximal solar mixing angle*, the predictions for the various neutrino mass spectra are very different. In particular, in the case of IO, the value of the solar mixing angle induces a significant lower bound on $|\langle m\rangle|$

$$|\langle m\rangle|^{\rm IO} \geq \sqrt{\Delta m^2_{32}}\cos 2\theta_{12} \simeq 16 \text{ meV}, \quad (25)$$

where we have used the current best-fit values of the oscillation parameters. In the case of NO, the effective Majorana mass can be zero even if neutrinos are Majorana particles. This cancellation is possible only for values of $m_{\rm MIN} \sim 0.005$ eV as shown in Fig. 6.

Due to these considerations, neutrinoless double beta decay can provide information on the neutrino mass spectrum.[72,75–77] In the ideal case of perfectly known NME, a measurement of $|\langle m\rangle| > 0.1$ eV would imply that the spectrum is QD and the ordering of the masses does not play a role in neutrinoless double beta decay. For values of $|\langle m\rangle|$ smaller than 16 meV, the ordering would necessarily be normal. For values in between, both orderings are allowed but with constraints on the values of the masses. For instance, for 16 meV $\leq |\langle m\rangle| \leq$ 49 meV, the neutrino mass spectrum could be IH or with NO and partial hierarchy with $m_1 >$ 15 meV. Similar, although somewhat weaker, conclusions can be drawn once the uncertainties on the NME and the experimental error on $|\langle m\rangle|$ are included.

In principle, in neutrinoless double beta decay, it might be possible to search for CPV due to Majorana phases.[72,75,76,78] A precise measurement of $|\langle m\rangle|$ combined with an accurate determination of the neutrino masses would open this possibility, for instance, for the QD spectrum with neutrino masses precisely determined by cosmological observations. However, this strategy is extremely challenging as it would require knowing the NME with a very small error, at most at the few 10% level.

4.5.2. *Experimental status*

Double beta decay is a process in which two nucleons simultaneously beta-decay and can be searched for in nuclei for which single beta decay is not allowed. These include, e.g., ^{76}Ge, ^{136}Xe, ^{130}Te, ^{100}Mo. The observation of neutrinoless double beta decay relies on the observation of the energy of the two electrons emitted in the double beta decay: the signal sits at the end point of the two-electron spectrum as there is no neutrino emission, contrary to 2DBD, and therefore the electrons carry away all the energy available in the process. These experiments are very challenging as they require very low backgrounds, large masses because of the very slow decay rate and excellent energy resolution in order to reduce the impact of the 2DBD events. After several years of development and construction, a new generation of experiments recently started and has provided new results.

The extraction of $|\langle m\rangle|$ from a limit on $T_{\beta\beta_{0\nu}}$ depends on the theoretical evaluation of the NME. At present, there are large uncertainties in their calculation and a strong theoretical effort is required. Limits on $|\langle m\rangle|$ are usually given as a range which takes into account the uncertainty on the NME present in the literature for a given nucleus.

We report here the current bounds and the future prospects.

- GERDA uses high-purity Ge detectors, looking for the decay of ^{76}Ge. This type of detectors have excellent energy resolution with no intrinsic background. Combining the phase I and phase II results, for a total of 35 kg, a limit on $T_{\beta\beta_{0\nu}} > 5.3 \times 10^{25}$ years has been reported corresponding to $|\langle m \rangle| < 160$–$260$ meV at 90% CL.[80] Slightly weaker bound has been recently reported by the Majorana collaboration using the first data from the Ge Majorana demonstrator.
- KamLAND-Zen and EXO-200 use ^{136}Xe which has the advantage of being available in large quantities. KamLAND-Zen Phase 2 has 383 kg of enriched Xe providing the current best bound on neutrinoless double beta decay: $T_{\beta\beta_{0\nu}} > 1.07 \times 10^{26}$ years and $|\langle m \rangle| < 61$–165 meV at 90% CL.[81] There are plans to increase the detector mass to 750 kg of Xe in KamLAND-Zen 800 and subsequently to 1 t in KL2-Zen which with better energy resolution plans to reach 20 meV and so get close to the lower value predicted for IO. EXO-200, with 110 kg active mass, reaches a bound of $T_{\beta\beta_{0\nu}} > 1.1 \times 10^{25}$ years and $|\langle m \rangle| < 190 - 450$ meV at 90% CL,[82] which is going to be improved by the currently running Phase II and subsequently by nEXO with 5 t of Xe.
- A different experimental strategy is employed in NEMO3 and his successor SuperNEMO in which the active material is located in thin sheets surrounded by a detector capable of full topological reconstruction of the events. These experiments have unique capabilities in terms of measuring the two-electron angular distribution. The current bounds from NEMO-3 are $T_{\beta\beta_{0\nu}} > 2.0 \times 10^{22}$ years and $|\langle m \rangle| < 1.6$–$5.3$ eV at 90% CL for ^{150}Nd[83] and $T_{\beta\beta_{0\nu}} > 1.1 \times 10^{24}$ years and $|\langle m \rangle| < 0.3$–$0.6$ eV at 90% CL for ^{100}Mo.[84] The SuperNEMO experiment plans to enhance the physics reach down to $|\langle m \rangle| < 0.04$–$0.1$ eV at 90% CL using 100 kg of ^{82}Se.
- Finally, CUORE is a ton-scale experiment searching for neutrinoless double beta decay in ^{130}Te using tellurium oxide bolometers. Its demonstrator, CUORE-0, has already obtained a bound of $T_{\beta\beta_{0\nu}} > 4.0 \times 10^{24}$ years at 90% CL when combined with the Cuoricino data.[85]

In addition to the experiments discussed above, there are other efforts at the R&D, prototype or initial stage such as COBRA with ^{116}Cd, CANDLES with ^{48}Ca, AMoRE and DCBA/MTD with ^{100}Mo, NEXT and PandaX III with ^{136}Xe and SNO+ with ^{130}Te. Current bounds on $|\langle m \rangle|$ correspond to quasidegenerate neutrino masses or with partial hierarchy for either ordering. The ultimate goal of the next generation of experiments is to cover the values of $|\langle m \rangle|$ predicted for the IO.

4.6. *Cosmology and astrophysics*

Additional information on the neutrino mass ordering may come from cosmology or from the detection of neutrinos emitted in supernovae explosions. Let us address the impact of neutrino masses in cosmology first (for some reviews, see Ref. 86, for example). In the early universe, neutrinos were in thermal equilibrium with the rest of the plasma and decoupled when the interaction rate due to their weak interactions was too slow compared to the expansion of the universe. This happened at a temperature of around 1 MeV. Since we know from beta-decay measurements that neutrino masses must be at (or below) the eV scale, this implies that relic neutrinos decoupled while they were still relativistic. After decoupling, their momentum redshifted with temperature so that they became colder and colder, turning nonrelativistic at late times. As a consequence massive neutrinos change the matter-radiation equality. This causes a small shift and an increase in the height of the peaks of the cosmic microwave background (CMB) power spectrum,[87] due to the Sachs–Wolfe effect.[88] However, the largest impact of neutrino masses in cosmology is due to their suppression of small-scale cosmological structures, such as galaxies and clusters of galaxies. These structures formed in the early universe starting from perturbations in the dark-matter density which grew due to the gravitational force. Cold, nonrelativistic dark matter falls into the gravitational wells due to overdensities, making them grow further and ultimately providing the seeds for the formation of galaxies and clusters of galaxies. Hot dark matter, such as neutrinos, is too fast to be trapped in the potential wells and streams freely out of it, suppressing the clustering at small scales. In more detail, the neutrino free-streaming length $\lambda_{\rm FS}$ relates to its thermal velocity $v_{\rm th}$ as[89]

$$\lambda_{\rm FS} = 2\pi\sqrt{\frac{2}{3}}\frac{v_{\rm th}(t)}{H(t)}\,,$$

where $H(t)$ is the Hubble rate at a given time t, and the thermal velocity is inversely proportional to the neutrino mass. Neutrinos are able to escape the potential wells with scales smaller than their free-streaming length, $\lambda < \lambda_{\rm FS}$. Thus, they do not cluster at those scales, producing a decay of the potential wells which eventually impacts the growth rate of both baryon and cold dark matter density perturbations. This leads to a suppression of the matter power spectrum P at small scales $\frac{\Delta P}{P} \propto 8\frac{\Omega_\nu}{\Omega_m}$. Here Ω_m is the matter density fraction, while Ω_ν is the neutrino density fraction and depends on the sum of neutrino masses, $\Omega_\nu = \sum_i m_{\nu_i}/(93\ {\rm eV}h^2)$. These effects are felt only gravitationally and therefore do not distinguish between different flavors or mass eigenstates but depend only on the total neutrino mass $\sum_i m_{\nu_i}$. The observation of successful structure formation at small scales eventually translates into an upper bound on the sum of neutrino masses. There are two main ways to test the distribution of matter at these scales: looking (i) at tracers of the matter distribution, e.g., galaxy surveys, such as SDSS,[90] BOSS,[91] HETDEX[92] and DES,[93] or low-density intergalactic gas for Ly-α surveys,[94] and (ii) directly at the impact of the dark-matter distribution in gravitational lensing experiments.[95] The former have been used extensively and have provided very strong bounds on neutrino masses but suffer from the problem of bias, i.e., the relation between the matter and the tracer power spectra. The latter are becoming more and more the tool of choice for constraining the matter power spectrum. It should be pointed out that bounds can change even very significantly when different sets of cosmological data are used. In fact, neutrino masses are degenerate with other cosmological parameters and different observables, for instance, high and low red-shift probes, allow to break these degeneracies. Moreover, these bounds rely on the assumption that the standard cosmological model ΛCDM holds; if this assumption is not true, for instance, a different explanation is invoked for the accelerated expansion of the universe, different limits will be inferred; see Ref. 96, for example. As a rather robust example, the combination of Planck CMB and BOSS BAO constraints currently sets the upper limit on the sum of neutrino masses at[97]

$$\sum_i m_{\nu_i} < 0.23 \text{ eV} \quad (95\% \text{ CI}).$$

Other recent bounds typically are around 0.15–0.3 eV.[98] From neutrino oscillation data, we know that at least two neutrinos are massive, while the third one could still be nearly massless. A lower limit on the sum of neutrino masses can then be derived by setting the lightest neutrino mass to vanish. Its value will depend on the ordering of neutrino masses:

$$\sum_i m_{\nu_i} > \sqrt{\Delta m^2_{21}} + \sqrt{\Delta m^2_{31}} \simeq 0.06 \text{ eV} \quad \text{for NO},$$

$$\sum_i m_{\nu_i} > \sqrt{\Delta m^2_{31}} + \sqrt{\Delta m^2_{32}} \simeq 0.10 \text{ eV} \quad \text{for IO}.$$

Therefore, as current and future cosmological measurements continue to improve their physics reach, they start becoming sensitive to the neutrino mass ordering[99,100] (under assumptions on the underlying cosmological model and of the treatment of different sets of data and degeneracies with other parameters). Current cosmological observations show a slight preference for NO,[100] although more data will be needed to confirm these results.

The ordering of neutrino masses can also have observable consequences on the evolution of neutrinos from core collapse supernovae (SN).[101] For a very detailed recent review we refer the reader to Ref. 102. During the SN explosion, 99% of the energy released is carried away by neutrinos as the latter have only weak interactions and therefore can escape from much deeper regions on the SN compared to photons. They are emitted in three phases: the neutronization peak, the accretion phase and the cooling phase. During the neutronization peak, which occurs in the first $\sim$25 ms, a large flux of ν_e is produced by electron capture on protons and nuclei. This is followed by the accretion phase, lasting tens to hundreds of milliseconds, when the shock stalls before the beginning of the explosion. In this phase, mainly electron neutrinos are emitted with a spectrum which reflects the complex nature of the standing accretion shock instability. The next phase, cooling, lasts few tens of seconds during which the neutron star born from the explosion gets colder and deleptonized. Most of the gravitational binding energy is released at this time with a flux of neutrinos and antineutrinos of all flavors and energies of tens of MeV. Because electron neutrinos interact more strongly with the background than electron antineutrinos which in turn have stronger interactions than muon and tau (anti)neutrinos, a hierarchy is expected to take place in their average energies $\langle E_{\nu_\mu,\nu_\tau}\rangle > \langle E_{\bar{\nu}_e}\rangle > \langle E_{\nu_e}\rangle$. The strength of the interaction determines how deep inside the SN the neutrinos decouple from the core and can escape freely. The weaker the interactions, the deeper is the resulting neutrinosphere and therefore the higher is the energy of the emitted neutrinos.

As the SN neutrinos propagate from the inner parts of the SN to the outer layers, their evolution is affected by oscillations in matter and consequently it is sensitive to the neutrino MO. Deep in the supernovae core, the neutrino density is so high that neutrinos become a background to themselves and a large matter potential due to neutrino–neutrino interactions is generated.[103] This potential can dominate the flavor transitions, causing a rapid conversion between different neutrino flavors. These oscillations are usually referred to as "collective oscillations" as the entire neutrino system evolves as a single collective mode. The most important consequence is the exchange between the ν_e ($\bar{\nu}_e$) and $\nu_{\mu,\tau}$ ($\bar{\nu}_{\mu,\tau}$) components of the total neutrino flux due to oscillation effects in certain energy ranges. These effects, called swaps,[103] depend on the MO and on the initial flux hierarchy among flavors. In the accretion phase, SN explosion studies suggest that the electron neutrino flux dominates over the antineutrino electron one, with the muon and tau ones being much smaller. Under these conditions, for NO, collective oscillations do not play a significant role. For IO, antineutrinos undergo a complete swap. In the neutrino channel instead, the swap happens only above a certain energy with characteristic spectral splits[104] in the energy distribution due to an abrupt change in the transition probability. More recently, detailed analysis of these effects have been performed showing that once a different hierarchy of the neutrino fluxes in the cooling phase, three-neutrino mixing effects, more realistic modelling of SN explosion are included, the evolution can be significantly more complex with multiple spectral splits and other features. These issues are still under intense investigation.

As SN neutrinos leave the inner parts of the SN, they will feel the effects of two matter resonances, see Section 4.1: the H resonance, associated with the atmospheric mass splitting at a density of $\rho_H \sim 10^3$ g/cm^3 and the L resonance due to the solar mass splitting at $\rho_L \sim 10$ g/cm^3. The L resonance always takes place for neutrinos. On the other hand, the H resonance happens in the neutrino (antineutrino) channel if the MO is normal (inverted). Therefore, for an NO of neutrino masses, the neutrino channel will be affected by a strong resonant flavor conversion and the neutronization peak will be severely reduced, while this will not be the case for IO. A final effect that needs to be considered is the oscillation of SN neutrinos as they cross the Earth on their way to the detector, which also depend on the neutrino MO.[105] The complex interplay of the different oscillation effects (collective, H resonance and oscillations in the Earth) needs to be carefully considered when predicting the observable neutrino spectrum at the detector.

SN neutrinos can be detected by several types of experiments. Large water Cherenkov detectors, such as SK, and liquid scintillators, e.g., KamLAND and the future SNO+ and JUNO, can detect mainly electron antineutrinos via inverse beta decay. The addition of gadolinium, which significantly improves the neutron capture detection efficiency, can further enhance the sensitivity by reducing the backgrounds for this channel. Elastic scattering can also give an important, although much smaller, sample of neutrinos mainly of the electron type. NC and other interactions on nuclei also contribute in small measure to the total number of events. LAr detectors, such as DUNE at SURF, have unique sensitivity to electron neutrinos via the reaction $\nu_e{}^{40}\mathrm{Ar} \to e^{-40}K^*$ and can also detect antineutrinos via inverse beta decay and NC interactions.

4.7. *Complementarity and synergy between different searches*

As it was discussed in the preceding sections, while to search for standard leptonic CPV the only foreseeable strategy is to use long-baseline neutrino oscillation experiments, there are different experimental techniques which are sensitive to the neutrino MO. This raises the issue of the complementarity and synergy between them. First of all, it should be pointed out that, being such an important theoretical question, it is of paramount importance to determine the MO using different physics principles, different type of experiments, different technologies. Moreover, its determination in one experiment will allow to solve degeneracies in others and test the standard three-neutrino scenario. Let's provide some important examples of these synergies.

- Long-baseline neutrino oscillation experiments. The sensitivity to CPV by the T2HK experiment is affected by the knowledge, or lack of, of the MO.[54] An independent determination of the MO would allow T2HK to fully exploit its great physics reach for δ. One possibility is that the neutrino MO is determined using atmospheric neutrino data collected by the HK detector itself. DUNE[53] is not affected by this issue as it can establish the MO at 5σ in a relatively short span of time.
- Neutrinoless double beta decay and neutrino oscillation experiments. As shown in Section 4.5, the predictions of $|\langle m\rangle|$ depend on the neutrino MO. The most important consequences would arise if neutrino oscillation experiments were to determine that the neutrino MO is inverted. In such case, a lower bound of 16 meV would hold on $|\langle m\rangle|$ providing a clear target for the next generation of experiments. Further conclusions could be derived depending on the results of experimental searches: (i) If a value $|\langle m\rangle| \geq 16$ meV is measured, one can conclude that neutrinos are Majorana particles. Moreover, if $|\langle m\rangle| > 49$ meV, both upper and lower bounds on m_3 can be deduced, given approximately by $|\langle m\rangle| \leq m_1 \leq |\langle m\rangle|/\cos 2\theta_{12}$. These bounds would also translate into a predicted range for the sum of neutrino masses relevant in cosmology. Instead for 16 meV $\leq |\langle m\rangle| \leq 49$ meV, the spectrum would be IH. If a precise measurement of the masses is obtained from cosmological observations, one might search for CPV due to Majorana phases, assuming $|\langle m\rangle|$ is determined precisely enough and the problem of NME is under control. (ii) Instead, if a value of $|\langle m\rangle| < 16$ meV is measured, one can conclude that neutrinos are Majorana particles but there must be some cancellation between the standard contribution and new physics. This could be the

case of a seesaw with heavy neutrino masses below 100 MeV, where an exact cancellation is at play between the contributions of standard and heavy neutrinos. (iii) Finally, if only an upper bound of 16 meV is set on $|\langle m\rangle|$, the simplest conclusion would be that neutrinos are Dirac particles, although a cancellation between the three-neutrino contribution and new physics could still be an option. It would be of particular importance to test this second hypothesis by looking for new particles and interactions which can give a sizable contribution to neutrinoless double beta decay.

Different considerations can be made if neutrino oscillation experiments determine the ordering to be normal. In this case, the predictions for $|\langle m\rangle|$ range from current bounds to the possibility of a complete cancellation if $m_1 \sim 5$ meV (see Fig. 6). For the limiting case of NH, we have 1 meV $\leq |\langle m\rangle| \leq 4$ meV. Measuring $|\langle m\rangle|$ would allow to constrain the allowed range of neutrino masses. (i) If it is found that $|\langle m\rangle| \gg 4$ meV, it is possible to conclude that neutrinos are Majorana and that their mass spectrum only has a partial hierarchy, with $|\langle m\rangle| \leq m_1 \leq |\langle m\rangle|/\cos 2\theta_{12}$. The latter constraint translates into a predicted range for $\Sigma_i m_i$ in cosmology. (ii) If instead $|\langle m\rangle| < 4$ meV, in case of a positive signal for the process, again it holds that neutrinos are Majorana particles and the spectrum is NH or with partial hierarchy. Finally, (iii) if the decay is not observed down to $|\langle m\rangle| < 1$ meV, neutrinos could be Dirac particles or Majorana ones but with values of the masses such that a strong cancellation is at play.

- Cosmology and terrestrial experiments (see Refs. 106–108, for example). If terrestrial experiments establish that the ordering is inverted, one would have $\Sigma_i m_i \geq 0.1$ eV. A precise measurement of $\Sigma_i m_i$, with $\Sigma_i m_i \geq 0.1$ eV would translate into an accurate determination of the values of neutrino masses. This would then be compared with the measurements of $|\langle m\rangle|$ in neutrinoless double beta decay. We have already mentioned that the possibility to search for Majorana CPV could open up. If, contrary to expectations, it is found that $\Sigma_i m_i < 0.1$ eV, then one would be forced to conclude that there are new effects reducing the impact of neutrino masses in the formation of large-scale structures. It should be pointed out that the determination of neutrino masses from cosmology and $|\langle m\rangle|$ are very difficult measurements and therefore the systematic and statistical errors, as well as the impact of NMEs, could significantly weaken the considerations made above.

5. Conclusions

Neutrino oscillations have been firmly established in the recent past, implying that neutrinos have mass and mix. Given two mass-squared differences, there are (at least) three massive neutrinos whose masses can be ordered in two ways: with NO for $m_1 < m_2 < m_3$ and IO for $m_3 < m_1 < m_2$. In order to establish the absolute values of neutrino masses, it is necessary to establish the type of ordering and measure or constrain the absolute neutrino mass scale set by the lightest neutrino mass. With three massive neutrinos, the mixing matrix $U_{\rm PMNS}$ is parameterised by three angles and one (three) CP-violating phases if neutrino are Dirac (Majorana) particles. Answering the questions related to the neutrino MO and leptonic CPV is crucial to hunt for the new physics responsible for neutrino masses and the leptonic mixing structure. Leptonic CPV can be searched for in long-baseline neutrino oscillation experiments measuring the appearance probabilities in the $\nu_\mu \to \nu_e$ and $\bar{\nu}_\mu \to \bar{\nu}_e$ channels. Current results from NOvA and T2K, combined with the determination of θ_{13} from reactor neutrino experiments, already provide some hints of possible CPV. These will be tested by future projects such as T2HK and DUNE which have a much better sensitivity. Various experimental strategies can be used to determine the MO. In this chapter, we have focussed on long-baseline and atmospheric neutrino experiments which exploit matter effects in neutrino oscillations, as well as medium-baseline reactor neutrino experiments and neutrinoless double beta decay. We have also mentioned briefly the impact of the MO on cosmological observables and SN neutrinos. Thanks to all these efforts, we can expect that the question of the neutrino MO will be solved in the near future. We have also discussed the synergies and complementarities between these searches pointing out how, by combining different sets of results, additional information on neutrino masses can be deduced and the standard three-neutrino mixing scenario can be tested.

Acknowledgments

This work has been partially supported by Elusives ITN (H2020- MSCA-ITN-2015, GA-2015-674896- ELUSIVES), InvisiblesPlus RISE (H2020-MSCARISE- 2015, GA-2015-690575-InvisiblesPlus). SP acknowledges partial support from ERC Grant "NuMass" (FP7- IDEAS-ERC ERC-CG 617143), the Wolfson Foundation and the Royal Society. SP thanks SISSA, Trieste for hospitality and the ERC Grant "DaMeSyFla" (FP7-IDEAS-ERC ERC-AG 267985) for their partial support. Fermilab is operated by Fermi Research Alliance, LLC under Contract No. DE-AC02-07CH11359 with the United States Department of Energy. The authors would like to thank C. Lunardini, S. Parke, S. Petcov for their useful comments.

References

1. B. Pontecorvo, *Zh. Eksp. Teor. Fiz.* **33**, 549 (1957) and **34**, 247 (1958); Z. Maki, M. Nakagawa, and S. Sakata, *Prog. Theor. Phys.* **28**, 870 (1962).
2. C. Jarlskog, *Phys. Rev. Lett.* **55**, 1039 (1985); C. Jarlskog, *Z. Phys.* **C29**, 491–497 (1985); O. W. Greenberg, *Phys. Rev. D* **32**, 1841 (1985); I. Dunietz, O. W. Greenberg, and D. d. Wu, *Phys. Rev. Lett.* **55**, 2935 (1985).
3. S. M. Bilenky, J. Hosek and S. T. Petcov, *Phys. Lett.* **B94**, 495 (1980).
4. J. Schechter and J. W. F. Valle, *Phys. Rev.* **D22**, 2227 (1980); M. Doi *et al.*, *Phys. Lett.* **B102**, 323 (1981).
5. I. Esteban, M. C. Gonzalez-Garcia, M. Maltoni, I. Martinez-Soler, and T. Schwetz, *JHEP* **1701**, 087 (2017) [arXiv:1611.01514 [hep-ph]] also available at NuFIT 3.0 (2016), www.nu-fit.org; F. Capozzi, E. Lisi, A. Marrone, D. Montanino, and A. Palazzo, *Nucl. Phys. B* **908**, 218 (2016) [arXiv:1601.07777 [hep-ph]].
6. L. J. Hall, H. Murayama, and N. Weiner, *Phys. Rev. Lett.* **84**, 2572 (2000) [hep-ph/9911341]; A. de Gouvea, and H. Murayama, *Phys. Lett. B* **747**, 479 (2015) [arXiv:1204.1249 [hep-ph]].
7. G. Altarelli, and F. Feruglio, *Rev. Mod. Phys.* **82**, 2701 (2010) [arXiv:1002.0211 [hep-ph]].
8. H. Ishimori, T. Kobayashi, H. Ohki, Y. Shimizu, H. Okada, and M. Tanimoto, *Prog. Theor. Phys. Suppl.* **183**, 1 (2010) [arXiv:1003.3552 [hep-th]].
9. S. F. King and C. Luhn, *Rept. Prog. Phys.* **76**, 056201 (2013) [arXiv:1301.1340 [hep-ph]].
10. R. Alonso, M. B. Gavela, G. Isidori, and L. Maiani, *JHEP* **1311**, 187 (2013) [arXiv:1306.5927 [hep-ph]].
11. S. F. King, *JHEP* **0508**, 105 (2005) [hep-ph/0506297].
12. C. S. Lam, *Phys. Lett. B* **656**, 193 (2007) [arXiv:0708.3665 [hep-ph]]; A. Blum, C. Hagedorn and M. Lindner, *Phys. Rev. D* **77**, 076004 (2008) [arXiv:0709.3450 [hep-ph]]; C. S. Lam, *Phys. Rev. D* **78**, 073015 (2008) [arXiv:0809.1185 [hep-ph]].
13. S. T. Petcov, *Phys. Lett.* **110B**, 245 (1982).
14. P. F. Harrison, D. H. Perkins, and W. G. Scott, *Phys. Lett. B* **530**, 167 (2002) [hep-ph/0202074]; P. F. Harrison and W. G. Scott, *Phys. Lett. B* **535**, 163 (2002) [hep-ph/0203209]; Z. Z. Xing, *Phys. Lett. B* **533**, 85 (2002) [hep-ph/0204049].
15. M. Fukugita, M. Tanimoto, and T. Yanagida, *Phys. Rev. D* **57**, 4429 (1998) [hep-ph/9709388]; V. D. Barger, S. Pakvasa, T. J. Weiler, and K. Whisnant, *Phys. Lett. B* **437**, 107 (1998) [hep-ph/9806387]; S. Davidson, and S. F. King, *Phys. Lett. B* **445**, 191 (1998) [hep-ph/9808296]; G. Altarelli, F. Feruglio, and L. Merlo, *JHEP* **0905**, 020 (2009) [arXiv:0903.1940 [hep-ph]]; D. Meloni, *JHEP* **1110**, 010 (2011) [arXiv:1107.0221 [hep-ph]].
16. A. Datta, F. S. Ling, and P. Ramond, *Nucl. Phys. B* **671**, 383 (2003) [hep-ph/0306002]; L. L. Everett, and A. J. Stuart, *Phys. Rev. D* **79**, 085005 (2009) [arXiv:0812.1057 [hep-ph]]; F. Feruglio, and A. Paris, *JHEP* **1103**, 101 (2011) [arXiv:1101.0393 [hep-ph]]; W. Rodejohann, *Phys. Lett. B* **671**, 267 (2009) [arXiv:0810.5239 [hep-ph]]; A. Adulpravitchai, A. Blum, and W. Rodejohann, *New J. Phys.* **11**, 063026 (2009) [arXiv:0903.0531 [hep-ph]].

17. S. F. King, *JHEP* **0209**, 011 (2002) [hep-ph/0204360]; P. H. Frampton, S. T. Petcov, and W. Rodejohann; *Nucl. Phys. B* **687**, 31 (2004) [hep-ph/0401206]; G. Altarelli, F. Feruglio, and I. Masina, *Nucl. Phys. B* **689**, 157 (2004) [hep-ph/0402155]; R. N. Mohapatra, and W. Rodejohann, *Phys. Rev. D* **72**, 053001 (2005) [hep-ph/0507312]; S. Antusch, and S. F. King, *Phys. Lett. B* **631**, 42 (2005) [hep-ph/0508044]; K. A. Hochmuth, S. T. Petcov, and W. Rodejohann, *Phys. Lett. B* **654**, 177 (2007) [arXiv:0706.2975 [hep-ph]].
18. S. F. King, *Phys. Lett. B* **659**, 244 (2008) [arXiv:0710.0530 [hep-ph]]; Y. Shimizu, M. Tanimoto, and A. Watanabe, *Prog. Theor. Phys.* **126**, 81 (2011) [arXiv:1105.2929 [hep-ph]]; S. F. King, and C. Luhn, *JHEP* **1109**, 042 (2011) [arXiv:1107.5332 [hep-ph]].
19. S. Antusch and S. F. King, *Phys. Lett. B* **591**, 104 (2004) [hep-ph/0403053]; I. Masina, *Phys. Lett. B* **633**, 134 (2006) [hep-ph/0508031].
20. S. Antusch, P. Huber, S. F. King, and T. Schwetz, *JHEP* **0704**, 060 (2007) [hep-ph/0702286 [HEP-PH]].
21. G. Altarelli, P. A. N. Machado, and D. Meloni, *PoS CORFU* **2014**, 012 (2015) [arXiv:1504.05514 [hep-ph]].
22. S. T. Petcov, *Nucl. Phys. B* **892**, 400 (2015) [arXiv:1405.6006 [hep-ph]]; D. Marzocca, S. T. Petcov, A. Romanino, and M. C. Sevilla, *JHEP* **1305**, 073 (2013) [arXiv:1302.0423 [hep-ph]]; P. Ballett, S. F. King, C. Luhn, S. Pascoli, and M. A. Schmidt, *JHEP* **1412**, 122 (2014) [arXiv:1410.7573 [hep-ph]].
23. F. Feruglio, C. Hagedorn, and R. Ziegler, *JHEP* **1307**, 027 (2013) [arXiv:1211.5560 [hep-ph]]; F. Feruglio, C. Hagedorn, and R. Ziegler, *Eur. Phys. J. C* **74**, 2753 (2014) [arXiv:1303.7178 [hep-ph]]; M. Holthausen, M. Lindner, and M. A. Schmidt, *JHEP* **1304**, 122 (2013) [arXiv:1211.6953 [hep-ph]]; C. Hagedorn, A. Meroni, and E. Molinaro, *Nucl. Phys. B* **891**, 499 (2015) [arXiv:1408.7118 [hep-ph]].
24. M. C. Chen, M. Fallbacher, K. T. Mahanthappa, M. Ratz, and A. Trautner, *Nucl. Phys. B* **883**, 267 (2014) [arXiv:1402.0507 [hep-ph]].
25. W. Grimus, and M. N. Rebelo, *Phys. Rept.* **281**, 239 (1997) [hep-ph/9506272].
26. A. D. Sakharov, *JTEP Lett.* **6**, 24 (1967).
27. M. Fukugita and T. Yanagida, *Phys. Lett. B* **174**, 45 (1986).
28. J. A. Harvey, M. S. Turner, *Phys. Rev. D* **42**, 3344 (1990).
29. P. Minkowski, *Phys. Lett. B* **67**, 421 (1977); M. Gell–Mann, P. Ramond, and R. Slansky in *Supergravity*, p. 315, edited by F. Nieuwenhuizen, and D. Friedman, North Holland, Amsterdam (1979); T. Yanagida, Proc. of the *Workshop on Unified Theories and the Baryon Number of the Universe*, edited by O. Sawada, and A. Sugamoto, KEK, Japan (1979); R. N. Mohapatra, G. Senjanovic, *Phys. Rev. Lett.* **44**, 912 (1980).
30. M. A. Luty, *Phys. Rev. D* **45**, 455 (1992); M. Flanz, E. A. Paschos, and U. Sarkar, *Phys. Lett. B* **345**, 248 (1995) [Erratum-ibid. B **382**, 447 (1996)] [hep-ph/9411366]; M. Plumacher, *Z. Phys. C* **74**, 549 (1997) [hep-ph/9604229].
31. L. Covi, E. Roulet, and F. Vissani, *Phys. Lett. B* **384**, 169 (1996) [hep-ph/9605319]; M. Flanz, E. A. Paschos, U. Sarkar, and J. Weiss, *Phys. Lett. B* **389**, 693 (1996) [hep-ph/9607310]; A. Pilaftsis, *Phys. Rev. D* **56**, 5431 (1997)

[hep-ph/9707235]; W. Buchmuller, and M. Plumacher, *Phys. Lett. B* **431**, 354 (1998) [hep-ph/9710460].

32. R. Barbieri, P. Creminelli, A. Strumia, and N. Tetradis, *Nucl. Phys. B* **575**, 61 (2000) [hep-ph/9911315].
33. A. Abada, S. Davidson, F. X. Josse-Michaux, M. Losada, and A. Riotto, *JCAP* **0604**, 004 (2006) [hep-ph/0601083]; E. Nardi, Y. Nir, E. Roulet, and J. Racker, *JHEP* **0601**, 164 (2006) [hep-ph/0601084]; A. Abada, S. Davidson, A. Ibarra, F.-X. Josse-Michaux, M. Losada, and A. Riotto, *JHEP* **0609**, 010 (2006) [hep-ph/0605281].
34. S. Davidson and A. Ibarra, *Nucl. Phys. B* **648**, 345 (2003) [hep-ph/0206304]; G. C. Branco, R. Gonzalez Felipe, F. R. Joaquim, and M. N. Rebelo, *Nucl. Phys. B* **640**, 202 (2002) [hep-ph/0202030]; T. Endoh, S. Kaneko, S. K. Kang, T. Morozumi, and M. Tanimoto, *Phys. Rev. Lett.* **89**, 231601 (2002) [hep-ph/0209020].
35. S. Pascoli, S. T. Petcov, and A. Riotto, *Phys. Rev. D* **75**, 083511 (2007) [hep-ph/0609125]; S. Pascoli, S. T. Petcov, and A. Riotto, *Nucl. Phys. B* **774**, 1 (2007) [hep-ph/0611338].
36. P. Langacker, J. P. Leveille, and J. Sheiman, *Phys. Rev. D* **27**, 1228 (1983). R. R. Lewis, *Phys. Rev. D* **21**, 663 (1980).
37. L. Wolfenstein, *Phys. Rev. D* **17**, 2369 (1978). L. Wolfenstein, *Phys. Rev. D* **20**, 2634 (1979). S. P. Mikheev, and A. Y. Smirnov, *Sov. J. Nucl. Phys.* **42**, 913 (1985) [Yad. Fiz. **42**, 1441 (1985)]; V. D. Barger, K. Whisnant, S. Pakvasa, and R. J. N. Phillips, *Phys. Rev. D* **22**, 2718 (1980).
38. K. Asano and H. Minakata, *JHEP* **1106**, 022 (2011) [arXiv:1103.4387 [hep-ph]]; P. B. Denton, H. Minakata, and S. J. Parke, *JHEP* **1606**, 051 (2016) [arXiv:1604.08167 [hep-ph]]; H. Minakata, and S. J. Parke, *JHEP* **1601**, 180 (2016) [arXiv:1505.01826 [hep-ph]].
39. A. Cervera, A. Donini, M. B. Gavela, J. J. Gomez Cadenas, P. Hernandez, O. Mena, and S. Rigolin, *Nucl. Phys. B* **579**, 17 (2000) Erratum: [*Nucl. Phys. B* **593**, 731 (2001)] [hep-ph/0002108]; M. Freund, *Phys. Rev. D* **64**, 053003 (2001) [hep-ph/0103300]; E. K. Akhmedov, R. Johansson, M. Lindner, T. Ohlsson, and T. Schwetz, *JHEP* **0404**, 078 (2004) [hep-ph/0402175].
40. J. Burguet-Castell, M. B. Gavela, J. J. Gomez-Cadenas, P. Hernandez, and O. Mena, *Nucl. Phys. B* **608**, 301 (2001) [hep-ph/0103258].
41. H. Minakata and H. Nunokawa, *JHEP* **0110**, 001 (2001) [hep-ph/0108085].
42. G. L. Fogli and E. Lisi, *Phys. Rev. D* **54**, 3667 (1996) [hep-ph/9604415]; V. Barger, D. Marfatia, and K. Whisnant, *Phys. Rev. D* **65**, 073023 (2002) [hep-ph/0112119].
43. P. Coloma and T. Schwetz, *Phys. Rev. D* **94**(5), 055005 (2016) [arXiv:1604.05772 [hep-ph]]; P. Bakhti, and Y. Farzan, *JHEP* **1407**, 064 (2014) [arXiv:1403.0744 [hep-ph]]; M. C. Gonzalez-Garcia, and M. Maltoni, *JHEP* **1309**, 152 (2013) [arXiv:1307.3092 [hep-ph]].
44. M. C. Gonzalez-Garcia, M. Maltoni, and J. Salvado, *JHEP* **1105**, 075 (2011) [arXiv:1103.4365 [hep-ph]].
45. H. Minakata and S. J. Parke, *Phys. Rev. D* **87**(11), 113005 (2013) [arXiv:1303.6178 [hep-ph]].

46. P. Coloma, H. Minakata, and S. J. Parke, *Phys. Rev. D* **90**, 093003 (2014) [arXiv:1406.2551 [hep-ph]].
47. D. Sgalaberna [NA61/SHINE Collaboration], *Nucl. Part. Phys. Proc.* **273–275**, 2672 (2016).
48. P. Coloma, and E. Fernandez-Martinez, *JHEP* **1204**, 089 (2012) [arXiv:1110.4583 [hep-ph]]. W. J. Marciano, hep-ph/0108181.
49. K. Abe *et al.* [T2K Collaboration], *Phys. Rev. Lett.* **107**, 041801 (2011) [arXiv:1106.2822 [hep-ex]]; K. Abe *et al.* [T2K Collaboration], arXiv:1701.00432 [hep-ex].
50. K. Abe *et al.* [T2K Collaboration], arXiv:1607.08004 [hep-ex].
51. D. S. Ayres *et al.* [NOvA Collaboration], hep-ex/0503053; P. Adamson *et al.* [NOvA Collaboration], [arXiv:1701.05891 [hep-ex]].
52. P. Huber, M. Lindner, T. Schwetz, and W. Winter, *JHEP* **0911**, 044 (2009) [arXiv:0907.1896 [hep-ph]]; S. K. Agarwalla, S. Prakash, S. K. Raut, and S. U. Sankar, *JHEP* **1212**, 075 (2012) [arXiv:1208.3644 [hep-ph]]; M. Ghosh, P. Ghoshal, S. Goswami, and S. K. Raut, *Nucl. Phys. B* **884**, 274 (2014) [arXiv:1401.7243 [hep-ph]].
53. C. Adams *et al.* [LBNE Collaboration], arXiv:1307.7335 [hep-ex]; R. Acciarri *et al.* [DUNE Collaboration], arXiv:1601.05471 [physics.ins-det]; R. Acciarri *et al.* [DUNE Collaboration], arXiv:1512.06148 [physics.ins-det].
54. K. Abe *et al.*, arXiv:1109.3262 [hep-ex]; K. Abe *et al.* [Hyper-Kamiokande Proto-Coll.], PTEP **2015**, 053C02 (2015) [arXiv:1502.05199 [hep-ex]].
55. M. Ishitsuka, T. Kajita, H. Minakata, and H. Nunokawa, *Phys. Rev. D* **72**, 033003 (2005) [hep-ph/0504026]; K. Abe *et al.* [Hyper-Kamiokande proto-Collaboration], arXiv:1611.06118 [hep-ex].
56. E. Baussan *et al.* [ESSnuSB Collaboration], *Nucl. Phys. B* **885**, 127 (2014) [arXiv:1309.7022 [hep-ex]]; E. Baussan, M. Dracos, T. Ekelof, E. F. Martinez, H. Ohman, and N. Vassilopoulos, arXiv:1212.5048 [hep-ex].
57. S. Geer, *Phys. Rev. D* **57**, 6989 (1998) Erratum: [*Phys. Rev. D* **59**, 039903 (1999)] [hep-ph/9712290]; A. De Rujula, M. B. Gavela, and P. Hernandez, *Nucl. Phys. B* **547**, 21 (1999) [hep-ph/9811390]; S. Geer, O. Mena, and S. Pascoli, *Phys. Rev. D* **75**, 093001 (2007) [hep-ph/0701258 [HEP-PH]]; A. D. Bross, M. Ellis, S. Geer, O. Mena, and S. Pascoli, *Phys. Rev. D* **77**, 093012 (2008) [arXiv:0709.3889 [hep-ph]].
58. S. Choubey *et al.* [IDS-NF Collaboration], arXiv:1112.2853 [hep-ex].
59. E. K. Akhmedov, S. Razzaque, and A. Y. Smirnov, *JHEP* **1302**, 082 (2013) [Erratum: [*JHEP* **1307**, 026 (2013)]] [arXiv:1205.7071 [hep-ph]].
60. S. Razzaque, and A. Y. Smirnov, *JHEP* **1505**, 139 (2015) [arXiv:1406.1407 [hep-ph]].
61. E. K. Akhmedov, A. Dighe, P. Lipari, and A. Y. Smirnov, *Nucl. Phys. B* **542**, 3 (1999) [hep-ph/9808270].
62. S. T. Petcov, *Phys. Lett. B* **434**, 321 (1998) [hep-ph/9805262]; M. Chizhov, M. Maris, and S. T. Petcov, hep-ph/9810501; M. V. Chizhov and S. T. Petcov, *Phys. Rev. Lett.* **83**, 1096 (1999) [hep-ph/9903399]; M. V. Chizhov and S. T. Petcov, *Phys. Rev. D* **63**, 073003 (2001) [hep-ph/9903424].

63. Y. Fukuda *et al.* [Super-Kamiokande Collaboration], *Phys. Rev. Lett.* **81**, 1562 (1998) [hep-ex/9807003].
64. M. G. Aartsen *et al.* [IceCube Collaboration], *Phys. Rev. Lett.* **111**(8), 081801 (2013) [arXiv:1305.3909 [hep-ex]].
65. S. Ahmed *et al.* [ICAL Collaboration], arXiv:1505.07380 [physics.ins-det].
66. M. M. Devi, T. Thakore, S. K. Agarwalla, and A. Dighe, *JHEP* **1410**, 189 (2014) [arXiv:1406.3689 [hep-ph]].
67. M. G. Aartsen *et al.* [IceCube PINGU Collaboration], arXiv:1401.2046 [physics.ins-det].
68. [Hyper-Kamiokande Collaboration], KEK-PREPRINT-2016-21, ICRR-REPORT-701-2016-1.
69. S. T. Petcov and M. Piai, *Phys. Lett. B* **533**, 94 (2002) [hep-ph/0112074]; S. Choubey, S. T. Petcov, and M. Piai, *Phys. Rev. D* **68**, 113006 (2003) [hep-ph/0306017].
70. Z. Djurcic *et al.* [JUNO Collaboration], arXiv:1508.07166 [physics.ins-det].
71. S. B. Kim, *Nucl. Part. Phys. Proc.* **265–266**, 93 (2015) [arXiv:1412.2199 [hep-ex]].
72. S. M. Bilenky, S. Pascoli, and S. T. Petcov, *Phys. Rev.* **D64**, 053010 (2001) [hep-ph/0102265]; S. Pascoli and S. T. Petcov, *Phys. Lett.* **B544**, 239 (2002) [hep-ph/0205022].
73. S. M. Bilenky *et al.*, *Phys. Rev.* **D54** 4432 (1996) [hep-ph/9604364]; S. M. Bilenky *et al.*, *Phys. Lett.* **B465** (1999) 193 [hep-ph/9907234]; V. Barger and K. Whisnant, *Phys. Lett.* **B456** (1999) 194 [hep-ph/9904281].
74. F. Vissani, *JHEP* **9906**, 022 (1999) [hep-ph/9906525]; W. Rodejohann, *Phys. Rev. D* **62**, 013011 (2000) [hep-ph/0003149]; H. V. Klapdor-Kleingrothaus, H. Pas, and A. Y. Smirnov, *Phys. Rev. D* **63**, 073005 (2001) [hep-ph/0003219]; F. Feruglio, A. Strumia, and F. Vissani, *Nucl. Phys. B* **637**, 345 (2002) Addendum: [*Nucl. Phys. B* **659**, 359 (2003)] [hep-ph/0201291]; S. Choubey and W. Rodejohann, *Phys. Rev. D* **72**, 033016 (2005) [hep-ph/0506102].
75. S. Pascoli, S. T. Petcov, and T. Schwetz, *Nucl. Phys. B* **734**, 24 (2006) [hep-ph/0505226].
76. H. Nunokawa, W. J. C. Teves, and R. Zukanovich Funchal, *Phys. Rev. D* **66**, 093010 (2002) [hep-ph/0206137].
77. S. Pascoli, S. T. Petcov, and W. Rodejohann, *Phys. Lett. B* **558**, 141 (2003) [hep-ph/0212113].
78. S. Pascoli, S. T. Petcov, and L. Wolfenstein, *Phys. Lett. B* **524**, 319 (2002) [hep-ph/0110287]; V. Barger, S. L. Glashow, P. Langacker, and D. Marfatia, *Phys. Lett. B* **540**, 247 (2002) [hep-ph/0205290]; H. Nunokawa, W. J. C. Teves, and R. Zukanovich Funchal, *Phys. Rev. D* **66**, 093010 (2002) [hep-ph/0206137];
79. S. Pascoli, S. T. Petcov, and W. Rodejohann, *Phys. Lett. B* **549**, 177 (2002) [hep-ph/0209059].
80. GERDA Coll., Talk presented at the XXVII International Conference on Neutrino Physics and Astrophysics (Neutrino 2016), Royal Geographical Society, London UK, 4–9 July 2016.

81. A. Gando *et al.* [KamLAND-Zen Collaboration], *Phys. Rev. Lett.* **117**(8), 082503 (2016) Addendum: [*Phys. Rev. Lett.* **117**(10), 109903 (2016)] [arXiv:1605.02889 [hep-ex]].
82. J. B. Albert *et al.* [EXO-200 Collaboration], *Nature* **510**, 229 (2014) [arXiv:1402.6956 [nucl-ex]].
83. R. Arnold *et al.* [NEMO-3 Collaboration], *Phys. Rev. D* **94**(7), 072003 (2016) [arXiv:1606.08494 [hep-ex]].
84. R. Arnold *et al.* [NEMO-3 Collaboration], *Phys. Rev. D* **92**(7), 072011 (2015) [arXiv:1506.05825 [hep-ex]].
85. A. D'Addabbo *et al.*, arXiv:1612.04276 [physics.ins-det].
86. J. Lesgourgues and S. Pastor, *Phys. Rept.* **429**, 307 (2006) [astro-ph/0603494]; K. N. Abazajian *et al.*, *Astropart. Phys.* **35**, 177 (2011) [arXiv:1103.5083 [astro-ph.CO]]; Y. Y. Y. Wong, *Ann. Rev. Nucl. Part. Sci.* **61**, 69 (2011) [arXiv:1111.1436 [astro-ph.CO]]; K. N. Abazajian *et al.* [Topical Conveners: K. N. Abazajian, J. E. Carlstrom, A. T. Lee Collaboration], *Astropart. Phys.* **63**, 66 (2015) [arXiv:1309.5383 [astro-ph.CO]].
87. S. Dodelson, E. Gates, and A. Stebbins, *Astrophys. J.* **467**, 10 (1996) [astro-ph/9509147]. W. Hu, and S. Dodelson, *Ann. Rev. Astron. Astrophys.* **40**, 171 (2002) [astro-ph/0110414].
88. R. K. Sachs and A. M. Wolfe, *Astrophys. J.* **147**, 73 (1967) [*Gen. Rel. Grav.* **39**, 1929 (2007)].
89. D. J. Eisenstein and W. Hu, *Astrophys. J.* **511**, 5 (1997) [astro-ph/9710252].
90. M. Tegmark *et al.* [SDSS Collaboration], *Astrophys. J.* **606**, 702 (2004) [astro-ph/0310725].
91. F. Beutler *et al.* [BOSS Collaboration], *Mon. Not. Roy. Astron. Soc.* **444**(4), 3501 (2014) [arXiv:1403.4599 [astro-ph.CO]].
92. G. J. Hill *et al.*, *ASP Conf. Ser.* **399**, 115 (2008) [arXiv:0806.0183 [astro-ph]].
93. T. Abbott *et al.*, [DES Collaboration], astro-ph/0510346.
94. N. Palanque-Delabrouille *et al.*, *JCAP* **1511**(11), 011 (2015) [arXiv:1506.05976 [astro-ph.CO]].
95. S. Hannestad, H. Tu, and Y. Y. Y. Wong, *JCAP* **0606**, 025 (2006) [astro-ph/0603019]; M. Kaplinghat, L. Knox, and Y. S. Song, *Phys. Rev. Lett.* **91**, 241301 (2003) [astro-ph/0303344]; R. A. Battye, and A. Moss, *Phys. Rev. Lett.* **112**(5), 051303 (2014) [arXiv:1308.5870 [astro-ph.CO]].
96. S. Hannestad, *Phys. Rev. Lett.* **95**, 221301 (2005) [astro-ph/0505551].
97. P. A. R. Ade *et al.*, [Planck Collaboration], *Astron. Astrophys.* **594**, A13 (2016) [arXiv:1502.01589 [astro-ph.CO]].
98. E. Giusarma, R. de Putter, S. Ho, and O. Mena, *Phys. Rev. D* **88**(6), 063515 (2013) [arXiv:1306.5544 [astro-ph.CO]]; A. B. Mantz *et al.*, *Mon. Not. Roy. Astron. Soc.* **446**, 2205 (2015) [arXiv:1407.4516 [astro-ph.CO]]; E. Giusarma, M. Gerbino, O. Mena, S. Vagnozzi, S. Ho, and K. Freese, *Phys. Rev. D* **94**(8), 083522 (2016) [arXiv:1605.04320 [astro-ph.CO]].
99. F. De Bernardis, T. D. Kitching, A. Heavens, and A. Melchiorri, *Phys. Rev. D* **80**, 123509 (2009) [arXiv:0907.1917 [astro-ph.CO]]; Q. G. Huang, K. Wang, and S. Wang, *Eur. Phys. J. C* **76**, 489 (2016) [arXiv:1512.05899 [astro-ph.CO]].

100. M. Gerbino, M. Lattanzi, O. Mena, and K. Freese, arXiv:1611.07847 [astro-ph.CO]; S. Hannestad, and T. Schwetz, *JCAP* **1611**(11), 035 (2016) [arXiv:1606.04691 [astro-ph.CO]]; S. Vagnozzi, E. Giusarma, O. Mena, K. Freese, M. Gerbino, S. Ho, and M. Lattanzi, arXiv:1701.08172 [astro-ph.CO]; F. Capozzi, E. Di Valentino, E. Lisi, A. Marrone, A. Melchiorri, and A. Palazzo, arXiv:1703.04471 [hep-ph].
101. A. S. Dighe and A. Y. Smirnov, *Phys. Rev. D* **62**, 033007 (2000) [hep-ph/9907423]; M. Kachelriess, R. Tomas, R. Buras, H.-T. Janka, A. Marek, and M. Rampp, *Phys. Rev. D* **71**, 063003 (2005) [astro-ph/0412082].
102. A. Mirizzi, *et al. Riv. Nuovo Cim.* **39**(1–2), 1 (2016) [arXiv:1508.00785 [astro-ph.HE]].
103. H. Duan, G. M. Fuller, J. Carlson, and Y. Z. Qian, *Phys. Rev. D* **74**, 105014 (2006) [astro-ph/0606616]; S. Hannestad, G. G. Raffelt, G. Sigl, and Y. Y. Y. Wong, *Phys. Rev. D* **74**, 105010 (2006) [Erratum: *Phys. Rev. D* **76**, 029901 (2007)] [astro-ph/0608695]; H. Duan, G. M. Fuller, and Y. Z. Qian, *Ann. Rev. Nucl. Part. Sci.* **60**, 569 (2010) [arXiv:1001.2799 [hep-ph]] .
104. G. G. Raffelt and A. Y. Smirnov, *Phys. Rev. D* **76**, 081301 (2007) [Erratum: *Phys. Rev. D* **77**, 029903 (2008)] [arXiv:0705.1830 [hep-ph]]; H. Duan, G. M. Fuller, and Y. Z. Qian, *Phys. Rev. D* **76**, 085013 (2007) [arXiv:0706.4293 [astro-ph]]; G. L. Fogli, E. Lisi, A. Marrone, and A. Mirizzi, *JCAP* **0712**, 010 (2007) [arXiv:0707.1998 [hep-ph]]. G. L. Fogli, E. Lisi, A. Marrone, A. Mirizzi, and I. Tamborra, *Phys. Rev. D* **78**, 097301 (2008) [arXiv:0808.0807 [hep-ph]]; B. Dasgupta, A. Dighe, G. G. Raffelt, and A. Y. Smirnov, *Phys. Rev. Lett.* **103**, 051105 (2009) [arXiv:0904.3542 [hep-ph]].
105. C. Lunardini and A. Y. Smirnov, *Phys. Rev. D* **63**, 073009 (2001) [hep-ph/0009356]; C. Lunardini, and A. Y. Smirnov, *Nucl. Phys. B* **616**, 307 (2001) [hep-ph/0106149]; K. Takahashi, and K. Sato, *Phys. Rev. D* **66**, 033006 (2002) [hep-ph/0110105].
106. V. D. Barger and K. Whisnant, *Phys. Lett. B* **456**, 194 (1999) [hep-ph/9904281].
107. G. L. Fogli, E. Lisi, A. Marrone, A. Melchiorri, A. Palazzo, P. Serra, and J. Silk, *Phys. Rev. D* **70**, 113003 (2004) [hep-ph/0408045]; G. L. Fogli, E. Lisi, A. Marrone, A. Melchiorri, A. Palazzo, P. Serra, J. Silk, and A. Slosar, *Phys. Rev. D* **75**, 053001 (2007) [hep-ph/0608060]; G. L. Fogli *et al.*, *Phys. Rev. D* **78**, 033010 (2008) [arXiv:0805.2517 [hep-ph]].
108. M. Gerbino, M. Lattanzi, and A. Melchiorri, *Phys. Rev. D* **93**(3), 033001 (2016) [arXiv:1507.08614 [hep-ph]].

Chapter 14

Beyond the Neutrino Standard Model

J. D. Lykken

Fermi National Accelerator Laboratory,
P.O. Box 500, Batavia 60510, IL, USA
lykken@fnal.gov

1. Introduction

Neutrinos, together with the Higgs boson, are the least understood particles that we have yet succeeded in detecting in laboratory experiments. Because they are so clearly and mysteriously different from the other particles of the standard model (SM), it seems plausible (so too for the Higgs) that they may also be a most promising window to a sector of new fundamental physics. Here we will review a number of ideas for how neutrinos might connect to new phenomena, via nonstandard interactions (NSIs), extra dimensions of space, exotic decays, violations of basic symmetry principles, or couplings to dark energy.

Our starting point is the "neutrino SM" that has emerged over the past few decades as the default framework for interpreting (as well as proposing) neutrino experiments. The neutrino SM begins with the particle content and interactions of the SM, including three flavors of "active" neutrinos related by the charged current (CC) weak interactions to three flavors of charged leptons. This is then supplemented by some unknown mechanism that gives the three flavors of neutinos mass, such that the three mass eigenstates are not the same as the three weak interaction eigenstates. While the mass generation mechanism is unspecified, it is assumed in the neutrino SM that neutrino mass is the only observable consequence of this mechanism. More generally, one might have imagined that neutrinos would experience new interactions or mix with additional species of neutral fermions.

2. NSIs in neutrino propagation through matter

Neutrino oscillation experiments as understood in the neutrino SM involve production of neutrinos via SM CC or neutral current (NC) interactions, propagation of the neutrinos over macroscopic distances during which they may interact with matter via CC or NC processes, and finally detection via CC or NC processes. It is useful to briefly review this standard description, to establish notation for the subsequent discussion of possible nonstandard effects. For recent reviews of nonstandard neutrino interactions, see Refs. 1 and 2; the early literature goes back to the seminal papers of Wolfenstein.[3–13]

Consider a process in which a neutrino ν_α^s of a particular flavor ($\alpha = \mathrm{e}, \mu, \tau$) is emitted by a source process (typically a radioactive decay with an associated charged lepton) and is later detected in flavor eigenstate β as indicated by an associated charged lepton in the detector. The propagation of the neutrinos between the source and the detector is described in terms of the neutrino mass eigenstates ν_i, $i = 1, 2, 3$, with masses m_i. For example, in the rest frame of ν_i, propagation in the vacuum is described by[14]

$$i\frac{d}{d\tau_i}|\nu_i(\tau_i)\rangle = m_i|\nu_i(\tau_i)\rangle, \tag{1}$$

where τ_i is the proper time of neutrino ν_i. In terms of lab frame quantities,

$$m_i\tau_i = E_i t - p_i L. \tag{2}$$

Since in experiments the neutrinos are always ultrarelativistic, we can write

$$p_i \simeq E_i - \frac{m_i^2}{2E_i}. \tag{3}$$

In this approximation, and ignoring overall phases, the probability for the process is given by the expression:

$$P_{\nu_\alpha^s \to \nu_\beta^d} = \left|\langle\nu_\beta^d|U_{\beta i}\mathrm{e}^{-iH_{ij}L}U_{\alpha j}^*|\nu_\alpha^s\rangle\right|^2, \tag{4}$$

where L is the lab frame distance between the source and the detector, and

$$H_{ij} = \frac{m_i^2}{2E}\delta_{ij} = \frac{\Delta m_{i1}^2}{2E}\delta_{ij} + \frac{m_1^2}{2E}\delta_{ij}, \tag{5}$$

where $\Delta m_{i1}^2 = m_i^2 - m_1^2$ and the second piece is often dropped since it just contributes an overall phase. The leptonic mixing matrix $U_{\alpha i}$ is known as the PMNS matrix in the neutrino SM, where it is a 3×3 unitary matrix. The alert reader may note that this description cannot be quite correct, since it assumes that the three neutrino mass eigenstates have exactly the same energy E. A more rigorous description of this process, and thus of neutrino oscillations, takes into account the quantum entanglement between the neutrino and other particles in the source interactions, the entanglement in the detector interactions, and the fact that the physical propagation is of a wave packet rather than a plane wave.[15,16] However for standard neutrino experiments this turns out to be a distinction without a difference.[17–19]

Ordinary matter such as comprises the Earth is almost completely transparent to neutrinos, however just as light waves can refract in a transparent medium, the wave function of a neutrino traveling through matter experiences a phase shift from the cumulative effect of multiple coherent forward scatterings.[3,4,20–22] This matter effect can also be expressed as an index of refraction,[3,23] or more commonly as a potential term added to the effective Hamiltonian of the neutrino:

$$H_{ij}^{\nu} = H_{ij}^{\text{vacuum}} + U_{i\alpha}^{\dagger} V_{\alpha\beta} U_{\beta j}, \tag{6}$$

where H_{ij}^{vacuum} is the Hamiltonian given in Eq. (5). In the neutrino SM, the matter potential arises from a combination of CC interactions between electron neutrinos and electrons and NC interactions between all three neutrino flavors and electrons and quarks in matter. Since the NC interactions are independent of neutrino flavor, they have no effect on neutrino oscillations, while the CC effect can be expressed simply as

$$V_{\alpha\beta} = \sqrt{2}\, G_F\, n_e \begin{pmatrix} 1 & 0 & 0 \\ 0 & 0 & 0 \\ 0 & 0 & 0 \end{pmatrix}, \tag{7}$$

where n_e is the number density of electrons in matter (assumed to be constant), and G_F is the Fermi constant: $2\sqrt{2}G_F \simeq (174\ \text{GeV})^2$.

If there are NSIs affecting neutrinos as they propagate through matter, then the matter potential may have additional contributions; generally one can write

$$V_{\alpha\beta} = \sqrt{2}\, G_F\, n_e \begin{pmatrix} 1 + \varepsilon_{ee} & \varepsilon_{e\mu} & \varepsilon_{e\tau} \\ \varepsilon_{e\mu}^* & \varepsilon_{\mu\mu} & \varepsilon_{\mu\tau} \\ \varepsilon_{e\tau}^* & \varepsilon_{\mu\tau}^* & \varepsilon_{\tau\tau} \end{pmatrix}, \tag{8}$$

where each of the NSI parameters $\varepsilon_{\alpha\beta}$ can in principle get contributions from neutrino interactions with electrons, up and down type quarks in ordinary matter such as the Earth. Since in Earth matter the number densities of electrons, protons, and neutrons are approximately equal, one can use $n_u \simeq n_d \simeq n_e$ to write

$$\varepsilon_{\alpha\beta} \simeq \varepsilon^e_{\alpha\beta} + 3\varepsilon^u_{\alpha\beta} + 3\varepsilon^d_{\alpha\beta}. \tag{9}$$

For antineutrinos, one should make the substitution $V_{\alpha\beta} \to -V^*_{\alpha\beta}$.

To understand how such NSIs might arise, it is useful to map the NSI parameters $\varepsilon_{\alpha\beta}$ into a four-fermion Lagrangian coupling neutrinos to fermions:

$$\mathcal{L}_{\rm NSI} = -\sum_{\alpha\beta f} 2\sqrt{2}\, G_F\, \varepsilon^{fC}_{\alpha\beta}\, (\bar{\nu}_\alpha \gamma^\mu P_L \nu_\beta)\, (\bar{f}\gamma_\mu P_C f), \tag{10}$$

where $f = e, u, d$, denote the relevant matter fermion fields and $C = L, R$ denotes the two possible chiralities, with P_L, P_R the chirality projection operators. The NSI parameters are regained as:

$$\varepsilon_{\alpha\beta} = \sum_f \left(\varepsilon^{fL}_{\alpha\beta} + \varepsilon^{fR}_{\alpha\beta}\right) \frac{n_f}{n_e}. \tag{11}$$

Note that the structure of the interactions is similar to the NCs of the SM.

Besides contributing nonstandard matter effects in neutrino propagation, four-fermion couplings as described by Eq. (10) will have effects on neutrino production and detection processes. Since neutrino production and detection in experiments typically involves the SM CC interaction (from W boson exchange), one can also address the possibility of additional four-fermion interactions that mimic the structure of CCs:

$$\mathcal{L}_{\rm NSI} = -\sum_{\alpha\beta f f'} 2\sqrt{2}\, G_F\, \varepsilon^{ff'C}_{\alpha\beta}\, (\bar{\ell}_\alpha \gamma^\mu P_L \nu_\beta)\, (\bar{f}\gamma_\mu P_C f'), \tag{12}$$

The most relevant of the resulting CC NSI parameters are $\varepsilon^{ud}_{\alpha\beta}$ and $\varepsilon^{\mu e}_{\alpha\beta}$; the former will cause what appear to be neutrino oscillation effects at zero baseline distance, for neutrinos produced from pion decays, while the latter will do the same for neutrinos produced by muon decays.[24]

Roughly speaking, if four-fermion interactions as described in Eq. (10) arise from tree-level exchange of some beyond-the-SM (BSM) heavy particle with mass Λ, one would expect the NSI parameters to be of order M_W^2/Λ^2; this is $\mathcal{O}(10^{-2})$ if we assume that the BSM particle must have mass $\gtrsim 1$ TeV to have escaped detection at the LHC. If the NSIs arise from loop effects, then one expects at least another order of magnitude suppression.

2.1. *Current direct experimental limits on NSIs*

Direct experimental bounds on the NSI parameters $\varepsilon^{f}_{\alpha\beta}$ derive from a variety of sources. These include $\nu_e e$ scattering data from LAMPF and LSND,[25,26] $\bar{\nu}_e e$ scattering data from reactor experiments,[27–31] $\nu_e q$ scattering data from CHARM,[32] $\nu_\mu e$ scattering data from CHARM II,[33,34] $\nu_\mu q$ scattering data from NuTeV,[35–37] $e^+e^- \to \nu\bar{\nu}\gamma$ data from the LEP experiments,[38–48] and neutrino oscillation data from Super-Kamiokande (SK),[49] IceCube and DeepCore,[50,51] KamLAND,[52] SNO,[53] and Borexino.[54,55] Collecting the results of a number of analyses[24,56–71] and following the discussions in[64,72,73] one can extract the following 90% CL bounds on the NSI parameters:

$$\begin{array}{lll} |\varepsilon^{e}_{ee}| < 0.1 & |\varepsilon^{e}_{\mu\mu}| < 0.04 & |\varepsilon^{e}_{\tau\tau}| < 0.6 \\ |\varepsilon^{e}_{e\mu}| < 0.33 & |\varepsilon^{e}_{e\tau}| < 3.0 & |\varepsilon^{e}_{\mu\tau}| < 0.33 \end{array} \tag{13}$$

$$\begin{array}{lll} |\varepsilon^{u/d}_{ee} - \varepsilon^{u/d}_{\mu\mu}| < 0.51 & |\varepsilon^{u/d}_{\mu\mu} - \varepsilon^{u/d}_{\tau\tau}| < 0.03 & \\ |\varepsilon^{u/d}_{e\mu}| < 0.09 & |\varepsilon^{u/d}_{e\tau}| < 0.14 & |\varepsilon^{u/d}_{\mu\tau}| < 0.006^{64} \end{array} \tag{14}$$

Analogous 90% CL bounds on the CC NSI parameters $\varepsilon^{ud}_{\alpha\beta}$ and $\varepsilon^{\mu e}_{\alpha\beta}$ are given in Ref. 24.

Less stringent bounds on the NSI parameters are often quoted in the literature and assumed in various analyses. As we shall see in the next section, it it already rather challenging to devise BSM models that are simultaneously consistent with precision data on charged leptons while also generating NSIs exceeding the limits (Eqs. (13) and (14)); thus entertaining looser bounds on NSIs may be a distinction without much of a difference.

Another caveat is that NSI effects can be mimicked by other sources. For example, if the three active neutrinos mix with a sterile neutrino that is too heavy for the oscillation effects to appear in short- or long-baseline experiments, the averaged-out oscillations are still in principle detectable as nonunitarity in the experimentally extracted values of the PMNS matrix. These departures from unitarity can be recast as nonzero values of the NSI parameters.[74]

2.2. *How large can NSIs be in BSM models?*

The Lagrangian Eq. (10) is still not especially enlightening since it is not gauge invariant. The three known active flavors of neutrinos reside in doublets of $SU(2)_L$ with their corresponding charged leptons. Thus any hypothesis about nonstandard neutrino interactions has some implication for NSIs of charged leptons as well.[75,76] These typically lead to much stronger experimental constraints on the $\varepsilon^f_{\alpha\beta}$ than the direct experimental limits.

For example, consider the $\varepsilon^e_{\alpha\beta}$. Following the treatment in Ref. 77, write down all of the possible gauge invariant four-fermion interactions that couple neutrinos to electrons:

$$\mathcal{L}_1^{\alpha\beta} = -c^1_{\alpha\beta}(\overline{L}_\alpha\gamma^\mu P_L L_\beta)(\overline{L}_e\gamma_\mu P_L L_e), \tag{15}$$

$$\mathcal{L}_2^{\alpha\beta} = -c^2_{\alpha\beta}(\overline{L}_\alpha\gamma^\mu P_L L_e)(\overline{L}_e\gamma_\mu P_L L_\beta), \tag{16}$$

$$\mathcal{L}_3^{\alpha\beta} = -c^3_{\alpha\beta}(\overline{L}_\alpha\gamma^\mu P_L L_\beta)(\bar{e}\gamma_\mu P_R e), \tag{17}$$

where the gauged $SU(2)_L$ indices are contracted to form a singlet within each bracket. In terms of these interactions, $\varepsilon^e_{\alpha\beta}$ can be written as

$$\varepsilon^e_{\alpha\beta} = \frac{1}{2\sqrt{2}G_F}(c_1^{\alpha\beta} + c_1^{ee}\delta_e^\alpha\delta_e^\beta + c_2^{e\beta}\delta_e^\alpha + c_2^{\alpha e}\delta_e^\beta + c_3^{\alpha\beta}). \tag{18}$$

These interactions lead to deviations from the SM in processes involving four charged leptons. The off-diagonal elements of $(c^1_{\alpha\beta} + c^2_{\alpha\beta})$ and $c^3_{\alpha\beta}$ are constrained by bounds on charged lepton flavor violating processes,[78]

$$\mathrm{Br}(\mu \to 3e) = \frac{|c_3^{e\mu}|^2 + 2|c_1^{e\mu} + c_2^{e\mu}|^2}{8G_F^2} < 1.0 \times 10^{-12}, \tag{19}$$

$$\frac{\mathrm{Br}(\tau \to 3e)}{\mathrm{Br}(\tau \to e\nu_\tau\bar{\nu}_e)} = \frac{|c_3^{e\tau}|^2 + 2|c_1^{e\tau} + c_2^{e\tau}|^2}{8G_F^2} < 2.0 \times 10^{-7}, \tag{20}$$

$$\frac{\mathrm{Br}(\tau \to \mu e^+e^-)}{\mathrm{Br}(\tau \to \mu\nu_\tau\bar{\nu}_\mu)} = \frac{|c_3^{\mu\tau}|^2 + 2|c_1^{\mu\tau} + c_2^{\mu\tau}|^2}{8G_F^2} < 1.6 \times 10^{-7}. \tag{21}$$

These typically constrain the off-diagonal $|\varepsilon^e_{\alpha\beta}|$ to be roughly in the range of 10^{-6} to a few $\times\, 10^{-4}$.

The diagonal elements of $(c^1_{\alpha\beta} + c^2_{\alpha\beta})$ and $c^3_{\alpha\beta}$ are all constrained by the LEP2 bounds[79] on four-lepton contact operators. These typically constrain the diagonal magnitudes $|\varepsilon^e_{\alpha\alpha}|$ to be less than a few $\times 10^{-3}$.

How can one avoid such stringent bounds? An obvious special case is a BSM model where $(c^1_{\alpha\beta} + c^2_{\alpha\beta}) = 0$ and $c^3_{\alpha\beta} = 0$ for all α and β. The simplest

BSM model with this feature[75–77,80,81] arises by adding to the SM an $SU(2)_L$ singlet scalar field S with electric charge +1 that has Yukawa-like couplings to lepton doublets:

$$\mathcal{L} = \lambda_{\alpha\beta}\overline{L}^c_\alpha(i\sigma_2)P_L L_\beta S + \text{h.c.}, \tag{22}$$

where the dimensionless couplings are antisymmetric $\lambda_{\alpha\beta} = -\lambda_{\beta\alpha}$. Assuming that the new scalar is very heavy and does not acquire a vacuum expectation value, one can integrate it out to get an effective low-energy NSI with the property that $(c^1_{\alpha\beta} + c^2_{\alpha\beta}) = 0$ and $c^3_{\alpha\beta} = 0$ for all α and β. More explicitly $\varepsilon^e_{ee} = \varepsilon^e_{e\mu} = \varepsilon^e_{e\tau} = 0$, and

$$\varepsilon^e_{\alpha\beta} = \frac{\lambda_{e\alpha}\lambda^*_{e\beta}}{\sqrt{2}G_F m_S^2}, \quad (\alpha, \beta = \mu, \tau). \tag{23}$$

Thus this BSM model of NSIs is free of strong constraints arising from interactions of four charged leptons. Unfortunately the model is still strongly constrained by BSM effects on the decays of muons and tau leptons that involve one charged lepton and two neutrinos in the final state; the resulting 90% CL bounds[75–77,81] on $\varepsilon^e_{\mu\mu}$, $\varepsilon^e_{\mu\tau}$, and $\varepsilon^e_{\tau\tau}$ are between 10^{-3} and 10^{-4}.

This discouraging result is rather generic for concrete BSM models that produce neutrino NSIs. At present, there appear to be three avenues for model building that hold some promise for producing NSIs that could be observable in planned or proposed neutrino oscillation experiments without, at the same time, coming into severe tension with other data:

2.2.1. *$SU(2)_L$ doublet leptoquark*

This is a simple model positing[77] a heavy (TeV mass) leptoquark scalar S with hypercharge −7/3 in a doublet under $SU(2)_L$. This doublet scalar field can couple the left-handed lepton doublets L_α of the SM to the right-handed $SU(2)_L$ singlet up quark field:

$$\mathcal{L} \supset \lambda_\alpha \bar{L}_\alpha u_R\, S + \text{h.c.}, \tag{24}$$

where the λ_α are dimensionless generation-dependent couplings. In terms of this coupling and the leptoquark mass M_S, the resulting NSI parameters are

$$\varepsilon^u_{\alpha\beta} = -\frac{\lambda_\alpha\lambda^*_\beta}{4\sqrt{2}G_F M_S^2}. \tag{25}$$

To avoid the strong constraints Eqs. (19)–(21) from charged lepton flavor violating processes, assume that all but one of the couplings λ_e, λ_μ, λ_τ is very small. In this case, current experimental constraints still allow ε_{ee} or

$\varepsilon_{\mu\mu}$ at the 10^{-2} level and $\varepsilon_{\tau\tau}$ approaching 0.1 or so.[77] One can make an analogous model with a hypercharge $-1/3$ doublet leptoquark that couples only to the right-handed down quark.

2.2.2. *Leptophilic $SU(2)_L$ doublet scalar*

In this approach,[77,82] one introduces an $SU(2)_L$ doublet scalar $S = (S^+, S^0)$ with the same quantum numbers as the Higgs doublet but arranged such that S does not acquire a vacuum expectation value. One further assumes that the new scalar has Yukawa interactions with electron fields:

$$\mathcal{L} \supset \lambda_\alpha \bar{L}_\alpha e_R S + \text{h.c.} \tag{26}$$

The tree-level exchange of the charged scalar S^+ induces an extra CC interaction between neutrinos and electrons; after a Fierz rearrangement, this can be written as

$$\mathcal{L} \supset -2\sqrt{2}\, G_F \varepsilon_{\alpha\beta} (\bar{\nu}_\alpha \gamma^\mu P_L \nu_\beta)(\bar{e}\gamma_\mu P_R e), \tag{27}$$

where the NSI parameters are

$$\varepsilon^e_{\alpha\beta} = -\frac{\lambda_\alpha \lambda^*_\beta}{4\sqrt{2} G_F M^2_{S^+}}. \tag{28}$$

This is already good enough to generate a flavor diagonal NSI parameter of order 0.01 without contradicting experimental bounds. As pointed out in Ref. 82, one can also obtain $\varepsilon_{e\tau}$ as large as $\sim$0.3 by mixing S^+ with a charged $SU(2)_L$ singlet scalar. In both of these realizations, there is a charged scalar particle just out of reach of current experimental limits, i.e., with mass $\sim\mathcal{O}(100)$ GeV.

2.2.3. *Neutrino NSIs mediated by a light Z' gauge boson*

In recent years, there has been considerable interest in BSM models with new light Z' gauge bosons in the mass range of roughly 1–100 MeV. The Z' is gauging a $U(1)$ global symmetry that is some linear combination of baryon number B and the three lepton numbers L_e, L_μ, and L_τ, which are separately conserved in the SM with massless neutrinos; the SM particle content is anomaly free under this new gauged $U(1)$ provided[83] that the linear combination can be be expressed as $yB - \sum_i x_i L_i$ with $3y = x_e + x_\mu + x_\tau$. The most popular example relevant to BSM neutrino physics[84] is a gauged $B-L$, but other examples are $L_\mu - L_\tau$[85–88] and $B-3L_\tau$.[89–91] Another potentially important possibility for BSM neutrinos is to gauge $U(1)_B$; this requires adding to the SM additional chiral fermions charged under the SM gauge group,[91–94] but such particles could be heavy enough to have so far escaped detection at the LHC.

Roughly speaking, if the Z' couples to both neutrinos and some combination of e, u, or d, then one expects to generate NSI parameters of order

$$|\varepsilon| \sim \frac{g^2}{2\sqrt{2}G_F M_{Z'}^2}, \tag{29}$$

where g is a $U(1)'$ gauge coupling, and $M_{Z'}$ is the mass of the Z' boson. If we assume, for example, $M_{Z'} \sim 10$ MeV, then $|\varepsilon| \gtrsim 0.01$ implies $g \gtrsim 10^{-5}$. Independent of neutrino physics, there are several intriguing motivations for considering BSM models with these general features. These include the long-standing 3.6σ discrepancy in the measured value[95] of the muon $g-2$, dark sectors,[96] and the 6.8σ anomaly in nuclear decays of excited Beryllium.[97]

Two recent studies[91,98] have focused on the question of how large neutrino NSIs can be in models of this type, consistent with all existing experimental constraints but not necessarily at the same time explaining other mysteries such as muon $g-2$. An example of a viable model with large NSIs is to gauge $U(1)_B$, adding extra heavy leptons to cancel anomalies. A sterile Dirac fermion Ψ is introduced that is assumed to carry baryon number, as well as a baryon number charged $SU(2)_L$ scalar doublet S that develops a vacuum expectation value (vev) much smaller than the vev of the Higgs field; this allows a mixing between neutrinos and the sterile fermion via the interaction

$$\mathcal{L} \supset y_\alpha \bar{L}_\alpha \Psi S + \text{h.c.}, \tag{30}$$

where the y_α are dimensionless couplings. Unlike the mixing with sterile Majorana fermions that generates the usual seesaw mechanism for neutrino masses, mixing with sterile Dirac fermions does not in itself generate neutrino mass; however, the mixing allows neutrinos to couple to the Z' gauge boson of $U(1)_B$, which also couples to quarks (in a generation-independent way, thus avoiding flavor-changing NCs).

The example of the NSI parameter $\varepsilon_{e\tau}$ is shown in Fig. 1. Here one sees that a model of this type could produce $\varepsilon_{e\tau} \sim 0.1$ for $M_{Z'} \sim 10$ MeV, sterile Dirac fermions with mass $\sim\mathcal{O}(\text{GeV})$, and a $U(1)_B$ gauge coupling to quarks of order 10^{-4}. The model predicts measurable NSI effects in future long-baseline experiments, and a heavy charged scalar boson and heavy exotic leptons that could be detected as the LHC, and possible signals in future dark-matter-direct-detection experiments as well as experiments looking for charged lepton violation. A fair conclusion is that, while viable models with

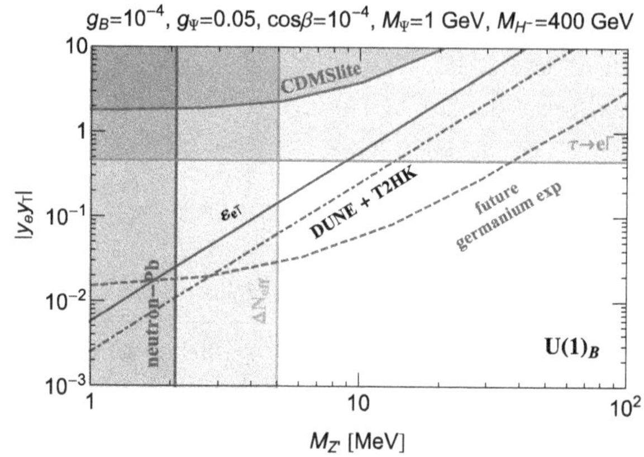

Fig. 1. Example of large NSIs produced by neutrinos and quarks interacting via a light Z' gauge boson. A contour of the NSI parameter $\varepsilon_{e\tau}$ is shown as a function of Z' boson mass and the couplings y_e, y_τ defined in Eq. (30). Current relevant experimental bounds are shown (shaded regions) as well as the sensitivity of planned or possible future long-baseline neutrino experiments (DUNE and T2HK), as well as planned future dark matter experiments. This figure has been taken from Ref. 91.

detectable NSIs are of necessity somewhat complicated, they lead to multiple correlated predictions for future experiments and may plausibly be motivated by other outstanding puzzles of particle physics.

2.3. *Future prospects for NSIs*

It is both important and challenging to continue to constrain, or find evidence for, the possibility of nonstandard neutrino interactions. Many studies have estimated the NSI sensitivity of future data from astrophysical neutrinos,[64,73] atmospheric neutrinos,[71,99–101] and future medium- to long-baseline experiments such as DUNE,[102–105] T2HK,[103] JUNO,[106] or a hypothetical neutrino factory.[107] As demonstrated in Fig. 2, significant improvements in sensitivity will eventually become available. It is important as well to explore new kinds of experiments with sensitivity to NSIs, such as those designed to detect low-energy neutrino coherent scattering.[108,109]

As has been studied by a number of researchers,[104,110–112] if either NSI or sterile neutrino effects are present in future long-baseline experiments, one could encounter a variety of interesting and challenging "confusion scenarios". In such scenarios, the extraction of the PMNS parameters, including the CP-violating phase δ, is confused by the presence of effects not included in the standard fitting. Of course, this is a challenge that would be most welcome.

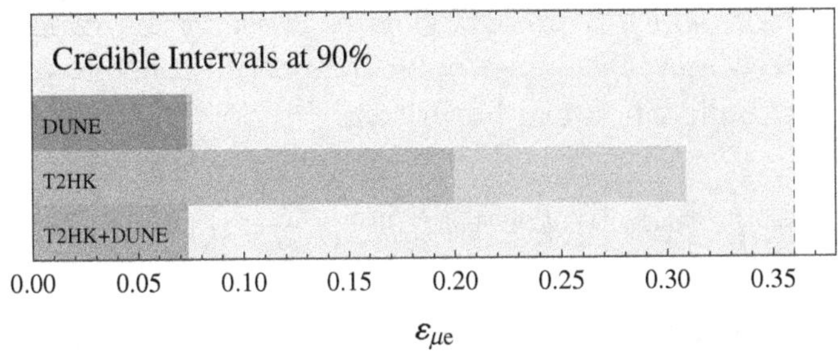

Fig. 2. Comparison of the expected sensitivites to the NSI parameter $\varepsilon_{e\mu}$ at DUNE and T2HK, before and after combining their respective data sets. Darker (lighter) bands show the results when prior constraints on NSI parameters are (not) included in the fit. This figure has been taken from Ref. 103.

2.4. *Large extra dimensions*

As clearly demonstrated by many examples from string theory, there is no reason to assume that the high-energy propagation of elementary particles is limited to three spatial dimensions. Indeed in braneworld examples[113–120] from string theory, one finds that some particles are confined to propagate on branes while other species of particles propagate in the full "bulk" of the higher-dimensional space. In the extreme version of this picture, the graviton would probe the full bulk space, while all of the SM fields might be restricted to a three-dimensional brane.

Neutrinos would seem to have a special role in such a scenario. Since the right-handed neutrino fields required to give Dirac masses to neutrinos are SM singlets, they are good candidates for probing a higher dimensionality than the fields of the SM. In that case what we naively label as dimensionless Yukawa couplings in a 4D description of neutrino Dirac masses is actually a composite of effects projected from higher dimensions, involving new mass/length scales such as the size of compactified extra dimensions. In this context, the smallness of neutrino masses could be explained by a volume suppression effect from large extra dimensions (LEDs).[121–124]

To see how this works in more detail, consider the case on a single extra spatial dimension, denoted by the coordinate y, compactified on a $\mathcal{Z}_2$ orbifold of a circle with radius R; the $\mathcal{Z}_2$ transformation takes $y \to -y$, and the orbifold can be thought of as a finite interval of length πR terminated on each end with a 3-brane. Let $\Psi(x^\mu, y)$ denote a 5D fermion that is a singlet under the SM gauge group. A 5D Dirac spinor has four complex components, like a 4D Dirac spinor; however, in 5D there is no analog of a Weyl or a Majorana spinor. The latter follows from the fact[125] that under charge conjugation in 5D, $(\Psi^c)^c = -\Psi$. It is possible to write the analog of a Majorana mass term for the 5D spinor, which if present violates conservation of lepton number and will lead to Majorana masses upon compactification to an effective 4D theory. The 5D bulk action can be written:[121,125,126]

$$S_{\text{bulk}} = \int \mathrm{d}^4x\mathrm{d}y \left[\mathrm{i}\bar{\Psi}\not{D}\Psi + M_D\bar{\Psi}\Psi - \frac{1}{2}M_M\left(\bar{\Psi}^c\Psi + \text{h.c.}\right)\right], \tag{31}$$

where M_D parameterizes a 5D Dirac mass and M_M is the analog of a Majorana mass.

It is also straightforward to write a 4D "brane" action, localized at $y = 0$, with a Yukawa coupling between the left-handed SM neutrino fields and Ψ:

$$S_{\text{brane}} = \int_{y=0} \mathrm{d}^4x \left(-\frac{1}{\sqrt{M_S}}\bar{\nu}_{\mathrm{L}}\hat{m}^c\Psi - \frac{1}{\sqrt{M_S}}\bar{\nu}^c_{\mathrm{L}}\hat{m}\Psi + \text{h.c.}\right), \tag{32}$$

where M_S represents some new mass scale of the higher-dimension theory, while $\hat{m}$, by gauge invariance, should be proportional to the SM Higgs vev v times some dimensionless Yukawa coupling.

The 5D bulk singlet Dirac fermion $\Psi(x^\mu, y)$ can be decomposed into two two-component objects: $\Psi = (\xi\,,\,\eta^c)^T$, where $\eta^c = \mathrm{i}\sigma^2\eta^*$. Then we can decompose $\xi(x^\mu, y)$ and $\eta(x^\mu, y)$ in a basis of Kaluza–Klein modes appropriate to the orbifold; each Kaluza–Klein mode will then be a 4D spinor. It is a feature of the orbifold compactification that we can consistently restrict ξ to have even parity under the $\mathcal{Z}_2$ and η to have odd parity; this will give us an effective theory in four dimensions that is chiral. The parity choices also ensure that only ξ, not η, couples to the left-handed neutrino in Eq. (32) and that the 5D Dirac mass term is absent, since it is odd under the $\mathcal{Z}_2$.

The Kaluza–Klein mode expansions are given by

$$\begin{aligned} \xi(x,y) &= \frac{1}{\sqrt{\pi R}}\xi_{(0)}(x) + \sqrt{\frac{2}{\pi R}}\sum_{n=1}^{N}\xi_{(n)}(x)\cos\left(\frac{ny}{R}\right), \\ \eta(x,y) &= \sqrt{\frac{2}{\pi R}}\sum_{n=1}^{N}\eta_{(n)}(x)\sin\left(\frac{ny}{R}\right), \end{aligned} \tag{33}$$

where the sums are cut off at some value N representing the short distance cutoff of this effective theory.

Inserting this expansion into Eqs. (31) and (32), and then integrating over the extra dimension, gives the following form of the four-dimensional action:

$$
\begin{aligned}
S = \int \mathrm{d}^4x \Bigg\{ & \xi_{(0)}^{\dagger} \mathrm{i}\bar{\sigma}^{\mu}\partial_{\mu}\xi_{(0)} + \sum_{n=1}^{N} \left(\xi_{(n)}^{\dagger} \mathrm{i}\bar{\sigma}^{\mu}\partial_{\mu}\xi_{(n)} + \eta_{(n)}^{\dagger} \mathrm{i}\bar{\sigma}^{\mu}\partial_{\mu}\eta_{(n)} \right) \\
& - \frac{\mathrm{i}}{2} \left[\sum_{n=1}^{N} \left({\xi_{(n)}}^{T} \; {\eta_{(n)}}^{T} \right) \sigma^2 \begin{pmatrix} 0 & n/R \\ n/R & 0 \end{pmatrix} \begin{pmatrix} \xi_{(n)} \\ \eta_{(n)} \end{pmatrix} + \text{h.c.} \right] \\
& - \mathrm{i} \left(\nu_{\mathrm{L}}^{T} \sigma^2 m_{\mathrm{D}} \xi_{(0)} + \sqrt{2} \sum_{n=1}^{N} \nu_{\mathrm{L}}^{T} \sigma^2 m_{\mathrm{D}} \xi_{(n)} + \text{h.c.} \right) \Bigg\},
\end{aligned} \tag{34}
$$

where $\bar{\sigma}_{\mu} = (\mathbf{1}, \sigma_i)$ with σ_i being the Pauli matrices. Here we have also taken the 5D Majorana-like mass parameter M_M to vanish, i.e., we assume lepton number is conserved. Note that the Kaluza–Klein modes of ξ and η are all left-handed 4D Weyl spinors. The Dirac mass parameter $m_{\mathrm{D}} = \hat{m}/\sqrt{2\pi M_S R}$.

The zero mode $\xi_{(0)}^{c}$ thus plays the role of the right-handed component of active neutrinos, while the heavy Kaluza–Klein modes $\xi_{(n)}$ and $\eta_{(n)}$ form Dirac pairs, i.e., $\nu_{s,n} = (\eta_{(n)}, \xi_{(n)}^{c})^{T}$ (for $n \geq 1$). In this notation, the neutrino mass term becomes

$$
\overline{(\nu, \nu_{s,1}, \nu_{s,2}, \cdots)_{\mathrm{L}}} \begin{pmatrix} m_{\mathrm{D}} & \sqrt{2}m_{\mathrm{D}} & \sqrt{2}m_{\mathrm{D}} & \cdots \\ 0 & 1/R & 0 & \cdots \\ 0 & 0 & 2/R & \cdots \\ \vdots & \vdots & \vdots & \ddots \end{pmatrix} \begin{pmatrix} \xi_{(0)}^{c} \\ \nu_{s,1} \\ \nu_{s,2} \\ \vdots \end{pmatrix}_{\mathrm{R}} . \tag{35}
$$

There is no seesaw mechanism here for the active neutrinos, which are purely Dirac. Their masses are roughly m_{D}, and the smallness of these neutrino masses is explained by the suppression factor $\sqrt{M_S R}$; for example if $M_S = 1$ TeV and $1/R = 1$ eV, we can understand neutrino masses without invoking Yukawa couplings much smaller than that of the electron. Expanding in powers of the small parameter $m_{\mathrm{D}}R$, the active neutrino masses are given by

$$
m_0 = m_{\mathrm{D}} \left(1 - \frac{\pi^2}{6} m_{\mathrm{D}}^2 R^2 + \cdots \right), \tag{36}
$$

while the heavier Kaluza–Klein tower of sterile neutrinos have masses

$$
m_{(n)} = \frac{n}{R} \left(1 + \frac{m_{\mathrm{D}}^2 R^2}{n^2} + \cdots \right). \tag{37}
$$

This type of model predicts mixing of the active neutrinos with the sterile neutrinos of the Kaluza–Klein tower. The mixing with the nth Kaluza–Klein sterile is proportional to $m_{\mathrm{D}}R/n \simeq m_0/m_{(n)}$. Such mixing could modify the

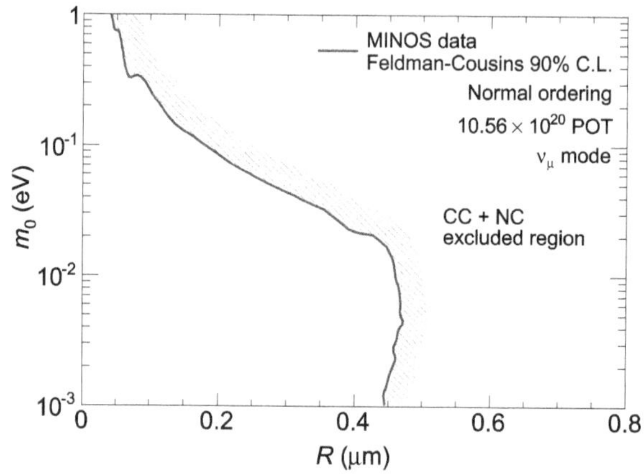

Fig. 3. The 90% CL data contour for the LED model, obtained using the Feldman–Cousins technique, based on 10.56×10^{20} POT MINOS data and assuming normal mass ordering. The large extra dimension size and the smallest neutrino mass are denoted as R and m_0, respectively. The shaded area indicates the excluded region to the right of the contour. This figure has been taken from Ref. 130.

neutrino oscillation patterns measured in experiments.[127,128] Currently the most stringent constraints on the orbifold model just described come from the MINOS experiment. As shown in Fig. 3, a model of this type with $R = 1$ $\text{eV}^{-1} \simeq 0.2$ μm is not yet excluded. Indeed it has even been suggested that bulk neutrinos could be an explanation of some of the anomalies still outstanding in short-baseline data from reactor and source experiments.[129]

It has been noted[125,131–133] that mixing of active neutrinos with bulk sterile neutrinos would alter the dynamics of core-collapse supernovae, by allowing significant energy loss in the form of sterile neutrinos. More detailed analysis[134,135] shows that the constraints arising from the observed neutrino signal from supernova 1987A are not as severe as first supposed but still pretty strong. Applied to our simple 5D orbifold model, they imply $m_{\text{D}}^2 R < 5 \times 10^{-5}$ eV.

2.5. *Neutrino decay*

As neutrino oscillations demonstrate, neutrino flavor is not conserved; thus there is no known mechanism forbidding heavier neutrinos from decaying into lighter ones. Indeed in the SM, the radiative decay $\nu_i \to \nu_j + \gamma$ and the three-body decay $\nu_i \to \nu_j \nu_k \nu_l$ are both predicted to occur, albeit at such a tiny rate that neutrino lifetimes should exceed 10^{37} years. BSM physics could make such decays more rapid, or open up new decay channels, e.g., $\nu_i \to \nu_j + X$, where X is a very light exotic pseudoscalar. One could also imagine an active neutrino decaying into a very light sterile fermion plus a pseudoscalar.

Neutrino survival probabilities can be characterized by the exponential factor

$$\exp(-t/\tau_i) = \exp\left(\frac{L}{E} \cdot \frac{m_i}{\tau_{i,0}}\right), \tag{38}$$

where τ_i is the lifetime of the neutrino mass eigenstate i in the observer's frame, $\tau_{i,0}$ is the rest frame lifetime, m_i is the neutrino mass, E is the energy, and t has been replaced by the distance traveled L using $c = 1$. Since the absolute neutrino masses are not known, neutrino decay constraints are usually expressed in terms of the quotient $\tau_{i,0}/m_i$ in units of seconds per eV.

The fact that neutrinos were detected from supernova 1987A implies that at least one neutrino mass eigenstate is rather stable; Fig. 4 shows a conservative estimate of the neutrino lifetime sensitivity (labeled "Gal. SN sens.") of a future detection of supernova neutrinos in the DUNE or Hyper-Kamiokande detectors. Very-high-energy astrophysical neutrinos, as

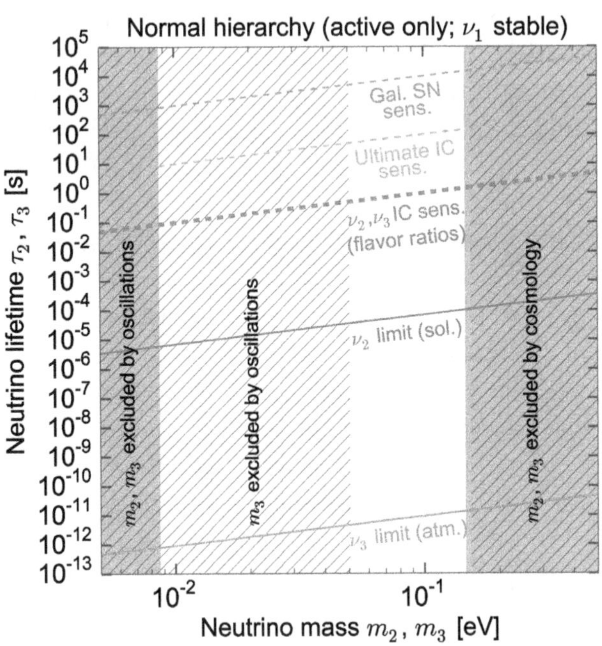

Fig. 4. Constraints on normal hierarchy neutrino masses and lifetimes, with hatched gray disallowed, hatched white allowed only for some eigenstates, and nonhatched white allowed for all. The assumption is that the sum of the neutrino masses is <0.3 eV. This figure has been taken from Ref. 140.

detected for example by IceCube, provide evidence of neutrino stability since the astrophysical sources are extremely distant; furthermore, if one knew the relative flavor composition of the neutrinos at the sources, then IceCube data on the relative flavor composition when these neutrinos reach the Earth might provide evidence of a particular pattern for neutrino decay.[136–140] An estimate of the ultimate reach of IceCube in this regard is shown in Fig. 4, labeled "Ulitmate IC sens."

Weaker but direct limits on neutrino lifetimes are provided by observations of solar[141] and atmospheric[142] neutrinos, as well as longer-baseline oscillation experiments with neutrinos from accelerators[142,143] and reactors.[144]

2.6. *Lorentz invariance violation and CPT violation*

Despite the spectacular successes of describing fundamental physical properties in terms of relativistic quantum field theories, this framework appears to fail when applied to quantum gravity. While very sophisticated statements can be made in the context of string theory, the mere existence of black holes is enough to show that there is a fundamental problem (for example, see Ref. 145). Even partial resolutions of this problem seem to imply that neither locality nor space-time itself are fundamental. If one could perform experiments sensitive to effects suppressed by powers of the Planck scale ($\simeq 10^{19}$ GeV), one might reasonably expect to observe explicit or spontaneous breakings of Lorentz invariance. More generally, if the building blocks of quantum field theory are not fundamental, one might encounter special physical systems where exotic deviations occur on more accessible length and energy scales (indeed black holes are believed to be an example of such systems).

It was suggested some time ago that neutrino experiments are an interesting arena to look for exotic violations of CPT symmetry.[146,147] Such speculations were motivated by early data that pointed towards different values for the mass differences of neutrinos versus antineutrinos.[146–151] The CPT theorem is so fundamental to quantum field theory that it is a challenge even to write down a framework for discussing the possibility of such gross violations. However, in string theory, it had been observed many years ago that string compactifications that do not fully respect the worldsheet symmetries of the superstring (namely modular invariance) produce particle spectra that violate CPT, and later work showed[152] that even in quantum field theory one can obtain CPT-violating spectra by abandoning the requirement of strict locality. Current data on neutrino versus antineutrino mass differences is completely consistent with CPT. Furthermore a careful re-examination of the famous CPT Theorem shows[153,154] that any violation of CPT must also imply violation of Lorentz invariance — although not neccessarily one that is detectable.

In any event it has become customary in recent years to focus on searching for violations of Lorentz invariance and CPT invariance together.[155] Generically violations of Lorentz invariance are assumed to appear as small deviations in the relativistic dispersion relations for neutrino propagation, i.e., the formula $E^2 = \vec{p}^2 + m^2$. The most gross violation of this dispersion relation would be neutrinos that travel faster than the speed of light, but this is difficult to constrain experimentally with much precision. More promising is to exploit the fact that neutrino oscillations over macroscopic distances depend on the dispersion relation, and thus the macroscopic baselines provide a lever arm for detecting small deviations. Small deviations could appear in the oscillation patterns or as differences between neutrino and antineutrino oscillations; they could also be dependent on the direction of the baseline.

Searches for such deviations have been carried out with data from a variety of experiements, including MINOS, IceCube, MiniBooNE, Double Chooz, and SK[156–162] While all the results so far are negative, they show remarkable sensitivity: as good as one part in 10^{22} for a certain type of potential deviation.

2.7. *Mass-varying neutrinos?*

The late time cosmic acceleration of our universe is usually attributed to dark energy, but the nature of dark energy remains a mystery. It is intriguing that the scale of dark energy density as extracted from cosmological data is about $(2 \times 10^{-3}\ \text{eV})^4$, which is similar to the measured mass-squared differences for neutrinos: $2.4 \times 10^{-3}\ \text{eV}^2$ and $7.5 \times 10^{-5}\ \text{eV}^2$. Could it be that neutrino mass generation is related to the existence of dark energy? Suppose that dark energy is due to the vev of a dynamical scalar field $\mathcal{A}$ (similar to, but not the same as, the Higgs field). We can suppose further that neutrino masses depend on the expectation value of $\mathcal{A}$; this could happen, for example, if the Majorana masses of the conventional neutrino seesaw mechanism depend on $\mathcal{A}$.

In this case, the scalar potential for $\mathcal{A}$ will acquire an additive contribution proportional to the background energy density from neutrinos[163,164]; changing variables from $\mathcal{A}$ to $m_\nu(\mathcal{A})$ for nonrelativistic neutrinos, one can write

$$V(m_\nu) = m_\nu n_\nu + V_0(m_\nu), \tag{39}$$

where V_0 is the scalar potential in the absence of neutrinos and n_ν is the total number density of neutrinos and antineutrinos. The additive contibution energetically disfavors a large mass for the neutrinos; thus as the early universe cools and the neutrino density dilutes, one expects that neutrino masses will increase. More generally, spatial variations in the neutrino density will also lead to spatial variations in neutrino masses. This general phenomenon is known as mass-varying neutrinos or MaVaNs.

There are serious challenges to making such a scenario work as a direct connection between neutrinos and dark energy.[165,166] Of course the basic idea could be correct even if the scalar field $\mathcal{A}$ has no connection to dark energy.[167]

2.8. *Other directions and final remarks*

This brief review of necessity has neglected a number of relevant topics. One large area of research is the possibility that neutrinos have nonstandard electromagnetic properties, such as neutrino magentic moments; for a recent review, see Ref. 168. Another prominent topic is grand unification, in particular the various schemes for unification based on $SO(10)$ gauge symmetry, which provide an appealing rationale for neutrino masses and perhaps leptogenesis as the origin of the matter vs. anitmatter excess in our universe. Any such scheme involves a great deal of physics beyond the SM and in principle relates neutrino properties to possibly detectable new experimental phenomena such as proton decay.

New physics involving neutrinos could involve ideas already well developed by theorists, such as extra dimensions and dark energy, or be something completely beyond our current imaginings. This presents an extraordinary challenge to neutrino experimentalists, since it implies that experiments, and the analysis of those experiments, should always keep in mind the unexpected and hold at bay any bias towards the attractive but very preliminary standard framework for neutrino physics. The possibility of the unexpected also underlines the importance of launching future experiments with the capability of *overconstraining* the measurements of neutrino properties and exploring novel experimental approaches. While enormous progress has been made in recent years towards revealing the surprising properties of neutrinos, it seems likely that many more surprises are yet to come.

Acknowledgments

Fermilab is operated by Fermi Research Alliance, LLC, under contract DE-AC02-07CH11359 with the United States Department of Energy.

References

1. T. Ohlsson, Status of non-standard neutrino interactions, *Rept. Prog. Phys.* **76**, 044201 (2013). doi: 10.1088/0034-4885/76/4/044201.
2. O. G. Miranda and H. Nunokawa, Non standard neutrino interactions: current status and future prospects, *New J. Phys.* **17**(9), 095002 (2015). doi: 10.1088/1367-2630/17/9/095002.
3. L. Wolfenstein, Neutrino oscillations in matter, *Phys. Rev.* **D17**, 2369–2374 (1978). doi: 10.1103/PhysRevD.17.2369.
4. L. Wolfenstein, Neutrino oscillations and stellar collapse, *Phys. Rev.* **D20**, 2634–2635 (1979). doi: 10.1103/PhysRevD.20.2634.
5. J. W. F. Valle, Resonant oscillations of massless neutrinos in matter, *Phys. Lett.* **B199**, 432–436 (1987). doi: 10.1016/0370-2693(87)90947-6.
6. E. Roulet, MSW effect with flavor changing neutrino interactions, *Phys. Rev.* **D44**, R935–R938 (1991). doi: 10.1103/PhysRevD.44.R935.
7. M. M. Guzzo, A. Masiero, and S. T. Petcov, On the MSW effect with massless neutrinos and no mixing in the vacuum, *Phys. Lett.* **B260**, 154–160 (1991). doi: 10.1016/0370-2693(91)90984-X.
8. Y. Grossman, Nonstandard neutrino interactions and neutrino oscillation experiments, *Phys. Lett.* **B359**, 141–147 (1995). doi: 10.1016/0370-2693(95)01069-3.
9. H. Nunokawa, A. Rossi, and J. W. F. Valle, Supernova bounds on supersymmetric R-parity violating interactions, *Nucl. Phys.* **B482**, 481–494 (1996). doi: 10.1016/S0550-3213(96)00541-X.
10. H. Nunokawa, Y. Z. Qian, A. Rossi, and J. W. F. Valle, Resonant conversion of massless neutrinos in supernovae, *Phys. Rev.* **D54**, 4356–4363 (1996). doi: 10.1103/PhysRevD.54.4356.
11. S. Bergmann, The solar neutrino problem in the presence of flavor changing neutrino interactions, *Nucl. Phys.* **B515**, 363–383 (1998). doi: 10.1016/S0550-3213(97)00794-3.
12. M. C. Gonzalez-Garcia, M. M. Guzzo, P. I. Krastev, H. Nunokawa, O. L. G. Peres, V. Pleitez, J. W. F. Valle, and R. Zukanovich Funchal, Atmospheric neutrino observations and flavor changing interactions, *Phys. Rev. Lett.* **82**, 3202–3205 (1999). doi: 10.1103/PhysRevLett.82.3202.
13. S. Bergmann, Y. Grossman, and E. Nardi, Neutrino propagation in matter with general interactions, *Phys. Rev.* **D60**, 093008 (1999). doi: 10.1103/PhysRevD.60.093008.
14. B. Kayser, Neutrino Oscillation Physics. In *Proceedings, 2011 European School of High-Energy Physics (ESHEP 2011): Cheile Gradistei, Romania, September 7-20, 2011*, pp. 107–117 (2014). doi: 10.5170/CERN-2014-003.107.
15. A. G. Cohen, S. L. Glashow, and Z. Ligeti, Disentangling neutrino oscillations, *Phys. Lett.* **B678**, 191–196 (2009). doi: 10.1016/j.physletb.2009.06.020.
16. E. K. Akhmedov and A. Y. Smirnov, Paradoxes of neutrino oscillations, *Phys. Atom. Nucl.* **72**, 1363–1381 (2009). doi: 10.1134/S1063778809080122.
17. C. Giunti and C. W. Kim, Quantum mechanics of neutrino oscillations, *Found. Phys. Lett.* **14**, 213–229 (2001). doi: 10.1023/A:1012230026160.

18. E. K. Akhmedov and A. Y. Smirnov, Neutrino oscillations: Entanglement, energy-momentum conservation and QFT, *Found. Phys.* **41**, 1279–1306 (2011). doi: 10.1007/s10701-011-9545-4.
19. B. Kayser, J. Kopp, R. G. H. Roberston, and P. Vogel, On a theory of neutrino oscillations with entanglement, *Phys. Rev.* **D82**, 093003 (2010). doi: 10.1103/PhysRevD.82.093003.
20. S. P. Mikheev and A. Y. Smirnov, Resonance amplification of oscillations in matter and spectroscopy of solar neutrinos, *Sov. J. Nucl. Phys.* **42**, 913–917 (1985). [*Yad. Fiz.* 42, 1441 (1985)].
21. S. P. Mikheev and A. Y. Smirnov, Resonant amplification of neutrino oscillations in matter and solar neutrino spectroscopy, *Nuovo Cim.* **C9**, 17–26 (1986). doi: 10.1007/BF02508049.
22. S. M. Bilenky, C. Giunti, and W. Grimus, Phenomenology of neutrino oscillations, *Prog. Part. Nucl. Phys.* **43**, 1–86 (1999). doi: 10.1016/S0146-6410(99)00092-7.
23. J. I. Peltoniemi and J. T. Peltoniemi, Coherence conditions for the forward scattering of neutrinos, *JHEP.* **08**, 008 (1999). doi: 10.1088/1126-6708/1999/08/008.
24. C. Biggio, M. Blennow, and E. Fernandez-Martinez, General bounds on non-standard neutrino interactions, *JHEP.* **08**, 090 (2009). doi: 10.1088/1126-6708/2009/08/090.
25. R. C. Allen *et al.*, Study of electron-neutrino electron elastic scattering at LAMPF, *Phys. Rev.* **D47**, 11–28 (1993). doi: 10.1103/PhysRevD.47.11.
26. L. B. Auerbach *et al.*, Measurement of electron-neutrino electron elastic scattering, *Phys. Rev.* **D63**, 112001 (2001). doi: 10.1103/PhysRevD.63.112001.
27. F. Reines, H. S. Gurr, and H. W. Sobel, Detection of $\bar{\nu}_e - e$ scattering, *Phys. Rev. Lett.* **37**, 315–318 (1976). doi: 10.1103/PhysRevLett.37.315.
28. G. S. Vidyakin, V. N. Vyrodov, I. I. Gurevich, Yu. V. Kozlov, V. P. Martemyanov, S. V. Sukhotin, V. G. Tarasenkov, E. V. Turbin, and S. K. Khakhimov, Limitations on the magnetic moment and charge radius of the electron-anti-neutrino, *JETP Lett.* **55**, 206–210 (1992). [*Pisma Zh. Eksp. Teor. Fiz.* 55, 212 (1992)].
29. A. I. Derbin, A. V. Chernyi, L. A. Popeko, V. N. Muratova, G. A. Shishkina, and S. I. Bakhlanov, Experiment on anti-neutrino scattering by electrons at a reactor of the Rovno nuclear power plant, *JETP Lett.* **57**, 768–772 (1993). [*Pisma Zh. Eksp. Teor. Fiz.* 57, 755 (1993)].
30. Z. Daraktchieva *et al.*, Limits on the neutrino magnetic moment from the MUNU experiment, *Phys. Lett.* **B564**, 190–198 (2003). doi: 10.1016/S0370-2693(03)00707-X.
31. M. Deniz *et al.*, Measurement of Nu(e)-bar -electron scattering cross-section with a CsI(Tl) scintillating crystal array at the Kuo-Sheng nuclear power reactor, *Phys. Rev.* **D81**, 072001 (2010). doi: 10.1103/PhysRevD.81.072001.
32. J. Dorenbosch *et al.*, Experimental verification of the universality of ν_e and ν_μ coupling to the neutral weak current, *Phys. Lett.* **B180**, 303–307 (1986). doi: 10.1016/0370-2693(86)90315-1.

33. P. Vilain *et al.*, Flavor universality of neutrino couplings with the Z, *Phys. Lett.* **B320**, 203–205 (1994). doi: 10.1016/0370-2693(94)90846-X.
34. P. Vilain *et al.*, Precision measurement of electroweak parameters from the scattering of muon-neutrinos on electrons, *Phys. Lett.* **B335**, 246–252 (1994). doi: 10.1016/0370-2693(94)91421-4.
35. G. P. Zeller *et al.*, A precise determination of electroweak parameters in neutrino nucleon scattering, *Phys. Rev. Lett.* **88**, 091802 (2002). doi: 10.1103/PhysRevLett.88.091802. [Erratum: *Phys. Rev. Lett.* 90, 239902 (2003)].
36. R. D. Ball, L. Del Debbio, S. Forte, A. Guffanti, J. I. Latorre, A. Piccione, J. Rojo, and M. Ubiali, Precision determination of electroweak parameters and the strange content of the proton from neutrino deep-inelastic scattering, *Nucl. Phys.* **B823**, 195–233 (2009). doi: 10.1016/j.nuclphysb.2009.08.003.
37. W. Bentz, I. C. Cloet, J. T. Londergan, and A. W. Thomas, Reassessment of the NuTeV determination of the weak mixing angle, *Phys. Lett.* **B693**, 462–466 (2010). doi: 10.1016/j.physletb.2010.09.001.
38. R. Barate *et al.*, Searches for supersymmetry in the photon(s) plus missing energy channels at $\sqrt{s} = 161\,\text{GeV}$ and $172\,\text{GeV}$, *Phys. Lett.* **B420**, 127–139 (1998). doi: 10.1016/S0370-2693(97)01555-4.
39. R. Barate *et al.*, Single photon and multiphoton production in e^+e^- collisions at a center-of-mass energy of 183 GeV, *Phys. Lett.* **B429**, 201–214 (1998). doi: 10.1016/S0370-2693(98)00468-7.
40. A. Heister *et al.*, Single photon and multiphoton production in e^+e^- collisions at $\sqrt{s}$ up to 209 GeV, *Eur. Phys. J.* **C28**, 1–13 (2003). doi: 10.1140/epjc/s2002-01129-7.
41. M. Acciarri *et al.*, Single and multiphoton events with missing energy in e^+ e^- collisions at $161\,\text{GeV} < \sqrt{s} < 172\,\text{GeV}$, *Phys. Lett.* **B415**, 299–310 (1997). doi: 10.1016/S0370-2693(97)01296-3.
42. M. Acciarri *et al.*, Single and multiphoton events with missing energy in e^+ e^- collisions at $\sqrt{s} = 183\,\text{GeV}$, *Phys. Lett.* **B444**, 503–515 (1998). doi: 10.1016/S0370-2693(98)01403-8.
43. M. Acciarri *et al.*, Single and multiphoton events with missing energy in e^+e^- collisions at $\sqrt{s}$ - 189-GeV, *Phys. Lett.* **B470**, 268–280 (1999). doi: 10.1016/S0370-2693(99)01286-1.
44. K. Ackerstaff *et al.*, Photonic events with large missing energy in e^+ e^- collisions at $\sqrt{s} = 161\,\text{GeV}$, *Phys. Lett.* **B391**, 210–220 (1997). doi: 10.1016/S0370-2693(96)01554-7.
45. K. Ackerstaff *et al.*, Search for anomalous production of photonic events with missing energy in e^+ e^- collisions at $\sqrt{s} = 130\,\text{GeV}$ to 172 GeV, *Eur. Phys. J.* **C2**, 607–625 (1998). doi: 10.1007/s100520050165.
46. G. Abbiendi *et al.*, Search for anomalous photonic events with missing energy in e^+ e^- collisions at $\sqrt{s} = 130\,\text{GeV}$, 136 GeV and 183 GeV, *Eur. Phys. J.* **C8**, 23–40 (1999). doi: 10.1007/s100520050442.
47. G. Abbiendi *et al.*, Photonic events with missing energy in e^+ e^- collisions at $\sqrt{s} = 189\,\text{GeV}$, *Eur. Phys. J.* **C18**, 253–272 (2000). doi: 10.1007/s100520000522.

48. J. Abdallah *et al.*, Photon events with missing energy in $e^+ e^-$ collisions at $\sqrt{s} = 130\,\text{GeV}$ to 209 GeV, *Eur. Phys. J.* **C38**, 395–411 (2005). doi: 10.1140/epjc/s2004-02051-8.
49. G. Mitsuka *et al.*, Study of non-standard neutrino interactions with atmospheric neutrino data in Super-Kamiokande I and II, *Phys. Rev.* **D84**, 113008 (2011). doi: 10.1103/PhysRevD.84.113008.
50. A. Gross, Atmospheric neutrino oscillations in IceCube, *Nucl. Phys. Proc. Suppl.* **237–238**, 272–274 (2013). doi: 10.1016/j.nuclphysbps.2013.04.107.
51. R. Abbasi *et al.*, The design and performance of IceCube DeepCore, *Astropart. Phys.* **35**, 615–624 (2012). doi: 10.1016/j.astropartphys.2012.01.004.
52. S. Abe *et al.*, Precision measurement of neutrino oscillation parameters with KamLAND, *Phys. Rev. Lett.* **100**, 221803 (2008). doi: 10.1103/PhysRevLett.100.221803.
53. B. Aharmim *et al.*, An independent measurement of the total active B-8 solar neutrino flux using an array of He-3 proportional counters at the sudbury neutrino observatory, *Phys. Rev. Lett.* **101**, 111301 (2008). doi: 10.1103/PhysRevLett.101.111301.
54. C. Arpesella *et al.*, Direct measurement of the Be-7 solar neutrino flux with 192 days of Borexino data, *Phys. Rev. Lett.* **101**, 091302 (2008). doi: 10.1103/PhysRevLett.101.091302.
55. G. Bellini *et al.*, Precision measurement of the ^{7}Be solar neutrino interaction rate in Borexino, *Phys. Rev. Lett.* **107**, 141302 (2011). doi: 10.1103/PhysRevLett.107.141302.
56. M. Maltoni, Standard and nonstandard physics in neutrino oscillations, *Nucl. Phys. Proc. Suppl.* **114**, 191–196 (2003). doi: 10.1016/S0920-5632(02)01904-7. [,191(2002)].
57. S. Davidson, C. Pena-Garay, N. Rius, and A. Santamaria, Present and future bounds on nonstandard neutrino interactions, *JHEP.* **03**, 011 (2003). doi: 10.1088/1126-6708/2003/03/011.
58. A. Friedland, C. Lunardini, and C. Pena-Garay, Solar neutrinos as probes of neutrino matter interactions, *Phys. Lett.* **B594**, 347 (2004). doi: 10.1016/j.physletb.2004.05.047.
59. A. Friedland, C. Lunardini, and M. Maltoni, Atmospheric neutrinos as probes of neutrino-matter interactions, *Phys. Rev.* **D70**, 111301 (2004). doi: 10.1103/PhysRevD.70.111301.
60. A. Friedland and C. Lunardini, A test of tau neutrino interactions with atmospheric neutrinos and K2K, *Phys. Rev.* **D72**, 053009 (2005). doi: 10.1103/PhysRevD.72.053009.
61. M. C. Gonzalez-Garcia, M. Maltoni, and J. Salvado, Testing matter effects in propagation of atmospheric and long-baseline neutrinos, *JHEP.* **05**, 075 (2011). doi: 10.1007/JHEP05(2011)075.
62. M. C. Gonzalez-Garcia, M. Maltoni, J. Salvado, and T. Schwetz, Global fit to three neutrino mixing: critical look at present precision, *JHEP.* **12**, 123 (2012). doi: 10.1007/JHEP12(2012)123.
63. M. C. Gonzalez-Garcia and M. Maltoni, Determination of matter potential from global analysis of neutrino oscillation data, *JHEP.* **09**, 152 (2013). doi: 10.1007/JHEP09(2013)152.

64. A. Esmaili and A. Y. Smirnov, Probing non-standard interaction of neutrinos with IceCube and DeepCore, *JHEP.* **06**, 026 (2013). doi: 10.1007/JHEP06(2013)026.
65. A. N. Khan, D. W. McKay, and F. Tahir, Sensitivity of medium-baseline reactor neutrino mass-hierarchy experiments to nonstandard interactions, *Phys. Rev.* **D88**, 113006 (2013). doi: 10.1103/PhysRevD.88.113006.
66. A. N. Khan, D. W. McKay, and F. Tahir, Short baseline reactor $\bar{\nu}-e$ scattering experiments and nonstandard neutrino interactions at source and detector, *Phys. Rev.* **D90**(5), 053008 (2014). doi: 10.1103/PhysRevD.90.053008.
67. I. Girardi and D. Meloni, Constraining new physics scenarios in neutrino oscillations from Daya Bay data, *Phys. Rev.* **D90**(7), 073011 (2014). doi: 10.1103/PhysRevD.90.073011.
68. I. Girardi, D. Meloni, and S. T. Petcov, The Daya Bay and T2K results on $\sin^2 2\theta_{13}$ and non-standard neutrino interactions, *Nucl. Phys.* **B886**, 31–42 (2014). doi: 10.1016/j.nuclphysb.2014.06.014.
69. S. K. Agarwalla, P. Bagchi, D. V. Forero, and M. Tortola, Probing non-standard interactions at Daya Bay, *JHEP.* **07**, 060 (2015). doi: 10.1007/JHEP07(2015)060.
70. M. C. Gonzalez-Garcia, M. Maltoni, and T. Schwetz, Global analyses of neutrino oscillation experiments, *Nucl. Phys.* **B908**, 199–217 (2016). doi: 10.1016/j.nuclphysb.2016.02.033.
71. S. Choubey, A. Ghosh, T. Ohlsson, and D. Tiwari, Neutrino physics with non-standard interactions at INO, *JHEP.* **12**, 126 (2015). doi: 10.1007/JHEP12(2015)126.
72. S. K. Agarwalla, Y. Kao, D. Saha, and T. Takeuchi, Running of oscillation parameters in matter with flavor-diagonal non-standard interactions of the neutrino, *JHEP.* **11**, 035 (2015). doi: 10.1007/JHEP11(2015)035.
73. M. C. Gonzalez-Garcia, M. Maltoni, I. Martinez-Soler, and N. Song, Non-standard neutrino interactions in the Earth and the flavor of astrophysical neutrinos, *Astropart. Phys.* **84**, 15–22 (2016). doi: 10.1016/j.astropartphys. 2016.07.001.
74. M. Blennow, P. Coloma, E. Fernandez-Martinez, J. Hernandez-Garcia, and J. Lopez-Pavon, Non-Unitarity, sterile neutrinos, and non-standard neutrino interactions, arXiv:1609.08637 [hep-ph].
75. S. Antusch, J. P. Baumann, and E. Fernandez-Martinez, Non-standard neutrino interactions with matter from physics beyond the standard model, *Nucl. Phys.* **B810**, 369–388 (2009). doi:10.1016/j.nuclphysb.2008. 11.018.
76. M. B. Gavela, D. Hernandez, T. Ota, and W. Winter, Large gauge invariant non-standard neutrino interactions, *Phys. Rev.* **D79**, 013007 (2009). doi: 10.1103/PhysRevD.79.013007.
77. M. B. Wise and Y. Zhang, Effective theory and simple completions for neutrino interactions, *Phys. Rev.* **D90**(5), 053005 (2014). doi: 10.1103/Phys-RevD.90.053005.
78. A. Celis, V. Cirigliano, and E. Passemar, Model-discriminating power of lepton flavor violating τ decays, *Phys. Rev.* **D89**(9), 095014 (2014). doi: 10.1103/PhysRevD.89.095014.

79. SLD Collaboration and LEP Collaborations, A combination of preliminary electroweak measurements and constraints on the standard model (2003).
80. A. Zee, A theory of lepton number violation, neutrino Majorana mass, and oscillation, *Phys. Lett.* **B93**, 389 (1980). doi: 10.1016/0370-2693(80)90349-4, 10.1016/0370-2693(80)90193-8. [Erratum: *Phys. Lett.* B95, 461 (1980)].
81. E. F. Martinez, Are there consistent models giving observable NSI? *J. Phys. Conf. Ser.* **408**, 012031 (2013). doi: 10.1088/1742-6596/408/1/012031.
82. D. V. Forero and W.-C. Huang, Sizable NSI from the $SU(2)_L$ scalar doublet-singlet mixing and the implications in DUNE, JHEP **1703**, 018 (2017) doi:10.1007/JHEP03(2017)018.
83. W. Altmannshofer, S. Gori, M. Pospelov, and I. Yavin, Neutrino trident production: a powerful probe of new physics with neutrino beams, *Phys. Rev. Lett.* **113**, 091801 (2014). doi: 10.1103/PhysRevLett.113.091801.
84. A. E. Nelson and J. Walsh, Short baseline neutrino oscillations and a new light gauge boson, *Phys. Rev.* **D77**, 033001 (2008). doi: 10.1103/PhysRevD.77.033001.
85. S. Baek, N. G. Deshpande, X. G. He, and P. Ko, Muon anomalous g–2 and gauged $L_\mu - L_\tau$ models, *Phys. Rev.* **D64**, 055006 (2001). doi: 10.1103/PhysRevD.64.055006.
86. E. Ma, D. P. Roy, and S. Roy, Gauged $L_\mu - L_\tau$ with large muon anomalous magnetic moment and the bimaximal mixing of neutrinos, *Phys. Lett.* **B525**, 101–106 (2002). doi: 10.1016/S0370-2693(01)01428-9.
87. J. Heeck and W. Rodejohann, Gauged $L_\mu - L_\tau$ and different muon neutrino and anti-neutrino oscillations: MINOS and beyond, *J. Phys.* **G38**, 085005 (2011). doi: 10.1088/0954-3899/38/8/085005.
88. K. Harigaya, T. Igari, M. M. Nojiri, M. Takeuchi, and K. Tobe, Muon g-2 and LHC phenomenology in the $L_\mu - L_\tau$ gauge symmetric model, *JHEP.* **03**, 105 (2014). doi: 10.1007/JHEP03(2014)105.
89. E. Ma, Gauged B - 3L(tau) and radiative neutrino masses, *Phys. Lett.* **B433**, 74–81 (1998). doi: 10.1016/S0370-2693(98)00599-1.
90. E. Ma and D. P. Roy, Phenomenology of the B - 3L(τ) gauge boson, *Phys. Rev.* **D58**, 095005 (1998). doi: 10.1103/PhysRevD.58.095005.
91. Y. Farzan and J. Heeck, Neutrinophilic non-standard interactions, *Phys. Rev. D* **94**(5) 053010 (2016) doi:10.1103/PhysRevD.94.053010.
92. P. Fileviez Perez and M. B. Wise, Baryon and lepton number as local gauge symmetries, *Phys. Rev.* **D82**, 011901 (2010). doi: 10.1103/PhysRevD.82.079901, 10.1103/PhysRevD.82.011901. [Erratum: *Phys. Rev.* **D82**, 079901 (2010)].
93. M. Duerr, P. Fileviez Perez, and M. B. Wise, Gauge theory for baryon and lepton numbers with leptoquarks, *Phys. Rev. Lett.* **110**, 231801 (2013). doi: 10.1103/PhysRevLett.110.231801.
94. P. Fileviez Perez, S. Ohmer, and H. H. Patel, Minimal theory for lepto-baryons, *Phys. Lett.* **B735**, 283–287 (2014). doi:10.1016/j.physletb.2014.06.057.
95. K. A. Olive *et al.*, Review of particle physics, *Chin. Phys.* **C38**, 090001 (2014). doi: 10.1088/1674-1137/38/9/090001.

96. R. Essig *et al.* Working Group Report: New Light Weakly Coupled Particles. In *Proceedings, Community Summer Study 2013: Snowmass on the Mississippi (CSS2013): Minneapolis, MN, USA, July 29-August 6, 2013*, arXiv: 1311.0029 [hep-ph].
97. J. L. Feng, B. Fornal, I. Galon, S. Gardner, J. Smolinsky, T. M. P. Tait, and P. Tanedo, Particle physics models for the 17 MeV anomaly in beryllium nuclear decays, *Phys. Rev. D* **95**(3), 035017 (2017) doi:10.1103/PhysRevD.95.035017.
98. Y. Farzan and I. M. Shoemaker, Lepton flavor violating non-standard interactions via light mediators, *JHEP.* **07**, 033 (2016). doi:10.1007/JHEP07(2016)033.
99. S. Fukasawa and O. Yasuda, Constraints on the nonstandard interaction in propagation from atmospheric neutrinos, *Adv. High Energy Phys.* **2015**, 820941 (2015). doi: 10.1155/2015/820941.
100. J. Salvado, O. Mena, S. Palomares-Ruiz, and N. Rius, Non-standard interactions with high-energy atmospheric neutrinos at IceCube, JHEP **1701**, 141 (2017) doi: 10.1007/JHEP01(2017)141.
101. S. Fukasawa and O. Yasuda, The possibility to observe the non-standard interaction by the Hyperkamiokande atmospheric neutrino experiment, *Nucl. Phys.* **B914**, 99–116 (2017). doi: 10.1016/j.nuclphysb.2016.11.004.
102. A. de Gouvea and K. J. Kelly, Non-standard neutrino interactions at DUNE, *Nucl. Phys.* **B908**, 318–335 (2016). doi: 10.1016/j.nuclphysb.2016.03.013.
103. P. Coloma, Non-standard interactions in propagation at the deep underground neutrino experiment, *JHEP.* **03**, 016 (2016). doi:10.1007/JHEP03(2016)016.
104. M. Blennow, S. Choubey, T. Ohlsson, D. Pramanik, and S. K. Raut, A combined study of source, detector and matter non-standard neutrino interactions at DUNE, *JHEP.* **08**, 090 (2016). doi: 10.1007/JHEP08(2016)090.
105. K. Huitu, T. J. Karkkainen, J. Maalampi, and S. Vihonen, Constraining the nonstandard interaction parameters in long baseline neutrino experiments, *Phys. Rev.* **D93** (5), 053016 (2016). doi: 10.1103/PhysRevD.93.053016.
106. F. An *et al.*, Neutrino physics with JUNO, *J. Phys.* **G43**(3), 030401 (2016). doi: 10.1088/0954-3899/43/3/030401.
107. J. Kopp, M. Lindner, and T. Ota, Discovery reach for non-standard interactions in a neutrino factory, *Phys. Rev.* **D76**, 013001 (2007). doi: 10.1103/PhysRevD.76.013001.
108. B. Dutta, R. Mahapatra, L. E. Strigari, and J. W. Walker, Sensitivity to Z-prime and nonstandard neutrino interactions from ultralow threshold neutrino-nucleus coherent scattering, *Phys. Rev.* **D93**(1), 013015 (2016). doi: 10.1103/PhysRevD.93.013015.
109. M. Lindner, W. Rodejohann, and X.-J. Xu, Coherent neutrino-nucleus scattering and new neutrino interactions, arXiv:1612.04150 [hep-ph].
110. M. Masud, A. Chatterjee, and P. Mehta, Probing CP violation signal at DUNE in presence of non-standard neutrino interactions, *J. Phys.* **G43**(9), 095005 (2016). doi: 10.1088/0954-3899/43/9/095005/meta, 10.1088/0954-3899/43/9/095005.

111. A. de Gouvea and K. J. Kelly, False signals of CP-invariance violation at DUNE, arXiv:1605.09376 [hep-ph].
112. M. Masud and P. Mehta, Non-standard interactions and the resolution of ordering of neutrino masses at DUNE and other long baseline experiments, *Phys. Rev. D* **94**(5), 053007 (2016). doi:10.1103/PhysRevD.94.053007.
113. P. Horava and E. Witten, Eleven-dimensional supergravity on a manifold with boundary, *Nucl. Phys.* **B475**, 94–114 (1996). doi: 10.1016/0550-3213(96)00308-2.
114. E. Witten, Strong coupling expansion of Calabi-Yau compactification, *Nucl. Phys.* **B471**, 135–158 (1996). doi: 10.1016/0550-3213(96)00190-3.
115. P. Horava, Gluino condensation in strongly coupled heterotic string theory, *Phys. Rev.* **D54**, 7561–7569 (1996). doi: 10.1103/PhysRevD.54.7561.
116. J. D. Lykken, Weak scale superstrings, *Phys. Rev.* **D54**, R3693–R3697 (1996). doi: 10.1103/PhysRevD.54.R3693.
117. A. Lukas, B. A. Ovrut, K. S. Stelle, and D. Waldram, The universe as a domain wall, *Phys. Rev.* **D59**, 086001 (1999). doi: 10.1103/PhysRevD.59.086001.
118. A. Lukas, B. A. Ovrut, K. S. Stelle, and D. Waldram, Heterotic M theory in five-dimensions, *Nucl. Phys.* **B552**, 246–290 (1999). doi: 10.1016/S0550-3213(99)00196-0.
119. I. Antoniadis, N. Arkani-Hamed, S. Dimopoulos, and G. R. Dvali, New dimensions at a millimeter to a Fermi and superstrings at a TeV, *Phys. Lett.* **B436**, 257–263 (1998). doi: 10.1016/S0370-2693(98)00860-0.
120. Z. Kakushadze and S. H. H. Tye, Brane world, *Nucl. Phys.* **B548**, 180–204 (1999). doi: 10.1016/S0550-3213(99)00082-6.
121. K. R. Dienes, E. Dudas, and T. Gherghetta, Neutrino oscillations without neutrino masses or heavy mass scales: a higher dimensional seesaw mechanism, *Nucl. Phys.* **B557**, 25 (1999). doi: 10.1016/S0550-3213(99)00377-6.
122. N. Arkani-Hamed, S. Dimopoulos, G. R. Dvali, and J. March-Russell, Neutrino masses from large extra dimensions, *Phys. Rev.* **D65**, 024032 (2001). doi: 10.1103/PhysRevD.65.024032.
123. G. R. Dvali and A. Y. Smirnov, Probing large extra dimensions with neutrinos, *Nucl. Phys.* **B563**, 63–81 (1999). doi: 10.1016/S0550-3213(99)00574-X.
124. R. N. Mohapatra, S. Nandi, and A. Perez-Lorenzana, Neutrino masses and oscillations in models with large extra dimensions, *Phys. Lett.* **B466**, 115–121 (1999). doi: 10.1016/S0370-2693(99)01119-3.
125. A. Lukas, P. Ramond, A. Romanino, and G. G. Ross, Neutrino masses and mixing in brane world theories, *JHEP.* **04**, 010 (2001). doi: 10.1088/1126-6708/2001/04/010.
126. W. Rodejohann and H. Zhang, Signatures of extra dimensional sterile neutrinos, *Phys. Lett.* **B737**, 81–89 (2014). doi: 10.1016/j.physletb.2014.08.035.
127. H. Davoudiasl, P. Langacker, and M. Perelstein, Constraints on large extra dimensions from neutrino oscillation experiments, *Phys. Rev.* **D65**, 105015 (2002). doi: 10.1103/PhysRevD.65.105015.
128. P. A. N. Machado, H. Nunokawa, and R. Zukanovich Funchal, Testing for large extra dimensions with neutrino oscillations, *Phys. Rev.* **D84**, 013003 (2011). doi: 10.1103/PhysRevD.84.013003.

129. P. A. N. Machado, H. Nunokawa, F. A. P. dos Santos, and R. Z. Funchal, Bulk neutrinos as an alternative cause of the gallium and reactor anti-neutrino anomalies, *Phys. Rev.* **D85**, 073012 (2012). doi: 10.1103/PhysRevD.85.073012.
130. P. Adamson *et al.*, Constraints on large extra dimensions from the MINOS experiment, *Phys. Rev. D* **94**(11), 111101 (2016). doi:10.1103/PhysRevD.94.111101.
131. R. Barbieri, P. Creminelli, and A. Strumia, Neutrino oscillations from large extra dimensions, *Nucl. Phys.* **B585**, 28–44 (2000). doi: 10.1016/S0550-3213(00)00348-5.
132. A. De Gouvea, G. F. Giudice, A. Strumia, and K. Tobe, Phenomenological implications of neutrinos in extra dimensions, *Nucl. Phys.* **B623**, 395–420 (2002). doi: 10.1016/S0550-3213(01)00621-6.
133. A. Strumia and F. Vissani, Neutrino masses and mixings and,... hep-ph/0606054.
134. G. Cacciapaglia, M. Cirelli, Y. Lin, and A. Romanino, Bulk neutrinos and core collapse supernovae, *Phys. Rev.* **D67**, 053001 (2003). doi: 10.1103/PhysRevD.67.053001.
135. G. Cacciapaglia, M. Cirelli, and A. Romanino, Signatures of supernova neutrino oscillations into extra dimensions, *Phys. Rev.* **D68**, 033013 (2003). doi: 10.1103/PhysRevD.68.033013.
136. J. F. Beacom, N. F. Bell, D. Hooper, S. Pakvasa, and T. J. Weiler, Decay of high-energy astrophysical neutrinos, *Phys. Rev. Lett.* **90**, 181301 (2003). doi: 10.1103/PhysRevLett.90.181301.
137. P. Baerwald, M. Bustamante, and W. Winter, Neutrino decays over cosmological distances and the implications for neutrino telescopes, *JCAP.* **1210**, 020 (2012). doi: 10.1088/1475-7516/2012/10/020.
138. L. Dorame, O. G. Miranda, and J. W. F. Valle, Invisible decays of ultra-high energy neutrinos, *Front. Phys.* **1**, 25 (2013). doi: 10.3389/fphy.2013.00025.
139. H. Nunokawa, B. Panes, and R. Zukanovich Funchal, How unequal fluxes of high energy astrophysical neutrinos and antineutrinos can fake new physics, *JCAP.* **1610**(10), 036 (2016). doi: 10.1088/1475-7516/2016/10/036.
140. M. Bustamante, J. F. Beacom, and K. Murase, Testing decay of astrophysical neutrinos with incomplete information, arXiv:1610.02096 [astro-ph.HE].
141. J. M. Berryman, A. de Gouvea, and D. Hernandez, Solar neutrinos and the decaying neutrino hypothesis, *Phys. Rev.* **D92**(7), 073003 (2015). doi: 10.1103/PhysRevD.92.073003.
142. M. C. Gonzalez-Garcia and M. Maltoni, Status of oscillation plus decay of atmospheric and long-baseline neutrinos, *Phys. Lett.* **B663**, 405–409 (2008). doi: 10.1016/j.physletb.2008.04.041.
143. R. A. Gomes, A. L. G. Gomes, and O. L. G. Peres, Constraints on neutrino decay lifetime using long-baseline charged and neutral current data, *Phys. Lett.* **B740**, 345–352 (2015). doi: 10.1016/j.physletb.2014.12.014.
144. T. Abrahao, H. Minakata, H. Nunokawa, and A. A. Quiroga, Constraint on neutrino decay with medium-baseline reactor neutrino oscillation experiments, *JHEP.* **11**, 001 (2015). doi: 10.1007/JHEP11(2015)001.

145. S. B. Giddings, Black holes, quantum information, and unitary evolution, *Phys. Rev.* **D85**, 124063 (2012). doi: 10.1103/PhysRevD.85.124063.
146. H. Murayama and T. Yanagida, LSND, SN1987A, and CPT violation, *Phys. Lett.* **B520**, 263–268 (2001). doi: 10.1016/S0370-2693(01)01136-4.
147. G. Barenboim, L. Borissov, J. D. Lykken, and A. Y. Smirnov, Neutrinos as the messengers of CPT violation, *JHEP.* **10**, 001 (2002). doi: 10.1088/1126-6708/2002/10/001.
148. G. Barenboim, L. Borissov, and J. D. Lykken, Neutrinos that violate CPT, and the experiments that love them, *Phys. Lett.* **B534**, 106–113 (2002). doi: 10.1016/S0370-2693(02)01597-6.
149. G. Barenboim, L. Borissov, and J. D. Lykken, CPT violating neutrinos in the light of KamLAND, hep-ph/0212116.
150. M. C. Gonzalez-Garcia, M. Maltoni, and T. Schwetz, Status of the CPT violating interpretations of the LSND signal, *Phys. Rev.* **D68**, 053007 (2003). doi: 10.1103/PhysRevD.68.053007.
151. G. Barenboim and J. D. Lykken, MINOS and CPT-violating neutrinos, *Phys. Rev.* **D80**, 113008 (2009). doi: 10.1103/PhysRevD.80.113008.
152. G. Barenboim and J. D. Lykken, A model of CPT violation for neutrinos, *Phys. Lett.* **B554**, 73–80 (2003). doi: 10.1016/S0370-2693(02)03262-8.
153. O. W. Greenberg, CPT violation implies violation of Lorentz invariance, *Phys. Rev. Lett.* **89**, 231602 (2002). doi: 10.1103/PhysRevLett.89.231602.
154. I. S. Tsukerman, CPT invariance and neutrino physics, arXiv:1006.4989 [hep-ph].
155. V. A. Kostelecky and M. Mewes, Lorentz and CPT violation in the neutrino sector, *Phys. Rev.* **D70**, 031902 (2004). doi: 10.1103/PhysRevD.70.031902.
156. P. Adamson *et al.*, Testing Lorentz invariance and CPT conservation with NuMI neutrinos in the MINOS near detector, *Phys. Rev. Lett.* **101**, 151601 (2008). doi: 10.1103/PhysRevLett.101.151601.
157. R. Abbasi *et al.*, Search for a Lorentz-violating sidereal signal with atmospheric neutrinos in IceCube, *Phys. Rev.* **D82**, 112003 (2010). doi: 10.1103/PhysRevD.82.112003.
158. A. A. Aguilar-Arevalo *et al.*, Test of Lorentz and CPT violation with short baseline neutrino oscillation excesses, *Phys. Lett.* **B718**, 1303–1308 (2013). doi: 10.1016/j.physletb.2012.12.020.
159. Y. Abe *et al.*, First test of Lorentz violation with a reactor-based antineutrino experiment, *Phys. Rev.* **D86**, 112009 (2012). doi: 10.1103/PhysRevD.86.112009.
160. B. Rebel and S. Mufson, The search for neutrino-antineutrino mixing resulting from lorentz invariance violation using neutrino interactions in MINOS, *Astropart. Phys.* **48**, 78–81 (2013). doi: 10.1016/j.astropartphys.2013.07.006.
161. T. Katori and J. Conrad, Beyond standard model searches in the MiniBooNE experiment, *Adv. High Energy Phys.* **2015**, 362971 (2015). doi: 10.1155/2015/362971.
162. K. Abe *et al.*, Test of Lorentz invariance with atmospheric neutrinos, *Phys. Rev.* **D91**(5), 052003 (2015). doi: 10.1103/PhysRevD.91.052003.

163. R. Fardon, A. E. Nelson, and N. Weiner, Dark energy from mass varying neutrinos, *JCAP.* **0410**, 005 (2004). doi: 10.1088/1475-7516/2004/10/005.
164. D. B. Kaplan, A. E. Nelson, and N. Weiner, Neutrino oscillations as a probe of dark energy, *Phys. Rev. Lett.* **93**, 091801 (2004). doi: 10.1103/PhysRevLett.93.091801.
165. N. Afshordi, M. Zaldarriaga, and K. Kohri, On the stability of dark energy with mass-varying neutrinos, *Phys. Rev.* **D72**, 065024 (2005). doi: 10.1103/PhysRevD.72.065024.
166. G. Y. Chitov, T. August, A. Natarajan, and T. Kahniashvili, Mass varying neutrinos, quintessence, and the accelerating expansion of the universe, *Phys. Rev.* **D83**, 045033 (2011). doi: 10.1103/PhysRevD.83.045033.
167. A. Ghalsasi, D. McKeen, and A. E. Nelson, Probing nonstandard neutrino cosmology with terrestrial neutrino experiments, arXiv:1609.06326 [hep-ph].
168. B. C. Canas, O. G. Miranda, A. Parada, M. Tortola and J. W. F. Valle, *J. Phys. Conf. Ser.* **761**(1), 012043 (2016). doi:10.1088/1742-6596/761/1/012043.

www.ingramcontent.com/pod-product-compliance
Lightning Source LLC
LaVergne TN
LVHW060618300926
849162LV00008B/1653
* 9 7 8 9 8 1 3 2 2 6 0 8 1 *